p163
170

col in A - rows in B

As many rows as columns

Applied linear algebra

Cover illustration:

The second X from the left results from superimposing a 20 × 20 grid on the first and coloring each square either all black or all white. The other two use the singular value decomposition of Section 8.4 to approximate the resulting 20 × 20 matrix by matrices of rank 1 or rank 2, using 41 and 82 data respectively instead of the original 400.

Third edition

Applied linear algebra

BEN NOBLE
UNIVERSITY OF WISCONSIN

and

JAMES W. DANIEL
UNIVERSITY OF TEXAS AT AUSTIN

PRENTICE-HALL / Englewood Cliffs, NJ 07632

Library of Congress Cataloging-in-Publication Data

Noble, Ben.
Applied linear algebra.

Bibliography: p.
Includes index.
1. Algebra Linear. I. Daniel, James W. II. Title.
QA184.N6 1988 512'.5 87–11511
ISBN 0-13-041260-0

Editorial/production: Nicholas C. Romanelli
Manufacturing buyer: Harry Baisley

Printed in the United States of America

10 9 8 7 6 5 4 3 2 1

ISBN 0-13-041260-0

Prentice-Hall International (UK) Limited, *London*
Prentice-Hall of Australia Pty Limited, *Sydney*
Prentice-Hall of Canada Inc., *Toronto*
Prentice-Hall Hispanoamericana, S.A., *Mexico*
Prentice-Hall of India Private Limited, *New Delhi*
Prentice-Hall of Japan, Inc., *Tokyo*
Prentice-Hall of Southeast Asia Pte., Ltd., *Singapore*
Editora Prentice-Hall do Brasil, Ltda., *Rio de Janeiro*

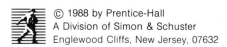

To Denise, Anna, and John Ben

To Adam and Joshua, and Ann and Susan

Contents

Preface

Overall approach

Linear algebra is an essential part of the mathematical toolkit required in the modern study of many areas in the behavioral, natural, physical, and social sciences, in engineering, in business, in computer science, and of course in pure and applied mathematics. Our aim in this book is to develop the fundamental concepts of linear algebra, emphasizing those concepts that are most important in applications, and to illustrate the applicability of these concepts through a variety of applications. Thus, while we present applications for illustrative and motivative purposes, our main goal is to present mathematics that can be applied.

We have taken great care in presenting the theory of linear algebra rather fully, although from as concrete a viewpoint as possible. We therefore begin with the concrete manipulative matrix and vector algebra (Chapter 1) and with Gauss elimination (Chapter 3) before the *theory* of linear equations (Chapter 4) or the abstract notions of vector spaces (Chapter 5) and linear transformations (Chapter 6). Similarly, eigensystems and various related decompositions and canonical forms are presented (Chapter 7 and, in more detail, Chapters 8 and 9) as tools for simplifying the study of linear transformations modeling the behavior of complicated systems.

In addition to presenting a number of applications briefly throughout the text, we have collected in Chapter 2 a special set of applications to motivate the later material; only familiarity with the basic matrix algebra of Chapter 1 is assumed, and we not only show how matrices arise in practice but also raise questions about matrices and their properties that stimulate interest in the ideas to come. Later chapters often refer to these examples for motivation, although the remainder of the book is in fact independent of Chapter 2 for those who prefer to omit it.

The entire book builds upon the notion of elementary row operations and Gauss elimination. The manipulative approach developed in Chapter 3 is fundamental to almost all the theoretical as well as applied topics. It is *absolutely essential* for the student to learn these techniques.

New features in the third edition

We created this new edition with the primary goal of making it easier to teach and to learn from the book. Except for some additions, the content is approximately the same as in the second edition, but the presentation has been reworked for a more consistent style and approach. Triangular forms produced by Gauss elimination often replace the (reduced-) row-echelon forms emphasized for theoretical work in the prior editions.

The problem sets have been expanded. This edition contains over 1100 problems, compared with just over 600 in the second edition. Problem sets appear at the end of each section instead of being scattered throughout the text as Exercises (a mix of examples and problems). An appendix contains answers or suggestions for over one-third of the problems (those indicated by a $\triangleright$ symbol preceding the problem numbers).

We now number all displayed material consecutively, so that displayed statement (4.15) falls immediately between Example 4.14 and **Key Theorem 4.16**, for example. Guidelines appear in most longer sections, indicating which of the problems can be solved using material presented by that point. A brief refresher on complex numbers appears early in Chapter 1; results are usually stated separately for the real and complex cases for those wishing to concentrate on one or the other.

In this edition many topics appear in the main text rather than merely in the problems. Greater emphasis is thus given to the Cayley-Hamilton theorem, Perron-Frobenius theorem, LU-decomposition, orthogonal projections and projection matrices, best approximation, fundamental frequencies of oscillating systems, and so on. New material has also been included, introducing such topics as adjoint transformations, Karmarkar-like methods for linear programming, theory of linear inequalities, convergence of sequences of vectors and matrices, and isomorphisms.

Examples and some problems (about 6%) have been added that make explicit use of computer software for matrix computations. The book can still be used completely independently of computers, but those wishing to take advantage of microcomputer and mainframe software will find it easy to do so. For more information on this, see the material following this Preface.

We realize that this text in its previous two editions has been used for a variety of courses. The new edition is designed to continue this option, which of course places a special responsibility on the instructor to select the appropriate material. Some guidelines on this now follow. Those interested in the possible support of this text through the use of computers should also examine the section after this Preface.

Use for elementary theoretical linear algebra courses

Those wishing to emphasize proofs in a classic theorem-proof approach can easily do so. Theory is fundamental to applied linear algebra and is therefore presented with great care (and affection). The standard one-semester sophomore course would cover Chapter 1 on matrix algebra, Sections 1 through 5 of Chapter 3 on techniques for equations and inverses, most of Chapter 4 on theory for equations and inverses (including determinants), Sections 1 through 5, 7, and 8 of Chapter 5 on vector spaces, the first two sections on linear transformations in Chapter 6, and possibly the first three sections on eigensystems in Chapter 7. We certainly hope that such courses will also include some application-oriented material: a section from Chapter 2 illustrating an application, the *LU*-decomposition of Section 3.7, and norms and least squares in Sections 5.6 and 5.9 are good possibilities.

Use for elementary applied linear algebra courses

Along with their intermediate/advanced versions, introductory applied linear algebra courses have been the main contexts in which the previous two editions of this book have been used for nearly twenty years. Many variations in these courses are possible, depending on such aspects as which application areas are to be emphasized (engineering? social science? et cetera) and how much proving is to be done (none? selected **Key Theorems**? et cetera). We sketch a basic syllabus that provides a good mix.

Chapter 1 on matrix algebra should be carefully covered, followed by two applications from Chapter 2 (we choose those that connect with topics emphasized later). We would then proceed carefully through Chapter 3 on methods for equations and inverses, followed by a light treatment (especially regarding determinants) of the theory in Chapter 4; most time would be spent explaining the theory, with an occasional instructive proof presented in detail. By emphasizing the facts rather than the proofs, we would cover vector spaces and linear transformations in Chapters 5 and 6 (most sections); plenty of time must be allowed for students to grasp the concepts of linear independence and dependence. Since Chapter 7 gives a broad overview of the theory and use of eigensystems, we would generally cover this rather than the more detailed and specialized Chapters 8 or 9; those chapters are good alternatives, however, for those wishing to stress particular types of applications. Depending on the availability of time, we might introduce some topics from later chapters: least squares (Section 8.5), differential equations (Section 9.5), quadratic forms (Section 10.3), or linear programs (Section 11.1).

This covers quite a bit of material. The key to doing so is to stress the *concepts* and *techniques*; with this approach material can be covered quite quickly.

Use for subsequent applied linear algebra courses

For many courses our text has been used by students who have already had an introduction to linear algebra—often in a theory-oriented course. In such courses, two or three topics from Chapter 2 can be used as early motivation, followed by a quick review of Chapters 1, 3, and 4 (how quick depends on how much one assumes the students recall); the later sections of Chapter 3 on the practical aspects of solving equations—particularly the *LU*-decomposition—are rarely familiar to students from traditional courses and should be covered thoroughly. After briefly reviewing the basic information about vector spaces in the first half of Chapter 5, we would concentrate on the material there regarding norms, inner products, projections, and least squares. Likewise for Chapter 6—material on norms and perturbations is likely to be new.

The heart of such a course should be constructed from Chapters 7 through 11, again depending strongly on the emphasis desired and whether the students learned anything about eigensystems in their previous linear algebra courses.

Ben Noble and James W. Daniel

On the use of computers

We do *not* assume that either students or instructors will use computers to support their learning and teaching of linear algebra. As has always been true of its various versions, this book can be used independently of computing support and independently of the reader's interest in computing. We note, however, that each year more of the students in our classes are acquainted with computing and often are accustomed to using computers—mainframes or microcomputers—to assist with their course work in various ways including word processing, calculating, data analysis, and the like. This third edition of *Applied Linear Algebra* has been written to make it easier *for those who wish to do so* to support their learning and teaching with active involvement with computing.

Examples and Problems

To reflect the fact that scientific computation is now done on computers, many of our numbered Examples in the text are computer-based. *Such material is self-contained: we assume no prior computing knowledge on the part of student or instructor.* Our using computers simply allows us to consider and present far more realistic and interesting examples than if we had to restrict ourselves to contrived problems that can be solved entirely by hand. Such Examples are flagged in the margin by the symbol $\mathfrak{M}$.

Roughly 6% of our more than 1100 Problems are likewise flagged by the symbol $\mathfrak{M}$. Solving these problems requires access to computing, particularly to software designed to facilitate matrix computations. *Students and instructors should not write their own programs for solving linear algebra problems—the problems involve subtle aspects that are not obvious to the inexpert.* Superb software is available essentially for the asking: see Sections 3.9 and 7.6 on how to obtain high-quality programs for matrix computations. Since so many educational institutions now provide microcomputer centers for student use, we discuss in a bit more detail some excellent software available for this environment.

MATLAB

Our symbol $\mathfrak{M}$ stands for the initial letter in MATLAB, which itself stands for matrix laboratory. Over a number of years, MATLAB was developed by Cleve Moler, initially for use in teaching linear algebra and numerical analysis courses, using programs based on those from the LINPACK and EISPACK projects described in the text sections referenced previously. MATLAB was eventually released and widely distributed in a FORTRAN version for mainframe computers. Several commercial organizations developed enhanced versions for special applications such as control theory.

PC-MATLAB was the next evolutionary step—a highly optimized second-generation version of MATLAB developed especially for use by the IBM PC microcomputer. This in turn is being followed by other microcomputer versions of the system for use with other operating systems and other machines. For simplicity, we refer to the entire spectrum of software—from mainframe to the latest microcomputer versions—by the generic term "MATLAB."

We have personally found MATLAB to be very easy to learn (especially for the computer-wary), easy to use, and extremely powerful. Moreover, *MATLAB is available in special versions at greatly reduced cost for use in instruction*: contact The MathWorks, Inc., Suite 250, 20 North Main St., Sherborn, MA 01770, (617) 653-1415. One need not learn computer programming with this highly flexible and interactive system. Matrices are easily entered; typing

A = [1 2 3; 4 5 6; 7 8 9]

creates precisely the 3×3 matrix one would expect. Typing

inv(A)

produces its inverse,

eig(A)

its eigenvalues,

lu(A)

its LU-decomposition, and so on. It is therefore so easy to use that it is excellent for association with even a first linear algebra course. On the other hand, it is so powerful and flexible that it is an important tool for anyone whose work requires matrix calculations, whether they involve equations, eigensystems, least squares, data analysis, signal processing, or whatever.

We like MATLAB. We recommend it to you. And we have no financial interest in its success.

Applied linear algebra

1

Matrix algebra

This first chapter is fundamental: its goal is to introduce matrices and those basic algebraic manipulations that the student must ***thoroughly*** *understand before proceeding. It is important to practice the addition and multiplication of matrices until these operations become automatic. Theorem 1.44 is a* ***Key Theorem****, providing the basis for later computational methods; the proof of Theorem 1.35 is a* ***Key Proof*** *because it illustrates a useful general argument.*

1.1 *INTRODUCTION*

Matrices provide convenient tools for systematizing laborious calculations by providing a compact notation for storing information and for describing complicated relationships.

(1.1) ***Definition.*** A $p \times q$ (read "p by q") *matrix* is a rectangular array $\mathbf{A}$ of pq numbers (or symbols representing numbers) enclosed in square brackets; the numbers in the array are called *entries* and are arranged in p horizontal *rows* and q vertical *columns*. The *(i, j)-entry* is denoted $\langle \mathbf{A} \rangle_{ij}$ and equals the entry in the ith row and jth column, numbering rows from top to bottom and columns from left to right; if $\mathbf{A}$ is $p \times 1$ (a *column matrix*) or $1 \times q$ (a *row matrix*), then $\langle \mathbf{A} \rangle_i$ is used instead of $\langle \mathbf{A} \rangle_{i1}$ or $\langle \mathbf{A} \rangle_{1i}$.

We denote matrices by boldface letters $\mathbf{A}$, $\mathbf{x}$, and so on. Using the helpful mnemonic device of denoting the (i, j)-entry of a matrix by a subscripted lowercase version of the boldface capital letter denoting the matrix, we write the general $p \times q$ matrix as

(1.2)
$$\mathbf{A} = \begin{bmatrix} a_{11} & a_{12} & \cdots & a_{1q} \\ a_{21} & a_{22} & \cdots & a_{2q} \\ & \cdots & & \\ a_{p1} & a_{p2} & \cdots & a_{pq} \end{bmatrix}.$$

Matrix entries may be real or complex numbers; readers unsure about the basic properties of complex numbers can consult the brief refresher at the end of this page. Since matrices with various special structures often arise, we introduce some terminology to describe them.

(1.3) ***Definition***

(a) A *square matrix* **A** is $p \times q$ with $p = q$; the (i, i)-entries for $1 \leq i \leq p$ form the *main diagonal* of **A**.

(b) A *diagonal matrix* is a square matrix all of whose entries *not* on the main diagonal equal zero. By $\mathbf{diag}(d_1, \ldots, d_p)$ is meant the $p \times p$ diagonal matrix whose (i, i)-entry equals d_i for $1 \leq i \leq p$.

(c) A $p \times q$ *lower-triangular matrix* **L** satisfies $\langle \mathbf{L} \rangle_{ij} = 0$ if $i < j$, for $1 \leq i \leq p$ and $1 \leq j \leq q$.

(d) A $p \times q$ *upper-triangular matrix* **U** satisfies $\langle \mathbf{U} \rangle_{ij} = 0$ if $i > j$, for $1 \leq i \leq p$ and $1 \leq j \leq q$.

(e) A *unit-lower* (or *-upper*)*-triangular matrix* **T** is a lower (or upper)-triangular matrix satisfying $\langle \mathbf{T} \rangle_{ii} = 1$ for $1 \leq i \leq \min(p, q)$.

The use of matrix notation allows us to consider an array of many numbers as a single object denoted by a single symbol. Relationships among the large sets of numbers often arising in applications can then be expressed in a concise fashion. The more complicated the problem, the more useful matrices become. Perhaps more important, the use of matrices often provides insights that could not be obtained easily—if at all—by other means.

A Brief Refresher on Complex Numbers. Recall that a *complex number* has the form $a + b\imath$, where a and b are real numbers and $\imath$ denotes $\sqrt{-1}$. Thus $3 + 2\imath$, $-6 + 4.7\imath$, and $e - \pi\imath$ are complex numbers. So also are all real numbers r—since they equal $r + 0\imath$—and all *pure imaginary numbers* $t\imath$ for real t—since they equal $0 + t\imath$. Complex numbers are added and multiplied much like real numbers, but you must keep in mind that $\imath \cdot \imath = -1$. Thus

$$(3 - 2\imath)(-5 + 7\imath) = -15 + 21\imath + 10\imath - 14\imath\imath = -15 + 31\imath - 14(-1)$$
$$= -1 + 31\imath.$$

The *conjugate* $\bar{z}$ of the complex number $z = a + b\imath$ is the complex number $\bar{z} = a - b\imath$. The *magnitude* $|z|$ of the complex number $z = a + b\imath$ is defined as

$$|z| = \sqrt{z\bar{z}} = \sqrt{a^2 + b^2}.$$

Thus for $z = 3 - 2\imath$ we have

$$\bar{z} = 3 + 2\imath \quad \text{and} \quad |z| = \sqrt{13}.$$

PROBLEMS 1.1

▷ **1.** Let $z_1 = 2 - 3i$, $z_2 = -1 + 5i$, $z_3 = 2i$, and $z_4 = -7$. Evaluate the following, writing each answer in the standard form $a + bi$ with a and b real.
(a) $z_1 + 2z_2$ (b) $(1 + i)\bar{z}_1$ (c) $z_3 z_4$
(d) $\bar{z}_1\bar{z}_3$ (e) $\overline{z_1 z_3}$ (f) $(z_2)^2$
(g) $(z_1)^3$ (h) $|z_1|$ (i) $|i\bar{z}_2|$
(j) $1/z_2$ (k) z_1/z_3

2. Prove that, for all complex numbers z and w, $\overline{zw} = \bar{z}\bar{w}$.

3. (a) How many entries are there in a $p \times q$ matrix?
(b) In its first row?
(c) In its last column?

▷ **4.** Let

$$\mathbf{A} = \begin{bmatrix} 1 & 2 \\ -2 & 3 \\ 4 & 0 \end{bmatrix} \quad \text{and} \quad \mathbf{B} = \begin{bmatrix} -1 & 6 & x \\ 2 & y & -3 \end{bmatrix}.$$

Give each of the following:
(a) the (1, 2)-entry of **A** (b) the (2, 2)-entry of **B**
(c) the (2, 2)-entry of **A** (d) $\langle \mathbf{A} \rangle_{21}$
(e) $\langle \mathbf{B} \rangle_{23}$ (f) $\langle \mathbf{A} \rangle_{32}$

5. Write the general 3×2 matrix **A** with entries a_{ij} in the form (1.2). Write out the general 2×3 matrix **B** with entries b_{ij} in the form (1.2).

6. Write the general 3×3 matrix of each type in Definition 1.3.

1.2 EQUALITY, ADDITION, AND MULTIPLICATION BY A SCALAR

In most applications, matrices must be combined in various ways much as numbers are combined through arithmetic. We in fact need concepts that correspond to the basic arithmetic operations on numbers; this section introduces addition, subtraction, negation, one form of multiplication, and—the key to them all—equality.

(1.4) ***Definition.*** Two matrices **A** and **B** are *equal* if and only if:
(a) **A** and **B** have the same number of rows and the same number of columns.
(b) All corresponding entries are equal—$\langle \mathbf{A} \rangle_{ij} = \langle \mathbf{B} \rangle_{ij}$ for all i and j.

This definition uses the phrase "if and only if," which you will see throughout this book. Recall that if P stands for some statement or condition and Q stands for another, then "P if and only if Q" means: if either of P and Q is true, then so is the other; if either is false, then so is the other. Like Siamese twins, P and Q always appear together.

We can now define matrix addition.

(1.5) ***Definition.*** Two matrices **A** and **B** can be added if and only if they have the same number p of rows and the same number q of columns. The *sum* $\mathbf{A} + \mathbf{B}$ of two $p \times q$ matrices is the $p \times q$ matrix obtained by adding corresponding entries of **A** and **B**. In symbols,

$$\langle \mathbf{A} + \mathbf{B} \rangle_{ij} = \langle \mathbf{A} \rangle_{ij} + \langle \mathbf{B} \rangle_{ij} \qquad \text{for } 1 \le i \le p \quad \text{and} \quad 1 \le j \le q.$$

(1.6) ***Example.*** Suppose that the 2×2 matrices **A**, **B**, and **C** are

$$\mathbf{A} = \begin{bmatrix} 9 & x+2 \\ -3 & 2 \end{bmatrix}, \qquad \mathbf{B} = \begin{bmatrix} y & -2 \\ 4 & 6 \end{bmatrix}, \qquad \mathbf{C} = \begin{bmatrix} 0 & 0 \\ 0 & 0 \end{bmatrix}.$$

Then

$$\mathbf{A} + \mathbf{B} = \mathbf{B} + \mathbf{A} = \begin{bmatrix} 9+y & x \\ 1 & 8 \end{bmatrix}, \qquad \mathbf{A} + \mathbf{C} = \mathbf{C} + \mathbf{A} = \mathbf{A}.$$

Properties of Matrix Addition

For the two matrices **A** and **B** above, you saw that $\mathbf{A} + \mathbf{B} = \mathbf{B} + \mathbf{A}$. Since you add matrices by adding their corresponding entries (which are numbers) and since $a + b = b + a$ for *numbers* a and b, you can expect $\mathbf{A} + \mathbf{B}$ to equal $\mathbf{B} + \mathbf{A}$ for *matrices* as well. A rigorous proof for this fact must be an argument that is valid for all $p \times q$ matrices **A** and **B**—for *all* positive integers p, for *all* positive integers q, for *all* $p \times q$ matrices **A**, and for *all* $p \times q$ matrices **B**. Example 1.6 is not a general proof; it holds only for $p = 2$, $q = 2$, and those specific matrices **A** and **B** given there. The argument

$$[u \quad v] + [x \quad y] = [u + x \quad v + y] = [x + u \quad y + v] = [x \quad y] + [u \quad v]$$

also is not a general proof; it only proves the result for arbitrary 1×2 matrices ("arbitrary" since u, v, x, y can assume all possible values). A general proof is given below. Although specific examples are often essential to our understanding, throughout mathematics we must keep in mind this distinction between a general proof and a specific example.

Important Notational Convention. We usually omit stating the numbers of rows and columns of matrices in theorems, assuming that they are such that all indicated operations make sense.

(1.7) ***Theorem*** (addition laws). Matrix addition is both *commutative* and *associative.* That is:

(a) $\mathbf{A} + \mathbf{B} = \mathbf{B} + \mathbf{A}$ (commutative law)
(b) $\mathbf{A} + (\mathbf{B} + \mathbf{C}) = (\mathbf{A} + \mathbf{B}) + \mathbf{C}$ (associative law)

PROOF

(a) The key is to show that $\langle \mathbf{A} + \mathbf{B} \rangle_{ij} = \langle \mathbf{B} + \mathbf{A} \rangle_{ij}$ for all i and j. To prove this we use the definition of matrix addition and the commutativity of addition *for numbers*:

$$\langle \mathbf{A} + \mathbf{B} \rangle_{ij} = \langle \mathbf{A} \rangle_{ij} + \langle \mathbf{B} \rangle_{ij} = \text{(why?)}\ \langle \mathbf{B} \rangle_{ij} + \langle \mathbf{A} \rangle_{ij} = \langle \mathbf{B} + \mathbf{A} \rangle_{ij}$$

for all i and j; we have proved that matrix addition is commutative.

(b) Problem 5. ■

You should now be able to solve Problems 1 to 6.

Zero and Negation for Matrices

There is more to learn from Example 1.6. Note that the matrix **C** there, all of whose entries equal 0, behaves much as the *number* zero does with respect to addition: $\mathbf{A} + \mathbf{C} = \mathbf{A}$ for all 2×2 matrices **A** just as $a + 0 = a$ for all numbers a. More generally, if **A** is an arbitrary $p \times q$ matrix and **0** is the $p \times q$ *zero* (or *null*) *matrix* each of whose entries is 0, then

(1.8) $$\mathbf{A} + \mathbf{0} = \mathbf{0} + \mathbf{A} = \mathbf{A}.$$

Problem 8 asks you to prove (1.8), and Problem 9 shows another way in which **0** behaves as the number 0.

Once you know about zero, you can view the negation $-a$ of a number a as that number for which $a + (-a) = 0$. Similarly, for a matrix **A** we want $-\mathbf{A}$ to indicate a matrix for which $\mathbf{A} + (-\mathbf{A}) = \mathbf{0}$; the definitions of matrix addition and of the zero matrix **0** show then that $-\mathbf{A}$ should be the matrix whose entries are just the negations of those of **A**: $\langle -\mathbf{A} \rangle_{ij} = -\langle \mathbf{A} \rangle_{ij}$. And from negation, we can define subtraction via $\mathbf{A} - \mathbf{B} = \mathbf{A} + (-\mathbf{B})$.

(1.9) ***Definition***

(a) The $p \times q$ *zero* (or *null*) *matrix* **0** is that $p \times q$ matrix with

$$\langle \mathbf{0} \rangle_{ij} = 0 \qquad \text{for } 1 \le i \le p \quad \text{and} \quad 1 \le j \le q.$$

(b) The *negation* $-\mathbf{A}$ of a $p \times q$ matrix **A** is that $p \times q$ matrix with

$$\langle -\mathbf{A} \rangle_{ij} = -\langle \mathbf{A} \rangle_{ij} \qquad \text{for } 1 \le i \le p \quad \text{and} \quad 1 \le j \le q.$$

(c) The *difference* $\mathbf{A} - \mathbf{B}$ of two matrices **A** and **B** is defined if and only if **A** and **B** are both $p \times q$, and then $\mathbf{A} - \mathbf{B}$ is that $p \times q$ matrix with $\mathbf{A} - \mathbf{B} = \mathbf{A} + (-\mathbf{B})$ and

$$\langle \mathbf{A} - \mathbf{B} \rangle_{ij} = \langle \mathbf{A} \rangle_{ij} - \langle \mathbf{B} \rangle_{ij} \qquad \text{for } 1 \le i \le p \quad \text{and} \quad 1 \le j \le q.$$

You should now be able to solve Problems 1 to 12.

Scalar Multiples of Matrices

It seems natural to denote $\mathbf{A} + \mathbf{A}$ by $2\mathbf{A}$, the product of the *number* 2 and the *matrix* $\mathbf{A}$. Of course,

$$\langle \mathbf{A} + \mathbf{A} \rangle_{ij} = \langle \mathbf{A} \rangle_{ij} + \langle \mathbf{A} \rangle_{ij} = 2\langle \mathbf{A} \rangle_{ij},$$

so it also seems natural to define the product $2\mathbf{A}$ by $\langle 2\mathbf{A} \rangle_{ij} = 2\langle \mathbf{A} \rangle_{ij}$. This motivates the general definition for the product $r\mathbf{A}$ of a number r—either real or complex—and a matrix $\mathbf{A}$.

(1.10) ***Definition.*** Let $\mathbf{A}$ be a $p \times q$ matrix and r be a real or complex number (often called a *scalar*). Then the *scalar multiple* $r\mathbf{A}$ of r and $\mathbf{A}$ is the $p \times q$ matrix with

$$\langle r\mathbf{A} \rangle_{ij} = r\langle \mathbf{A} \rangle_{ij} \qquad \text{for } 1 \le i \le p \quad \text{and} \quad 1 \le j \le q.$$

(1.11) ***Example.*** Let $\mathbf{E}$ and $\mathbf{F}$ be the 2×3 matrices

$$\mathbf{E} = \begin{bmatrix} -2 & 4 & 3 + 2i \\ x & 6 & -7 \end{bmatrix}, \qquad \mathbf{F} = \begin{bmatrix} 5 & y & -4 \\ 1 & 3 & -3i \end{bmatrix}.$$

Then

$$4\mathbf{E} = \begin{bmatrix} -8 & 16 & 12 + 8i \\ 4x & 24 & -28 \end{bmatrix},$$

$$2\mathbf{E} - 3\mathbf{F} = \begin{bmatrix} -19 & 8 - 3y & 18 + 4i \\ 2x - 3 & 3 & -14 + 9i \end{bmatrix}.$$

Addition, negation, and subtraction of matrices combine with multiplication by scalars in a simple way, as can be shown from the definitions of matrix equality and of the various operations. Problem 15 asks you to prove the following.

(1.12) ***Theorem*** (scalar multiple laws)

(a) $(r + s)\mathbf{A} = r\mathbf{A} + s\mathbf{A}$
(b) $r(s\mathbf{A}) = (rs)\mathbf{A}$
(c) $r(\mathbf{A} + \mathbf{B}) = r\mathbf{A} + r\mathbf{B}$
(d) $(-1)\mathbf{A} = -\mathbf{A}$
(e) $0\mathbf{A} = \mathbf{0}$
(f) $r\mathbf{0} = \mathbf{0}$

So far matrices have behaved in much the same way as symbols representing numbers. The surprises have been reserved for the next section, where we consider the multiplication of matrices with matrices.

PROBLEMS 1.2

▷ **1.** Interpret the statement $\mathbf{A} = \mathbf{B}$ entry by entry in order to find the values of x, y, z, and w, if

$$\mathbf{A} = \begin{bmatrix} x & -3 \\ 2 & y \end{bmatrix} \quad \text{and} \quad \mathbf{B} = \begin{bmatrix} 6 & w \\ z & -1 \end{bmatrix}.$$

2. Suppose that P and Q each stand for a statement, and you know for a fact that "P if and only if Q." What can you conclude from the additional fact that Q is false?

3. Let matrices **A**, **B**, **C**, **D**, **E**, and **F** be defined as

$$\mathbf{A} = \begin{bmatrix} 1 & 2 \\ -3 & 6 \end{bmatrix}, \quad \mathbf{B} = \begin{bmatrix} 4 & 7 \\ 2 & 0 \end{bmatrix}, \quad \mathbf{C} = \begin{bmatrix} 2 \\ -6 \\ 3 \end{bmatrix}, \quad \mathbf{D} = \begin{bmatrix} 0 \\ 14 \\ -2 \end{bmatrix},$$

$$\mathbf{E} = \begin{bmatrix} 4 \\ -3 \\ -2 \end{bmatrix}, \quad \mathbf{F} = \begin{bmatrix} 2+i & 6 \\ 3 & i \end{bmatrix}.$$

Evaluate (or answer "undefined").
(a) $\mathbf{A} + \mathbf{B}$ (b) $\mathbf{B} + \mathbf{F}$
(c) $\mathbf{C} + \mathbf{D}$ (d) $\mathbf{D} + \mathbf{E}$
(e) $\mathbf{A} + \mathbf{C}$
(f) $\mathbf{C} + \mathbf{E}$ and compare with $\mathbf{E} + \mathbf{C}$
(g) $(\mathbf{A} + \mathbf{B}) + \mathbf{F}$ and compare with $\mathbf{A} + (\mathbf{B} + \mathbf{F})$
(h) $\mathbf{F} + \mathbf{D}$

▷ **4.** (a) Prove that $\langle \mathbf{A} + \mathbf{A} \rangle_{ij} = 2\langle \mathbf{A} \rangle_{ij}$ for all 2×3 matrices **A**.
(b) Prove the same result for all $p \times q$ matrices.

5. Prove Theorem 1.7(b), the associative law of matrix addition.

6. The *trace* $\text{tr}(\mathbf{A})$ of a $p \times p$ matrix **A** is the sum of the entries on the main diagonal of **A**. Prove that $\text{tr}(\mathbf{A} + \mathbf{B}) = \text{tr}(\mathbf{A}) + \text{tr}(\mathbf{B})$.

7. Let the matrices **A** through **F** be those of Problem 3. Evaluate (or answer "undefined").
(a) $\mathbf{A} - \mathbf{D}$ (b) $-\mathbf{D}$
(c) $-\mathbf{F}$ (d) $\mathbf{B} - \mathbf{F}$
(e) $\mathbf{E} - \mathbf{C}$ (f) $-(\mathbf{A} - \mathbf{B})$
(g) $-\mathbf{B} - \mathbf{C}$ (h) $-\mathbf{B} - \mathbf{A}$
(i) $-(\mathbf{A} + \mathbf{B})$ and compare with $-\mathbf{A} - \mathbf{B}$

▷ **8.** Prove (1.8): $\mathbf{A} + \mathbf{0} = \mathbf{0} + \mathbf{A} = \mathbf{A}$ for all **A**.

9. For $p \times q$ matrices **A**, prove that $\mathbf{A} = -\mathbf{A}$ if and only if **A** is the zero matrix.

10. Prove that $\text{tr}(\mathbf{A} - \mathbf{B}) = \text{tr}(\mathbf{A}) - \text{tr}(\mathbf{B})$, where $\text{tr}(\mathbf{A})$ is the *trace* from Problem 6.

▷ **11.** By examining the entries, prove that $\mathbf{A} + (-\mathbf{A}) = \mathbf{0}$.

12. By examining the entries, prove that $\mathbf{A} - \mathbf{A} = \mathbf{0}$.

13. Let the matrices **A** through **F** be those of Problem 3. Evaluate (or answer "undefined").

(a) $5\mathbf{A}$

(b) $(2 - 3i)\mathbf{B}$

(c) $(-3)\mathbf{C}$ and compare with $-(3\mathbf{C})$

(d) $2\mathbf{D} - \mathbf{C}$ (e) $-10\mathbf{E}$

(f) $6\mathbf{F}$ (g) $(2 + i)\mathbf{F}$

(h) $0\mathbf{A}$ (i) $0\mathbf{B} + \mathbf{C}$

▷ **14.** Prove that $(-r)\mathbf{A} = -(r\mathbf{A})$.

15. Prove Theorem 1.12 on the laws of scalar multiples.

16. Prove that $\operatorname{tr}(r\mathbf{A}) = r\{\operatorname{tr}(\mathbf{A})\}$, where $\operatorname{tr}(\mathbf{A})$ is the *trace* defined in Problem 6.

1.3 MATRIX MULTIPLICATION

There are several different notions of the product of two matrices, depending on the use to which the concept is to be put. We present here the definition with the broadest range of applications; for other notions of product, see Problems 9, 10, 11, and 49. The product **AB** of two matrices will be defined in terms of the products of the *rows* of **A** and the *columns* of **B**, so we first define the product **uv** between a $1 \times q$ *row matrix* **u** and a $q \times 1$ *column matrix* **v**.

(1.13) ***Definition.*** Let **u** be $1 \times q$ and **v** be $q \times 1$. Then **uv** is the 1×1 matrix whose entry is

$$\langle \mathbf{u} \rangle_1 \langle \mathbf{v} \rangle_1 + \langle \mathbf{u} \rangle_2 \langle \mathbf{v} \rangle_2 + \cdots + \langle \mathbf{u} \rangle_q \langle \mathbf{v} \rangle_q = \sum_{i=1}^{q} \langle \mathbf{u} \rangle_i \langle \mathbf{v} \rangle_i.$$

For example,

$$[4 \quad -1 \quad 3] \begin{bmatrix} 2 \\ 1 \\ -5 \end{bmatrix} = [(4)(2) + (-1)(1) + (3)(-5)] = [-8].$$

Note that **uv** is a 1×1 *matrix*, not a number; in the example above, the product is the *matrix* $[-8]$, not the number -8. Upon this definition we build that for the general case.

(1.14) ***Definition.*** Let **A** be $p \times q$ and **B** be $q \times r$. Then the *product* **AB** is defined as the $p \times r$ matrix whose (i, j)-entry is the *entry* in the 1×1 matrix that is the product of the ith row of **A** and the jth column of **B**. That is,

$$\langle \mathbf{AB} \rangle_{ij} = \langle \mathbf{A} \rangle_{i1} \langle \mathbf{B} \rangle_{1j} + \langle \mathbf{A} \rangle_{i2} \langle \mathbf{B} \rangle_{2j} + \cdots + \langle \mathbf{A} \rangle_{iq} \langle \mathbf{B} \rangle_{qj}$$

$$= \sum_{k=1}^{q} \langle \mathbf{A} \rangle_{ik} \langle \mathbf{B} \rangle_{kj}.$$

Note that, in order to have **AB** be defined, the number of columns in **A** must equal the number of rows in **B**. You need not memorize this rule as it stands; just remember that you multiply the entries *across a row* of **A** times the entries *down a column* of **B**, and that you must run out of entries at the same time. You might also remember that $(p \times r)$ times $(r \times q)$ gives $(p \times q)$—the middle terms (r) must agree and "cancel out" in the result. A pictorial example of how to find the (3, 2)-entry of the product **C** of a 6×5 matrix **A** and a 5×4 matrix **B** is

$$\text{3rd row} \rightarrow \begin{bmatrix} x & x & x & x & x \\ x & x & x & x & x \\ \cdot & \cdot & \cdot & \cdot & \cdot \\ x & x & x & x & x \\ x & x & x & x & x \\ x & x & x & x & x \end{bmatrix} \begin{bmatrix} x & \cdot & x & x \\ x & \cdot & x & x \\ x & \cdot & x & x \\ x & \cdot & x & x \\ x & \cdot & x & x \end{bmatrix} = \begin{bmatrix} x & x & x & x \\ x & x & x & x \\ x & \cdot & x & x \\ x & x & x & x \\ x & x & x & x \\ x & x & x & x \end{bmatrix} \leftarrow \text{3rd row}$$

↑ *2nd column* ↑ *2nd column*

A × **B** = **C**

The procedure is to generate the product **AB** by multiplying each row of **A** times each column of **B**. One methodical approach is to get the first row of **AB** by multiplying the first row of **A** times each column of **B**; then get the second row of **AB** by multiplying the second row of **A** times each column of **B**; and so on, finally getting the last row of **AB** by multiplying the last row of **A** times each column of **B**.

(1.15) ***Example***

$$\begin{bmatrix} -1 & 5 \\ 2 & 1 \end{bmatrix} \begin{bmatrix} 4 & 3 & 6 \\ 0 & -1 & 2 \end{bmatrix} = \begin{bmatrix} -4 & -8 & 4 \\ 8 & 5 & 14 \end{bmatrix}$$

and

$$\begin{bmatrix} a & b \\ c & d \end{bmatrix} \begin{bmatrix} x \\ y \end{bmatrix} = \begin{bmatrix} ax + by \\ cx + dy \end{bmatrix}.$$

It is important to practice the row–column procedure for multiplying matrices until it becomes automatic. Also, you should be able to pick out immediately the row of **A** and the column of **B** that combine to give a particular entry in **AB**.

You should now be able to solve Problems 1 to 8.

Properties of Matrix Multiplication

We have defined what matrix multiplication *is*; the next task is to learn how it behaves—is it commutative, is it associative, and the like. Consider first commutativity: Does **AB** = **BA**? If

$$\mathbf{A} = [1 \quad 2] \quad \text{and} \quad \mathbf{B} = \begin{bmatrix} -3 \\ 4 \end{bmatrix}$$

then **AB** and **BA** are both defined:

$$\mathbf{AB} = [5] \quad \text{and} \quad \mathbf{BA} = \begin{bmatrix} -3 & -6 \\ 4 & 8 \end{bmatrix}.$$

The products **AB** and **BA** are quite different. In general, even if the product **AB** is defined, there is no reason why the product **BA** should be defined at all: **AB** makes sense if **A** is $p \times r$ and **B** is $r \times q$, but the product **BA** is nonsense unless $p = q$. When by chance **AB** and **BA** both make sense, as above, they need not both be $p \times q$: If **A** is $p \times r$ and **B** is $r \times p$, then **AB** is $p \times p$ while **BA** is $r \times r$. Finally—see Problems 12 and 13—even when **AB** and **BA** are both $p \times p$, they need not be equal (although they might be in some cases). Thus matrix multiplication is not commutative.

This means that when multiplying matrices, you need to be careful about the order of the terms in the product. To distinguish the order in the product **AB**, we say that **A** *premultiplies* **B** or multiplies **B** *from the left*; similarly, **B** *postmultiplies* **A** or multiplies **A** *from the right*. Then, if you want to multiply both sides of an equation **X** = **Y** by some matrix **P**, it is important that either you *premultiply both* sides by **P** or you *postmultiply both* sides by **P**; **PX** = **PY**, for example, will be valid since **X** = **Y** implies that **X** − **Y** = **0** and hence

$$\mathbf{0} = \mathbf{P0} = \mathbf{P}(\mathbf{X} - \mathbf{Y}) = \mathbf{PX} - \mathbf{PY},$$

which means that **PX** = **PY**. (We assumed here the distributive law, to be presented shortly.)

There is one special case when the order of multiplication is not important: when each matrix is some positive integral power of the same square matrix. Here we define powers in the natural way:

(1.16) $$\mathbf{A}^1 = \mathbf{A}, \qquad \mathbf{A}^2 = \mathbf{AA}, \qquad \mathbf{A}^3 = \mathbf{AAA}, \qquad \mathbf{A}^n = \mathbf{AA}\cdots\mathbf{A} \quad (n \text{ factors}).$$

Then it clearly follows that $\mathbf{A}^r\mathbf{A}^s = \mathbf{A}^{r+s} = \mathbf{A}^s\mathbf{A}^r$; that is, $\mathbf{A}^r$ and $\mathbf{A}^s$ commute.

𝔐 (1.17) ***Example.*** Even for relatively small square matrices—2×2 or 3×3, say—the calculation of more than small powers of the matrix can be tedious by hand; here computer software such as MATLAB can be extremely useful. In Sections 2.2 and 2.3 we will see that in many types of applications it is essential to learn how the powers $\mathbf{A}^r$ of a specific square matrix $\mathbf{A}$ behave for large integers r. For the 2×2 matrix $\mathbf{A}$ below, the powers $\mathbf{A}^2$ and $\mathbf{A}^3$ alone indicate little about how $\mathbf{A}^r$ behaves for large r. But MATLAB, for example, easily computes the higher powers for us and indicates that $\mathbf{A}^r$ tends to $\mathbf{0}$ as r tends to infinity. To *prove* this fact is a challenge that we will overcome in a later chapter.

$$\mathbf{A} = \begin{bmatrix} 0.6 & 0.5 \\ -0.18 & 1.2 \end{bmatrix}, \qquad \mathbf{A}^2 = \begin{bmatrix} 0.27 & 0.9 \\ -0.324 & 1.35 \end{bmatrix},$$

$$\mathbf{A}^3 = \begin{bmatrix} 0.0 & 1.215 \\ -0.437 & 1.458 \end{bmatrix},$$

$$\mathbf{A}^{25} = \begin{bmatrix} -0.527 & 0.997 \\ -0.359 & 0.670 \end{bmatrix}, \qquad \mathbf{A}^{50} = \begin{bmatrix} -0.081 & 0.143 \\ -0.052 & 0.09 \end{bmatrix},$$

$$\mathbf{A}^{100} = \begin{bmatrix} -0.0009 & 0.0015 \\ -0.0005 & 0.0009 \end{bmatrix}, \qquad \mathbf{A}^{200} = 10^{-7}\begin{bmatrix} -0.5 & 0.8 \\ -0.3 & 0.5 \end{bmatrix},$$

Although matrix multiplication is not commutative, it *is* distributive—

$$\mathbf{A}(\mathbf{B} + \mathbf{C}) = \mathbf{AB} + \mathbf{AC} \quad \text{and} \quad (\mathbf{B} + \mathbf{C})\mathbf{A} = \mathbf{BA} + \mathbf{CA}$$

—and associative—

$$\mathbf{A}(\mathbf{BC}) = (\mathbf{AB})\mathbf{C}$$

—whenever all the products make sense. Before proving these laws, we introduce a matrix that plays the same role in matrix multiplication as does the zero matrix in matrix addition. The number 1 has the special property that $1a = a1 = a$ for all numbers a; we seek a *matrix* with similar properties.

(1.18) ***Definition.*** The $r \times r$ *identity* (or *unit*) *matrix* is the diagonal matrix $\mathbf{I}_r = \mathbf{diag}(1, \ldots, 1)$:

$$\langle \mathbf{I}_r \rangle_{ij} = 1 \quad \text{if } i = j, \qquad \langle \mathbf{I}_r \rangle_{ij} = 0 \quad \text{if } i \neq j,$$

so that

$$\mathbf{I}_r = \begin{bmatrix} 1 & 0 & 0 & \cdots & 0 & 0 \\ 0 & 1 & 0 & \cdots & 0 & 0 \\ \vdots & \vdots & \vdots & & \vdots & \vdots \\ 0 & 0 & 0 & \cdots & 1 & 0 \\ 0 & 0 & 0 & \cdots & 0 & 1 \end{bmatrix}.$$

We drop the subscript r when it need not be emphasized.

It is simple to see that if $\mathbf{A}$ is $p \times q$, then $\mathbf{I}_p\mathbf{A} = \mathbf{A}$ and $\mathbf{A}\mathbf{I}_q = \mathbf{A}$. That is, matrices $\mathbf{I}_p$ and $\mathbf{I}_q$ play the role of the number 1 in multiplication.

We are now ready to state the key facts concerning matrix multiplication.

(1.19) ***Theorem*** (matrix multiplication laws)

(a) $\mathbf{A}(\mathbf{BC}) = (\mathbf{AB})\mathbf{C}$ (associative law)

(b) $\mathbf{A}(\mathbf{B} \pm \mathbf{C}) = \mathbf{AB} \pm \mathbf{AC}$ (distributive law)

$(\mathbf{A} \pm \mathbf{B})\mathbf{C} = \mathbf{AC} \pm \mathbf{BC}$ (distributive law)

(c) $\mathbf{AI} = \mathbf{A}$ ($\mathbf{I}$ is an identity for multiplication)

$\mathbf{IA} = \mathbf{A}$

(d) $c(\mathbf{AB}) = (c\mathbf{A})\mathbf{B} = \mathbf{A}(c\mathbf{B})$

(e) $\mathbf{A0} = \mathbf{0}$

$\mathbf{0B} = \mathbf{0}$

(f) For square $\mathbf{A}$:

$(\mathbf{A})^0 = \mathbf{I}$ (definition)

$(\mathbf{A})^1 = \mathbf{A}$ (definition)

$(\mathbf{A})^{n+1} = (\mathbf{A})(\mathbf{A})^n$ (definition); (n and m

$(\mathbf{A})^m(\mathbf{A})^n = (\mathbf{A})^{m+n}$ positive integers)

(g) Matrix multiplication is not commutative; $\mathbf{AB}$ does not generally equal $\mathbf{BA}$.

(h) Multiplicative cancellation is not in general valid; if $\mathbf{AB} = \mathbf{AC}$ (or $\mathbf{BA} = \mathbf{CA}$), it does not necessarily follow that $\mathbf{B} = \mathbf{C}$.

PROOF. Each proof involves showing that some matrix $\mathbf{L}$ equals some matrix $\mathbf{R}$ by Definition 1.4. We prove the equality of corresponding entries.

(a) To be definite, suppose that $\mathbf{A}$ is $p \times q$, $\mathbf{B}$ is $q \times r$, and $\mathbf{C}$ is $r \times s$. Then for $1 \le i \le p$ and $1 \le j \le s$, we have

$$\langle \mathbf{A}(\mathbf{BC})\rangle_{ij} = \sum_{k=1}^{q} \langle \mathbf{A}\rangle_{ik}\langle \mathbf{BC}\rangle_{kj} \qquad \text{(by Definition 1.14)}$$

$$= \sum_{k=1}^{q} \langle \mathbf{A}\rangle_{ik}\left\{\sum_{n=1}^{r} \langle \mathbf{B}\rangle_{kn}\langle \mathbf{C}\rangle_{nj}\right\} \qquad \text{(by Definition 1.14)}$$

$$= \sum_{k=1}^{q}\sum_{n=1}^{r} \langle \mathbf{A}\rangle_{ik}\langle \mathbf{B}\rangle_{kn}\langle \mathbf{C}\rangle_{nj} \qquad \text{(arithmetic)}$$

$$= \sum_{n=1}^{r}\sum_{k=1}^{q} \langle \mathbf{A}\rangle_{ik}\langle \mathbf{B}\rangle_{kn}\langle \mathbf{C}\rangle_{nj} \qquad \text{(reversing summation order)}$$

$$= \sum_{n=1}^{r}\left\{\sum_{k=1}^{q} \langle \mathbf{A}\rangle_{ik}\langle \mathbf{B}\rangle_{kn}\right\}\langle \mathbf{C}\rangle_{nj} \qquad \text{(arithmetic)}$$

$$= \sum_{n=1}^{r} \langle \mathbf{AB}\rangle_{in}\langle \mathbf{C}\rangle_{nj} \qquad \text{(by Definition 1.14)}$$

$$= \langle (\mathbf{AB})\mathbf{C}\rangle_{ij} \qquad \text{(by Definition 1.14)}$$

which completes the proof of (a).

(b) We prove only the first form, $\mathbf{A}(\mathbf{B} + \mathbf{C}) = \mathbf{AB} + \mathbf{AC}$, leaving the others for Problem 21. To be definite, suppose that $\mathbf{A}$ is $p \times r$ and $\mathbf{B}$ and $\mathbf{C}$

are $r \times q$. For equality of entries for $1 \le i \le p$ and $1 \le j \le q$, we have

$$\begin{aligned}
\langle \mathbf{A}(\mathbf{B}+\mathbf{C})\rangle_{ij} &= \sum_{k=1}^{r} \langle \mathbf{A}\rangle_{ik}\langle \mathbf{B}+\mathbf{C}\rangle_{kj} && \text{(by Definition 1.14)}\\
&= \sum_{k=1}^{r} \langle \mathbf{A}\rangle_{ik}\{\langle \mathbf{B}\rangle_{kj}+\langle \mathbf{C}\rangle_{kj}\} && \text{(by Definition 1.5)}\\
&= \sum_{k=1}^{r} \{\langle \mathbf{A}\rangle_{ik}\langle \mathbf{B}\rangle_{kj}+\langle \mathbf{A}\rangle_{ik}\langle \mathbf{C}\rangle_{kj}\} && \text{(arithmetic)}\\
&= \sum_{k=1}^{r} \langle \mathbf{A}\rangle_{ik}\langle \mathbf{B}\rangle_{kj}+\sum_{k=1}^{r} \langle \mathbf{A}\rangle_{ik}\langle \mathbf{C}\rangle_{kj} && \text{(arithmetic)}\\
&= \langle \mathbf{AB}\rangle_{ij}+\langle \mathbf{AC}\rangle_{ij} && \text{(by Definition 1.14)}\\
&= \langle \mathbf{AB}+\mathbf{AC}\rangle_{ij} && \text{(by Definition 1.5)}
\end{aligned}$$

which completes the proof of this part of (b).

(c), (d), (e), and (f) are left as problems.

(g) has already been demonstrated by example.

(h) By example, we show that the first cancellation law does not hold. Let

$$\mathbf{A} = \begin{bmatrix} 1 & 0 \\ 0 & 0 \end{bmatrix}, \quad \mathbf{B} = \begin{bmatrix} 0 & 0 \\ 1 & 0 \end{bmatrix}, \quad \text{and} \quad \mathbf{C} = \begin{bmatrix} 0 & 0 \\ 0 & 0 \end{bmatrix};$$

then **AB** and **AC** both equal **C**, so $\mathbf{AB} = \mathbf{AC}$ but $\mathbf{B} \ne \mathbf{C}$. This, together with the problems, completes our proof of the theorem. ■

A key point to remember in all this is to be careful with the order of matrix multiplication. Expressions such as $(\mathbf{A}+\mathbf{B})(\mathbf{A}-\mathbf{B})$ can of course be expanded as $\mathbf{A}^2 - \mathbf{AB} + \mathbf{BA} - \mathbf{B}^2$, but the middle terms do not cancel out.

You should now be able to solve Problems 1 to 24.

Transposes of Matrices

One matrix operation that has no analogue in arithmetic on numbers is that of taking the *transpose* or the *hermitian transpose* of a matrix. The basic idea is to create a new matrix whose rows are the columns of the original, which means that its columns are the rows of the original; for the hermitian transpose, complex conjugates are also taken. These operations will be very useful for us later.

(1.20) ***Definition.*** Let **A** be a $p \times q$ matrix.

(a) The *transpose* $\mathbf{A}^T$ of **A** is the $q \times p$ matrix obtained by interchanging the rows and columns of **A**—the first row becomes the first column, and so on. That is,

$$\langle \mathbf{A}^T\rangle_{ij} = \langle \mathbf{A}\rangle_{ji} \qquad \text{for } 1 \le i \le q \quad \text{and} \quad 1 \le j \le p.$$

(b) The *hermitian transpose* $\mathbf{A}^H$ of $\mathbf{A}$ is the $q \times p$ matrix obtained by taking the complex conjugates of the entries in $\mathbf{A}^T$—the first column of $\mathbf{A}^H$ consists of the complex conjugates of the first row of $\mathbf{A}$, and so on. That is,

$$\langle \mathbf{A}^H \rangle_{ij} = \overline{\langle \mathbf{A} \rangle_{ji}} \qquad \text{for } 1 \le i \le q \quad \text{and} \quad 1 \le j \le p.$$

(1.21) ***Example.*** Let $\mathbf{A}$ and $\mathbf{B}$ be as below; then $\mathbf{A}^T$, $\mathbf{B}^T$, $\mathbf{A}^H$, and $\mathbf{B}^H$ are as found below.

$$\mathbf{A} = \begin{bmatrix} -1 & 2 & 4 \\ 2 & 6 & 3 \end{bmatrix}, \qquad \mathbf{B} = \begin{bmatrix} 2 - 3i \\ 6 \end{bmatrix}.$$

By Definition 1.20,

$$\mathbf{A}^T = \begin{bmatrix} -1 & 2 \\ 2 & 6 \\ 4 & 3 \end{bmatrix}, \qquad \mathbf{B}^T = [2 - 3i \quad 6],$$

$$\mathbf{A}^H = \begin{bmatrix} -1 & 2 \\ 2 & 6 \\ 4 & 3 \end{bmatrix} \quad (= \mathbf{A}^T), \qquad \mathbf{B}^H = [2 + 3i \quad 6] \quad (\neq \mathbf{B}^T).$$

The transpose operations interact in a simple way with the arithmetic operations already introduced for matrices.

(1.22) ***Theorem*** (laws for transposes)
(a) $(\mathbf{A}^T)^T = \mathbf{A}$ and $(\mathbf{A}^H)^H = \mathbf{A}$
(b) $(\mathbf{A} \pm \mathbf{B})^T = \mathbf{A}^T \pm \mathbf{B}^T$ and $(\mathbf{A} \pm \mathbf{B})^H = \mathbf{A}^H \pm \mathbf{B}^H$
(c) $(c\mathbf{A})^T = c\mathbf{A}^T$ and $(c\mathbf{A})^H = \bar{c}\mathbf{A}^H$
(d) $(\mathbf{AB})^T = \mathbf{B}^T\mathbf{A}^T$ and $(\mathbf{AB})^H = \mathbf{B}^H\mathbf{A}^H$ (*Note reversed order.*)

PROOF. (a), (b), and (c) are straightforward and are left as Problems 28–30. We consider (d), a rather surprising result. For definiteness, suppose that $\mathbf{A}$ is $p \times r$ and $\mathbf{B}$ is $r \times q$, so that $\mathbf{AB}$ is defined and is $p \times q$. Then, by Definition 1.20, $(\mathbf{AB})^T$ and $(\mathbf{AB})^H$ are $q \times p$, $\mathbf{A}^T$ and $\mathbf{A}^H$ are $r \times p$, and $\mathbf{B}^T$ and $\mathbf{B}^H$ are $q \times r$; thus $\mathbf{B}^T\mathbf{A}^T$ and $\mathbf{B}^H\mathbf{A}^H$ are indeed defined and are also $q \times p$. We need only prove the equality of corresponding entries. We provide details only for the case of the transpose ^{T}. For $1 \le i \le q$ and $1 \le j \le p$,

$$\begin{aligned}
\langle (\mathbf{AB})^T \rangle_{ij} &= \langle \mathbf{AB} \rangle_{ji} && \text{(by Definition 1.20)} \\
&= \sum_{k=1}^{r} \langle \mathbf{A} \rangle_{jk} \langle \mathbf{B} \rangle_{ki} && \text{(by Definition 1.14)} \\
&= \sum_{k=1}^{r} \langle \mathbf{B}^T \rangle_{ik} \langle \mathbf{A}^T \rangle_{kj} && \text{(by Definition 1.20)} \\
&= \langle \mathbf{B}^T\mathbf{A}^T \rangle_{ij} && \text{(by Definition 1.14)}
\end{aligned}$$

and the desired equality of (i, j)-entries of $(\mathbf{AB})^T$ and $\mathbf{B}^T\mathbf{A}^T$ is proved. ■

Note that Theorem 1.22(d) can be extended to products of arbitrarily many matrices. For example,

$$(\mathbf{ABC})^T = \{(\mathbf{AB})\mathbf{C}\}^T = \mathbf{C}^T(\mathbf{AB})^T = \mathbf{C}^T(\mathbf{B}^T\mathbf{A}^T) = \mathbf{C}^T\mathbf{B}^T\mathbf{A}^T;$$

similarly, $(\mathbf{ABC})^H = \mathbf{C}^H\mathbf{B}^H\mathbf{A}^H$. See Problems 32 and 33.

You should now be able to solve Problems 1 to 33.

In terms of matrix transposes we can define two extremely important—as you will later see—classes of matrices.

(1.23) ***Definition.*** A (square) matrix $\mathbf{A}$ for which $\mathbf{A}^T = \mathbf{A}$, so that $\langle \mathbf{A} \rangle_{ij} = \langle \mathbf{A} \rangle_{ji}$ for all i and j, is said to be *symmetric*. A (square) matrix $\mathbf{B}$ for which $\mathbf{B}^H = \mathbf{B}$, so that $\langle \mathbf{B} \rangle_{ij} = \overline{\langle \mathbf{B} \rangle_{ji}}$ for all i and j, is said to be *hermitian*.

The reason for writing "(square) matrix" above is that a symmetric (or hermitian) matrix *must* be square: If $\mathbf{A}$ is $p \times q$, then $\mathbf{A}^T$ is $q \times p$, so q must equal p in order that $\mathbf{A} = \mathbf{A}^T$.

(1.24) ***Example.*** Consider the matrices

$$\mathbf{A} = \begin{bmatrix} 1 & 2 \\ 2 & 3 \end{bmatrix}, \quad \mathbf{B} = \begin{bmatrix} 1 & 2 \\ -2 & 3 \end{bmatrix}, \quad \mathbf{C} = \begin{bmatrix} 2 & 3 + 2i \\ 3 - 2i & -6 \end{bmatrix}.$$

Then $\mathbf{A}$ is both symmetric and hermitian; $\mathbf{B}$ is neither symmetric nor hermitian; and $\mathbf{C}$ is hermitian but not symmetric.

The product of two symmetric matrices is not in general symmetric, nor is the product of two hermitian matrices generally hermitian: If $\mathbf{A} = \mathbf{A}^T$ and $\mathbf{B} = \mathbf{B}^T$, then $(\mathbf{AB})^T = \mathbf{B}^T\mathbf{A}^T = \mathbf{BA} \neq \mathbf{AB}$ in general, so $(\mathbf{AB})^T \neq \mathbf{AB}$ and $\mathbf{AB}$ is not symmetric (see Problems 38 and 39).

If $\mathbf{A}$ is a $p \times p$ symmetric matrix and $\mathbf{B}$ is any $p \times q$ matrix, then $\mathbf{B}^T\mathbf{AB}$ *is* symmetric, since

(1.25) $$(\mathbf{B}^T\mathbf{AB})^T = \mathbf{B}^T\mathbf{A}^T(\mathbf{B}^T)^T = \mathbf{B}^T\mathbf{AB};$$

similarly, $\mathbf{B}^H\mathbf{AB}$ is hermitian if $\mathbf{A}$ is hermitian. Note that the proof (1.25) makes powerful use of the general properties of matrices already derived, without writing out in detail the complicated sums involved in the product $\mathbf{B}^T\mathbf{AB}$. This illustrates the convenience and power of matrix notation and the general results about matrices. As we continue to present material about matrices, we will take advantage of their notational power by avoiding—when possible—explicitly writing out the matrix entries. The student should follow this approach as well.

PROBLEMS 1.3

1. (a) Verify each product.

$$[1 \quad -3]\begin{bmatrix}-2\\4\end{bmatrix}=[-14]; \qquad [2 \quad -7 \quad 6]\begin{bmatrix}0\\-1\\2\end{bmatrix}=[19];$$

$$[0 \quad 0 \quad 0]\begin{bmatrix}a\\b\\c\end{bmatrix}=[0]; \qquad [3 \quad 2-4i]\begin{bmatrix}2\\-1\end{bmatrix}=[4+4i];$$

$$[i \quad 1]\begin{bmatrix}i\\1\end{bmatrix}=[0].$$

(b) Evaluate the products.

$$[-1 \quad -1 \quad 1 \quad 2]\begin{bmatrix}4\\-4\\6\\2\end{bmatrix}; \qquad [6][3]; \qquad [1 \quad 3 \quad -2]\begin{bmatrix}2\\0\\6\end{bmatrix};$$

$$[i \quad -i]\begin{bmatrix}2i\\i\end{bmatrix}.$$

2. Evaluate each product.

(a) $[1 \quad 6]\begin{bmatrix}2 & -1 & -4\\7 & 3 & 9\end{bmatrix}$ (b) $\begin{bmatrix}1 & 3\\-1 & 2\end{bmatrix}\begin{bmatrix}2 & -1 & 4\\7 & 2 & 0\end{bmatrix}$

(c) $\begin{bmatrix}4 & -1 & 2\\6 & 0 & 3\end{bmatrix}\begin{bmatrix}2\\-6\\9\end{bmatrix}$ (d) $\begin{bmatrix}2 & 1\\4 & 3\\1 & 2\end{bmatrix}\begin{bmatrix}-1 & 6 & 1 & 0 & 1\\3 & 2 & -1 & 2 & 0\end{bmatrix}$

(e) $\begin{bmatrix}2i & 1+i\\-3 & i\end{bmatrix}\begin{bmatrix}6\\2+3i\end{bmatrix}$ (f) $\begin{bmatrix}5\\3\end{bmatrix}[2 \quad -6]$

(g) $[2 \quad -6]\begin{bmatrix}5\\3\end{bmatrix}$

3. Compute **AB**, **AC**, **B** + **C**, and compare **AB** + **AC** with **A**(**B** + **C**) for

$$\mathbf{A}=\begin{bmatrix}1 & -1 & 2\\3 & 0 & 1\end{bmatrix}, \qquad \mathbf{B}=\begin{bmatrix}1 & 1\\2 & -1\\1 & 0\end{bmatrix}, \quad \text{and} \quad \mathbf{C}=\begin{bmatrix}2 & -2\\-1 & 3\\0 & 0\end{bmatrix}.$$

▷ **4.** Show that if the third row of **A** equals four times its first row, then the same holds for those rows of **AB** for all **B** for which the product is defined.

5. Show that if the third column of **A** equals two times its first column, then the same holds for those columns of **BA** for all **B** for which the product is defined.

6. Suppose that $\mathbf{D} = \mathbf{diag}(d_1, \ldots, d_p)$. Show that $\mathbf{DA}$ is obtained from $p \times q$ $\mathbf{A}$ just by replacing the ith row of $\mathbf{A}$ by d_i times it for all i. State and show the analogous result for the columns of $\mathbf{AD}$ for $r \times p$ $\mathbf{A}$.

▷ **7.** Suppose that $\mathbf{A}$ is $p \times q$ and that r is a number. While $r\mathbf{A}$ is of course defined, show that the matrix product $[r]\mathbf{A}$ makes sense if and only if $p = 1$; show that if $p = 1$, then $[r]\mathbf{A} = r\mathbf{A}$. State and verify the similar results concerning $\mathbf{A}[r]$. (*Note:* This shows that one must be careful about identifying r with $[r]$.)

8. Find *real* 2×2 matrices $\mathbf{X}$ and $\mathbf{Y}$—neither of which equals $\mathbf{0}$—for which $\mathbf{XX} + \mathbf{YY} = \mathbf{0}$.

9. A product different from that in Definition 1.14 is important in many areas of science and engineering; it is defined between any two real 1×3 row matrices

$$\mathbf{a} = [a_1 \quad a_2 \quad a_3] \quad \text{and} \quad \mathbf{b} = [b_1 \quad b_2 \quad b_3]$$

(although it could just as well be defined between two 3×1 column matrices). This product is called the *cross product* (or *vector product*), is denoted by $\mathbf{a} \times \mathbf{b}$, and is defined as

$$\mathbf{a} \times \mathbf{b} = [a_2b_3 - a_3b_2 \quad a_3b_1 - a_1b_3 \quad a_1b_2 - a_2b_1],$$

another 1×3 row matrix.

(a) Show that $\mathbf{a} \times \mathbf{b} = -(\mathbf{b} \times \mathbf{a})$, so that $\times$ is not commutative.
(b) Show that $\mathbf{a} \times \mathbf{a} = \mathbf{0}$ for all $\mathbf{a}$.
(c) Show by example that $\times$ is not associative.
(d) Show that there cannot exist an "identity matrix for $\times$," that is, that there cannot exist a special 1×3 matrix $\mathbf{e}$ for which $\mathbf{e} \times \mathbf{a} = \mathbf{a}$ for all $\mathbf{a}$.

10. A product different from that in Definition 1.14 is important in many areas of science and engineering; it is defined between any two matrices. If $\mathbf{A}$ is $p \times q$ and $\mathbf{B}$ is $r \times s$, then the *Kronecker product* (or *tensor product*) $\mathbf{A} \odot \mathbf{B}$ is defined as the $pr \times qs$ matrix containing all possible products of one entry of $\mathbf{A}$ with one entry of $\mathbf{B}$, arranged in a special way: denoting $\langle \mathbf{A} \rangle_{ij}$ by a_{ij}, the first r rows of $\mathbf{A} \odot \mathbf{B}$ are created by writing down $a_{11}\mathbf{B}$ followed by $a_{12}\mathbf{B}$ to its right followed by $a_{13}\mathbf{B}$ to its right followed by ... followed by $a_{1q}\mathbf{B}$ to its right; the second r rows are similarly generated from $a_{21}\mathbf{B}$, $a_{22}\mathbf{B}$, and so on; and this continues through the pth set of r rows. For example,

$$\begin{bmatrix} 1 & 2 \\ 3 & 4 \end{bmatrix} \odot \begin{bmatrix} 5 & 6 & 7 \\ 8 & 9 & 10 \end{bmatrix} = \begin{bmatrix} 5 & 6 & 7 & 10 & 12 & 14 \\ 8 & 9 & 10 & 16 & 18 & 20 \\ 15 & 18 & 21 & 20 & 24 & 28 \\ 24 & 27 & 30 & 32 & 36 & 40 \end{bmatrix}.$$

(a) Show by example that $\odot$ is not commutative.
(b) Show by example that $\odot$ is not associative.
(c) Show that $[1]$ is an "identity matrix" in the sense that $[1] \odot \mathbf{A} = \mathbf{A} \odot [1] = \mathbf{A}$ for all $\mathbf{A}$.

▷ **11.** A product different from that in Definition 1.14 that is useful primarily in connection with discrete Fourier analysis (see Problems 50 and 52) is defined entry-by-entry between any two $p \times q$ matrices **A** and **B** by defining

$$\langle \mathbf{A} \,\square\, \mathbf{B} \rangle_{ij} = \langle \mathbf{A} \rangle_{ij} \langle \mathbf{B} \rangle_{ij}.$$

(a) Show that $\square$ is commutative.
(b) Show that $\square$ is associative.
(c) Find the $p \times q$ "identity matrix" **E** so that $\mathbf{E} \,\square\, \mathbf{A} = \mathbf{A} \,\square\, \mathbf{E} = \mathbf{A}$ for all $p \times q$ **A**.

12. Find a matrix **A** and a matrix **B** so that **AB** and **BA** are both defined, so that **AB** and **BA** have the same shape, but so that $\mathbf{AB} \neq \mathbf{BA}$.

13. Find two 2×2 matrices **A** and **B** with $\mathbf{A} \neq \mathbf{B}$ but with $\mathbf{AB} = \mathbf{BA}$.

▷ **14.** Suppose that **A** is a 2×2 matrix that commutes with *every* 2×2 matrix. Show that **A** must be a scalar multiple of $\mathbf{I}_2$.

15. Compute $\mathbf{A}^2$ and $\mathbf{A}^3$ for

$$\mathbf{A} = \begin{bmatrix} 1 & -1 \\ 2 & 1 \end{bmatrix}.$$

16. Verify that $\mathbf{A}(\mathbf{BC}) = (\mathbf{AB})\mathbf{C}$ if

$$\mathbf{A} = [a_{11} \quad a_{12}], \qquad \mathbf{B} = \begin{bmatrix} b_{11} \\ b_{21} \end{bmatrix}, \quad \text{and} \quad \mathbf{C} = [c_{11} \quad c_{12} \quad c_{13}].$$

▷ **17.** (a) Explicitly write out $\mathbf{I}_2$.
(b) Show that $\mathbf{I}_2\mathbf{A} = \mathbf{A}$ for all 2×3 **A**.

18. (a) Explicitly write out $\mathbf{I}_3$.
(b) Show that $\mathbf{BI}_3 = \mathbf{B}$ for all 2×3 **B**.

𝔐 **19.** Use MATLAB or other software to discover experimentally what happens to $\mathbf{A}^n$ for large positive integers n if **A** is:

(a) $\begin{bmatrix} 0.6 & 0.5 \\ -0.2 & 1.2 \end{bmatrix}$ (b) $\begin{bmatrix} 0.6 & 0.5 \\ -0.16 & 1.2 \end{bmatrix}$

(c) $\begin{bmatrix} 0.6 & 0.5 \\ -0.1 & 1.2 \end{bmatrix}$ (d) $\begin{bmatrix} 0.9 & 1.0 \\ 0 & 0.9 \end{bmatrix}$

(e) $\begin{bmatrix} 0.99 & 1.0 \\ 0 & 0.99 \end{bmatrix}$ (f) $\begin{bmatrix} 1.0 & 1.0 \\ 0 & 1.0 \end{bmatrix}$

▷ **20.** Determine, in terms of the real number r, what happens to $\mathbf{A}^n$ for large positive integers n if

$$\mathbf{A} = \begin{bmatrix} r & 1 \\ 0 & r \end{bmatrix}.$$

21. Prove Theorem 1.19(b)—other than the one version proved in the text—on the distributive law.

22. Prove Theorem 1.19(c) on **I** as a multiplicative identity.

23. Prove Theorem 1.19(d) on scalar multiples of matrix products.

▷ **24.** Prove Theorem 1.19(e) on matrix multiplication with **0**.

25. Compute $\mathbf{A}^T$, $\mathbf{B}^T$, $\mathbf{AB}$, $(\mathbf{AB})^T$, and $\mathbf{B}^T\mathbf{A}^T$ and verify that $(\mathbf{AB})^T = \mathbf{B}^T\mathbf{A}^T$ if

$$\mathbf{A} = \begin{bmatrix} 1 & -2 \\ -2 & 3 \end{bmatrix} \quad \text{and} \quad \mathbf{B} = \begin{bmatrix} -2 & 1 \\ 1 & 1 \end{bmatrix}.$$

▷ **26.** Evaluate:
(a) $[1 \ \ 2 \ \ 3][-1 \ \ 3 \ \ 6]^T$ (b) $[1 \ \ 2 \ \ 3]^T[-1 \ \ 3 \ \ 6]$
(c) $[-1 \ \ 3 \ \ 6][1 \ \ 2 \ \ 3]^T$ (d) $[-1 \ \ 3 \ \ 6]^T[1 \ \ 2 \ \ 3]$

27. Suppose that $\mathbf{A}$ is $p \times p$ and $\mathbf{x}$ is $p \times 1$. Show that $\mathbf{x}^T\mathbf{A}\mathbf{x}$ is 1×1. If $\mathbf{x} = \mathbf{P}\mathbf{y}$, show that $\mathbf{x}^T\mathbf{A}\mathbf{x} = \mathbf{y}^T(\mathbf{P}^T\mathbf{A}\mathbf{P})\mathbf{y}$.

28. Prove Theorem 1.22(a) on $(\mathbf{A}^T)^T$ and $(\mathbf{A}^H)^H$.

29. Prove Theorem 1.22(b) on $(\mathbf{A} \pm \mathbf{B})^T$ and $(\mathbf{A} \pm \mathbf{B})^H$.

▷ **30.** Prove Theorem 1.22(c) on $(c\mathbf{A})^T$ and $(c\mathbf{A})^H$.

31. Prove the unproved part of Theorem 1.22(d), that $(\mathbf{AB})^H = \mathbf{B}^H\mathbf{A}^H$.

32. Prove that $(\mathbf{ABC})^H = \mathbf{C}^H\mathbf{B}^H\mathbf{A}^H$.

▷ **33.** Prove the extension of Theorem 1.22(d) to apply to a product of an arbitrary finite number k of matrices $\mathbf{A}_1\mathbf{A}_2 \cdots \mathbf{A}_k$.

34. Show that $\mathbf{I}_p$ and the $p \times p$ zero matrix $\mathbf{0}$ are symmetric and are hermitian.

35. (a) Show that every diagonal matrix is symmetric.
(b) Show that the diagonal matrix $\mathbf{D}$ is hermitian if and only if all of the entries in $\mathbf{D}$ are real.

36. Explicitly write down the general 3×3 symmetric matrix.

▷ **37.** Explicitly write down the general 2×2 hermitian matrix.

38. Show that $\mathbf{AB}$ need not be symmetric even though $\mathbf{A}$ and $\mathbf{B}$ are symmetric by using

$$\mathbf{A} = \begin{bmatrix} 1 & -2 \\ -2 & 3 \end{bmatrix} \quad \text{and} \quad \mathbf{B} = \begin{bmatrix} -2 & 1 \\ 1 & 1 \end{bmatrix}.$$

39. Show by example that $\mathbf{AB}$ can *possibly* be symmetric when $\mathbf{A}$ is symmetric and $\mathbf{B}$ is symmetric.

40. (a) Prove that $\mathbf{A}$ is symmetric if and only if $\mathbf{A}^T$ is symmetric.
(b) Prove that $\mathbf{A}$ is hermitian if and only if $\mathbf{A}^H$ is hermitian.

▷ **41.** (a) Prove that if $\mathbf{A}$ is symmetric, then $\mathbf{A}^2$ is symmetric.
(b) Prove that if $\mathbf{A}$ is symmetric, then $\mathbf{A}^n$ is symmetric for all positive integers n.
(c) What about the converse of (a)?

42. (a) Prove that if $\mathbf{A}$ is hermitian, then $\mathbf{A}^2$ is hermitian.
(b) Prove that if $\mathbf{A}$ is hermitian, then $\mathbf{A}^n$ is hermitian for all positive integers n.
(c) What about the converse of (a)?

43. Verify that $\mathbf{B}^T\mathbf{AB}$ is symmetric, in accordance with (1.25), if

$$\mathbf{A} = \begin{bmatrix} 1 & -1 \\ -1 & 1 \end{bmatrix} \quad \text{and} \quad \mathbf{B} = \begin{bmatrix} 1 & 0 & -2 \\ -1 & 3 & 0 \end{bmatrix}.$$

44. Prove the hermitian analogue of (1.25): $\mathbf{B}^H\mathbf{AB}$ is hermitian if $\mathbf{A}$ is hermitian.

▷ **45.** A matrix $\mathbf{A}$ is said to be *skew-symmetric* if $\mathbf{A}^T = -\mathbf{A}$.

(a) Show that a skew-symmetric matrix must be square and that its entries on the main diagonal must be zeros.

(b) Show that, given any square matrix $\mathbf{A}$, the matrix $\mathbf{A} - \mathbf{A}^T$ is skew-symmetric while the matrix $\mathbf{A} + \mathbf{A}^T$ is symmetric.

(c) By writing $\mathbf{A} = (\mathbf{A} + \mathbf{A}^T)/2 + (\mathbf{A} - \mathbf{A}^T)/2$, show that every square matrix can be *uniquely* written as the sum of a symmetric matrix and a skew-symmetric matrix.

46. If $\mathbf{x}$ is $p \times 1$, show that $\mathbf{x}^H\mathbf{x}$ is real (even if $\mathbf{x}$ has complex entries).

47. Generalize Problem 45 to the case of *skew-hermitian* matrices $\mathbf{A}$ for which $\mathbf{A}^H = -\mathbf{A}$.

▷ **48.** By considering its real and pure-imaginary parts, prove that every hermitian matrix is the sum of a real symmetric matrix and i times a real skew-symmetric matrix.

49. A product different from that in Definition 1.14 is important in many areas of science and engineering, especially in problems involving the processing of discrete signals; the *convolution* $\mathbf{x} * \mathbf{y}$ is defined between any two $1 \times p$ row matrices

$$\mathbf{x} = [x_1 \quad x_2 \quad \cdots \quad x_p] \quad \text{and} \quad \mathbf{y} = [y_1 \quad y_2 \quad \cdots \quad y_p]$$

and is itself a $1 \times p$ row matrix:

$$\begin{aligned}\langle \mathbf{x} * \mathbf{y} \rangle_i = {} & x_1 y_i + x_2 y_{i-1} + \cdots + x_i y_1 \\ & + x_{i+1} y_p + x_{i+2} y_{p-1} + \cdots + x_p y_{i+1}.\end{aligned}$$

For $p = 2$, this gives simply

$$[x_1 \quad x_2] * [y_1 \quad y_2] = [x_1 y_1 + x_2 y_2 \quad x_1 y_2 + x_2 y_1].$$

(a) Find the explicit formula for $[x_1 \quad x_2 \quad x_3] * [y_1 \quad y_2 \quad y_3]$.

(b) Show that $*$ is commutative for general p.

(c) Show that $*$ is associative for general p.

(d) Find a $1 \times p$ "identity matrix" $\mathbf{e}$ such that $\mathbf{e} * \mathbf{x} = \mathbf{x} * \mathbf{e} = \mathbf{x}$ for all $1 \times p$ $\mathbf{x}$.

(e) Find $[3 \quad 1 \quad 2 \quad 6] * [-2 \quad 1 \quad 3 \quad 1]$.

50. An important technique in applied mathematics is to approximate a given function $f(x)$ by a *Fourier series*: a sum of terms of the form $\sin kx$ or $\cos kx$ or, equivalently, of the complex form $e^{kix} = \cos kx + i \sin kx$. *Discrete Fourier analysis* chooses the coefficients c_k of e^{kix} by requiring (1) that all c_k except possibly some specific N coefficients are zero, and (2) that the series and the original function f are equal at N equally-spaced values of x. With

$N = 3$, for example we might approximate $f(x)$ for $0 \le x \le 2\pi$ by $s(x) = c_0 + c_1 e^{ix} + c_2 e^{2ix}$, with the coefficients c_k determined by requiring $f(x) = s(x)$ for $x = x_1 = 0$, $x = x_2 = 2\pi/3$, and $x = x_3 = 2(2\pi/3)$. Let f_i denote $f(x_i)$

(a) Show that these requirements on the c_k can be expressed as

$$f_1 = c_0 + c_1 + c_2$$

$$f_2 = c_0 + c_1 z + c_2 z^2$$

$$f_3 = c_0 + c_1 z^2 + c_2 z^4$$

where $z = e^{(2\pi/3)i} = \cos 2\pi/3 + i \sin 2\pi/3$
$= -1/2 + i\sqrt{3}/2.$

(b) Show that the equations above can be expressed as $\mathbf{f} = \mathbf{Zc}$, where $\mathbf{f} = [f_1 \quad f_2 \quad f_3]^T$, $\mathbf{c} = [c_0 \quad c_1 \quad c_2]^T$, and

$$\mathbf{Z} = \begin{bmatrix} 1 & 1 & 1 \\ 1 & z & z^2 \\ 1 & z^2 & z^4 \end{bmatrix}.$$

(c) Use the value of z above explicitly to find $\mathbf{Z}$.

(d) Let $\square$ and $*$ denote the products defined in Problems 11 and 49. Show that $\mathbf{Z}(\mathbf{x} * \mathbf{y})^T = \mathbf{Zx}^T \square \mathbf{Zy}^T$, first if $\mathbf{x} = [3 \quad -1 \quad 2]$ and $\mathbf{y} = [-2 \quad 2 \quad 3]$, and then for all 1×3 $\mathbf{x}$ and $\mathbf{y}$.

▷ **51.** Problem 50 examined discrete Fourier analysis with $N = 3$. Suppose more generally that we wish to approximate $f(x)$ for $0 \le x \le 2\pi$ by a finite Fourier series $s(x)$ using N terms:

$$s(x) = c_0 + c_1 e^{ix} + c_2 e^{2ix} + \cdots + c_{N-1} e^{(N-1)ix}$$

and that the c_k are determined by requiring $s(x_j) = f(x_j)$ where we take $x_j = (j-1)(2\pi/N)$ for $j = 1, 2, \ldots, N$. Show that the problem of finding the coefficients c_k from the given values $f_j = f(x_j)$ is the same as solving $\mathbf{Zc} = \mathbf{f}$, where $\mathbf{c} = [c_0 \quad c_1 \quad \cdots \quad c_{N-1}]^T$, $f = [f_1 \quad f_2 \quad \cdots \quad f_N]^T$, and $\mathbf{Z}$ is the $N \times N$ matrix with $\langle \mathbf{Z} \rangle_{jk} = z^{(j-1)(k-1)}$ and $z = e^{(2\pi/N)i}$. The product $\mathbf{Zc}$ is called the *discrete Fourier transform* of the row-matrix $\mathbf{c}$ and is extremely important in applied mathematics.

52. Let $\mathbf{Z}$ be the $N \times N$ matrix that generates the discrete Fourier transform as in Problem 51, and let $\square$ and $*$ denote the products defined in Problems 11 and 49. Show that $\mathbf{Z}(\mathbf{x} * \mathbf{y})^T = \mathbf{Zx}^T \square \mathbf{Zy}^T$ for all $1 \times N$ row matrices $\mathbf{x}$ and $\mathbf{y}$.

1.4 MATRIX INVERSES

The preceding sections extended to matrices many ideas associated with arithmetic on numbers; one idea omitted was the matrix analogue of inversion or reciprocation: computing $\frac{1}{3}$ from 3 or, more generally, computing $1/a$ from the nonzero number a.

One way to describe the inverse or reciprocal $1/a$ of a is to notice that $1/a$ is the solution x to the equation $ax = 1$, where of course 1 is the "identity number for number multiplication": $1b = b1 = b$ for all numbers b. Since identity matrices **I** (see Definition 1.18) are "identity matrices for matrix multiplication"—$\mathbf{IB} = \mathbf{BI} = \mathbf{B}$ for all matrices **B**—the natural analogue to seeking an x solving $ax = 1$ and $xa = 1$ for a given a is to seek a matrix **X** solving $\mathbf{AX} = \mathbf{I}$ and $\mathbf{XA} = \mathbf{I}$ for a given **A**.

(1.26) ***Definition.*** Let **A** be a given matrix.

(a) Any matrix **L** for which $\mathbf{LA} = \mathbf{I}$ is called a *left-inverse* of **A**.

(b) Any matrix **R** for which $\mathbf{AR} = \mathbf{I}$ is called a *right-inverse* of **A**.

(c) Any matrix **X** for which $\mathbf{XA} = \mathbf{I}$ and $\mathbf{AX} = \mathbf{I}$ is called an *inverse* of **A**. (To distinguish clearly, such an **X** is often called a *two-sided inverse of* **A**.)

We can determine which—if any—types of these inverses a matrix **A** has by solving sets of linear equations; and the solutions to the equations produce the inverses.

(1.27) ***Example.*** Consider the question of whether there is a right-inverse **R** for the matrix.

$$\mathbf{A} = \begin{bmatrix} 1 & -1 \\ 1 & 2 \end{bmatrix}.$$

Then we seek **R** with $\mathbf{AR} = \mathbf{I}$. For **R** to post-multiply the 2×2 **A**, **R** must be $2 \times q$ for some q, and then **AR** will be $2 \times q$. But we want **AR** to equal a (*square*) identity matrix, so $q = 2$ and **R** must be 2×2. So let us seek a 2×2 **R** of the form

$$\mathbf{R} = \begin{bmatrix} x & z \\ y & w \end{bmatrix}$$

by seeking x, y, z, w so that $\mathbf{AR} = \mathbf{I}$, that is,

$$\begin{bmatrix} 1 & -1 \\ 1 & 2 \end{bmatrix}\begin{bmatrix} x & z \\ y & w \end{bmatrix} = \mathbf{I} = \begin{bmatrix} 1 & 0 \\ 0 & 1 \end{bmatrix}.$$

Multiplying the two matrices on the left side together and equating the entries in the product with the corresponding entries in **I** produces two sets of linear equations, one involving only x and y and the other involving only z and w:

$$\begin{aligned} x - y &= 1 \\ x + 2y &= 0 \end{aligned} \quad \text{and} \quad \begin{aligned} z - w &= 0 \\ z + 2w &= 1. \end{aligned}$$

These are easily solved by the methods of high school algebra to produce exactly one solution $x = \frac{2}{3}$, $y = -\frac{1}{3}$, $z = \frac{1}{3}$, and $w = \frac{1}{3}$. This then says that there is exactly one right-inverse, namely

$$\mathbf{R} = \begin{bmatrix} \frac{2}{3} & \frac{1}{3} \\ -\frac{1}{3} & \frac{1}{3} \end{bmatrix}.$$

Our argument in Example 1.27 that determined that **R** is 2 × 2 works more generally (Problem 7):

(1.28) ***Theorem.*** Let **A** be $p \times q$. Then:
(a) Any possible right-inverse **R** of **A** must be $q \times p$.
(b) Any possible left-inverse **L** of **A** must be $q \times p$.

(1.29) ***Example.*** Consider whether there is a right-inverse **R** for the 2 × 3 matrix

$$\mathbf{A} = \begin{bmatrix} 1 & -1 & 1 \\ 1 & 1 & 2 \end{bmatrix}.$$

By Theorem 1.28, **R** would be 3 × 2, so we seek x, y, z, u, v, w so that

$$\begin{bmatrix} 1 & -1 & 1 \\ 1 & 1 & 2 \end{bmatrix} \begin{bmatrix} x & u \\ y & v \\ z & w \end{bmatrix} = \begin{bmatrix} 1 & 0 \\ 0 & 1 \end{bmatrix},$$

which translates to equations

$$\begin{aligned} x - y + z &= 1 \\ x + y + 2z &= 0 \end{aligned} \quad \text{and} \quad \begin{aligned} u - v + w &= 0 \\ u + v + 2w &= 1. \end{aligned}$$

Solving these shows that z and w (for example) can be given arbitrary values $z = \alpha$ and $w = \beta$; solving for the other variables in terms of α and β finally produces infinitely many different right-inverses, one for each choice of α and β:

$$\mathbf{R} = \begin{bmatrix} \frac{1}{2} - \frac{3\alpha}{2} & \frac{1}{2} - \frac{3\beta}{2} \\ -\frac{1}{2} - \frac{\alpha}{2} & \frac{1}{2} - \frac{\beta}{2} \\ \alpha & \beta \end{bmatrix}.$$

Example 1.27 presented a matrix with *exactly one* right-inverse, Example 1.29 one with *infinitely many* right-inverses. While demonstrating how to find *left*-inverses, we next examine a matrix that has *none.*

(1.30) ***Example.*** We seek a left-inverse **L** for

$$\mathbf{A} = \begin{bmatrix} 1 & -1 \\ -3 & 3 \end{bmatrix} \quad \text{in the form} \quad \mathbf{L} = \begin{bmatrix} x & y \\ z & w \end{bmatrix}.$$

LA = **I** translates to

$$\begin{aligned} x - 3y &= 1 \\ -x + 3y &= 0 \end{aligned} \qquad \text{and} \qquad \begin{aligned} z - 3w &= 0 \\ -z + 3w &= 1. \end{aligned}$$

Each set of equations is inconsistent: The first, for example, seeks x and y for which $1 = x - 3y$, which equals $-(-x + 3y) = -(0) = 0$, and it is certainly impossible to find x and y that make 1 equal 0.

𝔐 (1.31) ***Example.*** The preceding three examples have involved small matrices because it is so tedious to solve systems of equations by hand if they involve many variables. Many good computer programs exist, however, to make this an easy task; MATLAB, of course, contains this facility, and it is especially easy to use when the matrix is square. As an example, consider the 7×7 symmetric matrix **A** with $\langle \mathbf{A} \rangle_{ij} = 1/(i + j - 1)$:

$$\mathbf{A} = \begin{bmatrix} 1 & \frac{1}{2} & \frac{1}{3} & \frac{1}{4} & \frac{1}{5} & \frac{1}{6} & \frac{1}{7} \\ \frac{1}{2} & \frac{1}{3} & \frac{1}{4} & \frac{1}{5} & \frac{1}{6} & \frac{1}{7} & \frac{1}{8} \\ \frac{1}{3} & \frac{1}{4} & \frac{1}{5} & \frac{1}{6} & \frac{1}{7} & \frac{1}{8} & \frac{1}{9} \\ \frac{1}{4} & \frac{1}{5} & \frac{1}{6} & \frac{1}{7} & \frac{1}{8} & \frac{1}{9} & \frac{1}{10} \\ \frac{1}{5} & \frac{1}{6} & \frac{1}{7} & \frac{1}{8} & \frac{1}{9} & \frac{1}{10} & \frac{1}{11} \\ \frac{1}{6} & \frac{1}{7} & \frac{1}{8} & \frac{1}{9} & \frac{1}{10} & \frac{1}{11} & \frac{1}{12} \\ \frac{1}{7} & \frac{1}{8} & \frac{1}{9} & \frac{1}{10} & \frac{1}{11} & \frac{1}{12} & \frac{1}{13} \end{bmatrix}.$$

The MATLAB command **X** = inv(**A**) seeks a (two-sided) inverse of **A** and produces

$$\mathbf{X} = \begin{bmatrix} 49 & -1{,}176 & 8{,}820 & -29{,}400 & 48{,}510 & -38{,}808 & 12{,}012 \\ -1{,}176 & 37{,}632 & -317{,}520 & 1{,}128{,}960 & -1{,}940{,}400 & 1{,}596{,}672 & -504{,}504 \\ 8{,}820 & -317{,}520 & 2{,}857{,}680 & -10{,}584{,}000 & 18{,}711{,}000 & -15{,}717{,}240 & 5{,}045{,}040 \\ -29{,}400 & 1{,}128{,}960 & -10{,}584{,}000 & 40{,}320{,}000 & -72{,}765{,}000 & 62{,}092{,}800 & -20{,}180{,}160 \\ 48{,}510 & -1{,}940{,}400 & 18{,}711{,}000 & -72{,}765{,}000 & 133{,}402{,}500 & -115{,}259{,}760 & 37{,}837{,}800 \\ -38{,}808 & 1{,}596{,}672 & -15{,}717{,}240 & 62{,}092{,}800 & -115{,}259{,}760 & 100{,}530{,}336 & -33{,}297{,}264 \\ 12{,}012 & -504{,}504 & 5{,}045{,}040 & -20{,}180{,}160 & 37{,}837{,}800 & -33{,}297{,}264 & 11{,}099{,}088 \end{bmatrix}$$

These four examples illustrate how to find inverses as well as the fact that inverses—one or many—may not exist. Theory about inverses will be presented later; for the moment, we simply preview some of those later basic developments:

(1.32) ***Preview on Inverses***

(a) For a *square* $p \times p$ matrix **A**, either:

1. **A** has a unique (two-sided) inverse,

or

2. **A** has neither a left- nor a right-inverse. Indeed, if a square matrix **A** has either a left- or a right-inverse, then that is in fact a (two-sided) inverse.

(b) For a *nonsquare* $p \times q$ matrix **A** with $p \neq q$:
1. **A** cannot possibly have a (two-sided) inverse.
2. If $p < q$, then **A** cannot possibly have a left-inverse, but may possibly have a right-inverse.
3. If $p > q$, then **A** cannot possibly have a right-inverse, but may possibly have a left-inverse.

We *do* have enough tools at hand to prove one theorem about the different types of inverses.

(1.33) ***Theorem*** (two-sided inverses). Let **A** be a matrix.

(a) If both a left-inverse **L** and a right-inverse **R** exist for **A**, then they are equal and are a (two-sided) inverse.

(b) Any two (two-sided) inverses of **A** are identical.

PROOF.

(a) Let **L** and **R** be left- and right-inverses of **A**, so that $\mathbf{LA} = \mathbf{I}$ and $\mathbf{AR} = \mathbf{I}$. Then

$$\mathbf{L} = \mathbf{LI} = \mathbf{L}(\mathbf{AR}) = (\mathbf{LA})\mathbf{R} = (\mathbf{I})\mathbf{R} = \mathbf{R},$$

so $\mathbf{L} = \mathbf{R}$ as claimed and this one matrix serves as both a left- and a right-inverse—that is, it is a (two-sided) inverse.

(b) Suppose that **X** and **Y** are both (two-sided) inverses of **A**. Then **X** also is a left-inverse of **A** and **Y** also is a right-inverse of **A**; by part (a), we can conclude that $\mathbf{X} = \mathbf{Y}$. ■

You should now be able to solve Problems 1 to 9.

Nonsingular Matrices

The examples above and the preview in (1.32) show that (two-sided) inverses are the exception rather than the rule. The property of having such an inverse is so important that matrices with such inverses are distinguished by a special name, *nonsingular*.

(1.34) ***Definition***

(a) A *nonsingular matrix* is a (necessarily square) matrix **A** that possesses an inverse **X**: For nonsingular **A**, there is an **X** with $\mathbf{AX} = \mathbf{XA} = \mathbf{I}$. This inverse is denoted by $\mathbf{A}^{-1}$.

(b) A *singular matrix* is a square matrix that does *not* possess an inverse.

"Is it singular or nonsingular?" is often the first question we ask when confronted with a square matrix. Later we will develop theoretical tools for answering this question, but computationally it is usually answered by trying to *find* an inverse by the methods of Examples 1.27, 1.29, 1.30, and 1.31—remembering (1.32) (a). But if you somehow have a *candidate* $\mathbf{X}$ for the inverse $\mathbf{A}^{-1}$ of $\mathbf{A}$, you can test the candidate just by checking whether $\mathbf{AX} = \mathbf{I}$ and $\mathbf{XA} = \mathbf{I}$—which is the key tool used in the proof of the following theorem. In that proof we verify *both* equations $\mathbf{AX} = \mathbf{I}$ and $\mathbf{XA} = \mathbf{I}$, although Preview 1.32(a)—not yet proved—shows that we need check only *one* equation.

(1.35) ***Theorem.*** Let $\mathbf{A}$ and $\mathbf{B}$ be $p \times p$ nonsingular matrices. Then:
(a) $\mathbf{AB}$ is nonsingular and $(\mathbf{AB})^{-1} = \mathbf{B}^{-1}\mathbf{A}^{-1}$.
(b) $\mathbf{A}^{-1}$ is nonsingular and $(\mathbf{A}^{-1})^{-1} = \mathbf{A}$.
(c) $\mathbf{A}^T$ and $\mathbf{A}^H$ are nonsingular with $(\mathbf{A}^T)^{-1} = (\mathbf{A}^{-1})^T$ and $(\mathbf{A}^H)^{-1} = (\mathbf{A}^{-1})^H$.

Key Proof.
(a) We merely need to show that $\mathbf{AB}$ has a matrix $\mathbf{X}$ for which $(\mathbf{AB})\mathbf{X} = \mathbf{X}(\mathbf{AB}) = \mathbf{I}$. Since $\mathbf{A}$ and $\mathbf{B}$ are $p \times p$ and nonsingular, $\mathbf{B}^{-1}$ and $\mathbf{A}^{-1}$ both exist and are $p \times p$, so the product $\mathbf{B}^{-1}\mathbf{A}^{-1}$ exists and we take this as our candidate for $\mathbf{X}$. Checking whether $(\mathbf{AB})\mathbf{X}$ equals $\mathbf{I}$:

$$(\mathbf{AB})\mathbf{X} = (\mathbf{AB})(\mathbf{B}^{-1}\mathbf{A}^{-1}) = \mathbf{A}(\mathbf{BB}^{-1})\mathbf{A}^{-1} = \mathbf{AIA}^{-1} = \mathbf{AA}^{-1} = \mathbf{I},$$

as required. Checking whether $\mathbf{X}(\mathbf{AB})$ equals $\mathbf{I}$:

$$\mathbf{X}(\mathbf{AB}) = (\mathbf{B}^{-1}\mathbf{A}^{-1})(\mathbf{AB}) = \mathbf{B}^{-1}(\mathbf{A}^{-1}\mathbf{A})\mathbf{B} = \mathbf{B}^{-1}\mathbf{IB} = \mathbf{B}^{-1}\mathbf{B} = \mathbf{I},$$

also as required. Therefore, our candidate $\mathbf{X}$ is an inverse, so of course $\mathbf{AB}$ is nonsingular and its inverse is as stated.
(b) Since $\mathbf{A}$ is nonsingular, $\mathbf{A}^{-1}$ exists. Let $\mathbf{X} = \mathbf{A}$ be our candidate for the inverse of $\mathbf{A}^{-1}$; we need to check whether $\mathbf{X}(\mathbf{A}^{-1}) = \mathbf{I}$ and $(\mathbf{A}^{-1})\mathbf{X} = \mathbf{I}$. See Problem 14.
(c) See Problem 15. ■

The proof above is **key** because it is a model of proofs that a given matrix $\mathbf{G}$ is the inverse of a given matrix $\mathbf{H}$; we merely show that $\mathbf{GH} = \mathbf{I}$ and $\mathbf{HG} = \mathbf{I}$. In some cases we are given only $\mathbf{H}$ and part of the difficulty is to determine the formula for $\mathbf{G}$. In (a) of Theorem 1.35, for example, if we were only given $\mathbf{H} = \mathbf{AB}$ and had to guess the formula for $\mathbf{G}$, we could reason as follows: we need a $\mathbf{G}$ so that $\mathbf{G}(\mathbf{AB}) = \mathbf{I}$; postmultiplying this equation by $\mathbf{B}^{-1}$ changes it to $\mathbf{GA} = \mathbf{B}^{-1}$, and then postmultiplying this equation by $\mathbf{A}^{-1}$ changes it to $\mathbf{G} = \mathbf{B}^{-1}\mathbf{A}^{-1}$, giving us our *candidate* for the inverse of $\mathbf{AB}$. We would then proceed as in the proof of (a) to prove that this guess for the inverse is indeed correct.

You should now be able to solve Problems 1 to 23.

Matrices, Inverses, and Systems of Equations

Recall that we motivated our discussion of matrix inverses by analogy with number inverses $1/a$ for numbers a. One use you have often made of number inverses is to solve equations: You solve $3x = 7$ by multiplying both sides of the equation by $\frac{1}{3}$ to obtain $x = \frac{7}{3}$. We next examine solving matrix equations $\mathbf{Ax} = \mathbf{b}$ by using $\mathbf{A}^{-1}$.

Suppose that $\mathbf{A}$ is a $p \times q$ matrix of known numbers, $\mathbf{x}$ is a $q \times 1$ column matrix of unknowns, and $\mathbf{b}$ is a $p \times 1$ column matrix of known numbers. Then $\mathbf{Ax}$ is defined and is $p \times 1$, so the equation $\mathbf{Ax} = \mathbf{b}$ makes sense as a problem of finding an unknown matrix $\mathbf{x}$ to satisfy this equation. To see what this equation means, we write it out in terms of the entries of the matrices $\mathbf{A}$, $\mathbf{x}$, and $\mathbf{b}$. The definitions of matrix multiplication and matrix equality reveal that

(1.36) the equation $\mathbf{Ax} = \mathbf{b}$ with known $p \times q$ $\mathbf{A}$ having entries $\langle \mathbf{A} \rangle_{ij} = a_{ij}$, unknown $q \times 1$ $\mathbf{x}$ having entries $\langle \mathbf{x} \rangle_i = x_i$, and known $p \times 1$ $\mathbf{b}$ having entries $\langle \mathbf{b} \rangle_i = b_i$ is equivalent to

$$
\begin{aligned}
a_{11}x_1 + a_{12}x_2 + \cdots + a_{1q}x_q &= b_1 \\
a_{21}x_1 + a_{22}x_2 + \cdots + a_{2q}x_q &= b_2 \\
\cdots\cdots\cdots\cdots & \\
a_{p1}x_1 + a_{p2}x_2 + \cdots + a_{pq}x_q &= b_q
\end{aligned}
$$

a system of p linear equations in the q unknowns $x_1, x_2, \ldots, x_q$.

So $\mathbf{Ax} = \mathbf{b}$ is a compact notation for writing a system of p linear equations in q unknowns. The compact matrix notation also indicates a possible approach to solving such a system, by analogy with solving $3x = 7$ by multiplying both sides by $\frac{1}{3}$: Multiply both sides of $\mathbf{Ax} = \mathbf{b}$ by $\mathbf{A}^{-1}$ (if it exists) to obtain a solution $\mathbf{x} = \mathbf{A}^{-1}\mathbf{b}$. Of course, this works only if $\mathbf{A}$ is nonsingular; in particular, $\mathbf{A}$ must be square: the number of equations must equal the number of unknowns. Theorem 1.38 below explains what happens for non-square $\mathbf{A}$ having a left- or right-inverse.

(1.37) ***Example.*** Consider the system of two linear equations

$$
\begin{aligned}
2x - 3y &= -13 \\
x + 4y &= 10
\end{aligned}
$$

in the two unknowns x and y. By (1.36), this is equivalent to $\mathbf{Ax} = \mathbf{b}$ if

$$
\mathbf{A} = \begin{bmatrix} 2 & -3 \\ 1 & 4 \end{bmatrix}, \quad \mathbf{x} = \begin{bmatrix} x \\ y \end{bmatrix}, \quad \text{and} \quad \mathbf{b} = \begin{bmatrix} -13 \\ 10 \end{bmatrix}.
$$

In this case,

$$\mathbf{A}^{-1} = \begin{bmatrix} \frac{4}{11} & \frac{3}{11} \\ -\frac{1}{11} & \frac{2}{11} \end{bmatrix} \quad \text{and} \quad \begin{bmatrix} x \\ y \end{bmatrix} = \mathbf{x} = \mathbf{A}^{-1}\mathbf{b}$$

becomes

$$\begin{bmatrix} x \\ y \end{bmatrix} = \begin{bmatrix} \frac{4}{11} & \frac{3}{11} \\ -\frac{1}{11} & \frac{2}{11} \end{bmatrix} \begin{bmatrix} -13 \\ 10 \end{bmatrix} = \begin{bmatrix} -2 \\ 3 \end{bmatrix}.$$

You can check that $x = -2$ and $y = 3$ indeed solve the system of equations.

The general situation relating the solution of systems of linear equations to various matrix inverses is as follows.

(1.38) ***Theorem*** (equations and inverses). Let $\mathbf{Ax} = \mathbf{b}$ represent the system of p linear equations in q unknowns as in (1.36).

(a) Suppose that $\mathbf{A}$ has a left-inverse $\mathbf{L}$. If there are any solutions $\mathbf{x}$ at all to $\mathbf{Ax} = \mathbf{b}$, then they must all equal $\mathbf{Lb}$; thus there is *at most one solution*, namely $\mathbf{Lb}$, but possibly none.

(b) Suppose that $\mathbf{A}$ has a right-inverse $\mathbf{R}$. Then $\mathbf{Rb}$ is a solution to $\mathbf{Ax} = \mathbf{b}$; thus there is *at least one solution*, namely $\mathbf{Rb}$, but there may be more.

(c) Suppose that $\mathbf{A}$ is nonsingular. Then there is *exactly one solution* to $\mathbf{Ax} = \mathbf{b}$, namely $\mathbf{A}^{-1}\mathbf{b}$.

PROOF

(a) If there is some solution $\mathbf{x}_0$ to $\mathbf{Ax} = \mathbf{b}$, then of course $\mathbf{b} = \mathbf{Ax}_0$. Premultiplying this equation by $\mathbf{L}$ gives

$$\mathbf{Lb} = \mathbf{L}(\mathbf{Ax}_0) = (\mathbf{LA})\mathbf{x}_0 = \mathbf{Ix}_0 = \mathbf{x}_0,$$

which says that $\mathbf{x}_0$ must equal $\mathbf{Lb}$ if $\mathbf{x}_0$ is a solution.

(b) To show that $\mathbf{Rb}$ is a solution, we just check whether $\mathbf{A}(\mathbf{Rb})$ equals $\mathbf{b}$:

$$\mathbf{A}(\mathbf{Rb}) = (\mathbf{AR})\mathbf{b} = \mathbf{Ib} = \mathbf{b},$$

so $\mathbf{Rb}$ is a solution.

(c) If $\mathbf{A}$ is nonsingular, then $\mathbf{A}^{-1}$ exists and serves as both a left-inverse and a right-inverse for $\mathbf{A}$. Since $\mathbf{A}^{-1}$ is a right-inverse, part (b) shows that $\mathbf{A}^{-1}\mathbf{b}$ is one solution; since $\mathbf{A}^{-1}$ is also a left-inverse, part (a) shows that all other solutions must also equal $\mathbf{A}^{-1}\mathbf{b}$. ■

So we now know that we can solve systems of equations by using matrix inverses. But how do we obtain matrix inverses? By solving systems of equations—the very type of problem we wanted to solve in the first place! In practice, as we will later show in detail, it is usually far more efficient to solve the system of equations represented by $\mathbf{Ax} = \mathbf{b}$ directly than first to calculate $\mathbf{A}^{-1}$ and then calculate $\mathbf{A}^{-1}\mathbf{b}$; see Problem 27. However, the *concept* and *representation* of the solution as $\mathbf{x} = \mathbf{A}^{-1}\mathbf{b}$ can be very powerful in understanding certain aspects of problems.

PROBLEMS 1.4

▷ **1.** Directly determine whether there is a left-inverse $\mathbf{L}$ for the matrix $\mathbf{A}$ in Example 1.29; if so, find all such $\mathbf{L}$.

2. Directly determine whether there is a right-inverse $\mathbf{R}$ for the matrix $\mathbf{A}$ in Example 1.30; if so, find all such $\mathbf{R}$.

3. Show that there is neither a left- nor a right-inverse for

$$\mathbf{A} = \begin{bmatrix} 1 & -1 & 1 \\ -3 & 3 & -3 \end{bmatrix}.$$

▷ **4.** Find all possible left-inverses for the matrix $\mathbf{A}$ below, and show that there is no right-inverse for $\mathbf{A}$.

$$\mathbf{A} = \begin{bmatrix} 1 & -1 \\ 1 & 1 \\ 2 & 3 \end{bmatrix}.$$

𝔐 **5.** Use MATLAB or similar software to calculate both $\mathbf{XA}$ and $\mathbf{AX}$ using the matrices $\mathbf{X}$ and $\mathbf{A}$ in Example 1.31 to try to test whether $\mathbf{X}$ is the exact inverse of $\mathbf{A}$ or just a close numerical approximation. (Note that your computer is probably unable to represent the matrix $\mathbf{A}$ exactly because of the fractions $\frac{1}{3}$, $\frac{1}{7}$, and the like.)

𝔐 **6.** Use MATLAB or similar software to calculate the inverse of the matrix $\mathbf{A}$ in Example 1.31 and compare the result produced on your computer with the matrix $\mathbf{X}$ in Example 1.31.

7. Prove Theorem 1.28 on left- and right-inverses.

▷ **8.** (a) Prove that $\mathbf{L}$ is a left-inverse for a matrix $\mathbf{A}$ if and only if $\mathbf{L}^T$ is a right-inverse for $\mathbf{A}^T$.

(b) Prove that $\mathbf{R}$ is a right-inverse for a matrix $\mathbf{A}$ if and only if $\mathbf{R}^T$ is a left-inverse for $\mathbf{A}^T$.

9. Cancellation in matrix equations *is* valid if appropriate inverses exist.

(a) Suppose that $\mathbf{A}$ has a left-inverse and that $\mathbf{AB} = \mathbf{AC}$; prove that $\mathbf{B} = \mathbf{C}$.

(b) State and prove the analogous result for right-inverses.

▷ **10.** For each matrix below, determine whether it is nonsingular and—if it is—find its inverse.

(a) $\begin{bmatrix} 4 & -5 \\ -3 & 4 \end{bmatrix}$ (b) $\begin{bmatrix} -8 & -3 \\ 5 & 2 \end{bmatrix}$ (c) $\begin{bmatrix} 1 & -3 \\ 1 & 4 \end{bmatrix}$

(d) $\begin{bmatrix} 4 & -1 \\ -3 & 1 \end{bmatrix}$ (e) $\begin{bmatrix} -1 & 2 & 1 \\ 0 & 1 & -2 \\ 1 & -1 & -1 \end{bmatrix}$

11. Show that **A** is singular, where

$$\mathbf{A} = \begin{bmatrix} 4 & -3 \\ -8 & 6 \end{bmatrix}.$$

▷ **12.** Suppose that $\mathbf{D} = \mathbf{diag}(d_1 \ldots, d_p)$.
(a) If $d_i \neq 0$ for $1 \leq i \leq p$, show that **D** is nonsingular and that

$$\mathbf{D}^{-1} = \mathbf{diag}(1/d_1, \ldots, 1/d_p).$$

(b) If some $d_i = 0$, show that **D** is singular.

13. Show that the general 2×2 matrix **A** below is nonsingular if and only if its *determinant* Δ—defined as $\Delta = a_{11}a_{22} - a_{12}a_{21}$—is nonzero, and that $\mathbf{A}^{-1}$ is as given below if $\Delta \neq 0$.

$$\mathbf{A} = \begin{bmatrix} a_{11} & a_{12} \\ a_{21} & a_{22} \end{bmatrix}, \qquad \mathbf{A}^{-1} = \frac{1}{\Delta}\begin{bmatrix} a_{22} & -a_{12} \\ -a_{21} & a_{11} \end{bmatrix} \quad (\text{if } \Delta \neq 0).$$

14. Complete the proof of Theorem 1.35(b) on the inverse of $\mathbf{A}^{-1}$.

▷ **15.** Prove Theorem 1.35(c) on the inverse of $\mathbf{A}^T$ and of $\mathbf{A}^H$.

16. If **A** is nonsingular, we can define the matrix power $\mathbf{A}^r$ for *negative* integers r—although certainly not as the product of r factors of **A** as we did for *positive* integers r. If n is a negative integer and **A** is nonsingular, then we define $\mathbf{A}^n$ as $(\mathbf{A}^{-1})^{(-n)}$, where the exponent $-n$ of the inverse of **A** is now a *positive* integer for which our original definition is valid. Prove that, for nonsingular **A**, the statement $\mathbf{A}^r\mathbf{A}^s = \mathbf{A}^{r+s}$ is true for all integers r and s (regardless of their signs).

17. Suppose that **A**, **B**, and **A** + **B** are all nonsingular $p \times p$ matrices. Show that $\mathbf{A}^{-1} + \mathbf{B}^{-1}$ is also nonsingular, and that

$$(\mathbf{A}^{-1} + \mathbf{B}^{-1})^{-1} = \mathbf{A}(\mathbf{A} + \mathbf{B})^{-1}\mathbf{B}$$

and that this in turn equals $\mathbf{B}(\mathbf{A} + \mathbf{B})^{-1}\mathbf{A}$.

𝔐 **18.** Use MATLAB or similar software to verify the formula in Problem 17 when

$$\mathbf{A} = \begin{bmatrix} 2 & -1 & 0 & 0 \\ -1 & 2 & -1 & 0 \\ 0 & -1 & 2 & -1 \\ 0 & 0 & -1 & 2 \end{bmatrix} \quad \text{and} \quad \mathbf{B} = 3\mathbf{I}_4.$$

▷ **19.** Suppose that **A** is nonsingular.
(a) Prove that **A** is symmetric if and only if $\mathbf{A}^{-1}$ is symmetric.
(b) Prove that **A** is hermitian if and only if $\mathbf{A}^{-1}$ is hermitian.

20. Suppose that **A**, **B**, and **C** are $p \times p$ nonsingular matrices. Prove that **ABC** is nonsingular and that

$$(\mathbf{ABC})^{-1} = \mathbf{C}^{-1}\mathbf{B}^{-1}\mathbf{A}^{-1}.$$

▷ **21.** Suppose that $\mathbf{A}_1, \mathbf{A}_2, \ldots, \mathbf{A}_k$ are $p \times p$ nonsingular matrices. Prove that the product $\mathbf{A}_1\mathbf{A}_2 \cdots \mathbf{A}_k$ is nonsingular and that

$$(\mathbf{A}_1\mathbf{A}_2 \cdots \mathbf{A}_k)^{-1} = \mathbf{A}_k^{-1}\mathbf{A}_{k-1}^{-1} \cdots \mathbf{A}_1^{-1}.$$

𝔐 **22.** Let **B** be the 6×6 symmetric matrix formed from the 36 entries in the upper left corner of the matrix **A** of Example 1.31—that is, $\langle \mathbf{B} \rangle_{ij} = \langle \mathbf{A} \rangle_{ij}$ for $1 \le i \le 6$ and $1 \le j \le 6$. Use MATLAB or similar software to compute the inverse **X** of **B** and to evaluate both **XB** and **BX** to try to see whether **X** is an exact inverse of **B** or just a close numerical approximation.

𝔐 **23.** Use MATLAB or similar software to study the 5×5 symmetric matrices

$$\mathbf{A}_z = \begin{bmatrix} z & -1 & 0 & 0 & 0 \\ -1 & z & -1 & 0 & 0 \\ 0 & -1 & z & -1 & 0 \\ 0 & 0 & -1 & z & -1 \\ 0 & 0 & 0 & -1 & z \end{bmatrix}$$

for various real values of z as follows.

(a) By finding its inverse, show that $\mathbf{A}_5$ is nonsingular.
(b) By finding its inverse, show that $\mathbf{A}_2$ is nonsingular.
(c) By attempting to find its inverse, show that $\mathbf{A}_0$ is singular.
(d) By slowly decreasing z from the value 2 and attempting to find the inverses of the matrices $\mathbf{A}_z$, approximate the first value of z below 2 for which $\mathbf{A}_z$ is singular.

(*Note:* The $p \times p$ versions of matrices $\mathbf{A}_z$ are important in the numerical solution of certain types of differential equations, especially for very large p.)

▷ **24.** Find the matrices **A**, **x**, and **b** so that the equation $\mathbf{Ax} = \mathbf{b}$ is equivalent to the two linear equations in two unknowns x and y below, and use $\mathbf{A}^{-1}$ as given below to solve for x and y; show that $\mathbf{A}^{-1}$, x, and y are correct.

$$3x + 4y = -1 \qquad \mathbf{A}^{-1} = \begin{bmatrix} 3 & -4 \\ -2 & 3 \end{bmatrix}$$
$$2x + 3y = -1,$$

25. Find the matrices **A**, **x**, and **b** so that the equation $\mathbf{Ax} = \mathbf{b}$ is equivalent to the two linear equations in two unknowns x and y below; solve for x and y by computing and using $\mathbf{A}^{-1}$.

$$3x - y = 2$$
$$-5x + y = -4$$

▷ **26.** Calculate $\mathbf{A}^{-1}$ and use it to solve $\mathbf{AX} = \mathbf{B}$, where

$$\mathbf{A} = \begin{bmatrix} 4 & 1 \\ 3 & 1 \end{bmatrix}, \quad \mathbf{X} = \begin{bmatrix} x & u \\ y & v \end{bmatrix}, \quad \text{and} \quad \mathbf{B} = \begin{bmatrix} 1 & 2 \\ -1 & 3 \end{bmatrix}.$$

Write the systems of linear equations in x, y, u, and v equivalent to $\mathbf{AX} = \mathbf{B}$.

$\mathfrak{M}$ **27.** Consider the following system of 10 linear equations in the 10 unknowns x_1, $x_2, \ldots, x_{10}$:

$$\begin{array}{rcl}
2x_1 - x_2 & = & 1 \\
-x_1 + 2x_2 - x_3 & = & 1 \\
-x_2 + 2x_3 - x_4 & = & 1 \\
-x_3 + 2x_4 - x_5 & = & 1 \\
-x_4 + 2x_5 - x_6 & = & 1 \\
-x_5 + 2x_6 - x_7 & = & 1 \\
-x_6 + 2x_7 - x_8 & = & 1 \\
-x_7 + 2x_8 - x_9 & = & 1 \\
-x_8 + 2x_9 - x_{10} & = & 1 \\
-x_9 + 2x_{10} & = & 1.
\end{array}$$

(a) Find matrices **A**, **x**, and **b** so that the equation $\mathbf{Ax} = \mathbf{b}$ is equivalent to this system of linear equations.

(b) Use MATLAB to compute $\mathbf{A}^{-1}$ and the solution as the product of $\mathbf{A}^{-1}$ and **b**; by entering the MATLAB instruction *flops* both immediately before and immediately after this two-step process, count the number of floating-point operations involved in finding **x** by this method.

(c) Use MATLAB to find **x** directly by entering the MATLAB instruction $x = A\backslash b$ and use the MATLAB instruction *flops* as in (b) to count the number of floating-point operations involved in finding **x** by this method.

(d) Which is more efficient, the method in (b) or the method in (c)?

28. Consider the discrete Fourier analysis described for $N = 3$ in Problem 50 of Section 1.3. Determining the coefficients c_k from the given values f_j requires the solution of $\mathbf{Zc} = \mathbf{f}$ in the notation of that problem. Let $\bar{\mathbf{Z}}$ be the usual 3×3 matrix of complex conjugates of the entries of **Z**.

(a) Show that $\mathbf{Z}\bar{\mathbf{Z}} = \bar{\mathbf{Z}}\mathbf{Z} = 3\mathbf{I}_3$ and that $\mathbf{Z}^{-1} = \frac{1}{3}\bar{\mathbf{Z}}$.

(b) For the matrix **Z** explicitly found in Problem 50(c), use (a) explicitly to find $\mathbf{Z}^{-1}$.

(c) Use $\mathbf{Z}^{-1}$ to solve $\mathbf{Zc} = \mathbf{f}$ if $\mathbf{f} = [2 \quad -2 \quad 4]^T$.

▷ **29.** Problem 50(d) in Section 1.3 stated that $\mathbf{Z}(\mathbf{x} * \mathbf{y})^T = \mathbf{Zx}^T \square \mathbf{Zy}^T$ in the notation of that problem.

(a) Use Problem 28 to show that the convolution $\mathbf{x} * \mathbf{y}$ can be computed as

$$(\mathbf{x} * \mathbf{y})^T = \tfrac{1}{3}\bar{\mathbf{Z}}(\mathbf{Zx}^T \square \mathbf{Zy}^T).$$

(b) Verify that the formula in (a) works correctly when $\mathbf{x} = [2 \quad -4 \quad 2]$ and $\mathbf{y} = [-2 \quad 6 \quad 4]$.

30. Consider the general discrete Fourier transform $\mathbf{Zc}$ of $\mathbf{c}$ as described in Problem 51 of Section 1.3. As in Problem 28 for $N = 3$, show for general N that $\mathbf{Z}\bar{\mathbf{Z}} = \bar{\mathbf{Z}}\mathbf{Z} = (N)\mathbf{I}_N$ and $\mathbf{Z}^{-1} = (1/N)\bar{\mathbf{Z}}$. You will need to use the following facts: (1) $z^N = 1$; (2) $(\bar{z})^N = 1$; (3) $1 + r + r^2 \cdots + r^{N-1} = N$ if $r = 1$; and (4) $1 + r + r^2 + \cdots r^{N-1} = (r^N - 1)/(r - 1)$ if $r \neq 1$.

31. Use Problem 52 from Section 1.3 and Problem 30 to show that the convolution $\mathbf{x} * \mathbf{y}$ of two $1 \times N$ row matrices can be computed as

$$(\mathbf{x} * \mathbf{y})^T = \frac{1}{N}\bar{\mathbf{Z}}(\mathbf{Z}\mathbf{x}^T \square \mathbf{Z}\mathbf{y}^T).$$

▷ **32.** Use Problem 28 to find in explicit form the Fourier series

$$s(x) = c_0 + c_1 e^{ix} + c_2 e^{2ix}$$

as in Problem 50 of Section 1.3, using

$$e^{it} = \cos t + i \sin t,$$

if the values $f_j = f(x_j)$ are: (a) $f_1 = 0$, $f_2 = 4$, $f_3 = 16$; (b) $f_1 = 2$, $f_2 = \sqrt{3} - 1$, $f_3 = -(\sqrt{3} + 1)$.

1.5 PARTITIONED MATRICES

In Example 1.27 in the preceding section, the equations we needed to solve to find a right-inverse split into two sets, with each set involving only the variables forming one column of the right-inverse. The same thing happened in Example 1.29. In Example 1.30, on the other hand, the variables in each set of equations formed a row of the left-inverse—rather than a column of the right-inverse.

These examples show that it is sometimes useful to think of a matrix as being broken up or *partitioned* into its rows or into its columns; we saw this earlier in the definition of the matrix product $\mathbf{AB}$, where in effect we partition $\mathbf{A}$ into its rows and $\mathbf{B}$ into its columns in order to form the product. More generally, it is sometimes useful to think of a matrix as being made up of several smaller matrices of various shapes. In this way, we can think of the matrix

$$\mathbf{A} = \begin{bmatrix} 2 & 0 & -1 \\ 1 & 3 & 2 \end{bmatrix}$$

in terms of its columns

$$\mathbf{C}_1 = \begin{bmatrix} 2 \\ 1 \end{bmatrix}, \quad \mathbf{C}_2 = \begin{bmatrix} 0 \\ 3 \end{bmatrix}, \quad \text{and} \quad \mathbf{C}_3 = \begin{bmatrix} -1 \\ 2 \end{bmatrix}$$

by writing $\mathbf{A} = [\mathbf{C}_1 \quad \mathbf{C}_2 \quad \mathbf{C}_3]$. Or we can think of $\mathbf{A}$ in terms of its rows $\mathbf{R}_1 = [2 \quad 0 \quad -1]$ and $\mathbf{R}_2 = [1 \quad 3 \quad 2]$ by writing

$$\mathbf{A} = \begin{bmatrix} \mathbf{R}_1 \\ \mathbf{R}_2 \end{bmatrix}.$$

We could even write

$$\mathbf{A} = \begin{bmatrix} \mathbf{A}_{11} & \mathbf{A}_{12} \\ \mathbf{A}_{21} & \mathbf{A}_{22} \end{bmatrix},$$

where we let $\mathbf{A}_{11} = [2 \quad 0]$, $\mathbf{A}_{12} = [-1]$, $\mathbf{A}_{21} = [1 \quad 3]$, and $\mathbf{A}_{22} = [2]$ in order to think of $\mathbf{A}$ as partitioned as follows:

$$\mathbf{A} = \left[\begin{array}{cc:c} 2 & 0 & -1 \\ \hdashline 1 & 3 & 2 \end{array}\right].$$

Note the use of dashed lines to indicate the partitioning when the entries are numbers, not matrices.

(1.39) ***Definition.*** A matrix $\mathbf{A}$ is said to be *partitioned* when dashed vertical lines the full height of the matrix are drawn between selected columns and dashed horizontal lines the full width of the matrix are drawn between selected rows. The small matrices formed from the entries contained within the rectangles formed by these lines are called the *submatrices* of the partition of $\mathbf{A}$.

Suppose that two $p \times q$ matrices $\mathbf{A}$ and $\mathbf{B}$ are partitioned in the same way by

$$\mathbf{A} = [\mathbf{A}_1 \quad \mathbf{A}_2] \quad \text{and} \quad \mathbf{B} = [\mathbf{B}_1 \quad \mathbf{B}_2],$$

where $\mathbf{A}_1$ and $\mathbf{B}_1$ are both $p \times r$ and $\mathbf{A}_2$ and $\mathbf{B}_2$ are both $p \times (q - r)$; then it is obvious that

$$\mathbf{A} + \mathbf{B} = [\mathbf{A}_1 + \mathbf{B}_1 \quad \mathbf{A}_2 + \mathbf{B}_2].$$

Similarly, suppose that a $p \times r$ matrix $\mathbf{A}$ is partitioned as

$$\mathbf{A} = \begin{bmatrix} \mathbf{A}_{11} & \mathbf{A}_{12} \\ \mathbf{A}_{21} & \mathbf{A}_{22} \end{bmatrix}$$

while an $r \times q$ matrix $\mathbf{B}$ is partitioned as

$$\mathbf{B} = \begin{bmatrix} \mathbf{B}_1 \\ \mathbf{B}_2 \end{bmatrix}.$$

Then it is clear that the product $\mathbf{AB}$ can be partitioned as

$$\mathbf{AB} = \begin{bmatrix} \mathbf{A}_{11}\mathbf{B}_1 + \mathbf{A}_{12}\mathbf{B}_2 \\ \mathbf{A}_{21}\mathbf{B}_1 + \mathbf{A}_{22}\mathbf{B}_2 \end{bmatrix}$$

as long as the submatrices are such that each of the products and sums make sense; to check this, all you need do is write out the product in both its regular and partitioned form, displaying the individual entries, and see that the results are equal.

The general rule is as follows:

(1.40) Partitioned matrices can be added, subtracted, multiplied together, and multiplied by scalars just as though the submatrices were numbers, as long as the correct order in products is preserved and the partitions are such that the needed sums, differences, and products are well defined.

(1.41) ***Example.*** Suppose that we partition **A** into its columns as

$$\mathbf{A} = [\mathbf{a}_1 \quad \mathbf{a}_2 \quad \cdots \quad \mathbf{a}_q]$$

and **x** into its rows as

$$\mathbf{x} = \begin{bmatrix} x_1 \\ x_2 \\ \vdots \\ x_q \end{bmatrix}.$$

Then $\mathbf{Ax} = x_1\mathbf{a}_1 + x_2\mathbf{a}_2 + \cdots + x_q\mathbf{a}_q$.

(1.42) ***Example.*** We are accustomed to thinking of the product **AB** as being computed by multiplying the rows of **A** times the columns of **B**. It is also possible to view **AB** in terms of the *columns* of **A** and the *rows* of **B**. Suppose that **A** is partitioned into its columns and **B** into its rows:

$$\mathbf{A} = [\mathbf{a}_1 \quad \mathbf{a}_2 \quad \cdots \quad \mathbf{a}_r], \qquad \mathbf{B} = \begin{bmatrix} \mathbf{b}_1 \\ \mathbf{b}_2 \\ \vdots \\ \mathbf{b}_r \end{bmatrix}.$$

Each $\mathbf{a}_i$ is a column matrix, while each $\mathbf{b}_i$ is a row matrix. According to (1.40) we have

$$\mathbf{AB} = \mathbf{a}_1\mathbf{b}_1 + \mathbf{a}_2\mathbf{b}_2 + \cdots + \mathbf{a}_r\mathbf{b}_r,$$

which expresses the product **AB** in terms of the columns of **A** and the rows of **B** rather than vice versa.

Partitioned matrices are useful in a variety of ways. If a physical system of interest can be split into several interconnected subsystems, the behavior of the entire system can often be described by using a large matrix partitioned in such a way that the submatrices along the main diagonal describe the separate subsystems, while the off-diagonal submatrices describe the interconnections. This can help clarify the structure of a large and complicated system.

You should now be able to solve Problems 1 to 15.

Partitioned Matrices and Inverses

We began our discussion of partitioned matrices by noticing that the equations arising in finding inverses split into sets involving separate unknowns; we return now to this matter. For example, to seek a right-inverse **R** for a $p \times q$ matrix **A**, we attempt to solve $\mathbf{AX} = \mathbf{I}$ for a matrix $\mathbf{X}$ (to serve as $\mathbf{R}$). Partition $\mathbf{X}$ into its columns as $\mathbf{X} = [\mathbf{x}_1 \quad \mathbf{x}_2 \quad \cdots \quad \mathbf{x}_p]$ and partition $\mathbf{I}$ into its columns $[\mathbf{e}_1 \quad \mathbf{e}_2 \quad \cdots \quad \mathbf{e}_p]$. According to (1.40), the product $\mathbf{AX}$ can be computed as

$$\mathbf{AX} = \mathbf{A}[\mathbf{x}_1 \quad \mathbf{x}_2 \quad \cdots \quad \mathbf{x}_p] = [\mathbf{Ax}_1 \quad \mathbf{Ax}_2 \quad \cdots \quad \mathbf{Ax}_p];$$

since each $\mathbf{Ax}_i$ is just a column matrix, this representation partitions the product $\mathbf{AX}$ into its columns. But we want $\mathbf{AX}$ to equal $\mathbf{I}$, which we have partitioned into its columns $\mathbf{e}_i$. Thus $\mathbf{AX} = \mathbf{I}$ is equivalent to

$$\mathbf{Ax}_1 = \mathbf{e}_1, \mathbf{Ax}_2 = \mathbf{e}_2, \ldots, \mathbf{Ax}_p = \mathbf{e}_p.$$

According to (1.36), each of these equations $\mathbf{Ax}_i = \mathbf{e}_i$ is simply a set of p linear equations in q unknowns (the q entries in the column matrix $\mathbf{x}_i$); note that *the* ***coefficients*** *in each set of equations are the same—only the* ***right-hand sides*** *differ.* A similar situation holds for finding left-inverses.

Before stating these results formally, we need a definition.

(1.43) ***Definition.*** The ith *unit column matrix* $\mathbf{e}_i$ *of order* p is the $p \times 1$ column matrix that is the ith column of the $p \times p$ identity matrix $\mathbf{I}_p$: for $1 \le j \le p$, $\langle \mathbf{e}_i \rangle_j = 1$ if $i = j$ and $\langle \mathbf{e}_i \rangle_j = 0$ if $i \ne j$.

We can now summarize how the problem of finding matrix inverses can be reduced to the problem of solving several sets of linear equations, each with the same coefficient matrix. Later this will turn out to be extremely important for the development of efficient computational procedures for inverting matrices in practice.

(1.44) **Key Theorem.** Let $\mathbf{A}$ be a $p \times q$ matrix.

(a) The $q \times p$ matrix $\mathbf{R}$, partitioned into columns as

$$\mathbf{R} = [\mathbf{r}_1 \quad \mathbf{r}_2 \quad \cdots \quad \mathbf{r}_p],$$

is a right-inverse of $\mathbf{A}$ if and only if the $\mathbf{r}_i$ solve the equations $\mathbf{Ar}_i = \mathbf{e}_i$ for $1 \le i \le p$.

(b) The $q \times p$ matrix $\mathbf{L}$, with its $p \times q$ transpose $\mathbf{L}^T$ partitioned into its columns as

$$\mathbf{L}^T = [\mathbf{l}_1 \quad \mathbf{l}_2 \quad \cdots \quad \mathbf{l}_q],$$

is a left-inverse of $\mathbf{A}$ if and only if the $\mathbf{l}_i$ solve the equations $\mathbf{A}^T\mathbf{l}_i = \mathbf{e}_i$ for $1 \le i \le q$.

(c) If $p = q$ so that $\mathbf{A}$ is square, then the $p \times p$ matrix $\mathbf{X}$ is a (two-sided) inverse of $\mathbf{A}$ if and only if $\mathbf{X}$ satisfies the requirements both for $\mathbf{R}$ in (a) and for $\mathbf{L}$ in (b).

PROOF

(a) This was already proved in the material before Definition 1.43.

(b) We know that $\mathbf{L}$'s being a left-inverse of $\mathbf{A}$ means that $\mathbf{LA} = \mathbf{I}_q$. But then $\mathbf{I}_q = \mathbf{I}_q^T = (\mathbf{LA})^T = \mathbf{A}^T\mathbf{L}^T$, and the condition $\mathbf{A}^T\mathbf{L}^T = \mathbf{I}_q$ means that

$\mathbf{L}^T$ is a right-inverse of $\mathbf{A}^T$. Thus we can apply the already proved result (a) to $\mathbf{A}^T$, which proves (b).

(c) We know that $\mathbf{X}$ is a two-sided inverse if and only if $\mathbf{X}$ is both a left-inverse and a right-inverse. Since (a) and (b) have been proved, $\mathbf{X}$ is a two-sided inverse if and only if both (a) holds with $\mathbf{X} = \mathbf{R}$ and (b) holds with $\mathbf{X} = \mathbf{L}$. ■

PROBLEMS 1.5

1. Find the 3×2 matrix $\mathbf{A}_1$ and the 3×1 matrix $\mathbf{A}_2$ in order that $\mathbf{A}$ be partitioned as $\mathbf{A} = [\mathbf{A}_1 \quad \mathbf{A}_2]$, where

$$\mathbf{A} = \begin{bmatrix} 1 & 2 & -3 \\ 4 & 1 & 0 \\ 2 & 6 & -1 \end{bmatrix}.$$

▷ **2.** Find the submatrices $\mathbf{B}_1$ and $\mathbf{B}_2$ into which $\mathbf{B}$ below must be partitioned in order to use (1.40) to add it to the matrix $\mathbf{A}$ from Problem 1. Then calculate $[\mathbf{A}_1 + \mathbf{B}_1 \quad \mathbf{A}_2 + \mathbf{B}_2]$ and $\mathbf{A} + \mathbf{B}$ directly and show that the sums are equal.

$$\mathbf{B} = \begin{bmatrix} -1 & 6 & 3 \\ 2 & 4 & -8 \\ 7 & 4 & 0 \end{bmatrix}$$

3. Let $\mathbf{C} = [1 \quad -2 \quad 4]$ and let $\mathbf{A}$ be the matrix in Problem 1. Calculate $[\mathbf{CA}_1 \quad \mathbf{CA}_2]$ and $\mathbf{CA}$ directly and verify that the products are the same.

4. Let

$$\mathbf{A}_1 = \begin{bmatrix} 1 \\ -2 \\ 2 \end{bmatrix}, \qquad \mathbf{A}_2 = \begin{bmatrix} 3 & 1 \\ 0 & 2 \\ -1 & 4 \end{bmatrix},$$

$$\mathbf{B}_1 = [1 \quad 4], \quad \text{and} \quad \mathbf{B}_2 = \begin{bmatrix} -1 & 2 \\ 0 & 1 \end{bmatrix}.$$

Evaluate each of the expressions below both in partitioned form and then directly and verify that the same answer results.

(a) $[\mathbf{A}_1 \quad \mathbf{A}_2]^T$ (b) $[\mathbf{A}_1 \quad \mathbf{A}_2]\begin{bmatrix} \mathbf{B}_1 \\ \mathbf{B}_2 \end{bmatrix}$ (c) $\begin{bmatrix} \mathbf{B}_2 \\ \mathbf{B}_1 \end{bmatrix}\mathbf{A}_2^T$

▷ **5.** Evaluate $\mathbf{AB}$ both in partitioned form and directly and verify that the same answer results, where

$$\mathbf{A} = \left[\begin{array}{cc|ccc} 4 & 3 & -2 & 1 & 4 \\ 2 & -5 & 6 & 3 & -1 \end{array}\right] \quad \text{and} \quad \mathbf{B} = \left[\begin{array}{cc|c} 0 & -1 & 3 \\ 2 & -1 & 6 \\ \hline 5 & 2 & 1 \\ -3 & 4 & -1 \\ 2 & -1 & 2 \end{array}\right].$$

6. Show that, if $\mathbf{B}$ is as partitioned below, then $\mathbf{B}^T$ is also as given below.

$$\mathbf{B} = \begin{bmatrix} \mathbf{B}_{11} & \mathbf{B}_{12} \\ \mathbf{B}_{21} & \mathbf{B}_{22} \end{bmatrix}, \qquad \mathbf{B}^T = \begin{bmatrix} \mathbf{B}_{11}^T & \mathbf{B}_{21}^T \\ \mathbf{B}_{12}^T & \mathbf{B}_{22}^T \end{bmatrix}$$

▷ **7.** Verify (1.41) that $\mathbf{Ax} = x_1\mathbf{a}_1 + \cdots + x_q\mathbf{a}_q$ for

$$\mathbf{A} = \begin{bmatrix} 5 & -1 \\ 1 & 3 \\ 2 & -6 \end{bmatrix} \quad \text{and} \quad \mathbf{x} = \begin{bmatrix} 2 \\ -9 \end{bmatrix}.$$

8. Prove (1.41) that $\mathbf{Ax} = x_1\mathbf{a}_1 + \cdots + x_q\mathbf{a}_q$ in the general case by checking that the definition of equal matrices is satisfied.

9. Verify (1.42) on computing $\mathbf{AB}$ in terms of the *columns* of $\mathbf{A}$ and the *rows* of $\mathbf{B}$ for

$$\mathbf{A} = \begin{bmatrix} 0 & -1 & -3 \\ 2 & 1-2i & 6 \end{bmatrix} \quad \text{and} \quad \mathbf{B} = \begin{bmatrix} 2i & -1 \\ 3 & 6 \\ -1 & 2+i \end{bmatrix}.$$

10. Prove (1.42) on computing $\mathbf{AB}$ in terms of the *columns* of $\mathbf{A}$ and the *rows* of $\mathbf{B}$ in the general case by checking that the definition of equal matrices is satisfied.

▷ **11.** Suppose that $\mathbf{A}$, $\mathbf{B}$, and $\mathbf{C}$ are square matrices.

(a) If any of the three matrices is singular, show that the partitioned matrix $\mathbf{D} = \mathbf{diag}(\mathbf{A}, \mathbf{B}, \mathbf{C})$ below is singular.

(b) If all three matrices $\mathbf{A}$, $\mathbf{B}$, and $\mathbf{C}$ are nonsingular, show that the matrix $\mathbf{D}$ is nonsingular and that its inverse is

$$\mathbf{D}^{-1} = \mathbf{diag}(\mathbf{A}^{-1}, \mathbf{B}^{-1}, \mathbf{C}^{-1}),$$

as given below.

$$\mathbf{D} = \begin{bmatrix} \mathbf{A} & \mathbf{0} & \mathbf{0} \\ \mathbf{0} & \mathbf{B} & \mathbf{0} \\ \mathbf{0} & \mathbf{0} & \mathbf{C} \end{bmatrix} \quad \text{and} \quad \mathbf{D}^{-1} = \begin{bmatrix} \mathbf{A}^{-1} & \mathbf{0} & \mathbf{0} \\ \mathbf{0} & \mathbf{B}^{-1} & \mathbf{0} \\ \mathbf{0} & \mathbf{0} & \mathbf{C}^{-1} \end{bmatrix}$$

𝔐 **12.** Let $\mathbf{A}_3$ and $\mathbf{A}_4$ be the 5×5 matrices defined in Problem 23 of Section 1.4, and let

$$\mathbf{D} = \begin{bmatrix} \mathbf{A}_3 & \mathbf{0} \\ \mathbf{0} & \mathbf{A}_4 \end{bmatrix}.$$

(a) Use MATLAB to compute $\mathbf{D}^{-1}$ by the analogue of the method of Problem 11; by entering the MATLAB command *flops* immediately before and immediately after the computation, determine how many floating-point operations were used to invert $\mathbf{D}$ by this method.

(b) Use MATLAB to compute the inverse of $\mathbf{D}$ directly, again using *flops* to measure the work necessary.

(c) Which method is more efficient?

▷ **13.** Let $\mathbf{A}$ be a $p \times p$ nonsingular matrix, $\mathbf{u}$ and $\mathbf{v}$ $p \times 1$ column matrices, and d a number. Show that the *bordered matrix* $\hat{\mathbf{A}}$ defined below is nonsingular if and only if the number δ defined by $\delta = d - \mathbf{v}^T\mathbf{A}^{-1}\mathbf{u}$ is nonzero, and that the inverse of $\hat{\mathbf{A}}$ is then as given below.

$$\hat{\mathbf{A}} = \begin{bmatrix} \mathbf{A} & \mathbf{u} \\ \mathbf{v}^T & d \end{bmatrix}, \qquad \hat{\mathbf{A}}^{-1} = \begin{bmatrix} \mathbf{B} & \mathbf{w} \\ \mathbf{z}^T & \dfrac{1}{\delta} \end{bmatrix} \quad \text{if } \delta \neq 0,$$

where $\mathbf{B} = \mathbf{A}^{-1} + (1/\delta)\mathbf{A}^{-1}\mathbf{u}\mathbf{v}^T\mathbf{A}^{-1}$, $\mathbf{w} = -(1/\delta)\mathbf{A}^{-1}\mathbf{u}$, $\mathbf{z}^T = -(1/\delta)\mathbf{v}^T\mathbf{A}^{-1}$. In addition, show that $\mathbf{A}^{-1} = \mathbf{B} - \delta\mathbf{w}\mathbf{z}^T$. (Note that this shows that, once the inverse of a given matrix $\mathbf{F}$ is known, it is simple to compute the inverse of a matrix obtained from $\mathbf{F}$ either by adjoining a row and column to $\mathbf{F}$ or by deleting a row and column from $\mathbf{F}$.)

𝔐 **14.** Consider the 7×7 symmetric matrix $\mathbf{A}$ and its inverse given in Example 1.31 of Section 1.4.

(a) Use MATLAB or similar software with the method of Problem 13 to find the inverse of the 8×8 matrix obtained by adjoining a row and column to $\mathbf{A}$ that extend the obvious pattern in the entries of $\mathbf{A}$.

(b) Use MATLAB or similar software with the method of Problem 13 to find the inverse of the 6×6 matrix obtained by deleting the seventh row and seventh column of $\mathbf{A}$.

15. Consider the *Kronecker product* $\mathbf{A} \odot \mathbf{B}$ defined in Problem 10 of Section 1.3 for all $p \times q$ matrices $\mathbf{A}$ and all $r \times s$ matrices $\mathbf{B}$.

(a) Show that an equivalent way to define the product $\mathbf{C} = \mathbf{A} \odot \mathbf{B}$ is as that $pr \times qs$ matrix partitioned into pq submatrices $\mathbf{C}_{ij}$, each $r \times s$, where $\mathbf{C}_{ij} = \langle \mathbf{A} \rangle_{ij}\mathbf{B}$ and

$$\mathbf{C} = \begin{bmatrix} \mathbf{C}_{11} & \mathbf{C}_{12} & \cdots & \mathbf{C}_{1q} \\ \mathbf{C}_{21} & \mathbf{C}_{22} & \cdots & \mathbf{C}_{2q} \\ & \cdots\cdots & & \\ \mathbf{C}_{p1} & \mathbf{C}_{p2} & \cdots & \mathbf{C}_{pq} \end{bmatrix}.$$

(b) Show that, for all matrices $\mathbf{A}$,

$$[1] \odot \mathbf{A} = \mathbf{A} \odot [1] = \mathbf{A}.$$

(c) Show that, for all matrices $\mathbf{A}$, $\mathbf{B}$, and $\mathbf{D}$,

$$\mathbf{A} \odot (\mathbf{B} + \mathbf{D}) = \mathbf{A} \odot \mathbf{B} + \mathbf{A} \odot \mathbf{D}.$$

▷ **16.** Use the approach of Theorem 1.44(c) to find the inverse of the matrix $\mathbf{A}$ below, and verify that it is both a left- and a right-inverse.

$$\mathbf{A} = \begin{bmatrix} -1 & 8 \\ 1 & -7 \end{bmatrix}$$

17. Use the approach of Theorem 1.44(c) to find the inverse of the matrix **A** below, and verify that it is both a left- and a right-inverse.

$$\mathbf{A} = \begin{bmatrix} 4 & -1 \\ 3 & -1 \end{bmatrix}$$

18. (a) Find the second column of the inverse of the matrix **A** below.
(b) Find the third row of the inverse of **A**.

$$\mathbf{A} = \begin{bmatrix} 2 & -1 & 0 \\ -1 & 2 & -1 \\ 0 & -1 & 2 \end{bmatrix}$$

▷ **19.** Find the second row of the left-inverse of

$$\mathbf{A} = \begin{bmatrix} 2 & 1 \\ 1 & 2 \\ 3 & 3 \end{bmatrix}.$$

𝔐 **20.** Let **A** be the 10×10 matrix from Problem 27(a) of Section 1.4.
(a) Use MATLAB to compute $\mathbf{A}^{-1}$ directly; by entering the MATLAB command *flops* immediately before and immediately after this operation, count the number of floating-point operations used to compute $\mathbf{A}^{-1}$ this way.
(b) Use MATLAB to compute $\mathbf{A}^{-1}$ by solving the 10 systems $\mathbf{A}\mathbf{x}_i = \mathbf{e}_i$ as in Theorem 1.44(c) with the unit column matrices $\mathbf{e}_i$; use *flops* to measure the work done in computing $\mathbf{A}^{-1}$ by this method.
(c) Which method is more efficient?

1.6 MISCELLANEOUS PROBLEMS

PROBLEMS 1.6

1. Let $k \neq 0$ be a nonzero number; show by induction that for all positive integers n:

$$\begin{bmatrix} \cos\theta & k\sin\theta \\ -\dfrac{1}{k}\sin\theta & \cos\theta \end{bmatrix}^n = \begin{bmatrix} \cos n\theta & k\sin n\theta \\ -\dfrac{1}{k}\sin n\theta & \cos n\theta \end{bmatrix}.$$

{"Show by induction" means (a) verify for $n = 1$; (b) prove that if the result is true for any given n, then it is true for $n + 1$.}

▷ **2.** (a) Find all real matrices **A** for which $\mathbf{A}^T\mathbf{A} = \mathbf{0}$.
(b) Find all matrices **B** for which $\mathbf{B}^H\mathbf{B} = \mathbf{0}$.

▷ **3.** Suppose that **K** is a square matrix with $\mathbf{K} = -\mathbf{K}^T$ and that $\mathbf{I} - \mathbf{K}$ is nonsingular; define **B** as

$$\mathbf{B} = (\mathbf{I} + \mathbf{K})(\mathbf{I} - \mathbf{K})^{-1}.$$

Show that $\mathbf{B}^T\mathbf{B} = \mathbf{B}\mathbf{B}^T = \mathbf{I}$, so that $\mathbf{B}^{-1}$ is just $\mathbf{B}^T$ (such a $\mathbf{B}$ is said to be an *orthogonal matrix*).

▷ **4.** Prove that:

(a) If $\mathbf{A}$ has a full row of zeros, then $\mathbf{A}$ has no right-inverse.

(b) If $\mathbf{A}$ has a full column of zeros, then $\mathbf{A}$ has no left-inverse.

(c) If $\mathbf{A}$ is square and has either a full row or a full column of zeros, then $\mathbf{A}$ is singular.

5. If the entries $\langle \mathbf{A} \rangle_{ij}$ of a matrix $\mathbf{A}$ are functions of a variable t, then we define the derivative $d\mathbf{A}/dt$ as the obvious matrix of derivatives: $\langle d\mathbf{A}/dt \rangle_{ij} = d\langle \mathbf{A} \rangle_{ij}/dt$.

(a) If $\mathbf{AB}$ is defined, show that

$$\frac{d(\mathbf{AB})}{dt} = \mathbf{A}\left(\frac{d\mathbf{B}}{dt}\right) + \left(\frac{d\mathbf{A}}{dt}\right)\mathbf{B}.$$

(b) If $\mathbf{A}$ is nonsingular, differentiate $\mathbf{A}\mathbf{A}^{-1} = \mathbf{I}$ so as to obtain

$$\frac{d(\mathbf{A}^{-1})}{dt} = -\mathbf{A}^{-1}\left(\frac{d\mathbf{A}}{dt}\right)\mathbf{A}^{-1}.$$

▷ **6.** For a square matrix $\mathbf{A}$, define the *trace* of $\mathbf{A}$, $\mathrm{tr}(\mathbf{A})$, as the sum of the entries on the main diagonal of $\mathbf{A}$. (a) Prove that $\mathrm{tr}(\mathbf{AB}) = \mathrm{tr}(\mathbf{BA})$. (b) Prove that $\mathrm{tr}(\mathbf{S}^{-1}\mathbf{AS}) = \mathrm{tr}(\mathbf{A})$ if $\mathbf{S}$ is nonsingular.

7. If, in partitioned form,

$$\mathbf{A} = \begin{bmatrix} \mathbf{P} & \mathbf{Q} \\ \mathbf{R} & \mathbf{S} \end{bmatrix},$$

where $\mathbf{A}$ and $\mathbf{P}$ are nonsingular, prove that

$$\mathbf{A}^{-1} = \begin{bmatrix} \mathbf{X} & -\mathbf{P}^{-1}\mathbf{QW} \\ -\mathbf{WRP}^{-1} & \mathbf{W} \end{bmatrix},$$

where

$$\mathbf{W} = (\mathbf{S} - \mathbf{RP}^{-1}\mathbf{Q})^{-1}, \qquad \mathbf{X} = \mathbf{P}^{-1} + \mathbf{P}^{-1}\mathbf{QWRP}^{-1}.$$

Similarly, if $\mathbf{A}$ and $\mathbf{S}$ are nonsingular, prove that

$$\mathbf{A}^{-1} = \begin{bmatrix} \mathbf{X} & \mathbf{XQS}^{-1} \\ -\mathbf{S}^{-1}\mathbf{RX} & \mathbf{W} \end{bmatrix},$$

where

$$\mathbf{X} = (\mathbf{P} - \mathbf{QS}^{-1}\mathbf{R})^{-1}, \qquad \mathbf{W} = \mathbf{S}^{-1} + \mathbf{S}^{-1}\mathbf{RXQS}^{-1}.$$

If $\mathbf{P}$ and $\mathbf{S}$ are both nonsingular, prove directly that the forms are equivalent. Show that this proves that

$$(\mathbf{S} - \mathbf{RP}^{-1}\mathbf{Q})^{-1} = \mathbf{S}^{-1} + \mathbf{S}^{-1}\mathbf{RXQS}^{-1},$$

a result that is useful in applications. See the final subsection of Section 3.8, where the formula is derived from a different point of view.

2

Some simple applications and questions

This chapter has two goals: (1) to show the power of matrices in organizing and clarifying complicated sets of relationships; and (2) to raise some natural questions about matrices that arise from practical applications—as motivation for some of our later studies.

Since the chapter contains no theorems or proofs, it contains no **Key** *material; rather, it presents a wide variety of matrix applications in diverse fields, at least some of which should be meaningful and of interest to most students. The matrix information necessary to understand this chapter is just the matrix algebra presented in Chapter 1.*

The sections in this chapter are used as motivation and examples in later chapters. Questions raised in Sections 2.2, 2.3, and 2.5 motivate some of the material in Chapters 7, 8, and 9. Chapter 3 bears on the questions raised in the remaining sections of this chapter, and Chapter 4 does so for Sections 2.4 and 2.6. The issues of Section 2.6 are most explicitly addressed in Chapter 8, while Chapter 11 is devoted to the concerns of Section 2.7.

2.1 *INTRODUCTION*

In many applications, the usefulness of matrices arises from their representing an array of many numbers as a single object denoted by a single symbol, allowing relationships between variables to be expressed in a concise way. We do nothing that could not be done in terms of the entries of the matrices, but matrix notation often makes the essential relationships clearer.

In Section 2.2 we consider an application of matrices to a simple model of a marketing situation in which customers switch among suppliers; Section 2.3 studies the use of matrices in a model of competition between populations. Both applications, in matrix language, essentially reduce to the study of matrix multiplication and—more precisely—of the behavior of high positive powers of a fixed matrix; questions are raised about such powers that we later develop the tools to treat.

Section 2.4 examines an application that leads to questions about how to solve systems of linear equations; the setting is the study of static equilibrium in networks, especially pin-jointed frameworks.

We then examine simple oscillatory phenomena, an important topic in many areas of science and engineering. Section 2.5 studies the oscillation of two weights attached by springs; this raises questions about matrices that depend on parameters and how we can find values of those parameters that make the matrix singular. This motivates our later study of so-called *eigenvalues*.

A topic that pervades all of applied mathematics is that of modeling: How can we build mathematical models that sufficiently accurately reflect the properties of the real world? One aspect of this problem is: Given that a model of a real-world situation is to be of a certain form depending on certain parameters, how do we determine values for those parameters? The *least-squares method* is introduced in Section 2.6 to address this issue; questions are raised about whether the linear equations that arise have solutions and—if they do—about how to find them.

In Section 2.7 we consider a simple model of production planning; this leads to the need to solve what is called a *linear program*. Such problems provide an excellent example of the power of matrix methodology.

2.2 BUSINESS COMPETITION: MARKOV CHAINS

Many mathematical models in science, engineering, and business have to account for aspects of *chance* and *randomness*; we cannot be certain what our business competitors, the stock market, a subatomic particle, and so on, will do. Such models then use the mathematical tools of *probability theory* in order to, among other things, make predictions about *average* behavior. We lack the space to treat probability theory in this book, so we examine instead a simple deterministic—rather than probabilistic—model that leads to a key feature of probabilistic models: a *Markov chain*.

A Model of Competition Among Three Dairies

We illustrate the notion of a Markov chain by considering the competition among three dairies that supply all the milk consumed in a certain town. Over time, some customers switch among suppliers for various reasons: advertising, cost, convenience, and the like. We wish to model and analyze the movement of customers among suppliers, assuming for simplicity that the same *fractions* of customers switch from any one dairy to any other during each time period (one month, say).

Suppose that when we start our model—31 December, say—the three dairies named dairies 1, 2, and 3 control the fractions x_0, y_0, and z_0 (respectively) of the market; assuming for simplicity that these dairies are the only suppliers, we see that the fractions must sum to 1: $x_0 + y_0 + z_0 = 1$.

Suppose that after one month dairy 1 has managed to keep the fraction a_{11} of its own customers and has attracted the fractions a_{12} of dairy 2's customers and a_{13} of dairy 3's. Supposing that the total number of customers remains constant at, say, N, we can translate into a formula the fact that dairy 1's number of customers after one month equals the number it kept, plus the number it acquired from dairy 2, plus that from dairy 3: letting x_1 denote the *fraction* of the market held by dairy 1 after that month, we have

$$x_1N = a_{11}(x_0N) + a_{12}(y_0N) + a_{13}(z_0N).$$

Similar equations are obtained for the numbers of customers of dairies 2 and 3 after one month, in terms of their new shares y_1 and z_1. Dividing these equations through by N produces

(2.1)
$$\begin{aligned} x_1 &= a_{11}x_0 + a_{12}y_0 + a_{13}z_0 \\ y_1 &= a_{21}x_0 + a_{22}y_0 + a_{23}z_0 \\ z_1 &= a_{31}x_0 + a_{32}y_0 + a_{33}z_0, \end{aligned}$$

where a_{ii} = fraction of dairy i's customers kept by dairy i
a_{ij} = fraction of dairy j's customers lost to dairy i $(i \neq j)$.

Matrices express this concisely as

(2.2)
$$\mathbf{x}_1 = \mathbf{A}\mathbf{x}_0,$$

where $\mathbf{x}_r = [x_r \quad y_r \quad z_r]^T$ for $r = 0$ or 1 and $\langle \mathbf{A} \rangle_{ij} = a_{ij}$.

Since the a_{ij} are *fractions* of market shares, obviously $0 \leq a_{ij} \leq 1$. Note, too, that of dairy 1's original customers, the fraction a_{11} stays with dairy 1, the fraction a_{21} moves to dairy 2, and the fraction a_{31} moves to dairy 3; since these fractions must account for all of dairy 1's original customers, we find that $a_{11} + a_{21} + a_{31} = 1$. Applying the same argument to dairies 2 and 3 produces the fact that

(2.3) $0 \leq a_{ij} \leq 1$ for all i and j, and the sum of the entries in each column of $\mathbf{A}$ is 1:

$$a_{1i} + a_{2i} + a_{3i} = 1 \qquad \text{for } i = 1, 2, 3;$$

equivalently, $0 \leq \langle \mathbf{A} \rangle_{ij} \leq 1$ for all i and j, and

$$\langle \mathbf{A} \rangle_{1i} + \langle \mathbf{A} \rangle_{2i} + \langle \mathbf{A} \rangle_{3i} = 1 \qquad \text{for } i = 1, 2, 3.$$

The matrix $\mathbf{A}$ is known as the *transition matrix* for the model; note that in much of the literature, transition matrices are defined as the transpose of ours.

For simplicity, we assume that the *fractions* a_{ij} of customers that shift suppliers remains the same each month. If we use the symbols x_r, y_r, and z_r to denote the fractions of the total pool of customers held by each dairy at the end of r months, then we have not only (2.2) for the transition after one month but also

(2.4)
$$\mathbf{x}_{r+1} = \mathbf{A}\mathbf{x}_r \qquad \text{for } r = 0, 1, 2, \ldots,$$

where $\mathbf{x}_r = [x_r \quad y_r \quad z_r]^T$.

Since $\mathbf{x}_2 = \mathbf{A}\mathbf{x}_1 = \mathbf{A}(\mathbf{A}\mathbf{x}_0) = \mathbf{A}^2\mathbf{x}_0$ and similarly for $\mathbf{x}_r$ generally, we obtain

(2.5)
$$\mathbf{x}_r = \mathbf{A}^r\mathbf{x}_0 \qquad \text{for } r = 0, 1, 2, \ldots.$$

You should now be able to solve Problems 1 and 2.

It is intuitively obvious that the shares x_r, y_r, z_r of the market must always sum to 1; we show that this in fact follows from $x_0 + y_0 + z_0 = 1$ and the properties of $\mathbf{A}$. To do so, we use the power of matrix notation: we define $\mathbf{1}_3$ to be the 3×1 column matrix $\mathbf{1}_3 = [1 \quad 1 \quad 1]^T$, so that

$$\mathbf{1}_3^T\mathbf{x}_r = [1 \quad 1 \quad 1][x_r \quad y_r \quad z_r]^T = [x_r + y_r + z_r].$$

Thus our problem is to show $\mathbf{1}_3^T\mathbf{x}_r = [1]$ for all r; we proceed to do so. Since the sum of the entries in each column of $\mathbf{A}$ is 1, it is clear that

$$\mathbf{1}_3^T\mathbf{A} = [1 \quad 1 \quad 1] = \mathbf{1}_3^T.$$

Since $[1] = \mathbf{1}_3^T\mathbf{x}_0$, and since $\mathbf{1}_3^T\mathbf{A} = \mathbf{1}_3^T$, we can reason that

$$[1] = \mathbf{1}_3^T\mathbf{x}_0 = (\mathbf{1}_3^T\mathbf{A})\mathbf{x}_0 = \mathbf{1}_3^T(\mathbf{A}\mathbf{x}_0) = \mathbf{1}_3^T\mathbf{x}_1$$

since $\mathbf{A}\mathbf{x}_0 = \mathbf{x}_1$; looking at the far left and far right of these equalities, we see that $[1] = \mathbf{1}_3^T\mathbf{x}_1$. Continuing in this fashion—a precise argument must use induction—we conclude as desired that $[1] = \mathbf{1}_3^T\mathbf{x}_r$ for all r.

(2.6) ***Example.*** We consider a concrete numerical example to illustrate the ideas discussed above. Suppose that the initial market shares $\mathbf{x}_0$ and the transition matrix $\mathbf{A}$ are

$$\mathbf{x}_0 = \begin{bmatrix} 0.2 \\ 0.3 \\ 0.5 \end{bmatrix} \quad \text{and} \quad \mathbf{A} = \begin{bmatrix} 0.8 & 0.2 & 0.1 \\ 0.1 & 0.7 & 0.3 \\ 0.1 & 0.1 & 0.6 \end{bmatrix}.$$

Note that (2.3) holds. We easily—especially if we use MATLAB or similar software—obtain $\mathbf{x}_1 = \mathbf{A}\mathbf{x}_0$, $\mathbf{x}_2 = \mathbf{A}\mathbf{x}_1$, and so on, by matrix multiplication;

rounded to three decimals, some representative results are

$$\mathbf{x}_1 = \begin{bmatrix} 0.27 \\ 0.38 \\ 0.35 \end{bmatrix}, \qquad \mathbf{x}_3 = \begin{bmatrix} 0.327 \\ 0.398 \\ 0.275 \end{bmatrix}, \qquad \mathbf{x}_8 = \begin{bmatrix} 0.442 \\ 0.357 \\ 0.201 \end{bmatrix}, \qquad \mathbf{x}_{16} = \begin{bmatrix} 0.450 \\ 0.350 \\ 0.200 \end{bmatrix}.$$

On forming $\mathbf{x}_{17} = \mathbf{A}\mathbf{x}_{16}$, we find that $\mathbf{x}_{17} = \mathbf{x}_{16}$, which means that in fact all $\mathbf{x}_r$ for $r \geq 16$ are the same, namely $\mathbf{x}_{16}$, as given above. In other words, the market shares after 16 months become constant (at least to three decimals) at 45%, 35%, and 20%; the market is *in equilibrium.*

Limit Behavior and Equilibria

We want to explore whether the equilibrium reached in Example 2.6 occurs more generally than in that special case. We explore whether the column matrices $\mathbf{x}_r$ might possibly tend to some limit column matrix $\mathbf{x}_\infty$ as r grows. Since $\mathbf{x}_{r+1} = \mathbf{A}\mathbf{x}_r$, if $\mathbf{x}_k$ tends to $\mathbf{x}_\infty$, it follows that $\mathbf{x}_\infty = \mathbf{A}\mathbf{x}_\infty$; similarly, since we have $\mathbf{1}_3^T\mathbf{x}_r = [1]$ for all r, it follows that $\mathbf{1}_3^T\mathbf{x}_\infty = [1]$ also. In summary, we have the following:

(2.7) If the $\mathbf{x}_r$ tend to a limit $\mathbf{x}_\infty = [x_\infty \quad y_\infty \quad z_\infty]^T$, then

$$\mathbf{A}\mathbf{x}_\infty = \mathbf{x}_\infty \quad \text{and} \quad x_\infty + y_\infty + z_\infty = 1.$$

Since $\mathbf{A}$ is 3×3, the equation $\mathbf{A}\mathbf{x}_\infty = \mathbf{x}_\infty$ just represents a system of three equations in three unknowns; the final condition in (2.7) combines with this to give us a system of four linear equations in the three unknowns x_∞, y_∞, and z_∞. That is, *if* the $\mathbf{x}_r$ tend to a limit, the possible candidates for the limit can be found by solving a system of four equations in three unknowns.

(2.8) ***Example.*** To be concrete, consider again the model in Example 2.6. *If* the $\mathbf{x}_r$ tend to some $\mathbf{x}_\infty$—as they certainly appear to do—then $\mathbf{x}_\infty$ satisfies (2.7). Writing out those four equations in the three unknowns x_∞, y_∞, z_∞, we produce

$$\begin{aligned} 0.8x_\infty + 0.2y_\infty + 0.1z_\infty &= x_\infty \\ 0.1x_\infty + 0.7y_\infty + 0.3z_\infty &= y_\infty \\ 0.1x_\infty + 0.1y_\infty + 0.6z_\infty &= z_\infty \\ x_\infty + \quad y_\infty + \quad z_\infty &= 1. \end{aligned}$$

With some effort we can use high school algebra to solve this system of linear equations, discovering that it has exactly one solution, namely

$$x_\infty = 0.45, \qquad y_\infty = 0.35, \qquad z_\infty = 0.20,$$

which are precisely what we experimentally found the limit shares to be in Example 2.6. We still have not *proved* that the $\mathbf{x}_r$ actually do tend to some limit, but we *have* proved that the only *possible* limit is $[0.45 \quad 0.35 \quad 0.20]^T$—which in no way depends on the initial shares of the market!

You should now be able to solve Problems 1 to 18.

Returning to the general case, we recall that any possible limits of the $\mathbf{x}_r$ must satisfy (2.7). For the general case, we would like to know:

1. How can we determine whether—as in Example 2.8—the equations in (2.7) have any solutions?
2. How can we obtain these solutions?

Even if we can answer these questions, we still do not know whether the $\mathbf{x}_r$ actually tend to a limit. We still wonder:

3. How can we tell whether the $\mathbf{x}_r$ tend to a limit?

At the moment we lack the tools to answer these questions; Chapters 3 and 9 will develop the tools we need.

Limit Behavior of Matrix Powers

There is another approach to understanding the behavior of the $\mathbf{x}_r$ in general. Recall that $\mathbf{x}_r = \mathbf{A}^r\mathbf{x}_0$; if we can understand how $\mathbf{A}^r$ behaves for large r, then we should be able to deduce how $\mathbf{x}_r$ behaves for large r.

(2.9) ***Example.*** Consider the matrix $\mathbf{A}$ in our concrete model in Example 2.6. Calculation rounded to three places shows that

$$\mathbf{A}^2 = \begin{bmatrix} 0.67 & 0.31 & 0.20 \\ 0.18 & 0.54 & 0.40 \\ 0.15 & 0.15 & 0.40 \end{bmatrix}, \qquad \mathbf{A}^4 = \begin{bmatrix} 0.536 & 0.405 & 0.338 \\ 0.278 & 0.407 & 0.412 \\ 0.188 & 0.188 & 0.250 \end{bmatrix},$$

$$\mathbf{A}^8 = \begin{bmatrix} 0.462 & 0.445 & 0.432 \\ 0.339 & 0.356 & 0.365 \\ 0.199 & 0.199 & 0.203 \end{bmatrix}, \qquad \mathbf{A}^{16} = \begin{bmatrix} 0.450 & 0.450 & 0.450 \\ 0.350 & 0.350 & 0.350 \\ 0.200 & 0.200 & 0.200 \end{bmatrix},$$

and all higher powers of $\mathbf{A}$ equal $\mathbf{A}^{16}$ to three decimal places. In this particular case, it appears that $\mathbf{A}^r$ tends to a limit matrix $\mathbf{A}_\infty$.

Example 2.9 shows that we might succeed in studying $\mathbf{x}_r$ by studying $\mathbf{A}^r$ for large r. We add a fourth question to the three earlier ones we wanted to answer:

4. How can we tell whether the powers $\mathbf{A}^r$ tend to a limit $\mathbf{A}_\infty$, and how can we find the limit if there is one?

We will later learn the answer to this question as well as the previous three.

Markov Chains

The general model that we have been considering is a simple example of a (first-order) Markov chain. More generally, suppose that some system we wish to study —such as the dairy competition—is described at any point in time by the values of p variables—such as the $p = 3$ market shares. If we write these values as the entries in a $p \times 1$ matrix $\mathbf{x}_r$ at the rth moment in time, then these *state vectors* $\mathbf{x}_r$ describe the system. If $\mathbf{A}$ is then a $p \times p$ matrix, the equation $\mathbf{x}_{r+1} = \mathbf{A}\mathbf{x}_r$ describes how the state of the system evolves over time. One usually wants to learn how $\mathbf{x}_r$ or $\mathbf{A}^r$ behave for large r; under special assumptions on $\mathbf{A}$—such as (2.3) plus additional conditions—it is often possible to do so.

PROBLEMS 2.2

1. Show that the matrix equation (2.2) expresses the relationships written out in (2.1).

2. Use induction to prove that (2.5)'s $\mathbf{x}_r = \mathbf{A}^r\mathbf{x}_0$ holds, given (2.4).

3. Suppose that—rather than as given in Example 2.6—the initial shares $\mathbf{x}_0$ in Example 2.6 are $\mathbf{x}_0 = [0.5 \quad 0.5 \quad 0]^T$. Find what the market shares will be:
(a) After one month
(b) After two months

▷ **4.** Three companies, 1, 2, and 3, simultaneously introduce new brands of toothpaste on the market. At the start, their initial shares of the available market are 40%, 20%, and 40%, respectively. During the first year: company 1 kept 85% of its own customers, got 15% of company 2's, and got 5% of company 3's; company 2 got 5% of company 1's customers, kept 75% of its own, and got 5% of company 3's; and company 3 got 10% of company 1's customers, got 10% of company 2's, and kept 90% of its own. Assuming that the total market shared by these three firms does not change and that the same fractions shift among firms every year:
(a) Write down in the form (2.1) how the market shares shift from one year to the next.
(b) Find the transition matrix $\mathbf{A}$.

5. For the situation described in Problem 4, find what the market shares are:
(a) After one year
(b) After two years

▷ **6.** Suppose that each person's occupation can be classified as either professional, skilled, or unskilled. Assume that it is always true that: of the children of professionals, 70% are professional, 20% are skilled, and 10% are unskilled; of the children of the skilled, 20% are professional, 60% are skilled, and 20% are unskilled; and, of the children of the unskilled, 20% are professional, 30% are skilled, and 50% are unskilled. Assume that the total number of people with occupations is the same each generation, and that in the present generation 35% are professional, 35% are skilled, and 30% are unskilled.

(a) Write down in the form (2.1) how the percentage of each type evolves from year to year.
(b) Find the transition matrix **A**.

7. For the situation described in Problem 6, find the distribution of jobs:
(a) After one generation
(b) After two generations

8. For the general matrix **A** of (2.2), show that $\mathbf{1}_3^T\mathbf{A} = \mathbf{1}_3^T$.

9. Prove by induction that, for the model (2.3-4), $\mathbf{1}_3^T\mathbf{x}_r = [1]$ for all $r \geq 1$.

▷ **10.** By solving the equations in (2.7), find the equilibrium state—if there is one—of the market for the situation described in Problem 4.

11. By solving the equations in (2.7), find the equilibrium distribution—if there is one—among employee types for the situation in Problem 6.

𝔐 **12.** Use MATLAB or similar software to verify the behavior of $\mathbf{x}_r$ claimed in Example 2.6.

𝔐 ▷ **13.** Use MATLAB or similar software to check experimentally whether the market shares in Problem 4 converge to a limit set of shares as time goes on.

𝔐 **14.** Use MATLAB or similar software to check experimentally whether the employment distribution in Problem 6 converges to a limit distribution as time goes on.

𝔐 **15.** Use MATLAB or similar software to check experimentally whether the market shares in Problem 3 converge to a limit set of shares as time goes on.

16. Suppose that the matrix **A** below is the transition matrix for a problem of the type considered in this section.
(a) Show that it satisfies (2.3).
(b) By solving the equations in (2.7), show that there are infinitely many equilibrium positions $\mathbf{x}_\infty = [\alpha \quad 1 - 2\alpha \quad \alpha]^T$ for arbitrary α with $0 \leq \alpha \leq 0.5$.

$$\mathbf{A} = \begin{bmatrix} 0 & 0 & 1 \\ 0 & 1 & 0 \\ 1 & 0 & 0 \end{bmatrix}.$$

▷ **17.** Show for the situation in Problem 16 that, given $\mathbf{x}_0$, the sequence of states $\mathbf{x}_r$ does *not* tend to anything unless $\mathbf{x}_0$ is itself an equilibrium position.

𝔐 **18.** Use MATLAB or similar software to verify that the powers $\mathbf{A}^r$ behave as claimed in Example 2.9.

𝔐 **19.** For the matrix **A** in Problem 4, use MATLAB or similar software to determine how $\mathbf{A}^r$ behaves for large r.

𝔐 **20.** For the matrix **A** in Problem 6, use MATLAB or similar software to determine how $\mathbf{A}^r$ behaves for large r.

▷ **21.** For the matrix **A** in Problem 16, show by hand that the powers $\mathbf{A}^r$ do not converge to any matrix $\mathbf{A}_\infty$.

2.3 POPULATION GROWTH: POWERS OF A MATRIX

The preceding section considered a model of the evolution of a system; in that case we found that $\mathbf{x}_{r+1} = \mathbf{A}\mathbf{x}_r$ modeled the behavior of the system and that the matrix $\mathbf{A}$ satisfied some very special conditions (2.3) common to Markov chains. In many applications we can model the evolution of a system by $\mathbf{x}_{r+1} = \mathbf{A}\mathbf{x}_r$ but find that the special conditions (2.3) are not satisfied; such systems can exhibit very different behavior from that found in the preceding section. We consider a simple example to illustrate this.

Modeling Growth of a Population

In this section we create and study a simple model of the growth of populations over time. Imagine that we count the population at certain discrete points in time—such as every year, or every month, or every second—and that we let p_i denote the number of individuals in the population at the ith point in time. We are not concerned here with the *nature* of these individuals—people, chickens, bacteria, or whatever—just with their *number*. Our model supposes that the birth rate b and the death rate d are independent of time and that we know these numbers. Thus the number born between the ith and $(i + 1)$st times is just bp_i, while the number that die is dp_i. This gives $p_{i+1} - p_i = bp_i - dp_i$, that is,

(2.10) $$p_{i+1} = (1 + b - d)p_i \qquad \text{for } i = 1, 2, \ldots .$$

Since then

$$p_{i+1} = (1 + b - d)p_i = (1 + b - d)\{(1 + b - d)p_{i-1}\} = (1 + b - d)^2 p_{i-1},$$

and so on, we can conclude that

(2.11) $$p_{i+1} = (1 + b - d)^i p_1 \qquad \text{for } i = 0, 1, 2, \ldots ,$$

where p_1 is the population at the start of our consideration. From (2.11) we easily analyze the behavior of the population over time: (a) if the birth rate exceeds the death rate, then $1 + b - d > 1$ and p_i tends to infinity with i since $(1 + b - d)^i$ does so; (b) if the death rate exceeds the birth rate, then $1 + b - d < 1$ and p_i tends to zero since $(1 + b - d)^i$ does so.

(2.12) ***Example.*** Suppose that the birth rate $b = 0.2$ while the death rate $d = 0.1$; then $1 + b - d = 1.1$ and the population should explode. If the initial population is $p_1 = 10{,}000$, we find that (approximately) $p_2 = 11{,}000$, $p_3 = 12{,}100$, $p_4 = 13{,}331$, $p_5 = 14{,}641$, $p_{10} = 21{,}435$, $p_{60} = 970{,}137$, $p_{100} = 20{,}483{,}147$, and so on. On the other hand, if the birth rate $b = 0.1$ while the death rate $d = 0.2$, then $1 + b - d = 0.9$ and the population should die out; starting with $p_1 = 10{,}000$ produces (approximately) $p_2 = 9000$, $p_3 = 8100$, $p_4 = 7290$, $p_5 = 6561$, $p_{10} = 4304$, $p_{60} = 63$, $p_{100} = 2$, and so on.

Note that our model is just a model—we need not get precise quantitative results from it. For example, if $1 + b - d = 0.9$ and $p_1 = 10$, we quickly find that our model predicts a population of 8.1 individuals at the third time instant. The numbers are not perfect (real population counts are integers), but the qualitative behavior of decreasing population is correct.

You should now be able to solve Problems 1 to 5.

Competing Populations

The preceding simple model was just for background before considering the more complex situation in which two populations compete with one another. We denote the numbers in these populations by F_i and C_i, which you may view as counting *f*oxes and *c*hickens. Assume that the chickens, without any raiding foxes to harass them, would have a birth rate exceeding the death rate; to be specific, imagine that we would have $C_{i+1} = 1.2C_i$ in this situation. Without chickens to feed upon, we would expect the foxes to die out—$F_{i+1} = 0.6F_i$, say.

We want to model what happens when the foxes succeed in eating some number of chickens each time period, and we assume that this would allow an increase in the fox population proportional to the number of chickens killed. To be specific, we assume that $F_{i+1} = 0.6F_i + 0.5C_i$. The chicken population would, of course, decrease because of the raiding foxes, so we take $C_{i+1} = 1.2C_i - kF_i$, where k reflects the kill rate of chickens by foxes; we leave k as a variable in order to study the effect of different kill rates. Supposing that there are initially 1000 chickens and 100 foxes, we obtain as our model:

(2.13)
$$\begin{aligned} F_{i+1} &= 0.6F_i + 0.5C_i \quad \text{and} \\ C_{i+1} &= -kF_i + 1.2C_i \qquad \text{for } i = 1, 2, \ldots \end{aligned}$$

with $F_1 = 100$ and $C_1 = 1000$.

We use matrices to analyze the behavior of these populations as time passes. If we let

$$\mathbf{A} = \begin{bmatrix} 0.6 & 0.5 \\ -k & 1.2 \end{bmatrix}, \qquad \mathbf{x}_i = \begin{bmatrix} F_i \\ C_i \end{bmatrix}, \quad \text{so that} \quad \mathbf{x}_1 = \begin{bmatrix} 100 \\ 1000 \end{bmatrix},$$

our model (2.13) becomes

(2.14)
$$\mathbf{x}_{i+1} = \mathbf{A}\mathbf{x}_i \qquad \text{for } i = 1, 2, \ldots, \quad \text{with } \mathbf{x}_1 = [100 \quad 1000]^T.$$

From (2.14) follows $\mathbf{x}_{i+1} = \mathbf{A}\mathbf{x}_i = \mathbf{A}(\mathbf{A}\mathbf{x}_{i-1}) = \mathbf{A}^2\mathbf{x}_{i-1}$, and so on, so that, in fact,

(2.15)
$$\mathbf{x}_{i+1} = \mathbf{A}^i\mathbf{x}_1 \qquad \text{for } i = 0, 1, 2, \ldots.$$

According to (2.15), studying the behavior of $\mathbf{x}_i$ as i increases is essentially equivalent to studying the behavior of the powers $\mathbf{A}^i$ of $\mathbf{A}$, much as in Section 2.2. In the present case, however, we do *not* have the additional conditions (2.3) of Section 2.2 that were responsible for the simple behavior of $\mathbf{A}^i$ there. In fact, $\mathbf{A}^i$ can behave quite differently for different values of k.

(2.16) ***Example.*** Consider our model with $k = 0.1$, so that

$$\mathbf{A} = \begin{bmatrix} 0.6 & 0.5 \\ -0.1 & 1.2 \end{bmatrix}.$$

The table below shows approximately the numbers of foxes and chickens at the ith point in time for several values of i. The model experimentally verifies what we would expect: For a low kill rate k, the chicken population grows without bound and this allows the fox population also to grow without bound. The table indicates that the two populations eventually become equal.

i	1	2	3	4	5	6	8	12	16	20	30	100
F_i	100	560	931	1244	1523	1783	2292	3470	5107	7483	19,409	15,328,199
C_i	1000	1190	1372	1553	1739	1934	2367	3488	5111	7483	19,409	15,328,199

(2.17) ***Example.*** We consider a significantly higher kill rate $k = 0.18$; we will see that this causes the chicken population to be killed off, which in turn leads to the death of the foxes—both populations die out. With $k = 0.18$, we have

$$\mathbf{A} = \begin{bmatrix} 0.6 & 0.5 \\ -0.18 & 1.2 \end{bmatrix}.$$

The table below shows approximately the numbers of foxes and chickens at the ith point in time for several values of i. The model experimentally verifies what we expected: Both populations die out.

i	1	2	3	4	5	8	12	16	20	30	40	60	80	100
F_i	100	560	927	1214	1434	1808	1854	1654	1371	713	312	43	3	0
C_i	1000	1182	1317	1413	1477	1530	1400	1177	940	459	193	25	2	0

These two examples show that our model can produce radically different types of behavior for different values of k. We would like to know:

1. How can we determine the behavior of $\mathbf{x}_i$ as i grows for any specific value of k?

Since (2.15) tells us that the behavior of $\mathbf{x}_i$ is determined by that of the powers $\mathbf{A}^i$, we equally would like to know:

2. How can we determine the behavior of the powers $\mathbf{A}^i$ as i grows for any specific value of k?

At the moment we lack the tools to answer these questions; Chapters 7 and 9 will develop the tools we need.

PROBLEMS 2.3

1. Suppose that p_i is the population of a society at the ith instant, and suppose as in this section that the number of deaths over the next time period is dp_i—proportional to the size of the population. Suppose, however, that the society is "promiscuous" in that the number of births is proportional to the total number of possible pairings between two members of the society. Show that such an assumption leads to a model of the form

$$p_{i+1} = p_i - dp_i + b\left\{\frac{p_i(p_i - 1)}{2}\right\}.$$

2. Suppose in Problem 1 that $d = 0.3$, $b = 0.0008$, and $p_1 = 1000$. Determine experimentally what happens to the population.

▷ **3.** Somewhat as in Problem 1's "promiscuous society," find an interpretation for the model

$$p_{i+1} = p_i + bp_i - d\left\{\frac{p_i(p_i - 1)}{2}\right\}.$$

4. Suppose in Problem 3 that $b = 0.3$, $d = 0.0008$, and $p_1 = 1000$. Determine experimentally what happens to the population.

▷ **5.** Suppose that we are studying fungi of two types, A and B, and that their birth and death rates are, respectively, $b_A = 0.2$, $d_A = 0.1$, $b_B = 0.3$, and $d_B = 0.15$, and that the initial populations are $p_{A,1} = 1000$ and $p_{B,1} = 10$.

(a) Show that the population of type B fungi will eventually exceed that of type A.

(b) After approximately how many time periods will this first occur?

6. Use induction to prove (2.15): $\mathbf{x}_{i+1} = \mathbf{A}^i\mathbf{x}_1$.

7. Suppose in Example 2.16 that we change the initial populations to $C_1 = 100$ and $F_1 = 1000$. Find each population for $i = 1, 2, 3, 4$.

8. Explain whether the model (2.13) makes sense:

(a) If $C_1 = 0$

(b) If $F_1 = 0$

𝔐 ▷ **9.** Suppose that we take $k = 0.16$ in (2.13). Use MATLAB or similar software to determine what happens to F_i and C_i as i tends to infinity.

𝔐 **10.** Suppose that we take $k = 0.16$ in (2.13) as in Problem 9. Use MATLAB or similar software to determine what happens to $\mathbf{A}^i$ as i tends to infinity.

11. By examining (2.13), show that if the fox population dies out as i tends to infinity, then so does the chicken population.

𝔐 ▷ **12.** Use MATLAB or similar software to experiment with the model (2.13) in order to determine approximately the value of k at which the qualitative behavior switches from unbounded growth to convergence to zero.

2.4 *EQUILIBRIUM IN NETWORKS: LINEAR EQUATIONS*

The general notion *network* arises in a wide variety of areas: electrical circuits in electrical engineering; distribution systems in business or economics; pin-jointed frameworks in civil engineering; and so on. We consider a framework example to illustrate the usefulness of matrix notation in such problems and to see what questions about matrices arise naturally from networks.

A Planar Pin-Jointed Framework

A *pin-jointed framework* consists of inflexible members connected together in such a way that they are free to rotate at the joints somewhat like a tripod, say. For simplicity, we restrict ourselves to *planar* frameworks: All the members lie in one plane.

We consider a simple planar pin-jointed framework with five members, all of which are joined at the framework's single joint. The end opposite the one at the joint for each member is fastened to a vertical wall, and these ends all lie in a vertical straight line on the wall. (You can visualize this by considering the framework as a wall bracket from which you hang potted plants.) Figure 2.18 shows this framework.

(2.18)

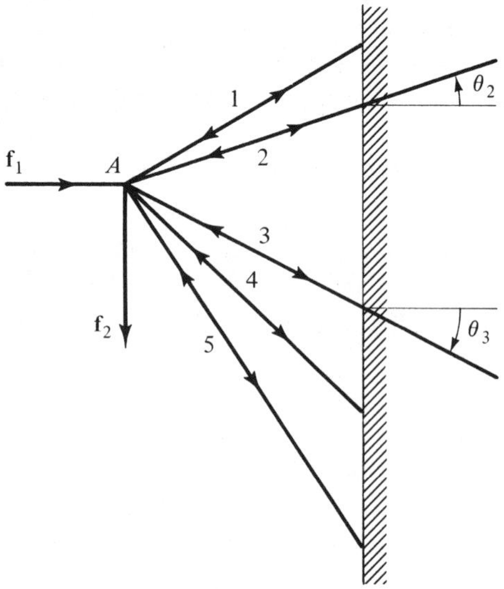

The double-arrowed lines numbered 1 through 5 in the figure represent the five members pin-jointed at the point indicated as A. The arrows labeled f_1 and f_2 represent forces applied at the joint A (from a hanging plant, for example). θ_2 and θ_3 measure the angles that members 2 and 3 make with the horizontal where they meet the vertical wall; θ_i measures the angle for member i, and we note that in the figure θ_1 and θ_2 are positive, while θ_3, θ_4, and θ_5 are negative. We want to be able to answer the question: When the forces f_1 and f_2 are applied at joint A, to what position will that joint move?

Let d_1 be the amount the joint moves in the direction of force f_1, and let d_2 be the amount in the direction of f_2. Our problem, then, is to determine $\mathbf{d}$ (defined as $[d_1 \quad d_2]^T$) from $\mathbf{f}$ (defined as $[f_1 \quad f_2]^T$). We assume that the weights of the members are negligible and that their lengths and the framework configuration are such that there are no forces on the members when f_1 and f_2 are zero. We also assume that the changes in the angles θ_i caused by the shifting of A under the forces $\mathbf{f}$ are negligible.

Mathematically, the model consists of three parts: (1) the forces $\mathbf{f}$ create compressive (or extensive) forces within the members, and these can be related by formulas; (2) the forces within each member cause extensions (or compressions) within that member, and these can be related by formulas; and (3) the extensions (or compressions) within each member cause movements $\mathbf{d}$ in the joint, and these can be related by formulas. For the first part of the model, we denote by t_i the extension force (tension) within the ith member, so that t_i is negative if the member is compressed. Each force t_i is in the direction along its member, and this can be resolved by trigonometry into the sum of a horizontal force and a vertical force; the sum of the horizontal forces for all the members balances the horizontal force f_1, and the sum of the vertical forces balances f_2. In equations, this says that

(2.19)
$$\begin{aligned} -t_1 \cos\theta_1 - t_2 \cos\theta_2 - t_3 \cos\theta_3 - t_4 \cos\theta_4 - t_5 \cos\theta_5 &= f_1 \\ t_1 \sin\theta_1 + t_2 \sin\theta_2 + t_3 \sin\theta_3 + t_4 \sin\theta_4 + t_5 \sin\theta_5 &= f_2. \end{aligned}$$

We already denote $[f_1 \quad f_2]^T$ by $\mathbf{f}$; let us denote $[t_1 \quad t_2 \quad t_3 \quad t_4 \quad t_5]^T$ by $\mathbf{t}$ and

$$\begin{bmatrix} -\cos\theta_1 & -\cos\theta_2 & -\cos\theta_3 & -\cos\theta_4 & -\cos\theta_5 \\ \sin\theta_1 & \sin\theta_2 & \sin\theta_3 & \sin\theta_4 & \sin\theta_5 \end{bmatrix} \text{ by } \mathbf{A}.$$

Then (2.19) can be expressed compactly as

(2.20)
$$\mathbf{At} = \mathbf{f}.$$

Now for the second part of the model: relating the internal forces $\mathbf{t}$ to the extensions or compressions they cause. We assume that the forces and extensions are small enough that Hooke's law applies: The extension e resulting from a force t is linearly proportional to t, with the constant k of proportionality called the *flexibility*: $e = kt$. We allow each member to have a different flexibility k_i, so that its extension $e_i = k_i t_i$. If we use $\mathbf{e}$ to denote

$$[e_1 \quad e_2 \quad e_3 \quad e_4 \quad e_5]^T$$

and **K** to denote the 5 × 5 diagonal matrix with $\langle \mathbf{K} \rangle_{ii} = k_i$, then the second part of our model ($e_i = k_i t_i$ for $i = 1, 2, 3, 4, 5$) can be expressed as

(2.21) $$\mathbf{e} = \mathbf{Kt}.$$

Only the third part of the model remains: relating the extensions **e** in the members to the displacement **d** they cause in the joint. The horizontal displacement d_1 and the vertical displacement d_2 in the joint must be equivalent to the displacement e_i at an angle θ_i for each i; interpreting this to first order in d_j yields $e_i = -d_1 \cos\theta_i + d_2 \sin\theta_i$ for $1 \le i \le 5$. If you write this as $\mathbf{e} = \mathbf{Bd}$ for a 5 × 2 matrix **B**, you will discover that **B** is just $\mathbf{A}^T$, where **A** is the same matrix that appears in (2.20). So the final part of the model is expressed as

(2.22) $$\mathbf{e} = \mathbf{A}^T\mathbf{d}.$$

Pulling together the three portions (2.20–2.22) of the model, we have

(2.23) $$\begin{aligned} \mathbf{At} &= \mathbf{f} \\ \mathbf{e} &= \mathbf{Kt} \\ \mathbf{e} &= \mathbf{A}^T\mathbf{d}. \end{aligned}$$

Recall that our objective was to find the displacement **d** of the joint in terms of the applied forces **f**. From (2.23) we see that

$$\mathbf{f} = \mathbf{At} = \mathbf{A}(\mathbf{K}^{-1}\mathbf{e}) = (\mathbf{AK}^{-1})\mathbf{e} = (\mathbf{AK}^{-1})(\mathbf{A}^T\mathbf{d}) = (\mathbf{AK}^{-1}\mathbf{A}^T)\mathbf{d};$$

here we have assumed that the flexibilities k_i are nonzero, so that the diagonal matrix **K** is nonsingular (see Problem 12 in Section 1.4). Recall that **A** is 2 × 5, **K** (and hence $\mathbf{K}^{-1}$) is 5 × 5, and $\mathbf{A}^T$ is 5 × 2; therefore, the product $\mathbf{AK}^{-1}\mathbf{A}^T$ is 2 × 2. So our model reduces to

(2.24) $$\mathbf{Hd} = \mathbf{f}, \qquad \text{where } \mathbf{H} = \mathbf{AK}^{-1}\mathbf{A}^T \text{ is } 2 \times 2,$$

which you should recognize as a system of two linear equations in two unknowns d_1 and d_2 that we want to solve in terms of the given forces in **f**.

(2.25) ***Example.*** To clarify (2.24), consider a specific example. Suppose that the angles θ_i of the members are: $\theta_1 = 30°$, $\theta_2 = 20°$, $\theta_3 = -30°$, $\theta_4 = -45°$, $\theta_5 = -60°$ (approximately as in Figure 2.18). Suppose also that all members are made from the same material and have the same flexibility $k_i = 0.01$. A little calculation then shows that in this case the matrix **H** of (2.24) is

$$\mathbf{H} = \mathbf{AK}^{-1}\mathbf{A}^T = \begin{bmatrix} 313.336 & 61.137 \\ 61.137 & 186.677 \end{bmatrix}.$$

According to Problem 17 of Section 1.4, this matrix **H** is nonsingular since

$$(313.336)(186.677) - (61.137)(61.137) \neq 0$$

(it equals 54,754.892). But when **H** is nonsingular so that $\mathbf{H}^{-1}$ exists, we can solve $\mathbf{Hd} = \mathbf{f}$ uniquely for **d** in terms of **f** no matter what the forces **f** are {see Theorem 1.38(c) in Section 1.4}. In this specific example, then, the displacements **d** are uniquely determined by the forces **f** (as intuition suggests) and we can calculate **d** easily from **f**.

In Example 2.25, the important matrix **H** in (2.24) turned out to be nonsingular. Does this always happen? That is, we want to be able to answer:

1. Is $\mathbf{AK}^{-1}\mathbf{A}^T$ always nonsingular?
2. If not, what are the conditions on the framework that make the matrix nonsingular?

At the moment we lack the tools to answer these questions; Chapters 3 and 4 will develop the tools we need.

You should now be able to solve Problems 1 to 5.

Networks More Generally

Although we treated only planar pin-jointed frameworks, the same ideas apply more generally. We can, for example, treat nonplanar frameworks, frameworks with several joints, frameworks with rigid rather than pin-jointed connections, and the like. In all cases we end up with three sets of equations analogous to (2.20–2.22) that can be combined into the fundamental relationship (2.24).

The same approach is useful for networks other than frameworks, as we mentioned at the start of this section; in other networks, concepts from the relevant disciplines replace the concepts of internal displacement **e**, internal force **t**, joint displacement **d**, and applied force **f**. In modeling the flow of goods between producers and consumers, economics concepts of course appear, while in modeling the flow of electricity we see concepts of electronics. These correspond to **e**, **t**, **d**, and **f** as follows:

Frameworks Concept	Economics Concept	Electronics Concept
Strain, **e**	Price difference	Voltage difference
Stress, **t**	Flow in branch	Current in branch
Displacement, **d**	Price	Voltage
Force, **f**	Flow at node	Current at node

In all these—and other—cases of networks, the final result is the fundamental relationship analogous to (2.24) that shows how the behavior (**d**) of the network is determined from the external actions (**f**) upon it. So, from a computational viewpoint, the main tasks in analyzing networks are the generation of the matrix **A**, the generation of the matrix **H**, and the solution of the system of equations defined by **H** (or the determination that **H** is singular and that the system has no solution). Matrix methods reveal the common nature of all network problems as well as how to solve them.

PROBLEMS 2.4

1. Verify that the equations (2.19) express the balance of horizontal forces and the balance of vertical forces.
2. Verify that (2.22)'s $\mathbf{e} = \mathbf{A}^T\mathbf{d}$ correctly represents the relation between member extensions and joint displacements.
3. ▷ Suppose in Example 2.25 that $f_1 = 1$ and $f_2 = 10$; find the displacements d_1 and d_2.
4. Suppose in Example 2.25 that you want the displacements to be $d_1 = 0.01$ and $d_2 = 0.1$. Find the forces f_1 and f_2 necessary to produce this.
5. In Example 2.25:
   (a) Find $\mathbf{H}^{-1}$.
   (b) Write out a formula expressing d_1 in terms of f_1 and f_2 and another expressing d_2 in terms of f_1 and f_2, for general f_1 and f_2.
6. A simple economic model assumes that a desired supply of goods to individuals can be maintained in equilibrium by an appropriate set of prices for the goods; given a desired supply, the problem is to determine the prices. Suppose that goods are produced and consumed by four individuals (I, II, III, IV in the illustration below); assume that individuals receive and send goods to others as indicated by the six directed paths (1, 2, 3, 4, 5, 6 below)—the arrows indicate the direction in which the goods flow. It is actually the *differences* in prices that lead to *differences* in supplies, so we can think of each price and each supply as a relative one—stated in comparison with those of individual IV,

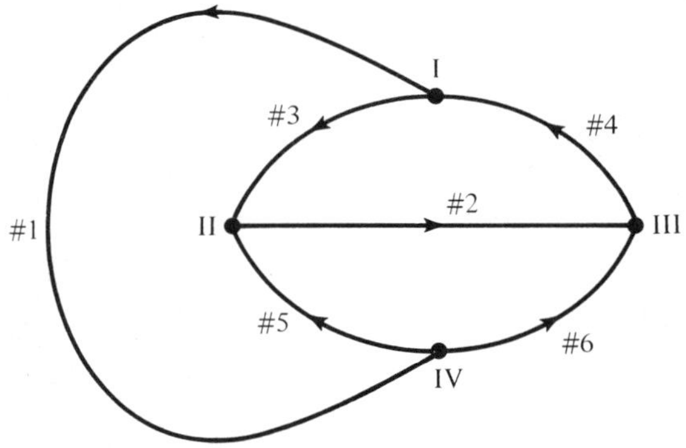

for example; mathematically, this means that we can take the supply and the price at IV to be zero. Let f_1, f_2, and f_3 denote the (relative) supply levels desired at I, II, and III, respectively, and let d_1, d_2, and d_3 denote the (relative) prices of goods there; our problem is to determine the d_i from the f_i. Let t_i be the amount of goods flowing along path i; then the supply for each individual equals the goods coming in minus the goods going out—$f_2 = t_3 + t_5 - t_2$, for example. (This will be expressed as $\mathbf{f} = \mathbf{At}$ below.) We make the simple economic assumption that the flow of goods along a path is directly proportional to the difference between the prices of the goods at the ends of the path; denoting these price differences by e_i for the ith path—so that $e_3 = d_2 - d_1$ and $e_5 = d_2 - 0$, for example—we then have $t_i = e_i/k_i$ for some known constants of proportionality k_i. Show that, just as in (2.23), this model is represented by $\mathbf{At} = \mathbf{f}$, $\mathbf{e} = \mathbf{Kt}$, $\mathbf{e} = \mathbf{A}^T\mathbf{d}$—so that $\mathbf{d}$ can be determined via $\mathbf{Hd} = \mathbf{f}$ for $\mathbf{H} = \mathbf{AK}^{-1}\mathbf{A}^T$—where in this case we take

$$\mathbf{f} = [f_1 \quad f_2 \quad f_3]^T, \qquad \mathbf{d} = [d_1 \quad d_2 \quad d_3]^T,$$

$$\mathbf{t} = [t_1 \quad t_2 \quad t_3 \quad t_4 \quad t_5 \quad t_6]^T, \qquad \mathbf{e} = [e_1 \quad e_2 \quad e_3 \quad e_4 \quad e_5 \quad e_6]^T,$$

$\mathbf{K}$ is the 6×6 diagonal matrix with main-diagonal entries $\langle\mathbf{K}\rangle_{ii} = k_i$, and

$$\mathbf{A} = \begin{bmatrix} -1 & 0 & -1 & 1 & 0 & 0 \\ 0 & -1 & 1 & 0 & 1 & 0 \\ 0 & 1 & 0 & -1 & 0 & 1 \end{bmatrix}.$$

▷ **7.** Suppose in Problem 6 that the desired relative supply levels are $f_1 = 10$, $f_2 = 4$, and $f_3 = 12$, and that the constants k_i all equal 1. Find the relative prices $\mathbf{d}$ needed to maintain these supply levels.

8. In the pin-jointed planar framework shown below, the support at joint A is fixed while the support at joint B can only move up or down the wall. Let $\mathbf{f}$, $\mathbf{t}$, $\mathbf{d}$, and $\mathbf{e}$ denote the applied forces, internal forces, joint displacements, and member extensions as in our model for (2.18). Show that (2.23) again relates these concepts, where now

$$\mathbf{A} = \begin{bmatrix} 0 & 0 & 0 & 1 & 0 & \frac{1}{\sqrt{2}} \\ 1 & 0 & 0 & 0 & 0 & \frac{1}{\sqrt{2}} \\ 0 & 0 & -1 & 0 & -\frac{1}{\sqrt{2}} & 0 \\ 0 & 0 & 0 & 1 & \frac{1}{\sqrt{2}} & 0 \\ 0 & -1 & 0 & 0 & 0 & -\frac{1}{\sqrt{2}} \end{bmatrix}.$$

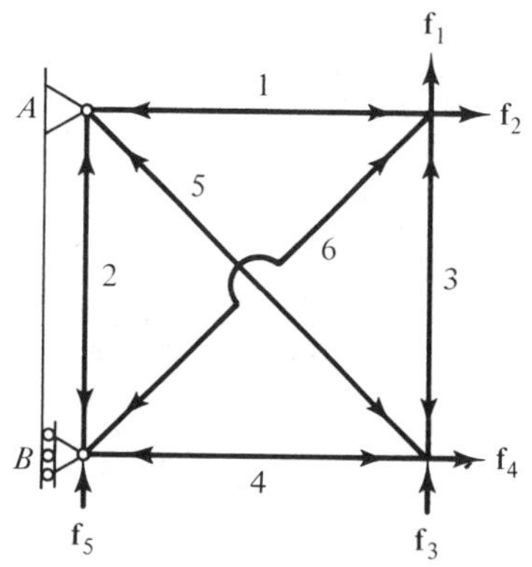

𝔐 9. Suppose in Problem 8 that the flexibilities k_i all equal 1 and that the applied forces are $f_1 = 0$, $f_2 = 1$, $f_3 = -10$, $f_4 = 1$, and $f_5 = -10$. Use MATLAB or similar software to solve the five linear equations (2.24) in the five unknowns **d** so as to obtain the displacements **d**.

2.5 *OSCILLATORY SYSTEMS: EIGENVALUES*

Many phenomena in various application areas exhibit oscillations: Airplane wings, bridges, and tall buildings oscillate in the wind; the economy oscillates (between inflation and deflation, for example); and so on. It is obviously important to understand the qualitative behavior of these oscillations: Will the wing fall off, the bridge collapse, the building topple, the economy crumble? The study of models of these phenomena usually leads to matrix problems in which we need to discover when certain matrices depending on parameters will be singular.

Two Masses Suspended and Coupled by Springs

As an illustration of oscillatory phenomena, we consider the motion of two masses coupled by one spring and suspended from the ceiling by another, as in the diagram below; the springs are assumed to have negligible weights. The two weights have masses m_1 and m_2. The two horizontal lines marked "Rest" indicate the position of the masses at rest, that is, where the restoring forces of the springs and the force of gravity are in perfect balance.

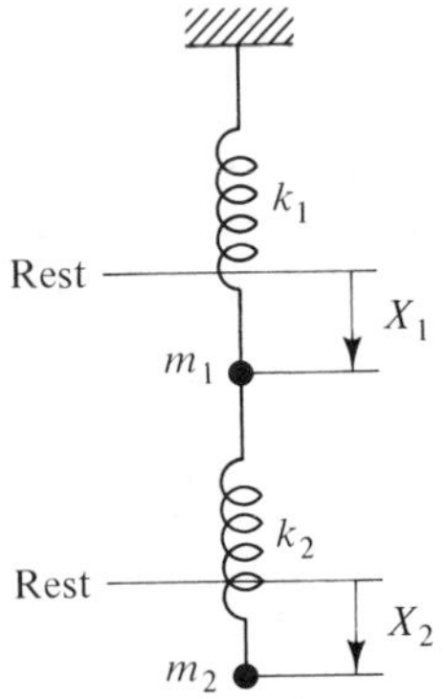

We want to model the vertical oscillations of the masses as time passes when they are not at rest. To do so, we introduce two functions of time t: $X_1(t)$ is the downward displacement at time t of the first mass from its rest position, and $X_2(t)$ is the downward displacement at time t of the second mass from its rest position. Then the forces acting on the first mass are (1) the upward force exerted by the first spring, which has been stretched $X_1(t)$ beyond equilibrium; and (2) the downward force exerted by the second spring, which has been stretched by $X_2(t) - X_1(t)$ beyond equilibrium; note that the force of gravity has been accounted for in the rest positions of the masses. The force on the second mass is the upward force exerted by the second spring.

We suppose that the displacements involved are small enough that Hooke's law is valid for the springs: The restoring force for each spring is equal to a constant k_i times the amount by which the spring is stretched. Recalling Newton's law that force equals mass times resulting acceleration and recalling that acceleration is the second derivative of displacement $X_i(t)$, we obtain the mathematical model of our system:

(2.26)
$$\begin{aligned} m_1X_1'' &= -k_1X_1 + k_2(X_2 - X_1) \\ m_2X_2'' &= -k_2(X_2 - X_1), \end{aligned}$$

where the primes denote differentiation with respect to t.

Our mathematical problem is to find functions X_1 and X_2 that satisfy (2.26). Experience and intuition tell us that such systems should be oscillatory: Both X_1 and X_2 should oscillate between positive and negative values much like the trigonometric functions sine and cosine. Therefore, we decide to see whether we can find solutions of (2.26) by using sines and cosines; more precisely, we seek solutions of the form

(2.27)
$$\begin{aligned} X_1(t) &= \xi_1 \sin \omega t + \eta_1 \cos \omega t \\ X_2(t) &= \xi_2 \sin \omega t + \eta_2 \cos \omega t, \end{aligned}$$

where $\xi_1, \eta_1, \xi_2, \eta_2$, and ω are constants to be determined in order that X_1 and X_2 solve (2.26).

If we substitute into (2.26) the expressions for X_1 and X_2 in (2.27), differentiate as required, and collect terms, we obtain

$$\{-m_1\omega^2\xi_1 + k_1\xi_1 + k_2\xi_1 - k_2\xi_2\} \sin \omega t + \{-m_1\omega^2\eta_1 + k_1\eta_1 + k_2\eta_1 - k_2\eta_2\} \cos \omega t = 0 \quad \text{for all } t$$

$$\{-m_2\omega^2\xi_2 + k_2\xi_2 - k_2\xi_1\} \sin \omega t + \{-m_2\omega^2\eta_2 + k_2\eta_2 - k_2\eta_1\} \cos \omega t = 0 \quad \text{for all } t.$$

Now, the only way that $A \sin \omega t + B \cos \omega t$ can equal zero for all t is for $A = B = 0$; this means that the expressions above inside braces $\{\cdot\}$ must all equal zero:

(2.28)
$$\begin{aligned} (-m_1\omega^2 + k_1 + k_2)\xi_1 + \qquad (-k_2)\xi_2 &= 0 \\ (-k_2)\xi_1 + (-m_2\omega^2 + k_2)\xi_2 &= 0 \end{aligned}$$

as well as

$$\begin{aligned}(-m_1\omega^2 + k_1 + k_2)\eta_1 + \qquad\qquad (-k_2)\eta_2 &= 0\\ (-k_2)\eta_1 + (-m_2\omega^2 + k_2)\eta_2 &= 0.\end{aligned}$$

Matrix notation will make it much clearer what (2.28) demands. We define

(2.29) $$\mathbf{K} = \begin{bmatrix} k_1 + k_2 & -k_2 \\ -k_2 & k_2 \end{bmatrix}, \quad \mathbf{M} = \begin{bmatrix} m_1 & 0 \\ 0 & m_2 \end{bmatrix}, \quad \boldsymbol{\xi} = \begin{bmatrix} \xi_1 \\ \xi_2 \end{bmatrix}, \quad \boldsymbol{\eta} = \begin{bmatrix} \eta_1 \\ \eta_2 \end{bmatrix}.$$

Note that $\mathbf{K}$ and $\mathbf{M}$ are *known* matrices, while $\boldsymbol{\xi}$ and $\boldsymbol{\eta}$ are the unknowns we seek to determine so that X_1 and X_2 of the form in (2.27) will solve (2.26). Using these matrices, we can rewrite the equations that we need to satisfy in (2.28) as

(2.30) $$(\mathbf{K} - \omega^2\mathbf{M})\boldsymbol{\xi} = \mathbf{0} \quad \text{and} \quad (\mathbf{K} - \omega^2\mathbf{M})\boldsymbol{\eta} = \mathbf{0}.$$

The systems of equations for $\boldsymbol{\xi}$ and for $\boldsymbol{\eta}$ are identical, both being represented by the coefficient matrix $\mathbf{K} - \omega^2\mathbf{M}$.

(2.31) ***Example.*** We consider a concrete example in order to clarify matters. Suppose that the masses are $m_1 = 10$ and $m_2 = 5$, while the spring constants are $k_1 = 80$ and $k_2 = 40$. Then the matrices in (2.29) are simply

$$\mathbf{K} = \begin{bmatrix} 120 & -40 \\ -40 & 40 \end{bmatrix}, \quad \mathbf{M} = \begin{bmatrix} 10 & 0 \\ 0 & 5 \end{bmatrix}.$$

The two sets of equations (2.30) each then have the form

$$\begin{aligned}(120 - 10\omega^2)x - \qquad\quad 40y &= 0\\ -40x + (40 - 5\omega^2)y &= 0.\end{aligned}$$

You should now be able to solve Problems 1 to 4.

Eigenvalues

Returning to the general problem, we note that each of the systems of equations in (2.30) is of the form $\mathbf{Ax} = \mathbf{0}$, where $\mathbf{A}$ is 2×2 and $\mathbf{x}$ is 2×1. It is clear that *one* solution to such a system is $\mathbf{x} = \mathbf{0}$; in our case this would make $\xi_1 = \xi_2 = \eta_1 = \eta_2 = 0$ and therefore $X_1(t) = X_2(t) = 0$, which describes the rest state of the system rather than an oscillatory state. *So we need solutions other than* $\mathbf{0}$. According to Section 1.4's Theorem 1.38(c) on equations and inverses, $\mathbf{Ax} = \mathbf{b}$ for square $\mathbf{A}$ will have exactly one solution $\mathbf{A}^{-1}\mathbf{b}$ *if* $\mathbf{A}$ *is nonsingular.* We want a solution to $\mathbf{Ax} = \mathbf{0}$ *other than* $\mathbf{x} = \mathbf{0}$ $(= \mathbf{A}^{-1}\mathbf{0}$ if $\mathbf{A}$ is nonsingular); for this to happen, then, we see that $\mathbf{A}$ *must be singular*. Recall that we are using $\mathbf{A}$ to stand for $\mathbf{K} - \omega^2\mathbf{M}$ in (2.30). We have deduced, then, that

(2.32) in order for (2.30) to have solutions $\boldsymbol{\xi}$ and $\boldsymbol{\eta}$ other than $\boldsymbol{\xi} = \boldsymbol{\eta} = \mathbf{0}$ {so that $X_1(t)$ and $X_2(t)$ are not identically 0}, the matrix $\mathbf{K} - \omega^2\mathbf{M}$ must be singular.

Remember that $\mathbf{K}$ and $\mathbf{M}$ are known—see (2.28)—but that ω is *un*known: We are trying to choose ω (along with $\boldsymbol{\xi}$ and $\boldsymbol{\eta}$) so as to have (2.27) provide solutions to (2.26). Thus our problem is:

(2.33) Choose ω^2 so that $\mathbf{K} - \omega^2\mathbf{M}$ is singular.

The problem of determining a number λ for which $\mathbf{K} - \lambda\mathbf{M}$ is singular is called an *eigenvalue problem*, and such a λ is called an *eigenvalue*.

(2.34) ***Example.*** Consider the concrete system from Example 2.31. According to what we have just done, our problem is to choose ω so that $\mathbf{K} - \omega^2\mathbf{M}$ is singular, where

$$\mathbf{K} - \omega^2\mathbf{M} = \begin{bmatrix} 120 - 10\omega^2 & -40 \\ -40 & 40 - 5\omega^2 \end{bmatrix}.$$

Problem 13 in Section 1.4 gives a necessary and sufficient condition on the entries in order that a 2×2 matrix be singular. In our case, this becomes

$$(120 - 10\omega^2)(40 - 5\omega^2) - (-40)(-40) = 0.$$

That is, $50(\omega^2)^2 - 1000(\omega^2) + 3200 = 0$; factoring this quadratic polynomial in ω^2 reduces the equation to $50(\omega^2 - 4)(\omega^2 - 16) = 0$. Therefore, $\omega^2 = 4$ and $\omega^2 = 16$ are the eigenvalues for this problem. This means that the only possible values for ω are $+2$, -2, $+4$, and -4; but these are just *possible* values as far as finding nonzero solutions $\boldsymbol{\xi}$ and $\boldsymbol{\eta}$ to (2.30) is concerned—we have to check to see whether we can actually solve for nonzero $\boldsymbol{\xi}$ and $\boldsymbol{\eta}$. Note that since it is ω^2 rather than ω that appears in the equations, we need only examine $\omega = +2$ and $\omega = +4$; also, changing the sign of ω is the same as changing the sign of ξ_1 and ξ_2 in (2.27), so we consider only the positive values of ω.

For $\omega = 2$, the equations for $\boldsymbol{\xi}$ are $80\xi_1 - 40\xi_2 = 0$ and $-40\xi_1 + 20\xi_2 = 0$, the solutions to which are $\xi_1 = \alpha$ arbitrary and $\xi_2 = 2\alpha$; similarly, the equations for $\boldsymbol{\eta}$ lead to $\eta_1 = \beta$ arbitrary and $\eta_2 = 2\beta$.

For $\omega = 4$, on the other hand, the equations for $\boldsymbol{\xi}$ are, instead,

$$-40\xi_1 - 40\xi_2 = 0 \quad \text{and} \quad -40\xi_1 - 40\xi_2 = 0,$$

the solutions to which are $\xi_1 = \gamma$ arbitrary and $\xi_2 = -\gamma$; similarly, the equations for $\boldsymbol{\eta}$ lead to $\eta_1 = \delta$ arbitrary and $\eta_2 = -\delta$.

Returning to the general situation, we see that in order to satisfy (2.33) for the 2×2 matrix $\mathbf{K} - \omega^2\mathbf{M}$, we will have to solve an equation for ω^2 just as in Example 2.34; and then we will have to see whether, for the specific values of ω we find, we can solve (2.30) for nonzero $\boldsymbol{\xi}$ and $\boldsymbol{\eta}$. For t measured in, say, seconds, solutions such as $\sin \omega t$ and $\cos \omega t$ repeat themselves every $2\pi/\omega$ seconds—that is, $\omega/2\pi$ times per second. The number $\omega/2\pi$ is thus the *frequency* of the oscillation. If we had p masses rather than just two as in our model, the matrices $\mathbf{K}$ and $\mathbf{M}$ would be $p \times p$ rather than 2×2. So, more generally, we want to know:

1. How do we know whether there are real numbers ω that make $\mathbf{K} - \omega^2\mathbf{M}$ singular and determine the frequencies of oscillation?
2. If we find such numbers ω, how do we know that we can then find nonzero $\mathbf{x}$ so that $(\mathbf{K} - \omega^2\mathbf{M})\mathbf{x} = \mathbf{0}$?

At the moment we lack the tools to answer these general questions; Chapters 7 and 8 will develop the tools we need.

You should now be able to solve Problems 1 to 6.

Solving the Oscillating-Masses Problem

We have not yet shown how this matrix analysis lets us solve the original problem of the coupled masses; for simplicity, we examine the concrete case treated in Examples 2.26 and 2.34.

In Example 2.34 we found nonzero solutions of (2.30) with $\omega = 2$, with ξ_1 and ξ_2 depending on an arbitrary number α and η_1 and η_2 depending on an arbitrary number β. Similarly, we found nonzero solutions with $\omega = 4$ depending on numbers γ and δ. If we substitute these into our expressions (2.27) for the functions $X_1(t)$ and $X_2(t)$ describing the behavior of the system of two masses, we obtain

$$X_1(t) = \alpha \sin 2t + \beta \cos 2t \quad \text{and} \quad X_2(t) = 2\alpha \sin 2t + 2\beta \cos 2t,$$

$$X_1(t) = \gamma \sin 4t + \delta \cos 4t \quad \text{and} \quad X_2(t) = -\gamma \sin 4t - \delta \cos 4t$$

as pairs of solutions to (2.26) for all values of α, β, γ, and δ. By combining these pairs, we can obtain solutions that describe the motions of the masses once we know how the oscillations begin. To see this, we write $\mathbf{X}(t)$ as the column matrix $\mathbf{X}(t) = [X_1(t) \quad X_2(t)]^T$; the solutions above can then be expressed as

$$\mathbf{X}(t) = \alpha\begin{bmatrix} \sin 2t \\ 2\sin 2t \end{bmatrix} + \beta\begin{bmatrix} \cos 2t \\ 2\cos 2t \end{bmatrix} + \gamma\begin{bmatrix} \sin 4t \\ -\sin 4t \end{bmatrix} + \delta\begin{bmatrix} \cos 4t \\ -\cos 4t \end{bmatrix} \tag{2.35}$$

for all arbitrary values of α, β, γ, and δ. We need only choose these values so as to match the starting conditions of the masses.

(2.36) ***Example.*** Suppose that the masses are started with the top mass placed one unit below its rest point and the bottom mass placed five units below its rest point; this requires $X_1(0) = 1$ and $X_2(0) = 5$. Suppose also that each mass is just released, so that no initial velocity is forced upon it; this requires that $X_1'(0) = X_2'(0) = 0$. If we apply these conditions at $t = 0$ to the solution represented by (2.35), we find that α, β, γ, and δ must satisfy

$$\begin{aligned} \beta + \delta &= 1 \\ 2\beta - \delta &= 5 \\ 2\alpha + 4\gamma &= 0 \\ 4\alpha - 4\gamma &= 0, \end{aligned}$$

from which we find that $\alpha = 0$, $\gamma = 0$, $\beta = 2$, and $\delta = -1$. This gives us the solution for the motion of the weights for all t:

$$\begin{aligned} X_1(t) &= 2 \cos 2t - \cos 4t \\ X_2(t) &= 4 \cos 2t + \cos 4t. \end{aligned}$$

PROBLEMS 2.5

1. Verify that the right-hand sides in (2.26) properly represent the forces on the masses.
2. Verify that (2.28) follows from substituting the expressions in (2.27) into the equations in (2.26) and using the fact that $A \sin \omega t + B \cos \omega t = 0$ for all t implies that $A = B = 0$.
3. Suppose that $m_1 = 12$, $m_2 = 16$, $k_1 = 36$, and $k_2 = 48$ in the model of this section. Find the matrices and equations analogous to those in Example 2.31.

▷ **4.** The behavior of the *linear triatomic molecule*—a molecule with three atoms arranged in a line—can be modeled by that of three masses connected by two springs as shown below, where the central mass has mass M and the others mass m. As in the basic model of this section, the functions X_i measure the displacements of the masses from rest positions; gravity is ignored, and the springs are assumed weightless.

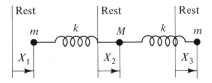

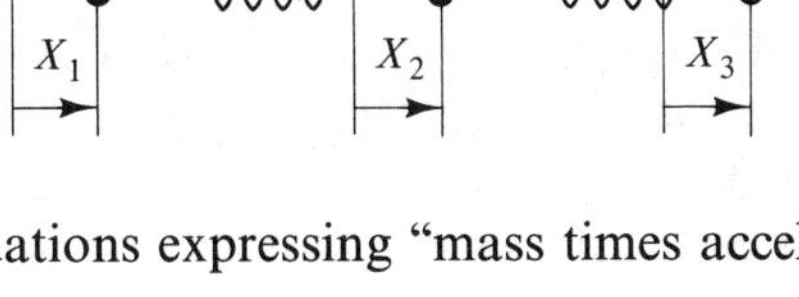

(a) Show that the equations expressing "mass times acceleration equals force" in this case are

$$mX_1'' = -k(X_1 - X_2)$$
$$MX_2'' = k(X_1 - 2X_2 + X_3)$$
$$mX_3'' = -k(X_3 - X_2).$$

(b) Substitute expressions *similar* to those in (2.27)—you must include an expression for X_3 using coefficients ξ_3 and η_3—into the equations above and derive the equations analogous to those in (2.28) and (2.30).

5. For the model in Problem 3, find the values of ω, ξ_i, and η_i as we did in Example 2.34.

▷ **6.** For the model in Problem 4, suppose that $k = 12$, $m = 2$, and $M = 6$. Find the values of ω, ξ_i, and η_i much as we did in Example 2.34.

7. For Example 2.36, verify:

(a) That the values of α, β, γ, and δ found there solve the equations given there for those numbers

(b) That the functions X_1 and X_2 found there satisfy the differential equations (2.26) using the relevant values of m_i and k_i

(c) That the functions X_1 and X_2 satisfy the conditions required at $t = 0$

8. For the model in Problem 3, find $X_1(t)$ and $X_2(t)$ explicitly in order that $X_1(0) = 0$, $X_2(0) = 1$, $X_1'(0) = X_2'(0) = 0$.

2.6 GENERAL MODELING: LEAST SQUARES

Central to applied mathematics is modeling, as the preceding sections have indicated. Often, one assumes a certain general form for the model and then must somehow determine the various parameters in that form so that the model conforms to reality as well as it can. A widely used approach to this is the *method of least squares*, in which one chooses the parameters so as to minimize the sum of the squares of the errors in the model's representation of known data; this approach is also central to the statistical analysis of data, in particular to what is called *regression analysis*.

Fitting Straight Lines to Data

Nearly the simplest form of a model is a straight line: We model the way that some variable y of interest depends on another variable t by the form

$$y \approx a + bt \tag{2.37}$$

where a and b are the parameters to be determined, and we write "$\approx$" rather than "$=$" because the model (2.37) is only an approximation. In order to choose a and b so as to reflect experience, we need some data; we therefore assume that we have p pairs of measurements of t and y values that we have found go together:

(2.38) (t_i, y_i) for $1 \le i \le p$ are the data from experience.

Our task is to select a and b so that (2.37) does as good a job as possible, in some sense, of matching the data in (2.38); once we determine a and b—if they match the data well—we can consider using the model (2.37) to predict what y will be for other values of t.

(2.39) ***Example.*** If a body moves in a perfectly straight line with perfectly constant velocity, then its position y along that line at time t is exactly given by $y = y_0 + vt$, where y_0 is the body's position at time $t = 0$ and v is the constant velocity. In the physical world, of course, the body will almost certainly have slight variations in its velocity and the model $y_0 + vt$ will not exactly describe the position y. Suppose that we have measured the position y of a particular body at five different moments and have obtained the following data:

t	0	3	5	8	10
y	2	5	6	9	11

By plotting these data on a graph with t and y axes we can see that the points lie nearly on a straight line as expected:

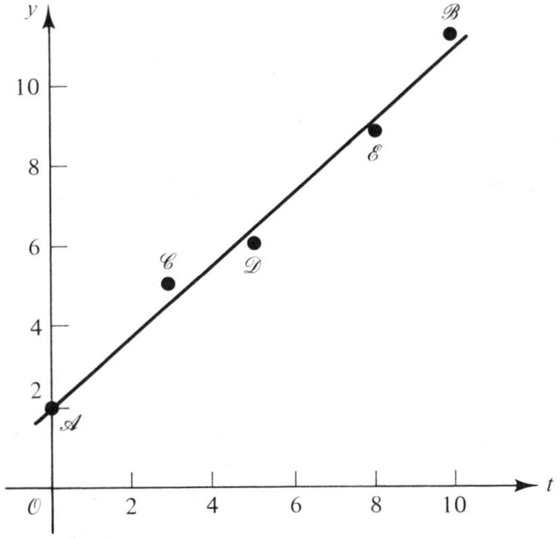

Our task is to choose y_0 and v so that $y \approx y_0 + vt$ at the five data points; since the equation $y = y_0 + vt$ between t and y has as its graph a straight line, geometrically we are trying to find a straight line that comes as close as possible to the five points on the graph above.

In the general case, we want to choose a and b so that $a + bt$ comes close to reproducing the data (2.38). For any choice of the parameters a and b, we measure their error in reproducing the data at the single data point (t_i, y_i) by the ith *residual*

(2.40)
$$r_i = y_i - (a + bt_i)$$

and we measure their overall error in reproducing the data by the sum of the squares of the residuals:

(2.41)
$$S(a, b) = \sum_{i=1}^{p} r_i^2 = \sum_{i=1}^{p} \{y_i - (a + bt_i)\}^2.$$

We use the notation $S(a, b)$ to emphasize that the error depends on the unknown parameters a and b. We now show that to determine a and b so as to make the overall error S as small as possible, we must find a and b satisfying

(2.42)
$$\alpha_{11}a + \alpha_{12}b = \beta_1 \quad \text{and} \quad \alpha_{21}a + \alpha_{22}b = \beta_2,$$

where

$$\alpha_{11} = \sum_{i=1}^{p} 1 = p, \qquad \alpha_{12} = \alpha_{21} = \sum_{i=1}^{p} t_i, \qquad \alpha_{22} = \sum_{i=1}^{p} t_i^2,$$

$$\beta_1 = \sum_{i=1}^{p} y_i, \quad \text{and} \quad \beta_2 = \sum_{i=1}^{p} t_i y_i.$$

By expanding the terms in (2.41), collecting terms, and using the notation α_{ij} and β_i in (2.42) for convenience, we find that

$$S(a, b) = \alpha_{11}a^2 - 2\beta_1 a + \alpha_{22}b^2 - 2\beta_2 b + (\alpha_{12} + \alpha_{21})ab + \sum_{i=1}^{p} y_i^2.$$

If we complete the square in the expression above with respect to the variable a, we get

$$S(a, b) = \alpha_{11}\left\{a + \frac{-2\beta_1 + 2\alpha_{12}b}{2\alpha_{11}}\right\}^2 + \text{terms independent of } a.$$

Since the only term above in $S(a, b)$ involving a is the one in braces, since we are to minimize $S(a, b)$ with respect to a, and since the coefficient α_{11} of $\{\cdot\}^2$ is positive, we minimize $S(a, b)$ with respect to a by making the term $\{\cdot\} = 0$. But $\{\cdot\} = 0$ converts to the first equation in (2.42); completing the square with respect to b in the same manner produces the second equation.

(2.43) ***Example.*** Consider again the problem in Example 2.39. Using the formulas in (2.42), we find that $\alpha_{11} = 5$, $\alpha_{12} = \alpha_{21} = 26$, $\alpha_{22} = 198$, $\beta_1 = 33$, and $\beta_2 = 227$. Minimizing the sum of the five squared errors $\{y_i - (y_0 + vt_i)\}^2$ is then equivalent to solving

$$5y_0 + 26v = 33$$
$$26y_0 + 198v = 227.$$

Solving this system gives $y_0 = 2.01$ and $v = 0.88$; the straight line determined by these values is graphed in the figure in Example 2.39 to show how well the model fits the data.

At this point we know that in order to fit a straight-line model $y \approx a + bt$ to the data (2.38), we need to solve the two linear equations in two unknowns a and b in (2.42). In the special case of Example 2.43 we found that these equations indeed had a solution; we do not, however, know that this is true in general. We would like to know:

1. Do the equations in (2.42) always have a solution? We will later learn that the answer is "yes."

You should now be able to solve Problems 1 to 3.

Since (2.42) gives two equations in two unknowns, the problem can be expressed in matrix notation; in fact, so can all the algebra that produced (2.42).

Define

$$\text{(2.44)} \qquad \mathbf{y} = \begin{bmatrix} y_1 \\ y_2 \\ \vdots \\ y_p \end{bmatrix}, \quad \mathbf{x} = \begin{bmatrix} a \\ b \end{bmatrix}, \quad \mathbf{A} = \begin{bmatrix} 1 & t_1 \\ 1 & t_2 \\ \vdots & \vdots \\ 1 & t_p \end{bmatrix}, \quad \text{and} \quad \mathbf{r} = \begin{bmatrix} r_1 \\ r_2 \\ \vdots \\ r_p \end{bmatrix}.$$

Then choosing a and b so that $a + bt_i \approx y_i$ for $1 \le i \le p$ is the same as choosing $\mathbf{x}$ so that $\mathbf{Ax} \approx \mathbf{y}$. In this notation, S—the sum of the squares of the residuals r_i—is the number with $[S] = \mathbf{r}^T\mathbf{r}$; for simplicity in notation, we write this as $S = \mathbf{r}^T\mathbf{r}$. That is, we seek to

$$\text{(2.45)} \qquad \text{minimize } S = \mathbf{r}^T\mathbf{r} = (\mathbf{y} - \mathbf{Ax})^T(\mathbf{y} - \mathbf{Ax}).$$

Surprisingly, the definitions of α_{11}, α_{12}, α_{21}, α_{22}, β_1, and β_2 show that

$$\text{(2.46)} \qquad \mathbf{A}^T\mathbf{A} = \begin{bmatrix} \alpha_{11} & \alpha_{12} \\ \alpha_{21} & \alpha_{22} \end{bmatrix} \quad \text{and} \quad \mathbf{A}^T\mathbf{y} = \begin{bmatrix} \beta_1 \\ \beta_2 \end{bmatrix}.$$

These α_{ij} and β_i matrices are precisely those in the system of equations (2.42). Thus, in matrix notation, the system of equations (2.42) that we need to solve to determine a and b is just

$$\text{(2.47)} \qquad \mathbf{A}^T\mathbf{Ax} = \mathbf{A}^T\mathbf{y},$$

where **A** and **y** are known and we need to find **x**. *Formally*—we repeat: *formally*—the equations (2.47) are obtained by premultiplying $\mathbf{Ax} \approx \mathbf{y}$ by $\mathbf{A}^T$ and replacing the $\approx$ by $=$.

You should now be able to solve Problems 1 to 8.

Fitting Data by More General Expressions

All of the development so far has addressed approximating data by straight lines. But the matrix notation reveals another way of visualizing what we are doing. In matrix notation, we have been given a matrix **A** and a matrix **y**—see (2.44)—and we want to find the matrix **x** so that $\mathbf{Ax} \approx \mathbf{y}$. Recall now Example 1.41 in Section 1.5; that example showed that in general

$$\mathbf{Ax} = x_1\mathbf{a}_1 + x_2\mathbf{a}_2 + \cdots + x_q\mathbf{a}_q,$$

where the numbers x_i are the entries in the $q \times 1$ matrix **x** and the $p \times 1$ column matrices $\mathbf{a}_i$ are the columns of the $p \times q$ matrix **A**. In our case $q = 2$ and we are trying, therefore, to choose x_1 (which equals a) and x_2 (which equals b) so that $\mathbf{y} \approx x_1\mathbf{a}_1 + x_2\mathbf{a}_2$ where

$$\mathbf{a}_1 = [1 \quad 1 \quad \cdots \quad 1]^T \quad \text{and} \quad \mathbf{a}_2 = [t_1 \quad t_2 \quad \cdots \quad t_p]^T.$$

This general problem arises quite often; we are required to

(2.48) approximate a given $p \times 1$ real column matrix **y** in terms of given $p \times 1$ real column matrices $\mathbf{a}_1, \mathbf{a}_2, \ldots, \mathbf{a}_q$ by finding numbers $x_1, x_2, \ldots, x_q$ so that $\mathbf{y} \approx x_1\mathbf{a}_1 + x_2\mathbf{a}_2 + \cdots + x_q\mathbf{a}_q$.

Written completely in matrix notation, (2.48) says:

(2.49) Given $p \times q$ real **A** and $p \times 1$ real **y**, find $q \times 1$ **x** so that $\mathbf{Ax} \approx \mathbf{y}$;

this is the general (linear) least-squares problem.

(2.50) ***Example.*** Suppose that we decide to approximate the data (t_i, y_i) in Example 2.39 by a quadratic rather than linear expression: $y \approx a + bt + ct^2$. This is equivalent to making $\mathbf{Ax} \approx \mathbf{y}$, where

$$\mathbf{A} = \begin{bmatrix} 1 & 0 & 0 \\ 1 & 3 & 9 \\ 1 & 5 & 25 \\ 1 & 8 & 64 \\ 1 & 10 & 100 \end{bmatrix}, \quad \mathbf{x} = \begin{bmatrix} a \\ b \\ c \end{bmatrix}, \quad \text{and} \quad \mathbf{y} = \begin{bmatrix} 2 \\ 5 \\ 6 \\ 9 \\ 11 \end{bmatrix}.$$

In the general case (2.48), just as in the case of straight-line fitting of data, we define the *residual* for any $\mathbf{x}$ as

(2.51) $$\mathbf{r} = \mathbf{y} - \mathbf{A}\mathbf{x}$$

and seek to find $\mathbf{x}$ to

(2.52) $$\text{minimize } \mathbf{r}^T\mathbf{r} = (\mathbf{y} - \mathbf{A}\mathbf{x})^T(\mathbf{y} - \mathbf{A}\mathbf{x}).$$

Later material will prove that the solution to (2.47) solves (2.52) just as in the special case of straight-line fitting except that now $\mathbf{A}$ is the $p \times q$ matrix of (2.48). With the information we have now, however, we need to know:

2. Why does the solution $\mathbf{x}$ to (2.47) solve (2.52)?

Moreover, there are the questions of whether (2.47) always has a solution and of how we find it, and of what happens when the data are complex:

3. Does $\mathbf{A}^T\mathbf{A}\mathbf{x} = \mathbf{A}^T\mathbf{y}$ always have a solution $\mathbf{x}$?
4. How can we efficiently and accurately solve for $\mathbf{x}$?
5. What if $\mathbf{A}$ or $\mathbf{y}$ or both are complex?

Chapters 5 and 8 will allow us to answer these questions.

PROBLEMS 2.6

1. For $y_0 = 2$ and $v = 1$, find the residuals r_i in Example 2.43.

2. There are situations in which some residuals should be given more weight in the overall error S than other residuals; this leads to *weighted least squares.* Suppose that we define—instead of $S(a, b)$ in (2.41)—

$$W(a, b) = \sum_{i=1}^{p} w_i r_i^2 \qquad \text{for given positive weights } w_j.$$

Show that the equations for determining a and b to minimize $W(a, b)$ are still those in (2.42), *except* that the α_{ij} and β_i are defined differently:

$$\alpha_{11} = \sum_{i=1}^{p} w_i, \qquad \alpha_{12} = \alpha_{21} = \sum_{i=1}^{p} w_i t_i, \qquad \alpha_{22} = \sum_{i=1}^{p} w_i t_i^2,$$

with analogous definitions for the β_j (inserting w_i into the sums).

▷ **3.** The table below gives the approximate U.S. population in millions every 10 years from 1900 through 1980. Find the equations (2.42) that will result from attempting to fit these data by the model

$$(\text{population in millions}) \approx a + b\,(\text{year}).$$

Yr.	1900	1910	1920	1930	1940	1950	1960	1970	1980
Pop.	76.0	92.0	105.7	123.2	131.7	150.7	179.3	203.2	226.5

4. For $\mathbf{A}$ and $\mathbf{y}$ as in Example 2.43, verify that (2.46) holds.
5. For the weighted least-squares method described in Problem 2, show that the matrix equation analogous to (2.47) is $\mathbf{A}^T\mathbf{W}\mathbf{A}\mathbf{x} = \mathbf{A}^T\mathbf{W}\mathbf{y}$, where $\mathbf{W}$ is the diagonal matrix with main-diagonal entries $\langle \mathbf{W} \rangle_{ii} = w_i$.

▷ 6. Suppose that the data in Example 2.39 are replaced by the t and y values below; fit a straight line to the data.

t	1	4	5	3	8
y	0	5	11	4	17

7. The relationship between Fahrenheit degrees F and Celsius degrees C is of the form $F = a + bC$ for appropriate a and b. Because of inaccurate measurements the table below does not perfectly reflect this relationship. Use the least-squares method to find approximate values for a and b from the data provided.

F	−1	2	10	15
C	32	36	51	57

𝔐 8. If $\mathbf{A}$ is $p \times q$ and $\mathbf{y}$ is $p \times 1$, the MATLAB command $x = A\backslash y$ actually computes the least-squares solution x to $\mathbf{A}\mathbf{x} \approx \mathbf{y}$ by a very accurate method.
   (a) Use this MATLAB command to solve for the least-squares model in Problem 3 *rather than solving* (2.47).
   (b) Use your model to predict the population in 1990.

9. By completing the square one variable x_i at a time, show that the least-squares solution for the general problem (2.48) can be obtained by solving $\mathbf{A}^T\mathbf{A}\mathbf{x} = \mathbf{A}^T\mathbf{y}$ ($\mathbf{A}$, $\mathbf{y}$ real).

▷ 10. Find the least-squares solution to $\mathbf{A}\mathbf{x} \approx \mathbf{y}$, where

$$\mathbf{A} = \begin{bmatrix} 1 & 2 \\ -1 & 0 \\ 3 & 7 \\ 2 & 1 \end{bmatrix}, \qquad \mathbf{y} = \begin{bmatrix} 4 \\ 0 \\ 8 \\ 1 \end{bmatrix}.$$

11. Find all least-squares solutions to $\mathbf{Ax} \approx \mathbf{y}$ and show that $\mathbf{A}^T\mathbf{A}$ is singular, where

$$\mathbf{A} = \begin{bmatrix} 1 & -2 \\ -1 & 2 \\ 2 & -4 \\ 1 & -2 \end{bmatrix}, \qquad \mathbf{y} = \begin{bmatrix} 1 \\ 2 \\ -1 \\ 1 \end{bmatrix}.$$

𝔐 **12.** Suppose that in the dairy-competition model in Example 2.6 in Section 2.2 we do not know how to compute dairy 1's new market fraction from the old ones but only know that $x_1 = \alpha x_0 + \beta y_0 + \gamma z_0$ for some unknown parameters α, β, and γ. Suppose that we have gathered data that show that initial market shares of $[0.2 \quad 0.3 \quad 0.5]^T$ lead to $x_1 = 0.3$, that $[0.3 \quad 0.2 \quad 0.5]^T$ lead to $x_1 = 0.3$, that $[0.4 \quad 0.3 \quad 0.3]^T$ lead to $x_1 = 0.4$, that $[0.1 \quad 0.1 \quad 0.8]^T$ lead to 0.2, and that $[0.7 \quad 0.2 \quad 0.1]^T$ lead to 0.6. Use MATLAB or similar software with the method of least squares to determine approximate values of α, β, and γ.

▷ **13.** Explicitly find the equations $\mathbf{A}^T\mathbf{Ax} = \mathbf{A}^T\mathbf{y}$ for the model in Example 2.50.

𝔐 **14.** Use MATLAB or similar software to solve for the quadratic fit in Example 2.50.

𝔐 **15.** Use MATLAB or similar software to find the least-squares quadratic fit to the population data in Problem 3; predict the population in 1990.

16. Suppose that $\mathbf{A}$ or $\mathbf{y}$ or both are possibly complex, so that the sum of the squares of the magnitudes of the residuals must be evaluated as $\mathbf{r}^H\mathbf{r}$ rather than $\mathbf{r}^T\mathbf{r}$. The equations that determine a least-squares solution $\mathbf{x}$ to $\mathbf{Ax} \approx \mathbf{y}$ then in fact become $\mathbf{A}^H\mathbf{Ax} = \mathbf{A}^H\mathbf{y}$ rather than $\mathbf{A}^T\mathbf{Ax} = \mathbf{A}^T\mathbf{y}$ as in (2.47). Prove this in the special case of fitting possibly complex data (2.38) by (2.37) with possibly complex a and b.

2.7 *PRODUCTION PLANNING: LINEAR PROGRAMS*

Many practical problems ask us to accomplish some task in an optimal (or nearly optimal) fashion: to minimize cost, to minimize effort, to maximize gain, and the like. Mathematical models of many such problems are called *mathematical programs*; here "program" is used in the sense of "plan of action" rather than of "computer program." The simplest such programs are *linear programs*; and although simple, they have had enormous application and impact in many areas of business and government. Presented here is a simple example of a linear program and of its solution.

A Production-Planning Problem

In trying to decide how to allocate its manufacturing capacity among various products, an industry is usually influenced by many factors, including, of course, the desire to make a reasonable profit. We now consider a very simplified model of such a situation in which we assume that the *sole* desire is to maximize profit.

Suppose that an industrial plant has three types (M_1, M_2, and M_3) of machines, each of which must be used in manufacturing the plant's products, of which there are two types (P_1 and P_2). The problem is to decide how many of each product to produce each week so as to maximize weekly profits. We assume that a fixed profit is made on each unit of each product manufactured, so that the total profit is simply the sum of the profits on each type P_1 and P_2, each of which profit is simply obtained by multiplying the profit per item by the number of items manufactured. Specifically, we assume that the profit per item made for product P_1 is \$40, while that for P_2 is \$60. Clearly, our manufacturer should simply produce as much as possible; the word "possible" here is the key, for the manufacturer obviously is limited by the capacities of the types M_1, M_2, M_3 of machines which must be used. Thus we must assume that we know the amount of time available on each machine and also the amount of time required on each machine to make each type of product. Specifically, suppose that an item of P_1 requires two hours on machines of type M_1 and one hour each on M_2 and M_3, while an item of P_2 requires one hour on each of M_1 and M_2 but three hours on M_3. In addition, suppose that the number of hours available each week on machines of types M_1, M_2, and M_3 are 70, 40, and 90, respectively. All these assumptions are summarized in (2.53).

(2.53)

Machine Type	Hours Needed by One Unit of P_1	Hours Needed by One Unit of P_2	Total Hours Available
M_1	2	1	70
M_2	1	1	40
M_3	1	3	90
Profit per unit of P_1 = \$40		Profit per unit of P_2 = \$60	

Next, let x_1 denote the number of units of P_1 to be produced each week, while x_2 denotes the number of units of P_2. Since each unit of P_1 requires two hours on machines of type M_1 while each unit of P_2 requires one such hour, we will require $2x_1 + x_2$ hours on machines of type M_1; since only 70 hours are available on these machines, we see that we must require

$$2x_1 + x_2 \leq 70.$$

Reasoning similarly from the limited time on machines of type M_2 and M_3 and from the time needed to produce x_1 of P_1 and x_2 of P_2, we see also that we must require

$$x_1 + x_2 \leq 40$$

$$x_1 + 3x_2 \leq 90.$$

Since it is impossible to produce a negative number of units, we also must require

$$x_1 \geq 0, \qquad x_2 \geq 0.$$

To compute the profit resulting from our production plan, recall that the profit on each unit of P_1 is 40, so that x_1 such units yield a profit of $40x_1$; similarly, x_2 units of P_2 yield a profit of $60x_2$, so that our total profit is $40x_1 + 60x_2$. Thus the mathematical version of our planning problem is as follows:

(2.54)
$$\text{maximize } M = 40x_1 + 60x_2$$

where x_1 and x_2 must satisfy the constraints

(2.55)
$$\begin{aligned} 2x_1 + x_2 &\leq 70 \\ x_1 + x_2 &\leq 40 \\ x_1 + 3x_2 &\leq 90 \\ x_1 &\geq 0 \\ x_2 &\geq 0. \end{aligned}$$

This is a typical *linear programming problem*, involving the optimization of a certain linear function of some unknowns which are subject to linear constraints restricting the permissible values of the unknowns; we will study such problems closely later and will learn how to solve them by matrix methods. For the present we limit ourselves to rewriting the linear program in matrix notation and to using graphical methods to solve the problem.

In order to express the equations above in matrix notation, we require the following concepts. A matrix $\mathbf{P}$ is said to be *greater than* $\mathbf{Q}$, written $\mathbf{P} > \mathbf{Q}$, when $\mathbf{P}$ and $\mathbf{Q}$ have the same numbers of rows and columns, and each entry of $\mathbf{P}$ is greater than the corresponding entry of $\mathbf{Q}$. Similar definitions hold for $\geq$, $<$, and $\leq$. If $\mathbf{P} > \mathbf{0}$, we say that $\mathbf{P}$ is *positive*. If $\mathbf{P} \geq \mathbf{0}$, we say that $\mathbf{P}$ is *nonnegative*.

If we introduce

(2.56)
$$\mathbf{A} = \begin{bmatrix} 2 & 1 \\ 1 & 1 \\ 1 & 3 \end{bmatrix}, \qquad \mathbf{b} = \begin{bmatrix} 70 \\ 40 \\ 90 \end{bmatrix}, \qquad \mathbf{c} = \begin{bmatrix} 40 \\ 60 \end{bmatrix}, \qquad \mathbf{x} = \begin{bmatrix} x_1 \\ x_2 \end{bmatrix},$$

then our problem described in (2.54) and (2.55) can be denoted easily as

(2.57)
$$\text{maximize } M = \mathbf{c}^T\mathbf{x},$$

where $\mathbf{x}$ must satisfy the constraints

$$\mathbf{Ax} \le \mathbf{b}, \qquad \mathbf{x} \ge \mathbf{0}. \tag{2.58}$$

(Strictly speaking, $\mathbf{c}^T\mathbf{x}$ is a 1×1 matrix, but we adopt the convention that $\mathbf{c}^T\mathbf{x}$ can also be used to denote the entry of the matrix.)

The same *form* of equations will hold if we have p different machines producing q products. Then $\mathbf{A}$ is a $p \times q$ matrix, and its (i, j)-entry represents the number of hours on machine i required to produce one unit of product j. The total hours available will form a $p \times 1$ column matrix, and the profit matrix $\mathbf{c}$ and the matrix $\mathbf{x}$ representing the numbers of units produced will be $q \times 1$ column matrices.

You should now be able to solve Problems 1 to 4.

Geometric Interpretation and Solution

Later we will learn matrix methods for solving general linear programs of the form in (2.57)–(2.58). For the present we describe the specific problem in (2.54)–(2.55) geometrically (or graphically) to visualize matters; we will be able to solve the problem by this approach as well.

To deal with the first inequality in (2.55), we note that $2x_1 + x_2 = 70$ is the equation of a straight line—see Figure 2.59. The inequality $2x_1 + x_2 \le 70$ means that the point (x_1, x_2) must lie *on or below* the straight line. Similarly, the other two inequalities in (2.55) define half-planes in which the permissible points (x_1, x_2) must lie. The inequalities $x_1 \ge 0$, $x_2 \ge 0$ mean that (x_1, x_2) must lie in the first quadrant. The net result is that the inequalities confine the point (x_1, x_2) to a polygonal region whose boundary is shaded in (2.59). For any given value of M, the equation (2.54), namely $M = 40x_1 + 60x_2$, defines a straight line. The three dashed lines in (2.59) represent this straight line for $M = 0, 1200, 2400$. The value of M corresponding to any point on a given line is a constant, and the lines are parallel. To maximize M we must go as far as possible in a direction perpendicular to these lines—in the direction of the arrow (near the origin) in (2.59)—without leaving the admissible region. In this way we reach the point $\mathscr{P}$, which is the intersection of

$$x_1 + 3x_2 = 90, \qquad x_1 + x_2 = 40,$$

that is, $x_1 = 15$, $x_2 = 25$. The corresponding value of M is 2100. From the geometrical picture it is obvious that two of the relations is (2.55) are now equalities and one is a strict inequality. In terms of our original production-planning problem, we obtain the maximum profit of \$2100 weekly by producing 15 units of P_1 and 25 units of P_2 each week; the machines of type M_2 and M_3 are used to fullest capacity each week, but only 55 hours of time on M_1 machines are used, 15 hours less than capacity. Our manufacturer should probably consider selling some of the M_1 machines since M_1 machine capacity can be reduced with no reduction in profit (see Problem 5).

(2.59)

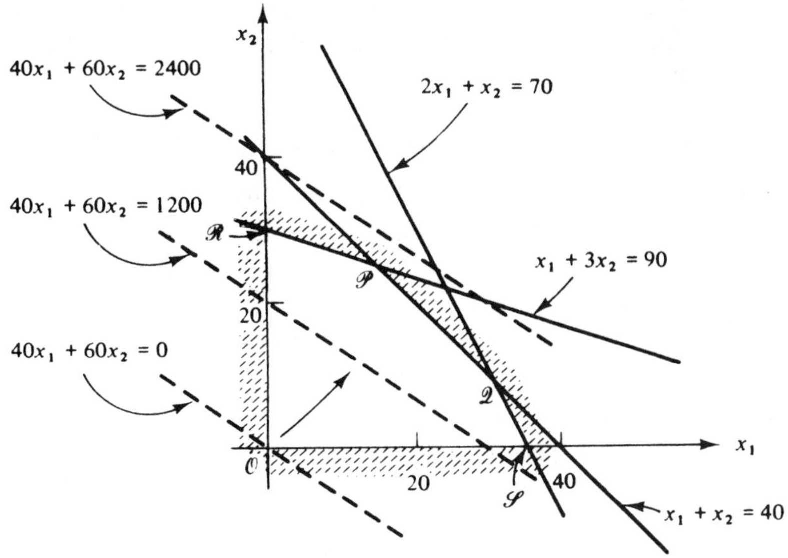

For problems like these, we need to know:

1. Are there always solutions?
2. How can we efficiently and accurately find solutions?

Such questions are answered in Chapter 11, using much of the information from Chapters 3 and 4.

PROBLEMS 2.7

1. Derive the second and third constraints of (2.55).
2. A balanced diet should consist of at least minimum amounts of certain nutrients. Suppose that we wish to determine, from a given collection of foods, the lowest-cost diet that satisfies the minimum requirements for certain nutrients; in particular, suppose that the table below gives the costs and nutrient levels in each of three foods and that we are to meet the minimum requirements listed there as cheaply as possible.

	Food 1	Food 2	Food 3	Minimum Number of Units Required
Nutrient 1	2	1	1	40
Nutrient 2	1	1	3	60
Cost	70	40	90	

Denote by y_1, y_2, and y_3 the amounts of each of the food types in the diet. Show that our problem is modeled by demanding that we find y_1, y_2, and y_3 to

$$\text{minimize } M = 70y_1 + 40y_2 + 90y_3$$

subject to the constraints

$$y_1 \geq 0, \quad y_2 \geq 0, \quad y_3 \geq 0,$$

$$2y_1 + y_2 + y_3 \geq 40,$$

$$y_1 + y_2 + 3y_3 \geq 60.$$

▷ **3.** Express the model in Problem 2 in matrix notation.

4. (a) By solving sets of two linear equations in two unknowns, find the x_1- and x_2-coordinates of each of the points $\mathcal{S}$, $\mathcal{Q}$, $\mathcal{P}$, $\mathcal{R}$, and $\mathcal{O}$ in (2.59).

(b) Evaluate M in (2.54) at each of these five points, and observe that the point that gives the largest value from these five is the same point found to solve the linear program (2.54)–(2.55).

▷ **5.** Suppose that the manufacturer in our main example decides to eliminate excess capacity by selling some type M_1 machines so that only 55 hours per week of M_1 machine time is available. Show that it is still possible to follow the production plan $x_1 = 15$ and $x_2 = 25$ and that this is still the optimal schedule.

6. Suppose in Problem 5 that the money gained from selling M_1 machines is used to obtain enough M_3 machines to increase the number of M_3 machine hours available each week to 100 hours instead of 90. Find the new optimal production schedule x_1, x_2 and the associated maximum profit; did this change produce increased profits?

7. Suppose in Problems 5 and 6 that instead the money from selling M_1 machines is used to buy M_2 machines so as to increase available M_2 machine time to 50 hours per week instead of 40. Find the new optimal production schedule x_1, x_2; did this change produce increased profits?

▷ **8.** Suppose that in (2.59) we change M to $M = 10x_1 + 40x_2$. Find the optimal production schedule x_1, x_2.

9. Suppose that in (2.59) we change M to $M = 90x_1 + 40x_2$. Find the optimal production schedule x_1, x_2.

▷ **10.** Find numbers c and d so that, if we change M to $M = cx_1 + dx_2$ in (2.59), the optimal production schedule will then occur at $\mathcal{Q}$, where $x_1 = 30$ and $x_2 = 10$.

11. Suppose that in (2.59) we change M to $M = 50x_1 + 50x_2$. Show that there are infinitely many optimal production schedules, namely $x_1 = \alpha$ and $x_2 = 40 - \alpha$ for all arbitrary α satisfying $15 \leq \alpha \leq 30$.

2.8 MISCELLANEOUS PROBLEMS

PROBLEMS 2.8

1. If a 3×3 matrix $\mathbf{A}$ satisfies appropriate conditions in addition to (2.3), it is possible to show that the powers $\mathbf{A}^r$ tend to a matrix $\mathbf{A}_\infty$ as r tends to infinity; in particular, suppose that the additional condition is that the matrix $\mathbf{A}$ is strictly positive: $\langle \mathbf{A} \rangle_{ij} \geq d$ for some number $d > 0$. (*Note:* d cannot possibly be more than $\frac{1}{3}$; why?)

(a) Let $a_{r,ij}$ denote the (i, j)-entry of $\mathbf{A}^r$: $\langle \mathbf{A}^r \rangle_{ij} = a_{r,ij}$; let m_r and M_r be the smallest and largest, respectively, of the entries $a_{r,1j}$ in the first row of $\mathbf{A}^r$. By writing out the entries of the first row of $\mathbf{A}^{r+1} = \mathbf{A}^r\mathbf{A}$ in terms of the entries of the first row of $\mathbf{A}^r$ and the various columns of $\mathbf{A}$, show that

$$m_{r+1} \geq M_r d + m_r(1 - d) \quad \text{and} \quad M_{r+1} \leq m_r d + M_r(1 - d).$$

(b) Deduce then that $(M_{r+1} - m_{r+1}) \leq (1 - 2d)(M_r - m_r)$.

(c) Deduce that $M_r - m_r$ tends to zero as r tends to infinity.

(d) Show that the numbers M_r decrease: $M_1 \geq M_2 \geq M_3 \cdots$, that the m_r increase: $m_1 \leq m_2 \leq m_3 \ldots$, and that the m_r and M_r both form bounded sequences.

(e) Conclude that both M_r and m_r converge, and to the same limit: that is, the first row of $\mathbf{A}^r$ converges to a row of the form $[\alpha \quad \alpha \quad \alpha]$ for some α.

(f) Show that applying the same type of argument to the second and third rows of $\mathbf{A}^r$ shows that $\mathbf{A}^r$ converges to some matrix of the form

$$\mathbf{A}_\infty = \begin{bmatrix} \alpha & \alpha & \alpha \\ \beta & \beta & \beta \\ \gamma & \gamma & \gamma \end{bmatrix} \qquad \text{with } \alpha, \beta, \text{ and } \gamma \text{ positive and } \alpha + \beta + \gamma = 1.$$

2. Generalize the theorem and proof in Problem 1 to apply to $p \times p$ matrices.

3. To explain the behavior of the fox and chicken populations in the model in Example 2.16 (where $k = 0.1$):

(a) Show that

$$F_{i+1} - C_{i+1} = 0.7(F_i - C_i) \quad \text{and} \quad 5C_{i+1} - F_{i+1} = 1.1(5C_i - F_i).$$

(b) Deduce that $(F_i - C_i)$ tends to zero while $5C_i - F_i$ tends to infinity when i tends to infinity.

(c) Conclude that F_i and C_i both tend to infinity.

4. To explain the behavior of the fox and chicken populations in the model in Example 2.17 (where $k = 0.18$):

(a) Show that

$$0.1C_{i+1} - 0.06F_{i+1} = 0.9(0.1C_i - 0.06F_i)$$

and that

$$2C_{i+1} - F_{i+1} = 0.9(2C_i - F_i) + (0.1C_i - 0.06F_i).$$

(b) Deduce that $0.1C_i - 0.06F_i$ tends to zero when i tends to infinity.
(c) Deduce that $2C_i - F_i$ tends to zero.
(d) Conclude that F_i and C_i both tend to zero.

5. Suppose that $\mathbf{x}$ is a $p \times 1$ column matrix with entries $\langle \mathbf{x} \rangle_i = x_i$ and that f is a real-valued function of the p variables $x_1, x_2, \ldots, x_p$. We denote $f(x_1, \ldots, x_p)$ by $f(\mathbf{x})$, and we define the $p \times 1$ column matrix $\partial f/\partial \mathbf{x}$ of partial derivatives of f with respect to its variables in the obvious way: $\langle \partial f/\partial \mathbf{x} \rangle_i = \partial f/\partial x_i$.
(a) If $f(\mathbf{x})$ is $\mathbf{c}^T\mathbf{x}$, where $\mathbf{c}$ is a constant $p \times 1$ column matrix, show that $\partial f/\partial \mathbf{x} = \mathbf{c}$.
(b) If $\mathbf{B}$ is a $p \times p$ symmetric matrix and $f(\mathbf{x})$ is (the entry in the 1×1 matrix) $\mathbf{x}^T\mathbf{A}\mathbf{x}$, show that $\partial f/\partial \mathbf{x} = 2\mathbf{A}\mathbf{x}$.
(c) Given $p \times q$ $\mathbf{A}$ and $p \times 1$ $\mathbf{y}$, for $\mathbf{x}$ to solve the least-squares problem (2.52) we need $\partial f/\partial \mathbf{x} = \mathbf{0}$, where $f(\mathbf{x})$ is (the entry in the 1×1 matrix) $(\mathbf{y} - \mathbf{A}\mathbf{x})^T(\mathbf{y} - \mathbf{A}\mathbf{x})$. Use (a) and (b) to show that this condition on the derivative just becomes $\mathbf{A}^T\mathbf{A}\mathbf{x} = \mathbf{A}^T\mathbf{y}$.

6. The following broad general topics—and associated bibliographic references—are closely related to the material covered in this chapter. It is possible to discuss the use of matrices in these problem areas either shallowly (based on the material in the first chapter and on a brief look into some of the references) or deeply (based on material in later chapters and on a detailed examination of the references and other sources). At a level as assigned by your teacher, use the suggested or other references to discuss the uses of matrices in one of the following areas.
(a) Markov chains—Karlin (17); and Kemeny and Snell (19 and 20).
(b) Equilibrium in networks—Martin (25); Robinson (29); and Strang (34).
(c) Electric circuits—Senturia and Wedlock (32); Strang (34); Tropper (35); and Weiss (37).
(d) Dynamics and differential equations—Bellman and Cooke (7); Braun (8); Frazer, Duncan, and Collar (10); Hurty and Rubinstein (16); and Strang (34).
(e) Least squares in modeling and statistics—Albert (5); Graybill (12); Linnik (22); Rao (28); Searle (31); and Walpole and Myers (36).
(f) Linear programming—Dantzig (9); Hadley (14); Karmarkar (18); and Simonnard (33).
(g) Economics—Karlin (17); Kemeny and Snell (19 and 20); Kemeny, Snell, and Thompson (21); Maki and Thompson (23); and Schwartz (30).
(h) Engineering and science—Amundsen (6); Haberman (13); Heading (15); Pipes (27); Searle (31); Strang (34); Tropper (35); and Weiss (37).
(i) Social and behavioral science—Haberman (13); Kemeny and Snell (19 and 20); Kemeny, Snell, and Thompson (21); and Maki and Thompson (23).

7. Once you have studied some of the later chapters, you will be able to discuss the use of matrices and linear algebra in areas in addition to those suggested in Problem 6; this problem is for assignment at such a later date. As in Problem 6, use the suggested references to discuss one of the following topics.

(a) Pseudoinverses—Albert (5); and Nashed (26).

(b) Nonlinear programming—Mangasarian (24).

(c) Theorems of the alternative for linear inequalities—Mangasarian (24).

(d) Computations with sparse matrices—Bjork, Plemmons, and Schneider (38); Bunch and Rose (39); Duff and Stewart (41); George and Liu (45); and Rose and Willoughby (51).

(e) Iterative methods for solving linear equations—Hageman and Young (47); and Varga (54).

(f) Solving least-squares problems—Golub and Van Loan (46); Lawson and Hanson (49); and Stewart (53).

(g) Computation of eigensystems—Golub and Van Loan (46); Parlett (50); Stewart (53); and Wilkinson (56).

(h) Nonnegative matrices—Varga (54).

3

Solving equations and finding inverses: methods

This chapter is the methodological heart of the book; the tools of Gauss elimination, elementary row operations, and back-substitution are used in virtually every subsequent chapter and must, therefore, be thoroughly mastered. Theorem 3.34 is a ***Key Theorem*** *because it is the basis for many of the theoretical arguments in Chapter 4. Sections 3.6, 3.8, and 3.9 will be of most interest to those with a computational orientation, while the material through Section 3.5 is central for those with theoretical, applied, or computational orientations. Section 3.7 falls somewhere in the middle; however,* ***Key Theorems 3.48*** *and* ***3.53*** *on the LU-decomposition should be known to all as an alternative view of Gauss elimination.*

3.1 *INTRODUCTION*

Chapter 2 provided ample evidence that systems of linear equations arise in the applications of mathematics to diverse areas; that chapter as well as Chapter 1—see Theorem 1.38 on equations and inverses, for example—showed too that matrices and their inverses are closely related to the problem of solving systems of linear equations. In this chapter we develop the matrix methodology for solving systems of equations and for finding inverses; the emphasis here is on *how*: *how* to discover whether a system has a solution or a matrix has an inverse, *how* to find all such solutions or inverses, and so on. Chapter 4 will then reexamine some of these procedures and problems from a theoretical viewpoint in order to develop a rigorous logical foundation for our practical approaches.

3.2 SOLVING EQUATIONS BY GAUSS ELIMINATION

The method of Gauss elimination is a straightforward but powerful procedure for reducing systems of linear equations to a simple *reduced form* easily solved by substitution. Note that many authors use "Gauss elimination" to refer to both the reduction and the following substitution, while we use it to refer to reduction only. To introduce it, we begin by ignoring matrix notation and writing out the equations in detail; later we will translate what we do into matrix language. Note that there are a number of equally effective variants of Gauss elimination; we present one in detail and mention some others later in problems.

Gauss Elimination for Equations

Consider the system of linear equations

$$\begin{aligned} -2x_1 + 2x_2 - 4x_3 - 6x_4 &= -4 \\ -3x_1 + 6x_2 + 3x_3 - 15x_4 &= -3 \\ 5x_1 - 8x_2 - x_3 + 17x_4 &= 9 \\ x_1 + x_2 + 11x_3 + 7x_4 &= 7. \end{aligned} \tag{3.1}$$

To apply *Gauss elimination* to (3.1), we use the first equation in (3.1) to eliminate the appearance of x_1 from the other three equations. To make the calculations simpler, first divide the first equation through by the coefficient -2 of x_1, thereby obtaining an equivalent equation in which the coefficient of x_1 is 1:

$$x_1 - x_2 + 2x_3 + 3x_4 = 2. \tag{3.2}$$

Next, use this equation to eliminate x_1 from the last three equations in (3.1).

> ***Note.*** This can be done in two *apparently* different ways: (1) by substituting, or (2) by combining equations. In substitution, you solve in (3.2) for x_1 in terms of the other variables and then substitute this expression in place of x_1 in each of the other equations; for example, the second equation of (3.1) then becomes
>
> $$-3(2 + x_2 - 2x_3 - 3x_4) + 6x_2 + 3x_3 - 15x_4 = -3,$$
>
> which, after combining terms, is just $3x_2 + 9x_3 - 6x_4 = 3$. With the combining approach, you add to each of the later equations an appropriate multiple of the first, so chosen that the coefficient of x_1 becomes zero in the new equation. For example, multiply the new first equation (3.2) in your head by 3 and add the result to the second equation in (3.1); you should get
>
> $$-3x_1 + 6x_2 + 3x_3 - 15x_4 + (3)(x_1 - x_2 + 2x_3 + 3x_4) = -3 + (3)2,$$

which, after combining terms, is just $3x_2 + 9x_3 - 6x_4 = 3$, exactly as in the first approach. The two approaches *always* give the same result. Since the combining-equations approach is easier to organize computationally, not requiring the exchange of terms across the equality sign, we will always use it. Thus,

(3.3) when we say "use equation A to eliminate variable B from equation C," we mean "replace equation C by itself plus that multiple of equation A that will produce a new equation not explicitly containing variable B."

Back now to Gauss elimination on (3.1): The task was to use equation (3.2) to eliminate x_1 from each of the other three equations. The result is

(3.4)
$$\begin{aligned} x_1 - x_2 + 2x_3 + 3x_4 &= 2 \\ 3x_2 + 9x_3 - 6x_4 &= 3 \\ -3x_2 - 11x_3 + 2x_4 &= -1 \\ 2x_2 + 9x_3 + 4x_4 &= 5. \end{aligned}$$

Note that the bottom three equations in (3.4) involve only the three unknowns x_2, x_3, and x_4, while the first equation allows the straightforward calculation of x_1 once x_2, x_3, and x_4 are found from the bottom equations.

The next step is to reduce the bottom three equations in three unknowns to two equations in two unknowns. First, divide the new second equation in (3.4) through by its coefficient 3 in order to simplify arithmetic; this replaces the second equation by

(3.5)
$$x_2 + 3x_3 - 2x_4 = 1.$$

Now use this new second equation to eliminate the second variable x_2 from the bottom two equations—for example, replace the third equation in (3.4) by itself plus 3 times the newest second equation (3.5). This replaces the set (3.4) of four equations by

(3.6)
$$\begin{aligned} x_1 - x_2 + 2x_3 + 3x_4 &= 2 \\ x_2 + 3x_3 - 2x_4 &= 1 \\ -2x_3 - 4x_4 &= 2 \\ 3x_3 + 8x_4 &= 3. \end{aligned}$$

Now we need to deal with the bottom two equations in two unknowns by the same process. We divide the third equation in (3.6) through by -2, replacing that equation by $x_3 + 2x_4 = -1$; we then use this equation to eliminate x_3 from the fourth equation in (3.6). After dividing this new fourth equation through by the new coefficient 2 found for x_4, we finally have the reduced system

$$\begin{aligned} x_1 - x_2 + 2x_3 + 3x_4 &= 2 \\ x_2 + 3x_3 - 2x_4 &= 1 \\ x_3 + 2x_4 &= -1 \\ x_4 &= 3. \end{aligned} \tag{3.7}$$

This completes Gauss elimination, having produced the *reduced form* (3.7) of the original equations (3.1). We define "reduced form" precisely in Chapter 4; the term describes a form produced by Gauss elimination.

We now use *back-substitution* to produce the solution to (3.7) as follows. The bottom equation in (3.7) clearly has exactly one solution: $x_4 = 3$. Now for back-substitution: Substituting $x_4 = 3$ into the third equation in (3.7) gives

$$x_3 = -1 - 2x_4 = -1 - 2(3) = -7;$$

substituting $x_4 = 3$ and $x_3 = -7$ into the second equation in (3.7) gives $x_2 = 28$; and one final back-substitution into the first equation of (3.7) gives $x_1 = 35$. If, as a check, you substitute these values into the *original* system (3.1), you will see that we do indeed have a solution.

You should now be able to solve Problems 1 and 2.

Gauss Elimination for Augmented Matrices

In performing Gauss elimination above, all the computation was done with the *numbers*—the coefficients and right-hand sides—rather than with the symbols x_i, although we certainly had to bother continuing to write down the x_i. Significant writing can be saved by working only with the numbers and not rewriting the variables all the time; we just need to be careful not to lose track of which numbers are coefficients of which unknowns, or are which right-hand sides. Matrices are ideally suited for this.

Instead of manipulating the equations, we operate on the *augmented matrix* $[\mathbf{A} \;\; \mathbf{b}]$ formed by appending the right-hand-side column $\mathbf{b}$ to the matrix $\mathbf{A}$ of coefficients to form a partitioned matrix; for the example worked above, the augmented matrix is

$$\left[\begin{array}{rrrr|r} -2 & 2 & -4 & -6 & -4 \\ -3 & 6 & 3 & -15 & -3 \\ 5 & -8 & -1 & 17 & 9 \\ 1 & 1 & 11 & 7 & 7 \end{array}\right]. \tag{3.8}$$

The first step in applying Gauss elimination for equations to (3.1) was to replace the first equation by itself divided by -2; in terms of the matrix of coefficients and right-hand sides, this just means that we replace the first *row*

$$[-2 \quad 2 \quad -4 \quad -6 \mid -4]$$

by itself divided by -2:

$$[1 \quad -1 \quad 2 \quad 3 \mid 2].$$

Note that this is just the matrix of numbers that appear in the new first equation (3.2). The next step using the *equations* approach was to use the new first *equation* to eliminate x_1 from the other three *equations*; for matrices, this means that we use the new first *row* to eliminate the entries in the first column from the other three *rows*. Note that

(3.9) when we say "use row A to eliminate the entry in column B from row C," we mean "replace row C by itself plus that multiple of row A that will produce a new row with a zero entry in column B";

note the similarity with the language (3.3) used to describe elimination for equations. Using the new first row $[1 \quad -1 \quad 2 \quad 3 \mid 2]$ to eliminate the entries in the first column from the other three rows of (3.8) replaces the augmented matrix (3.8) by

(3.10)
$$\left[\begin{array}{cccc|c} 1 & -1 & 2 & 3 & 2 \\ 0 & 3 & 9 & -6 & 3 \\ 0 & -3 & -11 & 2 & -1 \\ 0 & 2 & 9 & 4 & 5 \end{array}\right].$$

Note that (3.10) is precisely the augmented matrix for (3.4). Next, we replace the second row

$$[0 \quad 3 \quad 9 \quad -6 \mid 3]$$

in (3.10) by itself divided by 3—namely

$$[0 \quad 1 \quad 3 \quad -2 \mid 1]$$

—and then use that new second row to eliminate the entries in column 2 from the bottom two rows of (3.10); this produces

(3.11)
$$\left[\begin{array}{cccc|c} 1 & -1 & 2 & 3 & 2 \\ 0 & 1 & 3 & -2 & 1 \\ 0 & 0 & -2 & -4 & 2 \\ 0 & 0 & 3 & 8 & 3 \end{array}\right].$$

Note that (3.11) is precisely the augmented matrix for (3.6). Replace the third row in (3.11) by itself divided by -2, and use this new row to eliminate the entry in column 3 from the last row; this gives

(3.12)
$$\left[\begin{array}{cccc|c} 1 & -1 & 2 & 3 & 2 \\ 0 & 1 & 3 & -2 & 1 \\ 0 & 0 & 1 & 2 & -1 \\ 0 & 0 & 0 & 2 & 6 \end{array}\right].$$

Replacing the bottom row by itself divided by 2, we finally obtain the reduced matrix

(3.13)
$$\left[\begin{array}{cccc|c} 1 & -1 & 2 & 3 & 2 \\ 0 & 1 & 3 & -2 & 1 \\ 0 & 0 & 1 & 2 & -1 \\ 0 & 0 & 0 & 1 & 3 \end{array}\right].$$

This completes Gauss elimination, having produced the *reduced form* (3.13) of the original augmented matrix (3.8). *Note that (3.13) is precisely the augmented matrix for (3.7).* Had we not already done elimination on the equations, we would at this point interpret (3.13) as the augmented matrix for a system of equations—namely (3.7)—write that system down, and then use back-substitution to obtain the solution to the original system of equations.

The fact that we solved the system (3.1) in two different ways may have obscured the ease of using Gauss elimination on the augmented matrix. We consider another example as an illustration.

(3.14) ***Example.*** Consider the three linear equations in three unknowns

$$\begin{aligned} -3x_1 - 3x_2 - 3x_3 &= -3 \\ -2x_1 + 2x_2 + x_3 &= 0 \\ x_1 - 3x_2 + 3x_3 &= 0. \end{aligned}$$

We immediately write down the augmented matrix

(3.15)
$$\left[\begin{array}{ccc|c} -3 & -3 & -3 & -3 \\ -2 & 2 & 1 & 0 \\ 1 & -3 & 3 & 0 \end{array}\right].$$

Replacing the first row by itself divided by -3 and then using that new first row to eliminate the entries in the first column of the other two rows gives

$$\left[\begin{array}{ccc|c} 1 & 1 & 1 & 1 \\ 0 & 4 & 3 & 2 \\ 0 & -4 & 2 & -1 \end{array}\right].$$

Replacing the second row of this new augmented matrix by itself divided by 4 and then using that new row to eliminate the entry in the second column of the bottom row gives a matrix that—after dividing its bottom row by 5—in turn gives as the result of Gauss elimination the reduced matrix

$$\left[\begin{array}{ccc|c} 1 & 1 & 1 & 1 \\ 0 & 1 & 0.75 & 0.5 \\ 0 & 0 & 1 & 0.2 \end{array}\right].$$

If we interpret this final reduced matrix as the augmented matrix for a system of equations, then the last equation clearly gives $x_3 = 0.2$; back-substituting this into the second equation gives $x_2 = 0.35$; back-substituting both these values into the first equation gives $x_1 = 0.45$. If you substitute these into the original system of equations, you will see that Gauss elimination on the augmented matrix has indeed produced a solution.

You should now be able to solve Problems 1 to 5.

Interchanges in Gauss Elimination

In each case above we have proceeded systematically, using the rows (or equations) from the top down to eliminate in columns (the variables) from left to right. This is not necessary, in some cases it is impossible, and in practical computation—as you will see later—it is often unwise. If at the start we use a row other than the first to perform the elimination, it is simpler notationally to *interchange* this row with the first and then proceed as usual with this new first row; note that interchanging two rows is the same as writing the equations in a different order—which certainly has no impact on the solutions (see Theorem 4.12).

(3.16) ***Example.*** Consider again the system of three equations in three unknowns in Example 3.14 and its augmented matrix (3.15). Suppose that we decide to use the *third* row in (3.15) instead of the first for eliminating in the first column—perhaps because it already has a 1 as its first entry. We therefore interchange the first and third rows, obtaining

$$\left[\begin{array}{ccc|c} 1 & -3 & 3 & 0 \\ -2 & 2 & 1 & 0 \\ -3 & -3 & -3 & -3 \end{array}\right].$$

Replacing this new first row by itself divided by its first entry 1 of course changes nothing; using this first row to eliminate the entries in the first column from the other two rows gives

$$\left[\begin{array}{rrr|r} 1 & -3 & 3 & 0 \\ 0 & -4 & 7 & 0 \\ 0 & -12 & 6 & -3 \end{array}\right].$$

Simply in order to demonstrate that we can do so, suppose that we decide to use the third row for the elimination in the second column; we therefore interchange the present second and third rows to obtain

$$\left[\begin{array}{rrr|r} 1 & -3 & 3 & 0 \\ 0 & -12 & 6 & -3 \\ 0 & -4 & 7 & 0 \end{array}\right].$$

Replacing this second row by itself divided by -12 and then using the resulting new second row to eliminate the entry in the second column of the bottom row, we get a matrix that—after dividing its bottom row by 5—in turn gives the final reduced form

$$\left[\begin{array}{rrr|l} 1 & -3 & 3 & 0 \\ 0 & 1 & -0.5 & 0.25 \\ 0 & 0 & 1 & 0.2 \end{array}\right].$$

(Note that this is different from the reduced matrix obtained in Example 3.14; nonetheless, we will get the same solution to the equations.) If we interpret this reduced matrix as the augmented matrix for a system of equations, the bottom equation of course gives $x_3 = 0.2$; back-substitution produces $x_2 = 0.35$ and $x_1 = 0.45$—*exactly the same solution as obtained in Example 3.14 without interchanges.*

Example 3.16 illustrates that it is *possible* to interchange equations; sometimes it is *necessary*:

(3.17) ***Example.*** Consider the system of four linear equations in four unknowns

$$\begin{aligned} -2x_1 + 4x_2 - 2x_3 - 6x_4 &= 4 \\ 3x_1 - 6x_2 + 6x_3 + 10x_4 &= -1 \\ -2x_1 + 6x_2 - x_3 + x_4 &= 1 \\ 2x_1 - 5x_2 + 4x_3 + 8x_4 &= -3. \end{aligned}$$

The augmented matrix is

$$\left[\begin{array}{rrrr|r} -2 & 4 & -2 & -6 & 4 \\ 3 & -6 & 6 & 10 & -1 \\ -2 & 6 & -1 & 1 & 1 \\ 2 & -5 & 4 & 8 & -3 \end{array}\right].$$

Replacing the first row by itself divided by -2 and using the resulting first row to eliminate the entries in the first column from the bottom three rows gives

$$\left[\begin{array}{rrrr|r} 1 & -2 & 1 & 3 & -2 \\ 0 & 0 & 3 & 1 & 5 \\ 0 & 2 & 1 & 7 & -3 \\ 0 & -1 & 2 & 2 & 1 \end{array}\right].$$

Because the (2, 2)-entry in this matrix is zero, we cannot use the second row to eliminate in the second column: no matter what we multiply the second row by, when we add the result to a lower row the zero in the second entry of the second row leaves unchanged the second entry of the lower row. We are compelled to interchange the second row with the third or fourth.

Examples have so far demonstrated that it is always *possible* and sometimes *necessary* to interchange rows when performing Gauss elimination; we leave to Sections 3.6 and 3.9 the fact that it is often *wise* to do so in practical computations.

You should now be able to solve Problems 1 to 9.

Gauss Elimination on Augmented Matrices in General

Until now we have looked only at *examples* of systems of equations, augmented matrices, and Gauss elimination; we now examine the general situation and procedure. Look back at (1.36) in Section 1.4, where we showed that the general system

(3.18)
$$\begin{aligned} a_{11}x_1 + a_{12}x_2 + \cdots + a_{1q}x_q &= b_1 \\ a_{21}x_1 + a_{22}x_2 + \cdots + a_{2q}x_q &= b_2 \\ \cdots\cdots\cdots\cdots \\ a_{p1}x_1 + a_{p2}x_2 + \cdots + a_{pq}x_q &= b_p \end{aligned}$$

of p linear equations in q unknowns was equivalent to $\mathbf{Ax} = \mathbf{b}$ for $p \times q$ $\mathbf{A}$ with $\langle \mathbf{A} \rangle_{ij} = a_{ij}$, $q \times 1$ $\mathbf{x}$ with $\langle \mathbf{x} \rangle_j = x_j$, and $p \times 1$ $\mathbf{b}$ with $\langle \mathbf{b} \rangle_i = b_i$. Gauss elimination for matrices operates on the augmented matrix $[\mathbf{A} \quad \mathbf{b}]$:

(3.19) ***Definition.*** The partitioned matrix $[\mathbf{A} \quad \mathbf{b}]$ is the *augmented matrix* for the system of equations described by $\mathbf{Ax} = \mathbf{b}$.

Gauss elimination then proceeds to eliminate in the columns starting with column 1, then column 2, and so on; we use row r_j to eliminate in column j, and may well interchange a lower row with the r_jth row before elimination. We can describe the process using successive columns—(3.20)—or rows—see (4.1).

(3.20) Gauss elimination with interchanges on the $p \times (q + 1)$ augmented matrix $[\mathbf{A} \quad \mathbf{b}]$ proceeds as follows:

1. Set $j = 1$ and $r_1 = 1$ and use row r_j to eliminate in column j of the present augmented matrix as prescribed in steps 2 through 6.
2. Select one of the rows from among those numbered $r_j, r_j + 1, \ldots, p$ for use in eliminating in column j; call this row i, so that the (i, j)-entry—called the *pivot*—in the present augmented matrix must be nonzero. If there are no nonzero entries in this lower portion of column j, then no elimination is necessary: set $r_{j+1} = r_j$ to use the same row, and skip to step 6.
3. Interchange the ith and r_jth rows.
4. Replace this new r_jth row by itself divided by the pivot (its nonzero entry in its jth column).
5. Use this newest r_jth row to eliminate the entries in the jth column in the rows $r_j + 1, r_j + 2, \ldots, p$. Set $r_{j+1} = r_j + 1$ to use the next row.
6. (a) If $j \le q$ and $r_{j+1} \le p$, then further elimination is still possible: increase j by 1 and return to step 2.
   (b) Otherwise, Gauss elimination has been completed: go to step 7.
7. Interpret the final reduced matrix as the augmented matrix for a system of equations and proceed to find the solutions, if any, by back-substitution.

This outline applies to p equations in q unknowns, where p need not equal q. All the examples we have presented have had p equal to q; Section 3.3 will provide more general examples.

You should now be able to solve Problems 1 to 12.

Gauss–Jordan Elimination

Before looking at further examples of Gauss elimination, we consider a modification of this process that—although inefficient for practical calculations—is often useful for theoretical purposes.

In Gauss elimination, we use a particular row (equation) to eliminate *but only* in the *rows* (*equations*) ***below*** *that row* (*equation*). If we choose also to eliminate in the rows (equations) ***above*** that row (equation), the process is called *Gauss-Jordan elimination.* We will show later that Gauss–Jordan elimination involves more arithmetic than does Gauss elimination, so we are introducing it as a theoretical rather than computational tool. See Problem 14, however.

(3.21) ***Example.*** Consider using Gauss–Jordan elimination to solve (3.1); since the first step is identical with Gauss elimination, Gauss–Jordan elimination on the original augmented matrix (3.8) leads first to that in (3.10). After the second row in (3.10) is divided by 3, the resulting second row is used to eliminate *above* this row as well as below; this gives

$$\left[\begin{array}{cccc|c} 1 & 0 & 5 & 1 & 3 \\ 0 & 1 & 3 & -2 & 1 \\ 0 & 0 & -2 & -4 & 2 \\ 0 & 0 & 3 & 8 & 3 \end{array}\right].$$

We now replace the third row by itself divided by -2 and use the result to eliminate *above* this row as well as below; this yields

$$\left[\begin{array}{cccc|c} 1 & 0 & 0 & -9 & 8 \\ 0 & 1 & 0 & -8 & 4 \\ 0 & 0 & 1 & 2 & -1 \\ 0 & 0 & 0 & 2 & 6 \end{array}\right].$$

Dividing the bottom row by 2 and using the result to eliminate *above* it produces

$$\left[\begin{array}{cccc|c} 1 & 0 & 0 & 0 & 35 \\ 0 & 1 & 0 & 0 & 28 \\ 0 & 0 & 1 & 0 & -7 \\ 0 & 0 & 0 & 1 & 3 \end{array}\right].$$

Interpreting this as the augmented matrix for a system of equations produces a trivially–solved system with the same solution found earlier by Gauss elimination.

Gauss–Jordan elimination always leads to a trivially-solved system, for example as found in Example 3.21; in essence, Gauss–Jordan elimination performs the back-substitution of Gauss elimination as it goes along rather than waiting until the end. We repeat that Gauss–Jordan elimination turns out to involve more arithmetic for obtaining the solution to the system of equations than does Gauss elimination; see Problem 14, however.

PROBLEMS 3.2

▷ **1.** Solve each of the following systems of equations by Gauss elimination for equations.

(a)
$$\begin{aligned} 3x_1 - x_2 + x_3 &= -1 \\ 9x_1 - 2x_2 + x_3 &= -9 \\ 3x_1 + x_2 - 2x_3 &= -9 \end{aligned}$$

(b)
$$\begin{aligned} x_1 + 2x_2 - 5x_3 + x_4 &= 3 \\ 2x_1 - 3x_2 + 4x_3 - 2x_4 &= 2 \\ 4x_1 + x_2 - 6x_3 + 3x_4 &= 11 \\ 5x_1 + 9x_2 - 20x_3 + x_4 &= 10 \end{aligned}$$

2. Use Gauss elimination for equations to solve

$$\begin{aligned} 2x_1 + x_2 + 2x_3 + x_4 &= 5 \\ 4x_1 + 3x_2 + 7x_3 + 3x_4 &= 8 \\ -8x_1 - x_2 - x_3 + 3x_4 &= 4 \\ 6x_1 + x_2 + 2x_3 - x_4 &= 1. \end{aligned}$$

3. Each of the following matrices is an augmented matrix for a system of equations; write down the system.

(a) $\left[\begin{array}{rrr|r} 5 & 0 & 2 & -1 \\ -2 & 3 & 8 & 3 \\ 0 & 6 & 5 & -8 \\ 0 & 0 & 4 & 5 \end{array}\right]$ (b) $\left[\begin{array}{rr|r} 0 & 5 & 4 \\ 2 & 4 & 3 \\ 1 & 3 & 6 \end{array}\right]$

(c) $\left[\begin{array}{rrrr|r} 2 & 5 & 3 & 8 & 9 \\ 1 & 4 & 8 & 2 & 2 \end{array}\right]$

▷ **4.** Use Gauss elimination for augmented matrices to solve

$$\begin{aligned} -x_1 - x_2 + 2x_3 &= -5 \\ -3x_1 - x_2 + 7x_3 &= -22 \\ x_1 - 3x_2 - x_3 &= 10. \end{aligned}$$

5. Use Gauss elimination for augmented matrices to solve

$$\begin{aligned} -3x - 6y + 9z &= 0 \\ x + 4y + z &= 6 \\ 2x + 8y + 3z &= 13. \end{aligned}$$

6. Complete the solution of the system of equations in Example 3.17.

7. Interchange the second and third equations in Problem 4 before starting Gauss elimination for augmented matrices; verify that the solution you obtain solves both the original set of equations and the set with the two equations interchanged.

▷ **8.** (a) Use Gauss elimination for augmented matrices on

$$\begin{aligned} 3x - y - 2z &= 0 \\ -6x + 2y + 6z &= 4 \\ 2x + y + 6z &= 13 \end{aligned}$$

to obtain both the reduced form of the system and the solution; you should have to interchange the second and third equations *after* eliminating x.

(b) Interchange the second and third equations *before* using Gauss elimination for augmented matrices to obtain both the reduced form of the system and the solution; no interchange should be necessary during elimination.

(c) Compare the reduced forms of the equations and the solutions for the two methods (a) and (b).

9. Use Gauss elimination for augmented matrices to solve

$$\begin{aligned} 2x - y + 2z + 3w &= 3 \\ -4x + 6y - 3z - 6w &= 2 \\ 6x - y + 8z + 5w &= 9 \\ 4x - 2y + 6z + 12w &= 12. \end{aligned}$$

▷ **10.** Use Gauss elimination for augmented matrices to solve

$$\begin{aligned} -2x - 4y + 2z - 6w &= 0 \\ 3x + 6y - 2z + 13w &= 6 \\ 2x + 4y \qquad + 14w &= 12 \\ 4x + 8y - 7z \qquad &= -10. \end{aligned}$$

11. Use Gauss elimination for augmented matrices to solve

$$\begin{aligned} 2x + 6y - 4z &= 4 \\ -x - 3y + 5z &= 4 \\ -3x - 9y + 4z &= -11. \end{aligned}$$

▷ **12.** Find all 3×1 column matrices $\mathbf{b}$ for which there is at least one solution to $\mathbf{Ax} = \mathbf{b}$ and find all solutions $\mathbf{x}$ associated with that $\mathbf{b}$, where

$$\mathbf{A} = \begin{bmatrix} 4 & -1 & 2 & 6 \\ -1 & 5 & -1 & -3 \\ 3 & 4 & 1 & 3 \end{bmatrix}.$$

13. Solve the two systems of equations in Problem 1 by Gauss–Jordan elimination for augmented matrices.

14. Consider the following process for solving the equations $\mathbf{Ax} = \mathbf{b}$. First, perform Gauss elimination for augmented matrices. Then, starting at the *bottom right* and working toward the left, eliminate *above* the main diagonal as you would in Gauss–Jordan elimination.
(a) Show that this results in the same final augmented matrix as if Gauss–Jordan elimination had been performed at the outset.
(b) Show that the arithmetic involved is the same as in performing Gauss elimination followed by back-substitution.
(c) Explain why this approach involves less work than does performing Gauss–Jordan elimination from the outset.
(d) Use this on (3.1).

3.3 EXISTENCE OF SOLUTIONS TO SYSTEMS OF EQUATIONS: SOME EXAMPLES AND PROCEDURES

Each concrete system of equations considered in Section 3.2 involved as many equations as unknowns, and our elimination methods always produced a solution. Our methods work more generally, however.

The Number of Solutions

Consider first the apparently simple case of just one equation $ax = b$ in the one unknown x; we tend to say immediately that the solution of this equation is $x = b/a$, but in fact there are three possibilities:

1. If $a \neq 0$, then $x = b/a$ makes sense, and this is the *unique* solution.
2. If $a = 0$, then there are two possibilities:
   (a) If $b \neq 0$, then the equation asks us to find x so that $0x = b \neq 0$, and *no* solution x exists. We say that "no solution exists" or that "the equation is inconsistent" since it implies that $0 = b \neq 0$, a contradiction.
   (b) If $b = 0$, then there are *infinitely many* solutions: Every number x is a solution since $0x = 0 = b$ no matter what x is.

It is striking that precisely the same three possibilities—*exactly one* (*unique*) solution, *no* solutions, or *infinitely many* solutions—hold for two equations in two unknowns. For example the equations

$$x_1 + x_2 = 2$$

$$x_1 - x_2 = 0$$

have a unique solution,

$$x_1 + x_2 = 2$$
$$x_1 + x_2 = 1$$

are inconsistent, and

$$x_1 + x_2 = 2$$
$$2x_1 + 2x_2 = 4$$

have infinitely many solutions, namely $x_1 = k$, $x_2 = 2 - k$, for all k. Even more surprising is the fact that precisely the same three possibilities hold for p equations in q unknowns: *exactly one* (*unique*) solution, *no* solution, or *infinitely many* solutions. We will prove this later; for the moment, we can rely on Gauss elimination to discover all (if any) solutions to any concrete system of equations.

Gauss Elimination and the Existence of Solutions: Examples

(3.22) ***Example.*** Consider the three equations in three unknowns

$$\begin{aligned} x_1 + 2x_2 - 5x_3 &= 2 \\ 2x_1 - 3x_2 + 4x_3 &= 4 \\ 4x_1 + x_2 - 6x_3 &= 8. \end{aligned}$$

The augmented matrix is

$$\left[\begin{array}{ccc|c} 1 & 2 & -5 & 2 \\ 2 & -3 & 4 & 4 \\ 4 & 1 & -6 & 8 \end{array}\right].$$

Gauss elimination reduces this to

$$\left[\begin{array}{ccc|c} 1 & 2 & -5 & 2 \\ 0 & 1 & -2 & 0 \\ 0 & 0 & 0 & 0 \end{array}\right]$$

which we interpret as the augmented matrix for the system of equations

$$\begin{aligned} x_1 + 2x_2 - 5x_3 &= 2 \\ x_2 - 2x_3 &= 0 \\ 0x_3 &= 0. \end{aligned}$$

This bottom equation—which we usually rely on to determine x_3—is solved by $x_3 = k$ for all arbitrary numbers k. Back-substitution then gives $x_2 = 2k$,

$x_1 = 2 + k$ and we can write the solution as

$$\begin{bmatrix} x_1 \\ x_2 \\ x_3 \end{bmatrix} = \begin{bmatrix} 2 \\ 0 \\ 0 \end{bmatrix} + k \begin{bmatrix} 1 \\ 2 \\ 1 \end{bmatrix} \quad \text{for arbitrary } k.$$

Thus there are infinitely many solutions, and Gauss elimination found them for us.

(3.23) ***Example.*** Consider the three linear equations in four unknowns

$$\begin{aligned} x_1 + 2x_2 - x_3 + 2x_4 &= 4 \\ 2x_1 + 7x_2 + x_3 + x_4 &= 14 \\ 3x_1 + 8x_2 - x_3 + 4x_4 &= 17 \end{aligned}$$

whose augmented matrix is

$$\left[\begin{array}{rrrr|r} 1 & 2 & -1 & 2 & 4 \\ 2 & 7 & 1 & 1 & 14 \\ 3 & 8 & -1 & 4 & 17 \end{array}\right].$$

Gauss elimination on this matrix reduces it to

$$\left[\begin{array}{rrrr|r} 1 & 2 & -1 & 2 & 4 \\ 0 & 1 & 1 & -1 & 2 \\ 0 & 0 & 0 & 0 & 1 \end{array}\right].$$

We can interpret this as the augmented matrix for

$$\begin{aligned} x_1 + 2x_2 - x_3 + 2x_4 &= 4 \\ x_2 + x_3 - x_4 &= 2 \\ 0x_4 &= 1. \end{aligned}$$

The bottom equation—which we normally rely on to determine x_4—asks us to find x_4 so that $0x_4 = 1$. No such x_4 can exist, so the system of equations has no solution.

Our examples and the Problems show that it is not in general possible to determine whether equations have no, one, or infinitely many solutions merely from the *numbers* of equations or unknowns. Two equations in 10 unknowns can be inconsistent, whereas 10 equations in two unknowns can have a unique solution. Gauss elimination is the method for determining which situation prevails.

You should now be able to solve Problems 1 to 7.

Leading Variables and Leading Columns

In Example 3.22 it was possible to assign an arbitrary value to x_3, but then the remaining variables were completely determined by that value. This raises the question of how to tell, after Gauss elimination has been applied, which variables can be assigned arbitrary values and which are then completely determined. One approach is easily described in terms of *leading variables* (or *leading columns*).

(3.24) ***Definition.*** The *leading variable* in an equation is the first (reading from left to right) variable in that equation with a nonzero coefficient. The *leading column* for a row of a matrix is the column containing the first (reading from left to right) nonzero entry in that row.

In Chapter 4 we develop theory supporting the following rule:

(3.25) ***Preview on Solutions of Equations.*** After completing Gauss elimination on the augmented matrix $[\mathbf{A} \quad \mathbf{b}]$ of the system of equations $\mathbf{Ax} = \mathbf{b}$, find the leading variables (leading columns) of the reduced equations (reduced augmented matrix). Then:

1. There are no solutions if and only if the last column is a leading column for some row.
2. If the last column is not a leading column for any row:
   (a) There is a unique solution if and only if every variable is a leading variable for some equation.
   (b) There are infinitely many solutions if and only if there are some variables that are *not* leading variables; each such *nonleading variable* can be assigned completely arbitrary values, and then each leading variable is completely determined in terms of the values assigned the nonleading variables.

(3.26) ***Example.*** Consider the system of equations in Example 3.22, where we applied Gauss elimination to its augmented matrix. The leading columns in the resulting reduced matrix are numbers 1 and 2; the final column, number 4, is *not* a leading column, nor is column 3. Since x_3 is not a leading variable, according to (3.25) (2) (b) there are infinitely many solutions: we can assign an arbitrary value to x_3, in terms of which the others are completely determined. This agrees with what we found earlier in Example 3.22.

PROBLEMS 3.3

▷ **1.** Use Gauss elimination to solve

$$\begin{aligned} 2x - 3y &= -1 \\ 2x + \ \ y &= \ \ 3 \\ x - 3y &= -2. \end{aligned}$$

2. Use Gauss elimination to solve

$$\begin{aligned} 2x - 3y &= -1 \\ 2x + \ \ y &= \ \ 3 \\ x - \ \ y &= \ \ 2. \end{aligned}$$

3. Use Gauss elimination to solve

$$\begin{aligned} 2x - \ \ 3y &= -1 \\ -4x + \ \ 6y &= -2 \\ 12x - 18y &= -6. \end{aligned}$$

▷ **4.** Use Gauss elimination to solve

$$\begin{aligned} x_1 + 2x_2 - \ \ 3x_3 + \ \ x_4 &= \ \ 0 \\ 2x_1 + 5x_2 \qquad\qquad\qquad &= \ \ 8 \\ x_1 \qquad\ - 14x_3 + 8x_4 &= -15 \\ -2x_1 - 3x_2 + 14x_3 + 2x_4 &= \ \ 10. \end{aligned}$$

5. Use Gauss elimination to solve

$$\begin{aligned} -x_1 + \ \ x_2 &= \ \ 1 \\ 3x_1 - 2x_2 &= -1 \\ 2x_1 - \ \ x_2 &= \ \ 1. \end{aligned}$$

6. Give an example of (a) a system with fewer equations than unknowns but no solutions; (b) a system with fewer equations than unknowns and infinitely many solutions. (See Problem 7 also.)

7. Explain why you cannot possibly find two equations in three unknowns having exactly one solution.

▷ **8.** Each of the matrices below is the reduced form resulting from completing Gauss elimination on the augmented matrix of some system of equations. For each matrix: (1) identify its leading columns; (2) identify its leading variables;

(3) use the rule in (3.25) to conclude from (1) and (2) whether the original system has no, exactly one, or infinitely many solutions.

(a) $\left[\begin{array}{cc|c} 1 & -2 & 4 \\ 0 & 1 & 6 \\ 0 & 0 & 0 \end{array}\right]$ (b) $\left[\begin{array}{cc|c} 1 & -2 & 4 \\ 0 & 1 & 6 \\ 0 & 0 & 2 \end{array}\right]$ (c) $\left[\begin{array}{ccc|c} 0 & 1 & 2 & 3 \\ 0 & 0 & 0 & 0 \\ 0 & 0 & 0 & 0 \end{array}\right]$

9. For each of the reduced-form augmented matrices in Problem 8, write down the system of equations represented by the reduced-form augmented matrix and use back-substitution to solve the system if possible.

▷ **10.** Find all possible 3×4 matrices $\mathbf{X}$ for which $\mathbf{AX} = \mathbf{0}$, where $\mathbf{A}$ is

$$\mathbf{A} = \begin{bmatrix} 1 & -2 & 3 \\ -2 & 5 & -6 \\ 2 & -3 & 6 \end{bmatrix}.$$

11. Suppose that $\mathbf{A}$ is the matrix of Problem 10 and that $\mathbf{B}$ and $\mathbf{C}$ are 3×4 matrices such that $\mathbf{AB} = \mathbf{AC}$. What can you conclude about the relationship between $\mathbf{B}$ and $\mathbf{C}$? (See Problem 10.)

3.4 *FINDING INVERSES BY GAUSS ELIMINATION*

Chapter 1 demonstrated that matrix inverses and systems of equations were closely related in several ways; Theorem 1.38 in Section 1.4 explained how inverses provide information about the solution of equations, while Theorem 1.44 explained how to find inverses by solving several systems of equations. Since we have just developed Gauss elimination as a systematic approach to solving systems of equations, we can apply it to the problem of finding inverses—thanks to Theorem 1.44.

According to Theorem 1.44, we can construct a right-inverse $\mathbf{R}$ of $\mathbf{A}$ as

$$\mathbf{R} = [\mathbf{r}_1 \quad \mathbf{r}_2 \quad \cdots \quad \mathbf{r}_p]$$

if each $\mathbf{r}_i$ solves $\mathbf{Ar}_i = \mathbf{e}_i$. Similarly, we can construct a left-inverse $\mathbf{L}$ with

$$\mathbf{L}^T = [\mathbf{l}_1 \quad \mathbf{l}_2 \quad \cdots \quad \mathbf{l}_q]$$

if each $\mathbf{l}_j$ solves $\mathbf{A}^T\mathbf{l}_j = \mathbf{e}_j$. And, of course, if $\mathbf{A}$ is square ($p = q$) and nonsingular, then we can find the two-sided inverse $\mathbf{A}^{-1}$ as either $\mathbf{R}$ or $\mathbf{L}$.

For any of these inverses, the computational problem is similar:

(3.27) Given a general matrix $\mathbf{C}$ and n column matrices $\mathbf{b}_1, \mathbf{b}_2, \ldots, \mathbf{b}_n$, solve the n systems of equations $\mathbf{Cx}_i = \mathbf{b}_i$ for $1 \le i \le n$ for the n solutions $\mathbf{x}_1, \mathbf{x}_2, \ldots, \mathbf{x}_n$.

To find a right-inverse $\mathbf{R}$ of $\mathbf{A}$, take $\mathbf{C} = \mathbf{A}$; to find a left-inverse, take $\mathbf{C} = \mathbf{A}^T$; in either case, the $\mathbf{b}_i$ are unit column matrices $\mathbf{e}_i$ of appropriate order. The situation in (3.27) arises in other circumstances as well, however. For example, the equation $\mathbf{Cx} = \mathbf{b}$ may model some physical system, with $\mathbf{b}$ representing known inputs and $\mathbf{x}$ unknown outputs {see (2.24), where $\mathbf{b}$ describes external forces on a framework and $\mathbf{x}$ describes its resulting displacement}; (3.27) might then represent a design study in which we want to examine how the system responds to a variety of inputs.

We could attack (3.27) one system of equations at a time: Form the augmented matrix $[\mathbf{C} \quad \mathbf{b}_1]$ and use Gauss elimination to get $\mathbf{x}_1$; form $[\mathbf{C} \quad \mathbf{b}_2]$ and use Gauss elimination to get $\mathbf{x}_2$; . . . ; and finally form $[\mathbf{C} \quad \mathbf{b}_n]$ and use Gauss elimination to get $\mathbf{x}_n$. This is very inefficient for the following reason: *The operations performed in the elimination phase prior to back-substitution* (dividing a row by a constant, interchanging two rows, adding a multiple of one row to another) *are completely determined by the coefficients in the equations and are independent of the right-hand side.* The right-hand sides of course enter into the calculations, but they do not affect *which* operations are performed; this means that, given all the right-hand sides at the start as in (3.27), we can perform the entire elimination *with all the right-hand sides at the same time.* That is:

(3.28) To solve (3.27), perform Gauss elimination on the *multiply-augmented* matrix $[\mathbf{C} \quad \mathbf{b}_1 \quad \mathbf{b}_2 \quad \cdots \quad \mathbf{b}_n]$, interpret the reduced form as the multiply-augmented matrix for n reduced systems, and solve each reduced system by back-substitution.

(3.29) ***Example.*** Suppose that we need to solve $\mathbf{Cx}_i = \mathbf{b}_i$ for $i = 1, 2, 3$, where

$$\mathbf{C} = \begin{bmatrix} 2 & 1 \\ 1 & 1 \end{bmatrix}, \quad \mathbf{b}_1 = \begin{bmatrix} 3 \\ 2 \end{bmatrix}, \quad \mathbf{b}_2 = \begin{bmatrix} 5 \\ 3 \end{bmatrix}, \quad \mathbf{b}_3 = \begin{bmatrix} 3 \\ 1 \end{bmatrix}.$$

We form the multiply-augmented matrix $[\mathbf{C} \quad \mathbf{b}_1 \quad \mathbf{b}_2 \quad \mathbf{b}_3]$ and perform elimination as follows:

$$\left[\begin{array}{cc|c|c|c} 2 & 1 & 3 & 5 & 3 \\ 1 & 1 & 2 & 3 & 1 \end{array}\right]$$

then

$$\left[\begin{array}{cc|c|c|c} 1 & 0.5 & 1.5 & 2.5 & 1.5 \\ 1 & 1 & 2 & 3 & 1 \end{array}\right]$$

then

$$\left[\begin{array}{cc|c|c|c} 1 & 0.5 & 1.5 & 2.5 & 1.5 \\ 0 & 0.5 & 0.5 & 0.5 & -0.5 \end{array}\right]$$

and finally,

$$\left[\begin{array}{cc|c|c|c} 1 & 0.5 & 1.5 & 2.5 & 1.5 \\ 0 & 1 & 1 & 1 & -1 \end{array}\right].$$

We interpret this as the multiply-augmented matrix for three systems of equations and back-substitute for each solution. The first, for example, is

$$\begin{aligned} x_1 + 0.5x_2 &= 1.5 \\ x_2 &= 1, \end{aligned}$$

which easily gives the first solution as $\mathbf{x}_1 = [1 \quad 1]^T$. Handling the second system similarly gives $\mathbf{x}_2 = [2 \quad 1]^T$. The third solution is $\mathbf{x}_3 = [2 \quad -1]^T$.

The approach (3.28) can be quite cost-effective in solving (3.27); if $p = q = 50$ and $n = 10$, for example, (3.28) involves only about 15% of the work of solving each of the 10 sets of 50 equations in 50 unknowns by repeating the entire elimination process on one set of equations at a time.

PROBLEMS 3.4

▷ **1.** Find the right-inverse **R** of the matrix **A** below by applying Gauss elimination to the multiply-augmented matrix

$$[\mathbf{A} \quad \mathbf{e}_1 \quad \mathbf{e}_2 \quad \mathbf{e}_3] = [\mathbf{A} \quad \mathbf{I}_3]$$

and verify that **R** is a two-sided inverse.

$$\mathbf{A} = \begin{bmatrix} -1 & 2 & 1 \\ 0 & 1 & -2 \\ 1 & 4 & -1 \end{bmatrix}.$$

2. Use Gauss elimination on a multiply-augmented matrix to find the right-inverses of the matrices below.

(a) $\begin{bmatrix} 2 & 4 & 2 \\ 4 & 2 & -14 \\ 2 & 6 & 11 \end{bmatrix}$ (b) $\begin{bmatrix} 3 & 6 & -3 \\ 2 & 5 & 1 \end{bmatrix}$ (c) $\begin{bmatrix} 2 & -3 \\ -8 & 12 \end{bmatrix}$.

3. Use Gauss elimination on a multiply-augmented matrix to find the left-inverses of the matrices below.

(a) $\begin{bmatrix} 2 & -4 & -6 \\ 4 & -6 & -16 \\ -6 & 8 & 29 \end{bmatrix}$ (b) $\begin{bmatrix} 2 & 4 \\ -3 & -4 \\ 4 & 5 \end{bmatrix}$ (c) $\begin{bmatrix} 2 & -6 \\ -1 & 3 \\ 3 & -9 \end{bmatrix}$.

▷ **4.** Use Gauss elimination on the multiply-augmented matrix to find the three solutions $\mathbf{x}_1$, $\mathbf{x}_2$, $\mathbf{x}_3$ to $\mathbf{A}\mathbf{x}_i = \mathbf{b}_i$ for $i = 1, 2, 3$, where

$$\mathbf{A} = \begin{bmatrix} 2 & -8 & 4 \\ -4 & 19 & 10 \\ 6 & -22 & 15 \end{bmatrix}, \quad \mathbf{b}_1 = \begin{bmatrix} -2 \\ 25 \\ -1 \end{bmatrix}, \quad \mathbf{b}_2 = \begin{bmatrix} -10 \\ 44 \\ -23 \end{bmatrix}, \quad \mathbf{b}_3 = \begin{bmatrix} 8 \\ 23 \\ 32 \end{bmatrix}.$$

5. A health-food distributor wishes to create two powdered food supplements, one (for adults) containing 100% of the MDR (minimum daily requirement) of calcium, vitamin D, and phosphorus, and the second (for children) containing 75%, 100%, and 75%, respectively, of the MDRs. Three commercial compounds are available to create these special mixtures; 1 gram of these compounds contains that fraction of the MDR shown below.

	Compound 1	Compound 2	Compound 3
Calcium	0.15 MDR	0.23 MDR	0.26 MDR
Vitamin D	0.26 MDR	0.27 MDR	0 MDR
Phosphorus	0.15 MDR	0.28 MDR	0.08 MDR

Determine how many grams of each compound to blend for each of the food supplements in order that each exactly meets the MDR contents described above by setting up two systems of three equations in three unknowns and using a multiply-augmented matrix.

3.5 *ROW OPERATIONS AND ELEMENTARY MATRICES*

Gauss elimination was introduced in Section 3.2 as a way of manipulating equations; later it was viewed in matrix terms as a process for applying to augmented matrices. In fact, the *process itself* can be expressed in matrix terms as multiplication by a sequence of matrices; this turns out to be an essential tool in the study of some matrix problems.

Row Operations

Gauss elimination consists of a sequence of simple operations on the rows of an augmented matrix. These same simple operations can be used to learn a great deal about matrices other than augmented matrices as we will later see, and so are of interest in their own right.

(3.30) ***Definition.*** An *elementary row operation* (sometimes called just a *row operation*) on a matrix $\mathbf{A}$ is any one of the following three types of operations:
1. Interchange of two rows of $\mathbf{A}$.
2. Replacement of a row $\mathbf{r}$ of $\mathbf{A}$ by $c\mathbf{r}$ for some number $c \neq 0$.
3. Replacement of a row $\mathbf{r}_1$ of $\mathbf{A}$ by the sum $\mathbf{r}_1 + c\mathbf{r}_2$ of that row and a multiple of *another* row $\mathbf{r}_2$ of $\mathbf{A}$.

We have claimed that one of the uses of matrix notation is to express complicated relationships or operations in concise form; we do this for elementary row operations.

Elementary Matrices

(3.31)   **Definition.** A $p \times p$ *elementary matrix* is a matrix produced by applying exactly one elementary row operation to $\mathbf{I}_p$. $\mathbf{E}_{ij}$ is the elementary matrix obtained by interchanging the ith and jth rows of $\mathbf{I}_p$. $\mathbf{E}_i(c)$ is the elementary matrix obtained by multiplying the ith row of $\mathbf{I}_p$ by $c \neq 0$. $\mathbf{E}_{ij}(c)$ is the elementary matrix obtained by adding c times the jth row to the ith row of $\mathbf{I}_p$, where $i \neq j$.

The following are examples of 2×2 elementary matrices:

$$\begin{bmatrix} 0 & 1 \\ 1 & 0 \end{bmatrix}, \quad \begin{bmatrix} c & 0 \\ 0 & 1 \end{bmatrix} \text{ with } c \neq 0, \quad \begin{bmatrix} 1 & c \\ 0 & 1 \end{bmatrix}, \quad \text{and} \quad \begin{bmatrix} 1 & 0 \\ 0 & 1 \end{bmatrix}.$$

The first three of these are straightforward 2×2 illustrations of the three basic elementary matrices $\mathbf{E}_{12}$, $\mathbf{E}_1(c)$, and $\mathbf{E}_{12}(c)$. The fourth is, of course, just $\mathbf{I}_2$ itself, which can be thought of as $\mathbf{E}_{12}(0)$; because of this possibility, $\mathbf{I}_p$ is itself always an elementary matrix.

Consider now what happens when we premultiply any $2 \times q$ matrix $\mathbf{A}$ by one of these sample elementary matrices. For example, consider $\mathbf{E}_{12}(c)\mathbf{A}$:

$$\begin{bmatrix} 1 & c \\ 0 & 1 \end{bmatrix}\begin{bmatrix} x & y & z \\ u & v & w \end{bmatrix} = \begin{bmatrix} x + cu & y + cv & z + cw \\ u & v & w \end{bmatrix}.$$

This is the same as applying to $\mathbf{A}$ precisely the row operation that was applied to $\mathbf{I}_2$ to get the elementary matrix $\mathbf{E}_{12}(c)$ that premultiplied $\mathbf{A}$. These two operations *always* produce the same result.

by row op a matrix mult will get same answer

(3.32)   **Theorem** (elementary matrices and row operations). Suppose that $\mathbf{E}$ is a $p \times p$ elementary matrix produced by applying a particular elementary row operation to $\mathbf{I}_p$, and that $\mathbf{A}$ is a $p \times q$ matrix. Then $\mathbf{E}\mathbf{A}$ is the matrix that results from applying that same elementary row operation to $\mathbf{A}$.

PROOF. We prove the result for elementary matrices $\mathbf{E}_{ij}(c)$ and leave the proofs for the other two types of elementary matrices to the Problems. Observe that $\mathbf{E}_{ij}(c) = \mathbf{I}_p + c\mathbf{e}_i\mathbf{e}_j^T$, where $\mathbf{e}_i$ and $\mathbf{e}_j$ are the usual unit column matrices of order p. Therefore,

$$\mathbf{E}_{ij}(c)\mathbf{A} = (\mathbf{I}_p + c\mathbf{e}_i\mathbf{e}_j^T)\mathbf{A} = \mathbf{A} + c\mathbf{e}_i\mathbf{e}_j^T\mathbf{A}.$$

Now $\mathbf{e}_j^T\mathbf{A}$, by the definition of matrix multiplication, is just the jth row of $\mathbf{A}$; because of the same definition, $\mathbf{e}_i$ times this row is just a $p \times q$ matrix whose ith row equals this row and whose other entries are all zero. Thus $c\mathbf{e}_i\mathbf{e}_j^T\mathbf{A}$ is just that $p \times q$ matrix whose ith row is c times the jth row of $\mathbf{A}$ and whose other entries equal zero; adding this to $\mathbf{A}$ just performs the row operation we are discussing to the matrix $\mathbf{A}$, as asserted. ∎

This theorem tells us, for example, that Gauss elimination consists of a sequence of premultiplications by elementary matrices; we will make crucial use of this later together with the important fact that elementary matrices are nonsingular:

(3.33) ***Theorem*** (elementary matrices and nonsingularity). Each elementary matrix is nonsingular, and its inverse is itself an elementary matrix. More precisely:

(a) $\mathbf{E}_{ij}^{-1} = \mathbf{E}_{ji}\ (= \mathbf{E}_{ij})$
(b) $\mathbf{E}_i(c)^{-1} = \mathbf{E}_i(1/c)$ with $c \neq 0$
(c) $\mathbf{E}_{ij}(c)^{-1} = \mathbf{E}_{ij}(-c)$ with $i \neq j$

PROOF. Recall that in order to prove that a given square matrix $\mathbf{G}$ is nonsingular and has a particular matrix $\mathbf{H}$ as its inverse, what we need to show is $\mathbf{GH} = \mathbf{HG} = \mathbf{I}$. In the present case, the given matrices $\mathbf{G}$ are elementary matrices, as are their claimed inverses $\mathbf{H}$; we can therefore apply Theorem 3.32.

(a) To show that $\mathbf{E}_{ji}$ is the inverse of $\mathbf{E}_{ij}$, first consider $\mathbf{E}_{ij}\mathbf{E}_{ji}$. Since $\mathbf{E}_{ij}$ is an elementary matrix, Theorem 3.32 tells us that this product is the same as interchanging the ith and jth rows of $\mathbf{E}_{ji}$—which obviously gives us $\mathbf{I}$. A similar argument shows that $\mathbf{E}_{ji}\mathbf{E}_{ij}$ also equals $\mathbf{I}$, which proves (a).
(b) The proof, much like that for (a), is left for Problem 3.
(c) The proof can be modeled after that for (a), but we take another approach. Observe that $\mathbf{E}_{ij}(c) = \mathbf{I} + c\mathbf{e}_i\mathbf{e}_j^T$, where $\mathbf{e}_i$ and $\mathbf{e}_j$ are the usual unit column matrices. Then

$$\begin{aligned}\mathbf{E}_{ij}(c)\mathbf{E}_{ij}(-c) &= (\mathbf{I} + c\mathbf{e}_i\mathbf{e}_j^T)(\mathbf{I} - c\mathbf{e}_i\mathbf{e}_j^T) = \mathbf{I} + c\mathbf{e}_i\mathbf{e}_j^T - c\mathbf{e}_i\mathbf{e}_j^T - c^2\mathbf{e}_i\mathbf{e}_j^T\mathbf{e}_i\mathbf{e}_j^T \\ &= \mathbf{I} - c^2\mathbf{e}_i(\mathbf{e}_j^T\mathbf{e}_i)\mathbf{e}_j^T \\ &= \mathbf{I} - c^2\mathbf{e}_i[0]\mathbf{e}_j^T \\ &= \mathbf{I} - \mathbf{0} = \mathbf{I},\end{aligned}$$

since $i \neq j$. The product in the reverse order also produces $\mathbf{I}$, by the same argument. ∎

Now let's assemble several facts at our disposal:

1. Each elementary row operation is equivalent to premultiplication by an elementary matrix. (See Theorem 3.32.)
2. Any sequence of elementary row operations is equivalent to premultiplying by the product of a sequence of elementary matrices. {See (1) above.}
3. Each elementary matrix is nonsingular. {See Theorem 3.33.}
4. A product of nonsingular matrices is nonsingular. {See Theorem 1.35(a) and Problem 21, both in Section 1.4.}

Together these facts prove the following **Key Theorem**:

(3.34) **Key Theorem** (row operations). Suppose that **B** results from applying a sequence of elementary row operations to **A**. Then there exists a nonsingular **F** for which $\mathbf{B} = \mathbf{FA}$ and hence $\mathbf{F}^{-1}\mathbf{B} = \mathbf{A}$.

This result will be central to our analysis of the use of elementary row operations to solve a wide variety of problems.

PROBLEMS 3.5

1. Prove Theorem 3.32 for elementary matrices $\mathbf{E}_{ij}$.

▷ 2. Prove Theorem 3.32 for elementary matrices $\mathbf{E}_i(c)$, $c \neq 0$.

3. Prove Theorem 3.33(b).

4. If **B** is derived from **A** by applying a sequence of elementary row operations to **A**, then **B** is said to be *row-equivalent* to **A** and we write this as $\mathbf{B} \sim \mathbf{A}$. Prove that $\sim$ is a true *equivalence relation* on the set of all $p \times q$ matrices in the sense that:
   (a) $\mathbf{A} \sim \mathbf{A}$ for all **A**.
   (b) If $\mathbf{B} \sim \mathbf{A}$, then $\mathbf{A} \sim \mathbf{B}$.
   (c) If $\mathbf{A} \sim \mathbf{B}$ and $\mathbf{B} \sim \mathbf{C}$, then also $\mathbf{A} \sim \mathbf{C}$.

5. Suppose that **A** is a $p \times p$ matrix and that, when Gauss elimination is applied to reduce some augmented matrix $[\mathbf{A} \quad \mathbf{b}]$ to its reduced form, the process can be successfully carried to completion to produce a reduced matrix $[\mathbf{U} \quad \mathbf{b}']$ (which means that **U** is unit-upper-triangular). Show that such a matrix **U** is nonsingular, and use **Key Theorem 3.34** to show that **A** is therefore nonsingular.

▷ 6. Use Gauss elimination to reduce the augmented matrix **H** below to reduced form **G**. For each elementary row operation performed, find the corresponding elementary matrix. Find a nonsingular matrix **F** so that $\mathbf{FH} = \mathbf{G}$.

$$\mathbf{H} = \left[\begin{array}{cc|c} 0 & 1 & 2 \\ 2 & 3 & 4 \\ 4 & 6 & 8 \end{array}\right].$$

after Gauss that matrix G
$FH(H^{-1}) = GH^{-1}$
$F = GH^{-1}$

7. Let **A** be $p \times q$ and consider *post*multiplication by $q \times q$ elementary matrices. Show that:
   in worksheet see B for A
   (a) Postmultiplication by $\mathbf{E}_{ij}$ interchanges the ith and jth *columns* of **A**.
   (b) Postmultiplication by $\mathbf{E}_i(c)$ multiplies the ith *column* of **A** by $c \neq 0$.
   (c) Postmultiplication by $\mathbf{E}_{ij}(c)$ adds c times the ith *column* of **A** to the jth *column* of **A**.

▷ 8. Suppose that **A** is $p \times p$ and nonsingular. Use Problem 7 to show that:
   (a) if $\mathbf{A}'$ is obtained from **A** by interchanging two rows of **A**, then $\mathbf{A}'$ is non-

singular and its inverse can be obtained by interchanging the corresponding columns of $\mathbf{A}^{-1}$;

(b) if $\mathbf{A}'$ is obtained from $\mathbf{A}$ by multiplying the ith row of $\mathbf{A}$ by $c \neq 0$, then $\mathbf{A}'$ is nonsingular and its inverse can be obtained by dividing the ith column of $\mathbf{A}^{-1}$ by c; and

(c) if $\mathbf{A}'$ is obtained from $\mathbf{A}$ by adding c times the jth row of $\mathbf{A}$ to its ith row, then $\mathbf{A}'$ is nonsingular and its inverse can be obtained by subtracting c times $\mathbf{A}^{-1}$'s ith column from its jth column.

3.6 CHOOSING PIVOTS FOR GAUSS ELIMINATION IN PRACTICE

We consider now an important practical aspect of the solution of p linear equations in p unknowns by Gauss elimination. Previous examples—see Examples 3.16 and 3.17 in Section 3.2—have demonstrated that it is usually possible, and sometimes necessary, to interchange equations when performing Gauss elimination; the necessity arises, of course, when an entry that we intend to use as the next pivot is found to equal zero.

In practical computations, however, it is often *wise* to interchange equations even when not faced with a zero pivot. This claim should seem strange, since Gauss elimination always produces the (same) correct solution no matter what sequence of nonzero pivots is used; but *this assumes that arithmetic is carried out exactly, and the computers on which systems of equations are solved in practice rarely perform exact arithmetic.* This fact about inexact computer arithmetic dramatically affects Gauss elimination—and all other computational algorithms—and requires us to reexamine the choice of pivots. This section concentrates on this aspect of the practical solution of systems of linear equations when the arithmetic performed is not exact; note that arithmetic is usually inexact even for computations carried out by hand in decimal notation.

Floating-Point Arithmetic

Computers and calculators usually represent zero exactly as 0; nonzero numbers are usually represented as the product of a t-digit signed *fractional part* f and a *scale factor* 10^e, where $0.1 \leq |f| < 1$ and e is restricted to some range of integers. On many microcomputers with mathematics coprocessors, t is 16 and e is restricted to roughly $-308 \leq e \leq 308$. For simplicity, we will assume that e can take *any* integer value. The condition on f means that the first nonzero digit of f occurs immediately to the right of the decimal point; numbers represented in this form are called *t-digit floating-point numbers.* Thus, the 3-digit floating-point representation of 0.05 is 0.500×10^{-1}.

To deal with numbers such as $\frac{1}{3} = 0.333\ldots$ with unending decimal expansions or even with numbers requiring a very large number of digits in their decimal representation, in practice we must replace them with nearby numbers involving fewer digits. Although some computers and calculators just chop off the unwanted digits, most *round* the number (or *round* it *off*, as some say).

(3.35) ***Definition.*** To *round* a nonzero number x *to t floating-point digits*: (1) represent x in floating-point notation as $x = f \times 10^e$ with $0.1 \leq |f| < 1$, so that f's first digit (to the right of the decimal point) is nonzero; (2) retain as f_0 the first t digits of f; (3) if the $(t+1)$st digit of f is 0, 1, 2, 3, or 4, then define the rounded version $-\!\langle x \rangle\!-$ of x as $f_0 \times 10^e$; (4) otherwise, increase the tth digit of f_0 by 1 to produce f'_0, and define the rounded version $-\!\langle x \rangle\!-$ of x as $f'_0 \times 10^e$. Zero rounds to zero: $-\!\langle 0 \rangle\!- = 0$.

Having introduced this terminology, we can describe our model of *t-digit floating-point computer arithmetic.* In this model, by a *computer number* we mean either 0 or a (floating-point) number represented *exactly* by a fractional part of at most t digits with a scale factor. In this model, the result of any basic arithmetic operation $(+, -, /, \times)$ between two *computer numbers* u and v is defined as the result of rounding to t (floating-point) digits the result that you would get with perfect arithmetic:

(3.36) ***Definition.*** Let u and v be computer numbers, and let @ be any of the basic arithmetic operations $+$, $-$, $/$, or $\times$. Then the t-digit floating-point result u @ v is defined to be $-\!\langle u \mathbin{@} v \rangle\!-$.

(3.37) ***Example.*** Suppose that $t = 2$, so that we have just two digits, and that we evaluate $(\frac{2}{3} + \frac{2}{3}) - \frac{1}{3}$. First, $\frac{2}{3}$ will be evaluated as

$$-\!\langle 0.666\ldots \rangle\!- = 0.67;$$

then $0.67 + 0.67$ will be evaluated as

$$-\!\langle 0.67 + 0.67 \rangle\!- = -\!\langle 1.34 \rangle\!- = 1.3;$$

then $\frac{1}{3}$ will be evaluated as

$$-\!\langle 0.333\ldots \rangle\!- = 0.33;$$

and finally the overall expression $(\frac{2}{3} + \frac{2}{3}) - \frac{1}{3}$ will be given the value

$$-\!\langle 1.3 - 0.33 \rangle\!- = -\!\langle 0.97 \rangle\!- = 0.97.$$

Note that the "equivalent" expression $\frac{2}{3} + (\frac{2}{3} - \frac{1}{3})$ would be evaluated differently as 1.0—which is also the value in perfect arithmetic. MATLAB's 16-digit arithmetic evaluates the original expression as exactly 1; as a more realistic example of its arithmetic, it evaluates

$$\tfrac{1}{6} + \tfrac{1}{6} + \tfrac{1}{6} + \tfrac{1}{6} + \tfrac{1}{6} + \tfrac{1}{6}$$

as about $1 + 10^{-16}$ rather than 1.

You should now be able to solve Problems 1 and 2.

Gauss Elimination in Computer Arithmetic

We use our model computer arithmetic to illustrate the importance of the choice of pivots when Gauss elimination is performed in practice on computers. For simplicity, we suppose that $t = 2$, that is, that we have two-digit arithmetic; the same issues arise with a more realistic $t = 16$, but the assumption that $t = 2$ makes the issues more transparent.

Consider the system of equations

(3.38)
$$\begin{aligned} x_1 - x_2 &= 0 \\ 0.01x_1 + x_2 &= 1. \end{aligned}$$

Note that each coefficient and right-hand side in (3.38) is a two-digit computer number: 0.01, for example, is represented as 0.10×10^{-1}. The exact solution of (3.38) is $x_1^* = x_2^* = 1/1.01 = 0.990099\ldots$; the best that we could possibly ask Gauss elimination to do on our two-digit computer would be to obtain the closest two-digit computer numbers to x_1^* and x_2^*, namely $x_1' = x_2' = 0.99$. Let's see how well it does.

(3.39) ***Example*** (no interchanging). Suppose that we perform Gauss elimination without interchanges, using the first equation to eliminate the 0.01 in the second. Adding -0.01 times the first equation to the second produces $(1 + 0.01)$ as the new coefficient of x_2, *but we must evaluate this in two-digit arithmetic* as $\prec\!(1.01)\!\succ = 1.0$. Thus the reduced system is computed to be

$$\begin{aligned} x_1 - x_2 &= 0 \\ x_2 &= 1. \end{aligned}$$

Back-substitution (in computer arithmetic, remember) gives the computer-arithmetic solution $x_1 = x_2 = 1.0$—quite a good approximation to the best possible $x_1' = x_2' = 0.99$.

(3.40) ***Example*** (interchanging). Now suppose that we perform Gauss elimination after interchanging the equations to

$$\begin{aligned} 0.01x_1 + x_2 &= 1 \\ x_1 - x_2 &= 0. \end{aligned}$$

Adding -100 times the first equation to the second gives in the new second equation a coefficient for x_2 equal to $-100 + (-1)$ and a new right-hand side of -100, again to be evaluated in two-digit computer arithmetic; we obtain $\prec\!\!(-101)\!\!\succ = -100$, so that our computed reduced system is

$$\begin{aligned} 0.01x_1 + \quad x_2 &= \quad 1 \\ -100x_2 &= -100. \end{aligned}$$

Back-substitution gives first $x_2 = 1$ (still a good approximation to the best $x_2' = 0.99$) and then $x_1 = 0$ (an atrocious approximation to the best $x_1' = 0.99$).

These two examples illustrate that in floating-point arithmetic different pivots can produce dramatically different answers through Gauss elimination, that some of these answers can be quite acceptable, but that some can be totally unacceptable. Our problem is to understand this phenomenon so as to avoid pivots that lead to poor answers.

Backward Error Analysis

Consider again the application of Gauss elimination to (3.38) without interchanging, as in Example 3.39. If the second equation in (3.38) had been $0.01x_1 + 0.99x_2$, then *perfect* arithmetic on this changed system would have produced the same reduced form we obtained with computer arithmetic in Example 3.39. Therefore:

the effect of using two-digit computer arithmetic to implement Gauss elimination without interchanging in (3.38) is exactly the same as if we used exact arithmetic on the *slightly* different system

$$\begin{aligned} x_1 - \quad x_2 &= 0 \\ 0.01x_1 + 0.99x_2 &= 1. \end{aligned}$$

On the other hand, consider again the application of Gauss elimination to (3.38) with interchanging, as in Example 3.40. As above, we want to find a system that would be reduced in *perfect* arithmetic to what we obtained in Example 3.40 in two-digit arithmetic. Imagine that we replace the term $x_1 - x_2$ with $x_1 - \alpha x_2$; what would α have to be in order that perfect arithmetic in elimination as in Example 3.40 produces -100 as the coefficient of x_2 in the reduced equations? Perfect arithmetic produces as the second equation in the reduced set just $-(100 + \alpha)x_2 = -100$; we therefore need $-(100 + \alpha)$ to equal -100, so $\alpha = 0$. Therefore:

the effect of using two-digit computer arithmetic to implement Gauss elimination with interchanging in (3.38) is exactly the same as if we used exact arithmetic on the *greatly* changed system

$$\begin{aligned} x_1 \quad\quad &= 0 \\ 0.01x_1 + x_2 &= 1. \end{aligned}$$

Our analysis has shown that, for both ways of performing Gauss elimination on (3.38), the *computed* solution can be viewed as the *exact* solution to a *perturbed* (that is, changed) problem; this approach is called *backward error analysis*: "blame" for the errors in the result of a process on a computer is put *back* onto the data rather than onto the process or the computer arithmetic.

Backward error analysis is fundamental in applied mathematics generally and in the analysis of computational procedures particularly. The data used in applications are usually already inaccurate because of experimental and modeling error; if backward error analysis can show that the computed solution is the exact solution to a problem in which the data have been changed by amounts of about the same size as the errors already in the data, then the computer solution has not created errors larger than those already inherent in the problem. This allows the conclusion that the computer solution is as good as can reasonably be asked.

Note, however, that we have *not* shown that Gauss elimination always gives an exact solution to a slightly perturbed problem—on the contrary: The solution in Example 3.40 is the exact solution to a *greatly* perturbed problem. Our problem is to discover a strategy for interchanges—that is, for selecting the pivots—for which such a claim is valid. With a proper strategy, interchanging can improve matters rather than produce a disaster as in Example 3.40.

You may have the impression that the reason why the no-interchange solution was satisfactory while the interchange solution was unsatisfactory is related to the fact that the (1, 1)-entry—namely 1—used as the pivot to get the first solution is so much larger than the (2, 1)-entry—namely 0.01—used as the pivot to get the second solution. Thus you might well suggest the rule: "Avoid small pivots." This is not a bad rule, but it cannot be applied indiscriminately. To see this, suppose that we *rescale* (3.38) by multiplying the first equation by 10^{-2}, multiplying the second by 10^2, replacing x_1 by $x_1 = 10^2 z_1$, and replacing x_2 by $x_2 = 10^{-2} z_2$. This produces

(3.41)
$$\begin{aligned} z_1 - 0.0001z_2 &= 0 \\ 100z_1 + \qquad z_2 &= 100. \end{aligned}$$

If we believe that small pivots are bad and pivot on the large entry 100, in two-digit arithmetic we get $z_2 = 100$ and then $z_1 = 0$, which in turn gives the same poor solution $x_1 = 0$ and $x_2 = 1$ that we found in Example 3.40. The strategy of avoiding small pivots is not necessarily a successful one.

You should now be able to solve Problems 1 to 4.

Pivot-Selection Strategies

By considering the equivalent equations (3.38) and (3.41) that differ only by scaling, we showed that avoiding small pivots is not necessarily a good strategy: It gave good results in (3.38) but poor results in (3.41). Another way to view this is to say that avoiding small pivots might be a good strategy *if the equations are "properly" scaled.* By "scaling" we again mean multiplying each equation by a nonzero constant and replacing each variable by a nonzero multiple of a new variable. Many—but not all—good computer programs for solving systems of linear equations (see Section 3.9) scale the equations in some fashion before using Gauss elimination; small pivots are then avoided by one of two strategies:

1. *Partial pivoting,* in which the unknowns are eliminated in the usual order $x_1, x_2, \ldots, x_p$ and in which the pivot for the elimination with x_i is that coefficient of x_i in the equations numbered $i, i+1, \ldots, p$ that has the largest absolute value.
2. *Complete pivoting,* in which the unknowns are not necessarily eliminated in the usual order. The first variable eliminated is that variable x_i with the largest (in absolute value) coefficient, and that coefficient is used as the pivot. The rth variable eliminated is that variable with the largest (in absolute value) coefficient in the remaining $p - r + 1$ equations of all those $p - r + 1$ variables not yet eliminated, and that coefficient is used as the pivot.

Experience indicates that it is usually sufficient in practice to use partial pivoting; the possible advantages of complete pivoting tend to be overwhelmed by the disadvantages of increased "bookkeeping" in its implementation and of having to examine $(p - r + 1)^2$ coefficients to determine the pivot each time rather than just $p - r + 1$ as in partial pivoting. Systems of equations that arise in practice—unlike those created for textbooks in order to illustrate anomalies—seem often to have a built-in "natural" scaling that prevents partial pivoting from leading to disastrous pivot selection; in surveying, for example, we rarely measure some distances in miles and others in inches. Another factor is that it is unnecessary to find the *best* pivots; it is only necessary to avoid *very bad* pivots.

There are also theoretical reasons for using partial or complete pivoting. Backward error analysis can be used to prove that the computed solution $\mathbf{x}'$ exactly solves a perturbed system in which the size of the perturbations ("size" is measured in relation to that of the original coefficients) is related to the size of the pivots used during the elimination. In practice, the pivots rarely become large. In fact:

(3.42) Gauss elimination with partial pivoting performed in t-digit floating-point arithmetic produces a $p \times 1$ $\mathbf{x}'$ as an approximate solution to $\mathbf{Ax} = \mathbf{b}$ that is the exact solution to a perturbed problem $\mathbf{A}'\mathbf{x}' = \mathbf{b}$, where in practice it is usually the case that the largest entry of the perturbation $\mathbf{A} - \mathbf{A}'$ is no larger than about $p10^{1-t}$ times the largest entry in $\mathbf{A}$.

Thus, if $\mathbf{A}$ is scaled so that all its entries are of about the same size, then Gauss elimination with partial pivoting produces the exact solution to a system of equations each of whose coefficients usually is only slightly perturbed from its correct value.

The necessity of exhibiting the $\mathbf{A}'$ in (3.42) makes it difficult to illustrate (3.42) with a realistic example. There is a consequence of (3.42), however, that is interesting for its own sake and is easier to illustrate. If we calculate the residual $\mathbf{r} = \mathbf{b} - \mathbf{A}\mathbf{x}'$ to see how close $\mathbf{x}'$ comes to solving $\mathbf{A}\mathbf{x} = \mathbf{b}$, then because of (3.42) we can write

$$\mathbf{r} = \mathbf{b} - \mathbf{A}\mathbf{x}' = \mathbf{b} - \mathbf{A}'\mathbf{x}' + \mathbf{A}'\mathbf{x}' - \mathbf{A}\mathbf{x}' = (\mathbf{A}' - \mathbf{A})\mathbf{x}'.$$

From this we can conclude from (3.42) that

(3.43) Gauss elimination with partial pivoting performed in t-digit floating-point arithmetic produces a $p \times 1$ $\mathbf{x}'$ as an approximate solution to $\mathbf{A}\mathbf{x} = \mathbf{b}$ that is the exact solution to a perturbed problem $\mathbf{A}\mathbf{x}' = \mathbf{b} - \mathbf{r}$, where in practice it is usually the case that the largest entry in $\mathbf{r}$ is no larger than $p10^{1-t}$ times the largest entry in $\mathbf{A}$ times the largest entry in $\mathbf{x}'$.

𝔐 (3.44) ***Example.*** Consider solving $\mathbf{A}\mathbf{x} = \mathbf{b}$, where

$$\mathbf{A} = \begin{bmatrix} -4 & 7 & 1 & -3 & 71 & 6 & 5 & -2 & 9 & 8 \\ 2 & 1 & -6 & 3 & 5 & -60 & 9 & -8 & -4 & 7 \\ 3 & 53 & 2 & -7 & -6 & 4 & -9 & 8 & 5 & 1 \\ -9 & 3 & -5 & 8 & 2 & -1 & 7 & 4 & 6 & -61 \\ 7 & -5 & 4 & -6 & -1 & 2 & -8 & 51 & 9 & -3 \\ -5 & -8 & 2 & -50 & 9 & -7 & -3 & -1 & 6 & 4 \\ -48 & -1 & 3 & -7 & 9 & -2 & 4 & -6 & 8 & 5 \\ 1 & -9 & -63 & -2 & 8 & -3 & -6 & -7 & -4 & 5 \\ 8 & -6 & 1 & 4 & 7 & 3 & -46 & 2 & 9 & 5 \\ 2 & 1 & 9 & 5 & -3 & 8 & 6 & 4 & -73 & 7 \end{bmatrix} \quad \text{and} \quad \mathbf{b} = \begin{bmatrix} 10 \\ 4 \\ -6 \\ -3 \\ -2 \\ 9 \\ -8 \\ 1 \\ -5 \\ -7 \end{bmatrix}.$$

MATLAB solved this system and produced an $\mathbf{x}'$ whose largest entry is about 0.3; since $p = 10$ and the largest entry in $\mathbf{A}$ is 73, we expect from (3.43) that the largest entry of the residual $\mathbf{r} = \mathbf{b} - \mathbf{A}\mathbf{x}'$ should be less than about $(10)10^{1-16}(73)(0.3)$, which is about 2.2×10^{-13}. MATLAB calculated the residual $\mathbf{r}$—using computer arithmetic and therefore getting an imperfect result—as having a largest entry of about 2×10^{-15}, appreciably less than the upper bound from (3.43).

PROBLEMS 3.6

▷ **1.** Evaluate, in two-digit floating-point arithmetic:
(a) $(\frac{1}{3} + \frac{1}{3}) + \frac{1}{3}$.
(b) $(0.58 + 0.53) - 0.53$.
(c) $0.58 + (0.53 - 0.53)$—compare this result with that in (b).

2. Evaluate, in two-digit floating-point arithmetic:
   (a) $6(\frac{1}{6})$
   (b) $3(\frac{1}{3})$
3. Consider the system (3.41)—a scaled version of (3.38).
   (a) Find its exact solution.
   (b) Find the best possible two-digit approximation to this solution.
   (c) Find the solution obtained by Gauss elimination in two-digit floating-point arithmetic, assuming that the largest coefficient of any variable is used as the first pivot.

▷ 4. (a) Using two-digit floating-point arithmetic, apply Gauss elimination without interchanges to solve the system

$$0.98x_1 + 0.43x_2 = 0.91$$
$$-0.61x_1 + 0.23x_2 = 0.48.$$

   (b) Compare your answer with both the true solution $x_1^* = 0.005946\ldots$, $x_2^* = 2.102727\ldots$ and the best possible two-digit approximation $x_1' = 0.0059$, $x_2' = 2.1$;
   (c) Use backward error analysis and find a perturbed system from which Gauss elimination in perfect arithmetic would produce x_1' and x_2' as the true solution.
   (Note that the small absolute errors in the numerical answers actually represent large relative errors and may not be acceptable.)

5. Identify the entry that would be used as the first pivot in (1) partial pivoting and (2) complete pivoting for each of the augmented matrices below.

   (a) $\left[\begin{array}{ccc|c} -8 & 1 & 3 & 10 \\ 7 & 2 & 2 & 2 \\ 2 & 4 & 6 & 3 \end{array}\right]$

   (b) $\left[\begin{array}{ccc|c} 2 & 3 & 4 & 1 \\ 8 & 1 & 0 & 20 \\ -9 & 2 & 6 & 3 \end{array}\right]$

   (c) $\left[\begin{array}{ccc|c} 2 & 3 & 2 & 1 \\ 8 & 6 & 9 & 2 \\ 4 & 1 & 3 & 30 \end{array}\right]$

▷ 6. Show that the residual $\mathbf{b} - \mathbf{Ax}$ is small even for the atrocious approximate solution to (3.41) found in the text; show that the general rule (3.43) holds in this case.

7. Show that the residual $\mathbf{b} - \mathbf{Ax}$ is small for the approximate solution to (3.38) found in Example 3.39; verify that the general rule (3.43) holds in this case.

8. Show that each of the three pivots other than the largest in (3.41)—including the smallest one, 0.0001—gives a satisfactory approximation to x_1^* and x_2^* in two-digit floating-point arithmetic.

▷ **9.** The general rule (3.43) states that the computed solution produces a small residual; intuitively, this says that "the computed solution nearly solves $\mathbf{Ax} = \mathbf{b}$." This can be different from the statement "the computed solution is near the true solution." To see this, consider the system

$$0.89x_1 + 0.53x_2 = 0.36$$
$$0.47x_1 + 0.28x_2 = 0.19,$$

whose exact solution is $x_1^* = 1$, $x_2^* = -1$. Show that the "approximate solution" $x_1 = 0.47$, $x_2 = -0.11$ has a small residual and so "nearly solves" the equation, although the approximate solution is far from the true solution. This is an example of an *ill-conditioned* problem—see Section 6.4.

𝔐 **10.** (a) Use MATLAB or similar software to obtain an approximate solution $\mathbf{x}'$ to the system in Problem 9.

(b) Find the error $\mathbf{x}' - \mathbf{x}^*$ between the computed solution and the true solution.

(c) Compare the size of this error with the size of the residual $\mathbf{r} = \mathbf{b} - \mathbf{Ax}'$, but in order to calculate $\mathbf{r}$ accurately, note that

$$\mathbf{r} = \mathbf{b} - \mathbf{Ax}' = \mathbf{Ax}^* - \mathbf{Ax}' = \mathbf{A}(\mathbf{x}^* - \mathbf{x}')$$

and compute $\mathbf{r}$ as $\mathbf{A}(\mathbf{x}^* - \mathbf{x}')$. This again demonstrates that the system is *ill-conditioned.*

11. Neither partial pivoting nor complete pivoting is *guaranteed* to produce good results. To see this, show that both produce unsatisfactory results when Gauss elimination is applied in t-digit floating-point arithmetic to

$$2x_1 + x_2 + x_3 = 1$$
$$x_1 + \epsilon x_2 + \epsilon x_3 = 2\epsilon$$
$$x_1 + \epsilon x_2 - \epsilon x_3 = \epsilon$$

when ϵ is very small and known accurately. Show that rescaling via $z_1 = x_1/\epsilon$, $z_2 = x_2$, $z_3 = x_3$ and dividing the second and third equations by ϵ so as to obtain

$$2\epsilon z_1 + z_2 + z_3 = 1$$
$$z_1 + z_2 + z_3 = 2$$
$$z_1 + z_2 - z_3 = 1$$

allows both methods to work satisfactorily.

3.7 THE LU-DECOMPOSITION

Until now the emphasis in Gauss elimination has been on the *process*; the emphasis in this section will be more on the *result*—what it is that Gauss elimination accomplishes. The basic facts are summarized in (3.56) and are illustrated in Examples 3.45 and 3.55.

Recall from Example 3.17 that Gauss elimination *without interchanges* cannot always be successfully completed because zeros can appear on the main diagonal. Because it is simpler to understand, we first treat the case in which no such zeros appear and Gauss elimination can be successfully completed without interchanges.

LU and Gauss Elimination Without Interchanges

The main result is simple to state:

> If Gauss elimination on the $p \times p$ matrix **A** can be successfully completed as in (3.20) but without interchanges, then the process is equivalent to writing **A** as a product **LU** of a lower-triangular **L** and an upper-triangular **U** (see Definition 1.3).

Recall the process (3.20) of Gauss elimination, but in the special case when **A** *is* $p \times p$, *all pivots are nonzero, and no interchanges are used.* When eliminating in the ith column, we first divide the entries of the ith row by the pivot, the (i, i)-entry of the matrix in its partially reduced form; denote this ith pivot by α_{ii}. We then replace the jth row (for $j > i$) by the jth row plus a multiple m_{ji} of the ith row so as to produce a zero in the (j, i)-entry; the numbers m_{ji} are called *multipliers.* The triangular matrices involved in the product mentioned above are defined from the numbers produced during elimination.

(3.45) ***Example.*** Consider the 4×4 matrix **A** used to introduce Gauss elimination in Section 3.2; see (3.8). We form a lower-triangular matrix **L** and a unit-upper-triangular matrix **U**. **U** is simply the reduced form (3.13) of **A** produced during elimination. The main-diagonal entries of **L** are the pivots: $\langle \mathbf{L} \rangle_{ii} = \alpha_{ii}$; the subdiagonal entries of **L** are the negatives of the multipliers: $\langle \mathbf{L} \rangle_{ji} = -m_{ji}$. We can read off the multipliers m_{ji} and the pivots α_{ii} from the steps (3.8), (3.10), (3.11), (3.12) in the elimination: $\alpha_{11} = -2$, $m_{21} = 3$, $m_{31} = -5$, and $m_{41} = -1$ from (3.8); $\alpha_{22} = 3$, $m_{32} = 3$, and $m_{42} = -2$ from (3.10); $\alpha_{33} = -2$, and $m_{43} = -3$ from (3.11); and $\alpha_{44} = 2$ from (3.12). Forming **L** and **U** as described and then checking the multiplication reveals that the claimed decomposition of **A** as **A** = **LU** is correct:

(3.46)

$$\begin{matrix} \mathbf{A} & \text{does equal} & \mathbf{L} & \text{times} & \mathbf{U} & \text{since} \end{matrix}$$

$$\begin{bmatrix} -2 & 2 & -4 & -6 \\ -3 & 6 & 3 & -15 \\ 5 & -8 & -1 & 17 \\ 1 & 1 & 11 & 7 \end{bmatrix} = \begin{bmatrix} -2 & 0 & 0 & 0 \\ -3 & 3 & 0 & 0 \\ 5 & -3 & -2 & 0 \\ 1 & 2 & 3 & 2 \end{bmatrix} \begin{bmatrix} 1 & -1 & 2 & 3 \\ 0 & 1 & 3 & -2 \\ 0 & 0 & 1 & 2 \\ 0 & 0 & 0 & 1 \end{bmatrix}.$$

This is the *LU-decomposition* of $\mathbf{A}$. A related triangular decomposition can be derived from it. Let $\mathbf{D}_0$ be the diagonal matrix formed from the pivots; in our case

$$\mathbf{D}_0 = \mathbf{diag}(-2, 3, -2, 2).$$

Since $\mathbf{A} = \mathbf{LU} = \mathbf{LD}_0^{-1}\mathbf{D}_0\mathbf{U}$, we can also write $\mathbf{A} = \mathbf{L}_0\mathbf{U}_0$, where $\mathbf{L}_0 = \mathbf{LD}_0^{-1}$ and $\mathbf{U}_0 = \mathbf{D}_0\mathbf{U}$. $\mathbf{L}_0$ is found from $\mathbf{L}$ by dividing each column of $\mathbf{L}$ by the pivot in that column, while $\mathbf{U}_0$ is found from $\mathbf{U}$ by multiplying each row of $\mathbf{U}$ by the pivot from that row. This $\mathbf{A} = \mathbf{L}_0\mathbf{U}_0$ is the *L_0U_0-decomposition* of $\mathbf{A}$:

(3.47)

$$\begin{matrix} \mathbf{A} & \text{does equal} & \mathbf{L}_0 & \text{times} & \mathbf{U}_0 & \text{since} \end{matrix}$$

$$\begin{bmatrix} -2 & 2 & -4 & -6 \\ -3 & 6 & 3 & -15 \\ 5 & -8 & -1 & 17 \\ 1 & 1 & 11 & 7 \end{bmatrix} = \begin{bmatrix} 1 & 0 & 0 & 0 \\ \frac{3}{2} & 1 & 0 & 0 \\ -\frac{5}{2} & -1 & 1 & 0 \\ -\frac{1}{2} & \frac{2}{3} & -\frac{3}{2} & 1 \end{bmatrix} \begin{bmatrix} -2 & 2 & -4 & -6 \\ 0 & 3 & 9 & -6 \\ 0 & 0 & -2 & -4 \\ 0 & 0 & 0 & 2 \end{bmatrix}.$$

We described Gauss elimination as first dividing the row through by the pivot in order to obtain a 1 in that entry; this makes the calculations easier by hand. Computer programs, on the other hand, usually do *not* first divide through by this pivot but instead proceed to eliminate. The effect is that the reduced form obtained is the above $\mathbf{U}_0$ in this case rather than $\mathbf{U}$; the entries in $\mathbf{L}_0$ below the main diagonal are just the negatives of the multipliers used during this form of elimination. We showed above how to find $\mathbf{L}_0$ and $\mathbf{U}_0$ from $\mathbf{D}_0$, the diagonal matrix formed from the main-diagonal entries of $\mathbf{L}$. Conversely, having $\mathbf{L}_0$ and $\mathbf{U}_0$, we could find $\mathbf{L} = \mathbf{L}_0\mathbf{D}_0$ and $\mathbf{U} = \mathbf{D}_0^{-1}\mathbf{U}_0$, where $\mathbf{D}_0$ is the diagonal matrix formed from the main-diagonal entries of $\mathbf{U}$.

In Example 3.45 we were able to decompose $\mathbf{A}$ as $\mathbf{A} = \mathbf{LU}$ and as $\mathbf{A} = \mathbf{L}_0\mathbf{U}_0$, where $\mathbf{L}$ is lower-triangular with nonzero main-diagonal entries and $\mathbf{U}$ is unit-upper-triangular, and where $\mathbf{L}_0$ is unit-lower-triangular and $\mathbf{U}_0$ is upper-triangular with nonzero main-diagonal entries. The following theorem says that this can always be done when Gauss elimination can be completed without interchanges, and it characterizes when this occurs; the rather technical proof can be omitted by most readers.

(3.48) **Key Theorem** (*LU* without interchanges). Suppose that $\mathbf{A}$ is a $p \times p$ matrix. Then:

(a) Gauss elimination without interchanges can be carried to completion with nonzero pivots so as to produce a unit-upper-triangular reduced form $\mathbf{U}$ if and only if the *principal submatrices* $\mathbf{A}_k$ are nonsingular for $1 \le k \le p$, where $\mathbf{A}_k$ is the $k \times k$ upper-left corner of $\mathbf{A}$: $\langle \mathbf{A}_k \rangle_{ij} = \langle \mathbf{A} \rangle_{ij}$ for $1 \le i, j \le k$ (in particular, $\mathbf{A}$ is nonsingular).

(b) Gauss elimination without interchanges can be completed as in (a) if and only if $\mathbf{A} = \mathbf{LU}$, where $\mathbf{U}$ is the reduced unit-upper-triangular matrix from (a) and $\mathbf{L}$ is a lower-triangular matrix with the pivots α_{ii} from Gauss elimination on the main diagonal and the negatives of the multipliers m_{ji} from Gauss elimination on the subdiagonals; $\langle \mathbf{L} \rangle_{ii} = \alpha_{ii}$, $\langle \mathbf{L} \rangle_{ji} = -m_{ji}$. Equivalently,

(3.49)
$$\mathbf{A} = \mathbf{L}_0 \mathbf{U}_0$$

with $\mathbf{L}_0 = \mathbf{L}\mathbf{D}_0^{-1}$ unit-lower-triangular and $\mathbf{U}_0 = \mathbf{D}_0 \mathbf{U}$ upper-triangular with the pivots α_{ii} on the main diagonal; here

$$\mathbf{D}_0 = \mathbf{diag}(\alpha_{11}, \ldots, \alpha_{pp}) \quad \text{and} \quad \alpha_{ii} = \langle \mathbf{U}_0 \rangle_{ii} = \langle \mathbf{L} \rangle_{ii}.$$

PROOF. The proof is by induction on p. If $p = 1$ and $\mathbf{A} = [a]$, then $\mathbf{L} = [a]$, $\mathbf{U} = [1]$, $\mathbf{L}_0 = [1]$, $\mathbf{U}_0 = [a]$, and $a \neq 0$ if and only if the pivot(s)—that is, a—is nonzero. Suppose that the theorem is true for $p - 1$ and that $\mathbf{A}$ is $p \times p$. Implement Gauss elimination on $\mathbf{A}$ by eliminating only in the first $p - 1$ rows using the first $p - 1$ columns and then going back to eliminate the entries in the bottom row in each column from left to right; this is equivalent to normal Gauss elimination. The first part of this process involves regular Gauss elimination on $\mathbf{A}_{p-1}$, a $(p - 1) \times (p - 1)$ matrix for which the theorem is assumed to hold. So completing Gauss elimination on $\mathbf{A}$ is equivalent to completing it on the first $p - 1$ rows and then finishing up; completing it on $\mathbf{A}_{p-1}$ is equivalent to $\mathbf{A}_{p-1} = \mathbf{L}'\mathbf{U}'$ of the proper form. Consider trying to write $\mathbf{A} = \mathbf{LU}$ with $\mathbf{L}$ and $\mathbf{U}$ as follows:

$$\mathbf{A} = \begin{bmatrix} \mathbf{A}_{p-1} & \mathbf{w} \\ \mathbf{v}^T & a_{pp} \end{bmatrix} = \begin{bmatrix} \mathbf{L}' & \mathbf{0} \\ \mathbf{l}^T & \alpha \end{bmatrix} \begin{bmatrix} \mathbf{U}' & \mathbf{u} \\ \mathbf{0}^T & 1 \end{bmatrix},$$

where $a_{pp} = \langle \mathbf{A} \rangle_{pp}$. This requires $\mathbf{A}_{p-1} = \mathbf{L}'\mathbf{U}'$, which is true by construction; $\mathbf{L}'\mathbf{u} = \mathbf{w}$, which holds if and only if $\mathbf{u} = \mathbf{L}'^{-1}\mathbf{w}$ since $\mathbf{L}'$ is nonsingular (its main-diagonal entries are the nonzero pivots); $\mathbf{l}^T\mathbf{U}' = \mathbf{v}^T$, which holds if and only if $\mathbf{l}^T = \mathbf{v}^T\mathbf{U}'^{-1}$ since $\mathbf{U}'$ is nonsingular; and $a_{pp} = \mathbf{l}^T\mathbf{u} + \alpha$, which holds if and only if $\alpha = a_{pp} - \mathbf{l}^T\mathbf{u}$ with $\mathbf{l}^T$ and $\mathbf{u}$ as already determined. This gives the desired *LU*-decomposition.

Next we relate the decomposition to Gauss elimination. The elimination (as we modified it to deal only with the first $p-1$ rows first) reduced the first $p-1$ rows of $\mathbf{A}$ from $[\mathbf{A}_{p-1} \quad \mathbf{w}]$ to $[\mathbf{U}' \quad \mathbf{u}]$ without disturbing the last row $[\mathbf{v}^T \quad a_{pp}]$ of $\mathbf{A}$. Eliminating now in this bottom row merely adds to the bottom row some linear combination of the first $p-1$ new rows $[\mathbf{U}' \quad \mathbf{u}]$ so as to obtain $[\mathbf{0}^T \quad \alpha_{pp}]$:

$$\mathbf{z}^T[\mathbf{U}' \quad \mathbf{u}] + [\mathbf{v}^T \quad a_{pp}] = [\mathbf{0}^T \quad \alpha_{pp}],$$

which yields $\mathbf{z}^T = -\mathbf{v}^T\mathbf{U}'^{-1} = -\mathbf{l}^T$ (that is, the new row in the LU-decomposition already found is the negative of the row of multipliers, as claimed) and

$$\alpha_{pp} = a_{pp} + \mathbf{z}^T\mathbf{u} = a_{pp} - \mathbf{l}^T\mathbf{u},$$

which equals α, the new main-diagonal entry found for $\mathbf{L}$. This pivot $\alpha_{pp} = \alpha$ will be nonzero as required if and only if $\mathbf{L}$ is nonsingular; since $\mathbf{U}$ is nonsingular and $\mathbf{L} = \mathbf{AU}^{-1}$, $\mathbf{L}$ is nonsingular if and only if $\mathbf{A}$ is—which is the condition on $\mathbf{A}_p = \mathbf{A}$ that we needed. The construction of $\mathbf{L}_0$ and $\mathbf{U}_0$ from $\mathbf{L}$ and $\mathbf{U}$ demonstrates the equivalence of the two triangular decompositions. ■

You should now be able to solve Problems 1 to 10.

LU and Gauss Elimination with Interchanges

Example 3.17 revealed that not every matrix can be reduced by Gauss elimination without interchanges. If row interchanges are allowed, however, every *nonsingular* matrix can be reduced by Gauss elimination to a unit-upper-triangular matrix; this can be related as in **Key Theorem 3.48** to a triangular decomposition. If $\mathbf{A}$ is singular, however, our standard process (3.20) can be carried through but it *will not produce a **unit**-upper-triangular matrix.*

(3.50) ***Example.*** Consider the obviously singular matrix

$$\mathbf{A} = \begin{bmatrix} 0 & 2 \\ 0 & 2 \end{bmatrix}.$$

With or without interchanges, Gauss elimination reduces this to an upper-triangular matrix $\mathbf{R}$ with a *zero* on the main diagonal rather than the desired *one.* A simple argument (Problem 19) shows that $\mathbf{A}$ cannot possibly be written as a lower-triangular matrix times a unit-upper-triangular matrix. So $\mathbf{A}$ cannot be decomposed as $\mathbf{A} = \mathbf{LU}$ as in Theorem 3.48; but $\mathbf{A}$ *can be decomposed as* $\mathbf{A} = \mathbf{L}_0\mathbf{U}_0$ *of the second type in the theorem with* $\mathbf{L}_0$ *unit-lower-triangular if we allow zero entries on the main diagonal of* $\mathbf{U}_0$:

$$\mathbf{A} = \mathbf{L}_0\mathbf{U}_0 \qquad \text{with } \mathbf{L}_0 = \mathbf{I} \text{ and } \mathbf{U}_0 = \mathbf{A}$$

in this case.

Example 3.50 hints that the second decomposition—$\mathbf{A} = \mathbf{L}_0\mathbf{U}_0$ with $\mathbf{U}_0$ possibly singular (like **A**) and $\mathbf{L}_0$ unit-lower-triangular—allows the decomposition of some singular matrices (the reason for the "$_0$" on $\mathbf{L}_0$ and $\mathbf{U}_0$) without interchanges. And, of course, should **A** be nonsingular so that the diagonal entries of $\mathbf{U}_0$ are nonzero, we can construct **L** and **U** from $\mathbf{L}_0$ and $\mathbf{U}_0$ by using the diagonal matrix $\mathbf{D}_0$ (formed from the main-diagonal entries of $\mathbf{U}_0$ in this case) by using (3.49). But what if interchanges occur? Again, it is easy to state the basic facts:

> If Gauss elimination on the $p \times p$ matrix **A** can be successfully completed as in (3.20) with interchanges, then the process is equivalent to writing a modification of **A** called $\mathbf{A}'$ ($= \mathbf{A}$ with its rows reordered) as a product **LU** of a lower triangular **L** and an upper-triangular **U**.

To verify this claim for Gauss elimination with interchanges, imagine that somehow we know *in advance* of the elimination process what interchanges will have to be made and that we perform those interchanges among the rows *of the original matrix* **A**, *obtaining a matrix* $\mathbf{A}'$, *before beginning the elimination.* The fact is that Gauss elimination can then proceed on $\mathbf{A}'$ *without interchanges*, that *exactly the same numbers are used as pivots and multipliers for* $\mathbf{A}'$ *as for* **A**, and that *exactly the same reduced form results from* $\mathbf{A}'$ *as from* **A**. Therefore, we can use information from **Key Theorem 3.48** to understand the effects of Gauss elimination with interchanges by viewing it as Gauss elimination without interchanges on a matrix obtained from the original by *permuting*—that is, reordering—its rows. A definition helps make this reordering more precise.

(3.51) ***Definition.*** A $p \times p$ *permutation matrix* **P** is any $p \times p$ matrix that results from *permuting*—that is, rearranging the order of—the rows of $\mathbf{I}_p$. More precisely: Each row of **P** contains exactly one nonzero entry, namely 1; and each column of **P** contains exactly one nonzero entry, namely 1.

The very definition of matrix multiplication made it clear that premultiplying a matrix **A** by an elementary matrix produced the same result as applying the corresponding elementary row operation to **A**. Similarly it is clear that premultiplying **A** by a permutation matrix **P** produces the same result as permuting the rows of **A** in exactly the same way as the rows of $\mathbf{I}_p$ were permuted to produce **P**.

(3.52) ***Theorem*** (permutation matrices). Let **P** be a permutation matrix. Then:

(a) For any **A**, **PA** can be produced from **A** by permuting the rows of **A** exactly as the rows of **I** were permuted to obtain **P**.

(b) **P** is nonsingular, and $\mathbf{P}^{-1} = \mathbf{P}^T$: $\mathbf{P}\mathbf{P}^T = \mathbf{P}^T\mathbf{P} = \mathbf{I}$.

PROOF

(a) This follows easily from the definitions of matrix multiplication and of permutation matrices.

(b) Partition $\mathbf{P}$ into its rows $\mathbf{r}_1, \ldots, \mathbf{r}_p$, which are just the rows $\mathbf{e}_i^T$ of $\mathbf{I}$ in some order. $\mathbf{P}^T$ then has the $\mathbf{r}_i^T$ as its columns. The definition of matrix multiplication implies that the (i, j)-entry of $\mathbf{PP}^T$ is just (the entry in the 1×1 matrix) $\mathbf{r}_i\mathbf{r}_j^T$, and this is 1 if $i = j$ and 0 if $i \neq j$; that is, $\mathbf{PP}^T = \mathbf{I}$. A similar argument in terms of the columns of $\mathbf{P}$ shows that $\mathbf{P}^T\mathbf{P} = \mathbf{I}$ also. ■

We return once more to considering Gauss elimination with interchanges.

(3.53) **Key Theorem** (*LU* with interchanges). Suppose that $\mathbf{A}$ is $p \times p$. Then:

(a) $\mathbf{A}$ can be reduced by Gauss elimination with interchanges in the usual way to a reduced *unit*-upper-triangular matrix $\mathbf{U}$ if and only if $\mathbf{A}$ is nonsingular. When such $\mathbf{U}$ is produced, there exists a permutation matrix $\mathbf{P}$ and a nonsingular lower-triangular matrix $\mathbf{L}$ such that

$$\mathbf{PA} = \mathbf{LU} \quad \text{and} \quad \mathbf{A} = \mathbf{P}^T\mathbf{LU},$$

with $\mathbf{L}$ and $\mathbf{U}$ as in **Key Theorem 3.48(b)**.

(b) There exists a permutation matrix $\mathbf{P}$, a *unit*-lower-triangular matrix $\mathbf{L}_0$, and an upper-triangular matrix $\mathbf{U}_0$ such that

$$\mathbf{PA} = \mathbf{L}_0\mathbf{U}_0 \quad \text{and} \quad \mathbf{A} = \mathbf{P}^T\mathbf{L}_0\mathbf{U}_0.$$

$\mathbf{A}$ is nonsingular if and only if $\mathbf{U}_0$ is nonsingular; for nonsingular $\mathbf{A}$, this $\mathbf{L}_0$ and $\mathbf{U}_0$ relate to the $\mathbf{L}$ and $\mathbf{U}$ of (a) by

(3.54) $$\mathbf{U} = \mathbf{D}_0^{-1}\mathbf{U}_0 \quad \text{and} \quad \mathbf{L} = \mathbf{L}_0\mathbf{D}_0$$

where $\mathbf{D}_0 = \mathbf{diag}(\alpha_{11}, \ldots, \alpha_{pp})$ and $\alpha_{ii} = \langle \mathbf{U}_0 \rangle_{ii} = \langle \mathbf{L} \rangle_{ii}$.

PROOF. If $\mathbf{A}$ can be written $\mathbf{A} = \mathbf{P}^T\mathbf{LU}$ as in (a), then certainly $\mathbf{A}$ is nonsingular since $\mathbf{P}^T$, $\mathbf{L}$, and $\mathbf{U}$ are nonsingular. The proof of the rest of the theorem is by induction on p. For $p = 1$, the result is just as in **Key Theorem 3.48**.

(a) Suppose (a) to be true for $p - 1$. If $\mathbf{A}$ is nonsingular, then its first column is nonzero and the rows can be permuted so that the new $(1, 1)$-entry is nonzero, and then Gauss elimination can be carried out in the usual way in that column. The $(p - 1) \times (p - 1)$ matrix that remains to be reduced is also nonsingular since $\mathbf{A}$ is nonsingular and has been transformed to this partially reduced form by (nonsingular) elementary matrices. By the inductive hypothesis, the remainder of Gauss elimination with interchanges can be carried out, with the result being a unit-upper-triangular reduced form for $\mathbf{A}$. But this elimination process is equivalent (we omit the proof) to elimination without interchanges on $\mathbf{PA}$, and hence **Key Theorem 3.48** applies to $\mathbf{PA}$, giving the required result.

(b) Problem 18. ■

(3.55) ***Example.*** Gauss elimination, as indicated below, begins on the matrix $\mathbf{A}$ below with an interchange but shortly runs into a zero pivot:

$$\begin{bmatrix} 0 & 0 & 4 \\ 2 & 4 & 6 \\ -4 & -8 & -10 \end{bmatrix} \xrightarrow{\text{swap}} \begin{bmatrix} 2 & 4 & 6 \\ 0 & 0 & 4 \\ -4 & -8 & -10 \end{bmatrix} \xrightarrow[m_{31}=2]{m_{21}=0} \begin{bmatrix} 2 & 4 & 6 \\ 0 & 0 & 4 \\ 0 & 0 & 2 \end{bmatrix}.$$

In our standard Gauss elimination (3.20), we would next use the (2, 3)-entry 4 as a pivot and eliminate the 2; this will not, however, lead to an LU-decomposition. The modification needed to prove **Key Theorem 3.53(b)** requires that in case of encountering a zero lower column as in our present second column, we merely skip over that column (that is, take the multipliers to equal 0) and move on to the next column and next row. In our case, that is the (3, 3)-entry, so we have completed the elimination. We obtain the L_0U_0-decomposition $\mathbf{A} = \mathbf{P}^T\mathbf{L}_0\mathbf{U}_0$ by forming $\mathbf{P}$ as the permutation matrix that interchanges the top two rows, $\mathbf{U}_0$ as the upper-triangular form to which $\mathbf{A}$ has been reduced, and $\mathbf{L}_0$ as a unit-lower-triangular matrix with the negatives of the multipliers as entries. In our case, this gives $\mathbf{A} = \mathbf{P}^T\mathbf{L}_0\mathbf{U}_0$ as follows:

$$\overset{\mathbf{P}^T}{\begin{bmatrix} 0 & 1 & 0 \\ 1 & 0 & 0 \\ 0 & 0 & 1 \end{bmatrix}} \text{ times } \overset{\mathbf{L}_0}{\begin{bmatrix} 1 & 0 & 0 \\ 0 & 1 & 0 \\ -2 & 0 & 1 \end{bmatrix}} \text{ times } \overset{\mathbf{U}_0}{\begin{bmatrix} 2 & 4 & 6 \\ 0 & 0 & 4 \\ 0 & 0 & 2 \end{bmatrix}} \text{ does equal } \overset{\mathbf{A}:}{\begin{bmatrix} 0 & 0 & 4 \\ 2 & 4 & 6 \\ -4 & -8 & -10 \end{bmatrix}},$$

as can be checked by multiplication.

The details of the proofs obscure the basic ideas somewhat, so we summarize:

(3.56)
1. $\mathbf{A}$ is nonsingular if and only if $\mathbf{A}$ can be written in an LU-decomposition $\mathbf{A} = \mathbf{P}^T\mathbf{L}\mathbf{U}$ with $\mathbf{P}$ a permutation matrix, $\mathbf{U}$ unit-upper-triangular, and $\mathbf{L}$ lower-triangular with nonzero main-diagonal entries.
2. Every $p \times p$ matrix $\mathbf{A}$ can be written in an L_0U_0-decomposition $\mathbf{A} = \mathbf{P}^T\mathbf{L}_0\mathbf{U}_0$ with $\mathbf{P}$ a permutation matrix, $\mathbf{L}_0$ unit-lower-triangular (hence nonsingular), and $\mathbf{U}_0$ upper-triangular; $\mathbf{A}$ is nonsingular if and only if $\mathbf{U}_0$ is nonsingular.
3. For nonsingular $\mathbf{A}$, the LU- and L_0U_0-decompositions of (1) and (2) can be related by (3.54).
4. These decompositions can be found from Gauss elimination.

PROBLEMS 3.7

▷ **1.** For the matrix $\mathbf{A}$ below, perform Gauss elimination without interchanges and give the multipliers m_{21}, m_{31}, and m_{32}, and the pivots α_{11}, α_{22}, and α_{33}, as well as the elementary matrices that accomplish each row operation performed.

$$\mathbf{A} = \begin{bmatrix} 2 & 1 & -2 \\ 4 & -1 & 2 \\ 2 & -1 & 1 \end{bmatrix}$$

2. Verify (3.46).

3. Verify (3.47).

▷ **4.** Show that a lower-triangular matrix **L** is nonsingular if and only if its main-diagonal entries are all nonzero.

5. Show that every unit-upper-triangular matrix is nonsingular and that its inverse is unit-upper-triangular.

▷ **6.** Show that the product of arbitrarily many unit-upper (or -lower)-triangular matrices is unit-upper (or -lower)-triangular.

7. Show that the inverse of a nonsingular lower (or upper)-triangular matrix—see Problem 4—is lower (or upper)-triangular.

8. Show that the product of arbitrarily many upper (or lower)-triangular matrices is upper (or lower)-triangular.

9. For each matrix **A** below, find its *LU*-decomposition $\mathbf{A} = \mathbf{LU}$ and verify that $\mathbf{LU} = \mathbf{A}$.

(a) $\begin{bmatrix} 2 & 4 \\ 2 & 1 \end{bmatrix}$ (b) $\begin{bmatrix} -2 & -4 & 2 & -2 \\ 1 & 5 & 5 & -8 \\ -1 & 0 & 7 & -11 \\ 2 & 7 & 3 & -3 \end{bmatrix}$ (c) $\begin{bmatrix} 1 & 2 & -6 \\ -3 & 4 & 7 \\ 2 & 4 & 3 \end{bmatrix}$

▷ **10.** Suppose that $\mathbf{A} = \mathbf{L}_1\mathbf{U}_1 = \mathbf{L}_2\mathbf{U}_2$ are two *LU*-decompositions of **A**, with both $\mathbf{L}_i$ having nonzero main-diagonal entries. By examining $\mathbf{L}_2^{-1}\mathbf{L}_1$ and $\mathbf{U}_2\mathbf{U}_1^{-1}$, show that $\mathbf{L}_1 = \mathbf{L}_2$ and $\mathbf{U}_1 = \mathbf{U}_2$. In other words, the *LU*-decomposition $\mathbf{A} = \mathbf{LU}$ of a nonsingular matrix **A** is unique.

11. For each matrix **A** below, find its *LU*-decomposition $\mathbf{A} = \mathbf{P}^T\mathbf{LU}$ by performing Gauss elimination with interchanges.

(a) $\begin{bmatrix} 0 & 2 \\ 2 & 4 \end{bmatrix}$ (b) $\begin{bmatrix} 2 & 6 & -4 \\ -4 & -12 & 11 \\ 3 & 14 & -16 \end{bmatrix}$ (c) $\begin{bmatrix} 2 & 4 & 0 & -2 \\ -4 & -8 & 0 & 3 \\ 3 & 7 & 2 & 4 \\ 5 & 6 & 1 & -8 \end{bmatrix}$

12. The matrix **A** is defined by the *LU*-decomposition below. Perform Gauss elimination without interchanges and verify that the multipliers, pivots, and final reduced form are as contained in the *LU*-decomposition.

$$\mathbf{A} = \mathbf{LU} = \begin{bmatrix} -6 & 0 & 0 \\ 2 & 4 & 0 \\ -2 & 3 & 5 \end{bmatrix} \begin{bmatrix} 1 & -2 & 4 \\ 0 & 1 & -3 \\ 0 & 0 & 1 \end{bmatrix}$$

13. For each of the matrices in Problem 11 for which you found $\mathbf{A} = \mathbf{P}^T\mathbf{LU}$, compute the LU-decomposition without interchanges for the matrix $\mathbf{A}'$ defined as $\mathbf{A}' = \mathbf{PA}$ and verify that you obtain the same $\mathbf{L}$ and $\mathbf{U}$ matrices.

▷ **14.** Show that the LU-decomposition $\mathbf{A} = \mathbf{P}^T\mathbf{LU}$ is not unique even for nonsingular matrices by finding two different such decompositions for

$$\mathbf{A} = \begin{bmatrix} 2 & -4 \\ 2 & -7 \end{bmatrix}.$$

15. Many matrices in applications are *sparse*: Most of their entries are zero. In many cases they are *band matrices* in that their nonzero entries are in a band around the main diagonal. A matrix has *lower bandwidth* l if $\langle\mathbf{A}\rangle_{ij} = 0$ for $j < i - l$ and *upper bandwidth* u if $\langle\mathbf{A}\rangle_{ij} = 0$ for $j > i + u$. A matrix with upper and lower bandwidth 1 is called *tridiagonal.*

(a) Show that if Gauss elimination is performed to produce the LU-decomposition $\mathbf{A} = \mathbf{LU}$ *without interchanges* for a tridiagonal matrix $\mathbf{A}$, then $\mathbf{L}$ has lower bandwidth 1 and $\mathbf{U}$ has upper bandwidth 1.

(b) Show that if interchanges are required, then $\mathbf{U}$ can have upper bandwidth 2 while $\mathbf{L}$ can have lower bandwidth $p - 1$ for $p \times p$ $\mathbf{A} = \mathbf{P}^T\mathbf{LU}$.

(c) Generalize (a) and (b) to general band matrices.

16. Find the L_0U_0-decomposition of each matrix in Problem 11(a), (b), and (c).

▷ **17.** Find the L_0U_0-decomposition of $\mathbf{A}$ if $\mathbf{A}$ is:

(a) $\begin{bmatrix} 2 & 0 \\ 6 & 0 \end{bmatrix}$ (b) $\begin{bmatrix} 2 & 3 \\ 4 & 6 \end{bmatrix}$ (c) $\begin{bmatrix} 2 & 2 & 3 & 4 \\ -4 & -4 & 1 & 4 \\ -2 & -2 & 4 & 8 \\ 6 & 6 & 2 & 2 \end{bmatrix}$

18. Prove **Key Theorem 3.53(b)** as follows by using induction on p, where the result is clear for $p = 1$. Suppose that it is true for $p - 1$ and let $\mathbf{A}$ be $p \times p$.

(a) Show that elimination in the first column of $\mathbf{A}$ is equivalent to premultiplying $\mathbf{A}$ first by a permutation matrix $\mathbf{P}_1$ and then by a unit-lower-triangular matrix $\mathbf{L}_1$ whose only possibly nonzero subdiagonal entries are in the first column and are the negatives of the multipliers $\mathbf{m}$; since the remaining $(p - 1) \times (p - 1)$ matrix can be written as $\mathbf{P}'^T\mathbf{L}'\mathbf{U}'$ with unit-lower-triangular $\mathbf{L}'$ by the inductive hypothesis, show that this product can be split apart as $\mathbf{P}_2\mathbf{L}_2\mathbf{U}_2$, giving

$$\mathbf{L}_1\mathbf{P}_1\mathbf{A} = \mathbf{P}_2\mathbf{L}_2\mathbf{U}_2,$$

where

$$\mathbf{L}_1 = \begin{bmatrix} 1 & \mathbf{0}^T \\ \mathbf{m} & \mathbf{I} \end{bmatrix}, \quad \mathbf{P}_2 = \begin{bmatrix} 1 & \mathbf{0}^T \\ \mathbf{0} & \mathbf{P}'^T \end{bmatrix},$$

$$\mathbf{L}_2 = \begin{bmatrix} 1 & \mathbf{0}^T \\ \mathbf{0} & \mathbf{L}' \end{bmatrix}, \quad \mathbf{U}_2 = \begin{bmatrix} \alpha & \mathbf{w}^T \\ \mathbf{0} & \mathbf{U}' \end{bmatrix}.$$

Premultiplying by $\mathbf{P}_2^T\mathbf{L}_1^{-1}$ changes this equation to

$$\mathbf{P}_2^T\mathbf{P}_1\mathbf{A} = \mathbf{P}_2^T\mathbf{L}_1^{-1}\mathbf{P}_2\mathbf{L}_2\mathbf{U}_2.$$

(b) Show that $\mathbf{L}_1^{-1}$ looks like $\mathbf{L}_1$ except that the multipliers $\mathbf{m}$ are replaced by $-\mathbf{m}$.

(c) By taking advantage of the special structure of the terms $\mathbf{P}_2^T$, $\mathbf{L}_1^{-1}$, $\mathbf{P}_2$, and $\mathbf{L}_2$ on the right and using partitioned multiplication, show that this equation can be written as

$$(\mathbf{P}_2^T\mathbf{P}_1)\mathbf{A} = \begin{bmatrix} 1 & \mathbf{0}^T \\ -\mathbf{m}' & \mathbf{L}' \end{bmatrix}\begin{bmatrix} \alpha & \mathbf{w}^T \\ \mathbf{0} & \mathbf{U}' \end{bmatrix},$$

where $\mathbf{m}' = \mathbf{P}'\mathbf{m}$; this is the desired $\mathbf{PA} = \mathbf{L}_0\mathbf{U}_0$.

19. By multiplying out and equating entries, show that $\mathbf{A}$ below cannot be written as $\mathbf{A} = \mathbf{LU}$:

$$\begin{bmatrix} 0 & 2 \\ 0 & 2 \end{bmatrix} \neq \begin{bmatrix} a & 0 \\ b & c \end{bmatrix}\begin{bmatrix} 1 & 0 \\ u & 1 \end{bmatrix} \quad \text{for any } a, b, c, \text{ and } u.$$

3.8 *WORK MEASURES AND SOLVING SLIGHTLY MODIFIED SYSTEMS*

A given mathematical model of some system of interest is usually extensively used to explore properties of that system: The output of the system for several different inputs may be examined; the effects of small modifications in the model itself may be studied; and so on. If the model involves a system of linear equations, then such explorations usually require the solution of several systems of equations that differ only slightly from one another: Solutions may be required for several different right-hand sides but the same coefficient matrix; solutions may be required after a few changes in coefficients are made; and so on.

In such situations involving linear equations, it is usually possible to save quite a bit of work by making use of prior calculations in solving later, only slightly modified, systems of equations. In order to see what economies of effort are possible, it is important first to know what work is involved in solving a system of linear equations.

Work Measures for Elimination and Back-Substitution

Suppose that the problem is to solve for $\mathbf{x}$ in $\mathbf{Ax} = \mathbf{b}$, where $\mathbf{A}$ is $p \times p$, $\mathbf{b}$ is $p \times 1$, and $\mathbf{x}$ is $p \times 1$. How much effort is involved in Gauss elimination to produce a reduced system, and how much is involved in back-substitution to produce $\mathbf{x}$? The main effort is in the arithmetic, of course—the additions, subtractions, multiplications, and divisions essential to the process. For most computers and calculators as well as people, additions and subtractions are about equally time consuming, but are appreciably quicker than multiplications and divisions (which are of about equal cost). In measuring work, therefore, we will count addition/subtractions separately from multiplication/divisions but will lump all work into those two categories.

Consider Gauss elimination in column k of the $p \times p$ matrix $\mathbf{A}$; we will count the work of processing $\mathbf{b}$ later. The row containing the pivot must be divided by that pivot; we can avoid dividing the pivot by itself, since the answer is known to be 1. Therefore, $p - k$ divisions will be performed. For each row below, we must add to it a multiple m of our pivot row; the multiple m is just the negative of the coefficient we seek to eliminate (so getting m is free) and we need not calculate that eliminated entry since we know the result to be 0. Therefore, $p - k$ multiplications and $p - k$ additions must be performed for *each* of the $p - k$ rows below the pivot row.

The total cost of Gauss elimination is the sum of all these costs:

$$\sum_{k=1}^{p} \left\{ \{(p-k)^2 \text{ in } + \text{ and } -\} + \{(p-k) + (p-k)^2 \text{ in } \times \text{ and } /\} \right\}.$$

Using the facts that

$$1 + 2 + \cdots + n = \frac{n(n+1)}{2} \quad \text{and} \quad 1 + 4 + \cdots + n^2 = \frac{n(n+1)(2n+1)}{6},$$

we can calculate the total work to be $p^3/3 - p/3$ multiplication/divisions and $p^3/3 - p^2/2 + p/6$ addition/subtractions.

To solve $\mathbf{Ax} = \mathbf{b}$, of course, requires that $\mathbf{b}$ be processed also and that back-substitution be performed; this is easily expressed in terms of the LU-decomposition $\mathbf{A} = \mathbf{P}^T\mathbf{LU}$ of **Key Theorem 3.53**. $\mathbf{Ax} = \mathbf{b}$ means that $\mathbf{P}^T\mathbf{LUx} = \mathbf{b}$, so $\mathbf{LUx} = \mathbf{Pb}$; we write this as $\mathbf{Ly} = \mathbf{Pb}$ and $\mathbf{Ux} = \mathbf{y}$, each of which is simply a *triangular* system that can be solved immediately by forward- or back-substitution. The column matrix $\mathbf{y}$ is precisely what we would have obtained had we performed all the elimination steps on $\mathbf{b}$ during Gauss elimination.

(3.57) To solve $\mathbf{Ax} = \mathbf{b}$, given the LU-decomposition $\mathbf{A} = \mathbf{P}^T\mathbf{LU}$:

1. Permute the entries of $\mathbf{b}$ to obtain $\mathbf{b}' = \mathbf{Pb}$.
2. Solve the lower-triangular system $\mathbf{Ly} = \mathbf{b}'$ for $\mathbf{y}$ by forward-substitution.
3. Solve the unit-upper-triangular system $\mathbf{Ux} = \mathbf{y}$ for $\mathbf{x}$ by back-substitution.

(3.58) ***Example.*** Consider solving the system of equations in Example 3.17 by using (3.57); the relevant LU-decomposition $\mathbf{P}^T\mathbf{LU}$ of $\mathbf{A}$ is

$$\begin{bmatrix} 1 & 0 & 0 & 0 \\ 0 & 0 & 0 & 1 \\ 0 & 0 & 1 & 0 \\ 0 & 1 & 0 & 0 \end{bmatrix} \begin{bmatrix} -2 & 0 & 0 & 0 \\ 2 & -1 & 0 & 0 \\ -2 & 2 & 5 & 0 \\ 3 & 0 & 3 & -5.6 \end{bmatrix} \begin{bmatrix} 1 & -2 & 1 & 3 \\ 0 & 1 & -2 & -2 \\ 0 & 0 & 1 & 2.2 \\ 0 & 0 & 0 & 1 \end{bmatrix}.$$

Since $\mathbf{b} = [4 \quad -1 \quad 1 \quad -3]^T$, $\mathbf{b}' = \mathbf{Pb}$ is just $\mathbf{b}' = [4 \quad -3 \quad 1 \quad -1]^T$. The solution $\mathbf{y}$ to $\mathbf{Ly} = \mathbf{b}'$ is easily computed to be $\mathbf{y} = [-2 \quad -1 \quad -\frac{1}{5} \quad -1]^T$, and the solution $\mathbf{x}$ to $\mathbf{Ux} = \mathbf{y}$ is easily computed to be $\mathbf{x} = [1 \quad 1 \quad 2 \quad -1]$. This does indeed solve the system of equations in Example 3.17.

It is simple to count the work involved in (3.57) for a $p \times p$ matrix $\mathbf{A}$.

Step 1 involves interchanges but no arithmetic.

Step 2 requires finding $\langle \mathbf{y} \rangle_k$ in the order $k = 1, 2, \ldots, p$. In finding $\langle \mathbf{y} \rangle_k$, the $k - 1$ earlier $\mathbf{y}$-values are substituted in, multiplied by a coefficient, and subtracted from the right-hand side; one division then produces $\langle \mathbf{y} \rangle_k$. For each k, then, the cost is k multiplication/divisions and $k - 1$ addition/subtractions. This gives a total cost of $p(p + 1)/2$ multiplication/divisions and $p(p - 1)/2$ addition/subtractions for step 2.

Step 3 requires finding $\langle \mathbf{x} \rangle_k$ in the order $k = p, p - 1, \ldots, 1$. The work is similar to that in step 2 except that the equality $\langle \mathbf{U} \rangle_{kk} = 1$ eliminates the need for one division per variable. The total cost is thus $p(p - 1)/2$ multiplication/divisions and $p(p - 1)/2$ addition/subtractions.

The combined costs of the three steps in (3.57) can now easily be calculated.

(3.59) ***Theorem*** (work in eliminating and solving). Suppose that $\mathbf{A}$ is a $p \times p$ nonsingular matrix. Then:

(a) The cost of computing the LU-decomposition $\mathbf{A} = \mathbf{P}^T\mathbf{LU}$ of $\mathbf{A}$ is

$p^3/3 - p/3$ multiplication/divisions, and
$p^3/3 - p^2/2 + p/6$ addition/subtractions.

(b) Given the LU-decomposition, the cost of solving $\mathbf{Ax} = \mathbf{b}$ by the process in (3.57) is

p^2 multiplication/divisions, and
$p^2 - p$ addition/subtractions.

The key points to observe are that work in Gauss elimination to produce the LU-decomposition varies with the *cube* of p, while the cost of solving a system—once the LU-decomposition is available—varies with the *square* of p. The total cost for a single system thus varies with p^3: Doubling the number of equations and variables makes the work *eight* times as much. Observe, however, that the cost of simply multiplying together two $p \times p$ matrices is p^3 multiplications and $p^3 - p^2$ additions—both greater than in (3.59). Note, too, that solving a single system of 100 equations in 100 unknowns requires 343,300 multiplication/divisions and 338,250 addition/subtractions—a formidable task for most human beings; many microcomputers typically perform 10,000 arithmetic operations per second and could do this in about 70 seconds, while modern supercomputers would require only a tiny fraction of a second.

The same sort of analysis can be used to count the work in Gauss–Jordan elimination; the total for solving a single system of p linear equations in p unknowns is $p^3/2 + p^2/2$ multiplication/divisions and $p^3/2 - p/2$ addition/subtractions. The dominant terms here are $p^3/\underline{2}$ versus $p^3/\underline{3}$ for Gauss elimination; Gauss–Jordan elimination is about 50% more time consuming than is Gauss elimination for solving a system of equations. In fact, it is known that no method using only row (and column) operations can take less work on general matrices than Gauss elimination.

You should now be able to solve Problems 1 to 6.

Modifying the Right-hand Sides in Equations

Suppose now that it is necessary to solve $\mathbf{Ax}_i = \mathbf{b}_i$ with several different right-hand sides $\mathbf{b}_i$, $i = 1, 2, \ldots, n$; this might easily occur in a design study, as indicated earlier. If we know all the $\mathbf{b}_i$ at the start, then according to (3.28) we can simply use Gauss elimination and back-substitution on the multiply-augmented matrix

$$[\mathbf{A} \quad \mathbf{b}_1 \quad \mathbf{b}_2 \quad \cdots \quad \mathbf{b}_n].$$

In design studies and other applications, however, we often do not know the next right-hand side $\mathbf{b}_{i+1}$ until we have gotten the results $\mathbf{x}_i$ from the present right-hand side; therefore, we cannot set up the multiply-augmented matrix. Yet we want to avoid performing Gauss elimination all over again on each of the n different $p \times p$ systems, since that would cost us about $n(p^3/3)$ arithmetic operations according to Theorem 3.59.

The LU-decomposition of $\mathbf{A}$ provides the efficient resolution to this problem. We first compute the LU-decomposition $\mathbf{A} = \mathbf{P}^T\mathbf{LU}$ of $\mathbf{A}$, and then we use (3.57) on each right-hand side $\mathbf{b}_i$ as it becomes available. According to the work measures in Theorem 3.59:

(3.60) We can obtain the LU-decomposition $\mathbf{A} = \mathbf{P}^T\mathbf{LU}$ of $\mathbf{A}$ and solve n different sets of equations $\mathbf{Ax}_i = \mathbf{b}_i$ for $1 \le i \le n$ at a total cost of

$$p^3/3 - p/3 + np^2 \text{ multiplication/divisions, and}$$
$$p^3/3 - p^2/2 + p/6 + n(p^2 - p) \text{ addition/subtractions.}$$

One of the reasons for wanting to solve $\mathbf{Ax}_i = \mathbf{b}_i$ for several $\mathbf{b}_i$ could be to obtain the inverse $\mathbf{A}^{-1}$, since

$$\mathbf{A}^{-1} = [\mathbf{x}_1 \quad \mathbf{x}_2 \quad \cdots \quad \mathbf{x}_p]$$

if $\mathbf{b}_i = \mathbf{e}_i$, the unit-column matrix of order p—see **Key Theorem 1.44**. In this case $n = p$ and the work measures in (3.60) become roughly $4p^3/3$ operations. Each $\mathbf{e}_i$ is quite special, however, since it contains mainly zeros. If we avoid adding to zeros and avoid multiplying and dividing numbers into 0, we can greatly reduce the work in applying (3.57) to these special right-hand sides $\mathbf{e}_i$. The result is:

(3.61) The inverse of a nonsingular $p \times p$ matrix can be computed by Gauss elimination and back-substitution with

$$p^3 \text{ multiplication/divisions, and}$$
$$p^3 - 2p^2 + p \text{ addition/subtractions.}$$

Here is a surprising fact that follows from these work measures: No matter the number n of systems $\mathbf{Ax}_i = \mathbf{b}_i$ that you need to solve with the same matrix $\mathbf{A}$, it is always less effort for general $\mathbf{A}$ to find $\mathbf{A}$'s LU-decomposition and use (3.57) for each right-hand side than to compute $\mathbf{A}^{-1}$ by Gauss elimination and then compute each $\mathbf{x}_i$ as $\mathbf{A}^{-1}\mathbf{b}_i$. See Problem 7. This is one of the reasons why, in practice, it is rare that the actual inverse $\mathbf{A}^{-1}$ is really useful—most of what can be accomplished with $\mathbf{A}^{-1}$ can be accomplished more efficiently through other means, such as the LU-decomposition. There are exceptions to this, of course, such as when $\mathbf{A}$ has some very special structure or when you actually need to see the entries in $\mathbf{A}^{-1}$, but it is nonetheless a good general rule.

You should now be able to solve Problems 1 to 7.

Modifying the Coefficients in Equations

We now turn our attention to the case in which not $\mathbf{b}$ but $\mathbf{A}$ changes in some fairly simple way. If we change the role played by a particular variable or equation in a model, for example, then we might need to investigate how to solve systems after we have changed one row or one column of a matrix. It is easy to describe this situation in matrix language. To replace the ith column $\mathbf{c}_i$ of $\mathbf{A}$ by the column matrix $\mathbf{c}_i'$, we need only add to $\mathbf{A}$ the matrix $(\mathbf{c}_i' - \mathbf{c}_i)\mathbf{e}_i^T$; adding $\mathbf{e}_i(\mathbf{r}_i' - \mathbf{r}_i)^T$ instead replaces the ith row $\mathbf{r}_i$ by $\mathbf{r}_i'$.

(3.62) ***Example.*** Let

$$\mathbf{A} = \begin{bmatrix} 1 & 2 & -1 \\ 3 & 1 & 2 \\ 0 & -2 & 4 \end{bmatrix}, \qquad \mathbf{d} = \begin{bmatrix} 2 \\ 1 \\ 1 \end{bmatrix}.$$

Then

$$\mathbf{A} + \mathbf{de}_1^T = \begin{bmatrix} 1 & 2 & -1 \\ 3 & 1 & 2 \\ 0 & -2 & 4 \end{bmatrix} + \begin{bmatrix} 2 \\ 1 \\ 1 \end{bmatrix} [1 \quad 0 \quad 0]$$
$$= \begin{bmatrix} 1 & 2 & -1 \\ 3 & 1 & 2 \\ 0 & -2 & 4 \end{bmatrix} + \begin{bmatrix} 2 & 0 & 0 \\ 1 & 0 & 0 \\ 1 & 0 & 0 \end{bmatrix} = \begin{bmatrix} 3 & 2 & -1 \\ 4 & 1 & 2 \\ 1 & -2 & 4 \end{bmatrix}$$

while

$$\mathbf{A} + \mathbf{e}_2\mathbf{d}^T = \begin{bmatrix} 1 & 2 & -1 \\ 3 & 1 & 2 \\ 0 & -2 & 4 \end{bmatrix} + \begin{bmatrix} 0 \\ 1 \\ 0 \end{bmatrix} [2 \quad 1 \quad 1] = \begin{bmatrix} 1 & 2 & -1 \\ 5 & 2 & 3 \\ 0 & -2 & 4 \end{bmatrix}.$$

More generally, consider adding to $\mathbf{A}$ a matrix (constructed from a few column matrices and a few row matrices) of the form $\mathbf{CR}$, where $\mathbf{C}$ is $p \times n$ and $\mathbf{R}$ is $n \times p$, with n much less than p. The following result describes the inverse of the new matrix $\mathbf{A}' = \mathbf{A} + \mathbf{CR}$ and, perhaps more important, how to obtain the solution $\mathbf{x}'$ to $\mathbf{A}'\mathbf{x}' = \mathbf{b}$ from that to $\mathbf{Ax} = \mathbf{b}$.

(3.63) ***Theorem*** (inverses of modified matrices). Suppose that $\mathbf{A}$ is $p \times p$ and nonsingular, that $\mathbf{C}$ is $p \times n$ and $\mathbf{R}$ is $n \times p$ with $n \le p$ (usually n *much* less than p), and that the $n \times n$ matrix $\mathbf{K} = \mathbf{I}_n + \mathbf{RA}^{-1}\mathbf{C}$ is nonsingular. Then:

(a) $\mathbf{A}' = \mathbf{A} + \mathbf{CR}$ is nonsingular, and

$$\mathbf{A}'^{-1} = \mathbf{A}^{-1} - \mathbf{A}^{-1}\mathbf{CK}^{-1}\mathbf{RA}^{-1}.$$

(b) The solution $\mathbf{x}'$ to $\mathbf{A}'\mathbf{x}' = \mathbf{b}$ can be expressed in terms of the solution $\mathbf{x}$ to $\mathbf{Ax} = \mathbf{b}$ as

$$\mathbf{x}' = \mathbf{x} - \mathbf{A}^{-1}\mathbf{CK}^{-1}\mathbf{Rx}.$$

(c) Given $\mathbf{A}^{-1}$, $\mathbf{A}'^{-1}$ can be computed by (a) in approximately $3np^2 + 2n^2p + n^3/3$ multiplication/divisions and a comparable number of addition/subtractions.

(d) Given the LU-decomposition $\mathbf{A} = \mathbf{P}^T\mathbf{LU}$ and the solution $\mathbf{x}$ to $\mathbf{Ax} = \mathbf{b}$, $\mathbf{x}'$ can be computed by (b) in approximately $np^2 + n^2p + n^3/3$ multiplication/divisions and a comparable number of addition/subtractions.

PROOF

(a) Let $\mathbf{H}$ denote the claimed inverse of $\mathbf{A}'$; we need to show that $\mathbf{A}'\mathbf{H} = \mathbf{HA}' = \mathbf{I}_p$—this is Problem 8.

(b) $\mathbf{x}' = \mathbf{A}'^{-1}\mathbf{b} = (\mathbf{A}^{-1} - \mathbf{A}^{-1}\mathbf{CK}^{-1}\mathbf{RA}^{-1})\mathbf{b} = \mathbf{A}^{-1}\mathbf{b} - \mathbf{A}^{-1}\mathbf{CK}^{-1}\mathbf{RA}^{-1}\mathbf{b}$
$= \mathbf{x} - \mathbf{A}^{-1}\mathbf{CK}^{-1}\mathbf{Rx}$, as claimed.

(c) and (d) are both based on the work measures above and on the fact that it requires about rst multiplications (and similar additions) to compute the product of an $r \times s$ matrix times an $s \times t$ matrix—this is Problem 9. ■

(3.64) ***Example.*** We illustrate the point of the work measures in Theorem 3.63. Suppose that $p = 100$ and $n = 5$; suppose that we have already obtained the LU-decomposition of the 100×100 matrix $\mathbf{A}$. Now, suppose that we need to solve $\mathbf{A}'\mathbf{x}' = \mathbf{b}$ for some column matrix $\mathbf{b}$. If we ignore Theorem 3.63 and apply Gauss elimination to $\mathbf{A}'\mathbf{x}' = \mathbf{b}$, we will require on the order of $p^3/3 \approx$ 333,000 multiplication/divisions. Suppose that we instead use Theorem 3.63(b). We first need $\mathbf{x}$, the solution to $\mathbf{Ax} = \mathbf{b}$; since we have $\mathbf{A}$'s LU-decomposition, we can get this at a cost of about p^2 multiplication/divisions. We can then use the method in Theorem 3.63(d) so as to get $\mathbf{x}'$ in another roughly $np^2 + n^2p + n^3/3$ multiplication/divisions. The total effort with this approach is therefore about

$$p^2 + np^2 + n^2p + \frac{n^3}{3} \approx 62{,}500$$

multiplication/divisions—about an 80% saving over the first approach. For $p = 100$ and $n = 2$, for another example, the saving is about 90%: the direct approach takes 10 times as much time.

On first acquaintance it may appear that the circumstances of Theorem 3.63 would only rarely arise in practice, but this is not the case. Very often, extremely large models of complicated systems—large buildings, electrical-distribution networks, large economic systems, and the like—consist of a small number of very large and almost independent subsystems with a small number of interconnections among the subsystems. Theorem 3.63 can often be used in this case by letting $\mathbf{A}$ represent the situation in which the subsystems are truly independent and letting $\mathbf{C}$ and $\mathbf{R}$ provide the adjustments reflecting the interconnections. With just two subsystems, for example, you might need to solve a system of 2000 equations in 2000 unknowns with a coefficient matrix $\mathbf{A}'$ of the following non-standardly-partitioned form:

$$\mathbf{A}' = \begin{array}{c} \\ 1000 \\ 1000 \end{array}\!\!\begin{array}{cc} 1000 & 1000 \\ \mathbf{B} & \mathbf{0} \\ \mathbf{0} & \mathbf{D} \end{array} + \begin{array}{cccc} 990 & 10 & 10 & 990 \\ \mathbf{0} & \mathbf{0} & \mathbf{E} & \mathbf{0} \\ \mathbf{0} & \mathbf{F} & \mathbf{0} & \mathbf{0} \end{array},$$

with the dimensions of the blocks being as indicated. Let $\mathbf{A}$ be the first of the two partitioned matrices, and suppose that $\mathbf{B}$ and $\mathbf{D}$—and hence $\mathbf{A}$ itself—are nonsingular. We conceive of $\mathbf{B}$ as representing the internal relationships within one large subsystem, $\mathbf{D}$ the internal relationships within the other large subsystem, and $\mathbf{E}$ and $\mathbf{F}$ the interconnections between the two subsystems. We can write $\mathbf{A}'$ as in Theorem 3.63 with $\mathbf{A}' = \mathbf{A} + \mathbf{CR}$, where

$$\mathbf{C} = \begin{array}{c} \\ 1000 \\ 1000 \end{array}\!\!\begin{array}{cc} 10 & 10 \\ \mathbf{0} & \mathbf{E} \\ \mathbf{F} & \mathbf{0} \end{array} \quad \text{and} \quad \mathbf{R} = \begin{array}{c} \\ 10 \\ 10 \end{array}\!\!\begin{array}{cccc} 990 & 10 & 10 & 990 \\ \mathbf{0} & \mathbf{I} & \mathbf{0} & \mathbf{0} \\ \mathbf{0} & \mathbf{0} & \mathbf{I} & \mathbf{0} \end{array}.$$

In such circumstances the savings from using Theorem 3.63 can be enormous: Problems can sometimes be solved that might not be solved in more direct ways.

Various special techniques essentially based on this approach have been developed in a wide variety of application areas. This one basic idea leads, for example, to the *method of tearing* in network analysis, to *capacitance matrix methods* for the direct numerical solution of certain differential equations, to various *update methods* for both constrained and unconstrained nonlinear optimization, and to certain implementations of the *simplex method* in linear programming.

PROBLEMS 3.8

▷ **1.** Show that, in order to multiply an $r \times s$ matrix times an $s \times t$ matrix, at most rst multiplications and $rt(s-1)$ additions are required.

2. Use the formulas

$$1 + 2 + \cdots + n = \frac{n(n+1)}{2}$$

and

$$1 + 4 + \cdots + n^2 = \frac{n(n+1)(2n+1)}{6}$$

to prove Theorem 3.59(a).

3. Use the formulas given in Problem 2 to prove Theorem 3.59(b).

4. Derive the work measures stated in the text for Gauss–Jordan elimination.

▷ **5.** Use Theorem 3.59(a) to calculate the number of multiplication/divisions needed in order to find the LU-decomposition and solve one system of equations for a $p \times p$ matrix if:
(a) $p = 10$ (b) $p = 30$ (c) $p = 50$
(d) $p = 70$ (e) $p = 90$

6. Consider the LU-decomposition $\mathbf{A} = \mathbf{P}^T\mathbf{LU}$ found in Example 3.45 of Section 3.7. For that matrix $\mathbf{A}$, use (3.57) to solve

$$\mathbf{Ax} = \mathbf{b} = [-4 \quad 5 \quad 13 \quad 20]^T$$

and verify that the $\mathbf{x}$ found actually solves the original equations.

▷ **7.** Show that the work required to obtain the LU-decomposition $\mathbf{A} = \mathbf{P}^T\mathbf{LU}$ for general $\mathbf{A}$, and then use it to solve k systems $\mathbf{Ax}_i = \mathbf{b}_i$ for $1 \le i \le k$, is always less than that required to compute $\mathbf{A}^{-1}$ by Gauss elimination and then compute $\mathbf{x}_i$ as $\mathbf{x}_i = \mathbf{A}^{-1}\mathbf{b}_i$ for $1 \le i \le k$.

8. Complete the proof of Theorem 3.63(a).

9. Complete the proof of Theorem 3.63(c) and (d).

10. Suppose that $p = 1000$ and $n = 10$. As in Example 3.64, calculate the savings in using the approach of Theorem 3.63(d) to solve $\mathbf{A}'\mathbf{x}' = \mathbf{b}$.

▷ **11.** Let the nonsingular matrix $\mathbf{A}$ be $p \times p$ and $\mathbf{r}$ be $1 \times p$. Describe how to use Theorem 3.63(a) to find the inverse, if it exists, of:
(a) $\mathbf{A} + \mathbf{e}_i\mathbf{r}$ (b) $\mathbf{A} + \mathbf{r}^T\mathbf{e}_j^T$

12. Let the nonsingular matrix $\mathbf{A}$ be $p \times p$. Describe how to use Theorem 3.63(a) to find the inverse, if it exists, of the matrix formed by adding the number α to the (i, j)-entry of $\mathbf{A}$.

13. Use Theorem 3.63(a) in the fashion described in the paragraph following (3.64) to find the inverse of

$$\mathbf{A}' = \left[\begin{array}{cc|cc} 3 & 5 & 0 & 0 \\ 1 & 2 & 0 & 0 \\ \hline 0 & 0 & 2 & 3 \\ 0 & 0 & 1 & 2 \end{array}\right] + \left[\begin{array}{c|c|c|c} 0 & 0 & 0 & 0 \\ 0 & 0 & 1 & 0 \\ \hline 0 & 1 & 0 & 0 \\ 0 & 0 & 0 & 0 \end{array}\right].$$

𝔐 **14.** Consider the 50×50 matrix $\mathbf{A}'$ with the main-diagonal entries $\langle \mathbf{A}' \rangle_{ii} = 2$ for all i, the superdiagonal entries $\langle \mathbf{A}' \rangle_{i,i+1} = -1$ for all i, the subdiagonal entries $\langle \mathbf{A}' \rangle_{i-1,i} = -1$ for all i, and all other entires equal to zero. As in the paragraph following (3.64), view $\mathbf{A}'$ as the sum $\mathbf{A} + \mathbf{CR}$ in the form

$$\begin{bmatrix} \mathbf{B} & \mathbf{0} \\ \mathbf{0} & \mathbf{B} \end{bmatrix} + \begin{bmatrix} \mathbf{0} & \mathbf{0} & \mathbf{E} & \mathbf{0} \\ \mathbf{0} & \mathbf{F} & \mathbf{0} & \mathbf{0} \end{bmatrix},$$

where $\mathbf{B}$ is the 25×25 matrix that looks like a small version of $\mathbf{A}'$, where $\mathbf{F}$ and $\mathbf{E}$ are 25×1 with $\mathbf{F} = [-1 \quad 0 \quad \cdots \quad 0]^T$ and $\mathbf{E} = [0 \quad 0 \quad \cdots \quad 0 \quad -1]^T$.

(a) Use MATLAB to find the inverse of $\mathbf{A}'$ directly and—by entering the MATLAB command *flops* immediately before and after the computation—to find the number of floating-point operations required to do so.
(b) Use MATLAB to find the inverse of $\mathbf{A}'$ by means of Theorem 3.63(a)—computing the inverse of $\mathbf{A}$ by forming the obvious partitioned matrix using the inverse $\mathbf{B}^{-1}$—and compare the numbers of floating-point operations required by the two methods.

▷ **15.** Consider Problem 15 of Section 3.7, where Gauss elimination with and without interchanges was considered for tridiagonal matrices and for band matrices generally.
(a) Show that the work in finding the LU-decomposition of a $p \times p$ tridiagonal matrix *without interchanges* equals $(2p - 2)$ multiplication/divisions and $(p - 1)$ addition/subtractions.
(b) Show that, if interchanges are allowed, the work required is at most $(4p - 6)$ multiplication/divisions and $(p - 1)$ addition/subtractions.
(c) Generalize (a) and (b) to general band matrices.

3.9 COMPUTER SOFTWARE FOR GAUSS ELIMINATION

> **WARNING:** Always use a standard program rather than write your own program for solving linear equations or finding inverses!
>
> If you have no standard program, copy one from Forsythe, Malcolm, and Moler (42).

In practical applications, systems of linear equations are solved and inverses are found almost always with the use of computers—supercomputers, standard mainframe computers, or microcomputers. Over the years experts have developed excellent general-purpose computer programs for the efficient and accurate solution of systems of equations and for the calculation of inverses. In Section 3.6 we gave a slight indication of the subtleties involved in this task. Since the results of those efforts are generally available either free of charge or at low cost, there is no reason for nonexperts to create their own programs for these problems. And there are good reasons *not* to create your own programs—you are likely to produce inaccurate (and expensive) answers without any warning that the answers are inaccurate. Although you may find it instructive to write a simple program to implement Gauss elimination with or without interchanges as a pedagogic tool to assist in the understanding of exactly what the process is, do *not* use such a program on realistic problems whose answers are important to you.

What to Demand of Software

The least to demand is that the software be efficient, accurate, both easy and simple to use, and informative as to the confidence with which the results should be accepted in each specific case. Bonus features would include the ability to use special approaches for special problems (such as symmetric matrices, matrices all of whose nonzero entries are near the main diagonal, and so on), the ability to evaluate determinants (see Section 4.5), and so on. Not all software for equations and inverses meet these criteria, but much does; see the following subsection for specific suggestions.

What we mean by efficiency and accuracy probably seems clear, although there are in fact some subtle points here; these matters are beyond the scope of this book, however. Details may be found in Golub and Van Loan (46).

Software that is easy and simple to use should allow the data matrices to be entered as input in some natural and straightforward fashion and should not ask the user to provide much additional information.

An especially important feature is the providing of information about the reliability of the results. Many good programs in this area provide an indication of the so-called *condition number* $c(\mathbf{A})$ of the matrix **A** that is the coefficient matrix of the equations or whose inverse is being found; see Theorem 6.29. This number $c(\mathbf{A})$ measures the amount by which inaccuracies in the data of the problem are magnified in the answer; if the data contain errors of relative size 10^{-3}, for example, then the answers might contain errors of relative size as large as $c(\mathbf{A}) \times 10^{-3}$. These errors are not the "fault" of the program but rather are inherent in the answers because of the inaccuracies in the data. Since a careful implementation of Gauss elimination is capable of estimating this condition number cheaply, providing such an estimate is a feature that it is reasonable to demand.

What Software to Get

High-quality software meeting the general criteria above certainly is available commercially; the MATLAB system from The MathWorks is an obvious example, but there are excellent systems available elsewhere, such as from International Mathematical and Statistical Libraries {IMSL—see (57)} and from the Numerical Algorithms Group {NAG—see (59)}. For mainframe computers, state-of-the-art subroutines are available at no charge in the LINPACK system.

LINPACK was developed as a cooperative effort of many numerical analysts at many institutions; much of the work was concentrated at the Argonne National Laboratory, and financial support was provided by the National Science Foundation. The result was an excellent collection of easily portable FORTRAN programs that are available free from Argonne National Laboratory; details of the programs and their acquisition are contained in *LINPACK User's Guide* (40). LINPACK contains programs for most matrix computations related to equations, including of course the solution of systems of equations and the inversion of matrices.

There are actually a few dozen programs in LINPACK for solving systems or finding inverses; they vary by the special structures assumed for the matrix (general, symmetric, and so on), the type of numbers as matrix entries (real, complex, and so on), the way in which the matrix entries are stored in the computer, and the accuracy desired. Routines are of two basic types: those that compute LU-decompositions, and those that use LU-decompositions (to solve systems, find inverses, evaluate determinants, and so on). The basic methodology is Gauss elimination and back-substitution much as we have described; one variation is that **L** is *unit*-lower-triangular, while **U** has general entries on its main diagonal—our L_0U_0-decomposition (3.49), that is. The routines for finding LU-decompositions estimate the condition number and report this estimate as an output parameter accessible to the user; this lets the user assess the accuracy of any results to be generated. Anyone needing to solve systems of equations—or perform other matrix computations, for that matter—on a mainframe computer should consider obtaining the LINPACK system.

PROBLEMS 3.9

1. Determine whether the LINPACK collection of subroutines is available for your use.
2. Determine what programs are available to you for solving p linear equations in p unknowns and what special capabilities are available (such as for complex entries, special structures for the coefficient matrix, and so on).
3. Determine what programs are available to you for inverting $p \times p$ matrices and what special capabilities are available (such as complex entries, special structures for the matrix, and so on).
4. Determine whether the software available to you for linear systems and inverses provides an estimate of the condition number for the problem.

3.10 MISCELLANEOUS PROBLEMS

PROBLEMS 3.10

1. Suppose that $\mathbf{A}$ is $p \times p$, $\mathbf{B}$ is $p \times q$, and $\mathbf{AB} = \mathbf{0}$. Prove that either $\mathbf{A}$ is singular or $\mathbf{B} = \mathbf{0}$.

▷ 2. The matrix below is the augmented matrix for a system of three equations in three unknowns. Show that:
   (a) If $k = 0$, then the system has an infinite number of solutions.
   (b) For another specific value of k, which you must find, the system has no solutions.
   (c) For all other values of k, the system has a unique solution.

$$\mathbf{A} = \left[\begin{array}{ccc|c} 1 & -2 & 3 & 1 \\ 2 & k & 6 & 6 \\ -1 & 3 & k-3 & 0 \end{array}\right]$$

3. (a) Suppose that $\mathbf{A}$ is $p \times q$, $\mathbf{B}$ is $q \times q$, and $\mathbf{AB} = \mathbf{0}$. Prove that either $\mathbf{B}$ is singular or $\mathbf{A} = \mathbf{0}$.
   (b) Suppose that $\mathbf{A}$ and $\mathbf{B}$ are $p \times p$ and $\mathbf{AB} = \mathbf{0}$. Use (a) and Problem 1 to prove that either $\mathbf{A} = \mathbf{0}$ or $\mathbf{B} = \mathbf{0}$ or both $\mathbf{A}$ and $\mathbf{B}$ are singular.
4. Find the LU-decomposition of the matrix $\mathbf{A}$ below, not by Gauss elimination, but by writing $\mathbf{A} = \mathbf{LU}$ with $\mathbf{L}$ and $\mathbf{U}$ as below and then solving for the entries in $\mathbf{L}$ and $\mathbf{U}$.

$$\mathbf{A} = \begin{bmatrix} 2 & 2 \\ 4 & -1 \end{bmatrix}, \quad \mathbf{L} = \begin{bmatrix} l_{11} & 0 \\ l_{21} & l_{22} \end{bmatrix}, \quad \mathbf{U} = \begin{bmatrix} 1 & u_{12} \\ 0 & 1 \end{bmatrix}.$$

▷ 5. Generalize Problem 4 as follows. Suppose that $\mathbf{A}$ is $p \times p$ and has an LU-decomposition $\mathbf{A} = \mathbf{LU}$ for lower triangular $\mathbf{L}$ with nonzero main-diagonal entries and unit-upper-triangular $\mathbf{U}$. Write down the general form for $\mathbf{L}$ and $\mathbf{U}$ with entries l_{ij} and u_{ij} as in Problem 4. The (1, 1)-entry of $\mathbf{U}$ is known.

Show that this then lets you find the first column of $\mathbf{L}$, which lets you find the first row of $\mathbf{U}$, which lets you find the second column of $\mathbf{L}$, and so on, until $\mathbf{L}$ and $\mathbf{U}$ are both determined. Write down the equations for computing $\mathbf{L}$ and $\mathbf{U}$ in this fashion.

6. Our implementation of Gauss elimination is called the Crout variant. The Doolittle variant is similar except that it does not divide the kth row by the kth pivot before elimination; this produces our *unit*-lower-triangular $\mathbf{L}_0$ and a (*nonunit*) upper-triangular $\mathbf{U}_0$ with $\mathbf{A} = \mathbf{L}_0\mathbf{U}_0$. Use the Doolittle variant to perform Gauss elimination on the matrix $\mathbf{A}$ in Problem 4 and find this $\mathbf{A} = \mathbf{L}_0\mathbf{U}_0$ decomposition.

▷ **7.** Find the Doolittle variant—see Problem 6—of the LU-decomposition directly as in Problem 4 for the matrix $\mathbf{A}$ in Problem 4 by writing $\mathbf{A} = \mathbf{L}_0\mathbf{U}_0$ for $\mathbf{L}_0$ and $\mathbf{U}_0$ as below and solving for the entries in $\mathbf{L}_0$ and $\mathbf{U}_0$.

$$\mathbf{L}_0 = \begin{bmatrix} 1 & 0 \\ l_{21} & 1 \end{bmatrix}, \qquad \mathbf{U}_0 = \begin{bmatrix} u_{11} & u_{12} \\ 0 & u_{22} \end{bmatrix}.$$

8. Generalize the method of Problem 7 in the same way that you were asked in Problem 5 to generalize the method of Problem 4.

9. For the matrix $\mathbf{A}$ in Problem 4, show that $\mathbf{A} = \mathbf{L}_1\mathbf{D}_0\mathbf{U}_1$ with $\mathbf{L}_1$, $\mathbf{U}_1$, and $\mathbf{D}_0$ as given below. Note that $\mathbf{L}_1$ is unit-lower-triangular and equals the matrix $\mathbf{L}_0$ found in Problems 6 and 7, while $\mathbf{U}_1$ is unit-upper-triangular and equals the matrix $\mathbf{U}$ found in Problems 4 and 5.

$$\mathbf{L}_1 = \begin{bmatrix} 1 & 0 \\ 2 & 1 \end{bmatrix}, \qquad \mathbf{D}_0 = \begin{bmatrix} 2 & 0 \\ 0 & -5 \end{bmatrix}, \qquad \mathbf{U}_1 = \begin{bmatrix} 1 & 1 \\ 0 & 1 \end{bmatrix}.$$

▷ **10.** (a) Use **Key Theorem 3.48** to prove that, if $\mathbf{A}$ is a $p \times p$ matrix for which Gauss elimination can be completed without interchanges, then there is a unit-lower-triangular matrix $\mathbf{L}_1$, a nonsingular diagonal matrix $\mathbf{D}_0$, and a unit-upper-triangular matrix $\mathbf{U}_1$ for which $\mathbf{A} = \mathbf{L}_1\mathbf{D}_0\mathbf{U}_1$.

(b) State and prove the analogous generalization $\mathbf{PA} = \mathbf{L}_1\mathbf{D}_0\mathbf{U}_1$ of **Key Theorem 3.53**. (*Note:* This is the general *LDU-decomposition* of $\mathbf{A}$. Taking $\mathbf{L} = \mathbf{L}_1\mathbf{D}_0$ and $\mathbf{U} = \mathbf{U}_1$ gives the Crout variant we have used, while taking $\mathbf{L}_0 = \mathbf{L}_1$ and $\mathbf{U}_0 = \mathbf{D}_0\mathbf{U}_1$ gives the Doolittle variant of Problems 6–8.)

11. Find a Doolittle-variant L_0U_0-decomposition, as in Problem 10, of

$$\mathbf{A} = \begin{bmatrix} 2 & -1 & 3 & 4 \\ 2 & -1 & 4 & 2 \\ -4 & 2 & -7 & -7 \\ -2 & 1 & -1 & -1 \end{bmatrix}.$$

12. Prove that, if $\mathbf{A}$ is nonsingular, then its Doolittle-variant LU-decomposition—see Problem 10—is unique: If

$$\mathbf{L}_0\mathbf{U}_0 = \mathbf{L}'_0\mathbf{U}'_0, \quad \text{then} \quad \mathbf{L}_0 = \mathbf{L}'_0 \quad \text{and} \quad \mathbf{U}_0 = \mathbf{U}'_0.$$

▷ **13.** Prove that, if $\mathbf{A}$ is nonsingular, then its LDU-decomposition—see Problem 10—$\mathbf{A} = \mathbf{L}_1\mathbf{D}_0\mathbf{U}_1$ is unique.

14. To show that, for singular matrices, the Doolittle variant L_0U_0-decomposition—see Problem 10—is not necessarily unique, verify that the decomposition below is true for all numbers l.

$$\begin{bmatrix} 0 & 1 \\ 0 & 1 \end{bmatrix} = \begin{bmatrix} 1 & 0 \\ l & 1 \end{bmatrix} \begin{bmatrix} 0 & 1 \\ 0 & 1-l \end{bmatrix}.$$

▷ **15.** Suppose that $\mathbf{A}$ is symmetric and nonsingular and has the LDU-decomposition $\mathbf{A} = \mathbf{L}_1\mathbf{D}_0\mathbf{U}_1$ of Problem 10. Prove that $\mathbf{U}_1 = \mathbf{L}_1^T$.

4

Solving equations and finding inverses: theory

This chapter is the theoretical heart of the book; the concepts of Gauss-reduced form, rank, and row-echelon form are used in most subsequent chapters and must, therefore, be thoroughly understood. As you would expect in light of the theoretical importance of this chapter, it contains several ***Key Theorems****: **4.5**, **4.11**, **4.13**, **4.16**, **4.18** and its corollary, and **4.23**. In addition to these fundamental theoretical tools, the concepts of the determinant and of its use as a representational device are introduced.*

4.1 *INTRODUCTION*

In Chapter 3 we presented *methods* for solving systems of linear equations and for finding inverses of matrices. We ignored, however, the fundamental *theory* that supports those methods. For example, in Chapter 3 we simply assumed that Gauss elimination for equations did not change the set of solutions of the equations—that is, that the solutions of the original system of equations are exactly the same as those of the transformed system. Such an assumption is not necessarily true for all possible operations on equations; for example, squaring both sides of the equation $2x = 4$—whose only solution is $x = 2$—produces $4x^2 = 16$, which has the solution $x = -2$ in addition to the $x = 2$ of the original equation. So some theory is required to support Gauss elimination and show that it does not change the solution set.

In Chapter 3 we also ignored the theory that helps us *understand*—understand the nature of the set of solutions to a system of equations, or understand the circumstances in which matrices have right- or left-inverses, for example. The present chapter provides the missing theory.

4.2 GAUSS-REDUCED FORM AND RANK

The entire methodology of Chapter 3 was based on Gauss elimination: the systematic application of elementary row operations to reduce a matrix to a simple form. In solving the system of equations $\mathbf{Ax} = \mathbf{b}$, for example, Gauss elimination was applied to the augmented matrix $[\mathbf{A} \quad \mathbf{b}]$ to produce a reduced augmented matrix from which it was straightforward to find the solutions to the system of equations. We will find in later chapters that a reduced form such as produced by Gauss elimination is a powerful tool for problems other than solving equations, so we now develop the theory that supports this tool.

Gauss-Reduced Form

First, we reword our description (3.20) of Gauss elimination to apply to an arbitrary $p \times q$ matrix $\mathbf{A}$ without needing to consider any system of equations or augmented matrix. A row or column of a matrix is said to be *nonzero* if at least one entry in that row or column is nonzero; otherwise, it is a *zero row* or *zero column*, and all its entries are zero.

(4.1) ***Gauss Elimination.*** $\mathbf{A}$ is a given $p \times q$ matrix.

0. If $\mathbf{A} = \mathbf{0}$, no elimination is necessary—stop. Otherwise, set $r = 1$ (for "row $= 1$") and proceed with steps 1–3.
1. Find the first column (call it column c_r) of $\mathbf{A}$ that has a nonzero entry in a row i with $i \geq r$; interchange any such row with row r if necessary to make the rth entry of this column nonzero; divide this new row r by this nonzero entry to make the new rth entry of this column equal to 1.
2. Add suitable multiples of this new row r to all lower rows so that the $(r + 1)$st, $(r + 2)$nd, . . . , pth entries of this column become zeros. Column c_r now has a 1 as its rth entry and a 0 as its jth entry for $j > r$; also, the first nonzero entry in row r occurs in column c_r and equals 1.
3. If $r = p$ (we have used the bottom row of the matrix) or if rows $(r + 1)$, $(r + 2)$, . . . , p are all zero rows, then no further elimination is necessary—stop. Otherwise, increase r by 1 and return to the start of step 1.

Applying Gauss elimination as in (4.1) produces an upper-triangular reduced form $\mathbf{A}'$ from the original matrix $\mathbf{A}$; such a reduced form is given a special name.

(4.2) ***Definition.*** A $p \times q$ matrix $\mathbf{B}$ is said to be in *Gauss-reduced form* when there is an integer k with $0 \leq k \leq p$ for which the following are true:

(a) The first k rows of $\mathbf{B}$ are nonzero, while the remaining $p - k$ rows are zero.

(b) The first nonzero entry in each nonzero row equals 1, and the column in which this occurs is thus a leading column (see Definition 3.24).

(c) For the k leading columns, the leading column for each row is farther to the right than is the leading column for the row above.

(d) The ith entry in the ith leading column equals 1, while the jth entry equals 0 for $j > i$.

Using this language, the result of applying Gauss elimination as in (4.1) is to produce a matrix in Gauss-reduced form. Some authors refer to Gauss-reduced form as "row-echelon form," a term we reserve for a still-further-reduced form discussed later in this section.

(4.3) ***Example.*** The first three matrices below are in Gauss-reduced form; the second three are *not.*

$$\begin{bmatrix} 0 & 0 \\ 0 & 0 \end{bmatrix}, \quad \begin{bmatrix} 0 & 1 & -2 & 3 \\ 0 & 0 & 0 & 1 \\ 0 & 0 & 0 & 0 \end{bmatrix}, \quad \begin{bmatrix} 1 & 2 & 3 \\ 0 & 1 & 4 \\ 0 & 0 & 1 \end{bmatrix}$$

are Gauss-reduced;

$$\begin{bmatrix} 2 & 3 \\ 0 & 1 \\ 0 & 0 \end{bmatrix}, \quad \begin{bmatrix} 0 & 1 & 2 \\ 1 & 0 & 0 \\ 0 & 0 & 0 \end{bmatrix}, \quad \begin{bmatrix} 1 & 0 & 0 \\ 0 & 1 & 0 \\ 0 & 2 & 1 \end{bmatrix}$$

are not Gauss-reduced.

(4.4) ***Example.*** Consider the 2×2 matrix $\mathbf{A}$ below, which Gauss elimination easily reduces to the matrix $\mathbf{A}'$ below in Gauss-reduced form:

$$\mathbf{A} = \begin{bmatrix} 2 & 4 \\ 3 & 9 \end{bmatrix} \quad \text{is reduced by (4.1) to} \quad \mathbf{A}' = \begin{bmatrix} 1 & 2 \\ 0 & 1 \end{bmatrix}.$$

If, instead, we first interchange the two rows of $\mathbf{A}$ and then perform Gauss elimination on the result, we obtain the matrix $\mathbf{A}''$:

$$\mathbf{A}'' = \begin{bmatrix} 1 & 3 \\ 0 & 1 \end{bmatrix}.$$

Note that the matrix $\mathbf{A}''$ *also is in Gauss-reduced form.* Both $\mathbf{A}'$ and $\mathbf{A}''$ are in Gauss-reduced form and were produced from $\mathbf{A}$ by elementary row operations; therefore, *there can be more than one Gauss-reduced form produced from a given matrix.*

From Example 4.4 we see that there is not a unique Gauss-reduced form for each matrix $\mathbf{A}$; however, certain important aspects will be the same in all Gauss-reduced forms produced from a given matrix $\mathbf{A}$.

(4.5) **Key Theorem** (Gauss-reduced form). The number k of nonzero rows and the column numbers of the leading columns are the same in any Gauss-reduced form produced from a given matrix $\mathbf{A}$ by elementary row operations, regardless of the actual sequence of row operations used.

PROOF. Suppose that $\mathbf{A}$ is $p \times q$ and that two Gauss-reduced forms $\mathbf{A}'$ and $\mathbf{A}''$ are produced from $\mathbf{A}$ by sequences of elementary row operations. Let k' be the number of nonzero rows in $\mathbf{A}'$, and let the leading columns of $\mathbf{A}'$ be numbered $c_1', c_2', \ldots, c_{k'}'$. Similarly, let k'' be the number of nonzero rows in $\mathbf{A}''$, and let the leading columns of $\mathbf{A}''$ be numbered $c_1'', c_2'', \ldots, c_{k''}''$. What needs to be proved is that $k' = k''$ and that $c_j' = c_j''$ for $1 \le j \le k'$ $(=k'')$. Since elementary row operations were used to produce $\mathbf{A}'$ and $\mathbf{A}''$, by **Key Theorem 3.34** on row operations there are nonsingular matrices $\mathbf{F}'$ and $\mathbf{F}''$ with $\mathbf{F}'\mathbf{A} = \mathbf{A}'$ and $\mathbf{F}''\mathbf{A} = \mathbf{A}''$. Therefore,

$$\mathbf{A}'' = \mathbf{H}\mathbf{A}' \quad \text{and} \quad \mathbf{A}' = \mathbf{H}^{-1}\mathbf{A}'', \quad \text{where } \mathbf{H} = \mathbf{F}''\mathbf{F}'^{-1}.$$

This relationship is the key to the proof. If $\mathbf{A}' = \mathbf{0}$, then $\mathbf{A}'' = \mathbf{H}\mathbf{A}' = \mathbf{H}\mathbf{0} = \mathbf{0}$; also, if $\mathbf{A}'' = \mathbf{0}$, then $\mathbf{A}' = \mathbf{0}$. In other words, the theorem is true if either k' or k'' is 0. Next, consider when k' and k'' are both positive. Since column c_1' is the *first* leading column of $\mathbf{A}'$, the columns $\mathbf{a}_j'$ of $\mathbf{A}'$ for $1 \le j < c_1'$ must equal $\mathbf{0}$; by the rule for partitioned multiplication, the jth column $\mathbf{a}_j''$ of $\mathbf{A}''$ is $\mathbf{H}$ times that for $\mathbf{A}'$: $\mathbf{a}_j'' = \mathbf{H}\mathbf{a}_j'$. Therefore, since $\mathbf{H}\mathbf{0} = \mathbf{0}$, we find that $\mathbf{a}_j'' = \mathbf{0}$ for $1 \le j < c_1'$; but this says that the first leading column of $\mathbf{A}''$ must occur at or to the right of column c_1'—that is, $c_1'' \ge c_1'$. By applying the same argument to the relation $\mathbf{A}' = \mathbf{H}^{-1}\mathbf{A}''$ rather than to $\mathbf{A}'' = \mathbf{H}\mathbf{A}'$, we also obtain $c_1' \ge c_1''$. These two inequalities together give $c_1' = c_1''$. Now column c_1' is the first leading column of $\mathbf{A}'$ and, by Definition 4.2(d), it comes from the first row of $\mathbf{A}'$; thus, the first entry of that column is 1 and all the others are 0—that is, this column is just $\mathbf{e}_1$, the first unit column matrix of order p. The same argument applies to that first leading column of $\mathbf{A}''$. The rule for partitioned multiplication tells us that the columns of $\mathbf{A}'$ and $\mathbf{A}''$ are related by $\mathbf{a}'' = \mathbf{H}\mathbf{a}'$, so we obtain $\mathbf{e}_1 = \mathbf{H}\mathbf{e}_1$.

We next use induction. One of k' and k'' is the smaller; for convenience, call it k'. As an inductive hypothesis, suppose that we have proved that $c_i' = c_i''$ for $1 \le i \le n$, and

$$\mathbf{H}\mathbf{e}_i = \mathbf{e}_i + \sum_{j=1}^{i-1} \alpha_{ij}\mathbf{e}_j \qquad \text{for some numbers } \alpha_{ij}$$

for $1 \le i \le n$, where $n < k' \le k''$.

Note that we have already proved precisely this for $n = 1$. We next show that it is true for $n + 1$, which will then show it true for all n through k'. By Definition 4.2 on Gauss-reduced forms, the columns of $\mathbf{A}'$ numbered up through $(c_{n+1}' - 1)$ can have nothing but zero entries in rows $n + 1$, $n + 2, \ldots, p$; that is, each such column has nonzero entries only in rows n and above, so each such column can be written as a sum of multiples of $\mathbf{e}_1$, $\mathbf{e}_2, \ldots, \mathbf{e}_n$. Since each column $\mathbf{a}''$ of $\mathbf{A}''$ must equal $\mathbf{H}\mathbf{a}'$ for the corresponding column $\mathbf{a}'$ of $\mathbf{A}'$, we see that each column of $\mathbf{A}''$ numbered up through $(c_{n+1}' - 1)$ equals $\mathbf{H}$ times a sum of multiples of the $\mathbf{e}_i$ for $1 \le i \le n$. But our inductive hypothesis says that each $\mathbf{He}_i$ is itself a combination of multiples of the $\mathbf{e}_j$, so we conclude that each column of $\mathbf{A}''$ numbered up through $(c'_{n+1} - 1)$ equals a sum of multiples of $\mathbf{e}_i$ for $1 \le i \le n$; that is, each such column of $\mathbf{A}''$ has zeros in row $(n + 1)$ and below and hence the $(n + 1)$st leading column of $\mathbf{A}''$ cannot be any of these columns—so $c''_{n+1} \ge c'_{n+1}$. By using $\mathbf{H}^{-1}$ instead of $\mathbf{H}$ and reversing the roles of $\mathbf{A}'$ and $\mathbf{A}''$, we get by the same argument that $c'_{n+1} \ge c''_{n+1}$ as well, and hence $c'_{n+1} = c''_{n+1}$. Denoting these $(n + 1)$st leading columns of $\mathbf{A}'$ and $\mathbf{A}''$ by $\mathbf{a}'$ and $\mathbf{a}''$, we have that each equals $\mathbf{e}_{n+1}$ plus a sum of multiples of $\mathbf{e}_i$ for $1 \le i \le n$, as well as $\mathbf{a}'' = \mathbf{Ha}'$; this shows that our inductive hypothesis holds for $n + 1$.

We now know that the inductive hypothesis holds for $1 \le n \le k'$ $(\le k'')$. The bottom $p - k'$ rows of $\mathbf{A}'$ are zero rows, which says that all the columns of $\mathbf{A}'$ to the right of column $c'_{k'}$ are sums of multiples of the $\mathbf{e}_i$ for $1 \le i \le k'$; the same argument as above based on $\mathbf{a}'' = \mathbf{Ha}'$ and the inductive hypothesis for $n = k'$ then shows that the same must be true for all columns of $\mathbf{A}''$ to the right of column $c'_{k'}$—which says that there can be no further leading columns for $\mathbf{A}''$ beyond $c'_{k'}$. That is, $k'' = k'$. The theorem is proved. ■

You should now be able to solve Problems 1 to 6.

Rank

According to **Key Theorem 4.5** on Gauss-reduced form, the number of nonzero rows in every Gauss-reduced form of a given matrix is the same. This number plays a fundamental role in much of our subsequent work and is therefore given a special name.

(4.6) ***Definition.*** The number of nonzero rows in any Gauss-reduced form $\mathbf{A}'$ of a matrix $\mathbf{A}$ produced by elementary row operations on $\mathbf{A}$ is called the *rank* of $\mathbf{A}$.

(4.7) ***Example.*** The ranks of the first three matrices in Example 4.3 can immediately be read off from their being already in Gauss-reduced forms: 0, 2, and 3, respectively. The ranks of the second three can be found by performing Gauss elimination to get Gauss-reduced forms giving ranks of 2, 2, and 3, respectively. The rank of the matrix $\mathbf{A}$ in Example 4.4 is 2, as can be seen from either Gauss-reduced form $\mathbf{A}'$ or $\mathbf{A}''$. Note that those forms $\mathbf{A}'$ and $\mathbf{A}''$ illustrate **Key Theorem 4.5**.

Some authors refer to rank as "row rank," since its definition depends on the use of *row* operations and counts nonzero *rows*. A similar concept of "column rank" can be introduced by considering *column* operations and counting nonzero *columns* in a reduced form. It turns out that row rank and column rank are always equal, so we prefer to use the unadorned term "rank."

For some theoretical purposes it is occasionally useful to reduce a matrix by elementary row operations further than the Gauss-reduced form produced, for example, by Gauss elimination. One way to reach this form is through Gauss-Jordan elimination—see Example 3.21. A much more efficient method—see Problem 16—is as follows; see also Problem 14 in Section 3.2.

(4.8) ***Row-Echelon Reduction.*** **A** is a given $p \times q$ matrix.

1. Use Gauss elimination (4.1) or any other method to produce a Gauss-reduced form from **A**.
2. Use the bottom-most nonzero entry 1 in each leading column of the Gauss-reduced form—starting with the rightmost leading column and working to the left—so as to eliminate all nonzero entries in that column strictly above that entry 1.

This process produces, as does Gauss elimination, a special form with a special name.

(4.9) ***Definition.*** A $p \times q$ matrix **B** is said to be in *row-echelon form* when:

(a) It is in Gauss-reduced form (with k nonzero rows, say).

(b) The ith leading column equals $\mathbf{e}_i$, the ith unit column matrix of order p, for $1 \le i \le k$.

Note that some authors refer to row-echelon form as "reduced row-echelon form."

(4.10) ***Example.*** Consider the matrix **C** below; a Gauss-reduced form produced by Gauss elimination is given as **G** below, as developed in (3.8) through (3.13) of Section 3.2.

$$\mathbf{C} = \begin{bmatrix} -2 & 2 & -4 & -6 & -4 \\ -3 & 6 & 3 & -15 & -3 \\ 5 & -8 & -1 & 17 & 9 \\ 1 & 1 & 11 & 7 & 7 \end{bmatrix}, \qquad \mathbf{G} = \begin{bmatrix} 1 & -1 & 2 & 3 & 2 \\ 0 & 1 & 3 & -2 & 1 \\ 0 & 0 & 1 & 2 & -1 \\ 0 & 0 & 0 & 1 & 3 \end{bmatrix}.$$

The leading columns are columns 1, 2, 3, and 4. We use the fourth row to eliminate the nonzero entries in column 4 above the bottom entry, then likewise with the third row and column 3, and finally the second row and column 2, to get

$$\mathbf{R} = \begin{bmatrix} 1 & 0 & 0 & 0 & 35 \\ 0 & 1 & 0 & 0 & 28 \\ 0 & 0 & 1 & 0 & -7 \\ 0 & 0 & 0 & 1 & 3 \end{bmatrix}.$$

This matrix **R** is in row-echelon form, and is thus *a* row-echelon form of **C**. Unlike the situation with Gauss-reduced forms, it is in fact legitimate to refer to *the* row-echelon form of a matrix, as the next theorem shows.

(4.11) **Key Theorem** (row-echelon form). Each matrix has precisely one row-echelon form to which it can be reduced by elementary row operations, regardless of the actual sequence of operations used to produce it.

PROOF. Let $\mathbf{A}'$ and $\mathbf{A}''$ be two row-echelon forms of the same matrix **A**. Since $\mathbf{A}'$ and $\mathbf{A}''$ are also Gauss-reduced forms, we know from **Key Theorem 4.5** on Gauss-reduced forms that they have the same number k of nonzero rows and the same column numbers of their k leading columns. Also, the ith leading column of each matrix is just $\mathbf{e}_i$. In the notation of the proof of **Key Theorem 4.5**, we therefore have $\mathbf{He}_i = \mathbf{e}_i$ for $1 \leq i \leq k$. Any column $\mathbf{a}'$ of $\mathbf{A}'$ is just a sum of multiples of these first k unit column matrices $\mathbf{e}_i$:

$$\mathbf{a}' = \sum \alpha_i \mathbf{e}_i.$$

The corresponding column $\mathbf{a}''$ of $\mathbf{A}''$ is just $\mathbf{Ha}'$, so we have

$$\mathbf{a}'' = \mathbf{Ha}' = \mathbf{H}(\sum \alpha_i \mathbf{e}_i) = \sum \alpha_i \mathbf{He}_i = \sum \alpha_i \mathbf{e}_i = \mathbf{a}',$$

and thus $\mathbf{a}'' = \mathbf{a}'$ and corresponding columns of $\mathbf{A}'$ and of $\mathbf{A}''$ are equal. That is, $\mathbf{A}' = \mathbf{A}''$. ■

PROBLEMS 4.2

1. For each of the second three matrices in Example 4.3, state how it violates the definition of Gauss-reduced form and then reduce it to a Gauss-reduced form.

2. State which of the following matrices are in Gauss-reduced form, and for those that aren't, explain why not.

(a) $\begin{bmatrix} 2 & 1 & 3 & 2 \\ 0 & 0 & 1 & -2 \\ 0 & 0 & 0 & 0 \end{bmatrix}$ (b) $\begin{bmatrix} 0 & 1 & 2 \\ 0 & 0 & 0 \\ 0 & 0 & 0 \\ 0 & 0 & 0 \end{bmatrix}$

(c) $\begin{bmatrix} 1 & 3 & -1 & 2 \\ 0 & 0 & 0 & 0 \\ 0 & 1 & 1 & 3 \end{bmatrix}$

▷ 3. Find a Gauss-reduced form for each of the following matrices.

(a) $\begin{bmatrix} 1 & 1 & -8 & -14 \\ 3 & -4 & -3 & 0 \\ 2 & -1 & -7 & -10 \end{bmatrix}$ (b) $\begin{bmatrix} 1 & -2 & 1 & 4 \\ 2 & -3 & -1 & 2 \end{bmatrix}$

(c) $\begin{bmatrix} 1 & 2 & 3 & 4 & 7 \\ 0 & 5 & 7 & 6 & 8 \\ 0 & 5 & 2 & 4 & 7 \\ 0 & 0 & 2 & 1 & 0 \end{bmatrix}$ (d) $\begin{bmatrix} 4 & -1 & 2 & 6 \\ -1 & 5 & -1 & -3 \\ 3 & 4 & 1 & 3 \end{bmatrix}$

4. Use the approach of Section 3.8 to show that performing Gauss reduction as in (4.1) on a $p \times q$ matrix requires at most:
(a) N_{pq} addition/subtractions and $N_{pq} + q^2/2 - q/2$ multiplication/divisions if $p \geq q$, where

$$N_{pq} = \frac{q^2(3p - q)}{6} - \frac{q(3p - 1)}{6},$$

(b) N_{qp} addition/subtractions and $N_{qp} + p(2q - p - 1)/2$ multiplication/divisions if $p < q$.
Calculate these work measures for:
(c) $p = 50$, $q = 100$,
(d) $p = 100$, $q = 50$

▷ **5.** If $\mathbf{A}$ is $p \times q$, show that $[\mathbf{I}_p \quad \mathbf{A}]$ is in Gauss-reduced form.

6. Suppose that $\mathbf{A}$ is $p \times p$ and nonsingular. Use **Key Theorem 3.34** on row operations to show that any Gauss-reduced form $\mathbf{G}$ of $\mathbf{A}$ is unit-upper-triangular.

7. Find the rank of each of the second three matrices in Example 4.3.

▷ **8.** Suppose that $\mathbf{A}$ is $p \times q$ with rank p and $\mathbf{B}$ is $p \times r$; show that $[\mathbf{A} \quad \mathbf{B}]$ has rank p.

9. Find the rank for each of the matrices in Problem 3.

▷ **10.** Suppose that the $p \times q$ matrix $\mathbf{A}$ has n nonzero rows and $p - n$ zero rows. If k is the rank of $\mathbf{A}$, show that $k \leq n$.

11. Suppose that $\mathbf{A}$ is $p \times p$ and its rank k satisfies $k < p$. Show that $\mathbf{A}$ is singular.

12. Find the row-echelon form for each matrix in Example 4.3.

13. If $\mathbf{A}$ is $p \times q$, show that $[\mathbf{I}_p \quad \mathbf{A}]$ is in row-echelon form.

▷ **14.** Two different Gauss-reduced forms $\mathbf{A}'$ and $\mathbf{A}''$ were produced for one matrix $\mathbf{A}$ in Example 4.4. Produce the row-echelon form from each of $\mathbf{A}'$ and $\mathbf{A}''$ and verify that the same form results.

15. Suppose that $\mathbf{A}$ is $p \times q$ and that we reduce it to row-echelon form using (4.8). The work involved in doing this consists of, first, the work measured in Problem 4 to reduce $\mathbf{A}$ to Gauss-reduced form $\mathbf{G}$, and then the additional work to reduce $\mathbf{G}$ to row-echelon form. Suppose that the rank of $\mathbf{A}$ is k.
(a) Show that this additional work to reduce $\mathbf{G}$ requires at most

$$C_{kq} = \frac{k(k - 1)(q - k)}{2}$$

addition/subtractions and at most C_{kq} multiplication/divisions.

(b) Calculus can be used to show that C_{kq} is largest for k about $2q/3$; show that for such k the additional work C_{kq} is about

$$C_q = \frac{q^2(2q-3)}{27}.$$

(c) Find the additional work C_{kq} for a 50×100 matrix of rank 50.

(d) Find the largest amount C_q of additional work for a 90×100 matrix.

16. Suppose that $\mathbf{A}$ is $p \times p$ and that the $p \times (p+1)$ augmented matrix $[\mathbf{A} \quad \mathbf{b}]$ has rank p.

(a) Use the results of Problem 4 and Problem 15 to find the work necessary to reduce $[\mathbf{A} \quad \mathbf{b}]$ to row-echelon form by (4.8).

(b) Compare this with the $p^3/2 - p/2$ addition/subtractions and $p^3/2 + p^2/2$ multiplication/divisions necessary to produce the row-echelon form by Gauss-Jordan elimination.

(c) Evaluate each of the work measures from (a) and (b) when $p = 50$.

17. Show that the row-echelon form of a $p \times p$ nonsingular matrix $\mathbf{A}$ is $\mathbf{I}_p$. (See Problem 6.)

4.3 SOLVABILITY AND SOLUTION SETS FOR SYSTEMS OF EQUATIONS

In Chapter 2 we saw that it is often important in applied problems to know *whether there are solutions* to certain systems of linear equations and, if so, *how to find them.* For example, in our market-equilibrium problem we needed to solve $\mathbf{Ax} = \mathbf{x}$ for a nonzero $\mathbf{x}$, while in our least-squares problem we needed to solve $\mathbf{A}^T\mathbf{Ax} = \mathbf{A}^T\mathbf{y}$.

In Chapter 3 we addressed the *how to* issue: We introduced Gauss elimination as a method for manipulating systems of equations (and augmented matrices) so as to produce a system of equations that was straightforward to solve. As we observed in Section 4.1, however, even this methodological approach raises a theoretical imperative: It is necessary to show that this manipulation does not change the set of solutions. We now resolve this matter and then use the concept of rank to characterize the solvability of systems of equations; this allows us—see Problems 16 and 19—to understand *why* the examples in Chapter 2 turned out as they did.

Solvability of Equations

We will use a Gauss-reduced form of the augmented matrix to analyze the solvability of the system of equations $\mathbf{Ax} = \mathbf{b}$, so we must demonstrate that Gauss elimination leaves the solution set unchanged.

(4.12) ***Theorem*** (Gauss elimination and solution sets). Suppose that the system of equations $\mathbf{Ax} = \mathbf{b}$—or, equivalently, the augmented matrix $[\mathbf{A} \quad \mathbf{b}]$—is transformed by a sequence of elementary row operations into the system $\mathbf{A}'\mathbf{x} = \mathbf{b}'$—or, equivalently, into the augmented matrix $[\mathbf{A}' \quad \mathbf{b}']$. Then the solution sets are identical: $\mathbf{x}$ solves $\mathbf{A}\mathbf{x} = \mathbf{b}$ if and only if $\mathbf{x}$ solves $\mathbf{A}'\mathbf{x} = \mathbf{b}'$.

PROOF. By **Key Theorem 3.34** on row operations there is a nonsingular matrix $\mathbf{F}$ such that

$$[\mathbf{A}' \quad \mathbf{b}'] = \mathbf{F}[\mathbf{A} \quad \mathbf{b}] = [\mathbf{F}\mathbf{A} \quad \mathbf{F}\mathbf{b}],$$

where the latter equality follows from the rule for partitioned multiplication. This means that $\mathbf{A}' = \mathbf{F}\mathbf{A}$ and $\mathbf{b}' = \mathbf{F}\mathbf{b}$. Now, first suppose that $\mathbf{x}$ solves $\mathbf{A}\mathbf{x} = \mathbf{b}$. Premultiplication of this equation by $\mathbf{F}$ gives $\mathbf{F}\mathbf{A}\mathbf{x} = \mathbf{F}\mathbf{b}$, that is, $\mathbf{A}'\mathbf{x} = \mathbf{b}'$ and thus $\mathbf{x}$ solves the modified equations as well. Conversely, if $\mathbf{x}$ satisfies $\mathbf{A}'\mathbf{x} = \mathbf{b}'$, then premultiplication by $\mathbf{F}^{-1}$ similarly shows that $\mathbf{x}$ solves $\mathbf{A}\mathbf{x} = \mathbf{b}$. Thus, the two sets of solutions are identical. ■

Theorem 4.12 rigorously establishes that we can study the set of all solutions to $\mathbf{A}\mathbf{x} = \mathbf{b}$ by, instead, studying the set of solutions of the much simpler set $\mathbf{A}'\mathbf{x} = \mathbf{b}'$ obtained by reducing $[\mathbf{A} \quad \mathbf{b}]$ to a Gauss-reduced form $[\mathbf{A}' \quad \mathbf{b}']$. As was continually demonstrated by example throughout Chapter 3, three possible cases seem to occur:

1. No solutions exist.
2. Exactly one solution exists.
3. Infinitely many solutions exist.

In Section 3.2 (3.25) we previewed how these different cases could occur. To decide which case occurs in a particular situation, we write down the augmented matrix $[\mathbf{A} \quad \mathbf{b}]$, reduce it by Gauss elimination to a Gauss-reduced form $[\mathbf{A}' \quad \mathbf{b}']$, and then examine the simpler question of solvability of $\mathbf{A}'\mathbf{x} = \mathbf{b}'$. The resulting conclusions can be compactly stated in terms of rank.

(4.13) **Key Theorem** (rank and solvability). Consider the system of equations $\mathbf{A}\mathbf{x} = \mathbf{b}$. Exactly one of the following three possibilities must hold:

1. The rank of the augmented matrix $[\mathbf{A} \quad \mathbf{b}]$ is greater than that of $\mathbf{A}$, and no solution exists to $\mathbf{A}\mathbf{x} = \mathbf{b}$.
2. The rank of $[\mathbf{A} \quad \mathbf{b}]$ equals that of $\mathbf{A}$, which equals the number of unknowns, and the system $\mathbf{A}\mathbf{x} = \mathbf{b}$ has exactly one solution.
3. The rank of $[\mathbf{A} \quad \mathbf{b}]$ equals that of $\mathbf{A}$, which is strictly less than the number of unknowns, and the system $\mathbf{A}\mathbf{x} = \mathbf{b}$ has infinitely many solutions.

PROOF. We know from Theorem 4.12 on Gauss elimination and solution sets that the solution set of $\mathbf{A}\mathbf{x} = \mathbf{b}$ is identical with that for $\mathbf{A}'\mathbf{x} = \mathbf{b}'$, where $[\mathbf{A}' \quad \mathbf{b}']$ is a Gauss-reduced form of $[\mathbf{A} \quad \mathbf{b}]$. It is clear from Definition 4.2 on Gauss-reduced forms that, if $[\mathbf{A}' \quad \mathbf{b}']$ is in Gauss-reduced form, then so is $\mathbf{A}'$ itself. Let r be the rank of $\mathbf{A}$ and s be the rank of $[\mathbf{A} \quad \mathbf{b}]$; since r equals the number of nonzero rows in $\mathbf{A}'$ while s equals the number of nonzero rows in $[\mathbf{A}' \quad \mathbf{b}']$ and since any nonzero row of $\mathbf{A}'$ serves as the "start" of a nonzero row of $[\mathbf{A}' \quad \mathbf{b}']$, it follows that $r \leq s$. Note that the $p \times q$ matrix $\mathbf{A}'$ has q columns, and since r is the number of leading columns of $\mathbf{A}'$ it must be that $r \leq q$. There are then exactly three possibilities: (i) $r < s$; (ii) $r = s = q$; and finally, (iii) $r = s < q$.

In case (i), $[\mathbf{A}' \quad \mathbf{b}']$ must have leading columns beyond those of $\mathbf{A}'$; but $[\mathbf{A}' \quad \mathbf{b}']$ has only one column beyond those of $\mathbf{A}'$, namely the rightmost column $\mathbf{b}'$. Therefore, $s = r + 1$ and the $(r + 1)$st entry of $\mathbf{b}'$ equals 1, while the $(r + 1)$st row of $\mathbf{A}'$ equals $\mathbf{0}$. Then the $(r + 1)$st equation of the system $\mathbf{A}'\mathbf{x} = \mathbf{b}'$ asks that we find $\mathbf{x}$ that will make $\mathbf{0x} = [1]$, that is, make $0 = 1$—which is impossible. Thus in case (i) there are no solutions; this is case (1) of the theorem.

In case (ii), there are no nonzero entries in $\mathbf{b}'$ below the last nonzero row of $\mathbf{A}'$, so there is no equation demanding $0 = 1$. Since q is the total number of columns in $\mathbf{A}'$ while r is its number of leading columns and $r = q$, then every column is a leading column. Thus the qth equation $1\langle \mathbf{x} \rangle_q = \langle \mathbf{b}' \rangle_q$ determines $\langle \mathbf{x} \rangle_q$ uniquely, the $(q - 1)$st determines $\langle \mathbf{x} \rangle_{q-1}$ uniquely, and so on through the first equation determining $\langle \mathbf{x} \rangle_1$ uniquely. This is case (2) of the theorem and there is exactly one solution.

Finally, in case (iii), since r—the number of leading columns of $\mathbf{A}'$—is less than q—the total number of columns—there are some nonleading columns of $\mathbf{A}'$; that is, there are some nonleading variables. If each of these nonleading variables is assigned an arbitrary value and if these values are substituted for the variables and the results shifted to the right-hand side of the equations, then the remaining variables are all leading variables and their values are uniquely determined as in case (ii) by the right-hand sides—which now involve these $q - r$ arbitrary values. This is case (3) of the theorem, and there are infinitely many solutions—one for every choice of arbitrary values assigned to the nonleading variables. ■

(4.14) ***Example.*** Consider the system of equations $\mathbf{Ax} = \mathbf{b}$ below with α an arbitrary parameter.

$$\begin{aligned} x - 3y &= -2 \\ 2x + y &= 3 \\ 3x - 2y &= \alpha. \end{aligned}$$

The augmented matrix is easily reduced from

$$\left[\begin{array}{cc|c} 1 & -3 & -2 \\ 2 & 1 & 3 \\ 3 & -2 & \alpha \end{array}\right] \quad \text{to} \quad \left[\begin{array}{cc|c} 1 & -3 & -2 \\ 0 & 1 & 1 \\ 0 & 0 & \alpha - 1 \end{array}\right].$$

If $\alpha = 1$, then both $\mathbf{A}'$ and $[\mathbf{A}' \quad \mathbf{b}']$ are in Gauss-reduced form, and the ranks of each equal $2 = q$. We have case (2) of **Key Theorem 4.13** and there is a unique solution, namely $x = y = 1$. If $\alpha \neq 1$, however, then the third row can be divided by $\alpha - 1 \neq 0$ to produce a 1 at the bottom of the third column, giving the rank of $\mathbf{A}$ as 2 but the rank of $[\mathbf{A} \quad \mathbf{b}]$ as 3; this is case (1) and there are no solutions.

You should now be able to solve Problems 1 to 7.

Structure of the Solution Set

The structure of the set of solutions is quite clear in two cases: When the set is empty (there are no solutions), or when the set consists of one single solution. But what about the third case, when there are infinitely many solutions? If $\mathbf{y}$ and $\mathbf{z}$ both are solutions to $\mathbf{Ax} = \mathbf{b}$, so that $\mathbf{Ay} = \mathbf{b}$ and $\mathbf{Az} = \mathbf{b}$, then clearly

$$\mathbf{A}(\mathbf{y} - \mathbf{z}) = \mathbf{Ay} - \mathbf{Az} = \mathbf{b} - \mathbf{b} = \mathbf{0}.$$

That is, the column matrix $\mathbf{h} = \mathbf{y} - \mathbf{z}$ solves

(4.15) the so-called *homogeneous system* $\mathbf{Ah} = \mathbf{0}$ corresponding to the original system $\mathbf{Ax} = \mathbf{b}$.

Conversely, if $\mathbf{h}$ is any solution to the homogeneous system while $\mathbf{x}_0$ is some specific solution to $\mathbf{Ax} = \mathbf{b}$, then $\mathbf{A}(\mathbf{x}_0 + \mathbf{h}) = \mathbf{Ax}_0 + \mathbf{Ah} = \mathbf{b} + \mathbf{0} = \mathbf{b}$; that is, $\mathbf{x}_0 + \mathbf{h}$ is another solution to $\mathbf{Ax} = \mathbf{b}$.

(4.16) **Key Theorem** (solution sets). Suppose that $\mathbf{x}_0$ is some particular solution to the system of equations $\mathbf{Ax} = \mathbf{b}$. Then the set of all solutions $\mathbf{x}$ to $\mathbf{Ax} = \mathbf{b}$ is the same as the set of all matrices of the form $\mathbf{x}_0 + \mathbf{h}$, where $\mathbf{h}$ ranges over the set of all solutions of the homogeneous system $\mathbf{Ah} = \mathbf{0}$.

It follows from this that there being at most one solution to $\mathbf{Ax} = \mathbf{b}$ is equivalent to there being no nonzero solution $\mathbf{h}$ to $\mathbf{Ah} = \mathbf{0}$.

(4.17) ***Corollary*** (uniqueness of solutions). The system of equations $\mathbf{Ax} = \mathbf{b}$ has *at most* one solution (it may have none at all) if and only if the only solution to $\mathbf{Ah} = \mathbf{0}$ is $\mathbf{h} = \mathbf{0}$.

A far stronger result holds in the special case of $p = q$, that is, as many equations as variables.

(4.18) **Key Theorem** (existence = uniqueness when $p = q$). Suppose that the square matrix $\mathbf{A}$ is $p \times p$ and consider the system of equations $\mathbf{Ax} = \mathbf{b}$ for various $\mathbf{b}$. The following five conditions are equivalent:

1. The only solution to the homogeneous system $\mathbf{Ah} = \mathbf{0}$ is $\mathbf{h} = \mathbf{0}$.
2. The rank of $\mathbf{A}$ equals p.
3. For each $p \times 1$ column matrix $\mathbf{b}$, the system $\mathbf{Ax} = \mathbf{b}$ has exactly one solution.
4. A $p \times p$ right-inverse $\mathbf{R}$ exists for $\mathbf{A}$, that is, there is a $p \times p$ $\mathbf{R}$ with $\mathbf{AR} = \mathbf{I}_p$.
5. $\mathbf{A}$ is nonsingular. Moreover, $\mathbf{A}^{-1} = \mathbf{R}$ in (4).

PROOF. We prove that (1) implies (2), (2) implies (3), (3) implies (4), (4) implies (5), and (5) implies (1).

($1 \Rightarrow 2$) First, assume that (1) holds and apply **Key Theorem 4.13** on rank and solvability to the system $\mathbf{Ax} = \mathbf{0}$, which by assumption has exactly one solution $\mathbf{x} = \mathbf{0}$; so case (2) of **Key Theorem 4.13** must hold, and the rank of A equals p—that is, our (2) holds.

($2 \Rightarrow 3$) Second, assume that (2) holds and consider $\mathbf{Ax} = \mathbf{b}$ for some $\mathbf{b}$; if $[\mathbf{A}' \quad \mathbf{b}']$ is a Gauss-reduced form for $[\mathbf{A} \quad \mathbf{b}]$, then $\mathbf{A}'$ is a Gauss-reduced form for $\mathbf{A}$, and since the rank of $\mathbf{A}$ equals p, all p rows of $\mathbf{A}'$—and hence of $[\mathbf{A}' \quad \mathbf{b}']$—are nonzero. This says that the rank of $[\mathbf{A} \quad \mathbf{b}]$ also equals p; thus case (2) of **Key Theorem 4.13** holds and there is exactly one solution to $\mathbf{Ax} = \mathbf{b}$—that is, our case (3) holds.

($3 \Rightarrow 4$) Third, suppose that our (3) holds; by **Key Theorem 1.44(a)**, finding $\mathbf{R}$ for (4) is equivalent to solving systems of equations $\mathbf{Ar}_i = \mathbf{e}_i$, which, by our assumption of (3), can be done. Thus (4) follows from (3).

($4 \Rightarrow 5$) Fourth, suppose that (4) holds so that $\mathbf{A}$ has a right-inverse $\mathbf{R}$. We claim that this means that (1) holds for the matrix $\mathbf{A}^T$; for, if $\mathbf{A}^T\mathbf{y} = \mathbf{0}$, then

$$\mathbf{0} = \mathbf{R}^T\mathbf{0} = \mathbf{R}^T(\mathbf{A}^T\mathbf{y}) = (\mathbf{R}^T\mathbf{A}^T)\mathbf{y} = (\mathbf{AR})^T\mathbf{y} = \mathbf{I}_p\mathbf{y} = \mathbf{y},$$

so $\mathbf{y} = \mathbf{0}$, as asserted. But if (1) holds for $\mathbf{A}^T$, then by what we have already shown, so must (2), (3), and (4) hold for $\mathbf{A}^T$. From (4) for $\mathbf{A}^T$ we get a matrix $\mathbf{X}$ satisfying $\mathbf{A}^T\mathbf{X} = \mathbf{I}_p$, and therefore

$$\mathbf{X}^T\mathbf{A} = (\mathbf{A}^T\mathbf{X})^T = \mathbf{I}_p^T = \mathbf{I}_p,$$

which says that $\mathbf{X}^T$ is a left-inverse for $\mathbf{A}$. Theorem 1.33 then tells us that $\mathbf{R} = \mathbf{X}^T$ is actually a two-sided inverse $\mathbf{A}^{-1}$. That is, our (5) holds.

($5 \Rightarrow 1$) Fifth and finally, suppose that (5) holds and $\mathbf{Ah} = \mathbf{0}$ for some $\mathbf{h}$; then

$$\mathbf{0} = \mathbf{A}^{-1}\mathbf{0} = \mathbf{A}^{-1}(\mathbf{Ah}) = (\mathbf{A}^{-1}\mathbf{A})\mathbf{h} = \mathbf{I}_p\mathbf{h} = \mathbf{h},$$

so $\mathbf{h} = \mathbf{0}$ and (1) holds. ■

Key Theorem 4.18 contains a surprising and useful fact that we now extract for emphasis. It says that in order to determine whether the p equations $\mathbf{Ax} = \mathbf{b}$ in p unknowns have exactly one solution for every arbitrary right-hand side $\mathbf{b}$, all we need do is determine whether they have exactly one solution $\mathbf{h} = \mathbf{0}$ in the special case of $\mathbf{b} = \mathbf{0}$.

(4.19) **Key Corollary** (Fredholm Alternative). Let $\mathbf{Ax} = \mathbf{b}$ describe p equations in p unknowns. Then *either* (1) *or* (2) holds, but *not both*:
Either
1. There is precisely one solution for each arbitrary $\mathbf{b}$

or

2. There is a nonzero solution $\mathbf{h} \neq \mathbf{0}$ to $\mathbf{Ah} = \mathbf{0}$

but not both. That is, there is precisely one solution for each arbitrary $\mathbf{b}$ if and only if there is precisely one solution $\mathbf{h} = \mathbf{0}$ to $\mathbf{Ah} = \mathbf{0}$.

(4.20) ***Example*** (polynomial interpolation). Let $t_1, t_2, \ldots, t_{n+1}$ be given distinct numbers and let $b_1, b_2, \ldots, b_{n+1}$ be given numbers. Consider the problem of finding a polynomial

$$P(t) = x_1 + x_2 t + \cdots + x_{n+1} t^n$$

of degree n that *interpolates* the given data in the sense that $P(t_i) = b_i$ for $1 \leq i \leq n + 1$. The equations $P(t_i) = b_i$ are just equations for the unknown coefficients x_j and can be written as $\mathbf{Ax} = \mathbf{b}$, where

$$\mathbf{x} = \begin{bmatrix} x_1 \\ \vdots \\ x_{n+1} \end{bmatrix}, \qquad \mathbf{b} = \begin{bmatrix} b_1 \\ \vdots \\ b_{n+1} \end{bmatrix}, \qquad \mathbf{A} = \begin{bmatrix} 1 & t_1 & t_1^2 & t_1^3 & \cdots & t_1^n \\ 1 & t_2 & t_2^2 & t_2^3 & \cdots & t_2^n \\ & & \cdots & & & \\ 1 & t_{n+1} & t_{n+1}^2 & t_{n+1}^3 & \cdots & t_{n+1}^n \end{bmatrix}.$$

By **Key Corollary 4.19**, there is precisely one solution to $\mathbf{Ax} = \mathbf{b}$ for arbitrary $\mathbf{b}$ if and only if $\mathbf{h} = \mathbf{0}$ is the only solution to $\mathbf{Ah} = \mathbf{0}$—that is, if and only if the polynomial with all zero coefficients is the only one that equals zero at the $n + 1$ distinct points t_j. But we know that any *nonzero* polynomial of degree n has at most n points at which it is zero; to equal zero at $n + 1$ points, it must be the zero polynomial. Thus the only solution to $\mathbf{Ah} = \mathbf{0}$ is $\mathbf{h} = \mathbf{0}$, and hence there is a unique solution to $\mathbf{Ax} = \mathbf{b}$ for every $\mathbf{b}$. There is exactly one polynomial of degree at most n interpolating $n + 1$ pieces of data.

PROBLEMS 4.3

▷ **1.** Each of the following matrices is the augmented matrix for a system of linear equations. For each matrix, use **Key Theorem 4.13** on rank and solvability to determine whether the system has no, one, or infinitely many solutions. If there is one solution, find it. If there are infinitely many, find the general solution (in terms of arbitrary parameters).

(a) $\left[\begin{array}{cccc|c} 0 & 1 & 0 & 0 & 4 \\ 0 & 0 & 1 & 2 & 5 \\ 0 & 0 & 0 & 0 & 0 \end{array}\right]$ (b) $\left[\begin{array}{cccc|c} 0 & 1 & 0 & 0 & 4 \\ 0 & 0 & 1 & 0 & 5 \\ 0 & 0 & 0 & 1 & 6 \end{array}\right]$

(c) $\left[\begin{array}{cccc|c} 1 & 2 & 3 & 4 & 6 \\ 0 & 0 & 0 & 1 & 5 \\ 0 & 0 & 0 & 0 & 1 \end{array}\right]$ (d) $\left[\begin{array}{cccc|c} 1 & 0 & 0 & 0 & 4 \\ 0 & 1 & 2 & 0 & 5 \\ 0 & 0 & 0 & 1 & 6 \end{array}\right]$

(e) $\left[\begin{array}{ccc|c} 1 & 3 & 4 & 9 \\ 0 & 2 & 3 & 7 \\ 0 & 0 & 2 & 6 \\ 0 & 0 & 1 & 4 \end{array}\right]$ (f) $\left[\begin{array}{cccc|c} 1 & 2 & 3 & 4 & 7 \\ 0 & 5 & 7 & 6 & 8 \\ 0 & 5 & 2 & 4 & 7 \\ 0 & 0 & 2 & 1 & 0 \end{array}\right]$

2. Find the values of α for which the following equations possess solutions, and find the solutions.

$$\begin{aligned} x - 3y + 2z &= 4 \\ 2x + y - z &= 1 \\ 3x - 2y + z &= \alpha \end{aligned}$$

▷ **3.** Find all right-hand sides $\mathbf{b}$ for which $\mathbf{Ax} = \mathbf{b}$ has solutions, and find all solutions, where

$$\mathbf{A} = \begin{bmatrix} 4 & -1 & 2 & 6 \\ -1 & 5 & -1 & -3 \\ 3 & 4 & 1 & 3 \end{bmatrix}.$$

4. The matrix below is the augmented matrix for some system of equations. Find the values of the parameter k for which the system has no, one, and infinitely many solutions; when there are infinitely many, find the general form of the solution in terms of arbitrary parameters.

$$\left[\begin{array}{ccc|c} 1 & -2 & 3 & 1 \\ 2 & k & 6 & 6 \\ -1 & 3 & k-3 & 0 \end{array}\right]$$

5. Prove that $\mathbf{x} = \mathbf{0}$ is the only solution to $\mathbf{Ax} = \mathbf{0}$ if and only if the rank of the $p \times q$ matrix $\mathbf{A}$ equals q, the number of unknowns.

6. Suppose that $\mathbf{A}$ is $p \times p$ and nonsingular; prove that the rank of $\mathbf{A}$ equals p.

▷ **7.** The equation of a straight line in the x–y plane has the form $ax + by = c$. Consider p straight lines, with equations $a_i x + b_i y = c_i$ for $1 \le i \le p$. Prove that the p straight lines all pass through a common point if and only if the rank of $[\mathbf{a} \quad \mathbf{b}]$ equals the rank of $[\mathbf{a} \quad \mathbf{b} \quad \mathbf{c}]$, where $\mathbf{a}$, $\mathbf{b}$, and $\mathbf{c}$ are $p \times 1$ matrices with $\langle \mathbf{a} \rangle_i = a_i$, $\langle \mathbf{b} \rangle_i = b_i$, and $\langle \mathbf{c} \rangle_i = c_i$.

8. (a) Find the general 3×1 solution $\mathbf{h}$ to $\mathbf{Ah} = \mathbf{0}$, where

$$\mathbf{A} = \begin{bmatrix} 1 & 2 & 1 \\ 2 & -1 & 3 \end{bmatrix}.$$

(b) Given that $\mathbf{x}_0 = [2 \quad 1 \quad -1]^T$ solves $\mathbf{Ax} = \mathbf{b} = [3 \quad 0]^T$, use **Key Theorem 4.16** and the result of (a) to give the general solution to $\mathbf{Ax} = \mathbf{b}$ for this $\mathbf{b}$.

▷ **9.** Do the same things asked of you in Problem 8, only now

$$\mathbf{A} = \begin{bmatrix} 2 & 1 & 3 & -1 \\ 4 & 2 & 8 & -1 \\ -6 & -3 & -5 & 5 \end{bmatrix}, \quad \mathbf{x}_0 = \begin{bmatrix} 1 \\ 2 \\ 1 \\ -1 \end{bmatrix}, \quad \text{and} \quad \mathbf{b} = \begin{bmatrix} 8 \\ 17 \\ -22 \end{bmatrix}.$$

10. In Section 2.2 we saw that in models of how a system evolves over time we sometimes want to find whether there is an equilibrium state for the model; this involves solving $\mathbf{Ax} = \mathbf{x}$.
(a) For the matrix $\mathbf{A}$ in Example 2.6, find the general solution to $\mathbf{Ax} = \mathbf{x}$.
(b) Use this general solution to find all solutions that also have only nonnegative entries that sum to 1, as required in Section 2.2.

11. In Section 2.6 we saw that in applications it is often necessary to solve least-squares problems $\mathbf{Ax} \approx \mathbf{y}$ given real $\mathbf{A}$ and real $\mathbf{y}$ and that this can be done by solving $\mathbf{A}^T\mathbf{Ax} = \mathbf{A}^T\mathbf{y}$.
(a) For the matrices $\mathbf{A}$ and $\mathbf{y}$ in Problem 11 of Section 2.6, find the general solution to $\mathbf{A}^T\mathbf{Ax} = \mathbf{A}^T\mathbf{y}$.
(b) Write the general solution $\mathbf{x}$ in the form $\mathbf{x} = \mathbf{x}_0 + \mathbf{h}$ as in **Key Theorem 4.16.**

12. Find the polynomial $P(t)$ of degree 2 satisfying $P(2) = 4$, $P(-1) = 7$, and $P(1) = 1$.

▷ **13.** Use **Key Theorem 4.16** to describe the set of all right-inverses $\mathbf{R}$ of a $p \times q$ matrix $\mathbf{A}$ and use its corollary to describe when a right-inverse is unique.

14. Prove that the following sixth condition is equivalent to the other five in **Key Theorem 4.18**: (6) A $p \times p$ left-inverse $\mathbf{L}$ exists for $\mathbf{A}$; that is, there is a $p \times p$ $\mathbf{L}$ with $\mathbf{LA} = \mathbf{I}_p$.

15. Let $t_1, t_2, \ldots, t_r$ be distinct real numbers; let $b_1, b_2, \ldots, b_r$ and $b'_1, b'_2, \ldots, b'_r$ be real numbers. Prove that there exists precisely one polynomial P of degree less than or equal to $2r - 1$ such that

$$P(t_i) = b_i \quad \text{and} \quad P'(t_i) = b'_i$$

for $1 \le i \le r$, where P' denotes the derivative of P.

▷ **16.** In Section 2.2, we saw that in models of how a system evolves over time we sometimes want to find whether there is an equilibrium state for the model; this involves solving $\mathbf{Ax} = \mathbf{x}$ for nonzero $\mathbf{x}$, where $\mathbf{A}$ is a given $p \times p$ matrix.

Show that there exists such a nonzero $\mathbf{x}$ if and only if $\mathbf{A} - \mathbf{I}_p$ is singular; show that $\mathbf{A} - \mathbf{I}_3$ is singular for the matrix $\mathbf{A}$ in Example 2.6.

17. Suppose that the system $\mathbf{Ax} = \mathbf{b}$ of p equations in q unknowns has a solution $\mathbf{x}_0$; show that $\mathbf{x}_0$ is the only solution if and only if the rank of $\mathbf{A}$ equals q.

18. Let $\mathbf{A}$ be $p \times q$. Prove that $\mathbf{A}^T\mathbf{A}$ is singular if the rank of $\mathbf{A}$ is strictly less than q.

19. In Section 2.4 we saw that it is important in network-equilibrium problems to know whether a certain matrix $\mathbf{H}$ defined as $\mathbf{H} = \mathbf{AK}^{-1}\mathbf{A}^T$ is nonsingular, where $\mathbf{A}$ is nonsquare and real.

(a) Suppose that $\mathbf{A}$ is $r \times s$ and that $\mathbf{K} = \mathbf{I}_s$; show that $\mathbf{H}$ is nonsingular if the rank of $\mathbf{A}$ equals r.

(b) Suppose that $\mathbf{A}$ is $r \times s$ and that $\mathbf{K}$ is an $s \times s$ diagonal matrix with positive entries on the diagonal; show that $\mathbf{H}$ is nonsingular if the rank of $\mathbf{A}$ equals r.

(c) For the 2×5 matrix $\mathbf{A}$ immediately preceding (2.20) in Section 2.4, show that the rank of $\mathbf{A}$ equals 2; conclude that $\mathbf{H}$ in (2.24) is nonsingular, as shown for a special case in Example 2.25.

4.4 INVERSES AND RANK

Chapter 1 showed that the question of whether a matrix $\mathbf{A}$ has an inverse of some kind (left-, right-, or two-sided) can be resolved by considering systems of equations: **Key Theorem 1.44** spelled this out in detail. For example, the $p \times q$ matrix $\mathbf{A}$ has a $q \times p$ right-inverse $\mathbf{R}$ if and only if we can solve the equations $\mathbf{Ar}_i = \mathbf{e}_i$ for $1 \le i \le p$. Thanks to Section 4.3, we now have the theory necessary to understand the solvability of equations and therefore the existence of inverses.

Inverses and Rank

In order to characterize the existence of inverses in terms of rank, we only need to bring together two different sets of ideas: (1) the results of Theorem 1.38 and **Key Theorem 1.44** that relate various types of inverses to solvability of equations; and (2) **Key Theorem 4.13** that relates solvability of equations to rank. For simplicity we separate the results for each type of inverse; together, these theorems justify the statements in (1.32), the Preview on Inverses in Chapter 1.

(4.21) ***Theorem*** (right-inverses and rank). Let $\mathbf{A}$ be a $p \times q$ matrix and let k denote the rank of $\mathbf{A}$. Then $\mathbf{A}$ has a right-inverse $\mathbf{R}$ if and only if $k = p$ and $p \le q$.

PROOF. (rank $\Rightarrow$ inverse) Suppose first that $k = p \le q$; we must prove that an $\mathbf{R}$ exists. By **Key Theorem 1.44(a)**, $\mathbf{R}$ exists if we can solve $\mathbf{Ar}_i = \mathbf{e}_i$ for $1 \le i \le p$, where the $\mathbf{e}_i$ are the unit column matrices of order p. By **Key Theorem 4.13(2)** and **(3)**, we can solve $\mathbf{Ar}_i = \mathbf{e}_i$ if the rank of $\mathbf{A}$ equals that of $[\mathbf{A} \quad \mathbf{e}_i]$. Since the rank of **A** is p, all p rows of any Gauss-reduced form **G** of **A** are nonzero; if **F** is the nonsingular matrix of **Key Theorem 3.34** representing the row operations transforming **A** to **G**, then also

$$\mathbf{F}[\mathbf{A} \quad \mathbf{e}_i] = [\mathbf{FA} \quad \mathbf{Fe}_i] = [\mathbf{G} \quad \mathbf{Fe}_i],$$

which is in Gauss-reduced form since **G** is. Therefore, the rank of $[\mathbf{A} \quad \mathbf{e}_i]$—the number of nonzero rows in $[\mathbf{G} \quad \mathbf{Fe}_i]$—also equals p, the rank of **A**, so the systems are solvable and **R** exists.

(inverse ⇒ rank) Suppose next that **A** has a right-inverse **R**. Let **G** be a Gauss-reduced form of **A** and **F** be the nonsingular matrix as in the first part of the proof: $\mathbf{FA} = \mathbf{G}$. Consider the equations $\mathbf{Ax} = \mathbf{b}$, where $\mathbf{b} = \mathbf{F}^{-1}\mathbf{e}_p$. By Theorem 1.38(b), **Rb** solves these equations; therefore, by **Key Theorem 4.13(2)** and **(3)**, the ranks of **A** and of $[\mathbf{A} \quad \mathbf{b}]$ are equal. Now,

$$\mathbf{F}[\mathbf{A} \quad \mathbf{b}] = [\mathbf{FA} \quad \mathbf{Fb}] = [\mathbf{G} \quad \mathbf{e}_p]$$

is in Gauss-reduced form. Since the bottom row of this last matrix is not zero (because of $\mathbf{e}_p$), the rank of $[\mathbf{A} \quad \mathbf{b}]$—which equals that of **A**—must equal p. Since there are q columns in **G** and p leading columns, clearly $p \le q$. ■

(4.22) ***Theorem*** (left-inverses and rank). Let **A** be a $p \times q$ matrix and let k denote the rank of **A**. Then **A** has a left-inverse **L** if and only if $k = q$ and $q \le p$.

PROOF. This follows from Theorem 1.38 and **Key Theorems 1.44** and **4.13**, much as does Theorem 4.21. ■

(4.23) **Key Theorem** (nonsingularity and rank). Let **A** be a square $p \times p$ matrix. Then **A** is nonsingular if and only if the rank of **A** equals p.

PROOF. This follows immediately from the preceding two theorems and Theorem 1.33. ■

A consequence of the last result will be helpful later, so we state it separately.

(4.24) ***Corollary*** (nonsingularity and elementary matrices). Let **A** be a square $p \times p$ matrix. Then **A** is nonsingular if and only if **A** equals a product of elementary matrices.

PROOF. (product ⇒ nonsingular) If **A** equals such a product, since each elementary matrix is nonsingular and the product of nonsingular matrices is nonsingular, **A** is nonsingular.

(nonsingular ⇒ product) If **A** is nonsingular, by **Key Theorem 4.23** its rank is p. Let **R** be the row-echelon form of **A**; we know that $\mathbf{A} = \mathbf{E}_1\mathbf{E}_2 \cdots \mathbf{E}_r\mathbf{R}$ for some elementary matrices $\mathbf{E}_i$. We are done if $\mathbf{R} = \mathbf{I}_p$. By Definition

4.9 on row-echelon form, there are p leading columns in $\mathbf{R}$ and they equal the $\mathbf{e}_i$ in order for $1 \le i \le p$—which says that $\mathbf{R} = \mathbf{I}_p$, as needed. ∎

An immediate corollary of the proof just used and of the relationship between Gauss-reduced forms and row-echelon forms is the following.

(4.25) ***Corollary*** (nonsingularity and Gauss-reduced/row-echelon forms). Let $\mathbf{A}$ be square. Then the following three conditions are equivalent:
1. $\mathbf{A}$ is nonsingular.
2. Some (equivalently, every) Gauss-reduced form $\mathbf{G}$ of $\mathbf{A}$ is unit-upper-triangular.
3. The row-echelon form of $\mathbf{A}$ equals $\mathbf{I}$.

PROBLEMS 4.4

1. Provide the details of the proof of **Key Theorem 4.23**.

▷ **2.** (a) Prove that, if $\mathbf{A}$ is $p \times p$ and nonsingular, then the row-echelon form of $[\mathbf{A} \quad \mathbf{I}_p]$ is $[\mathbf{I}_p \quad \mathbf{A}^{-1}]$.
(b) Use this to find the inverse of

$$\mathbf{A} = \begin{bmatrix} 2 & 1 & -3 \\ 2 & 4 & -1 \\ -2 & 1 & 10 \end{bmatrix}.$$

(c) Explain how this result can be used to find the inverse of any nonsingular matrix.

3. (a) Suppose that $\mathbf{A}$ and $\mathbf{B}$ are $p \times p$ and that $\mathbf{AB}$ is nonsingular; prove that $\mathbf{A}$ and $\mathbf{B}$ both are nonsingular.
(b) Suppose that $\mathbf{A}$ and $\mathbf{B}$ are $p \times p$ and that one of them is nonsingular; show that the product is singular if and only if the other of the matrices is singular.

4. Suppose that $\mathbf{B}$ is $p \times q$ and that $\mathbf{A}$ is $q \times q$ and nonsingular. Prove that the rank of $\mathbf{BA}$ equals the rank of $\mathbf{B}$.

5. Suppose that $\mathbf{B}$ is $p \times q$ and that $\mathbf{A}$ is $p \times p$ and nonsingular. Prove that the rank of $\mathbf{AB}$ equals the rank of $\mathbf{B}$.

▷ **6.** Suppose that $\mathbf{A}$ is $p \times q$ and has rank k. Show that:
(a) $\mathbf{A}$ has no right-inverse if $p > q$.
(b) $\mathbf{A}$ has no right-inverse if $k < p < q$.
(c) $\mathbf{A}$ has infinitely many right-inverses if $k = p < q$.
(d) When does $\mathbf{A}$ have exactly one right-inverse?

7. Suppose that $\mathbf{A}$ is $p \times q$ and has rank k. Show that:
(a) $\mathbf{A}$ has no left-inverse if $p < q$.
(b) $\mathbf{A}$ has no left-inverse if $k < q < p$.

(c) **A** has infinitely many left-inverses if $k = q < p$.
(d) When does **A** has exactly one left-inverse?

▷ **8.** Suppose that **A** is $p \times q$, **B** is $q \times p$, and $q < p$. GOLD
(a) Show that **AB** is singular.
(b) Give an example to show that **BA** can be nonsingular.

9. Show that the rank of $\mathbf{AA}^T$ is less than or equal to the rank of **A**. (But see Problem 11, which is more difficult.)

10. Prove that the rank of **A** equals that of $\mathbf{A}^T$ by proving it first for Gauss-reduced **A** and then, using Problems 4 and 5, for general **A**.

▷ **11.** Prove that the rank of $\mathbf{AA}^T$ equals the rank of **A** for real **A**.

4.5 DETERMINANTS AND THEIR PROPERTIES

Many readers may have learned some years ago how to solve systems of equations by using determinants and may have wondered why we have not yet mentioned them, especially not in Chapter 3. The fact is that for practical and efficient computation of solutions of systems of equations larger than about 3×3, determinants are useless; they are, however, useful as a conceptual or descriptive device for representing solutions and inverses. For this reason, but *not* for computational purposes, we now take up determinants briefly.

Determinants

We are going to define a single number associated with a $p \times p$ matrix **A**, called the *determinant* of **A** and denoted by det **A** or det(**A**). The determinant for $p \times p$ matrices will be defined in terms of determinants for $(p-1) \times (p-1)$ matrices, and so on down to 1×1 matrices $[\alpha]$ for which we define det $[\alpha] = \alpha$. We need some terminology to make this precise.

(4.26) ***Definition.*** Let **A** be $p \times p$.
(a) The *(i, j)-minor* of **A**, denoted M_{ij}, is the determinant of the $(p-1) \times (p-1)$ matrix formed by deleting the ith row and jth column from **A**.
(b) The *(i, j)-cofactor* of **A**, denoted A_{ij}, is $(-1)^{i+j}M_{ij}$.

Note that the signs $(-1)^{i+j}$ in the definition of cofactor form a checkerboard pattern:

$$\begin{bmatrix} + & - & + & \cdots \\ - & + & - & \cdots \\ + & - & + & \cdots \\ & \cdots & & \end{bmatrix}.$$

We can now define determinants.

(4.27) ***Definition.***

(a) The *determinant* of the 1×1 matrix $[\alpha]$ is defined as $\det[\alpha] = \alpha$.

(b) The *determinant* of the $p \times p$ matrix $\mathbf{A}$ is defined as

$$\det \mathbf{A} = \sum_{j=1}^{p} \langle \mathbf{A} \rangle_{1j} A_{1j}.$$

In words: The determinant of $\mathbf{A}$ is the sum of the products of the entries of the first row and the cofactors of the first row.

(4.28) ***Example.*** Consider the determinant of the general 2×2 matrix

$$\mathbf{A} = \begin{bmatrix} a_{11} & a_{12} \\ a_{21} & a_{22} \end{bmatrix}.$$

By Definition 4.27, we have

$$\begin{aligned} \det \mathbf{A} &= a_{11}A_{11} + a_{12}A_{12} \\ &= a_{11} \det [a_{22}] - a_{12} \det [a_{21}] \\ &= a_{11}a_{22} - a_{12}a_{21}. \end{aligned}$$

For example,

$$\det \begin{bmatrix} -4 & -5 \\ 2 & 3 \end{bmatrix} = (-4)(3) - (-5)(2) = -12 - (-10) = -2.$$

(4.29) ***Example.*** Consider the determinant of the general 3×3 matrix

$$\mathbf{A} = \begin{bmatrix} a_{11} & a_{12} & a_{13} \\ a_{21} & a_{22} & a_{23} \\ a_{31} & a_{32} & a_{33} \end{bmatrix}.$$

By Definition 4.27, we have

$$\begin{aligned} \det \mathbf{A} &= a_{11}A_{11} + a_{12}A_{12} + a_{13}A_{13} \\ &= a_{11} \det \begin{bmatrix} a_{22} & a_{23} \\ a_{32} & a_{33} \end{bmatrix} - a_{12} \det \begin{bmatrix} a_{21} & a_{23} \\ a_{31} & a_{33} \end{bmatrix} \\ &\quad + a_{13} \det \begin{bmatrix} a_{21} & a_{22} \\ a_{31} & a_{32} \end{bmatrix} \\ &= a_{11}(a_{22}a_{33} - a_{23}a_{32}) - a_{12}(a_{21}a_{33} - a_{23}a_{31}) \\ &\quad + a_{13}(a_{21}a_{32} - a_{22}a_{31}) \\ &= a_{11}a_{22}a_{33} - a_{11}a_{32}a_{23} - a_{12}a_{21}a_{33} + a_{12}a_{31}a_{23} + a_{13}a_{21}a_{32} \\ &\quad - a_{13}a_{31}a_{22}. \end{aligned}$$

For example,

$$\det \begin{bmatrix} 2 & 1 & 3 \\ 1 & 2 & -1 \\ 2 & -1 & 3 \end{bmatrix} = 2\{(2)(3) - (-1)(-1)\} - 1\{(1)(3) - (-1)(2)\} + 3\{(1)(-1) - (2)(2)\}$$
$$= 2\{5\} - 1\{5\} + 3\{-5\} = -10.$$

If we examine any single product term in the expression for the determinant of 2×2 or 3×3 matrices, we see that each product involves an entry from each row and an entry from each column. In general, it turns out that det **A** is the sum of all possible products of p entries of **A**, with appropriate signs, where in each product there is exactly one entry from each row and exactly one from each column. Determinants can, in fact, be defined in this way, but we pursue this no further.

You should now be able to solve Problems 1 to 5.

Basic Properties of Determinants

The first row of **A** played a special role in the definition of det **A**: The entries of the *first row* are multiplied by their cofactors, the products summed, and the result called det **A**. The first important property is that "*first row*" may be replaced by "*any row or any column.*" Thus, in addition to the very first expression for det **A** for 3×3 **A** in Example 4.29, we have

$$\begin{aligned} \det \mathbf{A} &= a_{21}A_{21} + a_{22}A_{22} + a_{23}A_{23} \\ &= a_{11}A_{11} + a_{21}A_{21} + a_{31}A_{31} \\ &= a_{13}A_{13} + a_{23}A_{23} + a_{33}A_{33}. \end{aligned}$$

(4.30) ***Theorem*** (row/column expansion of det **A**). det **A** can be evaluated by expanding along any row or column; that is, det **A** equals the sum of the products of the entries along any row or column with their cofactors:

$$\det \mathbf{A} = \sum_{j=1}^{p} \langle \mathbf{A} \rangle_{rj} A_{rj} = \sum_{i=1}^{p} \langle \mathbf{A} \rangle_{is} A_{is} \qquad \text{for all } r \text{ and } s.$$

PROOF. The proof is more complicated than instructive and is omitted. ■

(4.31) ***Corollary.***

(a) $\det \mathbf{A} = \det \mathbf{A}^T$.

(b) If any row or column of **A** equals **0**, then

$$\det \mathbf{A} = 0.$$

(c) For any number c and for $p \times p$ **A**,

$$\det(c\mathbf{A}) = c^p \det(\mathbf{A}).$$

(d) If **A** has two rows (columns) that are equal, then det **A** = 0.

(e) If **A** and **B** are equal except for possibly the entries in their kth row (or column), and if **C** is defined as that matrix identical to **A** and **B** except that its kth row (column) is the sum of the kth rows (columns) of **A** and **B**, then

$$\det \mathbf{C} = \det \mathbf{A} + \det \mathbf{B}.$$

PROOF.

(a) Expanding det **A** by its first column is the same as expanding det $\mathbf{A}^T$ by its first row.

(b) Expanding det **A** along the zero row or column gives det **A** = 0.

(c) This is trivial for $p = 1$; for general p, it follows easily by induction on p and expanding along any row or column of $c\mathbf{A}$.

(d) The formula in Example 4.28 shows that this is true for 2×2 matrices. We then use induction on p; expanding along any row (column) other than the two in question gives 0 for det **A** since each cofactor is 0 by the inductive hypothesis that the result is true for $p - 1$.

(e) Expanding all three determinants along the kth row (column) easily gives the result. ■

We intend to use determinants to describe solutions of equations; since we presently describe solutions through row operations, we need to examine the relation between row operations and determinants.

(4.32) ***Theorem*** (determinants and row operations)

(a) If **A**′ is produced from **A** by interchanging two rows (or two columns) of **A**, then

$$\det(\mathbf{A}') = (-1)\det(\mathbf{A}).$$

(b) If **A**′ is produced from **A** by replacing one row (column) of **A** by the number c times that row (column) itself, then

$$\det(\mathbf{A}') = c\det(\mathbf{A}).$$

(c) If **A**′ is produced from **A** by replacing one row (column) of **A** by that row (column) plus some multiple of a *different row* (column), then

$$\det(\mathbf{A}') = \det(\mathbf{A}).$$

PROOF

(a) To be definite, suppose that we interchange the ith row $\mathbf{r}_i$ and the jth row $\mathbf{r}_j$ of **A** to get **A**′. Denote by $\{\mathbf{a}\,;\,\mathbf{b}\}$ the determinant of the matrix obtained from **A** by replacing the ith row of **A** by some row matrix **a** and the jth row of $\mathbf{A}$ by some row matrix $\mathbf{b}$; therefore, $\det(\mathbf{A})$ is just $\{\mathbf{r}_i\,;\mathbf{r}_j\}$ while $\det(\mathbf{A}')$ is $\{\mathbf{r}_j\,;\mathbf{r}_i\}$. By Corollary 4.31(d), we know that $\{\mathbf{a}\,;\mathbf{a}\} = 0$ for all $\mathbf{a}$. By Corollary 4.31(e), we know that

$$\{\mathbf{a} + \mathbf{b}\,;\mathbf{c}\} = \{\mathbf{a}\,;\mathbf{c}\} + \{\mathbf{b}\,;\mathbf{c}\}$$

and

$$\{\mathbf{a}\,;\mathbf{b} + \mathbf{c}\} = \{\mathbf{a}\,;\mathbf{b}\} + \{\mathbf{a}\,;\mathbf{c}\} \qquad \text{for all } \mathbf{a}, \mathbf{b}, \text{ and } \mathbf{c}.$$

Therefore,

$$\begin{aligned} 0 &= \{\mathbf{r}_i + \mathbf{r}_j\,;\mathbf{r}_i + \mathbf{r}_j\} = \{\mathbf{r}_i\,;\mathbf{r}_i + \mathbf{r}_j\} + \{\mathbf{r}_j\,;\mathbf{r}_i + \mathbf{r}_j\} \\ &= \{\mathbf{r}_i\,;\mathbf{r}_i\} + \{\mathbf{r}_i\,;\mathbf{r}_j\} + \{\mathbf{r}_j\,;\mathbf{r}_i\} + \{\mathbf{r}_j\,;\mathbf{r}_j\} \\ &= \quad 0 \quad + \det(\mathbf{A}) + \det(\mathbf{A}') + \quad 0 \\ &= \det(\mathbf{A}) + \det(\mathbf{A}'). \end{aligned}$$

This then says that $\det(\mathbf{A}') = -\det(\mathbf{A})$, as claimed.

(b) Expanding along the row (column) in question gives the result easily.

(c) Suppose that the ith row $\mathbf{r}_i$ is replaced by $\mathbf{r}_i + c\mathbf{r}_j$ where $\mathbf{r}_j$ is the jth row of $\mathbf{A}$ and c is a number. Using the notation of the proof of (a), we have

$$\begin{aligned} \det(\mathbf{A}') &= \{\mathbf{r}_i + c\mathbf{r}_j\,;\mathbf{r}_j\} = \{\mathbf{r}_i\,;\mathbf{r}_j\} + \{c\mathbf{r}_j\,;\mathbf{r}_j\} \\ &= \det(\mathbf{A}) + c\{\mathbf{r}_j; \mathbf{r}_j\} \qquad \text{by Corollary 4.31(b)} \\ &= \det(\mathbf{A}) + c(0) = \det(\mathbf{A}) \end{aligned}$$

as claimed. ■

The three basic row operations whose effects on determinants are described by Theorem 4.32 are of course equivalent to premultiplication by elementary matrices—see Definition 3.32 and Theorem 3.33. Each elementary matrix $\mathbf{E}$ results from applying its corresponding row operation to $\mathbf{I}$, whose determinant clearly equals 1; since the effects of row operations are described above, we can easily evaluate the determinants of the basic elementary matrices of Definition 3.31 and combine this information with Theorem 4.32 as follows.

(4.33) ***Corollary*** (elementary matrices and determinants). Let $\mathbf{E}_{ij}$, $\mathbf{E}_i(c)$, and $\mathbf{E}_{ij}(c)$ denote the elementary matrices of Definition 3.31. Then:

(a) $\det \mathbf{E}_{ij} = -1$.

(b) $\det \mathbf{E}_i(c) = c$.

(c) $\det \mathbf{E}_{ij}(c) = 1$.

(d) If $\mathbf{E}$ is any $p \times p$ elementary matrix and $\mathbf{A}$ is any $p \times p$ matrix, then

$$\det(\mathbf{EA}) = \det(\mathbf{E})\det(\mathbf{A}).$$

The basic properties of determinants contained in Theorems 4.30–4.32 provide the key to evaluating determinants. It is easiest to see how in an example.

(4.34) ***Example.*** Consider evaluating the determinant of the 4×4 matrix $\mathbf{A}$ below. Rather than immediately expand along some row or column, we first perform row operations so as to create a column with only one nonzero entry in it; expansion along that column will be easy. If we add multiples of row 2 to the other rows to create zeros in column 1, then by Theorem 4.32(c) we do not change the determinant. That is,

$$\det \mathbf{A} = \det \begin{bmatrix} 2 & -3 & 2 & 5 \\ 1 & -1 & 1 & 2 \\ 3 & 2 & 2 & 1 \\ 1 & 1 & -3 & -1 \end{bmatrix} = \det \begin{bmatrix} 0 & -1 & 0 & 1 \\ 1 & -1 & 1 & 2 \\ 0 & 5 & -1 & -5 \\ 0 & 2 & -4 & -3 \end{bmatrix},$$

which we then expand along its first column to get

$$\det \mathbf{A} = -(1) \det \begin{bmatrix} -1 & 0 & 1 \\ 5 & -1 & -5 \\ 2 & -4 & -3 \end{bmatrix}.$$

We could now evaluate as in Example 4.29, but it is easier to add -4 times the second row to the third—which does not change the determinant—so that the second column will have only one nonzero entry, again making for easy expansion along that column. This gives

$$\det \mathbf{A} = -(1) \det \begin{bmatrix} -1 & 0 & 1 \\ 5 & -1 & -5 \\ -18 & 0 & 17 \end{bmatrix} = -(1)(-1) \det \begin{bmatrix} -1 & 1 \\ -18 & 17 \end{bmatrix},$$

which equals

$$-(1)(-1)\{(-1)(17) - (1)(-18)\} = 1.$$

Thus $\det \mathbf{A} = 1$.

In Example 4.34 we never performed a row operation that changed the value of the determinant; had we interchanged two rows or replaced a row by a multiple of itself, we would have needed to keep track of the effect on the determinant.

You should now be able to solve Problems 1 to 12.

Determinants, Products, and Nonsingularity

The basic properties of determinants that we have presented have some surprising and far-from-obvious consequences; the next two theorems embody quite striking results.

(4.35) ***Theorem*** (determinants and nonsingularity).

(a) $\mathbf{A}$ is nonsingular if and only if $\det \mathbf{A} \neq 0$; equivalently, $\mathbf{A}$ is singular if and only if $\det \mathbf{A} = 0$.

(b) If $\mathbf{A}$ is nonsingular, then

$$\det(\mathbf{A}^{-1}) = 1/\det(\mathbf{A}).$$

PROOF

(a) (nonsingular $\Rightarrow$ det $\neq 0$) Suppose first that $\mathbf{A}$ is nonsingular; by Corollary 4.24, $\mathbf{A} = \mathbf{E}_1\mathbf{E}_2 \cdots \mathbf{E}_r$ for some elementary matrices $\mathbf{E}_i$. Applying Corollary 4.33(d) repeatedly gives

$$\det(\mathbf{A}) = \det(\mathbf{E}_1)\det(\mathbf{E}_2)\cdots\det(\mathbf{E}_r) \neq 0$$

since the determinant of each elementary matrix is nonzero by Corollary 4.33(a)–(c).

(det $\neq 0 \Rightarrow$ nonsingular) Let $\det(\mathbf{A}) \neq 0$. If, on the contrary, $\mathbf{A}$ *were* singular, then its rank would be less than p by **Key Theorem 4.18**; thus any Gauss-reduced form $\mathbf{G}$ would have a zero bottom row. Since row operations produce $\mathbf{G}$ from $\mathbf{A}$, we have

$$\mathbf{E}_1\mathbf{E}_2\cdots\mathbf{E}_r\mathbf{A} = \mathbf{G}$$

for some elementary matrices $\mathbf{E}_i$. Applying Corollary 4.33(d) repeatedly gives

$$\det(\mathbf{E}_1)\det(\mathbf{E}_2)\cdots\det(\mathbf{E}_r)\det(\mathbf{A}) = \det(\mathbf{G})$$

which would equal zero since $\mathbf{G}$'s bottom row would be zero. Since the determinant of each $\mathbf{E}_i$ is nonzero, this would mean that $\det(\mathbf{A}) = 0$ in contradiction to our assumption that $\det(\mathbf{A}) \neq 0$. Thus $\mathbf{A}$ could not in fact be singular.

(b) Since $\mathbf{A}$ is nonsingular, we can write $\mathbf{A} = \mathbf{E}_1\mathbf{E}_2\cdots\mathbf{E}_r$ for elementary matrices $\mathbf{E}_i$ by Corollary 4.24. Using $\det(\mathbf{I}_p) = 1$ and applying Corollary 4.33(d) repeatedly gives

$$\begin{aligned} 1 = \det(\mathbf{I}_p) &= \det(\mathbf{A}\mathbf{A}^{-1}) = \det(\mathbf{E}_1\mathbf{E}_2\cdots\mathbf{E}_r\mathbf{A}^{-1}) \\ &= \det(\mathbf{E}_1)\det(\mathbf{E}_2)\cdots\det(\mathbf{E}_r)\det(\mathbf{A}^{-1}) \\ &= \det(\mathbf{E}_1\mathbf{E}_2\cdots\mathbf{E}_r)\det(\mathbf{A}^{-1}) = \det(\mathbf{A})\det(\mathbf{A}^{-1}). \end{aligned}$$

From $1 = \det(\mathbf{A})\det(\mathbf{A}^{-1})$ and the fact that $\det(\mathbf{A}) \neq 0$ by (a), we get

$$\det(\mathbf{A}^{-1}) = 1/\det(\mathbf{A}). \quad \blacksquare$$

(4.36) ***Theorem*** (determinants and products)

$$\det(\mathbf{AB}) = \det(\mathbf{A})\det(\mathbf{B}).$$

PROOF. (**A** nonsingular $\Rightarrow$ result) First suppose that **A** is nonsingular. In this case, Corollary 4.24 says that $\mathbf{A} = \mathbf{E}_1\mathbf{E}_2 \cdots \mathbf{E}_r$ for elementary matrices $\mathbf{E}_i$ and Corollary 4.28(d) then gives

$$\begin{aligned}\det(\mathbf{AB}) &= \det(\mathbf{E}_1\mathbf{E}_2 \cdots \mathbf{E}_r\mathbf{B}) \\ &= \det(\mathbf{E}_1)\det(\mathbf{E}_2) \cdots \det(\mathbf{E}_r)\det(\mathbf{B}) \\ &= \det(\mathbf{E}_1\mathbf{E}_2 \cdots \mathbf{E}_r)\det(\mathbf{B}) = \det(\mathbf{A})\det(\mathbf{B})\end{aligned}$$

as asserted.

(**A** singular $\Rightarrow$ result) Now suppose that **A** is $p \times p$ and singular, so by **Key Theorem 4.18** it has rank less than p and therefore the bottom row of any Gauss-reduced form **G** must equal zero. Let **F** be the usual nonsingular matrix from **Key Theorem 3.34** with $\mathbf{FA} = \mathbf{G}$. Since the bottom row of **G** is zero, so also is the bottom row of **GB**; therefore, $\det(\mathbf{GB}) = 0$ by Corollary 4.31(b) and hence **GB** is singular by Theorem 4.35. Since **F** is nonsingular while $\mathbf{F}(\mathbf{AB}) = (\mathbf{FA})\mathbf{B} = \mathbf{GB}$ is singular, the matrix **AB** must be singular—otherwise, $\mathbf{F}(\mathbf{AB})$ would be nonsingular as the product of nonsingular matrices. Since **AB** is singular, its determinant equals zero. Thus

$$\det(\mathbf{AB}) = 0 = 0\det(\mathbf{B}) = \det(\mathbf{A})\det(\mathbf{B}),$$

as asserted, since $\det(\mathbf{A}) = 0$. ■

The definition of the product **AB**, the computation of the inverse $\mathbf{A}^{-1}$ from **A**, and the definition and evaluation of determinants are all quite complicated. It is quite striking that these complicated entities combine so simply that

$$\det(\mathbf{AB}) = \det(\mathbf{A})\det(\mathbf{B}) \quad \text{and} \quad \det(\mathbf{A}^{-1}) = \frac{1}{\det(\mathbf{A})}.$$

PROBLEMS 4.5

▷ **1.** Given the matrix **A**:

(a) Evaluate the minors M_{11}, M_{13}, M_{22}, M_{31}, and M_{33}.

(b) Evaluate the cofactors A_{12}, A_{21}, A_{23}, and A_{32}.

(c) Evaluate det **A**.

$$\mathbf{A} = \begin{bmatrix} -1 & 3 & 2 \\ 3 & 1 & -1 \\ -2 & 2 & -1 \end{bmatrix}.$$

2. Evaluate the determinants of the following matrices.

(a) $\begin{bmatrix} 4 & -6 \\ 2 & -3 \end{bmatrix}$ (b) $\begin{bmatrix} -3 \end{bmatrix}$ (c) $\begin{bmatrix} 0 & 0 \\ 0 & 0 \end{bmatrix}$ (d) $\begin{bmatrix} 1 & -1 & 2 \\ 2 & 1 & 3 \\ 0 & 2 & -1 \end{bmatrix}$

3. Evaluate the determinant of

$$\begin{bmatrix} 2 & 0 & 3 \\ 10 & 1 & 17 \\ 7 & 12 & -4 \end{bmatrix}.$$

▷ 4. Prove that det $\mathbf{I}_p = 1$ for all p.

5. Show that the determinant of a lower-triangular matrix $\mathbf{L}$ equals the product of its main-diagonal entries.

6. Verify that det $\mathbf{A}$ = det $\mathbf{A}^T$ for the general 2×2 matrix $\mathbf{A}$.

7. Provide the details of the proof of Corollary 4.33 on the determinants of elementary matrices.

▷ 8. Use the approach of Example 4.34 to evaluate the determinant of

$$\begin{bmatrix} 1 & 1 & 3 & 0 & 2 \\ 3 & 1 & 0 & 1 & 2 \\ 0 & 1 & 3 & 0 & 2 \\ 4 & -2 & 3 & 1 & 0 \\ 5 & 1 & 0 & 0 & 6 \end{bmatrix}.$$

9. Verify that the same result is produced by expanding the determinant of the matrix below along every row and along every column.

$$\begin{bmatrix} 2 & 0 & 3 \\ 10 & 1 & 17 \\ 7 & 12 & -4 \end{bmatrix}$$

▷ 10. If $\mathbf{A}$ is skew-symmetric—that is, $\mathbf{A}^T = -\mathbf{A}$—and $p \times p$, show that det $\mathbf{A} = (-1)^p$ det $\mathbf{A}$, and conclude that det $\mathbf{A} = 0$ if p is odd.

11. Let $\mathbf{P}$ be a permutation matrix as in Definition 3.51 in Section 3.7. Show that det $\mathbf{P} = \pm 1$.

▷ 12. Suppose that $\mathbf{T}$ is a square matrix and is either lower- or upper-triangular. Show that the determinant of $\mathbf{T}$ is the product of its main-diagonal entries.

13. The equation of a straight line in the x–y plane has the form $ax + by = c$. Consider three straight lines, with equations $a_i x + b_i y = c_i$ for $i = 1, 2, 3$. Prove that the three lines all pass through a common point if and only if det $[\mathbf{a} \;\; \mathbf{b} \;\; \mathbf{c}] = 0$, where $\mathbf{a}$, $\mathbf{b}$, and $\mathbf{c}$ are 3×1 and $\langle \mathbf{a} \rangle_i = a_i$, $\langle \mathbf{b} \rangle_i = b_i$, $\langle \mathbf{c} \rangle_i = c_i$.

14. Prove that the three points (x_i, y_i), $i = 1, 2, 3$, in the x–y plane are collinear (that is, lie on the same straight line) if and only if the determinant of the following matrix equals zero:

$$\begin{bmatrix} x_1 & y_1 & 1 \\ x_2 & y_2 & 1 \\ x_3 & y_3 & 1 \end{bmatrix}.$$

▷ **15.** Consider the straight line through the two distinct points (x_1, y_1) and (x_2, y_2) in the x–y plane. Show that the equation of this line is just

$$\det\begin{bmatrix} x & y & 1 \\ x_1 & y_1 & 1 \\ x_2 & y_2 & 1 \end{bmatrix} = 0.$$

16. Consider the circle through the three noncollinear points (x_1, y_1), (x_2, y_2), and (x_3, y_3) in the x–y plane. Show that the equation of this circle is just

$$\det\begin{bmatrix} x^2 + y^2 & x & y & 1 \\ x_1^2 + y_1^2 & x_1 & y_1 & 1 \\ x_2^2 + y_2^2 & x_2 & y_2 & 1 \\ x_3^2 + y_3^2 & x_3 & y_3 & 1 \end{bmatrix} = 0.$$

17. Suppose that $\mathbf{A} = \mathbf{P}^T\mathbf{LU}$ is the LU-decomposition of $\mathbf{A}$ as described in **Key Theorem 3.53** of Section 3.7. Show that the determinant of $\mathbf{A}$ equals plus or minus the product of the main-diagonal entries of $\mathbf{L}$.

▷ **18.** (a) Show that $\det(\mathbf{A}^n) = \{\det(\mathbf{A})\}^n$ for all nonnegative integers n.
(b) Show that if $\mathbf{A}$ is nonsingular, then $\det(\mathbf{A}^n) = \{\det(\mathbf{A})\}^n$ for all integers n.

19. In Section 2.2, we saw that in models of how a system evolves over time we sometimes want to find whether there is an equilibrium state for the model; this involves determining whether the equations $\mathbf{Ax} = \mathbf{x}$ have a nonzero solution, where $\mathbf{A}$ is a $p \times p$ matrix from the model.
(a) Show that $\mathbf{Ax} = \mathbf{x}$ has a solution $\mathbf{x} \neq \mathbf{0}$ if and only if $\det(\mathbf{A} - \mathbf{I}_p) = 0$.
(b) Verify that $\det(\mathbf{A} - \mathbf{I}_3) = 0$ for the matrix $\mathbf{A}$ in Example 2.6.

4.6 *DETERMINANTAL REPRESENTATION OF INVERSES AND SOLUTIONS*

Many of the uses of determinants derive from the fact that matrix inverses and solutions of equations can be represented compactly with determinants.

Representation of Inverses

Let $\mathbf{A}$ be a $p \times p$ matrix. Suppose for the moment that from $\mathbf{A}$ we produce a new matrix $\mathbf{A}'$ as follows: For some specific values of i and j with $i \neq j$, we replace the jth row of $\mathbf{A}$ by a copy of the ith row. $\mathbf{A}'$ is now identical with $\mathbf{A}$ except that the jth row of $\mathbf{A}'$ is different. Since $\mathbf{A}'$ has two equal rows—its ith and its jth—we know from Corollary 4.31(d) that $\det \mathbf{A}' = 0$. If we now evaluate $\det \mathbf{A}'$ by expanding along the jth row, the (j, k)-cofactors A'_{jk} of $\mathbf{A}'$ will be identical with the (j, k)-cofactors A_{jk} of $\mathbf{A}$ since we strike out the jth row—the one place where $\mathbf{A}'$ and $\mathbf{A}$ differ—when forming the cofactors. Recalling that the entries $\langle\mathbf{A}'\rangle_{jk}$ of the jth row of $\mathbf{A}'$ are just the entries $\langle\mathbf{A}\rangle_{ik}$ of the ith row of $\mathbf{A}$, we obtain from expanding along the jth row of $\mathbf{A}'$:

(4.37) If $i \neq j$, then

$$0 = \det \mathbf{A}' = \langle \mathbf{A}' \rangle_{j1} A'_{j1} + \langle \mathbf{A}' \rangle_{j2} A'_{j2} + \cdots + \langle \mathbf{A}' \rangle_{jp} A'_{jp}$$
$$= \langle \mathbf{A} \rangle_{i1} A_{j1} + \langle \mathbf{A} \rangle_{i2} A_{j2} + \cdots + \langle \mathbf{A} \rangle_{ip} A_{jp}.$$

In words: The sum of the products of the entries of *one* row of **A** and the cofactors of *another* row of **A** equals zero. We know, of course, that the sum of the products of the entries of one row of **A** and the cofactors of the *same* row of **A** equals det **A**. We can express the surprising relationship (4.37) in matrix notation by introducing the *adjoint matrix of* **A**.

(4.38) ***Definition.*** The *adjoint* (or *adjoint matrix*; sometimes also called the *adjugate*) of a $p \times p$ matrix **A** is that $p \times p$ matrix, denoted by **adj A**, defined by $\langle \mathbf{adj\ A} \rangle_{ij} = A_{ji}$, where A_{ji} denotes the (j, i)-cofactor of **A**.

Note that $\langle \mathbf{adj\ A} \rangle_{ij}$ equals the (j, i)-cofactor, not the (i, j)-cofactor.

(4.39) ***Example.*** Suppose that **A** is the 3×3 matrix

$$\mathbf{A} = \begin{bmatrix} 2 & 1 & -2 \\ 1 & -1 & 2 \\ 3 & 1 & 1 \end{bmatrix}.$$

Remembering to take into account the checkerboard pattern of signs that multiply the minors M_{ij} to produce the cofactors A_{ij}, we easily find that $A_{11} = -3$, $A_{12} = 5$, $A_{13} = 4$, $A_{21} = -3$, $A_{22} = 8$, $A_{23} = 1$, $A_{31} = 0$, $A_{32} = -6$, and $A_{33} = -3$. Definition 4.38 then gives

$$\mathbf{adj\ A} = \begin{bmatrix} -3 & -3 & 0 \\ 5 & 8 & -6 \\ 4 & 1 & -3 \end{bmatrix}.$$

The adjoint allows (4.37) to be expressed as

$$0 = \langle \mathbf{A} \rangle_{i1} \langle \mathbf{adj\ A} \rangle_{1j} + \langle \mathbf{A} \rangle_{i2} \langle \mathbf{adj\ A} \rangle_{2j} + \cdots + \langle \mathbf{A} \rangle_{ip} \langle \mathbf{adj\ A} \rangle_{pj},$$

which says that the (i, j)-entry of the product of **A** times **adj A** equals zero if $i \neq j$. Recalling that the sum in (4.37) equals det **A** rather than 0 when $i = j$ gives the first part of the following.

(4.40) ***Theorem*** (adjoints). $\mathbf{A}(\mathbf{adj\ A}) = (\det \mathbf{A})\mathbf{I} = (\mathbf{adj\ A})\mathbf{A}$.

PROOF. We have already proved the first part. The second follows by using the same approach, but this time producing $\mathbf{A}'$ by replacing the jth *column* of **A** by the ith *column* of **A**. ■

If det **A** is nonzero, then the equation in Theorem 4.40 can be divided through by det **A** to obtain, for example,

$$\mathbf{A}\{(\mathbf{adj\ A})/\det \mathbf{A}\} = \mathbf{I}_p.$$

That is, $(\mathbf{adj\ A})/(\det \mathbf{A})$ is just $\mathbf{A}^{-1}$.

(4.41) ***Corollary*** (adjoint formula for inverses). If **A** is nonsingular, then

$$\mathbf{A}^{-1} = \frac{\mathbf{adj\ A}}{\det \mathbf{A}}.$$

We cannot emphasize enough that this representation of $\mathbf{A}^{-1}$ is useful *theoretically* but *not useful computationally on large matrices* because of the work required to evaluate the determinants involved in **adj A**. A $p \times p$ determinant is defined in terms of $(p-1) \times (p-1)$ determinants, and so on, which implies that the work required to evaluate the determinant of a $p \times p$ matrix from its definition is proportional to

$$p! = p(p-1)(p-2)\cdots(2)(1).$$

Finding the inverse this way thus requires work like $p!$ (actually, nearly 2.7 times $p!$), while Gauss elimination produces the inverse with work like p^3. The table below should make clear why Gauss elimination is preferred.

(4.42)

$p =$	1	2	3	4	5	6	7	10
$p^3 \approx$	1	8	27	64	125	216	343	1000
$(2.7)p! \approx$	1	5	16	65	324	1944	13,608	9,797,760

(4.43) ***Example.*** Consider the matrix **A** whose adjoint was found in Example 4.39. Multiplication verifies that its inverse is correctly given, according to Corollary 4.41, as

$$\mathbf{A}^{-1} = \frac{\mathbf{adj\ A}}{\det \mathbf{A}} = -\frac{1}{9}\begin{bmatrix} -3 & -3 & 0 \\ 5 & 8 & -6 \\ 4 & 1 & -3 \end{bmatrix}.$$

You should now be able to solve Problems 1 to 11.

Representation of Solutions

Corollary 4.41 allows us to express $\mathbf{A}^{-1}$ in terms of determinants. Since we can express the solution of systems of equations (having nonsingular coefficient matrices) $\mathbf{Ax} = \mathbf{b}$ as $\mathbf{x} = \mathbf{A}^{-1}\mathbf{b}$, we can easily represent **x** with determinants. The result is known as *Cramer's Rule.*

(4.44) ***Theorem*** (Cramer's Rule). Let $\mathbf{A}$ be nonsingular. The entries $x_i = \langle \mathbf{x} \rangle_i$ in the solution to the equations $\mathbf{Ax} = \mathbf{b}$ are given by $x_i = \Delta_i/\Delta$, where Δ denotes the determinant of $\mathbf{A}$ and Δ_i denotes the determinant of the matrix $\mathbf{A}_i$ formed by replacing the ith column of $\mathbf{A}$ with $\mathbf{b}$.

PROOF. Problem 16. ■

We repeat what we emphasized after Corollary 4.41: These determinantal representations are *computationally inefficient*; Cramer's Rule is useful primarily theoretically as a way of representing, but not calculating, solutions.

(4.45) ***Example.*** Consider once again the matrix $\mathbf{A}$ of Example 4.39, whose determinant equals -9. The second entry $x_2 = \langle \mathbf{x} \rangle_2$ of the solution to $\mathbf{Ax} = [2 \quad 0 \quad 3]^T$ is given according to Cramer's Rule by

$$x_2 = \frac{\Delta_2}{\Delta} = \frac{\det \mathbf{A}_2}{\det \mathbf{A}} = \frac{\det \mathbf{A}_2}{-9}, \qquad \text{where } \mathbf{A}_2 = \begin{bmatrix} 2 & 2 & -2 \\ 1 & 0 & 2 \\ 3 & 3 & 1 \end{bmatrix}.$$

Straightforward evaluation gives $\det \mathbf{A}_2 = -8$ and hence $x_2 = -8/(-9) = \frac{8}{9}$.

PROBLEMS 4.6

▷ **1.** Find **adj A** and **A(adj A)** if see 4.40

$$\mathbf{A} = \begin{bmatrix} 4 & 5 & 6 \\ 1 & 2 & 3 \\ 7 & 8 & 9 \end{bmatrix}.$$

2. Provide details for the proof of the second part of Theorem 4.40 on the adjoint matrix.

3. Show that, if $\mathbf{A}$ is singular, then $\mathbf{A}(\mathbf{adj\,A}) = \mathbf{0}$ and $(\mathbf{adj\,A})\mathbf{A} = \mathbf{0}$, and give an example of a singular matrix $\mathbf{A}$ with $\mathbf{adj\,A} \neq \mathbf{0}$ (in order to show that the result of this problem is true for other than trivial reasons).

▷ **4.** Use the adjoint to find the inverse, if it exists, of each matrix below.

(a) $\begin{bmatrix} 4 & 2 \\ -1 & 3 \end{bmatrix}$ (b) $\begin{bmatrix} 2 & 1 & 1 \\ -1 & 0 & 3 \\ 1 & 2 & 0 \end{bmatrix}$ (c) $\begin{bmatrix} 2 & -1 & 0 \\ -1 & 2 & -1 \\ 0 & -1 & 2 \end{bmatrix}$

5. Use the adjoint to find a formula for the inverse of the general nonsingular 2×2 matrix.

6. Give an example of a matrix $\mathbf{A} \neq \mathbf{0}$ for which $\mathbf{adj\,A} = \mathbf{0}$.

▷ **7.** Prove that $\det(\mathbf{adj\,A}) = (\det \mathbf{A})^{p-1}$ if $\mathbf{A}$ is $p \times p$.

8. Determinants can be evaluated much more efficiently by using Gauss elimination than by using the definition of the determinant.

(a) Show that the determinant of a nonsingular matrix equals plus or minus the product of the pivots used in Gauss elimination to produce a Gauss-reduced form from the matrix.

(b) Use the measures in Section 3.8 ($p^3/3 - p/3$ multiplication/divisions and $p^3/3 - p^2/2 + p/6$ addition/subtractions) for producing a Gauss-reduced form of a $p \times p$ matrix to determine the work involved in evaluating the determinant of a $p \times p$ matrix using the approach of (a).

▷ **9.** Determinants are, in a sense, inherently difficult to evaluate because their values can be very large or very small for a matrix of moderate-sized numbers.

(a) If $\mathbf{A}$ is 100×100—not a "large" matrix in modern applications—show that $\det 2\mathbf{A} = 2^{100} \det \mathbf{A}$; although $\mathbf{A}$ and $2\mathbf{A}$ have entries of comparable size, the values of their determinants differ enormously ($2^{100} \approx 10^{30}$).

(b) The values of determinants can also be extremely sensitive to changes in the coefficients; to see this, evaluate the determinants of the two matrices below and state how the size of the change in the determinant compares with the change in the matrix:

$$\text{change} \begin{bmatrix} 1 & 1 \\ 10^{10} - 1 & 10^{10} \end{bmatrix} \quad \text{to} \quad \begin{bmatrix} 1.01 & 1 \\ 10^{10} - 1 & 10^{10} \end{bmatrix}.$$

10. Besides the difficulties mentioned in Problem 9, determinants often cause computational difficulties because of numerical *overflow* or *underflow*. Recall from Section 3.6 that computers store numbers as a fraction part times 10 to a power and that there is a limit on the size of this exponent; on many microcomputers, for example, it must be between -308 and $+308$. Suppose that $\mathbf{A}$ is a 100×100 matrix in Gauss-reduced form, that 50 of its main-diagonal entries equal 10^7, and that the other 50 equal 10^{-7}.

(a) Show that $\det \mathbf{A} = 1$.

(b) Show that, if the *first* 50 entries are the 10^7s, then attempting to multiply the diagonal entries (in order) so as to obtain the determinant will produce overflow—an exponent too large for the computer to store.

11. Prove that the adjoint of a singular matrix is singular.

▷ **12.** Use Cramer's Rule to solve

$$\begin{aligned} 2x_1 - 4x_2 &= 8 \\ 3x_1 + x_2 &= 5. \end{aligned}$$

13. Use Cramer's Rule to find x_2 solving

$$\begin{aligned} 2x_1 - x_2 + 3x_3 &= 5 \\ 3x_1 + 2x_2 - x_3 &= 0 \\ x_1 + 4x_2 + x_3 &= 6. \end{aligned}$$

▷ **14.** Suppose that $\mathbf{x}_0$ solves $\mathbf{Ax} = \mathbf{b}$ for some nonsingular matrix $\mathbf{A}$. If $\mathbf{b}$ was measured from an experiment and is subject to errors of size ϵ in its ith entry, we want to know the effect on the accuracy of $\mathbf{x}$. To do this, suppose that $\mathbf{b}$ changes to $\mathbf{b}' = \mathbf{b} + \epsilon\mathbf{e}_i$.

(a) Show that $\mathbf{x}_0$ changes by $\epsilon\mathbf{A}^{-1}\mathbf{e}_i$.

(b) Use Cramer's Rule or the adjoint matrix to find an expression in terms of ϵ and determinants for the change in $\langle\mathbf{x}_0\rangle_j$.

15. Problem 14 examined how a solution changed when the right-hand side changed; consider how the solution changes when the coefficient matrix changes, such as when it is subject to measurement error. Suppose that $\mathbf{A}$ is nonsingular and that $\mathbf{x}_0$ solves $\mathbf{Ax} = \mathbf{b}$. Suppose that $\langle\mathbf{A}\rangle_{11}$ changes to $\langle\mathbf{A}\rangle_{11} + \epsilon$.

(a) Use Cramer's Rule to express how $\langle\mathbf{x}_0\rangle_1$ changes in terms of ϵ.

(b) Use Cramer's Rule to express how $\langle\mathbf{x}_0\rangle_2$ changes in terms of ϵ.

16. Prove Cramer's Rule.

4.7 *MISCELLANEOUS PROBLEMS*

PROBLEMS 4.7

1. Suppose that $\mathbf{B}$ is formed in the obvious way from the entries in certain selected rows and columns of $\mathbf{A}$. Prove that the rank of $\mathbf{B}$ is less than or equal to that of $\mathbf{A}$.

▷ **2.** If $\mathbf{R}$ has rank r and $\mathbf{S}$ has rank s, show that, if $\mathbf{A}$ has the form below, then $\mathbf{A}$ has rank $r + s$.

$$\mathbf{A} = \begin{bmatrix} \mathbf{R} & \mathbf{0} \\ \mathbf{0} & \mathbf{S} \end{bmatrix}$$

3. Prove that every matrix of rank k can be written as a sum of k matrices of rank 1.

4. Suppose that $\mathbf{A}$ is a $p \times p$ matrix of rank 1. Show that $\mathbf{A}^2 = t\mathbf{A}$ for some number t.

▷ **5.** Suppose that $\mathbf{A}$ is a $p \times p$ matrix of rank 1. Show that there is a $p \times 1$ column matrix $\mathbf{c}$ and a $1 \times p$ row matrix $\mathbf{r}$ such that $\mathbf{A} = \mathbf{cr}$.

𝔐 **6.** Rank is a difficult concept computationally because small changes in a matrix—measurement errors, say—can change its rank (by a whole number, of course). Consider the matrix

$$\mathbf{A}_\epsilon = \begin{bmatrix} 10^7 + \epsilon & 10^7 - \epsilon \\ 1 & 1 \end{bmatrix}.$$

(a) Show that the rank of $\mathbf{A}_\epsilon$ equals 2 for $\epsilon \neq 0$, but that the rank of $\mathbf{A}_0$ equals 1.

(b) Suppose that—as on many microcomputers—calculations are performed in 16-digit floating-point arithmetic (see Section 3.6). Show that, as long as ϵ in magnitude equals at least about 10^{-8}, the *calculated* version of $\mathbf{A}_\epsilon$ also has rank 2 just as does the true $\mathbf{A}_\epsilon$. Show that, for ϵ in magnitude less than about 10^{-10}, the *calculated* $\mathbf{A}_\epsilon$ has rank 1 rather than 2 as for the true $\mathbf{A}_\epsilon$ with nonzero ϵ.

(c) Computer programs for calculating the rank of the calculated $\mathbf{A}_\epsilon$ may well have trouble correctly computing the rank to be 2 for ϵ larger than 10^{-8}. Use MATLAB or similar software to experiment with computing the rank of the calculated $\mathbf{A}_\epsilon$ for a range of values of ϵ from 10^{-1} through 10^{-9}.

▷ **7.** Consider the *block-diagonal* matrix $\mathbf{A} = \mathbf{diag}(\mathbf{A}_{11} \cdots \mathbf{A}_{rr})$:

$$\mathbf{A} = \begin{bmatrix} \mathbf{A}_{11} & \mathbf{0} \cdots & \mathbf{0} \\ \mathbf{0} & \mathbf{A}_{22} \cdots & \mathbf{0} \\ & \cdots & \\ \mathbf{0} & \mathbf{0} \cdots & \mathbf{A}_{rr} \end{bmatrix},$$

where the diagonal blocks $\mathbf{A}_{ii}$ are not necessarily square. Prove that $\mathbf{A}$ has a right-inverse $\mathbf{R}$ if and only if all of the blocks $\mathbf{A}_{ii}$ have right-inverses $\mathbf{R}_{ii}$ and that $\mathbf{R}$ can be formed from such $\mathbf{R}_{ii}$ in the obvious way.

8. State and prove the results for left-inverses and for two-sided inverses analogous to that in Problem 7.

9. Suppose that $\mathbf{A}$ is $p \times q$ and that $\mathbf{Ax} = \mathbf{Ay}$ for $q \times 1$ column matrices $\mathbf{x}$ and $\mathbf{y}$ implies that $\mathbf{x} = \mathbf{y}$. Prove that this is equivalent to the condition that the rank k of $\mathbf{A}$ satisfies $k = q \le p$.

▷ **10.** Define d_p as the determinant of the $p \times p$ matrix $\mathbf{A}_p$ such that $\langle \mathbf{A}_p \rangle_{ii} = a_i$ for all i, $\langle \mathbf{A}_p \rangle_{i,i-1} = 1$ for all i, $\langle \mathbf{A}_p \rangle_{i,i+1} = 1$ for all i, and all other $\langle \mathbf{A}_p \rangle_{ij} = 0$.

(a) Show that, for $p \ge 3$,

$$d_p = a_p d_{p-1} - d_{p-2}.$$

(b) Suppose that $a_i = 2 \cos\theta$ for all i; show that

$$d_p = \csc\theta \sin(p+1)\theta \qquad \text{for all } p.$$

▷ **11.** Show that the *Vandermonde determinant* $V(x_1, x_2, \ldots, x_p)$ equals the product of all terms of the form $(x_j - x_i)$ for $i < j$, that is,

$$V(x_1, \ldots, x_p) = \det \begin{bmatrix} 1 & x_1 & x_1^2 & \cdots & x_1^{p-1} \\ 1 & x_2 & x_2^2 & \cdots & x_2^{p-1} \\ & & \cdots & & \\ 1 & x_p & x_p^2 & \cdots & x_p^{p-1} \end{bmatrix}$$

$$= \prod_{i<j} (x_j - x_i).$$

12. Show that $\det \mathbf{A}$ is real if $\mathbf{A}$ is hermitian—$\mathbf{A}^H = \mathbf{A}$.

13. Show that the area of the triangle with vertices (x_i, y_i), $i = 1, 2, 3$, numbered counterclockwise equals $(\det \mathbf{A})/2$ if

$$\mathbf{A} = \begin{bmatrix} x_1 & y_1 & 1 \\ x_2 & y_2 & 1 \\ x_3 & y_3 & 1 \end{bmatrix}.$$

▷ **14.** Find a representation like those in Problems 15 and 16 of Section 4.5 for the curve (an ellipse, parabola, hyperbola, or straight line) with equation

$$ax^2 + bxy + cy^2 + dx + ey + f = 0$$

that passes through five given points (x_i, y_i), $i = 1, 2, 3, 4, 5$.

15. Bodies travel about the sun in orbits that approximate ellipses. In certain units with respect to certain coordinate axes whose origin is at the sun, a body is observed to pass through the points (0.6, 2.0), (1.0, 2.3), (1.4, 2.4), (2.0, 2.5), and (2.4, 2.4). Use Problem 14 to represent the equation of the ellipse through these five points.

𝔐 **16.** Use MATLAB or similar software to evaluate the cofactors of the first row of the determinant in Problem 15 and thus find the equation of the ellipse in more convenient form; graph the ellipse by plotting several points.

𝔐 **17.** As a hint of the difficulty of creating software that can evaluate determinants with excellent *relative* accuracy, consider the matrix

$$\mathbf{B}_\epsilon = \begin{bmatrix} 10^7 + \epsilon & 10^7 - \epsilon \\ 1/\epsilon & 1/\epsilon \end{bmatrix}$$

for various nonzero values of ϵ.

(a) Show that $\det \mathbf{B}_\epsilon = 2$ for all $\epsilon \neq 0$.

(b) Ideally, we would like software to compute $\det \mathbf{B}_\epsilon$ with small *relative* error, that is, with error that is small compared to the true value 2; on microcomputers with 16-digit arithmetic, we would hope to get $\det \mathbf{B}_\epsilon = 2$ correct to several decimal places. Use MATLAB or similar software to see what sort of accuracy it produces for $\det \mathbf{B}_\epsilon$ for a range of values of ϵ from 10^{-1} through 10^{-9}. The difficulty here, as often happens when evaluating determinants, is that we are attempting to produce a modest-sized number from very large and very small numbers.

▷ **18.** Prove that $\mathbf{A}^T\mathbf{A}$ is nonsingular if $\mathbf{A}$ is real and $p \times q$ and has rank q.

5

Vectors and vector spaces

In this chapter we continue the development of fundamental theoretical concepts and practical tools needed to understand how linear algebra is applied. Unlike the preceding chapters, however, this one has a distinctly geometric flavor. As with row operations and Gauss elimination in Chapters 3 and 4, a single fundamental concept dominates this chapter: linear dependence. The number of ***Key Theorems*** *reflects the fundamental nature of the material:* ***5.12, 5.43, 5.45, 5.50, 5.57, 5.75,*** *and* ***5.82****, as well as the corollaries* ***5.29, 5.31, 5.39,*** *and* ***5.86****.*

5.1 *INTRODUCTION; GEOMETRICAL VECTORS*

The presentation has so far been entirely *algebraic*: matrices have been added and multiplied, equations have been solved, and so on. Nothing of a *geometric* nature has been presented: no angles measured, no lengths calculated, and the like. Yet there is a natural geometric approach to matrices that is at least as important both theoretically and practically as the algebraic approach; and, not only can geometry be applied to the study of matrices, but matrices can be used to study other geometric objects.

The notion of there being geometric content to matrices may well seem quite strange—how, after all, can rectangular arrays of numbers have anything to do with lines and angles? In this section we first indicate the natural geometric aspect of certain matrices. In order to demonstrate the reverse approach—using matrices to study geometric objects—we then briefly examine some geometric ideas and show how matrices can be used to represent them; this second viewpoint is important because it exemplifies the approach that will be followed later when we use column matrices to represent abstract entities called *vectors*.

The Natural Geometry of Real 2 × 1 Matrices

After 1 × 1 matrices, the simplest matrices surely are those with two real entries; to be specific, we consider the 2 × 1 real matrices $\mathbf{u} = [u_1 \quad u_2]^T$. There is a geometric concept—a *vector*—naturally associated with $\mathbf{u}$. Consider a plane with a standard x–y coordinate system on it, with $\mathcal{O}$ designating the origin. Draw a line from the origin $\mathcal{O}$ to the point $\mathcal{P}(u_1, u_2)$ having x-coordinate u_1 and y-coordinate u_2, with an arrowhead at $\mathcal{P}$ pointing in the direction from $\mathcal{O}$ toward $\mathcal{P}$. We call this the *geometrical vector* associated with $\mathbf{u}$, and we denote it by $\vec{u}$. See (5.1).

(5.1)

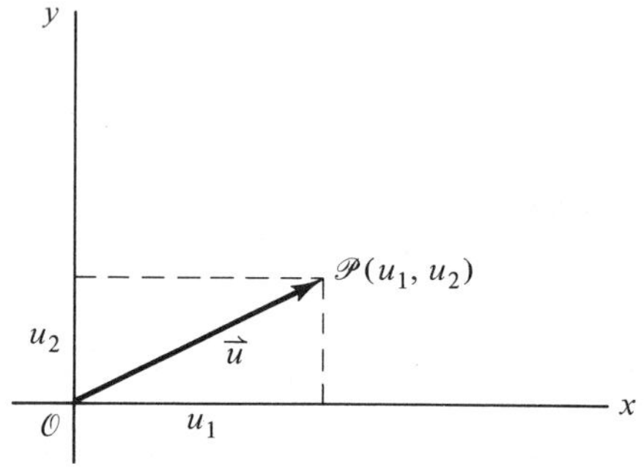

If this correspondence $\mathbf{u} \leftrightarrow \vec{u}$ between matrices and geometrical vectors is to be meaningful, our operations on matrices should have some geometric interpretation. Consider, for example, addition:

$$\mathbf{u} + \mathbf{v} = [u_1 \quad u_2]^T + [v_1 \quad v_2]^T = [u_1 + v_1 \quad u_2 + v_2]^T.$$

This has a simple geometric interpretation, as (5.2) reveals: To "add" the geometrical vectors $\vec{u}$ and $\vec{v}$ corresponding to $\mathbf{u}$ and $\mathbf{v}$, move a copy of $\vec{v}$ parallel to $\vec{v}$ so that its "tail" rests at the "head" of $\vec{u}$, and let the geometrical vector connecting $\mathcal{O}$ and the "head" of this shifted $\vec{v}$ be what we mean by $\vec{u} + \vec{v}$.

(5.2)

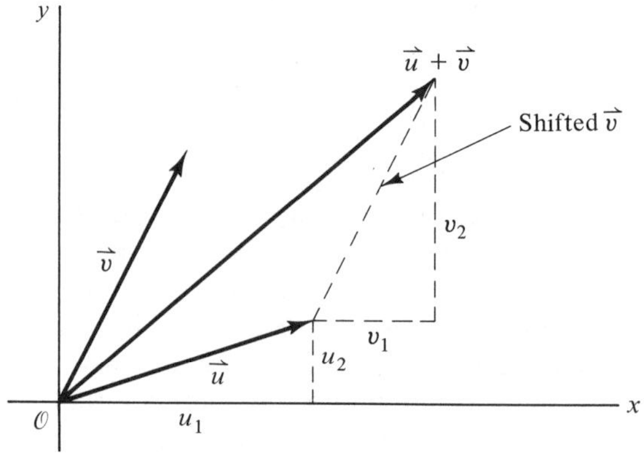

As (5.2) shows, this definition means that $\vec{u} + \vec{v}$ corresponds to $\mathbf{u} + \mathbf{v}$, so it certainly reflects the algebra of matrix addition. But it is also a geometrically natural definition for the addition of $\vec{u}$ and $\vec{v}$: If you think of geometrical vectors as representing displacements (that is, walking a certain distance in a certain direction), then the definition we have given means that $\vec{u} + \vec{v}$ represents following the displacement $\vec{u}$ by the displacement $\vec{v}$.

Other matrix operations also have a natural geometrical meaning, some of which are surprising. For example, the matrix product $\mathbf{u}^T\mathbf{v}$ is fundamentally related to *length* and *angle*. As can be seen from (5.1), the physical length of $\vec{u}$ can be calculated by the Pythagorean Theorem to be $(u_1^2 + u_2^2)^{1/2}$. Since $\mathbf{u}^T\mathbf{u} = [(u_1^2 + u_2^2)]$ we can (somewhat imprecisely) write the length of $\vec{u}$ as $(\mathbf{u}^T\mathbf{u})^{1/2}$. Also, the condition $\mathbf{u}^T\mathbf{v} = [0]$ means that $\vec{u}$ and $\vec{v}$ are perpendicular. To see this, note that the line through $\mathscr{O}$ and $\mathscr{P}(u_1, u_2)$ has slope u_2/u_1, and similarly for that through $\mathscr{O}$ and $\mathscr{P}(v_1, v_2)$; these lines will be perpendicular if the product $(u_2/u_1)(v_2/v_1)$ of their slopes equals -1, which easily gives $u_1v_1 + u_2v_2 = 0$. See (5.3).

(5.3)

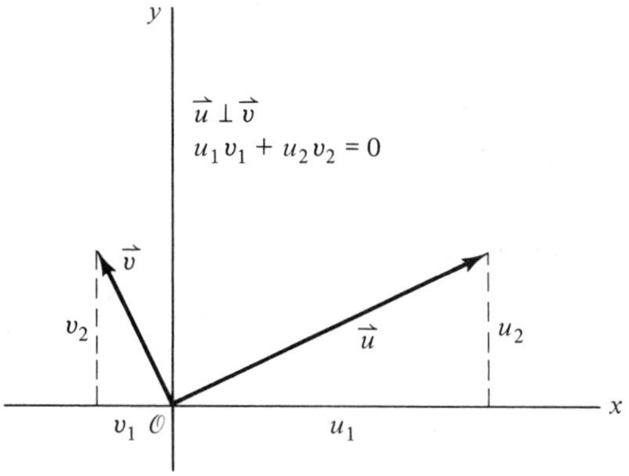

These few remarks have shown that there is natural geometric content to matrices and the operations we perform on them. We next take the reverse viewpoint: We introduce some geometric notions, and then use matrices to describe them—much as we will do more generally later in the chapter.

You should now be able to solve Problems 1 to 9.

Three-Dimensional Geometrical Vectors

The geometrical vectors we were just examining are easy to picture because they lie in a plane and can easily be drawn on graph paper. The physical world, however, is (at least) three-dimensional; we need to deal with geometrical vectors in this physical world.

Consider a point $\mathscr{O}$ fixed in physical space as a reference point. By a (three-dimensional) geometrical vector $\vec{u}$ we mean a (directed) line segment with its "tail" at $\mathscr{O}$ and its "head" at some point $\mathscr{P}$ in space; the direction is from $\mathscr{O}$ toward $\mathscr{P}$ and is usually indicated by an arrowhead in that direction at $\mathscr{P}$. See (5.4).

(5.4)

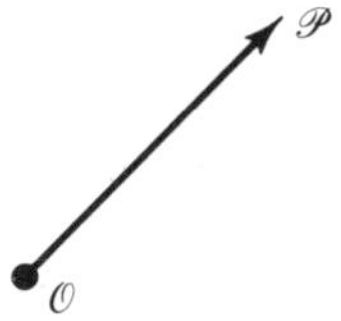

These geometrical vectors can be thought of as representing, for example, displacements of a certain distance in a certain direction from $\mathscr{O}$ to $\mathscr{P}$. When $\mathscr{P} = \mathscr{O}$, this geometrical zero vector is denoted by $\vec{0}$.

Two common operations on geometrical vectors are:

(5.5) 1. *Multiplication of a geometrical vector $\vec{u}$ by a scalar (a real number) k.* The result is denoted by $k\vec{u}$. If $k = 0$, then $k\vec{u}$ equals $\vec{0}$. If $k > 0$, then $k\vec{u}$ is in the same direction as $\vec{u}$ and its length equals k times that of $\vec{u}$. If $k < 0$, then $k\vec{u}$ is in the *opposite* direction from $\vec{u}$ and its length equals $|k|$ $(= -k)$ times that of $\vec{u}$.

(5.6) 2. *Addition of two geometrical vectors $\vec{u}$ and $\vec{v}$.* The result is denoted by $\vec{u} + \vec{v}$. We obtain $\vec{u} + \vec{v}$ just as we combine displacements: A copy of $\vec{v}$ is shifted parallel to $\vec{v}$ so that its "tail" rests at the "head" of $\vec{u}$, and then $\vec{u} + \vec{v}$ is that geometrical vector from $\mathscr{O}$ to the "head" of this shifted $\vec{v}$.

These operations on geometrical vectors have a number of important properties; for example:

$$\vec{u} + \vec{v} = \vec{v} + \vec{u} \qquad \text{(commutativity of addition)}$$

$$\vec{u} + \vec{0} = \vec{u} \qquad \text{(identity for addition)}$$

$$\vec{u} + (\vec{v} + \vec{w}) = (\vec{u} + \vec{v}) + \vec{w} \qquad \text{(associativity of addition)}$$

$$k(\vec{u} + \vec{v}) = k\vec{u} + k\vec{v} \qquad \text{(distributivity)}$$

and so on. The operations also allow us to describe some deeper concepts.

We have called our physical space "*three*-dimensional"; this is essentially because it requires *three* numbers (coordinates) to specify the location of any point in space with respect to some reference point. Similarly, a line is *one*-dimensional because *one* number is required, while a plane is *two*-dimensional because *two* numbers are required. (Later in the chapter there is a precise definition of dimension that formalizes this intuitive concept.)

If we consider all possible multiples $\alpha\vec{u}$ of a geometrical vector $\vec{u}$ by arbitrary scalars α, we have a line—a one-dimensional *subspace*—in three-dimensional space. Similarly, if $\vec{u}$ and $\vec{v}$ are not in the same (or opposite) direction, then all geometrical vectors of the form $\alpha\vec{u} + \beta\vec{v}$ for arbitrary scalars α and β form a plane—a two-dimensional *subspace*—in three-dimensional space. Thus our operations on geometrical vectors allow us to describe and discuss the deeper concepts of lines and planes. Geometrical vectors $\vec{x}$ in the plane just defined are of the form $\vec{x} = \alpha\vec{u} + \beta\vec{v}$, and we say that $\vec{x}$ is a *linear combination* of $\vec{u}$ and $\vec{v}$ and also that $\vec{x}$ is *linearly dependent on* $\vec{u}$ and $\vec{v}$.

Similarly, we say that one geometrical vector $\vec{x}$ is linearly dependent on another $\vec{y}$ when $\vec{x} = \alpha\vec{y}$ for some scalar α. We say that the set $\{\vec{x}, \vec{y}\}$ is a *linearly dependent set* when either $\vec{x}$ is linearly dependent on $\vec{y}$ or $\vec{y}$ is linearly dependent on $\vec{x}$; this simply means that $\vec{x}$ and $\vec{y}$ point in the same or opposite direction (or one of them equals $\vec{0}$). This terminology is useful in discussing lines and planes formed by all geometrical vectors of the form $\alpha\vec{u} + \beta\vec{v}$: All such geometrical vectors form a line if $\{\vec{u}, \vec{v}\}$ is linearly dependent but form a plane if $\{\vec{u}, \vec{v}\}$ is linearly independent (that is, not linearly dependent).

We will not pursue further the development of three-dimensional geometry; the ideas introduced should be kept in mind as motivation and examples for the concepts in the following sections. Before proceeding, however, we indicate how matrices can be used to study the geometric concepts we have been presenting.

Choose a system of three mutually perpendicular coordinate axes with their origin at our reference point $\mathcal{O}$; call these the x, y, and z axes. Then any point $\mathcal{P}$ in space can be specified by three coordinates, giving $\mathcal{P}(x, y, z)$. See (5.7).

(5.7)

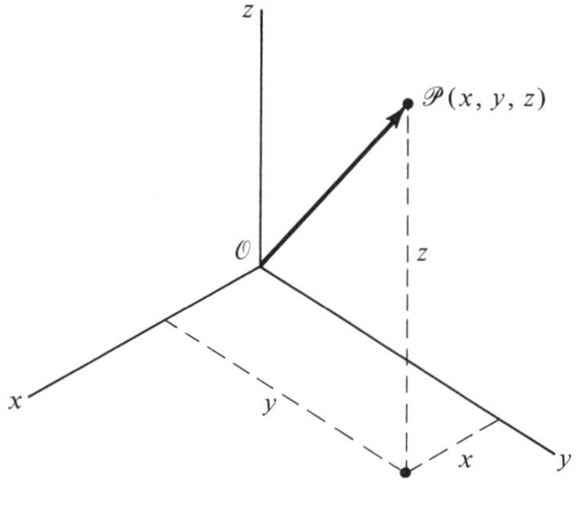

Any geometrical vector $\vec{u}$ is determined by the point $\mathcal{P}$ at its "head"; since $\mathcal{P}$ is determined by the three coordinates that specify it, we can say that $\vec{u}$ is described by those three coordinates x, y, z. If we place those three coordinates in a 3×1 column matrix $\mathbf{u} = [x \quad y \quad z]^T$, then the matrix $\mathbf{u}$ represents the geometrical vector $\vec{u}$. As we found at the start of this section for two-dimensional geometrical vectors, if **u** and **v** correspond to $\vec{u}$ and $\vec{v}$, then $\alpha\mathbf{u}$ corresponds to $\alpha\vec{u}$ and $\mathbf{u} + \mathbf{v}$ to $\vec{u} + \vec{v}$. This means that matrix expressions of the form $\alpha\mathbf{u} + \beta\mathbf{v}$ can be used to describe lines and planes, for example. By similar correspondences, questions about whether planes are parallel can be posed in terms of whether there exist solutions of systems $\mathbf{Ax} = \mathbf{b}$ of equations. Thus all the powerful matrix tools we have developed in the preceding chapters can be brought to bear on geometrical problems. This will be equally true for the geometric problems we encounter in subsequent sections as we develop the abstract—but extremely applicable—notions of general vectors and vector spaces.

PROBLEMS 5.1

1. For the geometrical vectors $\vec{u}$, $\vec{v}$, and $\vec{w}$ shown below, geometrically find:
(a) $\vec{u} + \vec{v}$ (b) $(\vec{u} + \vec{v}) + \vec{w}$
(c) $\vec{u} + (\vec{v} + \vec{w})$ (d) $\vec{u} + \vec{u} + \vec{u}$

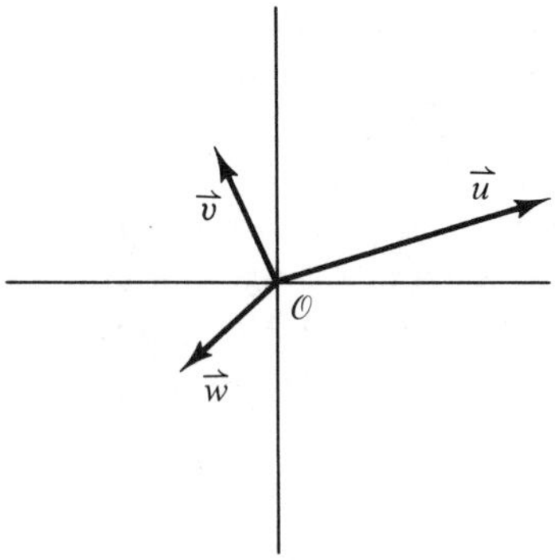

2. Let $\vec{u}$, $\vec{v}$, and $\vec{w}$ correspond to $[1 \quad 2]^T$, $[-1 \quad 3]^T$, and $[0 \quad 0]^T$, respectively. Find the 2×1 matrices corresponding to:
(a) $\vec{u} + \vec{v}$ (b) $\vec{u} + \vec{w}$
(c) $\vec{v} + \vec{w}$ (d) $\vec{u} + \vec{u} + \vec{u}$
3. Show geometrically that if $\vec{u}$, $\vec{v}$, and $\vec{w}$ are two-dimensional geometrical vectors, $\vec{u} + (\vec{v} + \vec{w}) = (\vec{u} + \vec{v}) + \vec{w}$.
4. ▷ A motorboat is pointing straight across a river, but is being swept along by the current. Without the current, the velocity of the boat would be 5 mph directly across the river; without the motor, the boat would be pushed by the river with a velocity of 3 mph downstream. Find the velocity of the powered boat crossing the river. (Assume that velocities may be represented by geometrical vectors).
5. Assume that you have a motorboat and river as in Problem 4. Find the angle at which the boat should aim upstream into the current so that the resulting actual motion of the boat will be straight across the river.

6. The *airspeed* of an airplane is its speed in *still* air, while the *groundspeed* is its actual speed that results from the combined effects of airspeed, the airplane's heading, and the wind velocity. Suppose that an airplane's airspeed is 140 mph. If the airplane is aiming due west and the wind is blowing due north at 20 mph, find the resulting groundspeed and direction of motion of the airplane.

▷ **7.** An airplane made three flights at full speed in three different directions. Ground observers noted that (1) it moved due east at a groundspeed (see Problem 6) of 160 mph; (2) it moved due west at a groundspeed of 120 mph; and (3) it moved due north at a groundspeed of 160 mph. Assuming that the wind's speed and direction were the same in all three cases, find the airspeed of the airplane as well as the direction and speed of the wind.

8. It seems reasonable for $2\vec{u}$ to mean $\vec{u} + \vec{u}$ for $\vec{u}$ a geometrical vector. Give a geometrical description of how to generate $2\vec{u}$ from $\vec{u}$.

▷ **9.** If $\vec{u}$ is a two-dimensional geometrical vector and corresponds to the 2×1 matrix $\mathbf{u}$, we could define $\alpha\vec{u}$ for real α as the geometrical vector corresponding to $\alpha\mathbf{u}$. Give a geometrical description of how to generate $\alpha\vec{u}$ from $\vec{u}$.

10. Use geometry to show that $\vec{u} + \vec{v} = \vec{v} + \vec{u}$ for all three-dimensional geometrical vectors $\vec{u}$, $\vec{v}$.

11. Suppose that $\vec{u}$ and $\vec{v}$ are collinear geometrical vectors in three dimensions (that is, they lie in the same straight line through the origin). Show that the set of all geometrical vectors of the form $\alpha\vec{u} + \beta\vec{v}$ as α and β range over all real numbers is a straight line.

12. Suppose that $\vec{u}$ and $\vec{v}$ are the geometrical vectors corresponding to

$$\mathbf{u} = [4 \quad -2 \quad -1]^T \quad \text{and} \quad \mathbf{v} = [-4 \quad 0 \quad 3]^T.$$

(a) Find the coordinates $\mathbf{w}$ of all vectors $\vec{w}$ of the form $\vec{w} = \alpha\vec{u} + \beta\vec{v}$ as α and β range over all real numbers.
(b) Find an equation of the form $ax + by + cz = d$ satisfied by the entries x, y, z in $\mathbf{w} = [x \quad y \quad z]^T$, where $\mathbf{w}$ corresponds to $\vec{w}$ as in (a).

13. Let $\vec{0}$ be the geometrical vector corresponding to $[0 \quad 0 \quad 0]^T$. Show that $\vec{u} + \vec{0} = \vec{u}$ for all geometrical vectors $\vec{u}$.

5.2 *GENERAL VECTOR SPACES*

The preceding section described (geometrical) vectors as geometric entities in physical space. The present section examines what really is at the heart of the concept of vectors and of spaces—like lines or planes—of vectors. To do this, we define an abstract model of a vector space and indicate how this abstract notion can be used to develop concepts and properties that are valid in all concrete instances of vectors spaces. From our present introductory viewpoint, the basic property of vectors is that we can form linear combinations of them—like $\alpha\vec{u} + \beta\vec{v}$ in the preceding section. There are important applications in which we need to form similar linear combinations of entities other than geometrical vectors $\vec{u}$; our approach will discover the properties in common to *all* such situations.

Vectors

We define a vector space in such a way as to reflect all the properties that appear essential and common to all the important special cases.

(5.8) ***Definition.*** A *vector space* is a nonempty set $\mathscr{V}$ of entities called *vectors* that, together with its associated numbers called *scalars*, satisfies the conditions below. A *real vector space* is a vector space in which the scalars comprise the set $\mathbb{R}$ of real numbers, and a *complex vector space* is a vector space in which the scalars comprise the set $\mathbb{C}$ of complex numbers.

1. There is an operation called *vector addition* that associates with each pair of vectors $\mathbf{x}$, $\mathbf{y}$ another vector, called their *sum* and denoted by $\mathbf{x} + \mathbf{y}$, and satisfying:
   (a) $\mathbf{x} + \mathbf{y} = \mathbf{y} + \mathbf{x}$ for all vectors $\mathbf{x}$, $\mathbf{y}$.
   (b) $\mathbf{x} + (\mathbf{y} + \mathbf{z}) = (\mathbf{x} + \mathbf{y}) + \mathbf{z}$ for all vectors $\mathbf{x}$, $\mathbf{y}$, $\mathbf{z}$.
   (c) There exists in $\mathscr{V}$ a unique vector, called zero and denoted by $\mathbf{0}$, such that $\mathbf{x} + \mathbf{0} = \mathbf{x}$ and $\mathbf{0} + \mathbf{x} = \mathbf{x}$ for all vectors $\mathbf{x}$.
   (d) For each vector $\mathbf{x}$ there is a unique vector in $\mathscr{V}$, called its negation and denoted by $-\mathbf{x}$, such that $\mathbf{x} + (-\mathbf{x}) = \mathbf{0}$ and $(-\mathbf{x}) + \mathbf{x} = \mathbf{0}$.
2. There is an operation called *multiplication by a scalar* that associates with each scalar α and each vector $\mathbf{x}$ in $\mathscr{V}$ a unique vector, called the *product* of α and $\mathbf{x}$ and denoted by $\alpha\mathbf{x}$ and $\mathbf{x}\alpha$, and satisfying:
   (a) $\alpha(\beta\mathbf{x}) = (\alpha\beta)\mathbf{x}$ for all scalars α, β, and all vectors $\mathbf{x}$.
   (b) $(\alpha + \beta)\mathbf{x} = \alpha\mathbf{x} + \beta\mathbf{x}$ for all scalars α, β and all vectors $\mathbf{x}$.
   (c) $\alpha(\mathbf{x} + \mathbf{y}) = \alpha\mathbf{x} + \alpha\mathbf{y}$ for all scalars α and all vectors $\mathbf{x}$, $\mathbf{y}$.
3. $1\mathbf{x} = \mathbf{x}$ for all vectors $\mathbf{x}$.

Some remarks are in order about Definition 5.8.

1. There is much redundancy in the definition: The second half of (1)(c) follows from the first half and from (1)(a), for example. We prefer to state such implications explicitly, at the expense of some mathematical elegance.
2. We have used familiar symbols (+, −, and 0) in new ways; we have even used + in more than one way in the same equation! Note in (2)(b), for example, that the "+" between α and β refers to addition as it is defined between *scalars*, while the "+" that appears between $\alpha\mathbf{x}$ and $\beta\mathbf{x}$ refers to addition as it is defined between *vectors* in $\mathscr{V}$. We prefer some possible confusion to a complicated proliferation of different symbols for +; the meaning should always be clear from the context. As the following examples show, the definition of + between vectors in applications is often so close to our usual concept of + between numbers that we hardly notice the difference. As with matrices, we define $\mathbf{x} - \mathbf{y} = \mathbf{x} + (-\mathbf{y})$.

(5.9) ***Example.*** Each of the following examples is a vector space with the appropriate set $\mathbb{R}$ or $\mathbb{C}$ of scalars, with the appropriate definitions of vector addition and of multiplication by a scalar in each case.

(a) $\mathscr{G}^3$, the real vector space of all geometrical vectors in three-dimensional physical space with vector addition and multiplication by a scalar as defined in the preceding section.

(b) $\mathbb{R}^p$, the real vector space of all real $p \times 1$ column matrices with vector addition being matrix addition, and multiplication by a scalar being as usual for matrices.

(c) $\mathbb{C}^p$, the complex vector space of all complex $p \times 1$ column matrices with the operations as in $\mathbb{R}^p$.

(d) $\mathscr{S}$, the real vector space of all real infinite sequences

$$\{x_i\} = x_1, x_2, \ldots,$$

with $\{x_i\} + \{y_i\}$ defined as $\{x_i + y_i\}$ and similarly for $r\{x_i\}$. This is not a particularly useful vector space unless restrictions are placed on the behavior of x_i as i tends to infinity.

(e) $C^{(k)}[a, b]$, the real vector space of all real-valued functions that are continuous, together with their first k derivatives, on the closed interval $[a, b]$. When $k = 0$, this space of continuous functions is often denoted by $C[a, b]$. Vector addition and multiplication by a scalar are defined in the obvious ways: If $\mathbf{x}$ and $\mathbf{y}$ are vectors (that is, functions) in $C^{(k)}[a, b]$, then $\mathbf{x} + \mathbf{y}$ is the function with

$$(\mathbf{x} + \mathbf{y})(t) = \mathbf{x}(t) + \mathbf{y}(t) \qquad \text{for } t \text{ in } [a, b].$$

(f) $\mathscr{P}$, the real vector space of all polynomials of arbitrary degree and having real coefficients. Operations are as defined in $C[a, b]$.

(g) $\mathscr{P}^k$, the real vector space of polynomials of degree less than or equal to $k - 1$ and having real coefficients. Operations are as defined in $\mathscr{P}$.

(h) The real vector space of all those functions $y(x)$ in $C^{(n)}[a, b]$ that solve the linear homogeneous differential equation

$$a_0(x)\frac{d^n y}{dx^n} + a_1(x)\frac{d^{n-1}y}{dx^{n-1}} + \cdots + a_n(x)y = 0.$$

Operations are as defined in $C[a, b]$.

(i) The complex vector space of all complex-valued continuous functions $y(x)$ that solve the integral equation.

$$\int_0^a K(x, t)y(t)\,dt + \lambda y(x) = 0, \qquad 0 \le x \le a,$$

where $K(x, t)$ is continuous and K and λ are given. Operations are the usual with functions.

(j) Suppose that $\mathscr{V}$ and $\mathscr{W}$ are both real or both complex vector spaces. The *product space* $\mathscr{V} \times \mathscr{W}$ is the vector space of ordered pairs $(\mathbf{v}, \mathbf{w})$ with

$\mathbf{v}$ in $\mathscr{V}$ and $\mathbf{w}$ in $\mathscr{W}$, where

$$(\mathbf{v}, \mathbf{w}) + (\mathbf{v}', \mathbf{w}') = (\mathbf{v} + \mathbf{v}', \mathbf{w} + \mathbf{w}') \quad \text{and} \quad \alpha(\mathbf{v}, \mathbf{w}) = (\alpha\mathbf{v}, \alpha\mathbf{w}),$$

using the same scalars as for $\mathscr{V}$.

There are vastly many more examples that are important in applications; some are introduced in the Problems.

From the basic properties of vector spaces follow some simple properties.

(5.10) ***Theorem***

(a) $0\mathbf{v} = \mathbf{0}$ and $\mathbf{r0} = \mathbf{0}$

(b) $(-1)\mathbf{v} = -\mathbf{v}$

PROOF. These statements may seem obvious, but the problem is to prove them using nothing but the defined properties of vector spaces.

(a) Note first that $0\mathbf{v} + 0\mathbf{v} = 0\mathbf{v}$, since $0\mathbf{v} + 0\mathbf{v} = (0 + 0)\mathbf{v} = 0\mathbf{v}$. Adding $-(0\mathbf{v})$ to both sides of the equation then gives $0\mathbf{v} = \mathbf{0}$ as claimed. Similarly, $r\mathbf{0} + r\mathbf{0} = (2r)\mathbf{0} = r(2\mathbf{0}) = r(\mathbf{0} + \mathbf{0}) = r\mathbf{0}$; adding $-(r\mathbf{0})$ to both sides gives $r\mathbf{0} = \mathbf{0}$.

(b) Note that $\mathbf{v} + (-1)\mathbf{v} = \mathbf{0}$, since

$$\mathbf{v} + (-1)\mathbf{v} = 1\mathbf{v} + (-1)\mathbf{v} = \{1 + (-1)\}\mathbf{v} = 0\mathbf{v} = \mathbf{0}$$

by (a). Adding $-\mathbf{v}$ to both sides then gives $(-1)\mathbf{v} = -\mathbf{v}$, as claimed. ■

Subspaces

It is obvious that many of the vector spaces in Example 5.9 are closely related. For example, every polynomial of degree at most $k - 1$ {see (g)} is also a polynomial of arbitrary degree {see (f)}, which is also a 423-times differentiable function {see (e)}, which is also a continuous function {see (e)}; *and the operations in each one of these vector spaces are the same*—simple addition of functions is vector addition, for example. This situation is very common and very important to our study of vector spaces.

(5.11) ***Definition.*** Suppose that $\mathscr{V}_0$ and $\mathscr{V}$ are both real or both complex vector spaces, that $\mathscr{V}_0$ is a ***subset*** of $\mathscr{V}$ (that is, each element of $\mathscr{V}_0$ is also an element of $\mathscr{V}$), and that the operations on elements of $\mathscr{V}_0$ as $\mathscr{V}_0$-vectors are the same as the operations on them as $\mathscr{V}$-vectors. Then $\mathscr{V}_0$ is said to be a ***subspace*** of $\mathscr{V}$.

Note that a sub*space* is much more than just a sub*set*; a subspace is a subset that is a vector space in its own right when the *same* operations are used as in the "parent" vector space.

Suppose that $\mathscr{V}$ is a vector space, that $\mathscr{V}_0$ is some sub*set* of $\mathscr{V}$, and that we want to know whether $\mathscr{V}_0$ is a sub*space* of $\mathscr{V}$. According to Definition 5.11, we have to check that the operations in $\mathscr{V}_0$ are the same as those in $\mathscr{V}$—no great problem—and then we have to check that $\mathscr{V}_0$ is itself a vector space in its own right by checking all the many conditions of Definition 5.8 on vector spaces. We will be presented with this situation so often in the future that it is quite useful to have a much simpler test to perform in order to check $\mathscr{V}_0$.

(5.12) **Key Theorem** (subspace theorem). Suppose that $\mathscr{V}$ is a vector space and that $\mathscr{V}_0$ is a subset of $\mathscr{V}$; define vector addition and multiplication by scalars for elements of $\mathscr{V}_0$ exactly as in $\mathscr{V}$—add, or multiply by scalars by thinking of elements of $\mathscr{V}_0$ as elements of $\mathscr{V}$. Then $\mathscr{V}_0$ is a subspace of $\mathscr{V}$ if and only if the following three conditions hold:

1. $\mathscr{V}_0$ is nonempty.
2. $\mathscr{V}_0$ is *closed under multiplication by scalars* in the sense that $\alpha\mathbf{v}_0$ is in $\mathscr{V}_0$ for all $\mathbf{v}_0$ in $\mathscr{V}_0$ and all scalars α.
3. $\mathscr{V}_0$ is *closed under vector addition* in the sense that $\mathbf{v}_0 + \mathbf{v}_0'$ is in $\mathscr{V}_0$ for all vectors $\mathbf{v}_0$ in $\mathscr{V}_0$ and $\mathbf{v}_0'$ in $\mathscr{V}_0$.

PROOF. (subspace $\Rightarrow$ three conditions) Suppose first that $\mathscr{V}_0$ is a subspace of $\mathscr{V}$. This means in particular that $\mathscr{V}_0$ is a vector space; the definition of vector spaces immediately shows that the three conditions here are satisfied.

(three conditions $\Rightarrow$ subspace) Suppose now that the three conditions above are satisfied; we have to show that $\mathscr{V}_0$ satisfies the conditions of Definition 5.11 for it to be a subspace of $\mathscr{V}$. By assumption, it is certainly a subset and certainly uses the same operations as in $\mathscr{V}$. All that remains is to verify that $\mathscr{V}_0$ is a vector space—that is, that $\mathscr{V}_0$ satisfies the many conditions of Definition 5.8. By condition (1), there is at least one $\mathbf{v}_0$ in $\mathscr{V}_0$; by condition (2), this means that $0\mathbf{v}_0$ is also in $\mathscr{V}_0$. But $0\mathbf{v}_0 = \mathbf{0}$, according to Theorem 5.10; thus $\mathbf{0}$ is in $\mathscr{V}_0$ and is unique as required for $\mathscr{V}_0$ to be a vector space (there cannot be a different zero, say $\mathbf{0}'$, in $\mathscr{V}_0$ because then $\mathbf{0} = \mathbf{0} + \mathbf{0}' = \mathbf{0}'$ and the two are really the same). For *any* $\mathbf{v}_0$ in $\mathscr{V}_0$, by condition (2), $(-1)\mathbf{v}_0$ is in $\mathscr{V}_0$. But $(-1)\mathbf{v}_0 = -\mathbf{v}_0$, according to Theorem 5.10; thus $-\mathbf{v}_0$ is in $\mathscr{V}_0$ and is unique (why?) as required for $\mathscr{V}_0$ to be a vector space. Now, all of the remaining requirements are simply properties of vector addition and scalar multiplication, the validity of which $\mathscr{V}_0$ inherits from their validity in $\mathscr{V}$ since the operations are the same. Thus $\mathscr{V}_0$ is a vector space and hence a subspace of $\mathscr{V}$. ■

(5.13) ***Example.*** Suppose that $\{\mathbf{v}_1, \mathbf{v}_2, \ldots, \mathbf{v}_r\}$ is some nonempty set of vectors from $\mathscr{V}$. Define $\mathscr{V}_0$ as the set of all linear combinations

$$\mathbf{v}_0 = \alpha_1\mathbf{v}_1 + \alpha_2\mathbf{v}_2 + \cdots + \alpha_r\mathbf{v}_r$$

where the scalars α_i are allowed to range over all arbitrary values. Then $\mathscr{V}_0$ is a subspace of $\mathscr{V}$. To demonstrate this, we use **Key Theorem 5.12**.

(1) Certainly, $\mathscr{V}_0$ is nonempty since

$$\mathbf{v}_1 = 1\mathbf{v}_1 + 0\mathbf{v}_2 + \cdots + 0\mathbf{v}_r$$

is in $\mathscr{V}_0$. (2) For any $\mathbf{v}_0$ in $\mathscr{V}_0$ and scalar α we have

$$\alpha\mathbf{v}_0 = \alpha(\alpha_1\mathbf{v}_1 + \alpha_2\mathbf{v}_2 + \cdots + \alpha_r\mathbf{v}_r) = (\alpha\alpha_1)\mathbf{v}_1 + (\alpha\alpha_2)\mathbf{v}_2 + \cdots + (\alpha\alpha_r)\mathbf{v}_r,$$

which is in $\mathscr{V}_0$ by the definition of $\mathscr{V}_0$. (3) is left for Problem 15.

(5.14) ***Example.*** Suppose that $\mathbf{A}$ is a fixed real $p \times q$ matrix and define $\mathscr{V}_0$ to be the set of all real $q \times 1$ column matrices $\mathbf{x}$ that satisfy $\mathbf{Ax} = \mathbf{0}$. Then $\mathscr{V}_0$ is a real vector space and is a subspace of $\mathbb{R}^q$. To demonstrate this, we use **Key Theorem 5.12**, the Subspace Theorem. (1) $\mathscr{V}_0$ is nonempty since $\mathbf{0}$ solves $\mathbf{A0} = \mathbf{0}$, so the matrix $\mathbf{0}$ is in $\mathscr{V}_0$. (2) If $\mathbf{v}_0$ is in $\mathscr{V}_0$, then $\mathbf{Av}_0 = \mathbf{0}$; therefore, for all real α we have

$$\mathbf{A}(\alpha\mathbf{v}_0) = \alpha(\mathbf{Av}_0) = \alpha\mathbf{0} = \mathbf{0},$$

so $\alpha\mathbf{v}_0$ is in $\mathscr{V}_0$. (3) If $\mathbf{v}_0$ and $\mathbf{v}'_0$ are in $\mathscr{V}_0$, then $\mathbf{Av}_0 = \mathbf{0}$ and $\mathbf{Av}'_0 = \mathbf{0}$; therefore,

$$\mathbf{A}(\mathbf{v}_0 + \mathbf{v}'_0) = \mathbf{Av}_0 + \mathbf{Av}'_0 = \mathbf{0} + \mathbf{0} = \mathbf{0},$$

so $\mathbf{v}_0 + \mathbf{v}'_0$ is in $\mathscr{V}_0$. Thus all three conditions are satisfied and $\mathscr{V}_0$ is a subspace of $\mathbb{R}^q$, so it certainly is a real vector space.

Note. In this book we will be concerned primarily with (subspaces of) the vector spaces $\mathbb{R}^p$ and $\mathbb{C}^p$ in Example 5.9(b) and (c). It is essential to be familiar with these vector spaces and with how to determine whether a subset of one is in fact a subspace.

Because we make so much use of $\mathbb{R}^p$ and $\mathbb{C}^p$, we restate their definitions.

(5.15) ***Definition.*** $\mathbb{R}^p$ (or $\mathbb{C}^p$) denotes the real (or complex) vector space of real (or complex) $p \times 1$ column matrices, with vector addition and multiplication by scalars being as for matrices.

PROBLEMS 5.2

1. Show that the spaces $\mathbb{R}^p$ and $\mathbb{C}^p$ of Example 5.9(b) and (c) are vector spaces.
2. Show that the space $\mathscr{P}$ of Example 5.9(f) is a vector space.

▷ 3. Show that the set $\mathscr{V} = \{0\}$ is a vector space—either real or complex.

4. Show that the space $\mathscr{V} \times \mathscr{W}$ of Example 5.9(j) is a vector space.

▷ 5. For each set below, determine whether it is a real vector space using the natural operations; if it isn't, explain why.
   (a) Two-dimensional geometrical vectors whose "heads" lie in the first quadrant.

(b) Ratios $P_m(x)/Q_n(x)$—with Q_n not the zero polynomial—of polynomials P_m of degree at most m and Q_n of degree at most n, where m and n are allowed to range over all nonnegative integers.

(c) As in (b), except that $m \leq M$ and $n \leq N$ for fixed M and N.

6. For each of the vector spaces in Example 5.9 for which it has not been done, define:

(a) Vector addition
(b) Multiplication by a scalar
(c) The zero vector
(d) $-\mathbf{v}$ for vectors $\mathbf{v}$

▷ **7.** Suppose that $\mathbf{A}$ is a real $p \times q$ matrix, that $\mathbf{b}$ is a real $p \times 1$ matrix with $\mathbf{b} \neq \mathbf{0}$, and that $\mathscr{V}_0$ is the set of $\mathbf{x}$ satisfying $\mathbf{Ax} = \mathbf{b}$. Show that $\mathscr{V}_0$ is not a subspace of $\mathbb{R}^q$.

8. Suppose that $\mathscr{V}_0$ is the set of all real 3×1 matrices $\mathbf{v}$ for which $\langle \mathbf{v} \rangle_1 = 0$. Show that $\mathscr{V}_0$ is a subspace of $\mathbb{R}^3$.

9. Define $\mathscr{V} = \mathbb{R}^{p \times q}$ (or $\mathbb{C}^{p \times q}$) as the set of all real (or complex) $p \times q$ matrices. Using the usual arithmetic of matrices, prove that $\mathscr{V}$ is a real (or complex) vector space.

▷ **10.** Explain why a subset of a vector space must be nonempty in order for it to be a subspace.

11. Show that the set $\mathscr{P}^k$ is a subspace of $\mathscr{P}$, where $\mathscr{P}^k$ and $\mathscr{P}$ are as defined in Example 5.9.

12. Prove that $\mathscr{V}_0$ defined as $\mathscr{V}_0 = \{\mathbf{0}\}$ is a subspace of $\mathscr{V}$.

13. Suppose that $\mathscr{V}$ and $\mathscr{W}$ are vector spaces and that $\mathscr{V} \times \mathscr{W}$ is as defined in Example 5.9. Let $\mathscr{V}'$ be the set of all vectors in $\mathscr{V} \times \mathscr{W}$ of the form $(\mathbf{v}, \mathbf{0})$, where $\mathbf{v}$ is in $\mathscr{V}$ and $\mathbf{0}$ is the zero vector of $\mathscr{W}$. Prove that $\mathscr{V}'$ is a subspace of $\mathscr{V} \times \mathscr{W}$.

▷ **14.** Section 2.2 showed that in applications that model the evolution of systems it is often important to solve $\mathbf{Ax} = \mathbf{x}$, where $\mathbf{A}$ is a given real $q \times q$ matrix. Show that the set of all such $\mathbf{x}$ is a subspace of $\mathbb{R}^q$.

15. Finish the proof in Example 5.13.

5.3 LINEAR DEPENDENCE AND LINEAR INDEPENDENCE

In applications we need to be able to calculate with vectors; this requires some sort of concrete representation of the vectors. For example, the subspace $\mathscr{V}$ of $C[0, 1]$ consisting of all functions f that satisfy

$$f(x + 3y) - 3f(x + 2y) + 3f(x + y) - f(x) = 0$$

identically in x and y is a vector space, but not one that we can easily manipulate with this definition; calculations in $\mathscr{V}$ become easy when we learn (and this is nontrivial) that every element of $\mathscr{V}$ can be written as $c_1 + c_2t + c_3t^2$ for some coefficients c_i—we can then, for example, add two vectors of $\mathscr{V}$ simply by adding their coefficients. For this reason we will find it important to be able to express the general element of a vector space as a combination (like $c_1 + c_2t + c_3t^2$) of special vectors (like 1, t, and t^2) in that space.

Spanning Sets

Some terminology is helpful in discussing the idea in the preceding sentence.

(5.16) ***Definition***

(a) A *linear combination* of the vectors $\mathbf{v}_1, \mathbf{v}_2, \ldots, \mathbf{v}_n$ is an expression of the form $\alpha_1\mathbf{v}_1 + \alpha_2\mathbf{v}_2 + \cdots + \alpha_n\mathbf{v}_n$, where the α_i are scalars.

(b) A vector $\mathbf{v}$ is said to be *linearly dependent* ***on*** the (set of) vectors $\mathbf{v}_1, \mathbf{v}_2, \ldots, \mathbf{v}_n$ if and only if $\mathbf{v}$ can be written as some linear combination of $\mathbf{v}_1, \mathbf{v}_2, \ldots, \mathbf{v}_n$; otherwise, $\mathbf{v}$ is said to be *linearly* ***independent of*** the (set of) vectors.

(c) A set S of vectors $\mathbf{v}_1, \mathbf{v}_2, \ldots, \mathbf{v}_n$ in $\mathscr{V}$ is said to *span* (or *generate*) some subspace $\mathscr{V}_0$ of $\mathscr{V}$ if and only if S is a subset of $\mathscr{V}_0$ and *every* vector $\mathbf{v}_0$ in $\mathscr{V}_0$ is linearly dependent on S; S is said to be a *spanning set* or *generating set* for $\mathscr{V}_0$.

In this language, then, the set $\{1, t, t^2\}$ spans the vector space $\mathscr{P}^3$ of polynomials of degree strictly less than 3; so also do $\{1, 3t, 6t^2\}$, $\{2, 3 + t, 2 - t^2\}$, $\{1, 2 + 2t, 1 - t + t^2, 2 - t^2\}$, $\{-2, -1 - t, t^2, 6 - 4t, 1 + t + t^2, t\}$, and others; $\{1, 3t\}$, $\{2, t, 4 + t\}$, $\{1 + t, t + t^2\}$ are *not* spanning sets for $\mathscr{P}^3$. The idea expressed at the start of this section is that we want to find spanning sets for vector spaces in order to calculate with the vectors.

Suppose, for example, that we need to work with $\mathbb{R}^3$—that is, with vectors like

$$\begin{bmatrix} 1 \\ -1 \\ 0 \end{bmatrix}, \quad \begin{bmatrix} 1 \\ 4 \\ -2 \end{bmatrix}, \quad \begin{bmatrix} 0 \\ 0 \\ 5 \end{bmatrix}, \quad \begin{bmatrix} 2 \\ 3 \\ -4 \end{bmatrix}. \tag{5.17}$$

One simple spanning set for $\mathbb{R}^3$ obviously is $S = \{\mathbf{e}_1, \mathbf{e}_2, \mathbf{e}_3\}$, where the $\mathbf{e}_i$ are the 3×1 unit column matrices. For example, the first vector in (5.17) equals $1\mathbf{e}_1 + (-1)\mathbf{e}_2 + 0\mathbf{e}_3$, the second is $1\mathbf{e}_1 + 4\mathbf{e}_2 + (-2)\mathbf{e}_3$, and the general $\mathbf{x} = [x_1 \quad x_2 \quad x_3]^T$ in $\mathbb{R}^3$ can be expressed easily as $x_1\mathbf{e}_1 + x_2\mathbf{e}_2 + x_3\mathbf{e}_3$. There are many other possible spanning sets for $\mathbb{R}^3$, of course; one such is $S' = \{\mathbf{e}'_1, \mathbf{e}'_2, \mathbf{e}'_3, \mathbf{e}'_4\}$, where

$$\mathbf{e}'_1 = \begin{bmatrix} 1 \\ 2 \\ 1 \end{bmatrix}, \quad \mathbf{e}'_2 = \begin{bmatrix} 1 \\ 0 \\ -1 \end{bmatrix}, \quad \mathbf{e}'_3 = \begin{bmatrix} 1 \\ -2 \\ 1 \end{bmatrix}, \quad \text{and} \quad \mathbf{e}'_4 = \begin{bmatrix} 1 \\ 4 \\ 3 \end{bmatrix}.$$

In order to express the first vector, for example, in (5.17) as a linear combination of the vectors in S', we need to find scalars $\alpha_1, \alpha_2, \alpha_3, \alpha_4$ such that

$$\begin{bmatrix} 1 \\ -1 \\ 0 \end{bmatrix} = \alpha_1 \begin{bmatrix} 1 \\ 2 \\ 1 \end{bmatrix} + \alpha_2 \begin{bmatrix} 1 \\ 0 \\ -1 \end{bmatrix} + \alpha_3 \begin{bmatrix} 1 \\ -2 \\ 1 \end{bmatrix} + \alpha_4 \begin{bmatrix} 1 \\ 4 \\ 3 \end{bmatrix}.$$

This gives three equations for the four unknowns α_i; the general solution is $\alpha_1 = -2k$, $\alpha_2 = 0.5 + k$, $\alpha_3 = 0.5$, and $\alpha_4 = k$ arbitrary.

This second spanning set S' is less convenient than S for at least two reasons: (1) S' has four vectors instead of three as in S; and (2) it is more difficult to find the coefficients to use to represent a vector via S' than via S because some nontrivial equations have to be solved.

The first objective can be met by dropping one of the vectors from S', but—*great care must be taken in selecting the one to drop.* If, for example, we drop $\mathbf{e}'_3$ and try to write the first vector in (5.17) as $\alpha_1\mathbf{e}'_1 + \alpha_2\mathbf{e}'_2 + \alpha_4\mathbf{e}'_4$ by solving for the α_i from

$$\begin{bmatrix} 1 \\ -1 \\ 0 \end{bmatrix} = \alpha_1 \begin{bmatrix} 1 \\ 2 \\ 1 \end{bmatrix} + \alpha_2 \begin{bmatrix} 1 \\ 0 \\ -1 \end{bmatrix} + \alpha_4 \begin{bmatrix} 1 \\ 4 \\ 3 \end{bmatrix}, \tag{5.18}$$

we find that the equations are inconsistent and there are no solutions. That is, the first vector in (5.17) is linearly independent of the vectors $\mathbf{e}'_1$, $\mathbf{e}'_2$, $\mathbf{e}'_4$. If we instead drop any one of $\mathbf{e}'_1$, $\mathbf{e}'_2$, and $\mathbf{e}'_4$ from S', then the remaining three vectors do in fact span $\mathbb{R}^3$ (see Problems 1 and 2). One way to see that we can drop $\mathbf{e}'_4$ from S' and still span $\mathbb{R}^3$ is to note that $\mathbf{e}'_4 = 2\mathbf{e}'_1 - \mathbf{e}'_2$, so that any vector $\mathbf{v}$ that is written as

$$\mathbf{x} = \alpha_1\mathbf{e}'_1 + \alpha_2\mathbf{e}'_2 + \alpha_3\mathbf{e}'_3 + \alpha_4\mathbf{e}'_4$$

can also immediately be written as

$$\begin{aligned} \mathbf{x} &= \alpha_1\mathbf{e}'_1 + \alpha_2\mathbf{e}'_2 + \alpha_3\mathbf{e}'_3 + \alpha_4(2\mathbf{e}'_1 - \mathbf{e}'_2) \\ &= (\alpha_1 + 2\alpha_4)\mathbf{e}'_1 + (\alpha_2 - \alpha_4)\mathbf{e}'_2 + \alpha_3\mathbf{e}'_3, \end{aligned}$$

which is a combination of the vectors with $\mathbf{e}'_4$ dropped.

You should now be able to solve Problems 1 to 10.

Linear Independence

We just saw in the preceding paragraph that we could produce from S' an easier spanning set to work with by deleting from S' a vector, such as $\mathbf{e}'_4$, that was linearly dependent on the others. More generally, we clearly can drop any vector from a spanning set that is linearly dependent on the other vectors in that spanning set. If we then drop from that smaller spanning set any vectors that are linearly dependent on the others in that smaller set, and so on, it seems intuitively clear that we will eventually obtain a set such that no member of the set can be written as a linear combination of the others in the set. We want to examine what properties such a set would have.

Suppose that the set L of vectors $\mathbf{v}_1, \mathbf{v}_2, \ldots, \mathbf{v}_k$ (with $k \geq 2$) is such that each vector $\mathbf{v}_i$ is linearly independent of the remaining vectors in L. Suppose now that there are some scalars α_i with

$$\alpha_1\mathbf{v}_1 + \alpha_2\mathbf{v}_2 + \cdots + \alpha_k\mathbf{v}_k = \mathbf{0}.$$

We claim that it must be the case that all of the scalars α_i equal 0. If not, suppose that $\alpha_j \neq 0$, move the other terms to the other side of the equation, and multiply by α_j^{-1}; this gives

$$\mathbf{v}_j = (-\alpha_1\alpha_j^{-1})\mathbf{v}_1 + \cdots + (-\alpha_k\alpha_j^{-1})\mathbf{v}_k$$

in contradiction to $\mathbf{v}_j$'s being linearly independent of the others. We will see in Theorem 5.23(a) that the implication goes the other direction as well: if the only way to write $\mathbf{0}$ as a linear combination of the $\mathbf{v}_i$ is with all the coefficients α_i equal to 0, then each of the $\mathbf{v}_i$ is linearly independent of the others.

This says that the property we are seeking in a set L—in order not to be able to drop a vector from L and still span the same space—can be stated in two equivalent ways:

> Each vector in L is linearly independent of the others in L;
> or: the only linear combination of the vectors in L that equals $\mathbf{0}$ has all the coefficients of the $\mathbf{v}_i$ equal 0.

Sets L with this special property are extremely important in the study of vector spaces and are given a special name. The definition is usually stated using the second form of the property.

(5.19) ***Definition.*** Let $L = \{\mathbf{v}_1, \mathbf{v}_2, \ldots, \mathbf{v}_k\}$ be a nonempty set of vectors.

(a) Suppose that

$$\alpha_1\mathbf{v}_1 + \alpha_2\mathbf{v}_2 + \cdots + \alpha_k\mathbf{v}_k = \mathbf{0}$$

implies that $\alpha_1 = \alpha_2 = \cdots = \alpha_k = 0$. L is then said to be *linearly independent*.

(b) A set that is not linearly independent is said to be *linearly dependent*; equivalently, L is linearly dependent if and only if there are scalars $\alpha_1, \alpha_2, \ldots, \alpha_k$, not *all* zero, with

$$\alpha_1\mathbf{v}_1 + \alpha_2\mathbf{v}_2 + \cdots + \alpha_k\mathbf{v}_k = \mathbf{0}.$$

(5.20) ***Example.*** Some sets of two-dimensional geometrical vectors:

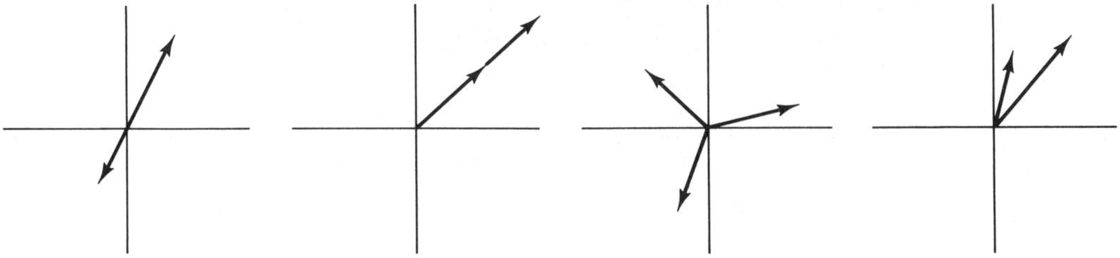

Dependent set — Dependent set — Dependent set — Independent set

(5.21) ***Example.*** The set $\{\mathbf{e}_1, \mathbf{e}_2\}$ is a linearly independent set in $\mathbb{R}^2$. To determine this, suppose that $\alpha_1\mathbf{e}_1 + \alpha_2\mathbf{e}_2 = \mathbf{0}$. Then

$$[0 \quad 0]^T = \mathbf{0} = \alpha_1[1 \quad 0]^T + \alpha_2[0 \quad 1]^T = [\alpha_1 \quad \alpha_2]^T,$$

so $\alpha_1 = \alpha_2 = 0$.

(5.22) ***Example.*** The set $\{1, 2 - 3t, 4 + t\}$ is linearly dependent in the vector space $\mathscr{P}^3$ of polynomials of degree strictly less than 3. To determine this, suppose that $\alpha_1(1) + \alpha_2(2 - 3t) + \alpha_3(4 + t)$ equals the zero polynomial $\mathbf{0}$—that is, that

$$(\alpha_1 + 2\alpha_2 + 4\alpha_3) + (-3\alpha_2 + \alpha_3)t = 0 \qquad \text{for all } t.$$

Then

$$\alpha_1 + 2\alpha_2 + 4\alpha_3 = 0 \quad \text{and} \quad -3\alpha_2 + \alpha_3 = 0,$$

a system of two equations for the three α_i. Gauss elimination produces the general solution $\alpha_1 = -10k/3$, $\alpha_2 = k/3$, and $\alpha_3 = k$ arbitrary. A solution with not all α_i zero comes, for example, from $k = 3$: $\alpha_1 = -10$, $\alpha_2 = 1$, and $\alpha_3 = 3$.

We conclude with a theorem stating, among other useful results, that our two different ways of stating that a set is linearly independent are in fact equivalent.

(5.23) ***Theorem*** (linear independence)
(a) Suppose that

$$L = \{\mathbf{v}_1, \mathbf{v}_2, \ldots, \mathbf{v}_k\}$$

with $k \geq 2$ with all the vectors $\mathbf{v}_i \neq \mathbf{0}$. Then L is linearly dependent if and only if at least one of the $\mathbf{v}_j$ is linearly dependent on the remaining vectors $\mathbf{v}_i$ $(i \neq j)$; in particular, L is linearly dependent if and only if at least one of the vectors $\mathbf{v}_j$ is linearly dependent on the *preceding* vectors $\mathbf{v}_1, \mathbf{v}_2, \ldots, \mathbf{v}_{j-1}$.

(b) Any set containing $\mathbf{0}$ is linearly dependent.
(c) $\{\mathbf{v}\}$ is linearly independent if and only if $\mathbf{v} \neq \mathbf{0}$.
(d) Suppose that $\mathbf{v}$ is linearly dependent on a set

$$L = \{\mathbf{v}_1, \mathbf{v}_2, \ldots, \mathbf{v}_k\},$$

and that $\mathbf{v}_j$ is linearly dependent on the others in L, namely

$$L'_j = \{\mathbf{v}_1, \mathbf{v}_2, \ldots, \mathbf{v}_{j-1}, \mathbf{v}_{j+1}, \ldots, \mathbf{v}_k\}.$$

Then $\mathbf{v}$ is linearly dependent on L'_j.
(e) Every subset of a linearly independent set is linearly independent.
(f) Suppose that L is a finite set of vectors and that some subset L_0 of L is linearly dependent. Then L is linearly dependent.

PROOF

(a) (dependent on $\Rightarrow$ dependent) See Definitions 5.16 and 5.19. Suppose that $\mathbf{v}_j$ is linearly dependent on the preceding vectors:

$$\mathbf{v}_j = \alpha_1\mathbf{v}_1 + \cdots + \alpha_{j-1}\mathbf{v}_{j-1}.$$

Then

$$\alpha_1\mathbf{v}_1 + \cdots + \alpha_{j-1}\mathbf{v}_{j-1} + (-1)\mathbf{v}_j + 0\mathbf{v}_{j+1} + \cdots + 0\mathbf{v}_k = \mathbf{0}$$

and hence L is linearly dependent.

(dependent $\Rightarrow$ dependent on) Suppose that L is linearly dependent. Define

$$L_j = \{\mathbf{v}_1, \mathbf{v}_2, \ldots, \mathbf{v}_j\} \qquad \text{for } 1 \leq j \leq k.$$

Since $\mathbf{v}_1 \neq \mathbf{0}$ by assumption, L_1 is linearly independent: see the proof of (c) below. Let L_i be the *first* of the L_j that is linearly dependent; such an i exists since L_k is linearly dependent. Also, $i \geq 2$ since L_1 is linearly independent. We claim that $\mathbf{v}_i$ is linearly dependent on the preceding vectors, that is, on L_{i-1}. To see this, note that there are α_r, not all zero, with

$$\alpha_1\mathbf{v}_1 + \cdots + \alpha_i\mathbf{v}_i = \mathbf{0};$$

were α_i to equal zero, then L_{i-1} would be linearly dependent, which it isn't. Since $\alpha_i \neq 0$, we can move the other terms in the equation to the right-hand side of the equation, multiply by α_i^{-1}, and thus express $\mathbf{v}_i$ in terms of the preceding $\mathbf{v}_r$, as claimed.

(b) Suppose that a set contains $\mathbf{0}$. Then we can certainly write $\mathbf{0}$ as 1 times $\mathbf{0}$ plus 0 times each of the other terms in the set, so the set is linearly dependent.

(c) If $\mathbf{v} = \mathbf{0}$, then the set is linearly dependent by (b). If the set is linearly dependent so that we have $\alpha\mathbf{v} = \mathbf{0}$ with $\alpha \neq 0$, then α^{-1} exists and we get

$$\mathbf{0} = \alpha^{-1}\mathbf{0} = \alpha^{-1}(\alpha\mathbf{v}) = 1\mathbf{v} = \mathbf{v},$$

so $\mathbf{v} = \mathbf{0}$.

(d) By assumption,

$$\mathbf{v} = \alpha_1\mathbf{v}_1 + \cdots + \alpha_k\mathbf{v}_k$$

and

$$\mathbf{v}_j = \beta_1\mathbf{v}_1 + \cdots + \beta_{j-1}\mathbf{v}_{j-1} + \beta_{j+1}\mathbf{v}_{j+1} + \cdots + \beta_k\mathbf{v}_k.$$

Then substituting the expression for $\mathbf{v}_j$ into the expression for $\mathbf{v}$ expresses $\mathbf{v}$ as linearly dependent on L_j', as claimed.

(e) Suppose that a linear combination of the vectors in some subset L_0 of L equals $\mathbf{0}$; by adding to this combination all terms of the form 0 times the vectors in L that are not in L_0, we obtain $\mathbf{0}$ as a linear combination of vectors in L. But L is linearly independent, so all the coefficients—including those of the vectors in L_0—must equal 0. Thus L_0 is linearly independent.

(f) Were L to be linearly independent, then by (e) L_0 would also be linearly independent, which it isn't. ■

Although it is not apparent so far, the concepts of linear independence and dependence are among the most crucial in linear algebra. We will see that they provide new ways of viewing the concepts of rank of matrices and of existence and uniqueness of solutions of systems of equations; they provide the key to understanding how to represent elements of vector spaces in a simple way; they help us understand the numerical difficulties that can arise in computing solutions to applied problems; and so on. *It is absolutely essential to grasp these concepts and to be able to determine whether a set is linearly independent.*

PROBLEMS 5.3

1. Show that the vectors $\mathbf{e}_1', \mathbf{e}_2', \mathbf{e}_3', \mathbf{e}_4'$ defined just after (5.17) span $\mathbb{R}^3$ as claimed.

2. Show that any one of $\mathbf{e}_1'$, $\mathbf{e}_2'$, or $\mathbf{e}_4'$ can be dropped in Problem 1 and the remaining vectors will still span $\mathbb{R}^3$.

3. Determine whether $\mathbb{R}^3$ is spanned by the vectors

$$\mathbf{v}_1 = [1 \quad -1 \quad 2]^T, \quad \mathbf{v}_2 = [-1 \quad 0 \quad 3]^T,$$
$$\mathbf{v}_3 = [0 \quad -1 \quad 5]^T, \quad \mathbf{v}_4 = [3 \quad -2 \quad 2]^T.$$

▷ **4.** Determine which of the vectors

$$\mathbf{u}_1 = [-3 \quad -1 \quad 15 \quad 6]^T, \quad \mathbf{u}_2 = [1 \quad 0 \quad -1 \quad 0]^T, \quad \mathbf{u}_3 = [1 \quad 1 \quad 1 \quad 0]^T$$

are in the subspace of $\mathbb{R}^4$ spanned by

$$\mathbf{v}_1 = [-1 \quad 3 \quad 5 \quad 2]^T, \quad \mathbf{v}_2 = [2 \quad -1 \quad 0 \quad 1]^T, \quad \mathbf{v}_3 = [1 \quad -8 \quad 5 \quad 3]^T.$$

5. Express the general vector $[x \quad y \quad z]^T$ as a linear combination of

$$\mathbf{v}_1 = [1 \quad 2 \quad 1]^T, \quad \mathbf{v}_2 = [1 \quad 0 \quad -1]^T, \quad \mathbf{v}_3 = [1 \quad -2 \quad 1]^T.$$

▷ **6.** Show that the space $\mathscr{P}^3$ of polynomials of degree strictly less than 3 is spanned by the set $\{2, 3 + t, 2 - t^2\}$.

7. Show that the space $\mathscr{P}^3$ of Problem 6 is spanned by the set

$$\{1, 2 + 2t, 1 - t + t^2, 2 - t^2\}.$$

8. Show that the space $\mathscr{P}^3$ of Problem 6 is *not* spanned by the set

$$\{1 + t, t + t^2\}.$$

▷ **9.** Suppose that the set $\{\mathbf{u}, \mathbf{v}, \mathbf{w}\}$ is linearly independent. Show that $\{\mathbf{u} + \mathbf{v}, \mathbf{v} + \mathbf{w}, \mathbf{w} + \mathbf{u}\}$ is also linearly independent.

10. Suppose that $\{\mathbf{u}, \mathbf{v}\}$ is linearly independent. Let $\mathbf{u}' = a\mathbf{u} + b\mathbf{v}$ and $\mathbf{v}' = c\mathbf{u} + d\mathbf{v}$. Show that $\{\mathbf{u}', \mathbf{v}'\}$ is linearly independent if and only if $ad - bc \neq 0$.

11. Show that the set of vectors $\mathbf{e}_1'$, $\mathbf{e}_2'$, $\mathbf{e}_3'$, $\mathbf{e}_4'$ defined just after (5.17) is linearly dependent.

▷ **12.** Show that the four vectors in Problem 3 form a linearly dependent set and determine which ones are linearly dependent on which others.

13. Show that in the space $\mathscr{P}^3$ of Problem 6, the set

$$\{1, 2 + 2t, 1 - t + t^2, 2 - t^2\}$$

is linearly dependent and determine which elements are linearly dependent on which others.

14. Show that in the space $\mathscr{P}^3$ of Problem 6, the set

$$\{1, t + 3, t + t^2, 2t - 1\}$$

is linearly dependent and determine which elements are linearly dependent on which others.

▷ **15.** Determine which of the following matrices cannot be written as a linear combination of the others.

$$\begin{bmatrix} 1 & -1 \\ -1 & 2 \end{bmatrix}, \quad \begin{bmatrix} -1 & 2 \\ 3 & 1 \end{bmatrix}, \quad \begin{bmatrix} 2 & -3 \\ -3 & 2 \end{bmatrix}, \quad \begin{bmatrix} 1 & 1 \\ 1 & 6 \end{bmatrix}$$

16. Determine the values of x for which the following set is linearly dependent.

$$\left\{ \begin{bmatrix} 1 \\ 2 \\ 1 \end{bmatrix}, \begin{bmatrix} 1 \\ x \\ -1 \end{bmatrix}, \begin{bmatrix} 1 \\ -2 \\ 1 \end{bmatrix} \right\}$$

17. Let $S = \{\mathbf{v}_1, \mathbf{v}_2, \ldots, \mathbf{v}_r\}$ be a nonempty set of vectors. Show that S is linearly independent if and only if every vector in the subspace $\mathscr{V}_0$ spanned by S has a *unique* representation in terms of S.

18. Let $S = \{\mathbf{v}_1, \mathbf{v}_2, \ldots, \mathbf{v}_r\}$ be a set of vectors in $\mathbb{R}^p$ with $p \geq r$ and suppose that $\langle \mathbf{v}_i \rangle_i \neq 0$ for all i, while $\langle \mathbf{v}_i \rangle_j = 0$ if $i > j$. Prove that S is linearly independent.

▷ **19.** An infinite set of vectors is said to be linearly dependent if and only if it has some finite subset that is linearly dependent; therefore, an infinite set is linearly independent if and only if every finite subset is linearly independent. State this definition in a form like that in Definition 5.19.

▷ **20.** (a) In the sense of Problem 19, show that $\{1, t, t^2, \ldots\}$ is linearly independent in the space $\mathscr{P}$ of all polynomials.
(b) For each $i \geq 1$, let $\mathbf{x}_i$ denote that infinite sequence whose ith term equals 1 and all of whose other terms equals zero. Show that $\{\mathbf{x}_1, \mathbf{x}_2, \ldots\}$ is a linearly independent set in the vector space $\mathscr{S}$ of all real sequences.

21. Suppose that $\mathbf{A}$ is a $p \times p$ real matrix that is unit-upper-triangular:

$$\langle \mathbf{A} \rangle_{ii} = 1 \qquad \text{for all } i, \text{ and } \langle \mathbf{A} \rangle_{ij} = 0 \text{ if } i > j.$$

(a) Show that the set of columns of $\mathbf{A}$ is linearly independent.
(b) Show that the set of rows of $\mathbf{A}$ is linearly independent.

5.4 *BASIS, DIMENSION, AND COORDINATES*

The definition of spanning sets in the preceding section was motivated by the need to find some concrete way to represent general elements of a vector space; the definition of a linearly independent set was motivated by an intuitive argument that we could continue to drop from a spanning set vectors that were linearly dependent on the others until we finally reach a set each of whose vectors is linearly independent of the others. We now make this argument rigorous.

Basis

(5.24) ***Theorem*** (reducing spanning sets). Suppose that $S = \{\mathbf{v}_1, \mathbf{v}_2, \ldots, \mathbf{v}_r\}$ spans $\mathscr{V}$.

(a) Suppose that one of these vectors $\mathbf{v}_j$ is linearly dependent on

$$S'_j = \{\mathbf{v}_1, \ldots, \mathbf{v}_{j-1}, \mathbf{v}_{j+1}, \ldots, \mathbf{v}_r\},$$

that is, on the remaining vectors. Then the reduced set S'_j also spans $\mathscr{V}$.

(b) If at least one vector in S is nonzero—that is, if $\mathscr{V}$ is more than just the trivial vector space $\{\mathbf{0}\}$—then there exists a subset S_0 of S that spans $\mathscr{V}$ and is linearly independent.

PROOF

(a) To show that S'_j spans $\mathscr{V}$, we must show that every $\mathbf{v}$ in $\mathscr{V}$ is linearly dependent on S'_j; suppose that $\mathbf{v}$ is an arbitrary vector. Since S spans

$\mathscr{V}$, $\mathbf{v}$ is linearly dependent on S. By Theorem 5.23(d) on linear independence, $\mathbf{v}$ must also be linearly dependent on S'_j, as required.

(b) If S itself is linearly independent, we are done; if not, then by (a) we can drop one of the vectors from S and still span $\mathscr{V}$, now with a set containing $r - 1$ vectors instead of r. We can continue dropping vectors in this fashion until we obtain either: (1) a linearly independent set S_0 containing at least two vectors and spanning $\mathscr{V}$, or (2) a set S_0 that spans $\mathscr{V}$ and contains just one vector $\mathbf{v}_s$, say. In case (1), S_0 is obviously the required set. In case (2), S_0 is again the required set; it certainly spans $\mathscr{V}$, and to see that it is linearly independent we only need note that $\mathbf{v}_s \neq \mathbf{0}$ since $\mathscr{V}$ is not just $\{\mathbf{0}\}$, and then use Theorem 5.23(c). ■

Theorem 5.24 says that we can reduce a spanning set to a linearly independent spanning set. Such a set is called a *basis* for the vector space.

(5.25) ***Definition.*** A *basis* for a vector space $\mathscr{V}$ is a linearly independent spanning set for $\mathscr{V}$.

(5.26) ***Example.*** We seek a basis for the subspace $\mathscr{V}_0$ of $\mathbb{R}^3$ consisting of all solutions to $x_1 + x_2 + x_3 = 0$; according to Example 5.14, this is indeed a subspace. Some caution is necessary here: we know that $\{\mathbf{e}_1, \mathbf{e}_2, \mathbf{e}_3\}$ is linearly independent and that every vector in $\mathbb{R}^3$—and thus certainly in $\mathscr{V}_0$—can be written as a linear combination of these three, yet this is *not* a basis for $\mathscr{V}_0$—the catch is that these vectors are *not in* $\mathscr{V}_0$, as elements of a basis must be. We must find a linearly independent spanning set of vectors that *are* in $\mathscr{V}_0$. To do so is simple: Note that the general solution to $x_1 + x_2 + x_3 = 0$ is $x_3 = \alpha$ arbitrary, $x_2 = \beta$ arbitrary, and $x_1 = -\alpha - \beta$. This says that the general vector $\mathbf{v}_0$ in the subspace $\mathscr{V}_0$ is

$$\mathbf{v}_0 = \begin{bmatrix} x_1 \\ x_2 \\ x_3 \end{bmatrix} = \begin{bmatrix} -\alpha - \beta \\ \beta \\ \alpha \end{bmatrix} = \alpha\mathbf{v}_1 + \beta\mathbf{v}_2 \quad \text{if} \quad \mathbf{v}_1 = \begin{bmatrix} -1 \\ 0 \\ 1 \end{bmatrix}, \quad \mathbf{v}_2 = \begin{bmatrix} -1 \\ 1 \\ 0 \end{bmatrix}.$$

The fact that every $\mathbf{v}_0$ in $\mathscr{V}_0$ can be written as above says that the set $B = \{\mathbf{v}_1, \mathbf{v}_2\}$ spans $\mathscr{V}_0$—note also that $\mathbf{v}_1$ and $\mathbf{v}_2$ are in $\mathscr{V}_0$; it is also easy to see that B is linearly independent: if $\alpha\mathbf{v}_1 + \beta\mathbf{v}_2 = \mathbf{0}$, then the $\mathbf{v}_0$ above equals $\mathbf{0}$ and we can see from the second and third entries of $\mathbf{v}_0$ that this forces $\alpha = \beta = 0$. Note that from the definition of $\mathscr{V}_0$ we know that $\mathscr{V}_0$ is a real vector space, but we have no concrete idea of what it really *is*. But now we know: it is simply everything of the form $\alpha\mathbf{v}_1 + \beta\mathbf{v}_2$—that is, $[-\alpha - \beta \quad \beta \quad \alpha]^T$—for arbitrary real numbers α and β. This is what finding a basis always does for us: it provides a concrete description of the vectors.

A basis has an important property that we will use a little later in this section.

(5.27) ***Theorem*** (unique basis representation). Let $B = \{\mathbf{v}_1, \ldots, \mathbf{v}_r\}$ be a basis. Then the representation of each $\mathbf{v}$ with respect to B is unique:

$$\text{if} \quad \mathbf{v} = \alpha_1\mathbf{v}_1 + \cdots + \alpha_r\mathbf{v}_r \quad \text{and also} \quad \mathbf{v} = \alpha'_1\mathbf{v}_1 + \cdots + \alpha'_r\mathbf{v}_r$$

then $\alpha_i = \alpha'_i$ for $1 \le i \le r$.

PROOF. $\mathbf{0} = \mathbf{v} - \mathbf{v} = (\alpha_1 - \alpha'_1)\mathbf{v}_1 + \cdots + (\alpha_r - \alpha'_r)\mathbf{v}_r$; since B is linearly independent, this means that $\alpha_i - \alpha'_i = 0$ for all i—that is, $\alpha_i = \alpha'_i$ for all i, as claimed. ■

You should now be able to solve Problems 1 to 8.

Dimension

Examples have shown that the set of vectors in a basis is not unique; many choices exist. We can show, however, that the *number* of vectors in a basis is unique; first, we need a preliminary result.

(5.28) ***Theorem.*** Suppose that $B = \{\mathbf{v}_1, \ldots, \mathbf{v}_p\}$ is a basis for the vector space $\mathscr{V}$ and that $D = \{\mathbf{u}_1, \ldots, \mathbf{u}_q\}$ is some linearly independent subset of $\mathscr{V}$ containing q vectors. Then $q \le p$.

PROOF. Suppose to the contrary that $q > p$; we will show that this produces a contradiction and therefore that $q \le p$. Let B_0 and B'_0 both equal B. Then the following statement is true for $i = 0$:

(#) B_i spans $\mathscr{V}$, where B_i consists of the first i vectors $\mathbf{u}_j$ from D followed by $p - i$ of the vectors $\mathbf{v}_j$ from B, and B'_i consists of these $p - i$ vectors from B.

We use induction on (#) to show it holds for $0 \le i \le p$. Suppose (#) holds for $i = r$ with $r < p$; we proceed to show that (#) holds for $i = r + 1$. Consider the set S consisting of $\mathbf{u}_{r+1}$ followed by the vectors of the spanning set B_r: that is, the first $r + 1$ vectors of D followed by $p - r$ vectors from B'_r (and hence from B). S is a linearly dependent spanning set since B_r spans $\mathscr{V}$. By Theorem 5.23(a) on linear independence, one vector in S is linearly dependent on the *preceding* vectors; this cannot occur for one of the $\mathbf{u}_j$ since they are preceded only by other $\mathbf{u}_i$ and D is linearly independent. Thus it is one of the $\mathbf{v}_j$ in B'_r that is linearly dependent on the preceding vectors in S. By Theorem 5.24(a) on reducing spanning sets, we can remove this $\mathbf{v}_j$ from S and still have a spanning set S_0. S_0 now consists of the first $r + 1$ vectors $\mathbf{u}_j$ from D followed by $p - (r + 1)$ of the vectors from B. If we let $B_{r+1} = S_0$ and let B'_{r+1} denote those $p - (r + 1)$ remaining vectors from B, then we have that (#) holds for $i = r + 1$. By induction, (#) must hold for $i = p$. That is, the first p of the $q > p$ vectors $\mathbf{u}_j$ in D (together with $p - p = 0$ of the vectors $\mathbf{v}_j$ from B) span $\mathscr{V}$; in particular, $\mathbf{u}_q$ can be written as a linear combination of the first p vectors $\mathbf{u}_j$, in contradiction to the linear independence of D. This contradiction completes the proof. ■

(5.29) **Key Corollary** (number of basis vectors). Suppose that the vector space $\mathscr{V}$ has some basis consisting of p vectors. Then every basis for $\mathscr{V}$ contains exactly p vectors.

PROOF. Suppose some other basis contains q vectors; since this other basis is linearly independent, Theorem 5.28 implies that $q \leq p$. By reversing the roles of these two bases, we also get $p \leq q$. So $p = q$. ■

The *two* geometrical vectors (see Section 5.1) that terminate at $\mathscr{P}(0, 1)$ and at $\mathscr{P}(1, 0)$ can easily be seen to form a basis for the *two*-dimensional plane of geometrical vectors; the *three* geometrical vectors that terminate at $\mathscr{P}(0, 0, 1)$, $\mathscr{P}(0, 1, 0)$, and at $\mathscr{P}(1, 0, 0)$ can easily be seen to form a basis for the *three*-dimensional physical space of geometrical vectors. By analogy, we call the number of vectors in a basis the *dimension* of the space.

(5.30) ***Definition.*** The number of vectors in a basis for a vector space is known as the *dimension* of the space. If the dimension of $\mathscr{V}$ is p, we also say that $\mathscr{V}$ is *p-dimensional.* When a vector space has a basis consisting of *some* finite number of vectors, we say that the space is *finite-dimensional.* The dimension of the (finite-dimensional) space $\{\mathbf{0}\}$ is defined to be 0.

The term "finite-dimensional" was introduced because not every vector space has a basis containing a finite number of vectors. In the vector space $\mathscr{P}$ of polynomials—see Example 5.9(f)—each of the sets $D_i = \{1, t, \ldots, t^{i-1}\}$ is linearly independent. Thus $\mathscr{P}$ contains linearly independent sets of arbitrarily many vectors and therefore cannot have a finite basis. It is also possible to introduce notions of linear independence and basis that allow for infinite sets (see Problem 19 in Section 5.3 and Problem 8), but we restrict ourselves—in our material on bases, at least—to finite-dimensional spaces.

This new terminology can be used to describe some useful results.

(5.31) **Key Corollary.** The real vector space $\mathbb{R}^p$ and the complex vector space $\mathbb{C}^p$ are p-dimensional.

PROOF. The p vectors $\mathbf{e}_1, \mathbf{e}_2, \ldots, \mathbf{e}_p$ form a basis for each space. ■

(5.32) ***Theorem.*** Let $\mathscr{V}$ be a p-dimensional vector space. Then:

(a) Every set containing strictly more than p vectors is linearly dependent.

(b) If $D = \{\mathbf{v}_1, \mathbf{v}_2, \ldots, \mathbf{v}_r\}$ is linearly independent and r is less than p, then there exist vectors $\mathbf{v}_{r+1}, \ldots, \mathbf{v}_p$ so that $\{\mathbf{v}_1, \mathbf{v}_2, \ldots, \mathbf{v}_p\}$ is a basis for $\mathscr{V}$.

(c) If a set of exactly p vectors *either* is a spanning set for $\mathscr{V}$ *or* is linearly independent, then it is *both* and is a basis for $\mathscr{V}$.

PROOF

(a) This follows immediately from Theorem 5.28.

(b) D cannot span $\mathscr{V}$ since $r < p$ means that D is not a basis; thus there is some vector, call it $\mathbf{v}_{r+1}$, that is linearly independent of D. The set of vectors $\mathbf{v}_1, \ldots, \mathbf{v}_{r+1}$ is also linearly independent; if not, one of the vectors would be linearly dependent on its predecessors, which is impossible by the assumptions on D and on $\mathbf{v}_{r+1}$. We can continue adjoining vectors to get $\mathbf{v}_1, \ldots, \mathbf{v}_s$ as long as $s \leq p$; we cannot adjoin a $\mathbf{v}_{p+1}$ since we would then violate (a). Thus the process stops at $\mathbf{v}_p$ with a linearly independent spanning set—that is, a basis as required.

(c) (spanning $\Rightarrow$ basis) Call the spanning set of p vectors S. Were S linearly dependent, by Theorem 5.24(b) on reducing spanning sets we could find a subset of S that is a basis but has fewer than p vectors, which is impossible. So S is linearly independent.
(independent $\Rightarrow$ basis) Let S be the linearly independent set of p vectors. If S were not spanning, then as in (b) we could adjoin a vector to S and obtain a linearly independent set of $p + 1$ vectors in contradiction to (a). ■

(5.33) ***Theorem.*** A vector space has finite dimension k if and only if k is the maximum number of vectors in a linearly independent set.

PROOF. (dimension $k \Rightarrow$ maximum k) If $\mathscr{V}$ has dimension k, then by definition it has some basis—hence a linearly independent set—containing exactly k vectors; by Theorem 5.32(a), there can be no greater number.

(maximum $k \Rightarrow$ dimension k) Suppose that k is the maximum and that D is such a linearly independent set of k vectors. Every vector $\mathbf{v}$ must be linearly dependent on D; otherwise, we could adjoin $\mathbf{v}$ to D and would then have a linearly independent set of more than k vectors, which is impossible. Thus D spans $\mathscr{V}$ and is a basis. ■

You should now be able to solve Problems 1 to 14.

Coordinates

We return to pursue an idea contained in Theorem 5.27, which stated that the representation of a vector with respect to a basis was unique. There is actually a technical, logical difficulty here. A set has no inherent order: $\{\mathbf{e}_1, \mathbf{e}_2\}$ and $\{\mathbf{e}_2, \mathbf{e}_1\}$ are identical as sets. There is no way, then, that we can precisely speak of the coefficient of the "first" vector in a basis; to resolve this, we define an *ordered basis.*

(5.34) ***Definition.*** An *ordered basis* $B = \{\mathbf{v}_1; \mathbf{v}_2; \ldots; \mathbf{v}_p\}$ is a basis $\{\mathbf{v}_1, \mathbf{v}_2, \ldots, \mathbf{v}_p\}$ in which there is a fixed and well-defined order as indicated by the relative positions of the vectors in B.

With this notation, $\{\mathbf{e}_1, \mathbf{e}_2\} = \{\mathbf{e}_2, \mathbf{e}_1\}$ but $\{\mathbf{e}_1; \mathbf{e}_2\} \neq \{\mathbf{e}_2; \mathbf{e}_1\}$. When B is an *ordered* basis, it makes sense to refer to the coefficient of the ith ordered-basis vector in the representation of a vector.

(5.35) ***Definition.*** Let $B = \{\mathbf{v}_1; \mathbf{v}_2; \ldots; \mathbf{v}_p\}$ be an ordered basis.

(a) The ith *coordinate* of a vector $\mathbf{v}$ *with respect to* B is the scalar α_i such that

$$\mathbf{v} = \alpha_1\mathbf{v}_1 + \cdots + \alpha_p\mathbf{v}_p.$$

(b) The *B-coordinates* or the *(B-)coordinate vector* of a vector $\mathbf{v}$ *with respect to* B is that $p \times 1$ column matrix $\mathbf{v}_B$ given by $\langle\mathbf{v}_B\rangle_i = \alpha_i$, where $\mathbf{v} = \alpha_1\mathbf{v}_1 + \cdots + \alpha_p\mathbf{v}_p$.

(c) The *coordinate mapping with respect to* B is that mapping (or *function*, or *correspondence*) c_B that assigns to each vector $\mathbf{v}$ in $\mathscr{V}$ its unique coordinate vector with respect to B:

$$c_B(\mathbf{v}) = \mathbf{v}_B = [\alpha_1 \quad \alpha_2 \quad \cdots \quad \alpha_p]^T, \quad \text{where} \quad \mathbf{v} = \alpha_1\mathbf{v}_1 + \cdots + \alpha_p\mathbf{v}_p.$$

(5.36) ***Example.*** Let $\mathscr{P}^3$ be the real vector space of polynomials of degree strictly less than 3, and let B' be the ordered basis

$$B' = \{1;\ 1 + t;\ 1 + t + t^2\}.$$

We tend to think of polynomials $\mathbf{v}$ in $\mathscr{P}^3$ in the form $a + bt + ct^2$. Trying to write this with respect to B' as

$$\alpha_1(1) + \alpha_2(1 + t) + \alpha_3(1 + t + t^2)$$

leads to the equations

$$\alpha_1 + \alpha_2 + \alpha_3 = a, \quad \alpha_2 + \alpha_3 = b, \quad \text{and} \quad \alpha_3 = c.$$

These are easily solved as $\alpha_3 = c$, $\alpha_2 = b - c$, and $\alpha_1 = a - b$. In the language of Definition 5.35, the coordinate vector with respect to B' of the vector

$$\mathbf{v} = a + bt + ct^2 \text{ is } \mathbf{v}_{B'} = [a - b \quad b - c \quad c]^T \text{ in } \mathbb{R}^3.$$

Another way to say this is that

$$c_{B'}(a + bt + ct^2) = [a - b \quad b - c \quad c]^T,$$

where $c_{B'}$ denotes the coordinate mapping with respect to B'. For example,

$$c_{B'}(2 - 3t + 6t^2) = [5 \quad -9 \quad 6]^T.$$

We can check that these really are the coordinates of the original polynomial $2 - 3t + 6t^2$ by using them as the coefficients of the ordered-basis vectors:

$$\begin{aligned}(5)(1) + (-9)(1 + t) + (6)(1 + t + t^2)\\ = (5 - 9 + 6) + (-9 + 6)t + 6t^2\\ = 2 - 3t + 6t^2,\end{aligned}$$

the original polynomial.

The importance of coordinates and coordinate vectors is that they provide a concrete way for us to calculate with vectors that have been defined in some abstract fashion.

(5.37) ***Example.*** Consider again the real vector space $\mathscr{V}_0$ from Example 5.26: all those vectors in $\mathbb{R}^3$ with $x_1 + x_2 + x_3 = 0$. From this definition alone it is hard to calculate with these vectors. But we found in Example 5.26 that—in present terminology—we can use $B = \{\mathbf{v}_1; \mathbf{v}_2\}$ as an ordered basis, where

$$\mathbf{v}_1 = [-1 \quad 0 \quad 1]^T, \quad \mathbf{v}_2 = [-1 \quad 1 \quad 0]^T.$$

The vectors in $\mathscr{V}_0$ no longer seem abstract; each is of the form $\alpha_1\mathbf{v}_1 + \alpha_2\mathbf{v}_2 = [-\alpha_1 - \alpha_2 \quad \alpha_2 \quad \alpha_1]^T$. Calculation is easy, too, if we use coordinates. We add $\mathbf{u}$ and $\mathbf{v}$ by adding their coordinates: if

$$\mathbf{u} = \beta_1\mathbf{v}_1 + \beta_2\mathbf{v}_2 \quad \text{and} \quad \mathbf{v} = \alpha_1\mathbf{v}_1 + \alpha_2\mathbf{v}_2,$$

then

$$\mathbf{u} + \mathbf{v} = (\beta_1 + \alpha_1)\mathbf{v}_1 + (\beta_2 + \alpha_2)\mathbf{v}_2.$$

In other language, we have $c_B(\mathbf{u} + \mathbf{v}) = c_B(\mathbf{u}) + c_B(\mathbf{v})$—the coordinates of a sum are the sum of the coordinates.

As in Example 5.37, calculations can always be done with the coordinates, thanks to the following important properties of coordinate mappings.

(5.38) ***Theorem*** (coordinate mappings). Let $\mathscr{V}$ be a p-dimensional real (or complex) vector space, and let c_B be the coordinate mapping with respect to some ordered basis B. Then:

(a) $c_B(\mathbf{u} + \mathbf{v}) = c_B(\mathbf{u}) + c_B(\mathbf{v})$ for all $\mathbf{u}$, $\mathbf{v}$ in $\mathscr{V}$
(b) $c_B(\alpha\mathbf{v}) = \alpha c_B(\mathbf{v})$ for all scalars α and all $\mathbf{v}$ in $\mathscr{V}$
(c) $c_B(\mathbf{v}) = \mathbf{0}$ in $\mathbb{R}^p$ (or $\mathbb{C}^p$) if and only if $\mathbf{v} = \mathbf{0}$ in $\mathscr{V}$
(d) For every $\mathbf{v}_B$ in $\mathbb{R}^p$ or ($\mathbb{C}^p$) there is exactly one $\mathbf{v}$ in $\mathscr{V}$ with $c_B(\mathbf{v}) = \mathbf{v}_B$

PROOF. Problem 17. ■

Representing abstract vectors by using a basis is an essential tool in applications. The fact that all computations on abstract vectors in $\mathscr{V}$ can equivalently be performed with their coordinates in $\mathbb{R}^p$ (or $\mathbb{C}^p$) says that $\mathscr{V}$ and $\mathbb{R}^p$ (or $\mathbb{C}^p$) are, for practical purposes, equivalent. Mathematically, any mapping c_B from $\mathscr{V}$ to $\mathbb{R}^p$ (or $\mathbb{C}^p$) satisfying (a)–(d) of Theorem 5.38 is said to be an *isomorphism* between those spaces, and the spaces are said to be *isomorphic*. We can therefore refer to the coordinate mapping c_B as the *coordinate isomorphism*.

(5.39) **Key Corollary.**
(a) Every p-dimensional real vector space is isomorphic to $\mathbb{R}^p$.
(b) Every p-dimensional complex vector space is isomorphic to $\mathbb{C}^p$.

This explains why the set of $p \times 1$ matrices is so important; it allows us to work with all finite-dimensional vector spaces. The isomorphism c_B not only allows us to perform simple calculations with the coordinates instead of with the abstract vectors but also allows us to verify linear independence, spanning, and so on with the coordinates; see Problems 18–21. The following general principle—although not a theorem—can usually be followed as a guideline, although its use in any specific case requires justification.

(5.40) ***Principle of Isomorphism.*** Suppose that c_B is the coordinate isomorphism with respect to the ordered basis B for a p-dimensional real or complex vector space $\mathscr{V}$. An algebraic property—essentially one involving the concepts of linear dependence and independence, basis, span, or dimension—holds for a set S in $\mathscr{V}$ if and only if it holds for the image $c_B(S)$, where $c_B(S)$ denotes the set of all those $\mathbf{v}_B$ in $\mathbb{R}^p$ or $\mathbb{C}^p$ satisfying $\mathbf{v}_B = c_B(\mathbf{v})$ for some $\mathbf{v}$ in S.

You should now be able to solve Problems 1 to 22.

Change of Basis

Examples have already appeared in which more than one basis was in use simultaneously; in Example 5.36, for example, we wrote polynomials in terms of both $\{1; t; t^2\}$ and $\{1; 1 + t; 1 + t + t^2\}$. That example was essentially an exercise in how to find the coordinates with respect to one ordered basis from those with respect to another. This "translation" problem is important, and we will make heavy use of the technique in later chapters. We therefore pursue it further now.

Suppose that

$$B = \{\mathbf{v}_1; \mathbf{v}_2; \ldots; \mathbf{v}_p\} \quad \text{and} \quad B' = \{\mathbf{v}'_1; \mathbf{v}'_2; \ldots; \mathbf{v}'_p\}$$

are two ordered bases for the same real or complex p-dimensional vector space $\mathscr{V}$. To see how to translate between B-coordinates and B'-coordinates, we first write each vector $\mathbf{v}'_i$ in the ordered basis B' as a combination of the vectors $\mathbf{v}_j$ in the ordered basis B as follows:

(5.41)
$$\mathbf{v}'_i = m_{1i}\mathbf{v}_1 + m_{2i}\mathbf{v}_2 + \cdots + m_{pi}\mathbf{v}_p,$$

that is,

$$c_B(\mathbf{v}'_i) = \mathbf{m}_i = [m_{1i} \quad m_{2i} \cdots m_{pi}]^T.$$

Suppose that we have the B'-coordinates $c_{B'}(\mathbf{v})$ and want the B-coordinates $c_B(\mathbf{v})$. That is, we have $\alpha'_1, \ldots, \alpha'_p$ such that $\mathbf{v} = \alpha'_1\mathbf{v}'_1 + \cdots + \alpha'_p\mathbf{v}'_p$ and we want to find $\alpha_1, \ldots, \alpha_p$ so that $\mathbf{v} = \alpha_1\mathbf{v}_1 + \cdots + \alpha_p\mathbf{v}_p$. From Theorem 5.38 on the isomorphism

c_B and from (5.41) we know that (using partitioned matrix notation)

(5.42)
$$\begin{aligned} c_B(\mathbf{v}) &= c_B(\alpha'_1\mathbf{v}'_1 + \cdots + \alpha'_p\mathbf{v}'_p) \\ &= \alpha'_1 c_B(\mathbf{v}'_1) + \cdots + \alpha'_p c_B(\mathbf{v}'_p) \\ &= \alpha'_1\mathbf{m}_1 + \alpha'_2\mathbf{m}_2 + \cdots + \alpha'_p\mathbf{m}_p \\ &= [\mathbf{m}_1 \quad \mathbf{m}_2 \quad \cdots \quad \mathbf{m}_p][\alpha'_1 \quad \alpha'_2 \quad \cdots \quad \alpha'_p]^T \\ &= \mathbf{M}c_{B'}(\mathbf{v}), \end{aligned}$$

where $\mathbf{M}$ is the $p \times p$ matrix $\mathbf{M} = [\mathbf{m}_1 \quad \mathbf{m}_2 \quad \cdots \quad \mathbf{m}_p]$.

This says how to get the B-coordinates from the B'-coordinates: Just multiply by the $p \times p$ matrix $\mathbf{M}$ whose entries $\langle\mathbf{M}\rangle_{ij}$ equal m_{ij} from (5.41). Conversely, we can get the B'-coordinates from the B-coordinates by multiplying by $\mathbf{M}^{-1}$, because $\mathbf{M}$ is nonsingular. To see that $\mathbf{M}$ is nonsingular, suppose that $\mathbf{Mx} = \mathbf{0}$; if we can show that $\mathbf{x} = \mathbf{0}$, then **Key Theorem 4.18** will tell us that $\mathbf{M}$ is nonsingular. Let $\mathbf{v}$ be the vector in $\mathscr{V}$ whose B'-coordinates α'_i are just the entries of some $\mathbf{x}$ for which $\mathbf{Mx} = \mathbf{0}$: $\alpha'_i = \langle\mathbf{x}\rangle_i$. By (5.42), the B-coordinates of $\mathbf{v}$ are just $\mathbf{Mx}$, which equals $\mathbf{0}$; therefore,

$$\mathbf{v} = 0\mathbf{v}_1 + 0\mathbf{v}_2 + \cdots + 0\mathbf{v}_p = \mathbf{0} \text{ in } \mathscr{V}.$$

But if $\mathbf{v} = \mathbf{0}$ in $\mathscr{V}$, then its B'-coordinates are all 0 since B' is linearly independent; since $\mathbf{v}$ was defined so that these coordinates are the $\langle\mathbf{x}\rangle_i$, we get $\mathbf{x} = \mathbf{0}$. Since $\mathbf{x} = \mathbf{0}$ if $\mathbf{Mx} = \mathbf{0}$, $\mathbf{M}$ is nonsingular. This proves the following extremely useful result.

(5.43) **Key Theorem** (change of basis). Let $B = \{\mathbf{v}_1; \ldots; \mathbf{v}_p\}$ and $B' = \{\mathbf{v}'_1; \ldots; \mathbf{v}'_p\}$ be two ordered bases for the same p-dimensional vector space $\mathscr{V}$. For any vector $\mathbf{v}$ in $\mathscr{V}$, the B-coordinates $\mathbf{v}_B$ of $\mathbf{v}$ and the B'-coordinates $\mathbf{v}_{B'}$ of $\mathbf{v}$ are related by

$$\mathbf{v}_B = \mathbf{M}\mathbf{v}_{B'} \quad \text{and} \quad \mathbf{v}_{B'} = \mathbf{M}^{-1}\mathbf{v}_B,$$

where $\mathbf{M}$ is the $p \times p$ nonsingular matrix with $\langle\mathbf{M}\rangle_{ij} = m_{ij}$, where $\mathbf{v}'_i = m_{1i}\mathbf{v}_1 + \cdots + m_{pi}\mathbf{v}_p$.

(5.44) ***Example.*** Consider again the polynomial space $\mathscr{P}^3$ and the two ordered bases $B = \{1; t; t^2\}$ and $B' = \{1; 1 + t; 1 + t + t^2\}$ for $\mathscr{P}^3$ as in Example 5.36. To translate between these, we need the matrix $\mathbf{M}$: we must write each $\mathbf{v}'_i$ in terms of the $\mathbf{v}_j$ as in (5.41). This is easy:

$$\mathbf{v}'_1 = 1 = 1(1) + 0(t) + 0(t^2) \quad \text{for the first column of } \mathbf{M};$$

$$\mathbf{v}'_2 = 1 + t = 1(1) + 1(t) + 0(t^2) \quad \text{for the second column;}$$

$$\mathbf{v}'_3 = 1 + t + t^2 = 1(1) + 1(t) + 1(t^2) \quad \text{for the third.}$$

Therefore,

$$\mathbf{M} = \begin{bmatrix} 1 & 1 & 1 \\ 0 & 1 & 1 \\ 0 & 0 & 1 \end{bmatrix}, \quad \text{and we find} \quad \mathbf{M}^{-1} = \begin{bmatrix} 1 & -1 & 0 \\ 0 & 1 & -1 \\ 0 & 0 & 1 \end{bmatrix}.$$

We can use $\mathbf{M}$ and $\mathbf{M}^{-1}$ to translate between coordinates. To write $7 - 3t + 4t^2$ in terms of B' we take its B-coordinates $[7 \quad -3 \quad 4]^T$ and multiply by $\mathbf{M}^{-1}$, getting $[10 \quad -7 \quad 4]^T$ as the B'-coordinates; as verification, we note that

$$\begin{aligned}(10)(1) + (-7)(1 + t) + (4)(1 + t + t^2) &= (10 - 7 + 4) + (-7 + 4)t + (4)t^2 \\ &= 7 - 3t + 4t^2,\end{aligned}$$

as required. More generally, the B'-coordinates of

$$a + bt + ct^2 \quad \text{are} \quad \mathbf{M}^{-1}[a \quad b \quad c]^T = [a - b \quad b - c \quad c]^T,$$

just as we found in Example 5.36.

PROBLEMS 5.4

▷ **1.** Find a subset of the following vectors that forms a basis for $\mathbb{R}^3$:

$$\mathbf{v}_1 = [1 \quad -1 \quad 2]^T, \quad \mathbf{v}_2 = [-1 \quad 0 \quad 3]^T,$$
$$\mathbf{v}_3 = [0 \quad -1 \quad 5]^T, \quad \mathbf{v}_4 = [3 \quad -2 \quad 2]^T.$$

2. Find a basis for the subspace $\mathscr{V}_0$ of $\mathbb{R}^3$ consisting of all those vectors

$$\mathbf{x} = [x_1 \quad x_2 \quad x_3]^T$$

satisfying $x_1 = 0$.

3. Find a basis for the subspace $\mathscr{V}_0$ of $\mathbb{R}^3$ consisting of all those vectors

$$\mathbf{x} = [x_1 \quad x_2 \quad x_3]^T$$

satisfying $x_1 + x_2 = 0$.

4. Find a basis for the subspace $\mathscr{V}_0$ of $\mathbb{R}^4$ consisting of all those vectors

$$\mathbf{x} = [x_1 \quad x_2 \quad x_3 \quad x_4]^T$$

satisfying $x_1 - x_2 = 0$ and $x_3 - x_4 = 0$.

▷ **5.** Find a basis for the subspace of $\mathbb{R}^4$ spanned by the vectors

$$\begin{bmatrix} 1 \\ 2 \\ -1 \\ 0 \end{bmatrix}, \quad \begin{bmatrix} 4 \\ 8 \\ -4 \\ -3 \end{bmatrix}, \quad \begin{bmatrix} 0 \\ 1 \\ 3 \\ 4 \end{bmatrix}, \quad \begin{bmatrix} 2 \\ 5 \\ 1 \\ 4 \end{bmatrix}.$$

6. From the set $\{1, 2 + 2t, 1 - t + t^2, 2 - t^2\}$ find a basis for $\mathscr{P}^3$, the real vector space of polynomials of degree strictly less than 3.

▷ **7.** Find a basis for the space $\mathscr{P}^3$ of Problem 6 from among the vectors in

$$\{1, t + 3, t + t^2, 2t - 1\}.$$

▷ **8.** An *infinite* set of vectors is said to be a basis for a vector space $\mathscr{V}$ if and only if it is linearly independent—see Problem 19 in Section 5.3—and every vector in the space can be written as a linear combination of some finite number of the vectors in the basis. Show that the set $\{1, t, t^2, \ldots\}$ is a basis for the real vector space $\mathscr{P}$ of all polynomials.

9. Definition 5.30 defined the dimension of $\{\mathbf{0}\}$ to equal 0. This is because this space has no basis—it is spanned by $\{\mathbf{0}\}$ but $\{\mathbf{0}\}$ is not a basis; explain why.

10. Adjoin as many vectors as necessary to the set $\{1 + t, t + t^2\}$ so as to obtain a basis for the space $\mathscr{P}^3$ of Problem 6.

▷ **11.** Show that the vectors below do *not* form a basis for $\mathbb{R}^3$.

$$\mathbf{v}_1 = \begin{bmatrix} 2 \\ -1 \\ 1 \end{bmatrix}, \quad \mathbf{v}_2 = \begin{bmatrix} 1 \\ 1 \\ 2 \end{bmatrix}, \quad \text{and} \quad \mathbf{v}_3 = \begin{bmatrix} 10 \\ -8 \\ 2 \end{bmatrix}.$$

12. Explain why you can be certain, without calculations, that the set below must be linearly dependent in $\mathbb{R}^2$:

$$\left\{\begin{bmatrix} 1 \\ -2 \end{bmatrix}, \begin{bmatrix} 17 \\ 36 \end{bmatrix}, \begin{bmatrix} -42.6 \\ 31.421 \end{bmatrix}\right\}.$$

13. Prove that the dimension of a finite-dimensional vector space equals the minimum of the number of vectors in all possible spanning sets for the space.

▷ **14.** Determine with reasons whether the following set spans $\mathbb{R}^3$:

$$\{[\pi^2 \quad 2.35 \quad \sqrt{\pi}]^T, [-3.289 \quad \pi + 2 \quad 1]^T\}.$$

15. Let $B = \{[1 \quad 1 \quad 0]^T; [1 \quad 0 \quad 1]^T; [0 \quad 1 \quad 1]^T\}$ be an ordered basis for $\mathbb{R}^3$.
(a) Find the second coordinate of $\mathbf{v} = [1 \quad 2 \quad 3]^T$ with respect to B.
(b) Find the coordinate vector of $\mathbf{v} = [a \quad b \quad c]^T$ with respect to B.
(c) Find a formula for the coordinate isomorphism $c_B(\mathbf{v})$ for general $\mathbf{v}$ in $\mathbb{R}^3$.

▷ **16.** Let $B = \{1 + t + t^2; 1 + t; 1\}$ be an ordered basis for the space $\mathscr{P}^3$ of Problem 6 (see Example 5.36).
(a) Find the coordinates of $2 - 3t + 6t^2$ with respect to B.
(b) Find the coordinate vector of $a + bt + ct^2$ with respect to B.
(c) Find a formula for the coordinate isomorphism $c_B(\mathbf{v})$ for a general vector $\mathbf{v}$ in $\mathscr{P}^3$.

17. Prove Theorem 5.38.

18. Let c_B be the coordinate isomorphism with respect to an ordered basis B from the real vector space $\mathscr{V}$ to $\mathbb{R}^p$. Prove that $\{\mathbf{v}_1, \ldots, \mathbf{v}_r\}$ is linearly independent in $\mathscr{V}$ if and only if $\{c_B(\mathbf{v}_1), \ldots, c_B(\mathbf{v}_r)\}$ is linearly independent in $\mathbb{R}^p$.

▷ **19.** Let c_B be as in Problem 18. Prove that a vector $\mathbf{v}$ in $\mathscr{V}$ is linearly dependent on $\mathbf{v}_1, \ldots, \mathbf{v}_r$ in $\mathscr{V}$ if and only if $c_B(\mathbf{v})$ is linearly dependent on $c_B(\mathbf{v}_1), \ldots, c_B(\mathbf{v}_r)$ in $\mathbb{R}^p$.

20. Let c_B be as in Problem 18. Prove that $\mathbf{u} = \mathbf{v}$ in $\mathscr{V}$ if and only if $c_B(\mathbf{u}) = c_B(\mathbf{v})$ in $\mathbb{R}^p$.

▷ **21.** Let c_B be as in Problem 18. Prove that $\{\mathbf{u}_1, \ldots, \mathbf{u}_r\}$ is a basis for $\mathscr{V}$ if and only if $\{c_B(\mathbf{u}_1), \ldots, c_B(\mathbf{u}_r)\}$ is a basis for $\mathbb{R}^p$.

22. By Theorem 5.38(d) on the isomorphism c_B, for each $\mathbf{v}_B$ there is a unique $\mathbf{v}$ in $\mathscr{V}$ with $c_B(\mathbf{v}) = \mathbf{v}_B$. This says that the *inverse isomorphism* c_B^{-1} exists mapping each $\mathbf{v}_B$ to that $\mathbf{v}$ that c_B maps back to $\mathbf{v}_B$. Show that

$$c_B^{-1}(\mathbf{u}_B + \mathbf{v}_B) = c_B^{-1}(\mathbf{u}_B) + c_B^{-1}(\mathbf{v}_B),$$

that

$$c_B^{-1}(\alpha\mathbf{v}_B) = \alpha c_B^{-1}(\mathbf{v}_B),$$

and that

$$c_B^{-1}(\mathbf{v}_B) = \mathbf{0} \text{ in } \mathscr{V}$$

if and only if $\mathbf{v}_B = \mathbf{0}$ in $\mathbb{R}^p$ or $\mathbb{C}^p$.

23. The sets

$$B = \{\mathbf{e}_1; \mathbf{e}_2; \mathbf{e}_3\} \quad \text{and} \quad B' = \{[1 \quad 2 \quad 3]^T; [1 \quad 0 \quad 1]^T; [3 \quad 4 \quad 6]^T\}$$

are both ordered bases for $\mathbb{R}^3$. Find the matrix $\mathbf{M}$ that translates between coordinates with respect to these two bases.

▷ **24.** The sets

$$B = \{\mathbf{e}_1; \mathbf{e}_2; \mathbf{e}_3\} \quad \text{and} \quad B' = \{\mathbf{v}_1; \mathbf{v}_2; \mathbf{v}_3\}$$

are both ordered bases for $\mathbb{R}^3$. In terms of the $\mathbf{v}_i$, find the matrix $\mathbf{M}$ that translates between coordinates with respect to these two bases.

25. The sets

$$B = \{1 + t; t + t^2; 1 + t^2\} \quad \text{and} \quad B' = \{1; 1 + t; 1 + t + t^2\}$$

are both ordered bases for the space $\mathscr{P}^3$ of Problem 6. Find the matrix $\mathbf{M}$ that translates between coordinates with respect to these two bases.

26. Suppose that

$$B = \{\mathbf{v}_1; \mathbf{v}_2; \mathbf{v}_3\} \quad \text{and} \quad B' = \{\mathbf{v}_3; \mathbf{v}_2; \mathbf{v}_1\}$$

are both ordered bases for $\mathbb{R}^3$. Find the matrix $\mathbf{M}$ that translates between coordinates with respect to these two bases.

27. Suppose that

$$B = \{\mathbf{v}_1; \mathbf{v}_2; \ldots; \mathbf{v}_p\} \quad \text{and} \quad B' = \{\mathbf{v}_p; \mathbf{v}_{p-1}; \ldots; \mathbf{v}_1\}$$

are two ordered bases for $\mathbb{R}^p$. Find the matrix $\mathbf{M}$ that translates between coordinates with respect to these two bases.

▷ **28.** The sets

$$B = \{[1 \quad 2 \quad 2 \quad 1]^T; [1 \quad 0 \quad 2 \quad 0]^T; [2 \quad 0 \quad 4 \quad -3]^T\}$$

and

$$B' = \{[0 \quad 2 \quad 0 \quad 1]^T; [2 \quad 1 \quad 4 \quad -1]^T; [1 \quad 2 \quad 2 \quad 4]^T\}$$

are bases for the same subspace of $\mathbb{R}^4$. Find the matrix $\mathbf{M}$ that translates between coordinates with respect to these two bases.

29. Suppose that B, B', and B'' are all bases for the real vector space $\mathcal{V}$; let $\mathbf{M}$ be the matrix that translates from B'-coordinates to B-coordinates, and let $\mathbf{M}'$ be the matrix that translates from B''-coordinates to B'-coordinates. Find the matrix that translates from B''-coordinates to B-coordinates.

5.5 *BASES AND MATRICES*

Most of the preceding section presents results about things that *can* be done—vectors *can* be dropped from linearly dependent spanning sets, linearly independent sets *can* be extended to form a basis, and so on. The section does not, however, address *how* to accomplish such tasks. Given a spanning set, *how* do we find a basis from within it? Given a linearly independent set, *how* do we find additional vectors so as to expand it to a basis?

The material on change of basis is an exception to this view of Section 5.4; that material explicitly showed *how* to "translate" between different bases for the same space. The key there was to use the matrix theory developed in the earlier chapters. In the present section, matrices again provide answers to the "how" questions. To use matrices, we restrict ourselves to $\mathbb{R}^p$, $\mathbb{C}^p$, and their subspaces; note, however, that this "restriction" still means that *our results apply to **all** p-dimensional real or complex vector spaces.* To see this, recall (5.40), the Principle of Isomorphism: In order to solve some problem about a set in $\mathcal{V}$, solve it instead for the set in $\mathbb{R}^p$ or $\mathbb{C}^p$ that consists of all the coordinate vectors of that set in $\mathcal{V}$. Thus the techniques of this chapter can actually be applied, through the coordinate vectors, to solve problems in all finite-dimensional real or complex vector spaces.

With this in mind, we now "restrict" ourselves to dealing with column matrices. Since $p \times 1$ matrices are vectors in $\mathbb{R}^p$ or $\mathbb{C}^p$, vector-space concepts such as linear dependence apply to them; since the set of all $1 \times q$ row matrices also is a vector space, the same concepts apply to them as well.

Row Operations and Linear Dependence

Two simple but fundamental facts about row operations and linear dependence provide answers to several questions.

(5.45) **Key Theorem** (row operations and linear dependence). Suppose that a sequence of elementary row operations transforms **A** to **B**. Then:

(a) a given collection of *columns* of **A** is linearly dependent (independent) if and only if the corresponding collection of columns of **B** is linearly dependent (independent);

(b) a row matrix can be written as a linear combination of—that is, is linearly dependent on—*all* the (nonzero) *rows* of **A** if and only if it can be written as a linear combination of *all* the (nonzero) rows of **B**.

PROOF. Let **F** be the nonsingular matrix that **Key Theorem 3.34** guarantees will satisfy $\mathbf{FA} = \mathbf{B}$.

(a) Let $\mathbf{a}_1, \ldots, \mathbf{a}_r$ denote some collection of r columns of **A**, and let $\mathbf{b}_1, \ldots, \mathbf{b}_r$ denote the corresponding columns of **B**; by the rule for multiplying partitioned matrices, $\mathbf{b}_i = \mathbf{F}\mathbf{a}_i$ for all i. Since **F** is nonsingular, $\mathbf{Fx} = \mathbf{0}$ if and only if $\mathbf{x} = \mathbf{0}$. Take **x** as a linear combination of the $\mathbf{a}_i$; **Fx** is that same combination, but of the $\mathbf{b}_i$. Therefore, a combination of the $\mathbf{a}_i$ is **0** if and only if that same combination of the $\mathbf{b}_i$ is **0**. This means that the one collection is linearly dependent if and only if the other is.

(b) A row matrix **y** is a linear combination of the rows of **A** if and only if $\mathbf{y} = \mathbf{xA}$ for some row matrix **x**; but $\mathbf{y} = \mathbf{xA}$ if and only if

$$\mathbf{y} = \mathbf{xF}^{-1}\mathbf{FA} = \mathbf{x}'\mathbf{B} \qquad \text{for } \mathbf{x}' = \mathbf{xF}^{-1},$$

since $\mathbf{FA} = \mathbf{B}$; and $\mathbf{y} = \mathbf{x}'\mathbf{B}$ if and only if **y** is a linear combination of the rows of **B**. ■

Note that (a) in the theorem refers to *columns*, and to *selected* columns from among all the columns; (b), in distinction, refers to *rows*, and to *all* rows *together*.

The basic idea for using this theorem to study questions of linear dependence about columns or rows of a matrix **A** is simple: create a Gauss-reduced or row-echelon form **B** from **A** and then study those questions on the much simpler matrix **B** where they are trivial to answer.

Dependence and Basis Problems

For the basic idea above to work, we have to understand linear dependence among columns of a matrix in Gauss-reduced or row-echelon form. The key for either form is that the ith leading column in either has a 1 as its ith entry and zeros below while every column to its left has zeros as the entries in row i as well as in the rows below. This makes it clear that the leading columns form a linearly independent set and that each column to the left of the ith leading column is linearly dependent on the first $i-1$ leading columns. Each nonzero row of a Gauss-reduced form starts with zeros until the 1 in that row's leading column is encountered; since those 1's move further to the right with each successive row, it is also clear that those rows form a linearly independent set. We summarize:

(5.46) ***Theorem*** (reduced matrices and dependence). Suppose that **G** is in Gauss-reduced form and has rank k. Then:

(a) The set of k leading columns of **G** is linearly independent.

(b) Any column to the left of the first leading column of **G** is a zero column. Any column to the left of the ith leading column of **G** (for $i > 1$) is linearly dependent on the first $i-1$ leading columns. Any column to the right of the kth leading column is linearly dependent on the k leading columns.

(c) The set of k nonzero rows of **G** is linearly independent.

The two preceding theorems are useful computational tools.

(5.47) ***Example.*** Consider the subspace $\mathscr{V}_0$ of $\mathbb{C}^4$ spanned by

$$\begin{bmatrix} 1 \\ -1 \\ -1 \\ 2 \end{bmatrix}, \quad \begin{bmatrix} -1 \\ 2 \\ 3 \\ 1 \end{bmatrix}, \quad \begin{bmatrix} 2 \\ -3 \\ -3 \\ 2 \end{bmatrix}, \quad \begin{bmatrix} 1 \\ 1 \\ 1 \\ 6 \end{bmatrix}.$$

We seek a basis for $\mathscr{V}_0$, from which we will discover its dimension. We use the four columns above as the columns of a matrix **A**, reduce **A** to Gauss-reduced form **G**, use Theorem 5.46 to identify the leading columns of **G** as linearly independent and generating all the other columns, and then use **Key Theorem 5.45** to deduce that the corresponding columns of **A** form a basis for $\mathscr{V}_0$ containing k—the rank of **G** and **A**—vectors. This process produces

$$\mathbf{A} = \begin{bmatrix} 1 & -1 & 2 & 1 \\ -1 & 2 & -3 & 1 \\ -1 & 3 & -3 & 1 \\ 2 & 1 & 2 & 6 \end{bmatrix} \quad \text{and then} \quad \mathbf{G} = \begin{bmatrix} 1 & -1 & 2 & 1 \\ 0 & 1 & -1 & 2 \\ 0 & 0 & 1 & -2 \\ 0 & 0 & 0 & 0 \end{bmatrix}.$$

The rank is 3, and the leading columns are the first three. Therefore, the first three columns of **A** are a basis for $\mathscr{V}_0$, and the dimension of $\mathscr{V}_0$ is 3.

This example shows the process for reducing a spanning set to a basis. A different problem is that of expanding a linearly independent set to obtain a basis.

(5.48) ***Example.*** Let $\mathscr{V}_0$ be the subspace of $\mathbb{R}^4$ of all solutions to

$$x_1 + x_2 = x_3 + x_4.$$

The two vectors $\begin{bmatrix} 1 & 0 & 1 & 0 \end{bmatrix}^T$ and $\begin{bmatrix} 0 & 1 & 0 & 1 \end{bmatrix}^T$ are in $\mathscr{V}_0$ and clearly form a linearly independent set. The problem is to expand this set (if needed) to get a basis for $\mathscr{V}_0$. Our approach is to form a matrix $\mathbf{A}$ with these two columns as its first columns and with the columns in *some* basis for $\mathscr{V}_0$ as the later columns, reduce $\mathbf{A}$ to Gauss-reduced form $\mathbf{G}$, use Theorem 5.46 to identify the leading columns of $\mathbf{G}$—which will include the first two columns since those two of $\mathbf{A}$ are independent of one another—as linearly independent and generating the others, and then take the corresponding columns of $\mathbf{A}$ as the expanded basis. To obtain *some* basis for $\mathscr{V}_0$, we proceed as in Example 5.26 to find the general solution to our equation and then write down the basis. The general solution is easily found to be

$$x_1 + x_2 = x_3 + x_4$$
$$x_1 = x_3 + x_4 - x_2$$

$$
\mathbf{v} = \begin{bmatrix} \alpha + \beta - \gamma \\ x_2 = \gamma \\ x_3 = \beta \\ x_4 = \alpha \end{bmatrix} = \alpha \begin{bmatrix} 1 \\ 0 \\ 0 \\ 1 \end{bmatrix} + \beta \begin{bmatrix} 1 \\ 0 \\ 1 \\ 0 \end{bmatrix} + \gamma \begin{bmatrix} -1 \\ 1 \\ 0 \\ 0 \end{bmatrix},
$$

so these three column matrices clearly form a basis for $\mathscr{V}_0$. We form the matrix $\mathbf{A}$ with the two originally given vectors as the first two columns and these as the last three, and then reduce $\mathbf{A}$ to Gauss-reduced form $\mathbf{G}$:

$$
\mathbf{A} = \begin{bmatrix} 1 & 0 & 1 & 1 & -1 \\ 0 & 1 & 0 & 0 & 1 \\ 1 & 0 & 0 & 1 & 0 \\ 0 & 1 & 1 & 0 & 0 \end{bmatrix} \quad \text{and then}
$$

$$
\mathbf{G} = \begin{bmatrix} 1 & 0 & 1 & 1 & -1 \\ 0 & 1 & 0 & 0 & 1 \\ 0 & 0 & 1 & 0 & -1 \\ 0 & 0 & 0 & 0 & 0 \end{bmatrix}.
$$

The leading columns of $\mathbf{G}$ are the first three, so the first three columns of $\mathbf{A}$ serve as our basis: We expand the original set of two vectors by appending to it the vector $\begin{bmatrix} 1 & 0 & 0 & 1 \end{bmatrix}^T$ to obtain a basis of three vectors, so the dimension of $\mathscr{V}_0$ is 3.

You should now be able to solve Problems 1 to 10.

Column Space and Row Space

The problems solved in the two examples above used the fact that the set of columns of a matrix $\mathbf{A}$ corresponding to the leading columns in a Gauss-reduced form $\mathbf{G}$ of $\mathbf{A}$ is linearly independent and that all the columns of $\mathbf{A}$ are linearly dependent on these. Therefore, any column matrix that is a linear combination of all the columns of **A** is a linear combination of these special columns of **A**. In other words, these special columns form a basis for the subspace of $\mathbb{R}^p$ or $\mathbb{C}^p$ spanned by all the columns of **A**. This space has a special name, the *column space* of **A**; by analogy, there is also a *row space.*

(5.49) ***Definition.*** Let **A** be a $p \times q$ matrix.

(a) The real (or complex) *column space* of **A** is the subspace of $\mathbb{R}^p$ (or $\mathbb{C}^p$) that is spanned by the set of columns of **A**.

(b) The real (or complex) *row space* of **A** is the subspace of the real (or complex) vector space of all real (or complex) $1 \times q$ matrices that is spanned by the set of rows of **A**.

In this terminology it is simple to state a fundamental result that follows immediately from **Key Theorems 5.45** and **4.23** and Theorem 5.46.

(5.50) **Key Theorem** (row/column spaces and rank). Let **A** have rank k. Then

(a) The column space of **A** has dimension k, and a basis for it consists of the set of columns of **A** *corresponding to* the leading columns in any Gauss-reduced form of **A**.

(b) The row space of **A** has dimension k, and a basis for it consists of the nonzero rows *in* any Gauss-reduced form of **A**.

(c) $p \times p$ **A** is nonsingular if and only if its columns form a linearly independent set, in which case the column space has dimension p.

(d) $p \times p$ **A** is nonsingular if and only if its rows form a linearly independent set, in which case the row space has dimension p.

According to this, the rank of **A** could have been *defined* as the dimension of the row space or as the dimension of the column space; since the column space of **A** is (isomorphic to) the row space of $\mathbf{A}^T$ (or $\mathbf{A}^H$), this says that the ranks of **A**, $\mathbf{A}^T$, and $\mathbf{A}^H$ are equal.

(5.51) ***Corollary*** (row rank = column rank). The ranks of **A**, $\mathbf{A}^T$, and $\mathbf{A}^H$ are equal.

Sometimes we are given two subsets S_1 and S_2 of $\mathbb{R}^p$ or of $\mathbb{C}^p$ and need to determine whether they span exactly the same space. If we take the *transposes* of the vectors in S_1 and use them as the *rows* of a matrix $\mathbf{A}_1$ and do likewise with S_2 to get an $\mathbf{A}_2$, then we can rephrase the question as whether $\mathbf{A}_1$ and $\mathbf{A}_2$ have the same row space. This can easily be answered.

(5.52) ***Theorem*** (equal row spaces). Let $\mathbf{A}_1$ be $r \times q$ and $\mathbf{A}_2$ be $s \times q$. Then the row space of $\mathbf{A}_1$ and the row space of $\mathbf{A}_2$ are the same if and only if the nonzero rows in the row-echelon form of $\mathbf{A}_1$ equal the nonzero rows in the row-echelon form of $\mathbf{A}_2$.

PROOF. (forms $\Rightarrow$ spaces) Since the nonzero rows in the row-echelon forms (which are Gauss-reduced forms) form bases for the row spaces, if the nonzero rows in the forms are equal, then the row spaces are equal.

(spaces $\Rightarrow$ forms) If the row spaces $\mathscr{V}$ are equal, then they certainly have the same dimension p, so there are p nonzero rows in the two row-echelon forms $\mathbf{R}_1$ and $\mathbf{R}_2$; let $\mathbf{R}'_1$ and $\mathbf{R}'_2$ denote the $p \times q$ matrices formed from the first p (nonzero) rows of $\mathbf{R}_1$ and $\mathbf{R}_2$. Let $\mathbf{v}_1, \mathbf{v}_2, \ldots, \mathbf{v}_p$ be the rows of $\mathbf{R}'_1$ in order, and let $\mathbf{v}'_1, \mathbf{v}'_2, \ldots, \mathbf{v}'_p$ be the rows of $\mathbf{R}'_2$ in order; each of these sets forms an ordered basis for $\mathscr{V}$. We use the material in Section 5.4 on change of basis by writing

$$\mathbf{v}'_i = m_{1i}\mathbf{v}_1 + m_{2i}\mathbf{v}_2 + \cdots + m_{pi}\mathbf{v}_p$$

exactly as in (5.41). We found in **Key Theorem 5.43** that the $p \times p$ matrix $\mathbf{M}$ formed as $\langle \mathbf{M} \rangle_{ij} = m_{ij}$ is nonsingular. The equation above, interpreted entry by entry, says

$$\langle \mathbf{v}'_i \rangle_k = \langle \mathbf{M}^T \rangle_{i1} \langle \mathbf{v}_1 \rangle_k + \langle \mathbf{M}^T \rangle_{i2} \langle \mathbf{v}_2 \rangle_k + \cdots + \langle \mathbf{M}^T \rangle_{ip} \langle \mathbf{v}_p \rangle_k.$$

Since the $\mathbf{v}'_i$ are the rows of $\mathbf{R}'_2$ while the $\mathbf{v}_i$ are the rows of $\mathbf{R}'_1$, this says that $\mathbf{R}'_2 = \mathbf{M}^T\mathbf{R}'_1$. Since $\mathbf{M}$ is nonsingular, so is $\mathbf{M}^T$; therefore, $\mathbf{M}^T$ can be written as a product of elementary matrices, each of which corresponds to performing an elementary row operation. In other words, $\mathbf{R}'_2$ can be produced from $\mathbf{R}'_1$ by elementary row operations; therefore, $\mathbf{R}'_2$ is the row-echelon form of $\mathbf{R}'_1$, which is itself in row-echelon form already. Since the row-echelon form of any matrix is unique, $\mathbf{R}'_1 = \mathbf{R}'_2$, as claimed. ∎

(5.53) ***Example.*** Suppose that we need to know whether the vectors

$$\begin{bmatrix} 1 \\ 3 \\ -7 \end{bmatrix}, \quad \begin{bmatrix} 2 \\ -1 \\ 0 \end{bmatrix}, \quad \begin{bmatrix} 3 \\ -1 \\ -1 \end{bmatrix}, \quad \begin{bmatrix} 4 \\ -3 \\ 2 \end{bmatrix}$$

span the same subspace of $\mathbb{C}^3$ as do

$$\begin{bmatrix} 1 \\ -1 \\ 1 \end{bmatrix}, \quad \begin{bmatrix} 1 \\ 1 \\ -3 \end{bmatrix}, \quad \begin{bmatrix} 1 \\ 2 \\ -5 \end{bmatrix}.$$

We form the 4×3 matrix $\mathbf{A}_1$ from the transposes of the first set and the 3×3 matrix $\mathbf{A}_2$ from the transposes of the second set:

$$\mathbf{A}_1 = \begin{bmatrix} 1 & 3 & -7 \\ 2 & -1 & 0 \\ 3 & -1 & -1 \\ 4 & -3 & 2 \end{bmatrix} \quad \text{and} \quad \mathbf{A}_2 = \begin{bmatrix} 1 & -1 & 1 \\ 1 & 1 & -3 \\ 1 & 2 & -5 \end{bmatrix}.$$

Reducing these to row-echelon forms $\mathbf{R}_1$ and $\mathbf{R}_2$ gives

$$\mathbf{R}_1 = \begin{bmatrix} 1 & 0 & -1 \\ 0 & 1 & -2 \\ 0 & 0 & 0 \\ 0 & 0 & 0 \end{bmatrix} \quad \text{and} \quad \mathbf{R}_2 = \begin{bmatrix} 1 & 0 & -1 \\ 0 & 1 & -2 \\ 0 & 0 & 0 \end{bmatrix}.$$

The nonzero rows in $\mathbf{R}_1$ and $\mathbf{R}_2$ are equal; therefore, the two original sets span the same subspace of $\mathbb{C}^3$.

PROBLEMS 5.5

▷ **1.** Find a basis for the subspace of $\mathbb{R}^4$ spanned by

$$S = \left\{ \begin{bmatrix} 1 \\ 2 \\ -1 \\ 3 \end{bmatrix}, \begin{bmatrix} 2 \\ -1 \\ 0 \\ 1 \end{bmatrix}, \begin{bmatrix} 1 \\ 7 \\ -3 \\ 8 \end{bmatrix}, \begin{bmatrix} 5 \\ 5 \\ -3 \\ 10 \end{bmatrix} \right\}.$$

2. Prove Theorem 5.46.

3. Find a basis for the subspace of $\mathbb{R}^4$ spanned by the vectors

$$\begin{bmatrix} 1 \\ -1 \\ 2 \\ 1 \end{bmatrix}, \begin{bmatrix} -1 \\ 2 \\ -3 \\ 1 \end{bmatrix}, \begin{bmatrix} -1 \\ 3 \\ -3 \\ 1 \end{bmatrix}, \begin{bmatrix} 2 \\ 1 \\ 2 \\ 6 \end{bmatrix}.$$

4. Find a basis for the subspace of $\mathbb{R}^4$ spanned by the vectors

$$\mathbf{v}_1 = \begin{bmatrix} 1 \\ 1 \\ 1 \\ 1 \end{bmatrix}, \quad \mathbf{v}_2 = \begin{bmatrix} 1 \\ 2 \\ 1 \\ 2 \end{bmatrix}, \quad \mathbf{v}_3 = \begin{bmatrix} 1 \\ 3 \\ 1 \\ 3 \end{bmatrix}, \quad \text{and} \quad \mathbf{v}_4 = \begin{bmatrix} 0 \\ 1 \\ 2 \\ 3 \end{bmatrix}.$$

▷ **5.** Find the values of k so that the vectors

$$[3-k \quad -1 \quad 0]^T, \quad [-1 \quad 2-k \quad -1]^T, \quad \text{and} \quad [0 \quad -1 \quad 3-k]^T$$

span a two-dimensional space.

6. Consider the subspace $\mathscr{V}_0$ of $\mathbb{R}^4$ consisting of all those vectors

$$\mathbf{x} = [x_1 \quad x_2 \quad x_3 \quad x_4]^T$$

satisfying $x_1 + x_2 = x_3 + x_4$. Determine whether S is a basis for $\mathscr{V}_0$, where

$$S = \left\{ \begin{bmatrix} 1 \\ -1 \\ 1 \\ -1 \end{bmatrix}, \begin{bmatrix} 1 \\ 0 \\ 1 \\ 0 \end{bmatrix}, \begin{bmatrix} 0 \\ 1 \\ 0 \\ 1 \end{bmatrix} \right\}.$$

▷ **7.** Extend $S = \{[1 \quad 1 \quad 2 \quad 3]^T\}$ to form a basis for the subspace of $\mathbb{R}^4$ consisting of all vectors of the form

$$[a \quad b-a \quad b \quad b+a]^T$$

for all real numbers a and b.

8. Extend $S = \{[3 \quad 1 \quad 1 \quad 1]^T, [3 \quad 0 \quad 2 \quad 1]^T\}$ to form a basis for the subspace of $\mathbb{R}^4$ consisting of all those vectors

$$\mathbf{x} = [x_1 \quad x_2 \quad x_3 \quad x_4]^T$$

satisfying $x_1 - x_2 = x_3 + x_4$.

9. Extend the set below to form a basis for $\mathbb{R}^4$:

$$\left\{\begin{bmatrix}1\\0\\-1\\0\end{bmatrix}, \begin{bmatrix}1\\1\\0\\0\end{bmatrix}, \begin{bmatrix}1\\2\\-1\\4\end{bmatrix}\right\}$$

10. Provide the details to prove **Key Theorem 5.50.**

▷ **11.** Find a basis for the column space of $\mathbf{A}$ from among its columns, where

$$\mathbf{A} = \begin{bmatrix}1 & 2 & 4 & 1\\2 & 1 & 5 & 5\\-1 & 2 & 0 & -5\end{bmatrix}.$$

12. Find a basis for the row space of the matrix $\mathbf{A}$ in Problem 11:

(a) By using **Key Theorem 5.50(b)** as usual.

(b) *From among the rows of* $\mathbf{A}$ by using **Key Theorem 5.50(a)** on $\mathbf{A}^T$ and then taking transposes of the resulting columns.

▷ **13.** Find the dimension of and a basis for the column space of

$$\mathbf{A} = \begin{bmatrix}1 & 4 & 0 & 2\\2 & 8 & 1 & 5\\-1 & -4 & 3 & 1\\0 & -3 & 4 & 4\end{bmatrix}.$$

▷ **14.** Find the dimension of and a basis for the row space of the matrix $\mathbf{A}$ in Problem 13:

(a) As in Problems 12(a)

(b) As in Problem 12(b)

15. Show that the two sets S_1 and S_2 below span the same subspace of $\mathbb{R}^3$.

$$S_1 = \left\{\begin{bmatrix}1\\0\\-1\end{bmatrix}, \begin{bmatrix}1\\1\\0\end{bmatrix}, \begin{bmatrix}0\\1\\1\end{bmatrix}\right\} \quad \text{and} \quad S_2 = \left\{\begin{bmatrix}2\\1\\-1\end{bmatrix}, \begin{bmatrix}1\\2\\1\end{bmatrix}\right\}$$

▷ **16.** Show that the two sets S_1 and S_2 below span the same subspace of $\mathbb{R}^3$.

$$S_1 = \left\{ \begin{bmatrix} 2 \\ -1 \\ 1 \end{bmatrix}, \begin{bmatrix} 1 \\ 2 \\ -3 \end{bmatrix} \right\} \quad \text{and} \quad S_2 = \left\{ \begin{bmatrix} 3 \\ 1 \\ -2 \end{bmatrix}, \begin{bmatrix} -1 \\ 3 \\ -4 \end{bmatrix} \right\}$$

17. Determine whether the two sets S_1 and S_2 below span the same subspace of $\mathbb{R}^3$.

$$S_1 = \left\{ \begin{bmatrix} 1 \\ 0 \\ -1 \end{bmatrix}, \begin{bmatrix} 1 \\ 1 \\ 0 \end{bmatrix}, \begin{bmatrix} 0 \\ 1 \\ 1 \end{bmatrix} \right\} \quad \text{and} \quad S_2 = \left\{ \begin{bmatrix} 2 \\ 1 \\ -1 \end{bmatrix}, \begin{bmatrix} 1 \\ -1 \\ 0 \end{bmatrix} \right\}$$

18. Since the row-echelon form of a matrix usually contains many zeros, its rows can be thought of as giving a basis of "simple" vectors for the row space of **A** or—after taking transposes—the column space of $\mathbf{A}^T$. By finding the row-echelon form of the transpose of the matrix **A** in Example 5.47, find a basis of "simple" vectors for the subspace $\mathscr{V}_0$ in that example.

▷ **19.** Use the approach of Problem 18 to find a basis of "simple" vectors for the *column* space—so you again have to work on $\mathbf{A}^T$—of

$$\mathbf{A} = \begin{bmatrix} 2 & -1 & 3 & -4 & 3 & 5 \\ -2 & 2 & -5 & 5 & -2 & -10 \\ 2 & 0 & 2 & -1 & 4 & 5 \\ 0 & 5 & -8 & 10 & 5 & -14 \\ 0 & -1 & 1 & -4 & -1 & -1 \\ 4 & 6 & -6 & 9 & 14 & -9 \end{bmatrix}.$$

20. Show that the column space of **AB** is a subspace of the column space of **A**.

21. Use Problem 20 to show that the rank of **AB** is less than or equal to that of **A**.

▷ **22.** (a) Use the fact that the rank of **C** equals that of $\mathbf{C}^T$ and the result of Problem 21 to show that the rank of **AB** is less than or equal to both the rank of **A** and the rank of **B**.

(b) Give an example to show that equality may hold in (a) and an example to show that strict inequality may hold in (a).

5.6 *LENGTH AND DISTANCE IN VECTOR SPACES: NORMS*

This chapter opened with the stated purpose of treating geometric aspects of vectors. Certainly, the material up to now on subspaces and the like has a strongly geometric flavor. Terms such as "length" and "angle," however, have been missing from all except Section 5.1 where geometrical vectors were discussed. We now take up both length and distance from **u** to **v** (defined as the length of $\mathbf{u} - \mathbf{v}$).

In Section 5.1 we saw that $(u_1^2 + u_2^2)^{1/2}$ gives the usual physical length of the two-dimensional geometrical vector $\vec{u}$ corresponding to the matrix $\mathbf{u} = [u_1 \quad u_2]^T$; in three dimensions, the Pythagorean Theorem again gives $(u_1^2 + u_2^2 + u_3^2)^{1/2}$ as the physical length. It seems natural, then, to measure the length, or magnitude, or size of a vector $\mathbf{u}$ *in* $\mathbb{R}^p$ by

$$(u_1^2 + u_2^2 + \cdots + u_p^2)^{1/2}.$$

This is, in fact, very often quite useful.

There are other situations, however, where some other measure of the size of a vector $\mathbf{u}$ is more meaningful. In our model of the fox and chicken populations in Section 2.3, for example, the matrix $\mathbf{x}_i = [F_i \quad C_i]^T$ was important, where F_i equals the number of foxes and C_i the number of chickens. Here $(F_i^2 + C_i^2)^{1/2}$ has little intuitive meaning. More meaningful is $|F_i| + |C_i|$—the total population—or perhaps $\max\{|F_i|, |C_i|\}$—the larger population.

In other applications, still other notions of the magnitude or length of a vector may be more appropriate than any of the above. We use the word *norm* to describe the general notion of length as long as it satisfies certain natural properties that most people intuitively associate with a sense of length.

(5.54) ***Definition.*** A *norm* (or *vector norm*) on $\mathscr{V}$ is a function that assigns to each vector $\mathbf{v}$ in $\mathscr{V}$ a nonnegative real number, called the *norm* of $\mathbf{v}$ and denoted by $\|\mathbf{v}\|$ (or sometimes $\|\mathbf{v}\|_{\mathscr{V}}$), satisfying:

(a) $\|\mathbf{v}\| > 0$ for $\mathbf{v} \neq \mathbf{0}$, and $\|\mathbf{0}\| = 0$

(b) $\|\alpha\mathbf{v}\| = |\alpha| \, \|\mathbf{v}\|$ for all scalars α and vectors $\mathbf{v}$

(c) $\|\mathbf{u} + \mathbf{v}\| \leq \|\mathbf{u}\| + \|\mathbf{v}\|$ (the triangle inequality) for all vectors $\mathbf{u}$ and $\mathbf{v}$.

The third condition, (c), is called the *triangle inequality* because it is a generalization—see (5.2) for motivation—of the fact that the length of any side of a triangle is less than or equal to the sum of the lengths of the other two sides.

There are three norms on $\mathbb{R}^p$ and $\mathbb{C}^p$ that are most commonly used in applications.

(5.55) ***Definition.*** For vectors $\mathbf{x} = [x_1 \quad x_2 \quad \cdots \quad x_p]^T$ in $\mathbb{R}^p$ or $\mathbb{C}^p$, the norms $\|\cdot\|_1$, $\|\cdot\|_2$, and $\|\cdot\|_\infty$—called the 1-norm, 2-norm, and ∞-norm—are defined as:

$$\|\mathbf{x}\|_1 = |x_1| + |x_2| + \cdots + |x_p|$$

$$\|\mathbf{x}\|_2 = (|x_1|^2 + |x_2|^2 + \cdots + |x_p|^2)^{1/2}$$

$$\|\mathbf{x}\|_\infty = \max\{|x_1|, |x_2|, \ldots, |x_p|\}.$$

(5.56) ***Example.*** Let $\mathbf{x} = [-5 \quad 3 \quad -2]^T$ and $\mathbf{y} = [2 + i \quad -1 - i \quad 2 \quad 1]^T$. Then

$$\|\mathbf{x}\|_1 = 10, \quad \|\mathbf{x}\|_2 = (38)^{1/2}, \quad \|\mathbf{x}\|_\infty = 5,$$

$$\|\mathbf{y}\|_1 = \sqrt{5} + \sqrt{2} + 3, \quad \|\mathbf{y}\|_2 = (12)^{1/2}, \quad \text{and} \quad \|\mathbf{y}\|_\infty = \sqrt{5}.$$

To verify that the quantities in Definition 5.55 actually are norms, we have to verify that they satisfy the conditions in Definition 5.54; see Problems 9–11. The only challenge in this is showing that $\|\cdot\|_2$ satisfies the triangle inequality; this follows from an important result known as the *Schwarz inequality*. To state it, we use the hermitian transpose ^{H} from Definition 1.20 and the expression $\mathbf{x}^H\mathbf{y}$ for two $p \times 1$ column matrices;

> *for notational convenience, we treat the 1×1 matrix $\mathbf{x}^H\mathbf{y}$ as equal to its sole entry.*

Thus, if $\mathbf{x} = [1 \quad 2 \quad -3]^T$ and $\mathbf{y} = [2 \quad 1 \quad 1]^T$, we write $\mathbf{x}^H\mathbf{y} = 1$ rather than the technically correct but notationally awkward $\mathbf{x}^H\mathbf{y} = [1]$. With this notation, we can write

$$\|\mathbf{x}\|_2 = (\mathbf{x}^H\mathbf{x})^{1/2}.$$

(5.57) **Key Theorem** (Schwarz inequality). Let $\mathbf{x}$ and $\mathbf{y}$ be $p \times 1$ column matrices. Then

$$|\mathbf{x}^H\mathbf{y}| \le \|\mathbf{x}\|_2\|\mathbf{y}\|_2.$$

PROOF. For all numbers α, we have

$$\begin{aligned} 0 \le \|\mathbf{x} + \alpha\mathbf{y}\|_2^2 &= \mathbf{x}^H\mathbf{x} + \alpha\mathbf{x}^H\mathbf{y} + \bar{\alpha}\mathbf{y}^H\mathbf{x} + \alpha\bar{\alpha}\mathbf{y}^H\mathbf{y} \\ &= \|\mathbf{x}\|_2^2 + \alpha\mathbf{x}^H\mathbf{y} + \bar{\alpha}\mathbf{y}^H\mathbf{x} + |\alpha|^2\|\mathbf{y}\|_2^2. \end{aligned}$$

Since the inequality is certainly true if $\mathbf{x}^H\mathbf{y} = 0$, we suppose that $\mathbf{x}^H\mathbf{y} \ne 0$ and let $\alpha = -\|\mathbf{x}\|_2^2/\mathbf{x}^H\mathbf{y}$ in the expression above. Combining terms above and using $\mathbf{y}^H\mathbf{x} = \overline{\mathbf{x}^H\mathbf{y}}$ produces

$$0 \le -\|\mathbf{x}\|_2^2 + \frac{\|\mathbf{x}\|_2^4\,\|\mathbf{y}\|_2^2}{|\mathbf{x}^H\mathbf{y}|^2},$$

from which the Schwarz inequality immediately follows. ■

You should now be able to solve Problems 1 to 11.

Convergence

In applications it is often important to know how some sequence of vectors $\mathbf{x}_i$ behaves as i tends to infinity. Section 2.2 presented a model of market competition in which the sequence $\mathbf{x}_i$ representing market shares tended to a limit $\mathbf{x}_\infty$; Section 2.3 presented a model of competing populations in which the sequence $\mathbf{x}_i$ representing populations sometimes tended to $\mathbf{0}$, sometimes exploded, and so on. Since norms measure the magnitude of vectors, they provide the natural tool for describing precisely what we mean by, for example, saying that the sequence of vectors $\mathbf{x}_i$ tends to $\mathbf{0}$: we should mean that their magnitudes $\|\mathbf{x}_i\|$ tend to 0.

(5.58) ***Definition.*** Let $\|\cdot\|$ be a norm on $\mathscr{V}$. A sequence of vectors $\mathbf{v}_i$ is said to *converge* to the vector $\mathbf{v}_\infty$ if and only if the sequence of real numbers $\|\mathbf{v}_i - \mathbf{v}_\infty\|$ converges to 0.

With this definition, we can now talk of a sequence of vectors in $\mathbb{R}^p$ or $\mathbb{C}^p$ converging by using norms. But which norm should we use? The interesting fact is: It doesn't matter—they are all "equivalent" as we now show.

From their very definitions, it is clear that $\|\mathbf{x}\|_\infty \le \|\mathbf{x}\|_1 \le p\|\mathbf{x}\|_\infty$ if $\mathbf{x}$ is $p \times 1$; similarly, $\|\mathbf{x}\|_2^2 \le \|\mathbf{x}\|_1^2$. If we apply the Schwarz inequality to the vectors $\mathbf{x}$ with $\langle\mathbf{x}\rangle_i = |u_i|$ and $\mathbf{y}$ with $\langle\mathbf{y}\rangle_i = 1$, we obtain $\|\mathbf{u}\|_1 \le p^{1/2}\|\mathbf{u}\|_2$ if $\mathbf{u}$ is $p \times 1$. Thus if any of the 1-norm, 2-norm, or ∞-norm of a sequence of $p \times 1$ matrices tends to zero, then they all do, and so do the entries of the matrices.

(5.59) ***Theorem*** (equivalence of norms). The 1-norm, 2-norm, and ∞-norm on $\mathbb{R}^p$ (or $\mathbb{C}^p$) are all equivalent in the sense that:

(a) If a sequence $\mathbf{x}_i$ of vectors converges to $\mathbf{x}_\infty$ as determined in *one* of the norms, then it converges to $\mathbf{x}_\infty$ as determined in *all three* norms and the entries $\langle\mathbf{x}_i\rangle_j$ converge to the entries $\langle\mathbf{x}_\infty\rangle_j$.

(b) For all $p \times 1$ matrices $\mathbf{x}$,

$$\frac{\|\mathbf{x}\|_2}{\sqrt{p}} \le \|\mathbf{x}\|_\infty \le \|\mathbf{x}\|_2$$

$$\|\mathbf{x}\|_2 \le \|\mathbf{x}\|_1 \le \|\mathbf{x}\|_2\sqrt{p}$$

$$\frac{\|\mathbf{x}\|_1}{p} \le \|\mathbf{x}\|_\infty \le \|\mathbf{x}\|_1.$$

This equivalence is actually a special case of a more general theorem that any two norms on a finite-dimensional space are equivalent, but we do not pursue this here.

(5.60) ***Example.*** By looking at the entries, it is intuitively clear that the sequence of vectors

$$\mathbf{x}_i = \begin{bmatrix} \dfrac{i}{(i+1)} & \dfrac{2}{i} \end{bmatrix}^T$$

converges to $\mathbf{x}_\infty = [1 \quad 0]^T$. According to Definition 5.58, however, convergence should be expressed in terms of norms of

$$\mathbf{x}_i - \mathbf{x}_\infty = \begin{bmatrix} \dfrac{-1}{(i+1)} & \dfrac{2}{i} \end{bmatrix}^T.$$

It is easy to see from this that $\|\mathbf{x}_i - \mathbf{x}_\infty\|_\infty = 2/i$, which certainly converges to 0; by Theorem 5.59, so do the other norms of the $\mathbf{x}_i - \mathbf{x}_\infty$, as you can easily verify directly. The formal definition of convergence agrees with our intuitive notion.

PROBLEMS 5.6

▷ **1.** For each of the following vectors $\mathbf{x}$, compute $\|\mathbf{x}\|_1$, $\|\mathbf{x}\|_2$, and $\|\mathbf{x}\|_\infty$:

(a) $[2 \quad -5]^T$ (b) $[2i \quad 0 \quad -3i]^T$

(c) $[2 \quad -5 \quad 5]^T$ (d) $[i \quad -i \quad 1 \quad -1]^T$

(e) $[0 \quad 0 \quad 0]^T$ (f) $[5i \quad -3 \quad 4]^T$

2. Let S be the set of points—a straight line—in $\mathbb{R}^2$ of the form $[1 \quad t]^T$ for all real numbers t. For each of the norms $\|\cdot\|_1$, $\|\cdot\|_2$, $\|\cdot\|_\infty$, find the point(s) in S that are closest to $\mathbf{0}$.

3. Compute $\mathbf{x}^T\mathbf{y}$, $\|\mathbf{x}\|_2$, and $\|\mathbf{y}\|_2$ and verify the Schwarz inequality for each of the following pairs of vectors.

$$\mathbf{x} = \begin{bmatrix} 1 \\ -3 \end{bmatrix}, \quad \mathbf{y} = \begin{bmatrix} 2 \\ 1 \end{bmatrix}; \quad \mathbf{x} = \begin{bmatrix} -3 \\ -4 \end{bmatrix}, \quad \mathbf{y} = \begin{bmatrix} 1 \\ -1 \end{bmatrix};$$

$$\mathbf{x} = \begin{bmatrix} 2 \\ -3 \end{bmatrix}, \quad \mathbf{y} = \begin{bmatrix} -6 \\ 9 \end{bmatrix}$$

4. Show graphically that the so-called $\|\cdot\|_2$-unit-circle in $\mathbb{R}^2$, namely the set of those vectors $\mathbf{x}$ with $\|\mathbf{x}\|_2 = 1$, is in fact the circle of radius 1 and center at the origin.

▷ **5.** Describe the so-called $\|\cdot\|_1$-unit-circle and $\|\cdot\|_\infty$-unit-circle (see Problem 4) in $\mathbb{R}^2$.

6. Suppose that $\|\cdot\|$ is a norm on $\mathbb{R}^p$ and that $\mathbf{A}$ is a real nonsingular $p \times p$ matrix, and define $\|\cdot\|_\mathbf{A}$ by $\|\mathbf{x}\|_\mathbf{A} = \|\mathbf{Ax}\|$. Prove that $\|\cdot\|_\mathbf{A}$ is a norm on $\mathbb{R}^p$.

7. In each of the norms $\|\cdot\|_1$, $\|\cdot\|_2$, and $\|\cdot\|_\infty$, calculate the norm of each of the columns of $\mathbf{A}' - \mathbf{A}$, where $\mathbf{A}$ and $\mathbf{A}'$ are as below. (Note that $\mathbf{A}$ is nonsingular while $\mathbf{A}'$ is singular.)

$$\mathbf{A} = \begin{bmatrix} 1 & 0 & 0.5 \\ 0 & 1 & 0.5 \\ 2 & 0.002 & 1.002 \end{bmatrix}, \quad \mathbf{A}' = \begin{bmatrix} 1 & 0 & 0.5005 \\ 0 & 1 & 0.5 \\ 2 & 0.002 & 1.002 \end{bmatrix}$$

▷ **8.** (a) By applying the Schwarz inequality to the expansion of

$$\|\mathbf{u} + \mathbf{v}\|_2^2 = (\mathbf{u} + \mathbf{v})^H(\mathbf{u} + \mathbf{v}),$$

prove the triangle inequality for $\|\cdot\|_2$.

(b) Conversely, use the triangle inequality for $\|\cdot\|_2$ to prove the Schwarz inequality.

9. Prove that $\|\cdot\|_1$ is a norm.

10. Prove that $\|\cdot\|_\infty$ is a norm.

11. Prove that $\|\cdot\|_2$ is a norm.

▷ **12.** For the sequence of vectors $\mathbf{x}_i$ below, find the limit $\mathbf{x}_\infty$. Use each of the norms $\|\cdot\|_1$, $\|\cdot\|_2$, and $\|\cdot\|_\infty$ to verify that $\mathbf{x}_i$ converges to $\mathbf{x}_\infty$.

$$\mathbf{x}_i = \begin{bmatrix} \dfrac{2i+3}{i+1} \\ \dfrac{i-2}{i^2+2} \\ \dfrac{i^2+1}{i^2-1} \end{bmatrix}$$

▷ **13.** Let $\mathscr{V}$ be a vector space with norm $\|\cdot\|$. Prove that

$$\big|\|\mathbf{u}\| - \|\mathbf{v}\|\big| \leq \|\mathbf{u} - \mathbf{v}\| \qquad \text{for all } \mathbf{u} \text{ and } \mathbf{v} \text{ in } \mathscr{V}.$$

14. (a) Use Problem 13 to show that, if $\mathbf{u}_i$ converges to $\mathbf{u}_\infty$ in the vector space $\mathscr{V}$ having norm $\|\cdot\|$, then $\|\mathbf{u}_i\|$ converges to $\|\mathbf{u}_\infty\|$.
(b) Give an example in $\mathbb{R}^2$ to show that the norms may converge without the vectors converging.

▷ **15.** In $\mathbb{R}^p$ (or $\mathbb{C}^p$), suppose that $\mathbf{u}_i$ converges to $\mathbf{u}_\infty$ and $\mathbf{v}_i$ converges to $\mathbf{v}_\infty$ as determined by one of the norms $\|\cdot\|_1$, $\|\cdot\|_2$, or $\|\cdot\|_\infty$. Show that $\mathbf{u}_i^T\mathbf{v}_i$ converges to $\mathbf{u}_\infty^T\mathbf{v}_\infty$ (or $\mathbf{u}_i^H\mathbf{v}_i$ to $\mathbf{u}_\infty^H\mathbf{v}_\infty$).

16. Suppose that B is an ordered basis for the p-dimensional real (or complex) vector space $\mathscr{V}$, and let c_B denote the coordinate isomorphism.
(a) Suppose that $\|\cdot\|$ is a norm on $\mathbb{R}^p$ (or $\mathbb{C}^p$). Show that $\|\cdot\|_{\mathscr{V}}$ is a norm on $\mathscr{V}$, where $\|\mathbf{v}\|_{\mathscr{V}}$ for $\mathbf{v}$ in $\mathscr{V}$ is defined to equal $\|c_B(\mathbf{v})\|$.
(b) Suppose that $\|\cdot\|_{\mathscr{V}}$ is a norm on $\mathscr{V}$. Show that $\|\cdot\|$ is a norm on $\mathbb{R}^p$ (or $\mathbb{C}^p$), where $\|\mathbf{v}_B\|$ for $\mathbf{v}_B$ in $\mathbb{R}^p$ (or $\mathbb{C}^p$) is defined as $\|c_B^{-1}(\mathbf{v}_B)\|_{\mathscr{V}}$ where $c_B^{-1}(\mathbf{v}_B)$ is the unique $\mathbf{v}$ in $\mathscr{V}$ for which $c_B(\mathbf{v}) = \mathbf{v}_B$.

17. Let $\mathscr{V}$ equal $\mathscr{P}^3$, the real vector space of polynomials of degree strictly less than 3, and let $B = \{1; 1+t; t+t^2\}$ be an ordered basis for $\mathscr{V}$. For each of the norms $\|\cdot\|_1$, $\|\cdot\|_2$, and $\|\cdot\|_\infty$ on $\mathbb{R}^3$, find a formula for the norm $\|a + bt + ct^2\|_{\mathscr{V}}$ on $\mathscr{V}$ as defined in Problem 16(a).

▷ **18.** Let $\mathscr{V}$ and B be as in Problem 17, and define $\|\cdot\|_{\mathscr{V}}$ on $\mathscr{V}$ as

$$\|\mathbf{v}\|_{\mathscr{V}} = \left\{\int_0^1 |\mathbf{v}(t)|^2\, dt\right\}^{1/2}.$$

Find a formula for the norm $\|[a \quad b \quad c]^T\|$ on $\mathbb{R}^3$ as defined in Problem 16(b).

5.7 ANGLE IN VECTOR SPACES: INNER PRODUCTS

Section 5.1 showed—see (5.3)—that the fact that two geometrical vectors $\vec{u}$ and $\vec{v}$ are perpendicular can be expressed in matrix language as $\mathbf{u}^T\mathbf{v} = [0]$, where $\mathbf{u}$ corresponds to $\vec{u}$ and $\mathbf{v}$ to $\vec{v}$ as described in that section. Thus the concept *right angle* can be expressed through the special product $\mathbf{u}^T\mathbf{v}$ between column matrices. We show that *angle* more generally can be expressed this way.

The Angle Between Geometrical Vectors

Consider the angle θ between the geometrical vectors $\vec{a}$ and $\vec{b}$ in (5.61).

(5.61)

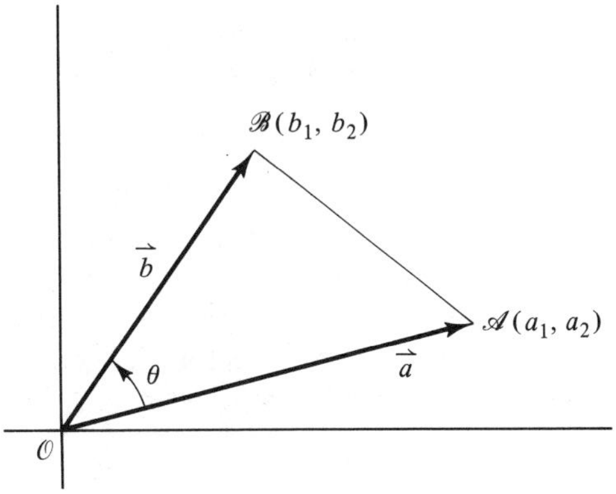

The law of cosines from trigonometry states that

$$|\mathscr{A}\mathscr{B}|^2 = |\mathscr{O}\mathscr{A}|^2 + |\mathscr{O}\mathscr{B}|^2 - 2|\mathscr{O}\mathscr{A}|\,|\mathscr{O}\mathscr{B}| \cos\theta$$

where $|\mathscr{P}\mathscr{Q}|$ denotes the distance between points $\mathscr{P}$ and $\mathscr{Q}$. The Pythagorean Theorem gives

$$|\mathscr{O}\mathscr{A}|^2 = a_1^2 + a_2^2$$
$$|\mathscr{O}\mathscr{B}|^2 = b_1^2 + b_2^2$$
$$|\mathscr{A}\mathscr{B}|^2 = (a_1 - b_1)^2 + (a_2 - b_2)^2.$$

Substituting these into the law of cosines and collecting terms gives

$$-2(a_1b_1 + a_2b_2) = -2(a_1^2 + a_2^2)^{1/2}(b_1^2 + b_2^2)^{1/2} \cos\theta.$$

If we let $\mathbf{a} = [a_1 \quad a_2]^T$ and $\mathbf{b} = [b_1 \quad b_2]^T$ and write $a_1b_1 + a_2b_2$ as $\mathbf{a}^T\mathbf{b}$, we can write the equation above as

$$-2\mathbf{a}^T\mathbf{b} = -2\|\mathbf{a}\|_2\|\mathbf{b}\|_2 \cos\theta$$

and therefore:

(5.62) If θ is the angle between the two nonzero geometrical vectors $\vec{a}$ and $\vec{b}$ that correspond to 2×1 matrices $\mathbf{a}$ and $\mathbf{b}$, then

$$\cos\theta = \frac{\mathbf{a}^T\mathbf{b}}{\|\mathbf{a}\|_2\|\mathbf{b}\|_2}.$$

Suppose now that we want to measure angles between two *three*-dimensional geometrical vectors; if we pass a plane through those vectors, then in that plane they appear as *two*-dimensional and we can measure angles as before. It turns out that the same formula as that in (5.62) results, except now $\mathbf{a}$ and $\mathbf{b}$ denote 3×1 matrices that contain the coordinates of the "heads" of the geometrical vectors. This approach works because the angle between two vectors really is a two-dimensional concept defined by the two vectors and the plane they determine; applying the same approach to the angle between two vectors $\mathbf{a}$ and $\mathbf{b}$ in $\mathbb{R}^p$ leads once again to the same formula in (5.62), where now $\mathbf{a}$ and $\mathbf{b}$ are $p \times 1$.

Inner Products

The discussion above shows that the concept of angle between two vectors in $\mathbb{R}^p$ is contained in the notions of the special product $\mathbf{a}^T\mathbf{b}$ and of the 2-norm, which itself can be defined through this product since $\|\mathbf{a}\|_2 = (\mathbf{a}^T\mathbf{a})^{1/2}$. To develop a concept of angle between two vectors in general vector spaces then requires a concept similar to this special product $\mathbf{a}^T\mathbf{b}$.

(5.63) ***Definition***

(a) (the *real* case) Let $\mathscr{V}$ be a real vector space. An *inner product* on $\mathscr{V}$ is a function that assigns to each ordered pair of vectors $\mathbf{u}$ and $\mathbf{v}$ in $\mathscr{V}$ a real number, denoted $(\mathbf{u}, \mathbf{v})$, satisfying:

1. $(\mathbf{u}, \mathbf{v}) = (\mathbf{v}, \mathbf{u})$ for all $\mathbf{u}$, $\mathbf{v}$ in $\mathscr{V}$.
2. $(\alpha\mathbf{u} + \beta\mathbf{v}, \mathbf{w}) = \alpha(\mathbf{u}, \mathbf{w}) + \beta(\mathbf{v}, \mathbf{w})$ and $(\mathbf{w}, \alpha\mathbf{u} + \beta\mathbf{v}) = \alpha(\mathbf{w}, \mathbf{u}) + \beta(\mathbf{w}, \mathbf{v})$ for all $\mathbf{u}$, $\mathbf{v}$, $\mathbf{w}$ in $\mathscr{V}$ and all real numbers α, β.
3. $(\mathbf{u}, \mathbf{u}) > 0$ if $\mathbf{u} \neq \mathbf{0}$, and $(\mathbf{u}, \mathbf{u}) = 0$ if and only if $\mathbf{u} = \mathbf{0}$.

The angle θ between two nonzero vectors $\mathbf{u}$ and $\mathbf{v}$ is defined by its cosine:

$$\cos\theta = \frac{(\mathbf{u}, \mathbf{v})}{(\mathbf{u}, \mathbf{u})^{1/2}(\mathbf{v}, \mathbf{v})^{1/2}}.$$

(b) (the *complex* case) Let $\mathscr{V}$ be a complex vector space. An *inner product* on $\mathscr{V}$ is a function that assigns to each ordered pair of vectors $\mathbf{u}$ and $\mathbf{v}$ in $\mathscr{V}$ a possibly complex number, denoted by $(\mathbf{u}, \mathbf{v})$, satisfying:

1. $(\mathbf{u}, \mathbf{v}) = \overline{(\mathbf{v}, \mathbf{u})}$ for all $\mathbf{u}$, $\mathbf{v}$ in $\mathscr{V}$.
2. $(\alpha\mathbf{u} + \beta\mathbf{v}, \mathbf{w}) = \bar{\alpha}(\mathbf{u}, \mathbf{w}) + \bar{\beta}(\mathbf{v}, \mathbf{w})$ and $(\mathbf{w}, \alpha\mathbf{u} + \beta\mathbf{v}) = \alpha(\mathbf{w}, \mathbf{u}) + \beta(\mathbf{w}, \mathbf{v})$ for all $\mathbf{u}$, $\mathbf{v}$, $\mathbf{w}$ in $\mathscr{V}$ and all complex numbers α, β.
3. $(\mathbf{u}, \mathbf{u}) > 0$ if $\mathbf{u} \neq \mathbf{0}$, and $(\mathbf{u}, \mathbf{u}) = 0$ if and only if $\mathbf{u} = \mathbf{0}$.

This definition of inner products is in two parts to simplify matters for those unfamiliar with complex numbers. The only differences in the cases involve the appearance of the complex conjugate in (1) and (2) and the omission of angle in the complex case.

(5.64) ***Example.*** By the *standard inner product* on $\mathbb{R}^p$ we mean $(\mathbf{u}, \mathbf{v})$ defined as (the entry in) $\mathbf{u}^T\mathbf{v}$:

$$(\mathbf{u}, \mathbf{v}) = u_1v_1 + \cdots + u_pv_p,$$

where $u_i = \langle\mathbf{u}\rangle_i$ and $v_i = \langle\mathbf{v}\rangle_i$. Unless explicitly stated otherwise, we always use this standard inner product in $\mathbb{R}^p$.

(5.65) ***Example.*** By the *standard inner product* on $\mathbb{C}^p$ we mean $(\mathbf{u}, \mathbf{v})$ defined as (the entry in) $\mathbf{u}^H\mathbf{v}$:

$$(\mathbf{u}, \mathbf{v}) = \bar{u}_1v_1 + \cdots + \bar{u}_pv_p,$$

where $u_i = \langle\mathbf{u}\rangle_i$ and $v_i = \langle\mathbf{v}\rangle_i$. Unless explicitly stated otherwise, we always use this standard inner product in $\mathbb{C}^p$.

(5.66) ***Example.*** Consider $\mathscr{P}^3$, the real vector space of polynomials of degree strictly less than 3. Define $(\mathbf{u}, \mathbf{v})$ for two polynomials $\mathbf{u}(t)$ and $\mathbf{v}(t)$ by

$$(\mathbf{u}, \mathbf{v}) = \int_0^1 \mathbf{u}(t)\mathbf{v}(t)\,dt.$$

It is easy to check that this defines an inner product, that is, that it satisfies Definition 5.63(a). Integration shows, for example, that $(1 + t, t^2) = 7/12$.

The standard inner products in $\mathbb{R}^p$ and $\mathbb{C}^p$ gave us the 2-norm of a vector as $\|\mathbf{v}\|_2 = (\mathbf{v}, \mathbf{v})^{1/2}$. The same is true for general inner products, but the Schwarz inequality is needed first.

(5.67) ***Theorem*** (Schwarz inequality). Let $(\mathbf{u}, \mathbf{v})$ be an inner product on the real or complex vector space $\mathscr{V}$. Then

$$|(\mathbf{u}, \mathbf{v})| \le (\mathbf{u}, \mathbf{u})^{1/2}(\mathbf{v}, \mathbf{v})^{1/2} \qquad \text{for all } \mathbf{u}, \mathbf{v} \text{ in } \mathscr{V}.$$

PROOF. The proof imitates that of **Key Theorem 5.57**; Problem 13. ∎

(5.68) ***Theorem*** (inner product norms). Let $(\mathbf{u}, \mathbf{v})$ be an inner product on $\mathscr{V}$, and define $\|\mathbf{v}\| = (\mathbf{v}, \mathbf{v})^{1/2}$. Then $\|\cdot\|$ is a norm on $\mathscr{V}$ (said to be the norm *induced* by the inner product).

PROOF. Problem 14. ∎

(5.69) ***Example.*** Consider a nonstandard inner product on $\mathbb{R}^2$ defined by

$$(\mathbf{u}, \mathbf{v})_0 = 3u_1v_1 + u_2v_2.$$

It is easy to show that this is an inner product, and therefore it defines an induced norm

$$\|\mathbf{u}\| = (3u_1^2 + u_2^2)^{1/2}.$$

You should now be able to solve Problems 1 to 15.

Orthogonality

In elementary geometry, an especially important role is played by right angles and by perpendicular (also called *orthogonal*) lines. So it is here. Since the cosine of a right angle equals zero, we have $(\mathbf{u}, \mathbf{v}) = 0$ for orthogonal vectors in $\mathbb{R}^p$. We use this same relationship to *define* orthogonality, even in complex vector spaces.

(5.70) ***Definition.*** Let $(\cdot, \cdot)$ be an inner product on $\mathcal{V}$ and (for the remainder of this chapter) let $\|\cdot\|$ be its induced norm from Theorem 5.68.

(a) Two vectors $\mathbf{u}$ and $\mathbf{v}$ are said to be *orthogonal* if and only if $(\mathbf{u}, \mathbf{v}) = 0$.
(b) A set of vectors is said to be *orthogonal* if and only if every two vectors from the set are orthogonal: $(\mathbf{u}, \mathbf{v}) = 0$ for all $\mathbf{u} \neq \mathbf{v}$ in that set.
(c) If a nonzero vector $\mathbf{u}$ is used to produce $\mathbf{v} = \mathbf{u}/\|\mathbf{u}\|$, so that $\|\mathbf{v}\| = 1$, then $\mathbf{u}$ is said to *have been normalized* to produce the *normalized vector* $\mathbf{v}$.
(d) A set of vectors is said to be *orthonormal* if and only if the set is orthogonal and $\|\mathbf{v}\| = 1$ for all $\mathbf{v}$ in the set.

By the properties of inner products, $(\mathbf{0}, \mathbf{v}) = (0\mathbf{v}, \mathbf{v}) = 0(\mathbf{v}, \mathbf{v}) = 0$. That is:

In any vector space with an inner product, $\mathbf{0}$ is orthogonal to every vector.

PROBLEMS 5.7

▷ **1.** Find the angle between the two two-dimensional geometrical vectors $\vec{u}$ and $\vec{v}$ corresponding to

$$\mathbf{u} = [1 \quad \sqrt{3}]^T \quad \text{and} \quad \mathbf{v} = [2\sqrt{3} \quad 2]^T.$$

2. Find the angle between the two vectors in $\mathbb{R}^5$,

$$\mathbf{u} = [1 \quad 2 \quad -1 \quad 0 \quad 1]^T \quad \text{and} \quad \mathbf{v} = [2 \quad 1 \quad -3 \quad 1 \quad -1]^T.$$

3. (a) Find the angle between the two vectors in $\mathbb{R}^2$ (using the standard inner product) $\mathbf{u} = [1 \quad 1]^T$ and $\mathbf{v} = [1 \quad -1]^T$.
(b) Find the "angle" between $\mathbf{u}$ and $\mathbf{v}$ when the nonstandard inner product of Example 5.69 is used to define the geometry.

▷ **4.** Find the general form of all those vectors $\mathbf{v}$ in $\mathbb{R}^3$ that are orthogonal to $\mathbf{n} = [1 \quad 2 \quad -1]^T$.

5. Suppose that $\mathscr{V}$ has an inner product, and that $\mathbf{n}$ is in $\mathscr{V}$. Show that the set $\mathscr{V}_0$ of all vectors in $\mathscr{V}$ that are orthogonal to $\mathbf{n}$ is a subspace.

6. Suppose that $\mathscr{V}$ has an inner product and that $\mathbf{n} \neq \mathbf{0}$ is in $\mathscr{V}$. Define $\mathscr{V}_1$ as the set of all vectors in $\mathscr{V}$ that satisfy $(\mathbf{n}, \mathbf{v}) = 1$.
(a) Show that $\mathscr{V}_1$ is not a subspace.
(b) Show that every vector $\mathbf{v}_1$ in $\mathscr{V}_1$ can be written as the sum $\mathbf{v}_0 + \mathbf{n}/\|\mathbf{n}\|^2$ of a vector $\mathbf{v}_0$ in the subspace $\mathscr{V}_0$ of Problem 5 and the multiple $\mathbf{n}/\|\mathbf{n}\|^2$ of $\mathbf{n}$, and that every vector $\mathbf{v}_1$ of this form is in fact in $\mathscr{V}_1$.

▷ **7.** Suppose that $\mathbf{n}$ is a nonzero vector in $\mathscr{V} = \mathbb{R}^p$. Show that the dimension of the subspace $\mathscr{V}_0$ of Problem 5 equals $p - 1$.

8. Let $\mathscr{P}^3$ be the real vector space of polynomials of degree strictly less than 3; define the inner product between the polynomials

$$\mathbf{f} = f_1 + f_2 t + f_3 t^2 \quad \text{and} \quad \mathbf{g} = g_1 + g_2 t + g_3 t^2$$

to be

$$(\mathbf{f}, \mathbf{g}) = f_1 g_1 + f_2 g_2 + f_3 g_3.$$

Find the angle between $\mathbf{f} = t$ and $\mathbf{g} = t^2 - t + 1$.

▷ **9.** Suppose that $\mathscr{V}$ is a real vector space with an inner product. Prove that

$$\|\mathbf{u} + \mathbf{v}\|^2 = \|\mathbf{u}\|^2 + \|\mathbf{v}\|^2$$

if and only if $\mathbf{u}$ and $\mathbf{v}$ are orthogonal, where $\|\ \|$ is the induced norm as usual.

10. We derived the definition of the induced norm from that of the inner product. Conversely, for a *real* vector space with inner product, show that

$$(\mathbf{u}, \mathbf{v}) = (\|\mathbf{u} + \mathbf{v}\|^2 - \|\mathbf{u} - \mathbf{v}\|^2)/4.$$

11. Let $C[0, 1]$ be the real vector space of continuous real-valued functions on $0 \leq t \leq 1$. For two functions f and g in $C[0, 1]$, define the inner product by

$$(f, g) = \int_0^1 f(t)g(t)\, dt.$$

(a) Verify that this is an inner product.
(b) Write out the Schwarz inequality in this case.

▷ **12.** (a) Show that the inner product on $C[0, 1]$ from Problem 11 defines an inner product on $\mathscr{P}^3$, the real vector space of polynomials of degree strictly less than 3; generalize this result to subspaces $\mathscr{V}_0$ of vector spaces $\mathscr{V}$ that have an inner product.
(b) Find the angle between t and $t^2 - t + 1$.

13. Prove Theorem 5.67 on the Schwarz inequality by imitating the proof of **Key Theorem 5.57**.

14. Prove Theorem 5.68 that the "norm" induced by an inner product is indeed a norm.

▷ **15.** Let B be an ordered basis for a p-dimensional real (or complex) vector space $\mathscr{V}$, and let c_B be the coordinate isomorphism.

(a) Suppose that $(\cdot, \cdot)$ is any inner product on $\mathbb{R}^p$ (or $\mathbb{C}^p$), and define

$$(\mathbf{u}, \mathbf{v})_{\mathscr{V}} = (c_B(\mathbf{u}), c_B(\mathbf{v})) \qquad \text{for } \mathbf{u}, \mathbf{v} \text{ in } \mathscr{V}.$$

Prove that $(\cdot, \cdot)_{\mathscr{V}}$ is an inner product on $\mathscr{V}$.

(b) Suppose that $(\cdot, \cdot)_{\mathscr{V}}$ is an inner product on $\mathscr{V}$. Define

$$(\mathbf{u}_B, \mathbf{v}_B) = (c_B^{-1}(\mathbf{u}_B), c_B^{-1}(\mathbf{v}_B))_{\mathscr{V}} \qquad \text{for } \mathbf{u}_B, \mathbf{v}_B \text{ in } \mathbb{R}^p \text{ (or } \mathbb{C}^p\text{)},$$

where $c_B^{-1}(\mathbf{u}_B)$ is the unique vector $\mathbf{u}$ in $\mathscr{V}$ whose B-coordinates are $\mathbf{u}_B$. Prove that $(\cdot, \cdot)$ is an inner product.

16. Determine whether the set below is orthogonal and whether it is orthonormal.

$$\left\{ \begin{bmatrix} 1 \\ 1 \\ 0 \end{bmatrix}, \begin{bmatrix} 0 \\ 0 \\ 1 \end{bmatrix}, \begin{bmatrix} 1 \\ -1 \\ 0 \end{bmatrix} \right\}$$

17. Produce an orthonormal set by normalizing the vectors in the orthogonal set

$$\left\{ \begin{bmatrix} 1 \\ 1 \\ 1 \end{bmatrix}, \begin{bmatrix} 1 \\ 2 \\ -3 \end{bmatrix} \right\}.$$

▷ **18.** Suppose that $\mathbf{v}_1, \ldots, \mathbf{v}_p$ form an orthogonal set of vectors in the vector space $\mathscr{V}$ with inner product. Show that

$$\|\alpha_1\mathbf{v}_1 + \cdots + \alpha_p\mathbf{v}_p\|^2 = |\alpha_1|^2\|\mathbf{v}_1\|^2 + \cdots + |\alpha_p|^2\|\mathbf{v}_p\|^2.$$

19. Show that $\{1, t - \frac{1}{2}\}$ is an orthogonal set in the space $\mathscr{P}^3$ of Problem 12, and normalize it to obtain an orthonormal set.

20. Show that the original set in Problem 19 is a basis for the subspace $\mathscr{P}^2$ of $\mathscr{P}^3$.

5.8 *ORTHOGONAL PROJECTIONS AND BASES: GENERAL SPACES AND GRAM–SCHMIDT*

Orthogonal sets as introduced in Section 5.7 are extremely important both theoretically and computationally. In this section and the next we present some indications of why this is so.

Orthogonal Projections

Our first result is essential for the derivation of the rest of the section; however, it is also important in its own right.

(5.71) ***Theorem*** (orthogonal projection). Let $\mathscr{V}$ be a vector space with an inner product. Let $\mathscr{V}_0$ be the subspace of $\mathscr{V}$ spanned by an orthogonal set

$$S = \{\mathbf{v}_1, \mathbf{v}_2, \ldots, \mathbf{v}_q\}$$

of *nonzero* vectors. Define the *orthogonal projection* P_0 *onto* $\mathscr{V}_0$ as follows: for any $\mathbf{v}$ in $\mathscr{V}$, set

$$P_0\mathbf{v} = \alpha_1\mathbf{v}_1 + \cdots + \alpha_q\mathbf{v}_q, \qquad \text{where } \alpha_i = \frac{(\mathbf{v}_i, \mathbf{v})}{(\mathbf{v}_i, \mathbf{v}_i)}.$$

Then:

(a) $\mathbf{v} - P_0\mathbf{v}$ is orthogonal to every vector $\mathbf{v}_0$ in $\mathscr{V}_0$.
(b) $P_0(\mathbf{u} + \mathbf{v}) = P_0\mathbf{u} + P_0\mathbf{v}$ for all $\mathbf{u}$, $\mathbf{v}$ in $\mathscr{V}$.
(c) $P_0(\alpha\mathbf{v}) = \alpha P_0\mathbf{v}$ for all scalars α and all $\mathbf{v}$ in $\mathscr{V}$.

PROOF

(a) Note first that $\mathbf{v} - P_0\mathbf{v}$ is orthogonal to each $\mathbf{v}_i$:

$$(\mathbf{v}_i, \mathbf{v} - P_0\mathbf{v}) = (\mathbf{v}_i, \mathbf{v}) - \alpha_1(\mathbf{v}_i, \mathbf{v}_1) - \cdots - \alpha_q(\mathbf{v}_i, \mathbf{v}_q) = (\mathbf{v}_i, \mathbf{v}) - \alpha_i(\mathbf{v}_i, \mathbf{v}_i) = 0.$$

Since each $\mathbf{v}_0$ in $\mathscr{V}_0$ is a linear combination of the $\mathbf{v}_i$, each of which satisfies $(\mathbf{v}_i, \mathbf{v} - P_0\mathbf{v}) = 0$, we get $(\mathbf{v}_0, \mathbf{v} - P_0\mathbf{v}) = 0$ as claimed.

(b) and (c) follow immediately from the definition of the coefficients α_i. For example, the coefficient α_i in $P_0(\alpha\mathbf{v})$ is

$$(\mathbf{v}_i, \alpha\mathbf{v})/(\mathbf{v}_i, \mathbf{v}_i) = \alpha(\mathbf{v}_i, \mathbf{v})/(\mathbf{v}_i, \mathbf{v}_i),$$

which is just α times the coefficient α_i in $P_0\mathbf{v}$. ■

(5.72) ***Example.*** Suppose that $\mathscr{V}$ is $\mathbb{R}^3$ and that $q = 2$; then $\mathscr{V}_0$ is a plane. The geometric interpretation of $\mathbf{v}' = P_0\mathbf{v}$ is shown below:

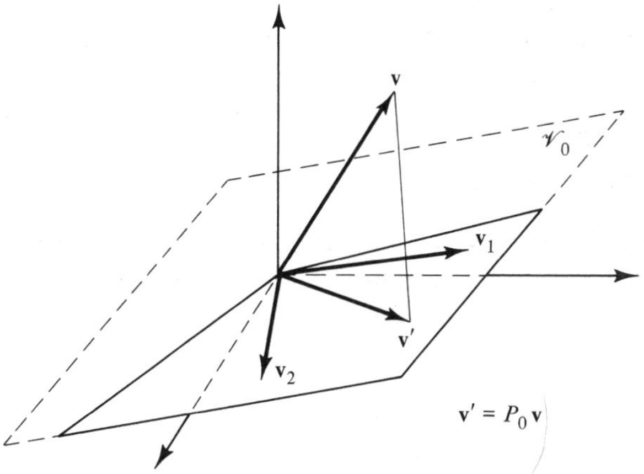

Best Approximations

Section 2.6 on least squares showed that in determining parameters in mathematical models one often has to approximate a given vector as a linear combination of other given vectors. If the vectors from which one forms linear combinations are mutually *orthogonal*, then the solution of this problem is straightforward.

(5.73) ***Theorem*** (best approximation). Let $\mathscr{V}$ be a vector space with an inner product and with induced norm $\|\cdot\|$, and let $\mathscr{V}_0$ be the subspace spanned by the *orthogonal* set of nonzero vectors $\mathbf{v}_1, \ldots, \mathbf{v}_q$. Then, for any $\mathbf{v}$, $P_0\mathbf{v}$ is the unique *closest point* in $\mathscr{V}_0$ to $\mathbf{v}$ and $\|\mathbf{v} - P_0\mathbf{v}\|$ is the *distance from* $\mathbf{v}$ *to* $\mathscr{V}_0$, in the sense that $P_0\mathbf{v}$ is in $\mathscr{V}_0$ and

$$\|\mathbf{v} - P_0\mathbf{v}\| < \|\mathbf{v} - \mathbf{v}_0\| \qquad \text{for all } \mathbf{v}_0 \neq P_0\mathbf{v} \text{ in } \mathscr{V}_0.$$

PROOF. For convenience, let $\hat{\mathbf{v}}_0$ denote $P_0\mathbf{v}$, which is clearly in $\mathscr{V}_0$ by the definition of $P_0\mathbf{v}$; for any $\mathbf{v}_0$ in $\mathscr{V}_0$, calculate $\|\mathbf{v} - \mathbf{v}_0\|$ as follows:

$$\begin{aligned}\|\mathbf{v} - \mathbf{v}_0\|^2 &= (\mathbf{v} - \mathbf{v}_0, \mathbf{v} - \mathbf{v}_0)\\ &= (\mathbf{v} - \hat{\mathbf{v}}_0 + \hat{\mathbf{v}}_0 - \mathbf{v}_0, \mathbf{v} - \hat{\mathbf{v}}_0 + \hat{\mathbf{v}}_0 - \mathbf{v}_0)\\ &= (\mathbf{v} - \hat{\mathbf{v}}_0, \mathbf{v} - \hat{\mathbf{v}}_0) + (\mathbf{v} - \hat{\mathbf{v}}_0, \hat{\mathbf{v}}_0 - \mathbf{v}_0)\\ &\quad + (\hat{\mathbf{v}}_0 - \mathbf{v}_0, \mathbf{v} - \hat{\mathbf{v}}_0) + (\hat{\mathbf{v}}_0 - \mathbf{v}_0, \hat{\mathbf{v}}_0 - \mathbf{v}_0).\end{aligned}$$

By Theorem 5.71(a), $\mathbf{v} - \hat{\mathbf{v}}_0$ is orthogonal to all vectors in $\mathscr{V}_0$, including $\hat{\mathbf{v}}_0 - \mathbf{v}_0$; the two middle terms of the four terms on the right above therefore equal zero. That equality then becomes

$$\|\mathbf{v} - \mathbf{v}_0\|^2 = \|\mathbf{v} - \hat{\mathbf{v}}_0\|^2 + \|\hat{\mathbf{v}}_0 - \mathbf{v}_0\|^2.$$

Therefore, $\|\mathbf{v} - \mathbf{v}_0\|^2 > \|\mathbf{v} - \hat{\mathbf{v}}_0\|^2$ unless $\mathbf{v}_0 = \hat{\mathbf{v}}_0$. ∎

According to Theorem 5.73, it is easy to find the best approximation in a given subspace $\mathscr{V}_0$ to a given vector $\mathbf{v}$ *as long as we have an orthogonal spanning set for* $\mathscr{V}_0$ *consisting of nonzero vectors.* Since this best-approximation problem arises so often in applications, we pursue this matter of orthogonal spanning sets.

Orthogonal and Orthonormal Bases

One important aspect of our main theorem on orthogonal projection is somewhat concealed by the notation: Every orthogonal set of nonzero vectors is linearly independent. To see this, suppose that $S = \{\mathbf{v}_1, \ldots, \mathbf{v}_q\}$ is such a set and that

$$c_1\mathbf{v}_1 + \cdots + c_q\mathbf{v}_q = \mathbf{0};$$

taking the inner product with each $\mathbf{v}_j$ gives

$$0 = (\mathbf{v}_j, \mathbf{0}) = (\mathbf{v}_j, c_1\mathbf{v}_1 + \cdots + c_q\mathbf{v}_q) = c_j(\mathbf{v}_j, \mathbf{v}_j),$$

which means that $c_j = 0$ since $(\mathbf{v}_j, \mathbf{v}_j) > 0$. Therefore, any orthogonal spanning set of nonzero vectors is an orthogonal linearly independent spanning set—that is, an *orthogonal basis* for $\mathscr{V}_0$. The orthogonal projection theorem says that orthogonal bases are extremely easy to use when we need to find a best approximation to a vector $\mathbf{v}$: We can immediately write it down as

$$\alpha_1\mathbf{v}_1 + \cdots + \alpha_q\mathbf{v}_q$$

with the α_i as defined in the theorem. If $\mathbf{v}$ happens itself to be in $\mathscr{V}_0$, then certainly $\mathbf{v}$ is the best approximation in $\mathscr{V}_0$ to itself and we have

$$\mathbf{v} = \alpha_1\mathbf{v}_1 + \cdots + \alpha_q\mathbf{v}_q$$

with the α_i as in the theorem. That is:

> We can express a vector $\mathbf{v}$ as a linear combination of vectors in an orthogonal basis without having to solve equations to determine the coefficients—we can simply evaluate some inner products to obtain the coefficients directly.

We summarize.

(5.74) ***Theorem*** (orthogonal bases). Let $B = \{\mathbf{v}_1, \mathbf{v}_2, \ldots, \mathbf{v}_q\}$ be an orthogonal (or orthonormal) basis. Then the representation of any vector $\mathbf{v}$ with respect to the orthogonal basis B can immediately be written down:

$$\mathbf{v} = \alpha_1\mathbf{v}_1 + \cdots + \alpha_q\mathbf{v}_q, \qquad \text{where } \alpha_i = \frac{(\mathbf{v}_i, \mathbf{v})}{(\mathbf{v}_i, \mathbf{v}_i)}.$$

We encountered a special case of this much earlier when we saw how easy it is to express any vector in $\mathbb{R}^p$ as a linear combination of $\mathbf{e}_1, \ldots, \mathbf{e}_p$: these vectors form an orthonormal basis for $\mathbb{R}^p$, so the coefficients in the representation of $\mathbf{v}$ are

$$\alpha_i = \frac{(\mathbf{e}_i, \mathbf{v})}{(\mathbf{e}_i, \mathbf{e}_i)} = \langle \mathbf{v} \rangle_i.$$

You should now be able to solve Problems 1 to 5.

Creating Orthogonal and Orthonormal Bases: Gram–Schmidt

Best-approximation problems are easy to solve *if we have an orthogonal basis for the approximating subspace.* This is not always the case in applications, however. Such a basis is so convenient that often we should attempt to *create* one for a space. This raises the following question:

> Given a basis or spanning set S for a vector space with an inner product, how can we find an orthogonal or orthonormal basis for the space?

In finite-dimensional spaces, this can be done in a straightforward manner. The process we use is called the *Gram–Schmidt* procedure. The theorem below appears complex but the idea is simple: the Gram–Schmidt process produces an orthogonal basis from a spanning set and detects whether the spanning set is linearly dependent.

(5.75) **Key Theorem** (Gram–Schmidt by projections). Let $S = \{\mathbf{v}_1, \ldots, \mathbf{v}_q\}$ span the vector space $\mathscr{V}$ having an inner product. The Gram–Schmidt process is as follows. Let $\mathscr{V}_i$ denote the subspace of $\mathscr{V}$ spanned by $\{\mathbf{v}_1, \ldots, \mathbf{v}_i\}$ and let P_i denote the orthogonal projection onto $\mathscr{V}_i$ (computed as in Theorem 5.71 using the orthogonal sets B_i below); if $\mathscr{V}_i = \{\mathbf{0}\}$, let $P_i\mathbf{v} = \mathbf{0}$ for all $\mathbf{v}$.

1. Define $\mathbf{u}_1 = \mathbf{v}_1$.
2. For $2 \le i \le q$, define $\mathbf{u}_i = \mathbf{v}_i - P_{i-1}\mathbf{v}_i$.

The following holds for the vectors produced by this process:

(a) $B = \{\mathbf{u}_1, \ldots, \mathbf{u}_q\}$ is an orthogonal set spanning $\mathscr{V}$.
(b) For $1 \le i \le q$, $B_i = \{\mathbf{u}_1, \ldots, \mathbf{u}_i\}$ is an orthogonal set spanning $\mathscr{V}_i$, the subspace spanned by $\{\mathbf{v}_1, \ldots, \mathbf{v}_i\}$.
(c) $\mathbf{u}_i = \mathbf{0}$ if and only if $\mathbf{v}_i$ is linearly dependent on the vectors $\mathbf{v}_1, \ldots, \mathbf{v}_{i-1}$.
(d) An orthogonal basis for $\mathscr{V}$ can be obtained from the orthogonal spanning set B by omitting the zero $\mathbf{u}_i$, if any.
(e) If S is a basis for $\mathscr{V}$, then B is an orthogonal basis for $\mathscr{V}$.

PROOF

(a) is just (b) with $i = q$.
(b) We use induction; since $\mathbf{u}_1 = \mathbf{v}_1$, (b) holds for $i = 1$ and P_1 is well defined. Suppose that (b) holds for $i = k < q$ so that P_k can be constructed using B_k. By definition, we have

$$(\#) \qquad \mathbf{u}_{k+1} = \mathbf{v}_{k+1} - P_k\mathbf{v}_{k+1}, \quad \text{so} \quad \mathbf{v}_{k+1} = \mathbf{u}_{k+1} + P_k\mathbf{v}_{k+1}.$$

Since P_k projects onto $\mathscr{V}_k$, $P_k\mathbf{v}_{k+1}$ is in $\mathscr{V}_k$ and, by the inductive hypothesis, is some linear combination of $\mathbf{u}_i, \ldots, \mathbf{u}_k$. Therefore, $(\#)$ expresses $\mathbf{v}_{k+1}$ as some linear combination of $\mathbf{u}_1, \ldots, \mathbf{u}_{k+1}$; since the preceding $\mathbf{v}_i$ are linear combinations of the preceding $\mathbf{u}_i$ (by the inductive hypothesis), (b) follows for $i = k + 1$ and hence for all i.
(c) Since $\mathbf{u}_i = \mathbf{v}_i - P_{i-1}\mathbf{v}_i$ and $P_{i-1}\mathbf{v}_i$ is linearly dependent on $\mathbf{v}_1, \ldots, \mathbf{v}_{i-1}$, the result holds.
(d) Omitting the zero vectors does not change the span of B, and an orthogonal set of nonzero vectors is linearly independent.
(e) follows from (d) since there can be no zero vectors $\mathbf{u}_i$. ■

We repeat for emphasis:

The Gram–Schmidt process produces an orthogonal spanning set from a spanning set and detects whether the original set is linearly dependent.

Notationally, the Gram–Schmidt process itself is simple:

$$\text{Compute} \quad \mathbf{u}_i = \mathbf{v}_i - P_{i-1}\mathbf{v}_i.$$

The question remains: How, in practice, do we compute $P_{i-1}\mathbf{v}_i$? There are several approaches to this, at least two of which—the "modified Gram–Schmidt process" and the "Householder process," mentioned in Problems 12–17—are numerically useful and are available in reliable computer software. The traditional approach below is less effective in practice (see Problem 11).

The problem is to compute $P_{i-1}\mathbf{v}_i$. We saw in Theorem 5.71 that it is easy to compute the orthogonal projection onto a subspace when we have an orthogonal spanning set for that subspace; our subspace is $\mathscr{V}_{i-1}$, and (b) of our Gram–Schmidt theorem says that $\mathbf{u}_1, \ldots, \mathbf{u}_{i-1}$ is just such an orthogonal spanning set. *At each step, the Gram–Schmidt process produces the orthogonal spanning set required for the next step*; therein lies its elegance. In light of this, the Gram–Schmidt process can be implemented—although it is not computationally sound—as

(5.76) ***The Traditional Gram–Schmidt Process.*** Given a spanning set $\mathbf{v}_1, \ldots, \mathbf{v}_q$:

1. Define $\mathbf{u}_1 = \mathbf{v}_1$.
2. For $2 \le i \le q$, define

$$\mathbf{u}_i = \mathbf{v}_i - \alpha_{1i}\mathbf{u}_1 - \cdots - \alpha_{i-1,i}\mathbf{u}_{i-1},$$

where $\alpha_{ji} = (\mathbf{u}_j, \mathbf{v}_i)/(\mathbf{u}_j, \mathbf{u}_j)$ if $\mathbf{u}_j \ne \mathbf{0}$ and $\alpha_{ji} = 0$ if $\mathbf{u}_j = \mathbf{0}$.
(See also Problem 9.)

(5.77) ***Example.*** Suppose that we want an orthogonal basis for the subspace of $\mathbb{R}^4$ spanned by

$$\mathbf{v}_1 = [1 \quad 1 \quad 1 \quad -1]^T, \qquad \mathbf{v}_2 = [2 \quad -1 \quad -1 \quad 1]^T,$$
$$\mathbf{v}_3 = [0 \quad 3 \quad 3 \quad -3]^T, \qquad \mathbf{v}_4 = [-1 \quad 2 \quad 2 \quad 1]^T.$$

The process in (5.76) gives

$$\mathbf{u}_1 = \mathbf{v}_1 = [1 \quad 1 \quad 1 \quad -1]^T;$$

$$\mathbf{u}_2 = \mathbf{v}_2 - \alpha_{12}\mathbf{u}_1 = \mathbf{v}_2 - \left(-\frac{1}{4}\right)\mathbf{u}_1 = \frac{[9 \quad -3 \quad -3 \quad 3]^T}{4};$$

$$\mathbf{u}_3 = \mathbf{v}_3 - \alpha_{13}\mathbf{u}_1 - \alpha_{23}\mathbf{u}_2 = \mathbf{v}_3 - (\tfrac{9}{4})\mathbf{u}_1 - (-1)\mathbf{u}_2$$
$$= [0 \quad 0 \quad 0 \quad 0]^T;$$

$$\mathbf{u}_4 = \mathbf{v}_4 - \alpha_{14}\mathbf{u}_1 - \alpha_{24}\mathbf{u}_2 - \alpha_{34}\mathbf{u}_3$$
$$= \mathbf{v}_4 - (\tfrac{1}{2})\mathbf{u}_1 - (-\tfrac{2}{3})\mathbf{u}_2 - (0)\mathbf{u}_3 = [0 \quad 1 \quad 1 \quad 2]^T.$$

Since $\mathbf{u}_3 = \mathbf{0}$, we see that the original set of four vectors is linearly dependent, with $\mathbf{v}_3$ linearly dependent on $\mathbf{v}_1$ and $\mathbf{v}_2$. The set $\{\mathbf{u}_1, \mathbf{u}_2, \mathbf{u}_4\}$ is an orthogonal basis for the subspace; we can obtain an orthonormal basis by replacing each of these three $\mathbf{u}_i$ by $\mathbf{u}_i/\|\mathbf{u}_i\|_2$.

One fact that follows immediately from the theorem on constructing an orthogonal basis of nonzero vectors merits separate mention.

(5.78) ***Corollary.*** Other than $\{\mathbf{0}\}$, every finite-dimensional vector space with an inner product has an orthonormal basis.

PROBLEMS 5.8

▷ **1.** Using $\{\mathbf{e}_2, \mathbf{e}_3\}$ as an orthogonal basis, find the closest point in the set of all vectors $[0 \quad \alpha \quad \beta]^T$ to
(a) $[1 \quad 2 \quad 3]^T$ (b) $[a \quad b \quad c]^T$

2. Using Problems 12, 19, and 20 from Section 5.7, find the point in the span of $\{1, t - \frac{1}{2}\}$ that is closest to t^2 in $\mathcal{P}^3$.

3. In $\mathcal{P}^3$ with the inner product as in Problem 12 of Section 5.7, show that

$$\{1, t - \tfrac{1}{2}, t^2 - t + \tfrac{1}{6}\}$$

is an orthogonal basis for $\mathcal{P}^3$.

▷ **4.** Show that if $\mathbf{v}$ is in the subspace $\mathcal{V}_0$ in Theorem 5.72, then the orthogonal projection $P_0\mathbf{v}$ of $\mathbf{v}$ onto $\mathcal{V}_0$ equals $\mathbf{v}$ itself.

5. Suppose that P_0 is the orthogonal projection onto $\mathcal{V}_0$ as in Theorem 5.72. Show that, for every $\mathbf{v}$ in $\mathcal{V}$,

$$\|\mathbf{v}\|^2 = \|P_0\mathbf{v}\|^2 + \|\mathbf{v} - P_0\mathbf{v}\|^2.$$

6. Verify that, in Example 5.77, $\{\mathbf{v}_1, \ldots, \mathbf{v}_i\}$ and $\{\mathbf{u}_1, \ldots, \mathbf{u}_i\}$ span the same space for each $i = 1, 2, 3, 4$.

▷ **7.** Apply the traditional Gram–Schmidt process (5.76) to the three vectors below, and make a valid statement about the general situation it represents.

$$\begin{bmatrix} 1 \\ 2 \\ 2 \\ 1 \end{bmatrix}, \quad \begin{bmatrix} 1 \\ 1 \\ -1 \\ -1 \end{bmatrix}, \quad \begin{bmatrix} 2 \\ -1 \\ -1 \\ 2 \end{bmatrix}.$$

8. Apply the traditional Gram–Schmidt process (5.76) to the three vectors

$$[1 \quad 1 \quad -1]^T, \quad [-1 \quad 2 \quad 2]^T, \quad \text{and} \quad [1 \quad 4 \quad 0]^T.$$

9. Our version (5.76) of the traditional Gram–Schmidt process is actually a slight variant of what is usually presented. In that other approach—which we call *normalized*—as soon as a $\mathbf{u}_i$ is computed, it is replaced by its normalized version $\mathbf{u}_i/\|\mathbf{u}_i\|$ if $\mathbf{u}_i \neq \mathbf{0}$. This avoids the calculation of the inner products $\{\mathbf{u}_i, \mathbf{u}_i\}$ in future steps since those will now equal 1. Apply this version to the vectors $\mathbf{v}_i$ in Example 5.77; the $\mathbf{u}_i$ you produce should just be normalized versions of the earlier $\mathbf{u}_i$—except for $\mathbf{u}_3 = \mathbf{0}$, of course.

▷ **10.** Using the inner product on $\mathscr{P}^3$ from Problem 8 of Section 5.7—taking the $\mathbb{R}^3$ inner product of the coefficients of the polynomial—apply the traditional Gram–Schmidt process (5.76) to the vectors $1 + t$, $t + t^2$, and $t^2 + 1$.

11. To see that the traditional Gram–Schmidt process (5.76)—or its normalized version as in Problem 9—is not necessarily a good way to compute the projections in **Key Theorem 5.75**, consider the following case. Let ϵ be a number such that the computer on which the calculations are done computes $1 + \epsilon^2$ to equal 1 but computes $2 + \epsilon$ and $\epsilon + \epsilon^2$ fairly accurately; $\epsilon = 10^{-6}$ works nicely on an 8- to 10-digit calculator, while $\epsilon = 10^{-10}$ works on most microcomputers having a math coprocessor. Apply the traditional Gram–Schmidt process (5.76) (or its normalized version) to the vectors

$$\begin{bmatrix} 1 \\ 1+\epsilon \\ 1 \\ 1 \end{bmatrix}, \quad \begin{bmatrix} 1 \\ 1 \\ 1+\epsilon \\ 1 \end{bmatrix}, \quad \begin{bmatrix} 1 \\ 1 \\ 1 \\ 1+\epsilon \end{bmatrix}.$$

The conditions on ϵ mean that the inner product of any different two of these will be accurately computed as $4 + 2\epsilon$, but that the norm squared of each will be computed as $4 + 2\epsilon$, rather than $4 + 2\epsilon + \epsilon^2$. Find the angles between the vectors you produce: $\mathbf{u}_1$ and $\mathbf{u}_2$, $\mathbf{u}_1$ and $\mathbf{u}_3$, and $\mathbf{u}_2$ and $\mathbf{u}_3$. You should find that $\mathbf{u}_1$ and $\mathbf{u}_2$ are within about ϵ of perpendicular, and likewise for $\mathbf{u}_1$ and $\mathbf{u}_3$; $\mathbf{u}_2$ and $\mathbf{u}_3$, however, make an angle of 60° rather than the desired right angle.

▷ **12.** The *modified Gram–Schmidt* process is one alternative for calculating the projections in **Key Theorem 5.75** that avoids the difficulties with traditional Gram–Schmidt as illustrated in Problem 11. In the traditional process, at the ith step only the vector $\mathbf{v}_i$ is altered: it is changed to $\mathbf{u}_i$ that is orthogonal to all the preceding $\mathbf{u}_j$. In the modified process, at the ith step, *all* of the remaining $\mathbf{v}_i$ are altered so as to be made orthogonal to the most recently computed $\mathbf{u}$-vector. More precisely:

Starting with the given $\mathbf{v}_1, \ldots, \mathbf{v}_q$, define $\mathbf{v}_1^0, \ldots, \mathbf{v}_q^0$ as

$$\mathbf{v}_1^0 = \mathbf{v}_1, \quad \ldots, \quad \mathbf{v}_q^0 = \mathbf{v}_q.$$

At the ith step, for $i = 1, \ldots, q$:

define $\mathbf{u}_i = \mathbf{v}_i^{i-1}$;

define $\mathbf{v}_j^i = \mathbf{v}_j^{i-1} - \dfrac{\mathbf{u}_i(\mathbf{u}_i, \mathbf{v}_j^{i-1})}{(\mathbf{u}_i, \mathbf{u}_i)}$ for $j = i + 1, \ldots, q$.

In perfect arithmetic, this produces the same vectors $\mathbf{u}_j$ as the traditional process. Apply this modified process to the vectors in Problem 11 and find the angles between the $\mathbf{u}_i$ you produce; you should get *much* better results than with the traditional process.

13. Apply the modified Gram–Schmidt process of Problem 12 to the vectors in Example 5.77.

𝔐 14. Apply MATLAB or similar software to find an orthogonal or orthonormal basis for the space spanned by the vectors in Problem 11 (with ϵ chosen appropriately for your computer) to see how well that software performs on this challenging problem.

15. Another alternative to the traditional process for calculating the projections in **Key Theorem 5.75** makes use of *Householder matrices* (see Problems 13–17 of Section 5.9); a (real) Householder matrix is a $p \times p$ matrix $\mathbf{H}$ of the form $\mathbf{H} = \mathbf{I}_p - 2\mathbf{w}\mathbf{w}^T/\mathbf{w}^T\mathbf{w}$ for a real nonzero $p \times 1$ column matrix $\mathbf{w}$.
    (a) Show that every Householder matrix is symmetric.
    (b) Show that every Householder matrix is nonsingular and that $\mathbf{H}^{-1} = \mathbf{H} = \mathbf{H}^T$.
    (c) Use (b) to show that the columns of a Householder matrix form an orthonormal set.

▷ 16. Given two $p \times 1$ real matrices $\mathbf{x}$ and $\mathbf{y}$, with $\mathbf{x} \neq \mathbf{y}$, define

$$\mathbf{w}_{\pm} = \mathbf{x} \pm \mathbf{y}\|\mathbf{x}\|_2/\|\mathbf{y}\|_2$$

and let $\mathbf{H}_{\pm}$ be the Householder matrices of Problem 15 defined by $\mathbf{w}_{\pm}$. Show that $\mathbf{H}_{\pm}$ transforms $\mathbf{x}$ into a multiple of $\mathbf{y}$; more precisely, show that

$$\mathbf{H}_{\pm}\mathbf{x} = \mp\,\mathbf{y}\|\mathbf{x}\|_2/\|\mathbf{y}\|_2.$$

17. (a) Let $\mathbf{x} = [2 \quad 2 \quad 1]^T$ and $\mathbf{y} = [1 \quad 0 \quad 0]^T$. Find the matrices $\mathbf{H}_{\pm}$ of Problem 16 and verify that $\mathbf{H}_{\pm}\mathbf{x} = \mp 3\mathbf{y}$.
    (b) Show that for any $\mathbf{x}$ in $\mathbb{R}^p$ there is a Householder matrix $\mathbf{H}$ so that $\mathbf{Hx}$ equals either plus or minus $\|\mathbf{x}\|_2\mathbf{e}_1$, where $\mathbf{e}_1$ is the usual unit column matrix.

5.9 *ORTHOGONAL PROJECTIONS AND BASES: $\mathbb{R}^p$, $\mathbb{C}^p$, QR, AND LEAST SQUARES*

Many applications make crucial use of the techniques and concepts of orthogonal projections and bases introduced in Section 5.8 for general vector spaces. In the special case of $\mathbb{R}^p$ (or $\mathbb{C}^p$), these results—when couched in matrix terminology—provide powerful computational tools.

Orthogonal Projections

Our first results—Theorems 5.72 and 5.73—in the preceding section involved orthogonal projections is general vector spaces with inner products; we now reformulate these in $\mathbb{R}^p$ and $\mathbb{C}^p$ using matrix terminology. To be specific, we consider $\mathcal{V} = \mathbb{R}^p$ in the discussion below; we only need change transposes ^{T} to hermitian transposes ^{H} to make the argument apply in $\mathbb{C}^p$.

Suppose that $\mathscr{V}_0$ is a subspace of $\mathbb{R}^p$ spanned by the orthogonal set of vectors $S = \{\mathbf{v}_1, \mathbf{v}_2, \ldots, \mathbf{v}_q\}$ and that $\mathbf{v}$ is another $p \times 1$ matrix; we wish to calculate the orthogonal projection $P_0\mathbf{v}$ of Theorem 5.72. We specialize slightly by assuming that S is an *orthonormal* set—that $\|\mathbf{v}_i\|_2 = 1$ in addition to the orthogonality conditions $0 = (\mathbf{v}_i, \mathbf{v}_j) = \mathbf{v}_i^T\mathbf{v}_j$ for $i \neq j$. Define the $p \times q$ matrix $\mathbf{Q}$ by

$$\mathbf{Q} = [\mathbf{v}_1 \quad \mathbf{v}_2 \quad \cdots \quad \mathbf{v}_q].$$

If we compute $\mathbf{Q}^T\mathbf{Q}$, we find that

$$\mathbf{Q}^T\mathbf{Q} = [\mathbf{v}_1 \quad \mathbf{v}_2 \quad \cdots \quad \mathbf{v}_q]^T[\mathbf{v}_1 \quad \mathbf{v}_2 \quad \cdots \quad \mathbf{v}_q]$$

has as its (i, j)-entry just $\mathbf{v}_i^T\mathbf{v}_j$—which equals 1 if $i = j$ and 0 if $i \neq j$. That is: $\mathbf{Q}^T\mathbf{Q} = \mathbf{I}_q$; this is our matrix reformulation of the fact that S is orthonormal. Now to the computation of $P_0\mathbf{v}$. According to Theorem 5.72,

$$P_0\mathbf{v} = \alpha_1\mathbf{v}_1 + \cdots + \alpha_q\mathbf{v}_q$$

for appropriate α_i; in matrix notation, this says that $P_0\mathbf{v} = \mathbf{Q}\boldsymbol{\alpha}$, where $\mathbf{Q}$ is the matrix above and

$$\boldsymbol{\alpha} = [\alpha_1 \quad \alpha_2 \quad \cdots \quad \alpha_q]^T.$$

Again according to Theorem 5.72, α_i is computed as

$$\alpha_i = \frac{(\mathbf{v}_i, \mathbf{v})}{(\mathbf{v}_i, \mathbf{v}_i)} = \mathbf{v}_i^T\mathbf{v}$$

in the present case; in matrix notation, this says that $\boldsymbol{\alpha} = \mathbf{Q}^T\mathbf{v}$. Putting these two facts together gives

$$P_0\mathbf{v} = \mathbf{Q}\mathbf{Q}^T\mathbf{v}.$$

Note here that $\mathbf{Q}\mathbf{Q}^T \neq \mathbf{I}_p$ in general; $\mathbf{Q}^T$ is a left-inverse of $\mathbf{Q}$ since $\mathbf{Q}^T\mathbf{Q} = \mathbf{I}_q$, but it is not generally a right-inverse since we are not assuming $\mathbf{Q}$ to be square. This formula for P_0 shows that orthogonal projection is nothing more than multiplication by a special matrix $\mathbf{Q}\mathbf{Q}^T$, called a *projection matrix*. This completes our reformulation of Theorems 5.72 and 5.73 in matrix terminology.

(5.79) ***Theorem*** (projection matrices). Let $\mathbf{Q}$ be a $p \times q$ matrix having orthonormal columns in $\mathbb{R}^p$ (or $\mathbb{C}^p$)—that is, let

$$\mathbf{Q}^T\mathbf{Q} = \mathbf{I}_q \qquad (\text{or } \mathbf{Q}^H\mathbf{Q} = \mathbf{I}_q),$$

and let $\mathscr{V}_0$ be the subspace spanned by the orthonormal basis for $\mathscr{V}_0$ formed by the columns of $\mathbf{Q}$. Then:

(a) The orthogonal projection P_0 onto $\mathscr{V}_0$ as described in Theorems 5.72 and 5.73 is computed with the *projection matrix* $\mathbf{Q}\mathbf{Q}^T$ (or $\mathbf{Q}\mathbf{Q}^H$) as

$$P_0\mathbf{v} = \mathbf{P}_0\mathbf{v}$$

where $\mathbf{P}_0$ is the $p \times p$ matrix $\mathbf{P}_0 = \mathbf{Q}\mathbf{Q}^T$ (or $\mathbf{P}_0 = \mathbf{Q}\mathbf{Q}^H$).

(b) The projection matrix $\mathbf{P}_0$ satisfies:
1. $\mathbf{P}_0$ is symmetric (or hermitian).
2. $\mathbf{P}_0^2 = \mathbf{P}_0$.
3. $\mathbf{P}_0(\mathbf{I}_p - \mathbf{P}_0) = (\mathbf{I}_p - \mathbf{P}_0)\mathbf{P}_0 = \mathbf{0}$.
4. $(\mathbf{I}_p - \mathbf{P}_0)\mathbf{Q} = \mathbf{0}$.

PROOF

(a) We have already proved this.

(b) 1. $\mathbf{P}_0^T = (\mathbf{Q}\mathbf{Q}^T)^T = (\mathbf{Q}^T)^T\mathbf{Q}^T = \mathbf{Q}\mathbf{Q}^T = \mathbf{P}_0$ (and similarly for the complex case).
2. $\mathbf{P}_0^2 = \mathbf{Q}\mathbf{Q}^T\mathbf{Q}\mathbf{Q}^T = \mathbf{Q}\mathbf{I}_q\mathbf{Q}^T = \mathbf{Q}\mathbf{Q}^T = \mathbf{P}_0$ (and similarly for the complex case).
3. $\mathbf{P}_0(\mathbf{I}_p - \mathbf{P}_0) = \mathbf{P}_0 - \mathbf{P}_0^2 = \mathbf{0}$, and similarly for the others.
4. $(\mathbf{I}_p - \mathbf{P}_0)\mathbf{Q} = \mathbf{Q} - \mathbf{Q}\mathbf{Q}^T\mathbf{Q} = \mathbf{Q} - \mathbf{Q}\mathbf{I}_q = \mathbf{0}$ (and similarly for the complex case). ■

(5.80) ***Example.*** Let $\mathbf{Q}$ be 4×3 and let its columns be normalized versions of the nonzero orthogonal vectors $\mathbf{u}_1$, $\mathbf{u}_2$, $\mathbf{u}_4$ found in Example 5.77:

$$\mathbf{Q} = \begin{bmatrix} \frac{1}{2} & \frac{\sqrt{3}}{2} & 0 \\ \frac{1}{2} & -\frac{\sqrt{3}}{6} & \frac{\sqrt{6}}{6} \\ \frac{1}{2} & -\frac{\sqrt{3}}{6} & \frac{\sqrt{6}}{6} \\ -\frac{1}{2} & \frac{\sqrt{3}}{6} & \frac{\sqrt{6}}{3} \end{bmatrix}. \quad \text{Then} \quad \mathbf{P}_0 = \begin{bmatrix} 1 & 0 & 0 & 0 \\ 0 & \frac{1}{2} & \frac{1}{2} & 0 \\ 0 & \frac{1}{2} & \frac{1}{2} & 0 \\ 0 & 0 & 0 & 1 \end{bmatrix}.$$

Therefore, the closest vector to $\mathbf{v} = [1 \quad 2 \quad 3 \quad 4]^T$ in the subspace $\mathscr{V}_0$ of $\mathbb{R}^4$ spanned by the columns of $\mathbf{Q}$ is

$$P_0\mathbf{v} = \mathbf{P}_0\mathbf{v} = [1 \quad 5/2 \quad 5/2 \quad 4]^T;$$

more generally, the closest point to $\mathbf{v} = [a \quad b \quad c \quad d]^T$ is

$$\mathbf{P}_0\mathbf{v} = \left[a \quad \frac{b+c}{2} \quad \frac{b+c}{2} \quad d\right]^T.$$

You should now be able to solve Problems 1 to 5.

QR-Decompositions

Our next task is to provide a matrix formulation for **Key Theorem 5.75** on how to compute an orthogonal spanning set from a general spanning set. Suppose that $\mathbf{v}_1, \mathbf{v}_2, \ldots, \mathbf{v}_q$ are $p \times 1$ column matrices and consider the traditional implementation (5.76) of the Gram–Schmidt process. From the definition of the column matrices $\mathbf{u}_j$ in (5.76)(2), we have

$$\mathbf{v}_j = \mathbf{u}_1\alpha_{1j} + \mathbf{u}_2\alpha_{2j} + \cdots + \mathbf{u}_{j-1}\alpha_{j-1,j} + \mathbf{u}_j.$$

In partitioned-matrix notation, this is just

(5.81a)
$$[\mathbf{v}_1 \quad \cdots \quad \mathbf{v}_q] = [\mathbf{u}_1 \quad \cdots \quad \mathbf{u}_q]\begin{bmatrix} 1 & \alpha_{12} & \cdots & \alpha_{1q} \\ 0 & 1 & \cdots & \alpha_{2q} \\ & & \cdots & \\ 0 & 0 & \cdots & 1 \end{bmatrix}.$$

Let $\mathbf{Q}_0$ denote the $p \times q$ matrix $[\mathbf{u}_1 \quad \cdots \quad \mathbf{u}_q]$ in (5.81a) and let $\mathbf{R}_0$ denote the $q \times q$ unit-upper-triangular matrix there; since $\mathbf{A} = [\mathbf{v}_1 \quad \cdots \quad \mathbf{v}_q]$ we can rewrite (5.81a) as

(5.81b)
$$\mathbf{A} = \mathbf{Q}_0\mathbf{R}_0.$$

The $\mathbf{u}_i$ in $\mathbf{Q}_0$ are constructed in (5.76), and are therefore mutually orthogonal; that is, $\mathbf{Q}_0$ has orthogonal columns, some of which may equal $\mathbf{0}$. Let $\mathbf{Q}$ and $\mathbf{R}$ be the matrices obtained by deleting the zero columns from $\mathbf{Q}_0$ and the corresponding rows from $\mathbf{R}_0$, and by dividing each nonzero column of $\mathbf{Q}_0$ by its 2-norm and multiplying each corresponding row of $\mathbf{R}_0$ by that same 2-norm. Then (5.81b) becomes

(5.81c)
$$\mathbf{A} = \mathbf{QR}$$

with $\mathbf{R}$ upper-triangular and $\mathbf{Q}$ having orthonormal columns. The form (5.81b) is called the *unnormalized QR-decomposition* of $\mathbf{A}$, while (5.81c) is the *normalized QR-decomposition.*

(5.82) **Key Theorem** (*QR*-decompositions). Let $\mathbf{A}$ be a $p \times q$ matrix of rank k. Then:

(a) $\mathbf{A}$ can be written in its *unnormalized QR-decomposition* as $\mathbf{A} = \mathbf{Q}_0\mathbf{R}_0$, where:
1. $\mathbf{Q}_0$ is $p \times q$ and has orthogonal columns (of which k are nonzero and $q - k$ are zero) that span the column space of $\mathbf{A}$.
2. $\mathbf{R}_0$ is $q \times q$, unit-upper-triangular, and nonsingular.
3. The 2-norm of the ith column of $\mathbf{Q}_0$ equals the distance from the ith column of $\mathbf{A}$ to the space spanned by $\mathbf{A}$'s first $i - 1$ columns.

(b) $\mathbf{A}$ can be written in its *normalized QR-decomposition* as $\mathbf{A} = \mathbf{QR}$, where:
1. $\mathbf{Q}$ is $p \times k$ and has ortho*normal* columns that span the column space of $\mathbf{A}$.
2. $\mathbf{R}$ is $k \times q$, upper-triangular, and has rank k.
3. If $k = q$, then $|\langle\mathbf{R}\rangle_{ii}|$ equals the distance from the ith column of $\mathbf{A}$ to the space spanned by A's first $i - 1$ columns.

PROOF. Since $\mathbf{A}$ has rank k, its column space has dimension k and there are k vectors in any basis.

(a) This follows from **Key Theorem 5.75**, (5.76), and (5.81a–c) above; for (3), note that $\mathbf{u}_i = \mathbf{v}_i - P_{i-1}\mathbf{v}_i$ and use Theorem 5.73.

(b) This follows from the constructions of $\mathbf{Q}$ and $\mathbf{R}$ from $\mathbf{Q}_0$ and $\mathbf{R}_0$; for (3), note that $k = q$ means that $\mathbf{Q}$ and $\mathbf{R}$ differ from $\mathbf{Q}_0$ and $\mathbf{R}_0$ only by scaling, and use (a) (3). ■

These decompositions are analogous to the LU-decompositions of Section 3.7 and are perhaps as useful computationally. As remarked in Section 5.8, the traditional Gram–Schmidt process is not computationally effective and should not be used in practice to find QR-decompositions; other methods are available, however—see Problems 11–17 in Section 5.8 and Problems 13–17 here.

(5.83) ***Example.*** Consider the 4 × 4 matrix $\mathbf{A}$ whose four columns are the original column matrices $\mathbf{v}_1, \mathbf{v}_2, \mathbf{v}_3, \mathbf{v}_4$ in Example 5.77:

$$\mathbf{A} = \begin{bmatrix} 1 & 2 & 0 & -1 \\ 1 & -1 & 3 & 2 \\ 1 & -1 & 3 & 2 \\ -1 & 1 & -3 & 1 \end{bmatrix}.$$

According to our proof above, we can construct the decomposition $\mathbf{Q}_0\mathbf{R}_0$ from the columns $\mathbf{u}_i$ and the coefficients α_{kj} computed during the Gram–Schmidt process in Example 5.77. This gives

$$\mathbf{Q}_0 = \begin{bmatrix} 1 & \frac{9}{4} & 0 & 0 \\ 1 & -\frac{3}{4} & 0 & 1 \\ 1 & -\frac{3}{4} & 0 & 1 \\ -1 & \frac{3}{4} & 0 & 2 \end{bmatrix} \quad \text{and} \quad \mathbf{R}_0 = \begin{bmatrix} 1 & -\frac{1}{4} & \frac{9}{4} & \frac{1}{2} \\ 0 & 1 & -1 & -\frac{2}{3} \\ 0 & 0 & 1 & 0 \\ 0 & 0 & 0 & 1 \end{bmatrix}.$$

It is easy to verify that indeed $\mathbf{A} = \mathbf{Q}_0\mathbf{R}_0$. To obtain $\mathbf{Q}$ and $\mathbf{R}$, according to the proof above we delete the third column of $\mathbf{Q}_0$ and the third row of $\mathbf{R}_0$ and scale the columns of $\mathbf{Q}_0$ and the rows of $\mathbf{R}_0$. For example, we divide the first column of $\mathbf{Q}_0$ by its 2-norm (2) and multiply the first row of $\mathbf{R}_0$ by the same number (2). Treating the second and third columns/rows similarly, we finally obtain

$$\mathbf{Q} = \begin{bmatrix} \frac{1}{2} & \frac{\sqrt{3}}{2} & 0 \\ \frac{1}{2} & -\frac{\sqrt{3}}{6} & \frac{\sqrt{6}}{6} \\ \frac{1}{2} & -\frac{\sqrt{3}}{6} & \frac{\sqrt{6}}{6} \\ -\frac{1}{2} & \frac{\sqrt{3}}{6} & \frac{\sqrt{6}}{3} \end{bmatrix} \quad \text{and} \quad \mathbf{R} = \begin{bmatrix} 2 & -\frac{1}{2} & \frac{9}{2} & 1 \\ 0 & \frac{3\sqrt{3}}{2} & \frac{-3\sqrt{3}}{2} & -\sqrt{3} \\ 0 & 0 & 0 & \sqrt{6} \end{bmatrix}.$$

It is again easy to verify that $\mathbf{A} = \mathbf{QR}$.

$\mathfrak{M}$ (5.84) ***Example.*** It is known that the 6×6 *Hilbert matrix* $\mathbf{H}_6$ is "nearly singular," in the sense that its columns are "nearly linearly dependent"; $\mathbf{H}_6$ is the matrix such that $\langle \mathbf{H}_6 \rangle_{ij} = 1/(i + j - 1)$. We can now give some meaning to the concept "nearly linearly dependent" since the QR-decompositions contain information about the distance of each column from the span of the preceding columns. We used the MATLAB function *qr* to compute a normalized QR-decomposition of $\mathbf{H}_6$. MATLAB found that $\mathbf{H}_6$ has rank 6, and MATLAB produced a 6×6 matrix $\mathbf{Q}$ with orthonormal columns and a 6×6 upper-triangular matrix $\mathbf{R}$. Since the rank equals 6, the diagonal entries of $\mathbf{R}$ measure the distance of each column of $\mathbf{H}_6$ from the span of the preceding columns; the magnitudes of these diagonal entries were computed by MATLAB as approximately 1.2, 0.14, 0.0096, 0.0048, 0.000017, and 0.00000040. If a vector $\mathbf{v}$ is at a distance d from a subspace $\mathscr{V}_0$, then $\alpha\mathbf{v}$ is αd away from $\mathscr{V}_0$; in other words, we should consider distance from $\mathscr{V}_0$ *relative to the size of* $\mathbf{v}$ *itself*. In our case, MATLAB computed the 2-norm of, for example, the sixth column of $\mathbf{H}_6$ as approximately 0.31; a measure of how nearly the sixth column of $\mathbf{H}_6$ is to being linearly dependent on the first five is then given by

$$0.00000040/0.31 \approx 0.0000013,$$

which indicates why $\mathbf{H}_6$ is called "nearly singular."

You should now be able to solve Problems 1 to 19.

Application: Least Squares

Section 2.6 on least squares explained how in mathematical modeling we often have the general form of a model and need to determine specific parameters in the model so as to approximate the actual behavior of the phenomenon being modeled. When the model is linear, this often means that we are given a $p \times q$ matrix $\mathbf{A}$ (the model) and a $p \times 1$ column matrix $\mathbf{y}$ (data on the real phenomenon) and we need to find a $q \times 1$ matrix $\mathbf{x}$ (the parameters) so that $\mathbf{Ax} \approx \mathbf{y}$. When we seek to determine $\mathbf{x}$ by making $\|\mathbf{Ax} - \mathbf{y}\|_2$ as small as possible, we have a *least-squares problem.*

One approach to finding $\mathbf{x}$ to minimize $\|\mathbf{Ax} - \mathbf{y}\|_2$ was stated in Section 2.6: $\mathbf{x}$ minimizes $\|\mathbf{Ax} - \mathbf{y}\|_2$ if and only if $\mathbf{x}$ solves $\mathbf{A}^T\mathbf{Ax} = \mathbf{A}^T\mathbf{y}$ (replace transpose ^{T} by hermitian transpose ^{H} if any of the data are complex); see Problem 26. We remarked, however, that actually computing $\mathbf{A}^T\mathbf{A}$ and then solving this system of equations often produces unacceptably large errors when the computations are done in floating-point arithmetic on computers; see Problem 24. The normalized QR-decomposition—which can be computed quite accurately and efficiently—provides a way to avoid the difficulty; see Problem 25.

Suppose that $\mathbf{A} = \mathbf{QR}$ is a normalized QR-decomposition of $\mathbf{A}$, so that—if $\mathbf{A}$ is $p \times q$ and has rank k—$\mathbf{Q}$ is $p \times k$ and has orthonormal columns, while $\mathbf{R}$ is

$k \times q$, is upper-triangular, and has rank k. Consider the equations

$$\mathbf{A}^T\mathbf{A}\mathbf{x} = \mathbf{A}^T\mathbf{y} \quad \text{when} \quad \mathbf{A} = \mathbf{Q}\mathbf{R}.$$

Substituting for $\mathbf{A}$ gives

$$(\mathbf{Q}\mathbf{R})^T(\mathbf{Q}\mathbf{R})\mathbf{x} = (\mathbf{Q}\mathbf{R})^T\mathbf{y}, \quad \text{that is,} \quad \mathbf{R}^T\mathbf{Q}^T\mathbf{Q}\mathbf{R}\mathbf{x} = \mathbf{R}^T\mathbf{Q}^T\mathbf{y}.$$

Recall that $\mathbf{Q}$ has orthonormal columns, so $\mathbf{Q}^T\mathbf{Q} = \mathbf{I}_k$. Therefore, the equations become

$$\mathbf{R}^T\mathbf{R}\mathbf{x} = \mathbf{R}^T\mathbf{Q}^T\mathbf{y}.$$

Corollary 5.51 states that the rank of $\mathbf{R}$ and of $\mathbf{R}^T$ are equal; therefore, the $q \times k$ matrix $\mathbf{R}^T$ has rank k. By Theorem 4.22 on left-inverses and rank, $\mathbf{R}^T$ has a left-inverse $\mathbf{L}$ with $\mathbf{L}\mathbf{R}^T = \mathbf{I}_k$. Multiplying both sides of the last equation above by $\mathbf{L}$ and using $\mathbf{L}\mathbf{R}^T = \mathbf{I}_k$ gives

$$\mathbf{R}\mathbf{x} = \mathbf{Q}^T\mathbf{y}$$

if $\mathbf{R}^T\mathbf{R}\mathbf{x} = \mathbf{R}^T\mathbf{Q}^T\mathbf{y}$; conversely, it is clear that if $\mathbf{R}\mathbf{x} = \mathbf{Q}^T\mathbf{y}$, then $\mathbf{R}^T\mathbf{R}\mathbf{x} = \mathbf{R}^T\mathbf{Q}^T\mathbf{y}$. Therefore, $\mathbf{x}$ solves the least-squares problem if and only if $\mathbf{R}\mathbf{x} = \mathbf{Q}^T\mathbf{y}$. Since $\mathbf{R}$ is $k \times q$, is upper-triangular, and has rank k, the equations $\mathbf{R}\mathbf{x} = \mathbf{Q}^T\mathbf{y}$ can immediately be solved for $\mathbf{x}$ by back-substitution. That is, once we find the normalized QR-decomposition of $\mathbf{A}$, the least-squares problem can be solved immediately. We summarize.

(5.85) ***Theorem*** (QR and least squares). Let $\mathbf{A} = \mathbf{Q}\mathbf{R}$ be a normalized QR-decomposition of the $p \times q$ matrix $\mathbf{A}$. Then all solutions to the least-squares problem of finding $\mathbf{x}$ to minimize $\|\mathbf{A}\mathbf{x} - \mathbf{y}\|_2$ can be obtained by applying back-substitution to solve $\mathbf{R}\mathbf{x} = \mathbf{Q}^T\mathbf{y}$ ($\mathbf{R}\mathbf{x} = \mathbf{Q}^H\mathbf{y}$ if $\mathbf{A}$ or $\mathbf{y}$ are complex).

(5.86) **Key Corollary** (least squares). Every least-squares problem $\mathbf{A}\mathbf{x} \approx \mathbf{y}$ has a solution, and all solutions can be found through Theorem 5.85 using the normalized QR-decomposition of $\mathbf{A}$.

(5.87) ***Example.*** Let $\mathbf{A}$ be the 4×4 matrix

$$\mathbf{A} = \begin{bmatrix} 1 & 2 & 0 & -1 \\ 1 & -1 & 3 & 2 \\ 1 & -1 & 3 & 2 \\ -1 & 1 & -3 & 1 \end{bmatrix}$$

whose normalized QR-decomposition $\mathbf{A} = \mathbf{Q}\mathbf{R}$ was found in Example 5.83. Suppose that we want to solve the least-squares problem

$$\mathbf{A}\mathbf{x} \approx \mathbf{y} = [1 \quad -1 \quad 2 \quad 1]^T.$$

According to Theorem 5.85, we merely need to solve $\mathbf{Rx} = \mathbf{Q}^T\mathbf{y}$. We find that

$$\mathbf{Q}^T\mathbf{y} = [1/2 \quad \sqrt{3}/2 \quad \sqrt{6}/2]^T.$$

The system to solve, therefore, is

$$\mathbf{Rx} = \mathbf{Q}^T\mathbf{y}, \quad \text{that is,} \quad \begin{bmatrix} 2 & -\frac{1}{2} & \frac{9}{2} & 1 \\ 0 & \frac{3\sqrt{3}}{2} & -\frac{3\sqrt{3}}{2} & -\sqrt{3} \\ 0 & 0 & 0 & \sqrt{6} \end{bmatrix} \mathbf{x} = \begin{bmatrix} \frac{1}{2} \\ \frac{\sqrt{3}}{2} \\ \frac{\sqrt{6}}{2} \end{bmatrix}.$$

This is easily solved immediately by back-substitution: we obtain $x_4 = \frac{1}{2}$, then $\mathbf{x}_3 = \alpha$ arbitrary, then $x_2 = \alpha + \frac{2}{3}$, and finally $x_1 = -2\alpha + \frac{1}{6}$. The general solution to the original least-squares problem is therefore

$$\mathbf{x} = \begin{bmatrix} \frac{1}{6} \\ \frac{2}{3} \\ 0 \\ \frac{1}{2} \end{bmatrix} + \alpha \begin{bmatrix} -2 \\ 1 \\ 1 \\ 0 \end{bmatrix} \quad \text{for arbitrary } \alpha.$$

PROBLEMS 5.9

▷ **1.** Suppose that $\mathbf{Q}$ is a $p \times q$ real matrix with orthogonal columns. What can be said about $\mathbf{Q}^T\mathbf{Q}$?

2. Suppose that $\mathbf{P}$ is a real $p \times p$ matrix, that $\mathbf{P}^T = \mathbf{P}$, and that $\mathbf{P}^2 = \mathbf{P}$; any such matrix is called a $p \times p$ *projection matrix*. Prove that, for each $\mathbf{v}$ in $\mathbb{R}^p$, $\mathbf{Pv}$ is the orthogonal projection of $\mathbf{v}$ onto the column space of $\mathbf{P}$.

3. Show that $\mathbf{P}$ is a projection matrix—see Problem 2—if and only if $\mathbf{I}_p - \mathbf{P}$ is a projection matrix.

▷ **4.** Find the 3×3 matrix $\mathbf{P}$ so that $\mathbf{Pv}$ is the orthogonal projection of $\mathbf{v}$ onto the subspace spanned by the orthogonal set $S = \{[1 \quad 2 \quad 2]^T, [-2 \quad 2 \quad -1]^T\}$.

5. Find the 4×4 matrix $\mathbf{P}$ so that $\mathbf{Pv}$ is the orthogonal projection of $\mathbf{v}$ onto the subspace spanned by the vectors

$$[0.5 \quad 0.5 \quad 0.5 \quad 0.5]^T, \qquad [-0.5 \quad 0.5 \quad -0.5 \quad 0.5]^T.$$

6. Find both the unnormalized and the normalized QR-decomposition of

$$\mathbf{A} = \begin{bmatrix} 1 & 2 \\ 0 & 1 \\ 1 & 4 \end{bmatrix}.$$

7. Find both the unnormalized and normalized QR-decomposition of

$$\mathbf{A} = \begin{bmatrix} 1 & 2 & 4 \\ 0 & 1 & 1 \\ 1 & 4 & 6 \end{bmatrix}.$$

▷ 8. Find both the unnormalized and normalized QR-decomposition of

$$\mathbf{A} = \begin{bmatrix} 1 & 2 & 3 \\ 0 & 1 & 1 \\ 1 & 4 & 6 \end{bmatrix}.$$

9. Find both the unnormalized and normalized QR-decomposition of

$$\mathbf{A} = \begin{bmatrix} 1 & 1 & 1 \\ 1 & 1 & 1 \end{bmatrix}.$$

10. Suppose that you are doing your computations on a computer and that the number ϵ is such that $1 + \epsilon$ is accurately evaluated but $1 + \epsilon^2$ is evaluated to equal 1; $\epsilon = 10^{-6}$ will work nicely on an 8- to 10-digit calculator, while $\epsilon = 10^{-10}$ works on most microcomputers having math coprocessors.
   (a) Apply the traditional Gram–Schmidt process on this "computer" in order to verify that you obtain the unnormalized QR-decomposition

$$\mathbf{A} = \begin{bmatrix} 1 & 1 \\ 1 & 1 \\ 1 & 1+\epsilon \end{bmatrix} = \mathbf{Q}_0\mathbf{R}_0 = \begin{bmatrix} 1 & -\frac{\epsilon}{3} \\ 1 & -\frac{\epsilon}{3} \\ 1 & \frac{2\epsilon}{3} \end{bmatrix} \begin{bmatrix} 1 & 1+\frac{\epsilon}{3} \\ 0 & 1 \end{bmatrix}.$$

   (Note that this is the correct decomposition—except of course that your computer will not divide perfectly by 3. The small ϵ did not cause difficulties. See Problems 24 and 25.)
   (b) Find the normalized QR-decomposition.

▷ 11. Suppose that $\mathbf{A}$ is real and $p \times q$, and that $\mathbf{A} = \mathbf{Q}_0\mathbf{R}_0$, where $\mathbf{Q}_0$ is $p \times q$ and has orthogonal columns while $\mathbf{R}_0$ is $q \times q$ and unit-upper-triangular. Prove that the columns of $\mathbf{Q}_0$ form an orthogonal spanning set for the column space of $\mathbf{A}$, and that the first i columns of $\mathbf{Q}_0$ span the space spanned by the first i columns of $\mathbf{A}$ for $1 \le i \le q$.

12. Suppose that $\mathbf{A}$ is real and $p \times q$ with rank k, and that $\mathbf{A} = \mathbf{QR}$, where $\mathbf{Q}$ is $p \times k$ and has orthonormal columns while $\mathbf{R}$ is $k \times q$, upper-triangular, and of rank k. Prove that the columns of $\mathbf{Q}$ form an orthonormal basis for the column space of $\mathbf{A}$ and that $\mathbf{P} = \mathbf{QQ}^T$ represents orthogonal projection onto the column space of $\mathbf{A}$.

13. Suppose that $\mathbf{A}$ is real and $p \times q$ with $p \geq q$; we describe how to use the *Householder matrices* of Problems 15–17 of Section 5.8 to compute a QR-decomposition of $\mathbf{A}$. Let $\mathbf{A}_0 = \mathbf{A}$; we compute $\mathbf{A}_1, \ldots, \mathbf{A}_q$ as follows. $\mathbf{A}_i$ has the special structure

$$\mathbf{A}_i = \begin{bmatrix} \mathbf{R}_i & \mathbf{B}_i \\ \mathbf{0} & \mathbf{C}_i \end{bmatrix}, \qquad \text{where } \mathbf{R}_i \text{ is } i \times i \text{ and upper-triangular;}$$

certainly, this holds for $i = 0$. Let $\mathbf{H}_i$ be a $(p - i) \times (p - i)$ Householder matrix such that $\mathbf{H}_i$ times the first column of $\mathbf{C}_i$ is a multiple of the $(p - i) \times 1$ unit column matrix $\mathbf{e}_1$, and define

$$\mathbf{Q}_i = \begin{bmatrix} \mathbf{I}_i & \mathbf{0} \\ \mathbf{0} & \mathbf{H}_i \end{bmatrix}, \qquad \text{where } \mathbf{I}_i \text{ is the } i \times i \text{ identity matrix.}$$

(a) Show that $\mathbf{Q}_i\mathbf{Q}_i^T = \mathbf{Q}_i^T\mathbf{Q}_i = \mathbf{I}_p$.

(b) Show that $\mathbf{Q}_i\mathbf{A}_i$ has the structure claimed for $\mathbf{A}_{i+1}$. We define $\mathbf{A}_{i+1} = \mathbf{Q}_i\mathbf{A}_i$ and continue until we have $\mathbf{A}_q$.

(c) Show that $\mathbf{Q}_{q-1}\mathbf{Q}_{q-2} \cdots \mathbf{Q}_1\mathbf{Q}_0\mathbf{A} = \mathbf{S}$, where $\mathbf{S}$ has the structure

$$\mathbf{S} = \begin{bmatrix} \mathbf{R} \\ \mathbf{0} \end{bmatrix}, \qquad \text{with } \mathbf{R} \; q \times q \text{ and upper-triangular.}$$

Therefore, $\mathbf{A} = \mathbf{PS}$, where $\mathbf{P} = \mathbf{Q}_0^T\mathbf{Q}_1^T \cdots \mathbf{Q}_{q-1}^T$.

(d) Show that $\mathbf{PP}^T = \mathbf{P}^T\mathbf{P} = \mathbf{I}_p$, so that $\mathbf{P}$ is nonsingular and has orthonormal columns.

(See also Problem 14.)

14. Continuing the development from Problem 13, define $\mathbf{Q}$ as the $p \times q$ matrix consisting of the first q columns of $\mathbf{P}$.

(a) Show that $\mathbf{A} = \mathbf{QR}$.

(b) *Assuming* that the rank of $\mathbf{A}$ is q and recalling that $\mathbf{P}$ is nonsingular, show that the rank of $\mathbf{S}$ is q. Show that the rank of $\mathbf{R}$ is q, so that $\mathbf{R}$ is nonsingular. Conclude that $\mathbf{A} = \mathbf{QR}$ is a normalized QR-decomposition of $\mathbf{A}$.

15. Use the Householder-matrix method of Problems 13 and 14 to find the normalized QR-decomposition of the matrix $\mathbf{A}$ from Problem 6 and verify that you obtain

$$\mathbf{A}_0 = \mathbf{A} = \begin{bmatrix} 1 & 2 \\ 0 & 1 \\ 1 & 4 \end{bmatrix} \quad \text{and} \quad \mathbf{Q}_0 = \begin{bmatrix} \frac{\sqrt{2}}{2} & 0 & \frac{\sqrt{2}}{2} \\ 0 & 1 & 0 \\ \frac{\sqrt{2}}{2} & 0 & -\frac{\sqrt{2}}{2} \end{bmatrix};$$

then

$$\mathbf{A}_1 = \begin{bmatrix} \sqrt{2} & 3\sqrt{2} \\ 0 & 1 \\ 0 & -\sqrt{2} \end{bmatrix} \quad \text{and} \quad \mathbf{Q}_1 = \begin{bmatrix} 1 & 0 & 0 \\ 0 & \frac{\sqrt{3}}{3} & -\frac{\sqrt{6}}{3} \\ 0 & -\frac{\sqrt{6}}{3} & -\frac{\sqrt{3}}{3} \end{bmatrix}.$$

Thus

$$\mathbf{A} = \mathbf{PS} = \begin{bmatrix} \frac{\sqrt{2}}{2} & -\frac{\sqrt{3}}{3} & -\frac{\sqrt{6}}{6} \\ 0 & \frac{\sqrt{3}}{3} & -\frac{\sqrt{6}}{3} \\ \frac{\sqrt{2}}{2} & \frac{\sqrt{3}}{3} & \frac{\sqrt{6}}{6} \end{bmatrix} \begin{bmatrix} \sqrt{2} & 3\sqrt{3} \\ 0 & \sqrt{3} \\ 0 & 0 \end{bmatrix},$$

and finally,

$$\mathbf{A} = \mathbf{QR} = \begin{bmatrix} \frac{\sqrt{2}}{2} & -\frac{\sqrt{3}}{3} \\ 0 & \frac{\sqrt{3}}{3} \\ \frac{\sqrt{2}}{2} & \frac{\sqrt{3}}{3} \end{bmatrix} \begin{bmatrix} \sqrt{2} & 3\sqrt{3} \\ 0 & \sqrt{3} \end{bmatrix}.$$

▷ **16.** Use the Householder-matrix method of Problems 13 and 14 to find a normalized QR-decomposition of the matrix $\mathbf{A}$ in Problem 8.

17. If $\mathbf{A}$ is $p \times q$ with $p \geq q$ and the rank of $\mathbf{A}$ is strictly less than q, then the Householder-matrix method of Problems 13 and 14 need not produce a normalized QR-decomposition. For example,

$$\mathbf{A} = \begin{bmatrix} 1 & 2 & 3 \\ 0 & 0 & 4 \\ 0 & 0 & 5 \end{bmatrix} = \mathbf{QR} \quad \text{with} \quad \mathbf{Q} = \mathbf{I}_3 \text{ and } \mathbf{R} = \mathbf{A}$$

is the decomposition that straightforward application of the Householder-matrix method to $\mathbf{A}$ would produce.

(a) Show that the rank of $\mathbf{A}$ equals 2.

(b) Show that no two of the columns of $\mathbf{Q}$ span the column space of $\mathbf{A}$.

(c) Conclude that we cannot obtain a normalized QR-decomposition of $\mathbf{A}$ *from among the columns of* $\mathbf{Q}$.

(d) Show that the traditional (or modified) Gram-Schmidt process when applied to this $\mathbf{A}$ produces both an unnormalized and a normalized QR-decomposition.

(When the rank of $\mathbf{A}$ is less than q, it is possible to implement the Householder-matrix method with what is called *column pivoting* so as to obtain the normalized QR-decomposition of a matrix with permuted columns; this method is widely used in computer software to obtain QR-decompositions.)

𝔐 **18.** Much as was done in Example 5.84 for the 6×6 Hilbert matrix $\mathbf{H}_6$, use MATLAB or similar software to measure how near the fifth column of the 5×5 Hilbert matrix $\mathbf{H}_5$ is to being linearly dependent on the first four columns.

𝔐 **19.** When the various powers t^i are graphed for $0 \le t \le 1$, their graphs seem quite similar; intuitively, it appears that these powers are nearly linearly dependent. For example, t^4, t^5, and t^6 appear so; we wish to measure how nearly linearly dependent these are when their values at 0.1, 0.2, 0.3, 0.4, 0.5, 0.6, 0.7, 0.8, 0.9, and 1.0 are considered. Set up an appropriate 10×3 matrix and use MATLAB or similar software to measure how nearly linearly dependent these are, much as in Example 5.84.

▷ **20.** Find a normalized QR-decomposition for $\mathbf{A}$ and use it to solve the least-squares problem $\mathbf{Ax} \approx \mathbf{y}$, where

$$\mathbf{A} = \begin{bmatrix} 1 & 1 \\ 2 & 2 \end{bmatrix} \quad \text{and} \quad \mathbf{y} = \begin{bmatrix} 2 \\ 8 \end{bmatrix}.$$

21. Find a normalized QR-decomposition for $\mathbf{A}$ and use it to solve the least-squares problem $\mathbf{Ax} \approx \mathbf{y}$, where

$$\mathbf{A} = \begin{bmatrix} 1 & 2 \\ 2 & 1 \\ 1 & -1 \end{bmatrix} \quad \text{and} \quad \mathbf{y} = \begin{bmatrix} 2 \\ 3 \\ 0 \end{bmatrix}.$$

22. In Section 2.3 we considered a growth model with competing populations; this led to a model of the form

$$\begin{bmatrix} F_{i+1} \\ C_{i+1} \end{bmatrix} = \begin{bmatrix} x_1 & x_2 \\ x_3 & x_4 \end{bmatrix} \begin{bmatrix} F_i \\ C_i \end{bmatrix},$$

where in Example 2.16 we assumed that $x_1 = 0.6$, $x_2 = 0.5$, $x_3 = -0.1$, and $x_4 = 1.2$. In practice, the problem could just as easily be to determine values of the model parameters x_j from actual population data. Use least squares to determine the parameters x_j so as to fit the population "data" from the table in Example 2.16 showing the evolution from time i to time $i + 1$ for $i = 1, 2, 3, 4, 5$, and see how the values obtained compare with those that generated the data. You may want to use MATLAB or similar software.

▷ **23.** Use a normalized QR-decomposition to solve the least-squares problem in Example 2.39.

24. Consider the matrix $\mathbf{A}$ of Problem 10 with ϵ as described there, and consider the least-squares problem $\mathbf{Ax} \approx \mathbf{y}$ where $\mathbf{y} = [2 \quad 3 \quad 2]^T$. The true least-squares solution is

$$\mathbf{x} = \frac{1}{\epsilon}\begin{bmatrix} \frac{1}{2} + \frac{5\epsilon}{2} \\ -\frac{1}{2} \end{bmatrix}.$$

(a) Remembering that ϵ is such that the computer evaluates $3 + 2\epsilon + \epsilon^2$ as equal to $3 + 2\epsilon$, find what will be *computed* as the *equations* $\mathbf{A}^T\mathbf{Ax} = \mathbf{A}^T\mathbf{y}$ for finding $\mathbf{x}$.

(b) Show that the unique solution to this computed system is

$$\frac{1}{\epsilon}\begin{bmatrix} -1 + 2\epsilon \\ 1 \end{bmatrix},$$

which is very far from the correct least-squares solution just given above. (This illustrates why using $\mathbf{A}^T\mathbf{Ax} = \mathbf{A}^T\mathbf{y}$ as a computational method for solving least-squares problems is unsatisfactory in practice.)

25. Problem 24 showed that computing $\mathbf{A}^T\mathbf{A}$ and then solving $\mathbf{A}^T\mathbf{Ax} = \mathbf{A}^T\mathbf{y}$ gave poor results on the least-squares problem there. Problem 10 found the *computed QR*-decomposition for the matrix $\mathbf{A}$. Use this computed QR-decomposition to solve the least-squares problem in Problem 24, and compare the answer you obtain with the true solution and with that obtained using $\mathbf{A}^T\mathbf{A}$.

26. The argument leading to Theorem 5.85 proved that there exists an $\mathbf{x}_0$ solving $\mathbf{A}^H\mathbf{Ax}_0 = \mathbf{A}^H\mathbf{b}$ (replace ^{H} by ^{T} in the real case).

(a) Show that any $\mathbf{x} = \mathbf{x}_0 + \mathbf{n}$ is also a solution if and only if $\mathbf{A}^H\mathbf{An} = \mathbf{0}$, and conclude from $\mathbf{n}^H(\mathbf{A}^H\mathbf{An}) = \|\mathbf{An}\|_2^2$ that this holds if and only if $\mathbf{An} = \mathbf{0}$.

(b) By substituting $\mathbf{A}^H\mathbf{Ax}_0$ for $\mathbf{A}^H\mathbf{b}$, show that

$$\|\mathbf{Ax} - \mathbf{b}\|_2^2 = \|\mathbf{b}\|_2^2 - \|\mathbf{Ax}_0\|_2^2 + \|\mathbf{A}(\mathbf{x} - \mathbf{x}_0)\|_2^2.$$

(c) Conclude from (b) that $\mathbf{x}$ minimizes $\|\mathbf{Ax} - \mathbf{b}\|_2$ if and only if $\mathbf{x} = \mathbf{x}_0 + \mathbf{n}$ where $\mathbf{An} = \mathbf{0}$.

(d) Conclude from (a) and (c) that $\mathbf{x}$ solves the least-squares problem $\mathbf{Ax} \approx \mathbf{b}$ if and only if $\mathbf{A}^H\mathbf{Ax} = \mathbf{A}^H\mathbf{b}$.

5.10 *MISCELLANEOUS PROBLEMS*

PROBLEMS 5.10

▷ **1.** Show that $\mathbb{C}$ can be viewed as a real vector space.

2. Show that $\mathbb{C}$ can be viewed as a complex vector space.

3. Let $\mathscr{V}$ be the real vector space of complex numbers over the field $\mathbb{R}$; show that the set of all real numbers is a subspace of $\mathscr{V}$.

▷ **4.** Suppose that $\mathbf{u}_1$, $\mathbf{u}_2$, $\mathbf{u}_3$ form a linearly independent set, and define

$$\mathbf{v}_1 = \mathbf{u}_1 + \mathbf{u}_2 + \mathbf{u}_3, \quad \mathbf{v}_2 = \mathbf{u}_1 + \alpha\mathbf{u}_2, \quad \text{and} \quad \mathbf{v}_3 = \mathbf{u}_2 + \beta\mathbf{u}_3.$$

Find conditions on α and β in order that the $\mathbf{v}_i$ form a linearly independent set.

▷ **5.** Let $\mathscr{V}_0$ be a subspace of $\mathbb{R}^p$. Define $\mathscr{V}_0^\perp$ as the set of all vectors $\mathbf{v}$ in $\mathbb{R}^p$ that are orthogonal to *all* the vectors in $\mathscr{V}_0$.

(a) Prove that $\mathscr{V}_0^\perp$ is a subspace of $\mathbb{R}^p$.

(b) Suppose that $p = 3$ and that $\mathscr{V}_0$ has as a basis the set of vectors

$$\mathbf{u}_1 = [2 \quad 1 \quad -3]^T, \qquad \mathbf{u}_2 = [-1 \quad 0 \quad -2]^T;$$

find a basis for $\mathscr{V}_0^\perp$.

6. Let $B = \{1 + t; t + t^2; t^2 + t^3; t^3\}$ be an ordered basis for the real vector space $\mathscr{P}^4$ of polynomials of degree strictly less than 4. By using the coordinate isomorphism c_B to convert to a problem in $\mathbb{R}^4$, extend the set $\{1 + 2t + t^2, t\}$ to form a basis for $\mathscr{P}^4$.

7. Consider the ordered basis B and space $\mathscr{P}^4$ of Problem 6. By using the coordinate isomorphism c_B to convert to a problem in $\mathbb{R}^4$, form a basis for $\mathscr{P}^4$ from among the vectors

$$1 + 2t + t^2, \quad t + 2t^2 + t^3, \quad 1 + t - t^2 - t^3, \quad t^2 + 2t^3, \quad t^3, \quad 1 + t + t^3.$$

8. Show that the rank k of a matrix $\mathbf{A}$ is the order of the largest nonsingular square submatrix of $\mathbf{A}$ by completing the details of the following outline of the proof. Suppose $\mathbf{A}$ has rank k and consider a $p \times p$ submatrix with $p > k$. Show that the p rows of the submatrix form a dependent set since the p rows of $\mathbf{A}$ do so (since $p > k =$ rank of $\mathbf{A}$); deduce that the submatrix is singular. Choose any k rows of $\mathbf{A}$ forming an independent subset; then the rank of the matrix of these k rows is k so there are some k columns from this matrix forming an independent set. Deduce that the resulting $k \times k$ submatrix is nonsingular. Conversely, suppose that k is the maximum order of nonsingular submatrices and choose a nonsingular $k \times k$ submatrix. Since the rows of the submatrix form a linearly independent set, show that the corresponding k rows of $\mathbf{A}$ form a linearly independent set and deduce then that the rank of $\mathbf{A}$ is not less than k. Next suppose that some subset of p rows of $\mathbf{A}$ is linearly independent; deduce as in the first part of the proof of this theorem that some $p \times p$ submatrix is nonsingular, so that $p \le k$. Since k is thus the maximum number of rows of $\mathbf{A}$ in any linearly independent set, deduce that k equals the rank of $\mathbf{A}$.

▷ **9.** By considering $\|\mathbf{u} - \alpha\mathbf{v}\|_2^2$ and seeking α so as to make this expression zero, prove that:

(a) $|\mathbf{u}^T\mathbf{v}| = \|\mathbf{u}\|_2\|\mathbf{v}\|_2$ for particular $\mathbf{u}$ and $\mathbf{v}$ in $\mathbb{R}^p$ (if and) only if $\mathbf{u}$ and $\mathbf{v}$ form a linearly dependent set;

(b) the same result holds with ^{T} replaced by ^{H} when $\mathbf{u}$ and $\mathbf{v}$ are in $\mathbb{C}^p$.

10. Prove that the normalized QR-decomposition $\mathbf{A} = \mathbf{QR}$ is unique if $p \times q$ $\mathbf{A}$ has rank q and if $\mathbf{R}$ is forced to have positive entries on its main diagonal.

▷ **11.** Suppose that $B = \{\mathbf{v}_1, \ldots, \mathbf{v}_q\}$ is a basis and that

$$\mathbf{v}_0 = \alpha_1 \mathbf{v}_1 + \cdots + \alpha_q \mathbf{v}_q.$$

(a) Show that any vector $\mathbf{v}_i$ whose coefficient α_i above is nonzero can be replaced in B by $\mathbf{v}_0$ and the resulting set will also be a basis.

(b) Show that if the coefficient α_i *is* zero, then the resulting set is *not* a basis.

𝔐 **12.** Use MATLAB or similar software to find a normalized QR-decomposition and use it to solve the least-squares problem in Problem 3 of Section 2.6 (on predicting the population of the United States in 1990).

13. Prove there is a (complex) Householder matrix

$$\mathbf{H}_{\mathbf{w}} = \mathbf{I} - (2/\mathbf{w}^H\mathbf{w})\mathbf{w}\mathbf{w}^H$$

such that $\mathbf{H}_{\mathbf{w}}\mathbf{x} = \mathbf{y}$ if and only if $\|\mathbf{x}\|_2 = \|\mathbf{y}\|_2$ and $\mathbf{x}^H\mathbf{y}$ is real.

6

Linear transformations and matrices

Much of the power of linear algebra and matrices stems from our ability not only to represent with vectors many objects of interest in applications but also to represent many of the operations or transformations performed on those objects — for example in models of how inputs to some process are transformed into outputs. This Chapter develops the basic theory of such transformations; the ***Key Theorems*** *are* ***6.14***, ***6.20***, ***6.23***, ***6.26***, *and* ***6.28***.

6.1 *INTRODUCTION; LINEAR TRANSFORMATIONS*

Many problems in applied mathematics involve the study of *transformations*—the way in which certain input data are transformed into output data. For example, the factors that produce certain profits for a business from a certain pricing and manufacturing structure can be viewed as a transformation of inputs (price and production data) into outputs (profit structure). In many mathematical models, it often turns out that the transformations are *linear* in the sense that the sum of two inputs is transformed into the sum of their individual outputs and a multiple of an input is transformed into that multiple of its output. The equation (2.4) represents such a linear transformation (of the market pattern one month into that for the next month), as does (2.13) (of the population pattern at one moment into that at a later moment). The applied mathematician attempts to deduce properties of the relevant transformations so as to learn about the properties of the real structures being modeled. In this chapter are presented some basic terminology and facts about linear transformations between vector spaces.

Linear Transformations

(6.1) ***Definition.*** Let $\mathscr{V}$ and $\mathscr{W}$ both be real (or both complex) vector spaces. A *linear transformation* $\mathscr{T}$ from $\mathscr{V}$—called the *domain* of $\mathscr{T}$—to $\mathscr{W}$—called the *range* of $\mathscr{T}$—is a correspondence that assigns to every vector $\mathbf{v}$ in $\mathscr{V}$ a vector $\mathscr{T}(\mathbf{v})$ in $\mathscr{W}$ in such a way that:
(a) $\mathscr{T}(\mathbf{v}_1 + \mathbf{v}_2) = \mathscr{T}(\mathbf{v}_1) + \mathscr{T}(\mathbf{v}_2)$ for all vectors $\mathbf{v}_1$, $\mathbf{v}_2$.
(b) $\mathscr{T}(\alpha\mathbf{v}) = \alpha\mathscr{T}(\mathbf{v})$ for all vectors $\mathbf{v}$ and scalars α.
{Equivalently, $\mathscr{T}(\alpha_1\mathbf{v}_1 + \alpha_2\mathbf{v}_2) = \alpha_1\mathscr{T}(\mathbf{v}_1) + \alpha_2\mathscr{T}(\mathbf{v}_2)$ for all vectors $\mathbf{v}_1$, $\mathbf{v}_2$, and scalars α_1, α_2.}

(6.2) ***Example***

(a) Let $\mathscr{V} = \mathbb{R}^q$ and $\mathscr{W} = \mathbb{R}^p$, and suppose that $\mathbf{A}$ is a $p \times q$ real matrix. Then $\mathscr{T}$ defined by $\mathscr{T}(\mathbf{v}) = \mathbf{A}\mathbf{v}$ is a linear transformation from $\mathscr{V}$ to $\mathscr{W}$. The same holds for $\mathbb{C}^q$ and $\mathbb{C}^p$, of course.

(b) Let $\mathscr{V} = \mathscr{W} = C[a, b]$, the space of real-valued continuous functions on $a \le t \le b$; suppose that $f(t)$ is a continuous function. Then $\mathscr{T}$ defined by letting $\mathscr{T}(\mathbf{x})$ be that function whose value at t is $f(t)\mathbf{x}(t)$ is a linear transformation from $\mathscr{V}$ to $\mathscr{W}$.

(c) If $\mathscr{V}$ is a real p-dimensional vector space and B is an ordered basis for $\mathscr{V}$, then the coordinate isomorphism c_B is a linear transformation from $\mathscr{V}$ to $\mathscr{W} = \mathbb{R}^p$.

(d) Let $\mathscr{V} = C^{(2)}[0, 1]$ be the space of twice-continuously differentiable functions on $[0, 1]$, and let $\mathscr{T}(\mathbf{v})$ be that function whose value at t equals

$$\mathscr{T}(\mathbf{v})(t) = \frac{d}{dt}(1 + t)\frac{d\mathbf{v}(t)}{dt} - (3 + e^t)\mathbf{v}(t).$$

$\mathscr{T}$ is a linear transformation from $\mathscr{V}$ to $\mathscr{W} = C[0, 1]$ of (b) above.

(e) Suppose that $\mathscr{V}$ and $\mathscr{W}$ are both real (or both complex) vector spaces, and let $\mathscr{L} = \mathscr{L}(\mathscr{V}, \mathscr{W})$ denote the set of all linear transformations from $\mathscr{V}$ to $\mathscr{W}$. We can define the sum $\mathscr{T}_1 + \mathscr{T}_2$ of two elements in $\mathscr{L}$ as that transformation with

$$(\mathscr{T}_1 + \mathscr{T}_2)(\mathbf{v}) = \mathscr{T}_1(\mathbf{v}) + \mathscr{T}_2(\mathbf{v});$$

$\mathscr{T}_1 + \mathscr{T}_2$ is a linear transformation from $\mathscr{V}$ to $\mathscr{W}$, that is, is in $\mathscr{L}$. Similarly, define $\alpha\mathscr{T}$ for scalars α by

$$(\alpha\mathscr{T})(\mathbf{v}) = \alpha\{\mathscr{T}(\mathbf{v})\};$$

$\alpha\mathscr{T}$ is in $\mathscr{L}$. The zero transformation $\mathscr{O}$ with $\mathscr{O}(\mathbf{v}) = \mathbf{0}$ in $\mathscr{W}$ for all $\mathbf{v}$ is in $\mathscr{L}$. The set $\mathscr{L}(\mathscr{V}, \mathscr{W})$ with operations as defined is a real (or complex) vector space.

To verify that each of the objects in Example 6.2 truly is a linear transformation, the two conditions in Definition 6.1 must be verified. In Example 6.2(a), for example, we have

$$\mathscr{T}(\mathbf{v}_1 + \mathbf{v}_2) = \mathbf{A}(\mathbf{v}_1 + \mathbf{v}_2) = \mathbf{A}\mathbf{v}_1 + \mathbf{A}\mathbf{v}_2 = \mathscr{T}(\mathbf{v}_1) + \mathscr{T}(\mathbf{v}_2)$$

and

$$\mathscr{T}(\alpha\mathbf{v}) = \mathbf{A}(\alpha\mathbf{v}) = \alpha\mathbf{A}\mathbf{v} = \alpha\mathscr{T}(\mathbf{v}).$$

See Problems 4–7.

Image Space and Null Space

Example 6.2(a) shows that linear transformations are generalizations of matrix multiplication with vectors—$\mathscr{T}(\mathbf{v}) = \mathbf{A}\mathbf{v}$. We know, of course, that equations $\mathbf{A}\mathbf{x} = \mathbf{b}$ are important as representations of systems of linear equations in unknowns $x_i = \langle\mathbf{x}\rangle_i$. Equations $\mathscr{T}(\mathbf{v}) = \mathbf{w}$, when $\mathbf{w}$ is given and $\mathbf{v}$ is to be found, are therefore generalizations of the familiar $\mathbf{A}\mathbf{x} = \mathbf{b}$. Such an equation with $\mathscr{T}$ as in Example 6.2(d), for example, is an *ordinary differential equation* and is very important in practical applications. We therefore want to consider the solvability of equations $\mathscr{T}(\mathbf{v}) = \mathbf{w}$.

It is, of course, obvious that $\mathscr{T}(\mathbf{v}) = \mathbf{w}$ has a solution $\mathbf{v}$ if and only if $\mathbf{w}$ is produced by $\mathscr{T}$ from some vector in $\mathscr{V}$. Consider the set $\mathscr{W}_0$ of all such vectors in $\mathscr{W}$: $\mathscr{W}_0$ equals the set of all vectors $\mathbf{w}_0$ in $\mathscr{W}$ such that $\mathbf{w}_0 = \mathscr{T}(\mathbf{v}_0)$ for some $\mathbf{v}_0$ in $\mathscr{V}$. It is easy to use the Subspace Theorem 5.12 to see that $\mathscr{W}_0$ is actually a subspace of $\mathscr{W}$. For example, if $\mathbf{w}_0$ is in $\mathscr{W}_0$ and α is a scalar, then $\alpha\mathbf{w}_0$ is in $\mathscr{W}_0$ since $\alpha\mathbf{w}_0 = \alpha\mathscr{T}(\mathbf{v}_0) = \mathscr{T}(\alpha\mathbf{v}_0)$, where $\mathbf{v}_0$ is a vector in $\mathscr{V}$ for which $\mathscr{T}(\mathbf{v}_0) = \mathbf{w}_0$.

(6.3) ***Definition.*** The *image space* of the linear transformation $\mathscr{T}$ from $\mathscr{V}$ to $\mathscr{W}$ is that subspace $\mathscr{W}_0$ of $\mathscr{W}$ consisting of all vectors $\mathbf{w}_0$ in $\mathscr{W}$ that equal $\mathscr{T}(\mathbf{v}_0)$ for some $\mathbf{v}_0$ in $\mathscr{V}$.

With this definition, it is obvious that the equation $\mathscr{T}(\mathbf{v}) = \mathbf{w}$ can be solved when $\mathbf{w}$ is chosen arbitrarily in $\mathscr{W}$ if and only if the image space of $\mathscr{T}$ is in fact all of $\mathscr{W}$; such a $\mathscr{T}$ is said to map $\mathscr{V}$ *onto* $\mathscr{W}$. Thus our generalization of the solvability question that concerned us so much with matrices—can $\mathbf{A}\mathbf{x} = \mathbf{b}$ always be solved?—is answered by studying the image space of $\mathscr{T}$ and checking whether it equals all of $\mathscr{W}$. In summary:

(6.4) ***Theorem*** (solvability of equations). Suppose that $\mathscr{T}$ is a linear transformation from $\mathscr{V}$ to $\mathscr{W}$.

(a) Given $\mathbf{w}$, the equation $\mathscr{T}(\mathbf{v}) = \mathbf{w}$ is solvable if and only if $\mathbf{w}$ is in the image space of $\mathscr{T}$.

(b) $\mathbf{v}$ can be found to solve $\mathscr{T}(\mathbf{v}) = \mathbf{w}$ for arbitrary $\mathbf{w}$ in $\mathscr{W}$ if and only if $\mathscr{T}$ maps $\mathscr{V}$ *onto* $\mathscr{W}$—that is, the image space of $\mathscr{T}$ equals $\mathscr{W}$.

Another question that concerned us in solving $\mathbf{A}\mathbf{x} = \mathbf{b}$ was *uniqueness*—is there at most one solution to $\mathbf{A}\mathbf{x} = \mathbf{b}$? In our more general setting, we ask whether there may be more than one $\mathbf{v}$ solving $\mathscr{T}(\mathbf{v}) = \mathbf{w}$. Suppose that $\mathbf{v}_1$ and $\mathbf{v}_2$ both solve this. Then $\mathbf{0} = \mathbf{w} - \mathbf{w} = \mathscr{T}(\mathbf{v}_1) - \mathscr{T}(\mathbf{v}_2) = \mathscr{T}(\mathbf{v}_1 - \mathbf{v}_2)$, which simply says that $\mathbf{v}_1 - \mathbf{v}_2$ solves the homogeneous equation $\mathscr{T}(\mathbf{v}) = \mathbf{0}$, just as in **Key Theorem 4.16**. Thus the question of uniqueness depends on the set $\mathscr{V}_0$ of all solutions to $\mathscr{T}(\mathbf{v}) = \mathbf{0}$. The Subspace Theorem 5.12 again easily shows that $\mathscr{V}_0$ is a subspace of $\mathscr{V}$; for example, if $\mathbf{v}_1$ and $\mathbf{v}_2$ are in $\mathscr{V}_0$, then so is $\mathbf{v}_1 + \mathbf{v}_2$ since

$$\mathscr{T}(\mathbf{v}_1 + \mathbf{v}_2) = \mathscr{T}(\mathbf{v}_1) + \mathscr{T}(\mathbf{v}_2) = \mathbf{0} + \mathbf{0} = \mathbf{0}.$$

(6.5) ***Definition.*** The *null space* (or *kernel*) $\mathscr{V}_0$ of the linear transformation $\mathscr{T}$ from $\mathscr{V}$ to $\mathscr{W}$ is the subspace $\mathscr{V}_0$ of $\mathscr{V}$ consisting of all those vectors $\mathbf{v}$ in $\mathscr{V}$ satisfying $\mathscr{T}(\mathbf{v}) = \mathbf{0}$ in $\mathscr{W}$.

The argument preceding Definition 6.5 shows that it is easy to generalize Corollary 4.17 as follows.

(6.6) ***Theorem*** (uniqueness of solutions). The equation $\mathscr{T}(\mathbf{v}) = \mathbf{w}$ has at most one solution $\mathbf{v}$ if and only if the null space of $\mathscr{T}$ is just $\{\mathbf{0}\}$.

The relationships among the domain $\mathscr{V}$, range $\mathscr{W}$, null space $\mathscr{V}_0$, and image space $\mathscr{W}_0$ for a transformation $\mathscr{T}$ are depicted schematically in (6.7).

(6.7)

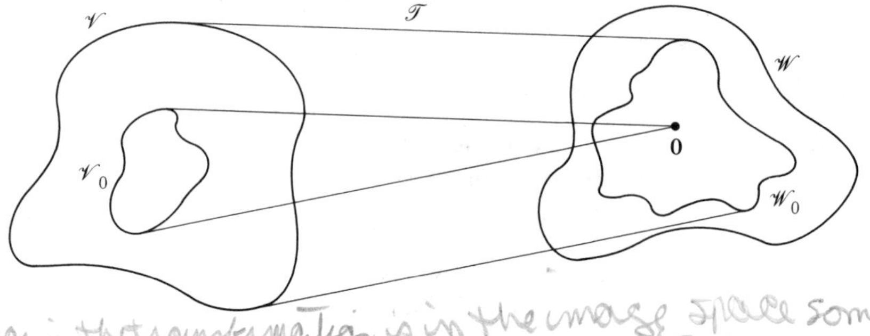

not every space in the transformation is in the image space some given the null space

The schematic above seems to suggest that the "shrinkage" of the null space $\mathscr{V}_0$ in $\mathscr{V}$ down to the vector $\mathbf{0}$ in $\mathscr{W}$ is somehow similar to the "shrinkage" of $\mathscr{V}$ down to the image space $\mathscr{W}_0$. The following theorem shows that this is correct in terms of dimensions.

(6.8) ***Theorem*** (domain, image, and null spaces). Suppose that $\mathscr{T}$ is a linear transformation from the finite-dimensional space $\mathscr{V}$ to $\mathscr{W}$. Then:

(a) The dimension of the domain $\mathscr{V}$ equals the sum of the dimension of the null space $\mathscr{V}_0$ and the dimension of the image space $\mathscr{W}_0$.

(b) In greater detail: If $\{\mathscr{T}(\mathbf{v}_1), \mathscr{T}(\mathbf{v}_2), \ldots, \mathscr{T}(\mathbf{v}_k)\}$ is a basis for $\mathscr{W}_0$ and $\{\mathbf{n}_1, \mathbf{n}_2, \ldots, \mathbf{n}_r\}$ is a basis for $\mathscr{V}_0$, then the set

$$\{\mathbf{v}_1, \mathbf{v}_2, \ldots, \mathbf{v}_k, \mathbf{n}_1, \mathbf{n}_2, \ldots, \mathbf{n}_r\}$$

is a basis for the domain $\mathscr{V}$.

PROOF. First, note that $\mathscr{V}$ has some finite basis $\{\mathbf{x}_1, \ldots, \mathbf{x}_p\}$ for some p. Since any $\mathbf{w}_0$ in $\mathscr{W}_0$ is of the form $\mathscr{T}(\mathbf{v})$ for some $\mathbf{v}$ in $\mathscr{V}$, and since any such $\mathbf{v}$ equals $\alpha_1\mathbf{x}_1 + \cdots + \alpha_p\mathbf{x}_p$ for appropriate scalars α_i, we have

$$\mathbf{w}_0 = \mathscr{T}(\mathbf{v}) = \mathscr{T}(\alpha_1\mathbf{x}_1 + \cdots + \alpha_p\mathbf{x}_p) = \alpha_1\mathscr{T}(\mathbf{x}_1) + \cdots + \alpha_p\mathscr{T}(\mathbf{x}_p),$$

which says that $\mathscr{T}(\mathbf{x}_1), \ldots, \mathscr{T}(\mathbf{x}_p)$ spans $\mathscr{W}_0$. Thus $\mathscr{W}_0$ is finite-dimensional; similarly, $\mathscr{V}_0$ is finite-dimensional since it is a subspace of the finite-dimensional $\mathscr{V}$ and cannot have more than p vectors in a linearly independent set. Thus all the dimensions in (a) of the theorem make sense. Part (a) follows immediately from (b), so we treat (b). Since $\mathscr{W}_0$ is finite-dimensional, it has a basis of, say, k vectors, each of which equals some $\mathscr{T}(\mathbf{v})$. That is, $\mathscr{W}_0$ has bases of the form $\{\mathscr{T}(\mathbf{v}_1), \ldots, \mathscr{T}(\mathbf{v}_k)\}$. Let $\{\mathbf{n}_1, \ldots, \mathbf{n}_r\}$ be a basis for $\mathscr{V}_0$. We have to prove that

$$B = \{\mathbf{v}_1, \ldots, \mathbf{v}_k, \mathbf{n}_1, \ldots, \mathbf{n}_r\}$$

is a basis for $\mathscr{V}$: a linearly independent spanning set.

(linear independence). Suppose that

$$\alpha_1\mathbf{v}_1 + \cdots + \alpha_k\mathbf{v}_k + \beta_1\mathbf{n}_1 + \cdots + \beta_r\mathbf{n}_r = \mathbf{0}.$$

Then

$$\begin{aligned}\alpha_1\mathscr{T}(\mathbf{v}_1) + \cdots + \alpha_k\mathscr{T}(\mathbf{v}_k) &= \mathscr{T}(\alpha_1\mathbf{v}_1 + \cdots + \alpha_k\mathbf{v}_k)\\ &= \mathscr{T}(\mathbf{0} - \beta_1\mathbf{n}_1 - \cdots - \beta_r\mathbf{n}_r)\\ &= \mathscr{T}(\mathbf{0}) - \beta_1\mathscr{T}(\mathbf{n}_1) - \cdots - \beta_r\mathscr{T}(\mathbf{n}_r) = \mathbf{0}\end{aligned}$$

since the $\mathbf{n}_i$ are in $\mathscr{V}_0$. But since the $\mathscr{T}(\mathbf{v}_i)$ form a basis for $\mathscr{W}_0$, the only combination of these vectors that equals $\mathbf{0}$ uses zero coefficients, so all the α_i must equal zero. But then we are left with a combination of the (linearly independent) $\mathbf{n}_i$ giving $\mathbf{0}$, which means that the β_i also must equal zero. Thus, B is linearly independent.

(spanning). For any $\mathbf{v}$ in $\mathscr{V}$, write $\mathscr{T}(\mathbf{v})$ as a linear combination of the $\mathscr{T}(\mathbf{v}_i)$:

$$\mathscr{T}(\mathbf{v}) = \alpha_1\mathscr{T}(\mathbf{v}_1) + \cdots + \alpha_k\mathscr{T}(\mathbf{v}_k).$$

Then $\mathbf{v}_0$ defined as

$$\mathbf{v}_0 = \mathbf{v} - \alpha_1\mathbf{v}_1 - \cdots - \alpha_k\mathbf{v}_k$$

is in $\mathscr{V}_0$ since

$$\mathscr{T}(\mathbf{v}_0) = \mathscr{T}(\mathbf{v}) - \alpha_1\mathscr{T}(\mathbf{v}_1) - \cdots - \alpha_k\mathscr{T}(\mathbf{v}_k) = \mathbf{0}.$$

Therefore, $\mathbf{v}_0$ can be written as a combination of the $\mathbf{n}_i$:

$$\mathbf{v}_0 = \beta_1\mathbf{n}_1 + \cdots + \beta_r\mathbf{n}_r.$$

But this says that

$$\mathbf{v} = \mathbf{v}_0 + \alpha_1\mathbf{v}_1 + \cdots + \alpha_k\mathbf{v}_k = \beta_1\mathbf{n}_1 + \cdots + \beta_r\mathbf{n}_r + \alpha_1\mathbf{v}_1 + \cdots + \alpha_k\mathbf{v}_k,$$

so B spans $\mathscr{V}$. ∎

You should now be able to solve Problems 1 to 20.

Inverses and Adjoints of Linear Transformations

Combining Theorems 6.4 and 6.6 shows that the equation $\mathscr{T}(\mathbf{v}) = \mathbf{w}$ has a unique solution $\mathbf{v}$ for every $\mathbf{w}$ if and only if the image space of $\mathscr{T}$ is $\mathscr{W}$ and the null space is $\{\mathbf{0}\}$. More can be said, however: the correspondence from $\mathbf{w}$ to that unique $\mathbf{v}$ solving $\mathscr{T}(\mathbf{v}) = \mathbf{w}$ is actually a linear transformation.

(6.9) ***Theorem*** (inverses). The equation $\mathscr{T}(\mathbf{v}) = \mathbf{w}$ has a unique solution $\mathbf{v}$ in $\mathscr{V}$ for each $\mathbf{w}$ in $\mathscr{W}$ if and only if (a) the image space of $\mathscr{T}$ equals $\mathbf{w}$, and (b) the null space of $\mathscr{T}$ equals $\{\mathbf{0}\}$. When (a) and (b) hold, the correspondence from $\mathbf{w}$ to the solution $\mathbf{v}$ is linear and is denoted by $\mathbf{v} = \mathscr{T}^{-1}(\mathbf{w})$, so that

$$\mathscr{T}\{\mathscr{T}^{-1}(\mathbf{w})\} = \mathbf{w} \quad \text{and} \quad \mathscr{T}^{-1}\{\mathscr{T}(\mathbf{v})\} = \mathbf{v}$$

for all $\mathbf{w}$ in $\mathscr{W}$ and $\mathbf{v}$ in $\mathscr{V}$.

PROOF. The first part follows immediately from Theorems 6.4 and 6.6; it remains to prove that the correspondence $\mathscr{T}^{-1}$ from $\mathbf{w}$ to $\mathbf{v}$ is a linear transformation—we must check against Definition 6.1. First, consider $\mathscr{T}^{-1}(\mathbf{w}_1 + \mathbf{w}_2)$; define $\mathbf{v}_1 = \mathscr{T}^{-1}(\mathbf{w}_1)$ and $\mathbf{v}_2 = \mathscr{T}^{-1}(\mathbf{w}_2)$ as the unique solutions to $\mathscr{T}(\mathbf{v}_1) = \mathbf{w}_1$ and $\mathscr{T}(\mathbf{v}_2) = \mathbf{w}_2$. Then, because $\mathscr{T}$ is linear,

$$\mathscr{T}(\mathbf{v}_1 + \mathbf{v}_2) = \mathscr{T}(\mathbf{v}_1) + \mathscr{T}(\mathbf{v}_2) = \mathbf{w}_1 + \mathbf{w}_2.$$

Since this says that $\mathbf{v}_1 + \mathbf{v}_2$ solves $\mathscr{T}(\mathbf{v}) = \mathbf{w}_1 + \mathbf{w}_2$ and since such solutions are unique, we deduce that

$$\mathbf{v}_1 + \mathbf{v}_2 = \mathscr{T}^{-1}(\mathbf{w}_1 + \mathbf{w}_2)$$

—that is, $\mathscr{T}^{-1}(\mathbf{w}_1 + \mathbf{w}_2) = \mathscr{T}^{-1}(\mathbf{w}_1) + \mathscr{T}^{-1}(\mathbf{w}_2)$. The argument treating $\mathscr{T}^{-1}(\alpha\mathbf{w}_1)$ is similar:

$$\mathscr{T}(\alpha\mathbf{v}_1) = \alpha\mathscr{T}(\mathbf{v}_1) = \alpha\mathbf{w}_1,$$

so $\mathscr{T}^{-1}(\alpha\mathbf{w}_1) = \alpha\mathbf{v}_1 = \alpha\mathscr{T}^{-1}(\mathbf{w}_1)$. Finally, the statements concerning $\mathscr{T}\mathscr{T}^{-1}$ and $\mathscr{T}^{-1}\mathscr{T}$ follow immediately from the definition of $\mathscr{T}^{-1}$. ■

The inverse $\mathscr{T}^{-1}$ is produced from the linear transformation $\mathscr{T}$ when the conditions above hold. There is another linear transformation that can be produced from $\mathscr{T}$ that turns out to be important in applications—the *adjoint transformation* (*unrelated* to the adjoint or adjugate matrix discussed in connection with determinants in Section 4.6).

(6.10) ***Definition.*** Let $\mathscr{T}$ be a linear transformation from $\mathscr{V}$ having inner product $(\cdot, \cdot)_{\mathscr{V}}$ to $\mathscr{W}$ having inner product $(\cdot, \cdot)_{\mathscr{W}}$. If there exists a linear transformation $\mathscr{T}^*$ from $\mathscr{W}$ to $\mathscr{V}$ such that

$$(\mathscr{T}(\mathbf{v}), \mathbf{w})_{\mathscr{W}} = (\mathbf{v}, \mathscr{T}^*(\mathbf{w}))_{\mathscr{V}} \qquad \text{for all } \mathbf{v} \text{ in } \mathscr{V} \text{ and } \mathbf{w} \text{ in } \mathscr{W}$$

then $\mathscr{T}^*$ is called the *adjoint transformation* of $\mathscr{T}$.

(6.11) ***Example.*** Suppose that the $p \times q$ real matrix $\mathbf{A}$ defines the linear transformation $\mathscr{T}$ from $\mathscr{V} = \mathbb{R}^q$ to $\mathscr{W} = \mathbb{R}^p$ in the usual way: $\mathscr{T}(\mathbf{v}) = \mathbf{A}\mathbf{v}$. Suppose that $\mathscr{V}$ and $\mathscr{W}$ have the standard inner products: $(\mathbf{x}, \mathbf{y}) = \mathbf{x}^T\mathbf{y}$. Then we have

$$(\mathscr{T}(\mathbf{v}), \mathbf{w})_{\mathscr{W}} = (\mathbf{A}\mathbf{v}, \mathbf{w})_{\mathscr{W}} = (\mathbf{A}\mathbf{v})^T\mathbf{w} = \mathbf{v}^T(\mathbf{A}^T\mathbf{w}) = (\mathbf{v}, \mathbf{A}^T\mathbf{w})_{\mathscr{V}},$$

which says that the adjoint transformation $\mathscr{T}^*$ is just defined by $\mathbf{A}^T$: $\mathscr{T}^*(\mathbf{w}) = \mathbf{A}^T\mathbf{w}$. Note that if we used $\mathbb{C}^q$ and $\mathbb{C}^p$, then $\mathscr{T}^*$ would be defined by $\mathbf{A}^H$ rather than $\mathbf{A}^T$. This shows that the adjoint transformation is a generalization of the extremely useful transpose—which gives some hint of why it is important.

See Problems 24–27 for some important properties of adjoint transformations.

PROBLEMS 6.1

▷ **1.** As asserted in Definition 6.1, show that conditions (a) and (b) of that definition are equivalent to the condition

$$\mathscr{T}(\alpha_1\mathbf{v}_1 + \alpha_2\mathbf{v}_2) = \alpha_1\mathscr{T}(\mathbf{v}_1) + \alpha_2\mathscr{T}(\mathbf{v}_2).$$

▷ **2.** The identity transformation $\mathscr{I}$ from $\mathscr{V}$ to $\mathscr{V}$ is defined by $\mathscr{I}(\mathbf{v}) = \mathbf{v}$ for all $\mathbf{v}$ in $\mathscr{V}$. Prove that $\mathscr{I}$ is a linear transformation.

3. The zero transformation $\mathscr{O}$ from $\mathscr{V}$ to $\mathscr{W}$ is defined by $\mathscr{O}(\mathbf{v}) = \mathbf{0}$ in $\mathscr{W}$ for all $\mathbf{v}$ in $\mathscr{V}$. Prove that $\mathscr{O}$ is a linear transformation.

4. Show that Example 6.2(a) defines a linear transformation.

5. Show that Examples 6.2(b) and (d) define linear transformations.

6. Show that Example 6.2(c) defines a linear transformation.

7. Show that $\mathscr{L}(\mathscr{V}, \mathscr{W})$ of Example 6.2(e) is a vector space.

▷ **8.** Suppose that $\mathscr{S}$ is a linear transformation from $\mathscr{V}$ to $\mathscr{W}$ and that $\mathscr{T}$ is a linear transformation from $\mathscr{W}$ to $\mathscr{Z}$. Prove that $\mathscr{T}\mathscr{S}$ is a linear transformation from $\mathscr{V}$ to $\mathscr{Z}$, where $\mathscr{T}\mathscr{S}$ is defined by $(\mathscr{T}\mathscr{S})(\mathbf{v}) = \mathscr{T}\{\mathscr{S}(\mathbf{v})\}$ for all $\mathbf{v}$ in $\mathscr{V}$; show that $\mathscr{S}\mathscr{T}$ is not defined unless $\mathscr{V} = \mathscr{Z}$.

9. Prove that the image space of a linear transformation $\mathscr{T}$ from $\mathscr{V}$ to $\mathscr{W}$ is a subspace of $\mathscr{W}$.

10. Find a formula for all vectors in the image space of $\mathscr{T}$, where $\mathscr{T}$ is from $\mathbb{R}^2$ to $\mathbb{R}^3$ and is defined by $\mathscr{T}(\mathbf{v}) = \mathbf{A}\mathbf{v}$ for

$$\mathbf{A} = \begin{bmatrix} 1 & 2 \\ 0 & -1 \\ 2 & 1 \end{bmatrix}.$$

▷ **11.** Let $\mathscr{T}$ be the linear transformation from $\mathbb{R}^q$ to $\mathbb{R}^p$ defined by $\mathscr{T}(\mathbf{v}) = \mathbf{A}\mathbf{v}$, where $\mathbf{A}$ is a $p \times q$ matrix. Prove that the image space of $\mathscr{T}$ equals the column space of $\mathbf{A}$.

12. Find the image space of the transformation $\mathscr{T}$ in Example 6.2(b) under the assumption that $f(t) \neq 0$ for all t.

13. $\mathscr{T}$ is a linear transformation from $C^{(1)}[0, 1]$—the space of all continuously differentiable functions on $[0, 1]$—to $C[0, 1]$—the space of continuous functions there. For any y in $C^{(1)}[0, 1]$, $\mathscr{T}(y)$ is that function whose value at t equals $y'(t) + y(t)$, where the prime denotes differentiation with respect to t. Show that the image space of $\mathscr{T}$ is all of $C[0, 1]$.

▷ **14.** Prove that the null space of a linear transformation $\mathscr{T}$ from $\mathscr{V}$ to $\mathscr{W}$ is a subspace of $\mathscr{V}$.

15. Prove Theorem 6.6 on uniqueness of solutions.

16. Find a basis for the null space of the transformation $\mathscr{T}$ from $\mathbb{R}^5$ to $\mathbb{R}^4$ defined by $\mathscr{T}(\mathbf{v}) = \mathbf{A}\mathbf{v}$, where

$$\mathbf{A} = \begin{bmatrix} 1 & 2 & 1 & 2 & -3 \\ 3 & 6 & 4 & -1 & 2 \\ 4 & 8 & 5 & 1 & -1 \\ -2 & -4 & -3 & 3 & -5 \end{bmatrix}.$$

17. Prove that $\mathscr{T}(\mathbf{x}) = \mathscr{T}(\mathbf{y})$ if and only if $\mathbf{x} - \mathbf{y}$ is in the null space of $\mathscr{T}$.

▷ **18.** Let $\mathbf{A}$ be $p \times q$ with rank k, and define $\mathscr{T}(\mathbf{v}) = \mathbf{A}\mathbf{v}$. Prove that the dimension of the null space of $\mathscr{T}$ equals $q - k$.

19. A transformation $\mathscr{T}$ is said to be *one-to-one* if and only if $\mathscr{T}(\mathbf{x}) = \mathscr{T}(\mathbf{y})$ implies $\mathbf{x} = \mathbf{y}$. Prove that a linear transformation is one-to-one if and only if the null space of $\mathscr{T}$ equals $\{\mathbf{0}\}$.

20. Find a basis for the domain of $\mathscr{T}$ of the type described in Theorem 6.8 if $\mathscr{T}(\mathbf{v}) = \mathbf{A}\mathbf{v}$ and

$$\mathbf{A} = \begin{bmatrix} 1 & 2 & 3 \\ 2 & 1 & -1 \\ 4 & 5 & 5 \end{bmatrix}.$$

▷ **21.** Determine whether the transformation defined in Problem 13 has an inverse.

22. Suppose that $\mathscr{T}$ is the transformation from $\mathbb{R}^p$ to $\mathbb{R}^p$ defined by $\mathscr{T}(\mathbf{v}) = \mathbf{A}\mathbf{v}$ for some $p \times p$ matrix $\mathbf{A}$. Prove that $\mathscr{T}$ has an inverse if and only if $\mathbf{A}$ is nonsingular, and that then $\mathscr{T}^{-1}(\mathbf{w}) = \mathbf{A}^{-1}\mathbf{w}$.

23. Suppose that $\mathscr{S}$ is a linear transformation from $\mathscr{V}$ to $\mathscr{W}$ and that $\mathscr{T}$ is a linear transformation from $\mathscr{W}$ to $\mathscr{Z}$; suppose that both $\mathscr{S}$ and $\mathscr{T}$ have inverses. Prove that $\mathscr{T}\mathscr{S}$—see Problem 8—has an inverse and that $(\mathscr{T}\mathscr{S})^{-1} = \mathscr{S}^{-1}\mathscr{T}^{-1}$.

▷ **24.** Suppose that $\mathscr{T}$ is the linear transformation from $\mathbb{C}^q$ to $\mathbb{C}^p$ defined by $\mathscr{T}(\mathbf{v}) =$

$\mathbf{Av}$ for some $p \times q$ matrix $\mathbf{A}$. Prove that $\mathscr{T}^*$ exists by showing that $\mathscr{T}^*(\mathbf{w}) = \mathbf{A}^H\mathbf{w}$.

25. Show that $(\mathscr{T}^*)^* = \mathscr{T}$.

26. Extend Problem 5 in Section 5.10 as follows. Suppose that $\mathscr{V}$ is a vector space with inner product and that $\mathscr{V}_0$ is a subspace. Define the *orthogonal complement* $\mathscr{V}_0{}^\perp$ of $\mathscr{V}_0$ as the set of vectors $\mathbf{v}$ orthogonal to all the vectors in $\mathscr{V}_0$. Prove that $\mathscr{V}_0{}^\perp$ is a subspace.

▷ **27.** Suppose that $\mathscr{T}$ is a linear transformation from $\mathscr{V}$ to $\mathscr{W}$, each with inner products, and that the adjoint transformation $\mathscr{T}^*$ exists. Let $\mathscr{N}$ be the null space of $\mathscr{T}$ and $\mathscr{R}^*$ be the image space of $\mathscr{T}^*$.

(a) Prove that $(\mathscr{R}^*)^\perp = \mathscr{N}$ (see Problem 26).

(b) Use orthogonal projection onto $\mathscr{R}^*$ to prove that if $\mathscr{R}^*$ is finite-dimensional, $\mathscr{N}^\perp = \mathscr{R}^*$.

6.2 *MATRIX REPRESENTATIONS OF LINEAR TRANSFORMATIONS*

We saw in Section 5.4 that every real or complex finite-dimensional vector space is essentially just $\mathbb{R}^p$ or $\mathbb{C}^p$. And we saw in Example 6.2(a) that $p \times q$ matrices define linear transformations between $\mathbb{R}^q$ and $\mathbb{R}^p$. These facts provide a hint about the fundamental fact regarding linear transformations defined between finite-dimensional vector spaces: Each is essentially nothing other than matrix multiplication.

Matrix Representations

Suppose that $B = \{\mathbf{v}_1; \ldots; \mathbf{v}_q\}$ is an ordered basis for $\mathscr{V}$, that $C = \{\mathbf{w}_1; \ldots; \mathbf{w}_p\}$ is an ordered basis for $\mathscr{W}$, and that $\mathscr{T}$ is a linear transformation from $\mathscr{V}$ to $\mathscr{W}$. For each $\mathbf{v}_j$ in B, since $\mathscr{T}(\mathbf{v}_j)$ is in $\mathscr{W}$ we can write it in terms of the ordered basis C:

(6.12)
$$\mathscr{T}(\mathbf{v}_j) = a_{1j}\mathbf{w}_1 + a_{2j}\mathbf{w}_2 + \cdots + a_{pj}\mathbf{w}_p.$$

In terms of the coordinate isomorphism c_C, this says that

$$c_C(\mathbf{v}_j) = [a_{1j} \quad \cdots \quad a_{pj}]^T.$$

Next, we write $\mathbf{v}$ in $\mathscr{V}$ in terms of its B-coordinates as

$$\mathbf{v} = x_1\mathbf{v}_1 + \cdots + x_q\mathbf{v}_q,$$

so that $c_B(\mathbf{v}) = [x_1 \quad \cdots \quad x_q]^T$. We then calculate $\mathscr{T}(\mathbf{v})$ as follows:

$$\begin{aligned}\mathscr{T}(\mathbf{v}) &= \mathscr{T}(x_1\mathbf{v}_1 + \cdots + x_q\mathbf{v}_q) = x_1\mathscr{T}(\mathbf{v}_1) + \cdots + x_q\mathscr{T}(\mathbf{v}_q)\\ &= x_1(a_{11}\mathbf{w}_1 + a_{21}\mathbf{w}_2 + \cdots + a_{p1}\mathbf{w}_p) + \cdots + x_q(a_{1q}\mathbf{w}_1 + \cdots + a_{pq}\mathbf{w}_p)\\ &= y_1\mathbf{w}_1 + \cdots + y_p\mathbf{w}_p,\end{aligned}$$

where the y_i are the C-coordinates of $\mathscr{T}(\mathbf{v})$ and are computed from

(6.13) $$y_i = a_{i1}x_1 + a_{i2}x_2 + \cdots + a_{iq}x_q.$$

If we now define the $p \times q$ matrix $\mathbf{A}$ by $\langle \mathbf{A} \rangle_{ij} = a_{ij}$ with the a_{ij} as defined in (6.12), we see that the C-coordinates $\mathbf{y} = c_C\{\mathscr{T}(\mathbf{v})\}$ of the output $\mathscr{T}(\mathbf{v})$ are related to the B-coordinates $\mathbf{x} = c_B(\mathbf{v})$ of the input $\mathbf{v}$ according to (6.13) by matrix multiplication: $\mathbf{y} = \mathbf{Ax}$. Thus, instead of applying $\mathscr{T}$ directly to $\mathbf{v}$, we can, instead, (1) find the B-coordinates $\mathbf{x}$ of $\mathbf{v}$ in $\mathscr{V}$, (2) calculate $\mathbf{y} = \mathbf{Ax}$, and (3) use the entries in $\mathbf{y}$ as the C-coordinates to form a vector $\mathbf{w}$ in $\mathscr{W}$; $\mathbf{w}$ will be the same as $\mathscr{T}(\mathbf{v})$. Thus all linear transformations between finite-dimensional real or complex vector spaces are equivalent to matrix multiplication on the coordinate vectors. We have proved:

(6.14) **Key Theorem** (matrix representations). Suppose that $\mathscr{T}$ is a linear transformation from the q-dimensional vector space $\mathscr{V}$ having ordered basis $B = \{\mathbf{v}_1; \ldots; \mathbf{v}_q\}$ to the p-dimensional vector space $\mathscr{W}$ having ordered basis $C = \{\mathbf{w}_1; \ldots; \mathbf{w}_p\}$. Define the $p \times q$ matrix $\mathbf{A}$ by $\langle \mathbf{A} \rangle_{ij} = a_{ij}$, where the a_{ij} are as in (6.12). Then $\mathbf{A}$ *represents* (or is *a representation of*) $\mathscr{T}$ in that:

(a) $\mathscr{T}(\mathbf{v}) = \mathbf{w}$ if and only if $\mathbf{Ax} = \mathbf{y}$, where $\mathbf{x} = c_B(\mathbf{v})$ are the B-coordinates of $\mathbf{v}$ and $\mathbf{y} = c_C(\mathbf{w})$ are the C-coordinates of $\mathbf{w}$.

(b) Stated symbolically, $\mathscr{T} = c_C^{-1}\mathbf{A}c_B$ in the sense that $\mathscr{T}(\mathbf{v}) = c_C^{-1}\{\mathbf{A}c_B(\mathbf{v})\}$.

Diagram 6.15 schematically represents **Key Theorem 6.14**.

(6.15)

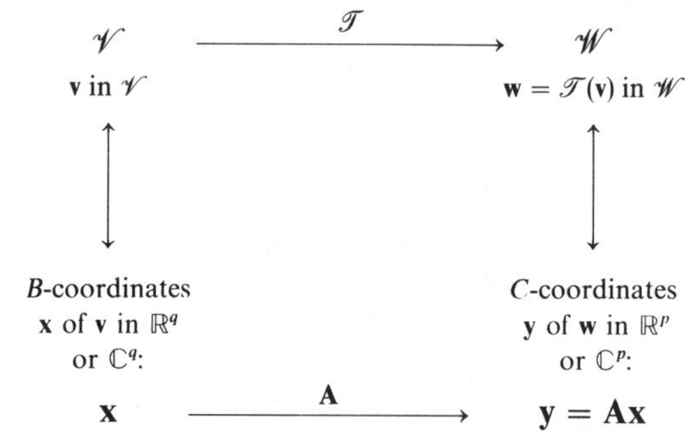

(6.16) ***Example.*** Let $\mathscr{V}$ be the real vector space $\mathscr{P}^3$ of polynomials of degree strictly less than 3 and let $\mathscr{W}$ be the analogous space $\mathscr{P}^2$. Define $\mathscr{T}$ from $\mathscr{V}$ to $\mathscr{W}$ by

$$\mathscr{T}(\mathbf{v}) = \frac{\mathbf{v}(t) - \mathbf{v}(0)}{t}.$$

Let the ordered basis for $\mathscr{V}$ be $B = \{1 + t; t + t^2; t^2 + 1\}$ and let the ordered basis C for $\mathscr{W}$ be $C = (1 + t; 1 - t\}$. In terms of the polynomials themselves, we can evaluate $\mathscr{T}$ by

$$\mathscr{T}(a + bt + ct^2) = b + ct.$$

To represent $\mathscr{T}$ by a matrix $\mathbf{A}$, we have to find a_{ij} from (6.12); we find

$$\mathscr{T}(1 + t) = 1 = 0.5(1 + t) + 0.5(1 - t),$$
$$\mathscr{T}(t + t^2) = 1 + t = 1(1 + t) + 0(1 - t)$$
$$\mathscr{T}(t^2 + 1) = t = 0.5(1 + t) - 0.5(1 - t).$$

This gives us

$$\mathbf{A} = \begin{bmatrix} 0.5 & 1 & 0.5 \\ 0.5 & 0 & -0.5 \end{bmatrix}$$

as the matrix representing $\mathscr{T}$ with respect to the ordered bases B and C. To check this we compute $\mathscr{T}(1 + t + t^2)$ in two ways: directly from $\mathscr{T}$ and indirectly through $\mathbf{A}$. To use $\mathbf{A}$, we first need the B-coordinates $\mathbf{x}$ of

$$1 + t + t^2 = 0.5(1 + t) + 0.5(t + t^2) + 0.5(t^2 + 1),$$

so $\mathbf{x}$ must be $\mathbf{x} = [0.5 \quad 0.5 \quad 0.5]^T$. Therefore, the C-coordinates $\mathbf{y}$ of $\mathscr{T}(1 + t + t^2)$ should be $\mathbf{Ax} = [1 \quad 0]^T$. Using these entries as C-coordinates gives

$$1(1 + t) + 0(1 - t) = 1 + t.$$

So, according to the calculation using $\mathbf{A}$,

$$\mathscr{T}(1 + t + t^2) = 1 + t;$$

it is easy to see that this is correct by using the definition of $\mathscr{T}$.

You should now be able to solve Problems 1 to 12.

Change of Basis

The representation $\mathbf{A}$ for a linear transformation $\mathscr{T}$ constructed above depends strongly on the ordered bases B and C used in $\mathscr{V}$ and $\mathscr{W}$. If the bases are changed, then the representing matrix $\mathbf{A}$ must be changed. The question is: how? **Key Theorem 5.43** explained how the coordinates of a vector with respect to two different ordered bases B and B' are related; we can easily use this to explain how the matrix representations of linear transformations change.

(6.17) ***Theorem*** (change of basis and representations). Suppose that a linear transformation $\mathscr{T}$ from $\mathscr{V}$ to $\mathscr{W}$ is represented by a matrix $\mathbf{A}$ with respect to the ordered bases B and C for $\mathscr{V}$ and $\mathscr{W}$ as in **Key Theorem 6.14**. Suppose also that new ordered bases B' and C' are introduced for $\mathscr{V}$ and $\mathscr{W}$. Let the B-coordinates $\mathbf{v}_B$ and the B'-coordinates $\mathbf{v}_{B'}$ of vectors $\mathbf{v}$ be related by the nonsingular matrix $\mathbf{P}$ (called $\mathbf{M}$ in **Key Theorem 5.43**) by $\mathbf{v}_B = \mathbf{P}\mathbf{v}_{B'}$. Let the C-coordinates $\mathbf{w}_C$ and the C'-coordinates $\mathbf{w}_{C'}$ of vectors $\mathbf{w}$ be related by the nonsingular matrix $\mathbf{S}$ by $\mathbf{w}_C = \mathbf{S}\mathbf{w}_{C'}$. Then the matrix $\mathbf{A}'$ representing $\mathscr{T}$ with respect to the ordered bases B' and C' is just $\mathbf{A}' = \mathbf{S}^{-1}\mathbf{AP}$.

PROOF. The problem is to find the matrix $\mathbf{A}'$ so that $\mathbf{w} = \mathscr{T}(\mathbf{v})$ is equivalent to $\mathbf{w}_{C'} = \mathbf{A}'\mathbf{v}_{B'}$. We already know that $\mathbf{w} = \mathscr{T}(\mathbf{v})$ is equivalent to $\mathbf{w}_C = \mathbf{A}\mathbf{v}_B$, and we have $\mathbf{v}_B = \mathbf{P}\mathbf{v}_{B'}$ and $\mathbf{w}_C = \mathbf{S}\mathbf{w}_{C'}$. Putting these together gives

$$\mathbf{w}_{C'} = \mathbf{S}^{-1}\mathbf{w}_C = \mathbf{S}^{-1}\mathbf{A}\mathbf{v}_B = \mathbf{S}^{-1}\mathbf{A}\mathbf{P}\mathbf{v}_{B'} = \mathbf{A}'\mathbf{v}_{B'} \quad \text{with} \quad \mathbf{A}' = \mathbf{S}^{-1}\mathbf{A}\mathbf{P},$$

as claimed. ∎

A common situation is that $\mathscr{V} = \mathscr{W}$, $B = C$, and $B' = C'$, in which case $\mathbf{S} = \mathbf{P}$.

(6.18) ***Example.*** Consider the matrix $\mathbf{A}$ in Example 2.17 on competing populations, where we found experimentally that the population distributions $\mathbf{x}_{i+1} = \mathbf{A}^i\mathbf{x}_1$ tended to zero as i became large, although we could not prove that this would always happen. $\mathbf{A}$ defines a linear transformation $\mathscr{T}$ from $\mathscr{V} = \mathbb{R}^2$ to $\mathscr{W} = \mathbb{R}^2$ by $\mathscr{T}(\mathbf{v}) = \mathbf{A}\mathbf{v}$, and in fact $\mathscr{T}$ is represented by $\mathbf{A}$ with respect to the ordered bases $B = C = \{\mathbf{e}_1; \mathbf{e}_2\}$—see Problem 10. Suppose that we take instead the new ordered bases

$$B' = C' = \{[5 \quad 3]^T; [0 \quad 1]^T\}.$$

According to **Key Theorem 5.43** the matrices $\mathbf{P} = \mathbf{S}$ translating between coordinates can easily be found to be

$$\mathbf{S} = \mathbf{P} = \begin{bmatrix} 5 & 0 \\ 3 & 1 \end{bmatrix}, \quad \text{so we find} \quad \mathbf{S}^{-1} = \mathbf{P}^{-1} = \begin{bmatrix} 0.2 & 0 \\ -0.6 & 1 \end{bmatrix}.$$

We next use Theorem 6.17 to find that $\mathscr{T}$ is represented with respect to the new ordered bases B' and C' by

$$\mathbf{A}' = \mathbf{S}^{-1}\mathbf{A}\mathbf{P} = \mathbf{S}^{-1}\mathbf{A}\mathbf{S} = \mathbf{A}' = \begin{bmatrix} 0.9 & 0.01 \\ 0 & 0.9 \end{bmatrix}.$$

This representation of $\mathscr{T}$ makes it far easier than with $\mathbf{A}$ to see what happens when $\mathscr{T}$ is applied repeatedly to an initial population distribution $\mathbf{x}_1$. Applying $\mathscr{T}$ repeatedly to $\mathbf{x}_1$ is equivalent to multiplying $\mathbf{A}'$ repeatedly times the B'-coordinates of $\mathbf{x}_1$. But it is easy to see from the special form of $\mathbf{A}'$ that repeated multiplication by $\mathbf{A}'$—that is, $(\mathbf{A}')^i$—is just

$$(\mathbf{A}')^i = \begin{bmatrix} (0.9)^i & 0.01(i)(0.9)^{i-1} \\ 0 & (0.9)^i \end{bmatrix},$$

each entry in which tends to zero as i becomes large. Thus the C'-coordinates of $\mathbf{x}_{i+1}$—which equal $(\mathbf{A}')^i$ times the B'-coordinates of $\mathbf{x}_1$—must tend to zero, which proves that the populations $\mathbf{x}_i$ tend to zero no matter what the initial population.

The special bases B' and C' in Example 6.18 were the key to obtaining such a simple representation of $\mathscr{T}$ that we could see what happened to the populations. Changing bases to obtain a simple representation of a linear transformation is often a key step in studying a transformation and the application it models. The problem still remains as to *how to choose a suitable ordered basis so that the representation is simple.* We will return to this difficulty in Chapter 7.

PROBLEMS 6.2

▷ **1.** Let $\mathscr{I}$ be the identity transformation $\mathscr{I}(\mathbf{v}) = \mathbf{v}$ from $\mathscr{V}$ to $\mathscr{V}$, where $\mathscr{V}$ is finite-dimensional and where the same ordered basis is used for $\mathscr{V}$ as domain and as range of $\mathscr{I}$. Find the matrix representation of $\mathscr{I}$.

2. Let $\mathscr{O}$ be the zero transformation $\mathscr{O}(\mathbf{v}) = \mathbf{0}$ from $\mathscr{V}$ to $\mathscr{W}$, both finite-dimensional. Find the matrix representation of $\mathscr{O}$.

▷ **3.** Let $\mathscr{V}$ equal $\mathscr{P}^3$, the space of polynomials of degree strictly less than 3, and let $\mathscr{W}$ equal the analogous space $\mathscr{P}^4$. Define $\mathscr{T}$ from $\mathscr{V}$ to $\mathscr{W}$ so that for each polynomial f in $\mathscr{V}$, $\mathscr{T}(f)$ is that polynomial in $\mathscr{W}$ whose value at t equals $tf(t) + \{f(t) - f(0)\}/t$. Find the matrix representation of $\mathscr{T}$ with respect to the ordered bases

$$B = \{1;\ 1 + t;\ 1 + t + t^2\} \qquad \text{for } \mathscr{V},$$

and

$$C = \{1;\ 1 - t;\ 1 + 2t + t^2;\ 1 - 3t + 3t^2 - t^3\} \qquad \text{for } \mathscr{W}.$$

Check that the representation is correct by computing $\mathscr{T}(-2 + 3t - t^2)$ two ways.

4. Let $\mathscr{V} = \mathscr{W}$ be p-dimensional, and let $\mathscr{I}$ be the identity transformation $\mathscr{I}(\mathbf{v}) = \mathbf{v}$ for all $\mathbf{v}$. Let B be the ordered basis used for $\mathscr{V}$ as the domain of $\mathscr{I}$, and let B' be a *different* ordered basis used for $\mathscr{V}$ as the range of $\mathscr{I}$. Compare the matrix that represents $\mathscr{I}$ with the matrices generated to change bases in **Key Theorem 5.43**.

5. Let $\mathscr{P}^3$ and $\mathscr{P}^4$ be the spaces from Problem 3, and let $\mathscr{T}$ be the linear transformation from $\mathscr{P}^4$ to $\mathscr{P}^3$ that takes each polynomial to its derivative. Find the matrix representation of $\mathscr{T}$ with respect to the ordered bases

$$B = \{1;\ t;\ t^2;\ t^3\} \qquad \text{for } \mathscr{P}^4$$

and

$$C = \{1;\ t;\ t^2\} \qquad \text{for } \mathscr{P}^3.$$

Check that the representation is correct by computing $\mathscr{T}(1 + 2t + t^2 + t^3\}$ in two ways.

▷ **6.** Let $\mathscr{P}^4$ be as in Problem 3, and let $\mathscr{T}$ be the linear transformation from $\mathscr{P}^4$ to $\mathscr{P}^4$ that takes each polynomial f to the derivative of $tf(t)$. Use the ordered basis $\{1;\ t;\ t^2;\ t^3\}$ for $\mathscr{P}^4$ as both domain and range to find a matrix representation for $\mathscr{T}$. Check that the representation is correct by computing $\mathscr{T}\{2 - 3t + t^2 - t^3\}$ two ways.

7. Suppose that $\mathbf{X}$ represents $\mathcal{T}$ from $\mathcal{V}$ to $\mathcal{W}$ with respect to ordered bases B for $\mathcal{V}$ and C for $\mathcal{W}$, and suppose that $\mathbf{Y}$ represents $\mathcal{S}$ from $\mathcal{W}$ to $\mathcal{X}$ with respect to ordered bases C for $\mathcal{W}$ and D for $\mathcal{X}$. Prove that $\mathbf{YX}$ represents $\mathcal{S}\mathcal{T}$ with respect to B for $\mathcal{V}$ and D for $\mathcal{X}$.

8. Suppose that $\mathbf{A}$ represents $\mathcal{T}$ from $\mathcal{V}$ to $\mathcal{V}$ with respect to some ordered basis used for $\mathcal{V}$ both as domain and range. Prove that $\mathbf{A}^i$ represents $\mathcal{T}^i$ for integers $i > 0$.

▷ **9.** Suppose that $\mathbf{A}$ represents $\mathcal{T}$ from $\mathcal{V}$ to $\mathcal{W}$. Prove that $\mathcal{T}^{-1}$ exists if and only if $\mathbf{A}$ is nonsingular, and that then $\mathbf{A}^{-1}$ represents $\mathcal{T}^{-1}$.

10. Suppose that $\mathbf{A}$ is a $p \times q$ matrix and that $\mathcal{T}(\mathbf{v}) = \mathbf{Av}$ as usual. Show that $\mathbf{A}$ represents $\mathcal{T}$ with respect to the ordered bases $\{\mathbf{e}_1; \ldots; \mathbf{e}_q\}$ for the domain and $\{\mathbf{e}_1; \ldots; \mathbf{e}_p\}$ for the range.

▷ **11.** Suppose that $\mathbf{A}$ represents $\mathcal{T}$ from $\mathcal{V}$ to $\mathcal{W}$ with respect to ordered bases B for $\mathcal{V}$ and C for $\mathcal{W}$. Show that $\mathcal{T}(\mathbf{v}) = \mathbf{0}$ if and only if $\mathbf{Av}_B = \mathbf{0}$, where $\mathbf{v}_B$ represents the B-coordinates of $\mathbf{v}$.

12. In the notation of Problem 11, show that $\mathbf{w}$ is in the image space of $\mathcal{T}$ if and only if $\mathbf{w}$'s C-coordinates $\mathbf{w}_C$ are in the column space of $\mathbf{A}$.

▷ **13.** Let the linear transformation $\mathbf{w} = \mathcal{T}(\mathbf{v})$ from $\mathbb{R}^2$ to $\mathbb{R}^3$ be defined by

$$w_1 = v_1 - v_2, \quad w_2 = 2v_1 + v_2, \quad \text{and} \quad w_3 = v_1 - 2v_2$$

where $\mathbf{v} = [v_1 \quad v_2]^T$ and $\mathbf{w} = [w_1 \quad w_2 \quad w_3]^T$. Find the matrix $\mathbf{A}$ that represents $\mathcal{T}$ with respect to the ordered bases $B = \{\mathbf{e}_1; \mathbf{e}_2\}$ for $\mathbb{R}^2$ and $C = \{\mathbf{e}_1; \mathbf{e}_2; \mathbf{e}_3\}$ for $\mathbb{R}^3$. Check by computing $\mathcal{T}([2 \quad 1]^T)$ two ways.

14. Let $B' = \{[1 \quad 0]^T; [1 \quad 1]^T\}$. Use Theorem 6.17 to find the matrix that represents $\mathcal{T}$ from Problem 13 with respect to B' and C. Check by computing $\mathcal{T}([2 \quad 1]^T)$ two ways.

15. Let $C' = \{[1 \quad 2 \quad 1]^T; [0 \quad 3 \quad -1]^T; [0 \quad 0 \quad 1]^T\}$. Use Theorem 6.17 to find the matrix that represents $\mathcal{T}$ from Problem 13 with respect to B and C'. Check by computing $\mathcal{T}([2 \quad 1]^T)$ two ways.

▷ **16.** Use Theorem 6.17 to find the matrix that represents $\mathcal{T}$ from Problem 13 with respect to B' from Problem 14 and C' from Problem 15. Check by computing $\mathcal{T}([2 \quad 1]^T)$ two ways.

17. Consider the population growth model of Section 2.3 with $k = 0.1$ as in Example 2.16, where the populations were shown experimentally to tend to infinity. Much as in Example 6.18, analyze the behavior of the populations as i tends to infinity by using the ordered basis $\{[5 \quad 1]^T; [1 \quad 1]^T\}$ for both the domain and range.

18. Consider the population growth model of Section 2.3 with $k = 0.16$. Much as in Example 6.18, analyze the behavior of the populations as i tends to infinity by using the ordered basis $\{[5 \quad 4]^T; [5 \quad 2]^T\}$ for both the domain and the range.

6.3 NORMS OF LINEAR TRANSFORMATIONS AND MATRICES

As indicated earlier, the applied mathematician often studies the properties of linear transformations in an attempt to understand their effects on inputs. One common problem is to understand the "size" of the transformation in the sense of the *size of its effects* on the magnitude of the inputs. For example, in Section 2.3 we might want to know by how much the total population of chickens and foxes can be amplified in a single time period. Since we have presented precise methods—norms—for measuring the magnitude of vectors, we can use these to help measure the size of the effects of a linear transformation; this in turn provides a means for measuring the size of matrices (as linear transformations).

Transformation Norms

Let $\mathscr{T}$ be a linear transformation from a vector space $\mathscr{V}$ having norm $\|\cdot\|_{\mathscr{V}}$ to a vector space $\mathscr{W}$ having norm $\|\cdot\|_{\mathscr{W}}$. For each nonzero $\mathbf{v}$ in $\mathscr{V}$, the quotient $\|\mathscr{T}(\mathbf{v})\|_{\mathscr{W}}/\|\mathbf{v}\|_{\mathscr{V}}$ measures the magnification caused by the transformation $\mathscr{T}$ on that specific vector $\mathbf{v}$; an upper bound on this quotient valid for *all* $\mathbf{v}$ would thus measure the overall effect of $\mathscr{T}$ on the size of vectors in $\mathscr{V}$. It can be shown that *in every finite-dimensional space* $\mathscr{V}$, the quotient above has a maximum value achieved with some special vector $\mathbf{v}$. Thus in all finite-dimensional real or complex spaces, there is a maximum value for the quotient $\|\mathscr{T}(\mathbf{v})\|_{\mathscr{W}}/\|\mathbf{v}\|_{\mathscr{V}}$ for all $\mathbf{v} \neq \mathbf{0}$, and this number measures the size of the *effect* of $\mathscr{T}$: If this number is small, then *every* vector is reduced in norm by $\mathscr{T}$; if this number is large, then *some* vectors are greatly increased in norm by $\mathscr{T}$. By analogy with the vector norms that measure the size of vectors, we call this number the norm of $\mathscr{T}$ *induced by* the given norms on $\mathscr{V}$ and $\mathscr{W}$.

In some infinite-dimensional spaces, the quotient $\|\mathscr{T}(\mathbf{v})\|_{\mathscr{W}}/\|\mathbf{v}\|_{\mathscr{V}}$ may not be bounded above; even if it is bounded, there may not be a maximum value achieved with some special $\mathbf{v}$. When the quotient is bounded, however, it will have a *least* upper bound called its *supremum*. The definition that follows therefore uses "supremum" rather than "maximum" to cover this possibility, even though this case does not occur in this book.

(6.19) ***Definition.*** Let $\mathscr{T}$ be a linear transformation from $\mathscr{V}$ to $\mathscr{W}$ having norms $\|\cdot\|_{\mathscr{V}}$ and $\|\cdot\|_{\mathscr{W}}$. The (transformation) *norm* $\|\cdot\|_{\mathscr{V},\mathscr{W}}$ *induced by* $\|\cdot\|_{\mathscr{V}}$ and $\|\cdot\|_{\mathscr{W}}$ is defined by

$$\|\mathscr{T}\|_{\mathscr{V},\mathscr{W}} = \underset{\mathbf{v}\neq\mathbf{0}\text{ in }\mathscr{V}}{\text{supremum}} \frac{\|\mathscr{T}(\mathbf{v})\|_{\mathscr{W}}}{\|\mathbf{v}\|_{\mathscr{V}}}$$

whenever this supremum is finite.

The norm of a transformation has most of the properties of vector norms as described in Definition 5.54; in fact—Problem 6—the transformation norm is a vector norm on the vector space $\mathscr{L}(\mathscr{V}, \mathscr{W})$ of all linear transformations from (finite-dimensional) $\mathscr{V}$ to $\mathscr{W}$. In addition, it satisfies some special properties because of its being computed from transformations.

(6.20) **Key Theorem** (transformation norms). Let $\mathscr{L}$, $\mathscr{M}$, and $\mathscr{N}$ be linear transformations from $\mathscr{V}$ to $\mathscr{W}$, $\mathscr{V}$ to $\mathscr{W}$, and $\mathscr{W}$ to $\mathscr{Z}$, respectively, and suppose that each has a finite transformation norm. Then:

(a) $\|\mathscr{L}\|_{\mathscr{V},\mathscr{W}} \geq 0$, and $\|\mathscr{L}\|_{\mathscr{V},\mathscr{W}} = 0$ if and only if $\mathscr{L}$ is the zero linear transformation with $\mathscr{L}(\mathbf{v}) = \mathbf{0}$ for all $\mathbf{v}$.

(b) $\|\alpha\mathscr{L}\|_{\mathscr{V},\mathscr{W}} = |\alpha|\,\|\mathscr{L}\|_{\mathscr{V},\mathscr{W}}$ for all scalars α.

(c) $\|\mathscr{L} + \mathscr{M}\|_{\mathscr{V},\mathscr{W}} \leq \|\mathscr{L}\|_{\mathscr{V},\mathscr{W}} + \|\mathscr{M}\|_{\mathscr{V},\mathscr{W}}$.

(d) $\|\mathscr{L}(\mathbf{v})\|_{\mathscr{W}} \leq \|\mathscr{L}\|_{\mathscr{V},\mathscr{W}}\|\mathbf{v}\|_{\mathscr{V}}$ for all $\mathbf{v}$ in $\mathscr{V}$.

(e) $\|\mathscr{I}\|_{\mathscr{V},\mathscr{V}} = 1$, where $\mathscr{I}$ is the identity linear transformation on $\mathscr{V}$ with $\mathscr{I}(\mathbf{v}) = \mathbf{v}$ for all $\mathbf{v}$ in $\mathscr{V}$.

(f) $\|\mathscr{N}\mathscr{L}\|_{\mathscr{V},\mathscr{Z}} \leq \|\mathscr{N}\|_{\mathscr{W},\mathscr{Z}}\,\|\mathscr{L}\|_{\mathscr{V},\mathscr{W}}$, where $\mathscr{N}\mathscr{L}$ is the linear transformation from $\mathscr{V}$ to $\mathscr{Z}$ defined by $\mathscr{N}\mathscr{L}(\mathbf{v}) = \mathscr{N}\{\mathscr{L}(\mathbf{v})\}$.

(g) If $\mathscr{V} = \mathscr{W}$, then $\|\mathscr{L}^i\|_{\mathscr{V},\mathscr{V}} \leq (\|\mathscr{L}\|_{\mathscr{V},\mathscr{V}})^i$.

PROOF. It is easy to see that each of the suprema is finite, since they are finite for $\mathscr{L}$, $\mathscr{M}$, and $\mathscr{N}$.

(a), (d), and (e) follow immediately from the definition of transformation norms.

(b) is obvious for $\alpha = 0$ and follows easily for nonzero α since $|\alpha|$ can be factored out of the supremum.

(c) From the triangle inequality for vector norms, we have

$$\begin{aligned}\|(\mathscr{L} + \mathscr{M})(\mathbf{v})\|_{\mathscr{W}} &= \|\mathscr{L}(\mathbf{v}) + \mathscr{M}(\mathbf{v})\|_{\mathscr{W}} \leq \|\mathscr{L}(\mathbf{v})\|_{\mathscr{W}} + \|\mathscr{M}(\mathbf{v})\|_{\mathscr{W}} \\ &\leq (\|\mathscr{L}\|_{\mathscr{V},\mathscr{W}} + \|\mathscr{M}\|_{\mathscr{V},\mathscr{W}})\|\mathbf{v}\| \qquad \text{by using (d).}\end{aligned}$$

This says that the quotient defining the norm of $\mathscr{L} + \mathscr{M}$ is bounded above by $\|\mathscr{L}\|_{\mathscr{V},\mathrm{W}} + \|\mathscr{M}\|_{\mathscr{V},\mathrm{W}}$, so the *least* upper bound is at most this.

(f) follows from an argument similar to that for (c), and then (g) follows by repeated application of (f). ■

You should now be able to solve Problems 1 to 8.

Matrix Norms

In this book we are concerned primarily with $\mathbb{R}^p$, $\mathbb{C}^p$, and their subspaces; the linear transformations of interest are defined by matrix multiplication: $\mathscr{T}(\mathbf{v}) = \mathbf{A}\mathbf{v}$. In these cases the most commonly used vector norms are $\|\cdot\|_1$, $\|\cdot\|_2$, and $\|\cdot\|_\infty$ from Definition 5.55. It is natural and useful to consider the transformation norms induced by these vector norms on linear transformations defined by matrices.

(6.21) ***Example.*** Suppose that $\mathbf{A}$ is a $p \times q$ matrix with $\langle \mathbf{A} \rangle_{ij} = a_{ij}$, and define α by

$$\alpha = \max_i \sum_{j=1}^{q} |a_{ij}|.$$

For any $\mathbf{v}$ in $\mathbb{R}^q$ (or $\mathbb{C}^q$), denote $\langle \mathbf{v} \rangle_i$ by v_i, and suppose that the maximum defining α above is attained for $i = i_0$. Then if $\mathscr{T}(\mathbf{v}) = \mathbf{A}\mathbf{v}$, we have

$$\|\mathscr{T}(\mathbf{v})\|_\infty = \|\mathbf{A}\mathbf{v}\|_\infty = \max_i |\langle \mathbf{A}\mathbf{v} \rangle_i| = \max_i \left| \sum_{j=1}^{q} a_{ij} v_l \right|$$

$$\leq \max_i \sum_{j=1}^{q} (|a_{ij}|\,|v_j|) \leq \max_i \sum_{j=1}^{q} (|a_{ij}| \max_k |v_k|)$$

$$\leq \alpha \|\mathbf{v}\|_\infty,$$

and so $\|\mathscr{T}(\mathbf{v})\|_\infty / \|\mathbf{v}\|_\infty \leq \alpha$ for all nonzero $\mathbf{v}$; this says that this transformation norm of $\mathscr{T}$ is at most α. In fact, it *equals* α; to show this, we must find a special $\mathbf{v}$ so that the usual quotient equals α. For each j, let v_j be chosen with absolute value 1 and such that $a_{i_0 j} v_j = |a_{i_0 j}|$. Clearly, $\|\mathbf{v}\|_\infty = 1$, so that $\|\mathscr{T}(\mathbf{v})\|_\infty \leq \alpha$; moreover,

$$|\langle \mathscr{T}(\mathbf{v}) \rangle_{i_0}| = \left| \sum_{j=1}^{q} a_{i_0 j} v_j \right| = \left| \sum_{j=1}^{q} |a_{i_0 j}| \right| = \alpha,$$

so that, in fact, $\|\mathscr{T}(\mathbf{v})\|_\infty = \alpha$ for this special $\mathbf{v}$. Thus this transformation norm of $\mathscr{T}$ equals α.

Since the transformation $\mathscr{T}$ in Example 6.21 is just multiplication by $\mathbf{A}$, it seems reasonable to call the norm that we computed a norm of $\mathbf{A}$ as well as of $\mathscr{T}$.

(6.22) ***Definition.*** For a $p \times q$ matrix $\mathbf{A}$, by the matrix norms $\|\mathbf{A}\|_1$, $\|\mathbf{A}\|_2$, and $\|\mathbf{A}\|_\infty$—and any other matrix norm $\|\mathbf{A}\|$—we mean the norms of the transformation $\mathscr{T}$ defined by $\mathscr{T}(\mathbf{v}) = \mathbf{A}\mathbf{v}$ that are induced when the corresponding vector norm is used on *both* the domain and range of $\mathscr{T}$. In particular,

$$\|\mathbf{A}\|_1 = \max_{\mathbf{x} \neq \mathbf{0}} \left\{ \frac{\|\mathbf{A}\mathbf{x}\|_1}{\|\mathbf{x}\|_1} \right\}$$

$$\|\mathbf{A}\|_2 = \max_{\mathbf{x} \neq \mathbf{0}} \left\{ \frac{\|\mathbf{A}\mathbf{x}\|_2}{\|\mathbf{x}\|_2} \right\}$$

$$\|\mathbf{A}\|_\infty = \max_{\mathbf{x} \neq \mathbf{0}} \left\{ \frac{\|\mathbf{A}\mathbf{x}\|_\infty}{\|\mathbf{x}\|_\infty} \right\}.$$

We found in Example 6.21 that it is possible to compute $\|\mathbf{A}\|_\infty$ by a straightforward calculation with the entries of $\mathbf{A}$; a similar argument—Problem 10—leads to a related formula for $\|\mathbf{A}\|_1$. The computation of $\|\mathbf{A}\|_2$ unfortunately is much more complicated and involves a concept that we will not encounter until the next chapter, although for completeness we state the formula here.

(6.23) **Key Theorem** (matrix norms). Let $\mathbf{A}$ be $p \times q$. Then:

(a) $\|\mathbf{A}\|_1 = \max_j \sum_{i=1}^{p} |a_{ij}|$ *(maximum absolute column sum).*

(b) $\|\mathbf{A}\|_\infty = \max_i \sum_{j=1}^{q} |a_{ij}|$ *(maximum absolute row sum).*

(c) $\|\mathbf{A}\|_2 =$ *(maximum eigenvalue of* $\mathbf{A}^H\mathbf{A})^{1/2}$

$=$ *maximum singular value of* $\mathbf{A}$.

(6.24) ***Example.*** In Section 2.3 we considered—in our present language—a linear transformation modeling the growth of competing populations of chickens and foxes from one time period to the next. In particular, Example 2.17 treated the model with $\mathscr{T}(\mathbf{x}) = \mathbf{A}\mathbf{x}$ for

$$\mathbf{A} = \begin{bmatrix} 0.6 & 0.5 \\ -0.18 & 1.2 \end{bmatrix}.$$

If we are interested in the *total* number of animals in the model, then $\|\mathbf{x}\|_1$ computes this figure for each pair $\mathbf{x}$ of populations. Using this norm and the formula above gives $\|\mathbf{A}\|_1 = 1.7$, so we know that the total population can be magnified by a factor of 1.7 in any one time period. If we are, instead, interested in the *maximum* number of animals of one type, then $\|\mathbf{x}\|_\infty$ computes this figure. Using this norm and the formula above gives $\|\mathbf{A}\|_\infty = 1.38$, so we know that the maximum population can be magnified by a factor of 1.38 in any one period. Note also that from **Key Theorem 6.20(g)** we can conclude that, for example, $\|\mathbf{A}^i\|_\infty \leq (1.38)^i$, which might give the impression that we should expect $\mathbf{A}^i$ to contain large numbers for large i; the inequality gives just an *upper bound* on $\|\mathbf{A}^i\|_\infty$, however—Example 6.18 in the preceding section showed that $\mathbf{A}^i$ actually converges to zero as i tends to infinity. Norms provide useful information about matrices, but they are a rather imprecise tool—at least for studying the behavior of powers of a matrix.

Recall that Theorem 5.59 on the equivalence of norms showed that the vector 1-norm, 2-norm, and ∞-norm are all equivalent. We can easily use the inequalities in (b) of that theorem to develop similar facts for the induced matrix norms. For example, for $p \times q$ $\mathbf{A}$ we have

$$\|\mathbf{A}\mathbf{x}\|_1 \leq p\|\mathbf{A}\mathbf{x}\|_\infty \leq p\|\mathbf{A}\|_\infty\|\mathbf{x}\|_\infty \leq p\|\mathbf{A}\|_\infty\|\mathbf{x}\|_1.$$

which says that $\|\mathbf{A}\mathbf{x}\|_1/\|\mathbf{x}\|_1$ is bounded above by $p\|\mathbf{A}\|_\infty$. Since $\|\mathbf{A}\|_1$ is the *least* such upper bound for this quotient, we conclude that $\|\mathbf{A}\|_1 \leq p\|\mathbf{A}\|_\infty$. Similar arguments lead to similar inequalities among the other matrix norms. Moreover, we can use the concept of *convergence* from Definition 5.58 in considering sequences $\mathbf{A}_i$ of matrices, so that $\mathbf{A}_i$ converges to $\mathbf{A}_\infty$ if and only if $\|\mathbf{A}_i - \mathbf{A}_\infty\|$ converges to zero for whatever norm we are using. Since $|\langle\mathbf{A}\rangle_{jk}| \le \|\mathbf{A}\|_\infty$ for all j and k, convergence in the sense above means that the entries of $\mathbf{A}_i$ converge to the corresponding entries of $\mathbf{A}_\infty$. Thus we have a direct analogue of Theorem 5.59 on the equivalence of vector norms.

(6.25) ***Theorem*** (matrix-norm equivalence). The 1-norm, 2-norm, and ∞-norm on matrices are all equivalent in the sense that:

(a) If a sequence of matrices $\mathbf{A}_i$ converges to $\mathbf{A}_\infty$ as determined in one of the norms, then it converges as determined in all three norms and the individual entries $\langle\mathbf{A}_i\rangle_{jk}$ converge to the entries $\langle\mathbf{A}_\infty\rangle_{jk}$.

(b) For all $p \times q$ matrices $\mathbf{A}$,

$$\frac{\|\mathbf{A}\|_2}{\sqrt{p}} \le \|\mathbf{A}\|_\infty \le \|\mathbf{A}\|_2\sqrt{q}$$

$$\frac{\|\mathbf{A}\|_2}{\sqrt{q}} \le \|\mathbf{A}\|_1 \le \|\mathbf{A}\|_2\sqrt{p}$$

$$\frac{\|\mathbf{A}\|_1}{p} \le \|\mathbf{A}\|_\infty \le \|\mathbf{A}\|_1 q.$$

PROOF. Problem 14. ∎

PROBLEMS 6.3

▷ **1.** Suppose that $\mathcal{T}$ is a linear transformation and that for all $\mathbf{v}$ in $\mathcal{V}$ we have

$$\|\mathcal{T}(\mathbf{v})\|_{\mathcal{W}} \le k\|\mathbf{v}\|_{\mathcal{V}}$$

for some fixed k. Prove that $\|\mathcal{T}\|_{\mathcal{V},\mathcal{W}} \le k$.

2. Show that each supremum defining a transformation norm is finite in **Key Theorem 6.20(b)**, **(c)**, **(f)**, and **(g)**.

3. Provide the details to prove **Key Theorem 6.20(a)**, **(b)**, and **(d)–(g)**.

▷ **4.** Suppose that $\mathcal{I}$ is the identity transformation with $\mathcal{I}(\mathbf{v}) = \mathbf{v}$ for all $\mathbf{v}$ in $\mathcal{V}$, and that $\|\cdot\|$ is a norm on $\mathcal{V}$. Find the transformation norm of $\mathcal{I}$.

5. Suppose that $\mathcal{O}$ is the zero transformation with $\mathcal{O}(\mathbf{v}) = \mathbf{0}$ in $\mathcal{W}$ for all $\mathbf{v}$ in $\mathcal{V}$, and that $\mathcal{V}$ and $\mathcal{W}$ both have norms. Find the transformation norm of $\mathcal{O}$.

6. Suppose that $\mathcal{V}$ and $\mathcal{W}$ are finite-dimensional vector spaces, each with a norm. Show that $\|\cdot\|_{\mathcal{V},\mathcal{W}}$ is a vector norm on the vector space $\mathcal{L}(\mathcal{V}, \mathcal{W})$ of all linear transformations from $\mathcal{V}$ to $\mathcal{W}$. See Example 6.2(e).

▷ **7.** For each continuous function x in the real vector space $C[0, 1]$, define $\|x\|$ as the maximum value of $|x(t)|$ for $0 \le t \le 1$. Find the norm of the linear transformation $\mathcal{T}$ defined in Example 6.2(b).

8. Suppose that $\mathscr{T}$ is a linear transformation from $\mathscr{V}$ to $\mathscr{V}$ with $\|\mathscr{T}\|_{\mathscr{V},\mathscr{V}} < 1$. Prove that $\mathscr{T}^i(\mathbf{v})$ converges to $\mathbf{0}$ as i tends to infinity for each vector $\mathbf{v}$.

▷ **9.** Let $k = 0.1$ in the model in Section 2.3 of competing populations; see Example 2.16. For the matrix $\mathbf{A}$ there, find $\|\mathbf{A}\|_1$ and $\|\mathbf{A}\|_\infty$.

10. Prove that the formula in Theorem 6.23(a) correctly evaluates $\|\mathbf{A}\|_1$.

11. For each of the following matrices $\mathbf{A}$, evaluate $\|\mathbf{A}\|_1$ and $\|\mathbf{A}\|_\infty$ and verify the inequality in Theorem 6.25(b).

(a) $\begin{bmatrix} -3 & 2 & 1 \end{bmatrix}$ (b) $\begin{bmatrix} -3 \\ 2 \\ 1 \end{bmatrix}$

(c) $\begin{bmatrix} 4 & -7 \\ -6 & 1 \end{bmatrix}$ (d) $\begin{bmatrix} i & 3 & 2 \\ 1 & -i & 2 \end{bmatrix}$.

▷ **12.** For each of the following matrices $\mathbf{A}$, evaluate $\|\mathbf{A}\|_1$ and $\|\mathbf{A}\|_\infty$ and verify the inequality in Theorem 6.25(b).

(a) $\begin{bmatrix} -6 \\ 1 \\ 3 \end{bmatrix}$ (b) $\begin{bmatrix} -8 & 1 & 2 & 8 \end{bmatrix}$

(c) $\begin{bmatrix} 0 & 0 \\ 0 & 0 \end{bmatrix}$ (d) $\begin{bmatrix} 8 & -3 \\ -6 & 6 \\ -2 & -6 \end{bmatrix}$

13. Suppose that $\mathbf{A}$ is a square matrix for which $\|\mathbf{A}\| < 1$. Prove that $\mathbf{A}^i$ tends to $\mathbf{0}$ as i tends to infinity.

14. Use Theorem 5.59 on the equivalence of vector norms to prove Theorem 6.25 on the equivalence of matrix norms.

▷ **15.** (a) Use the formulas in **Key Theorem 6.23** to show that $\|\mathbf{I}\|_1$ and $\|\mathbf{I}\|_\infty$ both equal 1.
(b) Prove that $\|\mathbf{I}\| = 1$ for any matrix norm by using the definition of transformation norms.

16. In Section 2.2 we discussed special transition matrices arising in models of business competition; see (2.3) for the properties of these matrices $\mathbf{A}$. Use norms to show that, for such matrices $\mathbf{A}$, the powers $\mathbf{A}^i$ must remain bounded as i tends to infinity.

6.4 *INVERSES OF PERTURBED MATRICES; CONDITION OF LINEAR EQUATIONS*

As often mentioned before, in practical applications it is important to realize that inaccuracies in measuring and modeling force us to deal with vectors, matrices, equations, and the like that are slightly perturbed (changed) from the ideal values that would result from perfect measurements and models. It is essential to understand what effects these perturbations in the data create in the fundamental issues of linear algebra: nonsingularity of matrices, solvability of equations, solutions of equations, and the like.

Inverses

One issue of this type involves nonsingularity. Some "ideal" matrix **A** might be nonsingular, but the perturbed approximation to it that we measure might be singular. We can show, however, that this cannot happen if the perturbations are small enough. We first treat a special case that makes the general case simple.

(6.26) **Key Lemma** (Banach Lemma). Let **P** be a $p \times p$ matrix, and let $\|\cdot\|$ denote any vector norm and its corresponding matrix norm. Suppose that $\|\mathbf{P}\| < 1$. Then $\mathbf{I}_p + \mathbf{P}$ is nonsingular, and

$$\frac{1}{1 + \|\mathbf{P}\|} \leq \|(\mathbf{I}_p + \mathbf{P})^{-1}\| \leq \frac{1}{1 - \|\mathbf{P}\|}.$$

PROOF. According to **Key Theorem 4.18**, $\mathbf{I}_p + \mathbf{P}$ is nonsingular if and only if the only solution to $(\mathbf{I}_p + \mathbf{P})\mathbf{x} = \mathbf{0}$ is $\mathbf{x} = \mathbf{0}$. Suppose then that $(\mathbf{I}_p + \mathbf{P})\mathbf{x} = \mathbf{0}$, so that $\mathbf{x} = -\mathbf{P}\mathbf{x}$. Then we have

$$\|\mathbf{x}\| = \|-\mathbf{P}\mathbf{x}\| = \|\mathbf{P}\mathbf{x}\| \leq \|\mathbf{P}\|\,\|\mathbf{x}\|,$$

and since $\|\mathbf{P}\| < 1$ this is a contradiction unless $\mathbf{x} = \mathbf{0}$, as we sought to prove. Thus $(\mathbf{I}_p + \mathbf{P})^{-1}$ exists; denote it by **B**. From $\mathbf{I}_p = \mathbf{B}(\mathbf{I}_p + \mathbf{P}) = \mathbf{B} + \mathbf{BP}$, we have $1 = \|\mathbf{I}_p\| = \|\mathbf{B}(\mathbf{I}_p + \mathbf{P})\| \leq \|\mathbf{B}\|\,\|\mathbf{I}_p + \mathbf{P}\| \leq \|\mathbf{B}\|(1 + \|\mathbf{P}\|)$, which yields $\|\mathbf{B}\| \geq 1/(1 + \|\mathbf{P}\|)$, as claimed. Also, $\mathbf{B} = \mathbf{I}_p - \mathbf{BP}$, which therefore gives $\|\mathbf{B}\| = \|\mathbf{I}_p - \mathbf{BP}\| \leq 1 + \|\mathbf{BP}\| \leq 1 + \|\mathbf{B}\|\,\|\mathbf{P}\|$, which yields $\|\mathbf{B}\| \leq 1/(1 - \|\mathbf{P}\|)$, completing the proof. ■

(6.27) ***Example.*** Consider the matrix

$$\mathbf{A} = \begin{bmatrix} 1.1 & -0.6 \\ 0.8 & 0.9 \end{bmatrix}.$$

We can write **A** as $\mathbf{A} = \mathbf{I} + \mathbf{P}$, where

$$\mathbf{P} = \begin{bmatrix} 0.1 & -0.6 \\ 0.8 & -0.1 \end{bmatrix}.$$

Since $\|\mathbf{P}\|_\infty = 0.9 < 1$, the Banach Lemma says that $\mathbf{A} = \mathbf{I} + \mathbf{P}$ is nonsingular and that $1/1.9 \leq \|\mathbf{A}^{-1}\|_\infty \leq 1/0.1$.

The Banach Lemma makes the general result simple to prove.

(6.28) **Key Theorem** (perturbed inverses). Let **A** and **R** be $p \times p$ with **A** being nonsingular, and let $\|\cdot\|$ denote any vector norm and its corresponding matrix norm. Define $\alpha = \|\mathbf{A}^{-1}\mathbf{R}\|$ or $\alpha = \|\mathbf{R}\mathbf{A}^{-1}\|$. If $\alpha < 1$—so in particular if $\|\mathbf{R}\| < 1/\|\mathbf{A}^{-1}\|$—then $\mathbf{A} + \mathbf{R}$ is also nonsingular, and

$$\frac{\|\mathbf{A}^{-1}\|}{1 + \alpha} \leq \|(\mathbf{A} + \mathbf{R})^{-1}\| \leq \frac{\|\mathbf{A}^{-1}\|}{1 - \alpha}.$$

PROOF. We take the case $\alpha = \|\mathbf{A}^{-1}\mathbf{R}\| < 1$; the other case is similar. Since $\mathbf{A}^{-1}$ exists, we can write

$$\mathbf{A} + \mathbf{R} \quad \text{as} \quad \mathbf{A}(\mathbf{I}_p + \mathbf{A}^{-1}\mathbf{R}) = \mathbf{A}(\mathbf{I}_p + \mathbf{P})$$

if we define $\mathbf{P} = \mathbf{A}^{-1}\mathbf{R}$. By assumption, $\|\mathbf{P}\| = \alpha < 1$, so the Banach Lemma applies and $\mathbf{I}_p + \mathbf{P}$ is nonsingular, as is **A**. Therefore, the product $\mathbf{A}(\mathbf{I}_p + \mathbf{P})$—which equals $\mathbf{A} + \mathbf{R}$—is also nonsingular and

$$(\mathbf{A} + \mathbf{R})^{-1} = \{\mathbf{A}(\mathbf{I}_p + \mathbf{P})\}^{-1} = (\mathbf{I}_p + \mathbf{P})^{-1}\mathbf{A}^{-1}.$$

This gives

$$\|(\mathbf{A} + \mathbf{R})^{-1}\| \leq \|(\mathbf{I}_p + \mathbf{P})^{-1}\| \, \|\mathbf{A}^{-1}\| \leq \|\mathbf{A}^{-1}\|/(1 - \alpha),$$

by the Banach Lemma; this is the desired upper bound. To get the lower bound, we write

$$\mathbf{A}^{-1} = (\mathbf{I}_p + \mathbf{P})(\mathbf{A} + \mathbf{R})^{-1}$$

and deduce that

$$\|\mathbf{A}^{-1}\| \leq \|\mathbf{I}_p + \mathbf{P}\| \, \|(\mathbf{A} + \mathbf{R})^{-1}\| \leq (1 + \alpha)\|(\mathbf{A} + \mathbf{R})^{-1}\|.$$

Division by $(1 + \alpha)$ gives the claimed lower bound. ■

The theorem in essence says that if **A** is nonsingular and if $\|\mathbf{R}\|$ is small enough, then $\mathbf{A} + \mathbf{R}$ is also nonsingular and the norm of its inverse is comparable to the norm of $\mathbf{A}^{-1}$. That is, *all matrices sufficiently near a nonsingular matrix are also nonsingular.* Inaccuracies in data and modeling—*if* they are small enough—will not perturb a nonsingular matrix enough to make it singular; note that how *small* the perturbations $\|\mathbf{R}\|$ need to be for nonsingularity depends on how *large* $\|\mathbf{A}^{-1}\|$ is: $\|\mathbf{R}\| < 1/\|\mathbf{A}^{-1}\|$ is sufficient, in particular. For example, in order for perturbations $\mathbf{A} + \mathbf{R}$ of

$$\mathbf{A} = \begin{bmatrix} 1 + 10^{-10} & 1 \\ 1 & 1 \end{bmatrix}$$

to be guaranteed to be nonsingular by **Key Theorem 6.28**, we need $\|\mathbf{R}\|_\infty$ smaller than about $10^{-10}/2$, since $\|\mathbf{A}^{-1}\|_\infty$ is about $2(10^{10})$.

Condition

The notion of *condition* is important in all of applied mathematics. If small changes in the data of some problem always lead to reasonably small changes in the answer to the problem, the problem is said to be *well-conditioned.* If small changes in the data of some problem can *sometimes* lead to unacceptably large changes in the answer to the problem, the problem is said to be *ill-conditioned.* The reason for the importance of this concept should be obvious: In applied problems, data are almost always inaccurate because of measuring and modeling errors, and it is crucial for us to know what effect such inaccuracies in the data have on the answer to the problem.

We consider here the condition of the solution $\mathbf{x}$ to the system of equations $\mathbf{Ax} = \mathbf{b}$ in terms of the data $\mathbf{b}$ and $\mathbf{A}$. We want to know by how much the solution $\mathbf{x}$ changes—say from $\mathbf{x}$ to $\mathbf{x} + \boldsymbol{\delta}\mathbf{x}$—when the data $\mathbf{b}$ and $\mathbf{A}$ change—say to $\mathbf{b} + \boldsymbol{\delta}\mathbf{b}$ and $\mathbf{A} + \boldsymbol{\delta}\mathbf{A}$.

(6.29) ***Theorem*** (condition of equations). Let $\mathbf{A}$ be nonsingular and let $\|\cdot\|$ denote any vector norm and its corresponding matrix norm. Suppose that $\mathbf{x}$ solves $\mathbf{Ax} = \mathbf{b}$ while

$$\mathbf{x} + \boldsymbol{\delta}\mathbf{x} \quad \text{solves} \quad (\mathbf{A} + \boldsymbol{\delta}\mathbf{A})(\mathbf{x} + \boldsymbol{\delta}\mathbf{x}) = \mathbf{b} + \boldsymbol{\delta}\mathbf{b}$$

for some perturbations $\boldsymbol{\delta}\mathbf{A}$ and $\boldsymbol{\delta}\mathbf{b}$ in the data. Suppose that the perturbation $\boldsymbol{\delta}\mathbf{A}$ is small enough that $\alpha < 1$, where $\alpha = \|(\boldsymbol{\delta}\mathbf{A})\mathbf{A}^{-1}\|$ or $\alpha = \|\mathbf{A}^{-1}(\boldsymbol{\delta}\mathbf{A})\|$. Then the change $\boldsymbol{\delta}\mathbf{x}$ in the solution satisfies

$$\frac{\|\boldsymbol{\delta}\mathbf{x}\|}{\|\mathbf{x}\|} \leq M \cdot c(\mathbf{A}) \cdot \left(\frac{\|\boldsymbol{\delta}\mathbf{b}\|}{\|\mathbf{b}\|} + \frac{\|\boldsymbol{\delta}\mathbf{A}\|}{\|\mathbf{A}\|} \right),$$

where $M = 1/(1 - \alpha)$ and where $c(\mathbf{A}) = \|\mathbf{A}\|\,\|\mathbf{A}^{-1}\|$ is the so-called *condition number* of $\mathbf{A}$.

PROOF. Since $\alpha < 1$, **Key Theorem 6.28** implies that $\mathbf{A} + \boldsymbol{\delta}\mathbf{A}$ is nonsingular and gives a bound on the norm of its inverse. Since $\mathbf{A} + \boldsymbol{\delta}\mathbf{A}$ is nonsingular, the solution $\mathbf{x} + \boldsymbol{\delta}\mathbf{x}$ to the perturbed problem exists. In fact, $\boldsymbol{\delta}\mathbf{x}$ itself solves

$$(\mathbf{A} + \boldsymbol{\delta}\mathbf{A})\,\boldsymbol{\delta}\mathbf{x} = \mathbf{b} + \boldsymbol{\delta}\mathbf{b} - \mathbf{Ax} - \boldsymbol{\delta}\mathbf{Ax} = \boldsymbol{\delta}\mathbf{b} - \boldsymbol{\delta}\mathbf{Ax},$$

so that

$$\boldsymbol{\delta}\mathbf{x} = (\mathbf{A} + \boldsymbol{\delta}\mathbf{A})^{-1}(\boldsymbol{\delta}\mathbf{b} - \boldsymbol{\delta}\mathbf{Ax}).$$

Applying the upper bound in **Key Theorem 6.28** with $\mathbf{R} = \boldsymbol{\delta}\mathbf{A}$ in this formula for $\boldsymbol{\delta}\mathbf{x}$ immediately gives

$$\begin{aligned} \|\boldsymbol{\delta}\mathbf{x}\| &\leq M \cdot \|\mathbf{A}^{-1}\| \cdot \|\boldsymbol{\delta}\mathbf{b} - \boldsymbol{\delta}\mathbf{Ax}\| \\ &\leq M \cdot \|\mathbf{A}^{-1}\| (\|\boldsymbol{\delta}\mathbf{b}\| + \|\boldsymbol{\delta}\mathbf{A}\|\,\|\mathbf{x}\|). \end{aligned}$$

Therefore,

$$\frac{\|\boldsymbol{\delta}\mathbf{x}\|}{\|\mathbf{x}\|} \leq M \cdot \|\mathbf{A}^{-1}\| \left(\frac{\|\boldsymbol{\delta}\mathbf{b}\|}{\|\mathbf{x}\|} + \|\boldsymbol{\delta}\mathbf{A}\| \right)$$

$$\leq M \cdot \|\mathbf{A}^{-1}\| \cdot \left(\frac{\|\boldsymbol{\delta}\mathbf{b}\|}{\|\mathbf{b}\|/\|\mathbf{A}\|} + \|\boldsymbol{\delta}\mathbf{A}\| \right)$$

since $\mathbf{b} = \mathbf{A}\mathbf{x}$ implies that $\|\mathbf{b}\| \leq \|\mathbf{A}\|\,\|\mathbf{x}\|$. Simplifying, we obtain

$$\frac{\|\boldsymbol{\delta}\mathbf{x}\|}{\|\mathbf{x}\|} \leq M \cdot c(\mathbf{A}) \cdot \left(\frac{\|\boldsymbol{\delta}\mathbf{b}\|}{\|\mathbf{b}\|} + \frac{\|\boldsymbol{\delta}\mathbf{A}\|}{\|\mathbf{A}\|} \right),$$

as required, completing the proof. ■

To see what this theorem says about condition, suppose that we decide to measure the size of the changes $\boldsymbol{\delta}\mathbf{y}$ in quantities $\mathbf{y}$ by examining $\|\boldsymbol{\delta}\mathbf{y}\|/\|\mathbf{y}\|$—that is, by looking at the overall size of the changes relative to the overall size of the original quantities. Then the theorem compares the changes—measured in this sense—in the solution $\mathbf{x}$ to those in the data $\mathbf{b}$ and $\mathbf{A}$. If the change $\boldsymbol{\delta}\mathbf{A}$ in $\mathbf{A}$ is small enough, the constant M in the theorem is near 1; in this case, the bound on the change $\|\boldsymbol{\delta}\mathbf{x}\|/\|\mathbf{x}\|$ will not be much larger than the changes in the data *if the condition number*

$$c(\mathbf{A}) = \|\mathbf{A}\|\,\|\mathbf{A}^{-1}\|$$

is not too large. This means that a moderate condition number $c(\mathbf{A})$ guarantees that the equations are well-conditioned: Small changes in the data produce reasonably small changes in the solution. If $c(\mathbf{A})$ is large, however, the changes in $\mathbf{x}$ caused by changes in the data *may* be much larger than the changes in the data (whether this actually occurs depends on the specific $\mathbf{b}$, but it is usually wise to view such a system as ill-conditioned).

> *Note.* We can combine Theorem 6.29 with the results (3.42) and (3.43) that say that Gauss elimination with partial pivoting in floating-point arithmetic usually gives an exact solution to a slightly perturbed problem. The result is a generally valid bound on the error in the solution thus produced; the bound is primarily in terms of the condition number of the matrix and the nature of the computer arithmetic—see Problem 14.

The meaning of the condition number $c(\mathbf{A})$ can be examined from another viewpoint. If $\mathbf{A}$ is nonsingular, we know that $\mathbf{A} + \mathbf{R}$ is nonsingular for all $\mathbf{R}$ satisfying $\|\mathbf{R}\| < 1/\|\mathbf{A}^{-1}\|$, according to **Key Theorem 6.28**. We can restate this requirement on $\mathbf{R}$ as

$$\|\mathbf{R}\|/\|\mathbf{A}\| < 1/c(\mathbf{A}).$$

That is,

$$\mathbf{A} + \delta\mathbf{A} \quad \text{is nonsingular as long as} \quad \frac{\|\delta\mathbf{A}\|}{\|\mathbf{A}\|} < \frac{1}{c(\mathbf{A})}.$$

Put differently, if $\mathbf{A} + \delta\mathbf{A}$ *is* singular, then $\|\delta\mathbf{A}\|/\|\mathbf{A}\| \geq 1/c(\mathbf{A})$ and hence

(6.30) $$c(\mathbf{A}) \geq \frac{\|\mathbf{A}\|}{\|\delta\mathbf{A}\|} \qquad \text{whenever } \mathbf{A} + \delta\mathbf{A} \text{ is singular.}$$

In fact, $c(\mathbf{A})$ actually equals the supremum of $\|\mathbf{A}\|/\|\delta\mathbf{A}\|$ taken over all $\delta\mathbf{A}$ for which $\mathbf{A} + \delta\mathbf{A}$ is singular. Thus the condition number—more precisely, its reciprocal—measures how far $\mathbf{A}$ is from the nearest singular matrix.

It is unfortunate for practical purposes that it is not at all easy to recognize an ill-conditioned matrix $\mathbf{A}$—that is, one with large condition number—just by looking at $\mathbf{A}$. As we mentioned in Section 3.9, good computer programs for solving $\mathbf{Ax} = \mathbf{b}$ usually provide the user with an estimate of $c(\mathbf{A})$ from information gathered during the computation, but it is rarely possible to estimate $c(\mathbf{A})$ without such computational assistance except in simple cases. The 7×7 matrix $\mathbf{A}$ in Example 1.31, for example, seems perfectly innocent, yet its inverse found there has ∞-norm about $110{,}000{,}000 \approx 10^8$. Thus the condition number of $\mathbf{A}$ is about 10^8, so there is a singular matrix within about 10^{-8} of $\mathbf{A}$—not something we would easily guess from looking at $\mathbf{A}$.

𝔐 (6.31) ***Example.*** In the analysis of chemical reactions it turns out that the number of truly independent factors in a process can be computed as the rank of a certain matrix of measurements of the concentrations of the chemicals in the reaction. Suppose that there are three chemicals involved and that we measure the concentration matrix $\mathbf{C}$ to be

$$\mathbf{C} = \begin{bmatrix} 1.02 & 2.03 & 4.20 \\ 0.25 & 0.51 & 1.06 \\ 1.74 & 3.46 & 7.17 \end{bmatrix},$$

where each of our measurements may be subject to experimental errors of size 0.015. To determine the rank of $\mathbf{C}$, we find a Gauss-reduced form; working first with the first column produces

$$\begin{bmatrix} 1.00 & 1.99 & 4.12 \\ 0 & 0.013 & -0.03 \\ 0 & 0.003 & 0.005 \end{bmatrix}.$$

If we accept these numbers we can go on to find that the rank is 3; however, given the fact that our numbers contain experimental errors, these last two rows look suspiciously like zeros. In fact, the slightly perturbed matrix

$$\mathbf{C}' = \begin{bmatrix} 1.02 & 2.03 & 4.20 \\ 0.26 & 0.517451\ldots & 1.070588\ldots \\ 1.74 & 3.462941\ldots & 7.164706\ldots \end{bmatrix}$$

reduces exactly to

$$\begin{bmatrix} 1.00 & 1.99 & 4.12 \\ 0 & 0 & 0 \\ 0 & 0 & 0 \end{bmatrix}.$$

Since the largest change in any element from $\mathbf{C}$ to $\mathbf{C}'$ is $0.010588\ldots < 0.015$, the bound on our experimental error, the data in $\mathbf{C}'$ are just as good as those in $\mathbf{C}$ for which the rank was 3 rather than 1. We probably should conclude that there is only one independent factor in the reaction. From the viewpoint of condition number, since $\|\mathbf{C}\|_\infty = 12.37$ while $\mathbf{C}'$ is singular and

$$\|\mathbf{C} - \mathbf{C}'\|_\infty = 0.028\ldots$$

we know that the condition number $c(\mathbf{C})$ for the norm $\|\cdot\|_\infty$ is at least $12.37/0.028 \geq 441$. In fact, by similarly perturbing only the third row of $\mathbf{C}$, we can find an even closer singular matrix and deduce that $c(\mathbf{C}) \geq 1502$. By using MATLAB or similar software to find the inverse of $\mathbf{C}$, we can discover that $c(\mathbf{A}) \approx 6885$ and that the ∞-norm distance to the nearest singular matrix is about 0.0018 and to the nearest matrix of rank 1 is about 0.006. In light of the measurement errors of size 0.015, it certainly seems that we can only be *confident* of there being one independent factor in the reaction, although there *may* be more.

PROBLEMS 6.4

▷ **1.** Suppose that $\mathbf{P}$ is a $q \times q$ matrix whose rows satisfy

$$|\langle\mathbf{P}\rangle_{i1}| + |\langle\mathbf{P}\rangle_{i2}| + \cdots + |\langle\mathbf{P}\rangle_{iq}| < 1 \qquad \text{for each } i.$$

Prove that $\mathbf{I}_q + \mathbf{P}$ is nonsingular.

2. Suppose that $\mathbf{P}$ is a $q \times q$ matrix whose columns satisfy

$$|\langle\mathbf{P}\rangle_{1j}| + |\langle\mathbf{P}\rangle_{2j}| + \cdots + |\langle\mathbf{P}\rangle_{qj}| < 1 \qquad \text{for each } j.$$

Prove that $\mathbf{I}_q + \mathbf{P}$ is nonsingular.

3. Suppose that the $q \times q$ matrix $\mathbf{A}$ is *strictly row-diagonally dominant* in the sense that the rows satisfy

$$|\langle\mathbf{A}\rangle_{ii}| > |\langle\mathbf{A}\rangle_{i1}| + \cdots + |\langle\mathbf{A}\rangle_{i,i-1}| + |\langle\mathbf{A}\rangle_{i,i+1}| + \cdots + |\langle\mathbf{A}\rangle_{iq}| \qquad \text{for each } i.$$

Prove that $\mathbf{A}$ is nonsingular.

▷ **4.** Suppose that the $q \times q$ matrix $\mathbf{A}$ is *strictly column-diagonally dominant* in the sense that the columns satisfy

$$|\langle \mathbf{A} \rangle_{jj}| > |\langle \mathbf{A} \rangle_{1j}| + \cdots + |\langle \mathbf{A} \rangle_{j-1,j}| + |\langle \mathbf{A} \rangle_{j+1,j}| + \cdots + |\langle \mathbf{A} \rangle_{qj}| \qquad \text{for each } j.$$

Prove that $\mathbf{A}$ is nonsingular.

5. Prove **Key Theorem 6.28** for the case $\alpha = \|\mathbf{R}\mathbf{A}^{-1}\| < 1$.

▷ **6.** Prove that, no matter what norm is used, the condition number

$$c(\mathbf{A}) = \|\mathbf{A}\| \, \|\mathbf{A}^{-1}\|$$

satisfies $c(\mathbf{A}) \geq 1$.

7. For both the 1-norm and the ∞-norm, find 2×2 examples to show that the condition number $c(\mathbf{A})$ can be arbitrarily large.

8. Given the matrix $\mathbf{A}$ below:
(a) Find its condition number.
(b) Find a nearby singular matrix.
(c) Verify (6.30) in this case.

$$\mathbf{A} = \begin{bmatrix} 1 & 0 \\ 0 & \dfrac{1}{k} \end{bmatrix}.$$

9. Let $\mathbf{A}$ be the matrix

$$\mathbf{A} = \begin{bmatrix} 1 & k \\ 0 & 1 \end{bmatrix},$$

so that

$$\mathbf{A}^{-1} = \begin{bmatrix} 1 & -k \\ 0 & 1 \end{bmatrix}.$$

In either the norm $\|\cdot\|_1$ or the norm $\|\cdot\|_\infty$,

$$\|\mathbf{A}\| = \|\mathbf{A}^{-1}\| = 1 + k \text{ for } k \geq 0,$$

so that the condition number $c(\mathbf{A}) = (1 + k)^2$, which is large for large k. However, if we consider the system of equations $\mathbf{A}\mathbf{x} = \mathbf{b}$ with

$$\mathbf{b} = \begin{bmatrix} 1 \\ 1 \end{bmatrix},$$

the solution is

$$\mathbf{x} = \begin{bmatrix} 1 - k \\ 1 \end{bmatrix},$$

while if we perturb only $\mathbf{b}$ via nonzero δ_1, δ_2 to

$$\mathbf{b} + \boldsymbol{\delta}\mathbf{b} = \begin{bmatrix} 1 + \delta_1 \\ 1 + \delta_2 \end{bmatrix},$$

we find the change $\boldsymbol{\delta}\mathbf{x}$ in the solution to be

$$\boldsymbol{\delta}\mathbf{x} = \begin{bmatrix} \delta_1 - k\delta_2 \\ \delta_2 \end{bmatrix}.$$

Find a bound on $\|\boldsymbol{\delta}\mathbf{x}\|/\|\mathbf{x}\|$ in terms of $\|\boldsymbol{\delta}\mathbf{b}\|/\|\mathbf{b}\|$ using either the 1-norm or the ∞-norm to show that this problem is well-conditioned, despite the large condition number of $\mathbf{A}$.

10. Perform similar calculations to those in Problem 9, but this time let $\mathbf{b} = [1 \quad -1/k]^T$ and show that in this case the problem is ill-conditioned.

▷ **11.** Using the 1-norm, show that $c(\mathbf{A}) \geq 600$, where

$$\mathbf{A} = \begin{bmatrix} 1.1 & 2.1 & 3.1 \\ 1.0 & -1.0 & 2.0 \\ 0.2 & 3.3 & 1.4 \end{bmatrix}.$$

12. Let $\mathbf{A}$ be $p \times p$ and let $c(\mathbf{A})$ be its condition number using any norm you wish. Prove that there exists some $\mathbf{b}$ and some perturbation $\boldsymbol{\delta}\mathbf{b}$ so that the solutions $\mathbf{x}$ and $\mathbf{x} + \boldsymbol{\delta}\mathbf{x}$ have $\|\boldsymbol{\delta}\mathbf{x}\|/\|\mathbf{x}\|$ approximately equal to $c(\mathbf{A})\|\boldsymbol{\delta}\mathbf{b}\|/\|\mathbf{b}\|$.

13. By exhibiting a small perturbation $\boldsymbol{\delta}\mathbf{b}$ that produces a large change $\boldsymbol{\delta}\mathbf{x}$, show that the equations

$$x_1 + x_2 = 2.0000$$

$$1.00001x_1 + x_2 = 2.00001$$

are ill-conditioned; find $c(\mathbf{A})$ using the ∞-norm.

▷ **14.** Use Theorem 6.29 with each of (3.42) and (3.43) to obtain generally valid bounds on the error in the solution $\mathbf{x}$ to $\mathbf{Ax} = \mathbf{b}$ produced by applying Gauss elimination with partial pivoting on a computer using t-digit floating-point arithmetic.

15. Use Theorem 6.29 and the size of the residual found in Example 3.44 to bound the error—as in Problem 14—in the numerical solution found in that example.

▷ **16.** Consider the system of equations

$$0.89x_1 + 0.53x_2 = 0.36$$

$$0.47x_1 + 0.28x_2 = 0.19$$

with exact solution $x_1 = 1$, $x_2 = -1$.

(a) Find $\boldsymbol{\delta}\mathbf{b}$ so that if you replace the right-hand side $\mathbf{b}$ by $\mathbf{b} + \boldsymbol{\delta}\mathbf{b}$, the exact solution will be $x_1 = 0.47$, $x_2 = -0.11$.

(b) Is the system ill-conditioned or well-conditioned?

(c) Find the condition number of the matrix for the system using the ∞-norm.

𝔐 **17.** (a) Use MATLAB or similar software to obtain an approximate solution $\mathbf{x}'$ to the equations in Problem 16.
(b) Compare the error $\mathbf{x} - \mathbf{x}'$ to the bound obtained by using Problem 14 in this case.

𝔐 **18.** Use MATLAB or similar software to compute the condition number of the 5×5 matrix $\mathbf{H}$ formed from the 25 entries in the upper-left-hand corner of the matrix in Example 1.31, and deduce how far it is from $\mathbf{H}$ to the nearest singular matrix. (Compare Problem 18 in Section 5.9.)

6.5 *MISCELLANEOUS PROBLEMS*

PROBLEMS 6.5

▷ **1.** Find a basis for the null space of the usual transformation $\mathscr{T}$ with $\mathscr{T}(\mathbf{v}) = \mathbf{A}\mathbf{v}$ generated:
(a) By the matrix

$$[1 \quad 1 \quad -1 \quad -1].$$

(b) By the matrix

$$\begin{bmatrix} 1 & -1 & 0 & 0 \\ 0 & 1 & -1 & 0 \\ 1 & -2 & 1 & 0 \end{bmatrix}.$$

2. Let $\mathscr{V}_0$ be the subspace of $C^{(1)}[0, 1]$—see Example 5.9(e)—consisting of those functions y that satisfy $y(0) = 0$. Define $\mathscr{T}$ from $\mathscr{V}_0$ to $\mathscr{W} = C[0, 1]$ so that $\mathscr{T}(y)$ is that continuous function whose value at t equals $y'(t) + y(t)$, where the prime denotes differentiation.
(a) Show that the image space of $\mathscr{T}$ is $\mathscr{W}$ and that $\mathscr{T}$ has an inverse.
(b) Show how to find $\mathscr{T}^{-1}(f)$ for f in $\mathscr{W}$.

3. Define the inner product of h and g in $C[0, 1]$ as the definite integral from 0 to 1 of the product hg. Show that the adjoint transformation $\mathscr{T}^*$ of the transformation $\mathscr{T}$ in Example 6.2(b) is just $\mathscr{T}$ itself: $\mathscr{T}^* = \mathscr{T}$.

▷ **4.** Let B and C be ordered bases for q-dimensional $\mathscr{V}$ and p-dimensional $\mathscr{W}$, respectively. Let $[\cdot]_{B,C}$ denote the transformation that takes each linear transformation $\mathscr{T}$ from $\mathscr{V}$ to $\mathscr{W}$ into its matrix representation $\mathbf{A}$: $[\mathscr{T}]_{B,C} = \mathbf{A}$ if and only if $\mathbf{A}$ represents $\mathscr{T}$ with respect to the ordered bases B and C. Prove that the transformation $[\cdot]_{B,C}$ is an isomorphism from $\mathscr{L}(\mathscr{V}, \mathscr{W})$—see Example 6.2(e)—to the vector space of all $p \times q$ matrices.

5. Suppose that $\mathbf{v}_0$ solves $\mathscr{T}(\mathbf{v}_0) = \mathbf{w}$, where $\mathscr{T}$ is a linear transformation from $\mathscr{V}$ to $\mathscr{W}$. Prove that $\mathbf{v}$ also solves $\mathscr{T}(\mathbf{v}) = \mathbf{w}$ if and only if $\mathbf{v} = \mathbf{v}_0 + \mathbf{n}$ for some $\mathbf{n}$ in the null space of $\mathscr{T}$.

6. Suppose that $\mathscr{T}$ is a linear transformation. Prove that, given $\mathbf{w}$, the equation $\mathscr{T}(\mathbf{v}) = \mathbf{w}$ has either no, exactly one, or infinitely many solutions $\mathbf{v}$.

▷ **7.** Suppose that $\mathscr{T}$ is a linear transformation from $\mathscr{V}$ to $\mathscr{W}$, both being p-dimensional. Prove that the following are equivalent:
1. The null space of $\mathscr{T}$ equals $\{\mathbf{0}\}$.
2. A linear transformation $\mathscr{X}$ exists from $\mathscr{W}$ to $\mathscr{V}$ such that $\mathscr{T}\mathscr{X}$ is the identity transformation on $\mathscr{W}$.
3. $\mathscr{T}^{-1}$ exists and equals $\mathscr{X}$ from (2).
4. The equation $\mathscr{T}(\mathbf{v}) = \mathbf{w}$ has exactly one solution for each $\mathbf{w}$ in $\mathscr{W}$.

8. Suppose that $\mathscr{T}$ is a linear transformation from $\mathscr{V}$ to $\mathscr{W}$, both p-dimensional. Prove that the null space of $\mathscr{T}$ equals $\{\mathbf{0}\}$ if and only if the image space of $\mathscr{T}$ equals $\mathscr{W}$.

9. Assume as in Problem 8. Prove that $\mathscr{T}^{-1}$ exists if and only if the null space of $\mathscr{T}$ equals $\{\mathbf{0}\}$.

▷ **10.** Assume as in Problem 8. Prove that $\mathscr{T}^{-1}$ exists if and only if the image space of $\mathscr{T}$ equals $\mathscr{W}$.

11. Suppose that $\mathscr{V}$ and $\mathscr{W}$ are finite-dimensional and have inner products, and that $\mathscr{T}$ has an adjoint transformation $\mathscr{T}^*$. Prove that the union of a basis for the null space of $\mathscr{T}$ and a basis for the image space of $\mathscr{T}^*$ is a basis for $\mathscr{V}$.

12. Assume as in Problem 11. Prove that the dimension of the image space of $\mathscr{T}^*$ equals the dimension of the image space of $\mathscr{T}$.

13. Assume as in Problem 11. Prove that $\mathscr{T}$ has an inverse if and only if $\mathscr{T}^*$ has an inverse, and that then $(\mathscr{T}^*)^{-1}$ equals $(\mathscr{T}^{-1})^*$ (which exists).

▷ **14.** Suppose that $\mathscr{T}$ is a linear transformation from $\mathscr{V}$ to $\mathscr{W}$ and that $\mathscr{S}$ is a linear transformation from $\mathscr{W}$ to $\mathscr{Z}$, and that $\mathscr{T}^*$ and $\mathscr{S}^*$ both exist for the inner products on $\mathscr{V}$, $\mathscr{W}$, and $\mathscr{Z}$. Prove that $(\mathscr{S}\mathscr{T})^*$ exists and equals $\mathscr{T}^*\mathscr{S}^*$.

15. Suppose that $\mathbf{P}$ is $q \times q$ and that $\|\mathbf{P}\|_\infty < 1$. Prove that the infinite series

$$\mathbf{I} - \mathbf{P} + \mathbf{P}^2 - \mathbf{P}^3 + \mathbf{P}^4 - \cdots$$

converges to some matrix $\mathbf{B}$, and that

$$(\mathbf{I} + \mathbf{P})\mathbf{B} = \mathbf{B}(\mathbf{I} + \mathbf{P}) = \mathbf{I},$$

so that the series in fact converges to $(\mathbf{I} + \mathbf{P})^{-1}$, which must therefore exist.

7

Eigenvalues and eigenvectors: an overview

One of the most powerful approaches to analyzing the behavior of some interesting real-world system is to determine the so-called eigensystem of matrices involved in modeling the system. An overview of this topic appears in this chapter both as a preparation for the following two chapters and as a survey for those who do not read those chapters. The ***Key Theorems*** *are* ***7.9****,* ***7.10****,* ***7.14****,* ***7.18****, and* ***7.38****; previews—see* ***(7.13)****,* ***(7.26)****,* ***(7.42)****, and* ***(7.43)****—of* ***Key Theorems*** *from Chapters 8 and 9 are also included.*

7.1 *INTRODUCTION*

Studying so-called *eigenvalues* and *eigenvectors* of matrices and of linear transformations generally is of fundamental importance in applied mathematics. One reason for this is that these basic concepts arise when models are studied from a surprisingly wide variety of viewpoints. This section considers some examples that illustrate four fundamental viewpoints stressed throughout Chapters 7, 8, and 9: (1) singularity of $\mathbf{A} - \lambda\mathbf{I}$, where λ is a parameter; (2) invariant subspaces; (3) simple representations of transformations; and (4) decompositions of matrices.

Singularity of Parameter-Dependent Matrices

It is often the case in applications that the mathematical model leads to a matrix that depends on an unknown but important parameter whose value must be chosen so as to make the matrix singular. An important class of such problems is those that arise from modeling oscillatory phenomena (such as the motion of airplane wings under various aerodynamic forces, the swings in the economy under various market forces, and the like).

As a concrete example, consider the motion of the two masses suspended and coupled by springs in Section 2.5. Mathematical analysis of that situation leads to a 2×2 matrix $\mathbf{K} - \omega^2\mathbf{M}$, where $\mathbf{K}$ and $\mathbf{M}$ are known matrices and ω is an unknown parameter. The problem is to determine ω so that $\mathbf{K} - \omega^2\mathbf{M}$ is singular. Example 2.34 shows that this is equivalent to having ω^2 be the root of a polynomial of exact degree 2. More generally, if $\mathbf{M}$ is nonsingular, the problem is equivalent to finding a parameter λ for which $\mathbf{A} - \lambda\mathbf{I}$ is singular, where $\mathbf{A} = \mathbf{M}^{-1}\mathbf{K}$ and $\lambda = \omega^2$.

(7.1) ***Example.*** In the specific case in Example 2.34, we have

$$\mathbf{A} - \lambda\mathbf{I} = \begin{bmatrix} 12 & -4 \\ -8 & 8 \end{bmatrix} - \lambda\begin{bmatrix} 1 & 0 \\ 0 & 1 \end{bmatrix} = \begin{bmatrix} 12-\lambda & -4 \\ -8 & 8-\lambda \end{bmatrix}.$$

From Theorem 4.35 we know that a matrix is singular if and only if its determinant equals zero, so λ must be chosen so that

$$\begin{aligned} 0 = \det(\mathbf{A} - \lambda\mathbf{I}) &= (12-\lambda)(8-\lambda) - (-4)(-8) \\ &= \lambda^2 - 20\lambda + 64 = (\lambda - 4)(\lambda - 16). \end{aligned}$$

Thus the values of λ $(= \omega^2)$ must be $\lambda = 4$ or $\lambda = 16$.

As defined in the next section, such a value of λ that makes $\mathbf{A} - \lambda\mathbf{I}$ singular is called an *eigenvalue* of $\mathbf{A}$.

This illustrates one way in which eigenvalues arise in practice: in oscillation and other problems in which a parameter must be determined so as to make a parameter-dependent matrix singular.

Invariant Subspaces

We have seen that mathematical models of applied problems often lead to the study of linear transformations defined by matrices: $\mathscr{T}(\mathbf{x}) = \mathbf{A}\mathbf{x}$, where $\mathbf{A}$ is $p \times p$ and $\mathbf{x}$ is in $\mathbb{R}^p$. Section 2.2, for example, modeled the shifting patterns of market shares among dairies from month to month in terms of a matrix $\mathbf{A}$ so that the shares $\mathbf{x}_i$ at one month were transformed into the shares $\mathbf{x}_{i+1}$ the next month by $\mathbf{x}_{i+1} = \mathbf{A}\mathbf{x}_i$. In real situations, it is very often the case that the vectors analogous to $\mathbf{x}_i$ have an extremely large number p of entries. In such cases a useful approach is to attempt to break the large system being modeled into a collection of much smaller subsystems that can be dealt with more easily; that is, we seek certain variables or combinations of variables whose values after being transformed can be expressed in terms of just those variables or combinations. Algebraically, this corresponds to finding a low-dimensional subspace $\mathscr{V}_0$ of $\mathbb{R}^p$ so that $\mathbf{A}\mathbf{x}$ is also in $\mathscr{V}_0$ whenever $\mathbf{x}$ is in $\mathscr{V}_0$. Such a subspace is called an *invariant subspace* of $\mathbf{A}$ (and of $\mathscr{T}$).

The simplest possible such invariant subspace would, of course, be one whose dimension equals 1 so that $\mathscr{V}_0$ is spanned by some single vector $\mathbf{x} \neq \mathbf{0}$. For $\mathbf{Ax}$ also to be in $\mathscr{V}_0$ would require $\mathbf{Ax}$ to be a multiple of $\mathbf{x}$, that is, $\mathbf{Ax} = \lambda\mathbf{x}$ for some scalar λ. But this would mean that $(\mathbf{A} - \lambda\mathbf{I})\mathbf{x} = \mathbf{0}$ while $\mathbf{x} \neq \mathbf{0}$, so $\mathbf{A} - \lambda\mathbf{I}$ *must be singular*. That is, λ must be an eigenvalue of $\mathbf{A}$; the nonzero vectors $\mathbf{x}$ associated in this way with λ are called *eigenvectors* of $\mathbf{A}$.

(7.2) ***Example.*** Consider the dairy-competition model of Example 2.6 mentioned above. In this case, we find

$$\mathbf{A} - \lambda\mathbf{I} = \begin{bmatrix} 0.8 - \lambda & 0.2 & 0.1 \\ 0.1 & 0.7 - \lambda & 0.3 \\ 0.1 & 0.1 & 0.6 - \lambda \end{bmatrix}.$$

In order that $\mathbf{A} - \lambda\mathbf{I}$ be singular, we require that $\det(\mathbf{A} - \lambda\mathbf{I}) = 0$. Direct calculation gives

$$\det(\mathbf{A} - \lambda\mathbf{I}) = -\lambda^3 + 2.1\lambda^2 - 1.4\lambda + 0.3 = -(\lambda - 0.5)(\lambda - 0.6)(\lambda - 1.0)$$

so that the eigenvalues λ must be

$$\lambda = 0.5, \qquad \lambda = 0.6, \qquad \lambda = 1.0.$$

To find the associated eigenvectors, consider, for example, the equation $(\mathbf{A} - \lambda\mathbf{I})\mathbf{x} = \mathbf{0}$ for $\lambda = 0.5$. In terms of the entries x_1, x_2, x_3 of $\mathbf{x}$, this is

$$\begin{aligned} 0.3x_1 + 0.2x_2 + 0.1x_3 &= 0 \\ 0.1x_1 + 0.2x_2 + 0.3x_3 &= 0 \\ 0.1x_1 + 0.1x_2 + 0.1x_3 &= 0. \end{aligned}$$

Our usual techniques for linear systems of equations reduce this to

$$\begin{aligned} x_1 + 0x_2 - x_3 &= 0 \\ 0x_1 + x_2 + 2x_3 &= 0 \\ 0x_1 + 0x_2 + 0x_3 &= 0; \end{aligned}$$

hence we can let x_3 equal an arbitrary α, $x_1 = \alpha$, and $x_2 = -2\alpha$, so that

$$\mathbf{x} = \alpha\begin{bmatrix} 1 \\ -2 \\ 1 \end{bmatrix}$$

is an eigenvector associated with the eigenvalue $\lambda = 0.5$ for every nonzero α. By similar methods we find that arbitrary multiples of the vectors

$$\begin{bmatrix} 1 \\ -1 \\ 0 \end{bmatrix}, \quad \begin{bmatrix} 9 \\ 7 \\ 4 \end{bmatrix}$$

are eigenvectors associated with the eigenvalues $\lambda = 0.6$ and $\lambda = 1.0$, respectively. The three one-dimensional subspaces of $\mathbb{R}^3$, each spanned by one of these three eigenvectors, are thus invariant subspaces of **A**.

This illustrates a second way in which eigenvalues and eigenvectors arise naturally: in seeking invariant subspaces for linear transformations in order to simplify the study of the transformations by breaking the overall system into smaller subsystems.

Simple Representations of Transformations

Besides seeking invariant subspaces, there is another approach to simplifying the study of linear transformations $\mathcal{T}$ defined by $p \times p$ matrices $\mathbf{A}$ with $\mathcal{T}(\mathbf{x}) = \mathbf{Ax}$: Find different bases in terms of which $\mathcal{T}$ is represented by a simpler matrix that is easier to study. $\mathcal{T}$ is represented by $\mathbf{A}$ with respect to the standard ordered basis $B = \{\mathbf{e}_1; \ldots; \mathbf{e}_p\}$ for $\mathbb{R}^p$ as both domain and range. Suppose that we use a different ordered basis $B' = \{\mathbf{x}_1; \ldots; \mathbf{x}_p\}$. According to **Key Theorem 5.43** on change of basis, the B'-coordinates $\mathbf{x}_{B'}$ of a vector $\mathbf{x}$ and the B-coordinates $\mathbf{x}_B$ are related by $\mathbf{x}_B = \mathbf{M}\mathbf{x}_{B'}$, where $\mathbf{M}$ is the $p \times p$ nonsingular matrix whose ith column contains the coefficients used to write $\mathbf{x}_i$ in terms of the $\mathbf{e}_j$—that is, because of the special form of the $\mathbf{e}_j$,

$$\mathbf{M} = [\mathbf{x}_1 \cdots \mathbf{x}_p].$$

According to Theorem 6.17 on change of basis and representations, the matrix that represents $\mathcal{T}$ with respect to B' is $\mathbf{A}' = \mathbf{M}^{-1}\mathbf{AM}$. Our problem, then, is to find vectors $\mathbf{x}_1, \ldots, \mathbf{x}_p$ so that $\mathbf{M}^{-1}\mathbf{AM}$ is simple.

(7.3) ***Example.*** Consider the matrix **A** of Example 7.2, and suppose that as the vectors $\mathbf{x}_i$ we take *eigenvectors of* **A**:

$$\mathbf{x}_1 = [1 \quad -2 \quad 1]^T, \quad \mathbf{x}_2 = [1 \quad -1 \quad 0]^T, \quad \mathbf{x}_3 = [9 \quad 7 \quad 4]^T,$$

for example. A little calculation gives $\mathbf{M}^{-1}$ and then

$$\begin{aligned}\mathbf{A}' &= \mathbf{M}^{-1}\mathbf{AM}\\ &= \begin{bmatrix} -0.20 & -0.20 & 0.20 \\ 0.75 & -0.25 & -1.25 \\ 0.05 & 0.05 & 0.05 \end{bmatrix} \begin{bmatrix} 0.8 & 0.2 & 0.1 \\ 0.1 & 0.7 & 0.3 \\ 0.1 & 0.1 & 0.6 \end{bmatrix} \begin{bmatrix} 1 & 1 & 9 \\ -2 & -1 & 7 \\ 1 & 0 & 4 \end{bmatrix}\\ &= \begin{bmatrix} 0.5 & 0 & 0 \\ 0 & 0.6 & 0 \\ 0 & 0 & 1.0 \end{bmatrix},\end{aligned}$$

a diagonal matrix with *the eigenvalues of* **A** on the diagonal.

Thus we have found another viewpoint on eigenvectors and eigenvalues of $p \times p$ matrices: providing extremely simple representations of linear transformations.

Decompositions of Matrices

One final viewpoint on eigenvalues and eigenvectors of a $p \times p$ matrix $\mathbf{A}$ remains to be illustrated. We saw in Example 7.3 that $\mathbf{M}^{-1}\mathbf{A}\mathbf{M} = \mathbf{A}'$ is a diagonal matrix. Premultiplying by $\mathbf{M}$ and postmultiplying by $\mathbf{M}^{-1}$ rewrites this as $\mathbf{A} = \mathbf{M}\mathbf{A}'\mathbf{M}^{-1}$, that is, $\mathbf{A}$ has been *decomposed* into a special type of product with $\mathbf{A}'$ having a special form.

This decomposition can be surprisingly useful; it was this that was in the background of Example 6.18, where we analyzed the behavior of large powers $\mathbf{A}^i$ of the matrix $\mathbf{A}$ from Example 2.17. Example 6.18 found a matrix $\mathbf{M}$ (denoted by $\mathbf{S}$ and $\mathbf{P}$ there) for which $\mathbf{A}'$, although not diagonal, was simple enough that we could show easily that its powers tend to zero. And since

$$\mathbf{A}^i = (\mathbf{M}\mathbf{A}'\mathbf{M}^{-1})(\mathbf{M}\mathbf{A}'\mathbf{M}^{-1})\cdots(\mathbf{M}\mathbf{A}'\mathbf{M}^{-1}) = \mathbf{M}(\mathbf{A}')^i\mathbf{M}^{-1}$$

and $(\mathbf{A}')^i$ tends to zero, so does $\mathbf{A}^i$. What has this to do with eigenvalues and eigenvectors since $\mathbf{A}'$ is not diagonal? The first column $\mathbf{x}_1 = [5 \quad 3]^T$ of $\mathbf{M}$ satisfies $\mathbf{A}\mathbf{x}_1 = 0.9\mathbf{x}_1$ as can easily be checked; that is, $\mathbf{x}_1$ is an eigenvector of $\mathbf{A}$ associated with the eigenvalue 0.9. It is easy to see that in this case $\det(\mathbf{A} - \lambda\mathbf{I}) = (\lambda - 0.9)^2$, so there is only one (but double) eigenvalue $\lambda_1 = \lambda_2 = 0.9$.

Thus we have still another use for eigenvalues and eigenvectors: as a means to obtain useful decompositions of matrices.

The variety of ways in which eigenvalues and eigenvectors arise indicates why they are so important both theoretically and practically. But the illustrations above glossed over many crucial questions: Do eigenvalues always exist? Do eigenvectors always exist? How simple a representation can be found for a linear transformation by proper choice of ordered bases? How simple a decomposition can be found for a given matrix? And how, in practice, can we actually find these eigenvalues and eigenvectors? Our next task is to address these and other fundamental issues.

PROBLEMS 7.1

▷ **1.** Let $m_1 = 12$, $m_2 = 16$, $k_1 = 36$, and $k_2 = 48$ in the oscillating-masses model of Section 2.5. Find the matrices $\mathbf{K}$ and $\mathbf{M}$ from (2.29), find $\mathbf{A} = \mathbf{M}^{-1}\mathbf{K}$, and then find the eigenvalues $\lambda = \omega^2$ of $\mathbf{A}$ as in Example 7.1.

2. Find the eigenvalues of the matrix $\mathbf{A}$ in Example 2.16 arising from the competing-populations model with $k = 0.1$.

▷ **3.** Find the eigenvalues of

$$\mathbf{A} = \begin{bmatrix} 3.5 & 2 \\ -6 & -3.5 \end{bmatrix}.$$

4. Find the eigenvalues (see Problem 2) and eigenvectors of the matrix $\mathbf{A}$ in Example 2.16 arising from the competing-populations model with $k = 0.1$.

5. Find the eigenvalues and eigenvectors of the matrix $\mathbf{A}$ in Problem 3.

▷ **6.** Use the eigenvectors of $\mathbf{A}$ found in Problem 5 as an ordered basis for $\mathbb{R}^2$ as domain and range of $\mathscr{T}(\mathbf{v}) = \mathbf{Av}$, and find the matrix $\mathbf{A}'$ representing $\mathscr{T}$ with respect to that basis by directly using **Key Theorem 6.14**.

7. Consider the transition matrix $\mathbf{A}$ for the dairy-competition model of Example 2.6 and its eigenvalues λ found in Example 7.2. When $\mathbf{A} - \lambda\mathbf{I}$ is singular, so is $\mathbf{A}^T - \lambda\mathbf{I}$; thus there is a $\mathbf{y}$ with $\mathbf{A}^T\mathbf{y} = \lambda\mathbf{y}$, that is,

$$\mathbf{y}^T\mathbf{A} = \lambda\mathbf{y}^T.$$

($\mathbf{y}^T$ is called a *left-eigenvector* of $\mathbf{A}$.)

(a) Show, for any market distribution $\mathbf{x}_i$ and the following $\mathbf{x}_{i+1} = \mathbf{Ax}_i$, that

$$\mathbf{y}^T\mathbf{x}_{i+1} = \lambda\mathbf{y}^T\mathbf{x}_i,$$

which tells how the linear combination $\mathbf{y}^T\mathbf{x}$ of the market shares $\langle\mathbf{x}\rangle_i$ changes from month to month.

(b) Find the left-eigenvector $\mathbf{y}_i$ associated with each eigenvalue λ_i.

(c) Use (a) and the left-eigenvector associated with $\lambda = 1$ to show that the sum of the market shares never changes.

8. Use the eigenvectors and eigenvalues of the matrix $\mathbf{A}$ in Problem 3 to decompose $\mathbf{A}$ as $\mathbf{A} = \mathbf{MA}'\mathbf{M}^{-1}$ with a diagonal matrix $\mathbf{A}'$.

9. Decompose the matrix $\mathbf{A}$ of Problem 4 as $\mathbf{A} = \mathbf{MA}'\mathbf{M}^{-1}$ with $\mathbf{A}'$ a diagonal matrix.

7.2 DEFINITIONS AND BASIC PROPERTIES

The concepts of eigenvectors and eigenvalues make sense for a linear transformation from any vector space $\mathscr{V}$ to itself; the theory for infinite-dimensional spaces is quite complicated and subtle, however. We therefore restrict ourselves to finite-dimensional $\mathscr{V}$. In this case, it follows from the fact that all linear transformations can be represented by matrices that we can further restrict ourselves to the study of eigenvalues and eigenvectors of matrices.

Eigensystems and Characteristic Polynomials

(7.4) ***Definition.*** The *eigenvalues* of a $p \times p$ real or complex matrix $\mathbf{A}$ are the real or complex numbers λ for which there is a *nonzero* $\mathbf{x} \neq \mathbf{0}$ with $\mathbf{Ax} = \lambda\mathbf{x}$. The *eigenvectors* of $\mathbf{A}$ are the *nonzero* vectors $\mathbf{x} \neq \mathbf{0}$ for which there is a number λ with $\mathbf{Ax} = \lambda\mathbf{x}$. If $\mathbf{Ax} = \lambda\mathbf{x}$ for $\mathbf{x} \neq \mathbf{0}$, then $\mathbf{x}$ is an eigenvector *associated with* the eigenvalue λ, and *vice versa.* The associated eigenvalues and eigenvectors together make up the *eigensystem* of $\mathbf{A}$. An *invariant substance* $\mathscr{V}_0$

of **A** is a subspace of $\mathbb{R}^p$ or $\mathbb{C}^p$ for which **Ax** is in $\mathscr{V}_0$ whenever **x** is in $\mathscr{V}_0$.

If **A** is 2 × 2, the equation $\mathbf{Ax} = \lambda\mathbf{x}$ is a system of two equations; however, (a) it involves *three* unknowns x_1, x_2, and λ; and (b) the equations are *nonlinear* because the unknown λ multiplies each of the unknowns x_1 and x_2. Our first task is to show that we can in fact "decouple" the variables: We can first find λ from a nonlinear equation and then, once λ is known, solve the *linear* equations $\mathbf{Ax} = \lambda\mathbf{x}$ for **x**.

(7.5) ***Theorem*** (eigenvalues and singularity). λ is an eigenvalue of **A** if and only if $\mathbf{A} - \lambda\mathbf{I}$ is singular, which in turn holds if and only if the determinant of $\mathbf{A} - \lambda\mathbf{I}$ equals zero: $\det(\mathbf{A} - \lambda\mathbf{I}) = 0$ (the so-called *characteristic equation* of **A**).

PROOF. λ being an eigenvalue means that there is a nonzero **x** with $\mathbf{Ax} = \lambda\mathbf{x}$, that is, $(\mathbf{A} - \lambda\mathbf{I})\mathbf{x} = \mathbf{0}$ and $\mathbf{x} \neq \mathbf{0}$. Application of **Key Theorem 4.18** to $\mathbf{A} - \lambda\mathbf{I}$ shows that this is equivalent to the singularity of $\mathbf{A} - \lambda\mathbf{I}$, which—by Theorem 4.35—is equivalent to $\det(\mathbf{A} - \lambda\mathbf{I}) = 0$. ■

Thus we can first find those values λ that make $\det(\mathbf{A} - \lambda\mathbf{I}) = 0$, and then for each such known λ solve $\mathbf{Ax} = \lambda\mathbf{x}$ for $\mathbf{x} \neq \mathbf{0}$. We need to understand the nature of the expression $\det(\mathbf{A} - \lambda\mathbf{I})$ in the characteristic equation of **A**.

(7.6) ***Theorem*** (characteristic polynomials). Suppose that **A** is $p \times p$. Then:

(a) $\det(\mathbf{A} - \lambda\mathbf{I})$ is a polynomial of exact degree p in the variable λ; this is called the *characteristic polynomial* $f(\lambda)$ of **A**.
(b) The coefficient of λ^p in $f(\lambda)$ equals $(-1)^p$.
(c) The coefficient of λ^{p-1} in $f(\lambda)$ equals $(-1)^{p-1}$ tr **A**, where tr **A** (the *trace* of **A**) equals the sum of the entries on the main diagonal of **A**.
(d) The constant term of $f(\lambda)$ equals det **A**.

PROOF

(a) We remarked immediately after Example 4.29 in Section 4.5 on determinants that the determinant of a matrix **B** equals the sum of all possible products—with appropriate signs—of entries from **B**, where in each product there must be exactly one entry from each row and exactly one from each column of **B**. This shows that the characteristic polynomial of **A** indeed is a polynomial and that its exact degree is p.
(b) The only product of the form described in (a) that involves p entries, each including λ, equals the product of the diagonal entries of $\mathbf{A} - \lambda\mathbf{I}$. By expanding the determinants repeatedly along the first row it is easy to see that the sign given this product is positive, so that the highest power of λ appears as $(-\lambda)^p$.

(c) Arguing much as in (b), we see that the only products that involve λ in $p-1$ factors must use $p-1$ of the p entries on the main diagonal: since each product uses one term from each row and one from each column, the remaining pth term must just be the remaining diagonal entry. Thus the product is the same as in (b), namely

$$+(a_{11}-\lambda)(a_{22}-\lambda)\cdots(a_{pp}-\lambda), \qquad \text{where } a_{ii} = \langle \mathbf{A} \rangle_{ii}.$$

The coefficient of λ^{p-1} in this term is as asserted.

(d) The constant term in any polynomial $f(\lambda)$ can be found as $f(0)$. In our case, this is $\det(\mathbf{A} - 0\mathbf{I}) = \det \mathbf{A}$. ■

The eigenvalues of a $p \times p$ matrix $\mathbf{A}$ are therefore the roots of a polynomial of exact degree p. From the theory of polynomials we know that such a polynomial has p real or complex roots $\lambda_1, \ldots, \lambda_p$ in the sense that we can factor the polynomial as

(7.7) $\det(\mathbf{A} - \lambda\mathbf{I}) = (\lambda_1 - \lambda)(\lambda_2 - \lambda)\cdots(\lambda_p - \lambda)$, where the λ_i need not be *distinct*—that is, some of the λ_i may well equal one another.

Thus every $p \times p$ matrix has p real or complex eigenvalues, some of which may well be equal to one another. For example, the eigenvalues of some 7×7 matrix might be 3, 3, 3, 3, 2, 5, 5; there are seven eigenvalues, but only three *distinct* eigenvalues 3, 2, and 5. According to (7.7), the characteristic polynomial of $\mathbf{A}$ in this case is

$$\det(\mathbf{A} - \lambda\mathbf{I}) = (3-\lambda)(3-\lambda)(3-\lambda)(3-\lambda)(2-\lambda)(5-\lambda)(5-\lambda),$$

which can be written more compactly as

$$\det(\mathbf{A} - \lambda\mathbf{I}) = (3-\lambda)^4(2-\lambda)^1(5-\lambda)^2$$

by writing a term for each distinct eigenvalue but raising it to the power equal to the number of times it repeats. This can of course be done in general.

(7.8) ***Definition.*** When the characteristic polynomial of a $p \times p$ matrix $\mathbf{A}$ is written in the form

$$\det(\mathbf{A} - \lambda\mathbf{I}) = (\lambda_1 - \lambda)^{m_1}(\lambda_2 - \lambda)^{m_2}\cdots(\lambda_r - \lambda)^{m_r}$$

with $\lambda_i \neq \lambda_j$ for $1 \leq i \neq j \leq r$ and $m_1 + m_2 + \cdots + m_r = p$, the positive integer m_i is called the *algebraic multiplicity* of the eigenvalue λ_i. An eigenvalue of algebraic multiplicity 1 is called a *simple* eigenvalue.

Algebraic multiplicity is distinguished from *geometric* multiplicity defined in Definition 7.12 below.

In the notation of Definition 7.8, the totality of eigenvalues of $\mathbf{A}$ consists of λ_1 counted m_1 times, λ_2 counted m_2 times, ..., and λ_r counted m_r times. The mythical 7×7 matrix we discussed prior to this definition therefore has eigenvalues $\lambda_1 = 3$ of algebraic multiplicity $m_1 = 4$, $\lambda_2 = 2$ of algebraic multiplicity $m_2 = 1$ (a simple eigenvalue), and $\lambda_3 = 5$ of algebraic multiplicity $m_3 = 2$. Using Definition 7.8, we have developed a fairly simple description of the structure of the set of eigenvalues:

(7.9) **Key Theorem** (eigenvalues' structure). The set of eigenvalues of a $p \times p$ matrix consists of $r \le p$ distinct numbers $\lambda_1, \lambda_2, \ldots, \lambda_r$ with each λ_i of algebraic multiplicity m_i, where $m_1 + m_2 + \cdots + m_r = p$—that is, there are p eigenvalues if each distinct λ_i is counted m_i times, where m_i equals λ_i's multiplicity as a root of the characteristic polynomial of **A**.

PROOF. This follows immediately from Theorems 7.5 and 7.6 and Definition 7.8. ■

The situation regarding the structure of the set of eigen*vectors* is rather more subtle. If λ is a root of the characteristic polynomial—that is, an eigenvalue—then we know from Theorem 7.5 that there is at least one associated eigenvector. Although there is a great deal more to the story, we can use this simple fact to prove quite a bit about the structure of the set of all associated eigenvectors.

(7.10) **Key Theorem** (eigenvectors). Let **A** be $p \times p$. Then:

(a) There exists at least one eigenvector $\mathbf{x}$ associated with each distinct eigenvalue λ; if **A** and λ are real, then $\mathbf{x}$ can be taken to be real.

(b) The set $\mathscr{V}_0$ of all eigenvectors associated with a given eigenvalue λ forms an invariant subspace of **A** (if we join $\mathbf{0}$ to the set, since $\mathbf{0}$ is not an eigenvector).

(c) If $\lambda_1, \ldots, \lambda_r$ is a collection of distinct eigenvalues (that is, $\lambda_i \ne \lambda_j$ for $1 \le i \ne j \le r$) and if $\mathbf{x}_i$ is an eigenvector associated with λ_i for each i, then $\{\mathbf{x}_1, \ldots, \mathbf{x}_r\}$ is linearly independent.

(d) If $p \times p$ **A** has p distinct eigenvalues, then any set of p eigenvectors—one of which is associated with each eigenvalue—is linearly independent, and every eigenvector of **A** is a multiple of one of these p eigenvectors.

(e) If λ is an eigenvalue and $\|\cdot\|$ is a matrix (transformation) norm, then $|\lambda| \le \|\mathbf{A}\|$.

PROOF

(a) The first part follows from Theorem 7.5. If λ and **A** are real and $\mathbf{x} = \mathbf{u} + i\mathbf{v}$ for real $\mathbf{u}$ and $\mathbf{v}$, then $\mathbf{Ax} = \lambda\mathbf{x}$ gives

$$\mathbf{A}(\mathbf{u} + i\mathbf{v}) = \lambda(\mathbf{u} + i\mathbf{v})$$

which implies that $\mathbf{Au} = \lambda\mathbf{u}$ and $\mathbf{Av} = \lambda\mathbf{v}$. At least one of $\mathbf{u}$ and $\mathbf{v}$ must be nonzero since $\mathbf{x}$ is nonzero, and this will be the real eigenvector required.

(b) $\mathscr{V}_0$ is nonempty. If $\mathbf{x}$ and $\mathbf{y}$ are in $\mathscr{V}_0$, so that $\mathbf{Ax} = \lambda\mathbf{x}$ and $\mathbf{Ay} = \lambda\mathbf{y}$, then

$$\mathbf{A}(\mathbf{x} + \mathbf{y}) = \mathbf{Ax} + \mathbf{Ay} = \lambda\mathbf{x} + \lambda\mathbf{y} = \lambda(\mathbf{x} + \mathbf{y}) \quad \text{and}$$

$$\mathbf{A}(\alpha\mathbf{x}) = \alpha(\mathbf{Ax}) = \alpha(\lambda\mathbf{x}) = \lambda(\alpha\mathbf{x})$$

for all scalars α. Thus $\mathbf{x} + \mathbf{y}$ and $\alpha\mathbf{x}$ are in $\mathscr{V}_0$, which must be a subspace by the Subspace Theorem. Since $\mathbf{Ax} = \lambda\mathbf{x}$ for all $\mathbf{x}$ in $\mathscr{V}_0$, it clearly is an *invariant* subspace.

(c) Suppose that the set is linearly dependent; by using Theorem 5.23(a) on linear independence we can consider the *first* vector $\mathbf{x}_s$ that is linearly dependent on the preceding $\mathbf{x}_1, \ldots, \mathbf{x}_{s-1}$:

$$\mathbf{x}_s = \alpha_1\mathbf{x}_1 + \cdots + \alpha_{s-1}\mathbf{x}_{s-1}.$$

Multiplying by λ_s gives

$$\lambda_s\mathbf{x}_s = \alpha_1\lambda_s\mathbf{x}_1 + \cdots + \alpha_{s-1}\lambda_s\mathbf{x}_{s-1},$$

while multiplying instead by $\mathbf{A}$ and using $\mathbf{Ax}_i = \lambda_i\mathbf{x}_i$ gives

$$\lambda_s\mathbf{x}_s = \alpha_1\lambda_1\mathbf{x}_1 + \cdots + \alpha_{s-1}\lambda_{s-1}\mathbf{x}_{s-1}.$$

Subtracting these last two equations gives

$$\mathbf{0} = \alpha_1(\lambda_s - \lambda_1)\mathbf{x}_1 + \cdots + \alpha_{s-1}(\lambda_s - \lambda_{s-1})\mathbf{x}_{s-1}.$$

The set $\{\mathbf{x}_1, \ldots, \mathbf{x}_{s-1}\}$ is linearly independent, since otherwise $\mathbf{x}_s$ would not be the *first* of the vectors $\mathbf{x}_i$ dependent on its predecessors; therefore, the coefficients $\alpha_i(\lambda_s - \lambda_i)$ in the foregoing linear combination giving $\mathbf{0}$ must all equal zero for $1 \le i \le s - 1$. Since $\lambda_s - \lambda_i \neq 0$ for distinct eigenvalues, this says that $\alpha_1 = \cdots = \alpha_{s-1} = 0$. But then also $\mathbf{x}_s$ would be zero, which is impossible for an eigenvector; thus our assumption that the original set was linearly dependent must have been false.

(d) The first parts follow immediately from (a) and (c); we need to show that every eigenvector $\mathbf{x}$ must be a multiple of one of $\mathbf{x}_1, \ldots, \mathbf{x}_p$. Since we have a linearly independent set of p vectors in $\mathbb{C}^p$, they span $\mathbb{C}^p$ by Theorem 5.32(c); thus

$$\mathbf{x} = \alpha_1\mathbf{x}_1 + \cdots + \alpha_p\mathbf{x}_p \qquad \text{for some } \alpha_i.$$

Suppose that λ_r is the eigenvalue with which $\mathbf{x}$ is associated, and consider $\mathbf{v} = \mathbf{x} - \alpha_r\mathbf{x}_r$. By (b), $\mathbf{v}$ either is $\mathbf{0}$ or is an eigenvector associated with λ_r; if $\mathbf{v}$ equals $\mathbf{0}$, then $\mathbf{x} = \alpha_r\mathbf{x}_r$, as asserted; if $\mathbf{v}$ is instead an eigenvector associated with λ_r, then by (c),

$$\{\mathbf{v}, \mathbf{x}_1, \ldots, \mathbf{x}_{r-1}, \mathbf{x}_{r+1}, \ldots, \mathbf{x}_p\}$$

would be linearly independent in contradiction to the fact that we know

$$\mathbf{v} = \mathbf{x} - \alpha_r\mathbf{x}_r = \alpha_1\mathbf{x}_1 + \cdots + \alpha_{r-1}\mathbf{x}_{r-1} + \alpha_{r+1}\mathbf{x}_{r+1} + \cdots + \alpha_p\mathbf{x}_p.$$

(e) We know that $\|\mathbf{Ax}\| \le \|\mathbf{A}\|\,\|\mathbf{x}\|$ for any $\mathbf{x}$, so it certainly holds for this eigenvector. But

$$\|\mathbf{Ax}\| = \|\lambda\mathbf{x}\| = |\lambda|\,\|\mathbf{x}\|;$$

since $\mathbf{x} \neq \mathbf{0}$, we can divide this last equation by $\|\mathbf{x}\|$ and use the inequality above to obtain $|\lambda| \le \|\mathbf{A}\|$. ■

The examples in Section 6.1 illustrate Theorem 7.10. In Example 7.1 the 2×2 matrix $\mathbf{A}$ has two distinct simple eigenvalues $\lambda_1 = 4$ and $\lambda_2 = 16$; it is easy to solve $\mathbf{Ax} = \lambda\mathbf{x}$ to find associated eigenvectors

$$\mathbf{x}_1 = [1 \quad 2]^T \quad \text{and} \quad \mathbf{x}_2 = [1 \quad -1]^T,$$

say, which form a linearly independent set. The 3×3 matrix $\mathbf{A}$ of Example 7.2 was there found to have three distinct simple eigenvalues and three eigenvectors that can easily be shown to form a linearly independent set. Each of these cases involved only eigenvalues of algebraic multiplicity 1; the situation regarding the number of eigenvectors associated with a nonsimple eigenvalue is appreciably more complicated, as we next illustrate.

You should now be able to solve Problems 1 to 13.

Eigenvectors of Multiple Eigenvalues

We want to examine the question of the number of eigenvectors associated with an eigenvalue of algebraic multiplicity greater than 1. We know of course that if $\mathbf{x}$ is an eigenvector of $\mathbf{A}$ associated with λ, then so is $\alpha\mathbf{x}$ for all $\alpha \neq 0$; thus we have found infinitely many eigenvectors associated with λ, *but they all are linearly dependent on the one eigenvector* $\mathbf{x}$ *with which we started.* What we want to learn is whether there are any "really different" eigenvectors—eigenvectors that are linearly independent of $\mathbf{x}$.

(7.11) ***Example.*** Consider the simple 5×5 matrix

$$\mathbf{A} = \begin{bmatrix} 7 & 0 & 0 & 0 & 0 \\ 0 & 4 & 1 & 0 & 0 \\ 0 & 0 & 4 & 0 & 0 \\ 0 & 0 & 0 & 7 & 0 \\ 0 & 0 & 0 & 0 & 4 \end{bmatrix}.$$

It is easy to find that the characteristic polynomial of $\mathbf{A}$ is $(7 - \lambda)^2(4 - \lambda)^3$, so that the eigenvalues of $\mathbf{A}$ are $\lambda_1 = 7$ of algebraic multiplicity $m_1 = 2$ and $\lambda_2 = 4$ of algebraic multiplicity $m_2 = 3$. We consider the eigenvectors associated with these eigenvalues.

($\lambda_1 = 7$). Consider the equation $(\mathbf{A} - 7\mathbf{I})\mathbf{x} = \mathbf{0}$ that we must solve to find eigenvectors $\mathbf{x}$ associated with $\lambda_1 = 7$. The augmented matrix of this system is

$$\left[\begin{array}{ccccc|c} 0 & 0 & 0 & 0 & 0 & 0 \\ 0 & -3 & 1 & 0 & 0 & 0 \\ 0 & 0 & -3 & 0 & 0 & 0 \\ 0 & 0 & 0 & 0 & 0 & 0 \\ 0 & 0 & 0 & 0 & -3 & 0 \end{array}\right],$$

from which it is easy to see that the general solution

$$\mathbf{x} = [x_1 \quad x_2 \quad x_3 \quad x_4 \quad x_5]^T$$

must be $x_5 = 0$, $x_4 = k$ arbitrary, $x_3 = 0$, $x_2 = 0$, and $x_1 = h$ arbitrary. This gives

$$\mathbf{x} = [h \quad 0 \quad 0 \quad k \quad 0]^T = h\mathbf{e}_1 + k\mathbf{e}_4$$

for arbitrary h and k as the set of all eigenvectors associated with $\lambda_1 = 7$. Therefore, $\mathbf{e}_1$ and $\mathbf{e}_4$ are eigenvectors associated with $\lambda_1 = 7$, and all other such eigenvectors are linearly dependent on $\mathbf{e}_1$ and $\mathbf{e}_4$. *We have found a linearly independent set of* ***two*** *eigenvectors associated with this eigenvalue of algebraic multiplicity* ***two***.

($\lambda_2 = 4$). Similarly, consider $(\mathbf{A} - 4\mathbf{I})\mathbf{x} = \mathbf{0}$ and the augmented matrix of this system that we must solve to find all eigenvectors associated with $\lambda_2 = 4$:

$$\left[\begin{array}{ccccc|c} 3 & 0 & 0 & 0 & 0 & 0 \\ 0 & 0 & 1 & 0 & 0 & 0 \\ 0 & 0 & 0 & 0 & 0 & 0 \\ 0 & 0 & 0 & 3 & 0 & 0 \\ 0 & 0 & 0 & 0 & 0 & 0 \end{array}\right].$$

We can again easily see that the general solution must be $x_5 = k$ arbitrary, $x_4 = 0$, $x_3 = 0$, $x_2 = h$ arbitrary, and $x_1 = 0$. This gives

$$\mathbf{x} = [0 \quad h \quad 0 \quad 0 \quad k]^T = h\mathbf{e}_2 + k\mathbf{e}_5$$

as the set of all eigenvectors associated with $\lambda_2 = 4$. Therefore, $\mathbf{e}_2$ and $\mathbf{e}_5$ are eigenvectors associated with $\lambda_2 = 4$, and all other such eigenvectors are linearly dependent on $\mathbf{e}_2$ and $\mathbf{e}_5$. This appears similar to what happened with $\lambda_1 = 7$, but note that in the present case for $\lambda_2 = 4$ *we have found a linearly independent set of* ***only two*** *eigenvectors associated with an eigenvalue of algebraic multiplicity* ***three***. We cannot find a third eigenvector associated with $\lambda_2 = 4$ that is linearly independent of the two already found. Note, however, that although we cannot find a third such *eigenvector*, we can find a third vector—$\mathbf{e}_3$, say—so that it and the two eigenvectors form a basis $\{\mathbf{e}_3, \mathbf{e}_2, \mathbf{e}_5\}$ for an *invariant subspace* of $\mathbf{A}$ of dimension 3, since it is easy to see that

$$\mathbf{A}(\alpha\mathbf{e}_3 + \beta\mathbf{e}_2 + \gamma\mathbf{e}_5) = 4\alpha\mathbf{e}_3 + (4\beta + \alpha)\mathbf{e}_2 + 4\gamma\mathbf{e}_5.$$

The situation illustrated in Example 7.11 is typical: An eigenvalue λ_i of algebraic multiplicity m_i may or may not have a linearly independent set of m_i associated eigenvectors, although it will always have an invariant subspace of dimension m_i. Algebraic multiplicity was called "algebraic" because it came from the algebraic concept of λ_i as a multiple root of the characteristic polynomial; the concept of dimension of the space of associated eigenvectors is geometric, and leads to the notion of *geometric multiplicity* of an eigenvalue.

(7.12) ***Definition.*** The *geometric multiplicity* μ_i of an eigenvalue λ_i of $\mathbf{A}$ is the maximum number of eigenvectors associated with λ_i in a linearly independent set of such eigenvectors; that is, μ_i is the dimension of the subspace (when $\mathbf{0}$ is adjoined) of all eigenvectors associated with λ_i.

In this notation, the matrix $\mathbf{A}$ in Example 7.11 has $m_1 = \mu_1 = 2$ but $m_2 = 3$ while $\mu_2 = 2$. It is possible to create examples in which μ_i equals any integer with $1 \le \mu_i \le m_i$. Not until Chapter 9 will we have the tools to prove:

(7.13) **Preview of Key Theorem 9.11.** The geometric multiplicity μ_i of an eigenvalue λ_i of algebraic multiplicity m_i can be any integer satisfying $1 \le \mu_i \le m_i$; that is, the dimension of the subspace (with $\mathbf{0}$ adjoined) of eigenvectors associated with λ_i may be any number from 1 through the number of times λ_i is counted as a root of the characteristic polynomial. However, there always is an invariant subspace of dimension m_i associated with λ_i, and every vector in $\mathbb{R}^p$ or $\mathbb{C}^p$ (for a $p \times p$ matrix) can be uniquely written as a linear combination of vectors $\mathbf{u}_j$, where each $\mathbf{u}_i$ is in the m_i-dimensional invariant subspace associated with λ_i.

PROBLEMS 7.2

1. Show that the matrix

$$\begin{bmatrix} 2 & -1 & 0 \\ -1 & 2 & -1 \\ 0 & -1 & 2 \end{bmatrix}$$

has eigenvalues 2, $2 \pm \sqrt{2}$, and find the corresponding eigenvectors.

▷ **2.** Find a linearly independent set of two eigenvectors of the matrix

$$\begin{bmatrix} 2 & 2 & -6 \\ 2 & -1 & -3 \\ -2 & -1 & 1 \end{bmatrix}$$

corresponding to the eigenvalue $\lambda = -2$. Find the other eigenvalue and eigenvector.

3. Prove that $\lambda = 0$ is an eigenvalue of $\mathbf{A}$ if and only if $\mathbf{A}$ is singular.

▷ **4.** Find the characteristic polynomial of the general 2×2 matrix.

5. Prove that the eigenvalues of a triangular (upper- or lower-) matrix $\mathbf{T}$ are the entries $\langle \mathbf{T} \rangle_{ii}$, and find the associated eigenvectors.

6. For the matrix $\mathbf{A}$ in the competing-populations model of Example 2.13 with $k = 0.16$, find the characteristic polynomial, eigenvalues, and eigenvectors.

▷ **7.** Find the characteristic polynomial of the transition matrix $\mathbf{A}$ requested in Problem 6 of Section 2.2.

𝔐 **8.** Use MATLAB or similar software to find the eigenvalues and eigenvectors of the matrix in Problem 7.

9. Generalize the oscillating-masses model in Section 2.5 to the case of p masses $m_1, m_2, \ldots, m_p$ and p springs with spring constants $k_1, k_2, \ldots, k_p$ by showing that the equations generalizing (2.28) are again represented by (2.30), where now $\mathbf{K}$ and $\mathbf{M}$ are $p \times p$,

$$\mathbf{M} = \mathbf{diag}(m_1, \ldots, m_p),$$

and $\mathbf{K}$ is *tridiagonal*: $\langle \mathbf{K} \rangle_{ii} = k_i + k_{i+1}$ (defining $k_{p+1} = 0$ for convenience) for $1 \le i \le p$, $\langle \mathbf{K} \rangle_{i,i-1} = \langle \mathbf{K} \rangle_{i-1,i} = -k_i$ for $2 \le i \le p$, and all other $\langle \mathbf{K} \rangle_{ij} = 0$.

𝔐 **10.** By defining $\mathbf{A} = \mathbf{M}^{-1}\mathbf{K}$ and $\lambda = \omega^2$ as in Example 7.1, use MATLAB or similar software to find the eigenvalues λ and corresponding ω if, in Problem 9, $p = 10$, masses $m_i = 5$, and $k_i = 40$ for all i. Also find the oscillation frequencies $\omega/(2\pi)$ for the system of coupled masses.

11. Let $\mathbf{A}$ have the eigenvalue λ_i. Prove:
(a) The transpose of $\mathbf{A}$ has the same eigenvalues as $\mathbf{A}$.
(b) The matrix $k\mathbf{A}$ has the eigenvalue $k\lambda_i$.
(c) The matrix $\mathbf{A}^r$, where r is a positive integer, has the eigenvalue λ_i^r.
(d) If $\mathbf{A}$ is nonsingular, $\mathbf{A}^{-1}$ has the eigenvalue $1/\lambda_i$.
(e) The matrix $\mathbf{A} + k\mathbf{I}$ has the eigenvalue $\lambda_i + k$.

▷ **12.** If $f(x)$ is a polynomial in x and $\mathbf{A}$ is a square matrix, then $f(\mathbf{A})$ denotes the matrix obtained by replacing x^i by $\mathbf{A}^i$ (and x^0 by $\mathbf{I}$) in the formula for $f(x)$. Prove that $f(\lambda)$ is an eigenvalue of $f(\mathbf{A})$ associated with the eigenvector $\mathbf{x}$ if λ is an eigenvalue of $\mathbf{A}$ associated with $\mathbf{x}$.

13. (a) Show that if $\mathbf{A}$ is a hermitian matrix, then $\mathbf{A}$ can be written as $\mathbf{A} = \mathbf{B} + i\mathbf{C}$, where $\mathbf{B}$ and $\mathbf{C}$ are real, $\mathbf{B} = \mathbf{B}^T$, and $\mathbf{C} = -\mathbf{C}^T$.

(b) Describe how to derive the eigensystem of **A** in (a) from that of the real symmetric matrix

$$\begin{bmatrix} \mathbf{B} & -\mathbf{C} \\ \mathbf{C} & \mathbf{B} \end{bmatrix}.$$

14. The eigenvalues of a 10×10 matrix **A** are 3, 2, 2, 2, 2, 6, 6, 6, 12, 12.
(a) Find the characteristic polynomial of **A**.
(b) Find the algebraic multiplicity of each distinct eigenvalue.

15. Find the characteristic polynomial and the algebraic and geometric multiplicity of each distinct eigenvalue of the $p \times p$ identity matrix **I** and of the $p \times p$ zero matrix **0**.

16. Given that $\lambda = -2$ is one eigenvalue of

$$\mathbf{A} = \begin{bmatrix} 2 & 2 & -6 \\ 2 & -1 & -3 \\ -2 & -1 & 1 \end{bmatrix},$$

find the eigenvalues of **A** together with their algebraic and geometric multiplicities, and find as many eigenvectors (in a linearly independent set) as possible.

▷ **17.** Find the eigenvalues of **A** together with their algebraic and geometric multiplicities, and find as many eigenvectors (in a linearly independent set) as possible, where

$$\mathbf{A} = \begin{bmatrix} 2 & -1 & 1 \\ 0 & 2 & 1 \\ 0 & 0 & 3 \end{bmatrix}.$$

18. Do what is asked of you in Problem 17, but for

$$\mathbf{A} = \begin{bmatrix} 7 & 1 & 2 \\ -1 & 7 & 0 \\ 1 & -1 & 6 \end{bmatrix}.$$

7.3 EIGENSYSTEMS, DECOMPOSITIONS, AND TRANSFORMATION REPRESENTATIONS

We began the chapter by motivating the study of eigensystems from four different viewpoints. Section 7.2 discussed the first two of these: the singularity of $\mathbf{A} - \lambda\mathbf{I}$, and eigenvectors and more general invariant subspaces. We now turn to the remaining viewpoints.

Matrix Decompositions

Consider again the matrix

$$\mathbf{A} = \begin{bmatrix} 0.8 & 0.2 & 0.1 \\ 0.1 & 0.7 & 0.3 \\ 0.1 & 0.1 & 0.6 \end{bmatrix}$$

that arose originally in the dairy-competition model of Example 2.6 and whose eigenvalues and eigenvectors were found in Example 7.2 to be $\lambda_1 = 0.5$, $\lambda_2 = 0.6$, and $\lambda_3 = 1.0$ with

$$\mathbf{x}_1 = [1 \quad -2 \quad 1]^T, \quad \mathbf{x}_2 = [1 \quad -1 \quad 0]^T, \quad \mathbf{x}_3 = [9 \quad 7 \quad 4]^T$$

forming a linearly independent set of eigenvectors. Writing the equations $\mathbf{Ax}_i = \lambda_i \mathbf{x}_i$ in the block form

$$\mathbf{A}[\mathbf{x}_1 \quad \mathbf{x}_2 \quad \mathbf{x}_3] = [\lambda_1\mathbf{x}_1 \quad \lambda_2\mathbf{x}_2 \quad \lambda_3\mathbf{x}_3]$$

reveals that, if we define

$$\mathbf{P} = [\mathbf{x}_1 \quad \mathbf{x}_2 \quad \mathbf{x}_3] \quad \text{and} \quad \mathbf{\Lambda} = \mathbf{diag}(\lambda_1, \lambda_2, \lambda_3),$$

the block equation above can be written as $\mathbf{AP} = \mathbf{P\Lambda}$. Since the set of columns of $\mathbf{P}$ is linearly independent, by **Key Theorem 5.50(c)**, $\mathbf{P}$ is nonsingular; we thus can rewrite $\mathbf{AP} = \mathbf{P\Lambda}$ as $\mathbf{A} = \mathbf{P\Lambda P}^{-1}$, which can easily be checked numerically ($\mathbf{P}^{-1}$ was exhibited in Example 7.3, where it was denoted $\mathbf{M}^{-1}$). Thus the existence of a linearly independent set of three eigenvectors allows us to *decompose* the 3×3 matrix $\mathbf{A}$ into a special product $\mathbf{A} = \mathbf{P\Lambda P}^{-1}$ for diagonal $\mathbf{\Lambda}$. This can be done for $p \times p$ $\mathbf{A}$ more generally and is in fact *equivalent* to the existence of a linearly independent set of p eigenvectors of $\mathbf{A}$.

(7.14) **Key Theorem** (decompositions and eigenvectors). A $p \times p$ matrix $\mathbf{A}$ has a linearly independent set of p eigenvectors if and only if there exists a nonsingular matrix $\mathbf{P}$ and a diagonal matrix $\mathbf{\Lambda}$ for which

$$\mathbf{A} = \mathbf{P\Lambda P}^{-1} \quad (\text{and equivalently, } \mathbf{\Lambda} = \mathbf{P}^{-1}\mathbf{AP});$$

these decompositions hold if and only if the columns

$$\mathbf{x}_1, \ldots, \mathbf{x}_p \quad \text{of} \quad \mathbf{P} = [\mathbf{x}_1 \quad \cdots \quad \mathbf{x}_p]$$

form a linearly independent set of eigenvectors of $\mathbf{A}$ associated with the eigenvalues $\lambda_1, \ldots, \lambda_p$, which are the diagonal entries of $\mathbf{\Lambda} = \mathbf{diag}(\lambda_1, \ldots, \lambda_p)$.

PROOF. We proceed as in the material preceding the theorem.

(eigensystem $\Rightarrow$ decomposition) Form a matrix $\mathbf{P} = [\mathbf{x}_1 \quad \cdots \quad \mathbf{x}_p]$ from a linearly independent set of eigenvectors and form a diagonal matrix $\mathbf{\Lambda} =$

diag$(\lambda_1, \ldots, \lambda_p)$ from the associated eigenvalues. $\mathbf{Ax}_i = \lambda_i\mathbf{x}_i$ means that $\mathbf{AP} = \mathbf{P\Lambda}$; since $\mathbf{P}$ is nonsingular by **Key Theorem 5.50(c)**, we have $\mathbf{A} = \mathbf{P\Lambda P}^{-1}$ and $\mathbf{\Lambda} = \mathbf{P}^{-1}\mathbf{AP}$, as asserted.

(decomposition $\Rightarrow$ eigensystem) If $\mathbf{A} = \mathbf{P\Lambda P}^{-1}$ or $\mathbf{\Lambda} = \mathbf{P}^{-1}\mathbf{AP}$ with $\mathbf{\Lambda}$ diagonal, the rules for partitioned-matrix multiplication show that $\mathbf{Ax}_i = \lambda_i\mathbf{x}_i$ where the $\mathbf{x}_i$ are the columns of $\mathbf{P}$ and λ_i the diagonal entries of $\mathbf{\Lambda}$ (both taken in order). Since $\mathbf{P}$ is nonsingular, the $\mathbf{x}_i$ form a linearly independent set and are nonzero; thus they form a linearly independent set of p eigenvectors, as asserted. ■

(7.15) ***Example.*** Consider the simple matrix

$$\mathbf{A} = \begin{bmatrix} 1 & 1 & 0 \\ 0 & 2 & 1 \\ 0 & 0 & 3 \end{bmatrix},$$

whose characteristic polynomial clearly equals

$$(1 - \lambda)(2 - \lambda)(3 - \lambda);$$

thus the (simple) eigenvalues are $\lambda_1 = 1$, $\lambda_2 = 2$, and $\lambda_3 = 3$. The corresponding eigenvectors are easily found; used as columns of $\mathbf{P}$ they form

$$\mathbf{P} = \begin{bmatrix} 1 & 1 & 1 \\ 0 & 1 & 2 \\ 0 & 0 & 2 \end{bmatrix}.$$

A straightforward computation leads to

$$\mathbf{P}^{-1} = \frac{1}{2}\begin{bmatrix} 2 & -2 & 1 \\ 0 & 2 & -2 \\ 0 & 0 & 1 \end{bmatrix}.$$

You can easily verify that

$$\mathbf{P}^{-1}\mathbf{AP} = \begin{bmatrix} 1 & 0 & 0 \\ 0 & 2 & 0 \\ 0 & 0 & 3 \end{bmatrix} = \mathbf{\Lambda} \quad \text{and} \quad \mathbf{P\Lambda P}^{-1} = \mathbf{A}.$$

You should now be able to solve Problems 1 to 6.

Transformation Representations

Recall from Section 6.2 how matrices represent linear transformations with respect to ordered bases; according to **Key Theorem 6.14** the linear transformation $\mathscr{T}$ is represented by the matrix whose ith column is formed from the coefficients used to write $\mathscr{T}(\mathbf{v}_j)$ in terms of the $\mathbf{w}_i$, where the $\mathbf{v}_j$ form the ordered basis for the domain and the $\mathbf{w}_j$ form the ordered basis for the range of $\mathscr{T}$.

Suppose now that $\mathscr{T}$ is defined by the $p \times p$ matrix $\mathbf{A}$ as $\mathscr{T}(\mathbf{v}) = \mathbf{Av}$ and that the ordered basis used both for the domain and range $\mathbb{R}^p$ or $\mathbb{C}^p$ is the standard $\{\mathbf{e}_1; \ldots; \mathbf{e}_p\}$. Then, according to the recipe above, the matrix representing $\mathscr{T}$ has as its ith column the coefficients of $\mathbf{e}_1, \ldots, \mathbf{e}_p$ in the representation of $\mathscr{T}(\mathbf{e}_i) = \mathbf{Ae}_i$, which is just the ith column of $\mathbf{A}$ itself. Thus:

(7.16) $p \times p$ $\mathbf{A}$ itself represents the linear transformation $\mathscr{T}$ defined by $\mathscr{T}(\mathbf{v}) = \mathbf{Av}$ with respect to the ordered basis $\{\mathbf{e}_1; \ldots; \mathbf{e}_p\}$ used both for the domain and range of $\mathscr{T}$.

Recall also from Section 6.2 how the matrix representation of a linear transformation changes when we change ordered bases. According to Theorem 6.17, the matrix $\mathbf{A}$ representing $\mathscr{T}$ with respect to one pair of ordered bases is replaced by $\mathbf{A}' = \mathbf{S}^{-1}\mathbf{AP}$ with respect to another pair, where the matrices $\mathbf{S}$ and $\mathbf{P}$ translate between C- and C'-coordinates in the range and B- and B'-coordinates in the domain, respectively. In the notation of Theorem 6.2, suppose that the original ordered bases B and C are

$$B = C = \{\mathbf{e}_1; \ldots; \mathbf{e}_p\}$$

used for both domain and range, and that the new ordered bases are

$$B' = C' = \{\mathbf{x}_1; \ldots; \mathbf{x}_p\}, \quad \text{say.}$$

Then the translation matrices $\mathbf{S}$ and $\mathbf{P}$ are equal; we use $\mathbf{P}$ to denote the common matrix. What is this matrix $\mathbf{P}$? According to **Key Theorem 5.43**, we obtain the ith column of this matrix as the coefficients used to write the ith vector in $B' = C'$ as a linear combination of those in $B = C$. In our case, that means the ith column of $\mathbf{P}$ comes from writing $\mathbf{x}_i$ as a combination of $\mathbf{e}_1, \ldots, \mathbf{e}_p$—and these coefficients are nothing other than the entries in $\mathbf{x}_i$ itself. That is, the translation matrix $\mathbf{P}$ equals $[\mathbf{x}_1 \quad \mathbf{x}_2 \quad \cdots \quad \mathbf{x}_p]$, and the new matrix $\mathbf{A}'$ representing $\mathscr{T}$ as in (7.16) is $\mathbf{A}' = \mathbf{P}^{-1}\mathbf{AP}$. We summarize what has been proved.

(7.17) ***Theorem.*** Suppose that $\mathbf{A}$ is $p \times p$ and that the linear transformation $\mathscr{T}$ is defined by $\mathscr{T}(\mathbf{v}) = \mathbf{Av}$. Suppose that $B = \{\mathbf{e}_1; \ldots; \mathbf{e}_p\}$ is the standard ordered basis while $B' = \{\mathbf{x}_1; \ldots; \mathbf{x}_p\}$ is another ordered basis. Then:

(a) $\mathbf{A}$ itself represents $\mathscr{T}$ with respect to the ordered basis B used both for the domain and range of $\mathscr{T}$.

(b) With respect to the ordered basis B' used for both the domain and range of $\mathscr{T}$, $\mathscr{T}$ is represented by $\mathbf{A}' = \mathbf{P}^{-1}\mathbf{AP}$, where $\mathbf{P} = [\mathbf{x}_1 \quad \cdots \quad \mathbf{x}_p]$.

Since $\mathbf{A}' = \mathbf{P}^{-1}\mathbf{A}\mathbf{P}$ and $\mathbf{A} = \mathbf{P}\mathbf{A}'\mathbf{P}^{-1}$ are equivalent, Theorem 7.17 combines with **Key Theorem 7.14** to describe our fourth viewpoint on eigensystems.

(7.18) **Key Theorem** (transformation representations and eigenvectors). A $p \times p$ matrix $\mathbf{A}$ has a linearly independent set of p eigenvectors if and only if the linear transformation $\mathscr{T}$ with $\mathscr{T}(\mathbf{v}) = \mathbf{A}\mathbf{v}$ can be represented by a diagonal matrix $\mathbf{\Lambda}$ with respect to some single ordered basis used for both the domain and range of $\mathscr{T}$; this representation holds if and only if the vectors $\mathbf{x}_1, \ldots, \mathbf{x}_p$ in the ordered basis form a linearly independent set of eigenvectors of $\mathbf{A}$ associated with the eigenvalues $\lambda_1, \ldots, \lambda_p$, which are the diagonal entries of $\mathbf{\Lambda}$ in the same order as the $\mathbf{x}_i$.

For an illustration of this theorem, see Example 7.3 and the material leading up to it in Section 7.1.

Eigensystems have now been discussed from four viewpoints, although one of these—the singularity of $\mathbf{A} - \lambda\mathbf{I}$—actually refers only to eigen*values*. Eigen*vectors* have actually been discussed from three viewpoints: eigenvectors and invariant subspaces, decompositions of matrices, and representations of linear transformations. In this section, for example, we have expressed a single idea in each of these three fashions:

1. $p \times p$ $\mathbf{A}$ has a linearly independent set of p eigenvectors.
2. $\mathbf{A}$ can be decomposed as $\mathbf{A} = \mathbf{P}\mathbf{\Lambda}\mathbf{P}^{-1}$—so that $\mathbf{\Lambda} = \mathbf{P}^{-1}\mathbf{A}\mathbf{P}$—with diagonal $\mathbf{\Lambda}$.
3. $\mathscr{T}(\mathbf{v}) = \mathbf{A}\mathbf{v}$ can be represented by a diagonal matrix $\mathbf{\Lambda}$.

Similarly, we will find that these various viewpoints are useful when $\mathbf{A}$ does *not* have such a simply structured set of eigenvectors; depending on the situation, any one of these three viewpoints might be the most help in understanding the nature of a particular matrix or linear transformation. It is important, therefore, to grasp the equivalence of the viewpoints as indicated in (7.19).

(7.19)

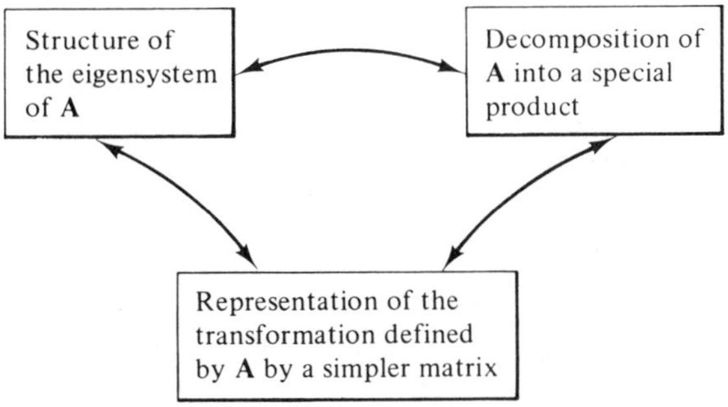

PROBLEMS 7.3

▷ **1.** Find a decomposition $\mathbf{A} = \mathbf{P\Lambda P}^{-1}$ with diagonal $\mathbf{\Lambda}$ for

$$\mathbf{A} = \begin{bmatrix} 2 & 1 \\ 1 & 2 \end{bmatrix}.$$

2. Find a decomposition $\mathbf{A} = \mathbf{P\Lambda P}^{-1}$ with diagonal $\mathbf{\Lambda}$ for

$$\mathbf{A} = \begin{bmatrix} 1 & -1 \\ 2 & 4 \end{bmatrix}.$$

3. Find a decomposition $\mathbf{A} = \mathbf{P\Lambda P}^{-1}$ with diagonal $\mathbf{\Lambda}$ for the matrix $\mathbf{A}$ in the competing-populations model of Example 2.13 when $k = 0.16$.

▷ **4.** Determine whether the matrix

$$\mathbf{A} = \begin{bmatrix} 2 & 1 \\ 0 & 2 \end{bmatrix}$$

can be decomposed as $\mathbf{A} = \mathbf{P\Lambda P}^{-1}$ with diagonal $\mathbf{\Lambda}$. If so, do so; if not, explain why not.

𝔐 **5.** Verify $\mathbf{P}^{-1}\mathbf{AP} = \mathbf{\Lambda}$ and $\mathbf{A} = \mathbf{P\Lambda P}^{-1}$ with appropriate $\mathbf{P}$ and diagonal $\mathbf{\Lambda}$ for the matrix $\mathbf{A}$ in Problem 10 of Section 7.2.

6. For the matrix in each of the following Problems from Section 7.2, determine whether there is a decomposition $\mathbf{A} = \mathbf{P\Lambda P}^{-1}$ with diagonal $\mathbf{\Lambda}$. If so, find one; if not, explain why not.
(a) Problem 1 (b) Problem 2 (c) Problem 16
(d) Problem 17 (e) Problem 18

▷ **7.** For the linear transformation in the competing-populations model in Example 2.13 with $k = 0.16$, find an ordered basis and a corresponding diagonal representation of the transformation.

8. Find an ordered basis and corresponding diagonal matrix representing the linear transformation $\mathscr{T}$ defined by $\mathscr{T}(\mathbf{v}) = \mathbf{Av}$ for the matrix $\mathbf{A}$ in Problem 2.

9. Do as asked in Problem 8, but for the matrix from Problem 1.

10. For the matrix in each of the following problems from Section 7.2, determine whether there is an ordered basis and a corresponding diagonal representation for the transformation $\mathscr{T}(\mathbf{v}) = \mathbf{Av}$. If so, find them; if not, explain why not.
(a) Problem 1 (b) Problem 2 (c) Problem 16
(d) Problem 17 (e) Problem 18

▷ **11.** Do as asked in Problem 10, but for the matrix in Problem 4 above.

12. Consider the linear transformation $\mathscr{T}$ taking $\mathbf{x}$ in $\mathbb{R}^2$ to $\mathscr{T}(\mathbf{x}) = \mathbf{y}$ in $\mathbb{R}^2$, where $y_1 = 2x_1 + x_2$ and $y_2 = x_1 + 2x_2$. Find an ordered basis so that the corresponding representation of $\mathscr{T}$ is diagonal, and find the representation.

Know (will not go over in class)

7.4 SIMILARITY TRANSFORMATIONS; JORDAN FORM

Evidence has been accumulating in prior sections that the relationships

$$\mathbf{A}' = \mathbf{P}^{-1}\mathbf{A}\mathbf{P} \text{ and } \mathbf{A} = \mathbf{P}\mathbf{A}'\mathbf{P}^{-1}$$

between two $p \times p$ matrices $\mathbf{A}$ and $\mathbf{A}'$ are important and useful; see, for instance, Example 6.18, **Key Theorem 7.14**, and Theorem 7.17.

Similarity Transformations

(7.20) ***Definition.*** If there exists a nonsingular matrix $\mathbf{P}$ such that $\mathbf{P}^{-1}\mathbf{A}\mathbf{P} = \mathbf{B}$, then $\mathbf{B}$ is said to be *similar* to $\mathbf{A}$ and to be obtained from $\mathbf{A}$ by means of a *similarity transformation.*

(7.21) ***Example***

$$\mathbf{B} = \begin{bmatrix} 0 & -1 \\ 1 & 1 \end{bmatrix} \quad \text{is similar to} \quad \mathbf{A} = \begin{bmatrix} 2 & -3 \\ 1 & -1 \end{bmatrix}$$

since

$$\mathbf{B} = \mathbf{P}^{-1}\mathbf{A}\mathbf{P} \quad \text{with} \quad \mathbf{P} = \begin{bmatrix} 2 & 1 \\ 1 & 1 \end{bmatrix} \quad \text{so that} \quad \mathbf{P}^{-1} = \begin{bmatrix} 1 & -1 \\ -1 & 2 \end{bmatrix},$$

as you can readily verify.

Similarity has some simple and obvious properties:

(7.22) ***Theorem*** (similarity as an equivalence). Similarity is an *equivalence relation* in the sense that:

(a) $\mathbf{A}$ is similar to itself.
(b) If $\mathbf{B}$ is similar to $\mathbf{A}$, then $\mathbf{A}$ is similar to $\mathbf{B}$.
(c) If $\mathbf{C}$ is similar to $\mathbf{B}$ and $\mathbf{B}$ is similar to $\mathbf{A}$, then $\mathbf{C}$ is similar to $\mathbf{A}$.

PROOF.
(a) Use $\mathbf{P} = \mathbf{I}$ so that $\mathbf{P}^{-1}\mathbf{A}\mathbf{P} = \mathbf{A}$.
(b) If $\mathbf{B} = \mathbf{P}^{-1}\mathbf{A}\mathbf{P}$, then $\mathbf{A} = \mathbf{Q}^{-1}\mathbf{B}\mathbf{Q}$ with $\mathbf{Q} = \mathbf{P}^{-1}$.
(c) If $\mathbf{C} = \mathbf{Q}^{-1}\mathbf{B}\mathbf{Q}$ and $\mathbf{B} = \mathbf{P}^{-1}\mathbf{A}\mathbf{P}$, then $\mathbf{C} = \mathbf{S}^{-1}\mathbf{A}\mathbf{S}$ with $\mathbf{S} = \mathbf{P}\mathbf{Q}$. ■

(7.23) ***Theorem*** (similarity and eigensystems).

(a) Similar matrices have the same characteristic polynomial and the same eigenvalues.
(b) Suppose that $\mathbf{B}$ is similar to $\mathbf{A}$ with $\mathbf{B} = \mathbf{P}^{-1}\mathbf{A}\mathbf{P}$. Then $\mathbf{x}$ is an eigenvector of $\mathbf{A}$ associated with the eigenvalue λ if and only if $\mathbf{P}^{-1}\mathbf{x}$ is an eigenvector of $\mathbf{B}$ associated with the eigenvalue λ.

PROOF

(a) Since $\det \mathbf{P}^{-1} = 1/(\det \mathbf{P})$, we have

$$\begin{aligned}\det(\mathbf{B} - \lambda\mathbf{I}) &= \det\{\mathbf{P}^{-1}(\mathbf{A} - \lambda\mathbf{I})\mathbf{P}\} \\ &= \det(\mathbf{P}^{-1}) \det(\mathbf{A} - \lambda\mathbf{I}) \det(\mathbf{P}) \\ &= \det(\mathbf{A} - \lambda\mathbf{I})\end{aligned}$$

and the characteristic polynomials are identical. Since the eigenvalues are just the roots of the characteristic polynomials, (a) follows.

(b) $\mathbf{x}$ is an eigenvector of $\mathbf{A}$ associated with λ if and only if $\mathbf{A}\mathbf{x} = \lambda\mathbf{x}$—that is, $(\mathbf{P}\mathbf{B}\mathbf{P}^{-1})\mathbf{x} = \lambda\mathbf{x}$, which is equivalent to $\mathbf{B}(\mathbf{P}^{-1}\mathbf{x}) = \lambda(\mathbf{P}^{-1}\mathbf{x})$, as claimed. Note also that $\mathbf{x}$ is zero if and only if $\mathbf{P}^{-1}\mathbf{x}$ is zero. ■

(7.24) ***Example.*** For the similar matrices $\mathbf{A}$ and $\mathbf{B}$ of Example 7.21, we verify that

$$\det(\mathbf{A} - \lambda\mathbf{I}) = \det\left(\begin{bmatrix} 2-\lambda & -3 \\ 1 & -1-\lambda \end{bmatrix}\right) = \lambda^2 - \lambda + 1,$$

while

$$\det(\mathbf{B} - \lambda\mathbf{I}) = \det\left(\begin{bmatrix} 0-\lambda & -1 \\ 1 & 1-\lambda \end{bmatrix}\right) = \lambda^2 - \lambda + 1$$

as well.

(7.25) ***Theorem*** (similarity and powers). Suppose that $\mathbf{B}$ is similar to $\mathbf{A}$ with $\mathbf{B} = \mathbf{P}^{-1}\mathbf{A}\mathbf{P}$. Then:

(a) For each positive integer k, $\mathbf{B}^k$ is similar to $\mathbf{A}^k$ with $\mathbf{B}^k = \mathbf{P}^{-1}\mathbf{A}^k\mathbf{P}$.

(b) $\det \mathbf{B} = \det \mathbf{A}$.

(c) $\mathbf{B}$ is nonsingular if and only if $\mathbf{A}$ is nonsingular.

(d) If $\mathbf{A}$ and $\mathbf{B}$ are nonsingular, then $\mathbf{B}^k$ is similar to $\mathbf{A}^k$ with $\mathbf{B}^k = \mathbf{P}^{-1}\mathbf{A}^k\mathbf{P}$ for negative integers k as well, so in particular $\mathbf{B}^{-1} = \mathbf{P}^{-1}\mathbf{A}^{-1}\mathbf{P}$.

(e) If f is a polynomial with $f(x) = a_0x^m + \cdots + a_m$ and if $f(\mathbf{X})$ for a square matrix $\mathbf{X}$ denotes

$$a_0\mathbf{X}^m + \cdots + a_m\mathbf{I},$$

then $f(\mathbf{B})$ is similar to $f(\mathbf{A})$ with $f(\mathbf{B}) = \mathbf{P}^{-1}f(\mathbf{A})\mathbf{P}$.

PROOF

(a) $\mathbf{B}^k = (\mathbf{P}^{-1}\mathbf{A}\mathbf{P})(\mathbf{P}^{-1}\mathbf{A}\mathbf{P}) \cdots (\mathbf{P}^{-1}\mathbf{A}\mathbf{P})$—with k terms—and the associative law allows the removal of parentheses so that $\mathbf{P}\mathbf{P}^{-1}$ repeatedly gives $\mathbf{I}$ and the product collapses to $\mathbf{P}^{-1}\mathbf{A}^k\mathbf{P}$.

(b) $\det \mathbf{B} = \det(\mathbf{P}^{-1}\mathbf{A}\mathbf{P}) = \det(\mathbf{P}^{-1}) \det(\mathbf{A}) \det(\mathbf{P}) = \det \mathbf{A}$.

(c) This follows from (b) and the fact that a matrix is nonsingular if and only if its determinant is nonzero.

(d) $\mathbf{B}^{-1} = (\mathbf{P}^{-1}\mathbf{A}\mathbf{P})^{-1} = \mathbf{P}^{-1}\mathbf{A}^{-1}(\mathbf{P}^{-1})^{-1} = \mathbf{P}^{-1}\mathbf{A}^{-1}\mathbf{P}$. The result for general negative k follows from applying (a) to $\mathbf{B}^{-1}$ and $\mathbf{A}^{-1}$ with $|k|$.

(e)
$$f(\mathbf{B}) = a_0\mathbf{P}^{-1}\mathbf{A}^m\mathbf{P} + a_1\mathbf{P}^{-1}\mathbf{A}^{m-1}\mathbf{P} + \cdots + a_m\mathbf{P}^{-1}\mathbf{P}$$
$$= \mathbf{P}^{-1}(a_0\mathbf{A}^m + a_1\mathbf{A}^{m-1} + \cdots + a_m\mathbf{I})\mathbf{P} = \mathbf{P}^{-1}f(\mathbf{A})\mathbf{P},$$

as claimed. ■

Note that the relation $\mathbf{B}^k = \mathbf{P}^{-1}\mathbf{A}^k\mathbf{P}$ can be extremely useful if we need to calculate or analyze $\mathbf{B}^k$ and if $\mathbf{A}^k$ is much more easily analyzed; see Example 6.18 on our competing-populations model in this regard.

You should now be able to solve Problems 1 to 8.

Jordan Form

In Chapter 9 we will exploit similarity transformations systematically so as to transform any $p \times p$ matrix into a similar one of special form. Thanks to the relationship in (7.19) between information about decompositions described by similarity transformations and information about the structure of eigensystems, we will be able to use this special similar form to describe the structure of eigensystems generally as previewed in (7.13).

(7.26) **Preview of Key Theorem 9.4.** Every $p \times p$ matrix $\mathbf{A}$ is similar to a matrix $\mathbf{J}$—its *Jordan form*—having the following special properties, among others:
(a) $\mathbf{J}$ is upper-triangular.
(b) The p entries on the main diagonal of $\mathbf{J}$ equal the p eigenvalues of $\mathbf{A}$ (repeated according to their algebraic multiplicities).
(c) Each of the $p - 1$ $(i, i + 1)$-entries on the first super-diagonal of $\mathbf{J}$ equals either 0 or 1; the number of zeros is one less than the maximum number of eigenvectors of $\mathbf{A}$ in a linearly independent set.
(d) Each of the $(i, i + s)$-entries for $s > 1$—that is, above the first superdiagonal—equals zero.

Note. This theorem says only that such a form exists; it does *not* describe a constructive approach for finding $\mathbf{J}$ in practice.

(7.27) ***Example.*** The matrix $\mathbf{J}$ is in Jordan form:

$$\mathbf{J} = \begin{bmatrix} 4 & 1 & 0 & 0 & 0 \\ 0 & 4 & 0 & 0 & 0 \\ 0 & 0 & 7 & 1 & 0 \\ 0 & 0 & 0 & 7 & 0 \\ 0 & 0 & 0 & 0 & 4 \end{bmatrix}.$$

The eigenvalues are $\lambda_1 = 4$ of algebraic multiplicity 3 and geometric multiplicity 2, and $\lambda_2 = 7$ of algebraic multiplicity 2 and geometric multiplicity 1.

Thus there are three eigenvectors in a linearly independent set, as indicated—according to (7.26)—by the presence of two zeros in the first superdiagonal of **J**.

To relate the preview (7.26) to what we know when there is a linearly independent set of p eigenvectors for $p \times p$ **A**, note that (7.26) states that there will be $p - 1$ zeros in that superdiagonal, so that in fact **J** will be diagonal (as we know from **Key Theorem 7.14**).

(7.28) **Partial Restatement of Key Theorem 7.14.** A $p \times p$ matrix **A** has a linearly independent set of p eigenvectors if and only if **A** has a diagonal Jordan form.

Similarity is a powerful theoretical tool for simplifying the study of square matrices; unfortunately, similarity transformations in general often create serious numerical difficulties in practical computations (see Problems 14 and 15). A special *subclass* of similarity transformations is very useful computationally, however, and is thus extremely important in applications for this and other reasons. The next section examines these *unitary* and *orthogonal* similarity transformations in detail.

PROBLEMS 7.4

▷ **1.** **A** is known to be similar to the matrix **B** below; find the eigenvalues of **A**.

$$\mathbf{B} = \begin{bmatrix} 4 & 1 & 2 \\ 0 & -2 & 1 \\ 0 & 0 & 3 \end{bmatrix}.$$

2. Using the similar matrices **A** and **B** from Example 7.21, verify:
(a) Theorem 7.25(a) for $k = 2$
(b) Theorem 7.25(b)
(c) Theorem 7.25(d) for $k = -1$ and $k = -2$

▷ **3.** Suppose that **A** is similar to an upper-triangular matrix **U**; describe the eigenvalues of **A**.

4. Show that two diagonal matrices are similar if and only if the diagonal entries of one are just those of the other but possibly rearranged.

5. Find explicit formulas for $\mathbf{A}^k$ in terms of the positive integer k if:

(a) $\mathbf{A} = \begin{bmatrix} 1 & -1 \\ 2 & 4 \end{bmatrix}$ (b) $\mathbf{A} = \begin{bmatrix} 2 & 1 \\ 1 & 2 \end{bmatrix}$

▷ **6.** Use Theorem 7.25 to analyze the behavior of the population vectors $\mathbf{x}_i$ in the competing-populations model as i tends to infinity in Example 2.13 with:
(a) $k = 0.1$ (b) $k = 0.16$

𝔐 **7.** *Without experimenting with* $\mathbf{A}^k$, *for various* k, use MATLAB or similar software to determine what happens to the powers $\mathbf{A}^k$ as the positive integers k tend to infinity if **A** is the matrix in Problem 10 of Section 7.2.

8. We have stated that the geometric multiplicity μ_i of an eigenvalue λ_i of **A** cannot exceed the algebraic multiplicity m_i. Prove this with the following approach.

(a) Show that you can expand a set of μ_i eigenvectors to obtain a nonsingular matrix **P** so that $\mathbf{A}' = \mathbf{P}^{-1}\mathbf{A}\mathbf{P}$ has the form

$$\mathbf{A}' = \begin{bmatrix} \lambda_i \mathbf{I}_{\mu_i} & \mathbf{B} \\ \mathbf{0} & \mathbf{C} \end{bmatrix}$$

with **C** square.

(b) Prove that $\mu_i \le m_i$ by considering the characteristic polynomials of **A** and of **A**′.

9. Prove that **A** and **B** are similar if and only if they possess a common Jordan form **J**.

▷ **10.** Show that **A** and **B** are not similar if

$$\mathbf{A} = \begin{bmatrix} 1 & -1 \\ 2 & 4 \end{bmatrix} \quad \text{and} \quad \mathbf{B} = \begin{bmatrix} 2 & 1 \\ 1 & 2 \end{bmatrix}.$$

11. Show that **A** and **B** are similar if

$$\mathbf{A} = \begin{bmatrix} 3 & 4 \\ 1 & 3 \end{bmatrix} \quad \text{and} \quad \mathbf{B} = \begin{bmatrix} 4 & 3 \\ 1 & 1 \end{bmatrix}.$$

12. Verify directly the assertions about the eigenvalues and eigenvectors of the matrix **J** in Example 7.27.

▷ **13.** For the Jordan form **J** below, find the eigenvalues and eigenvectors and verify (7.26)(c) in this case.

$$\mathbf{J} = \begin{bmatrix} 3 & 1 & 0 & 0 & 0 & 0 & 0 \\ 0 & 3 & 0 & 0 & 0 & 0 & 0 \\ 0 & 0 & 5 & 1 & 0 & 0 & 0 \\ 0 & 0 & 0 & 5 & 1 & 0 & 0 \\ 0 & 0 & 0 & 0 & 5 & 0 & 0 \\ 0 & 0 & 0 & 0 & 0 & 3 & 0 \\ 0 & 0 & 0 & 0 & 0 & 0 & 3 \end{bmatrix}$$

14. We have asserted that similarity transformations can be difficult computationally. Consider the matrices **A**, **P**, and $\mathbf{B} = \mathbf{P}^{-1}\mathbf{A}\mathbf{P}$ below.

$$\mathbf{A} = \begin{bmatrix} 2-\epsilon & 1-\epsilon \\ -2+2\epsilon & -1+2\epsilon \end{bmatrix}, \qquad \mathbf{P} = \begin{bmatrix} \epsilon & 0 \\ 1-\epsilon & 1 \end{bmatrix}$$

$$\mathbf{B} = \mathbf{P}^{-1}\mathbf{A}\mathbf{P} = \begin{bmatrix} \dfrac{1}{\epsilon} & \dfrac{1-\epsilon}{\epsilon} \\ \dfrac{-1+\epsilon^2}{\epsilon} & \dfrac{-1+\epsilon+\epsilon^2}{\epsilon} \end{bmatrix} \qquad \text{for } \epsilon \text{ nonzero but small.}$$

A (and therefore **B**) is nonsingular, has eigenvalues $\lambda_1 = 1$, $\lambda_2 = \epsilon$, and has determinant ϵ. Suppose that ϵ is of such size that $1 + \epsilon$ and similar expressions are accurately evaluated on a computer, while $-1 + \epsilon^2$ is evaluated as -1. Show that the *computed* version of **B** in this case will be singular, and compare its eigenvalues to those of **A**.

𝔐 **15.** Use MATLAB or similar software to carry out the computations of **A**, $\mathbf{P}^{-1}$, **B**, the eigenvalues of **A** and **B**, the determinants of **A** and **B**, and the ranks of **A** and **B** for the matrices from Problem 14; use ϵ as small as needed—for example, $\epsilon = 10^{-9}$ for the 16-digit decimal arithmetic on many microcomputers with math coprocessors.

7.5 *UNITARY MATRICES AND UNITARY SIMILARITY; SCHUR AND DIAGONAL FORMS*

The preceding section treated similarity: the relationship $\mathbf{B} = \mathbf{P}^{-1}\mathbf{AP}$, where **P** is an arbitrary nonsingular matrix. As remarked there and indicated in that section's Problems 14 and 15, general similarity transformations can be difficult to work with computationally. We introduce a special type of similarity transformation that avoids such difficulty and is extremely important in practical work with matrices.

Unitary Matrices and Transformations

Similarity transformations $\mathbf{B} = \mathbf{P}^{-1}\mathbf{AP}$ were introduced in part because they describe how the matrix **A** representing a linear transformation is changed when the ordered bases change from $\{\mathbf{e}_1; \ldots; \mathbf{e}_p\}$ to $\{\mathbf{x}_1; \ldots; \mathbf{x}_p\}$, where $\mathbf{P} = [\mathbf{x}_1 \quad \mathbf{x}_2 \quad \cdots \quad \mathbf{x}_p]$; see Theorem 7.17 in Section 7.3. We showed much earlier—see Sections 5.8 and 5.9—that *orthogonal* and *orthonormal* bases were especially useful and convenient computationally. We consider the use of such bases in similarity transformations.

Suppose that $\{\mathbf{x}_1, \ldots, \mathbf{x}_p\}$ is an orthonormal set in $\mathbb{R}^p$ (or $\mathbb{C}^p$) with the standard inner product; then

$$\mathbf{x}_i^T\mathbf{x}_i = 1 \quad (\text{or } \mathbf{x}_i^H\mathbf{x}_i = 1)$$

for all i and

$$\mathbf{x}_i^T\mathbf{x}_j = 0 \quad (\text{or } \mathbf{x}_i^H\mathbf{x}_j = 0)$$

for all i and j with $i \neq j$, where as usual we treat the 1×1 matrix $\mathbf{x}^T\mathbf{y}$ (or $\mathbf{x}^H\mathbf{y}$) as equal to its sole entry. If we then form the $p \times p$ matrix

$$\mathbf{P} = [\mathbf{x}_1 \quad \mathbf{x}_2 \quad \cdots \quad \mathbf{x}_p],$$

it follows from partitioned multiplication that

$$\mathbf{P}^T\mathbf{P} = \mathbf{I} \quad (\text{or } \mathbf{P}^H\mathbf{P} = \mathbf{I}).$$

That is, $\mathbf{P}^T = \mathbf{P}^{-1}$ (or $\mathbf{P}^H = \mathbf{P}^{-1}$).

(7.29) ***Definition.*** A $p \times p$ matrix $\mathbf{P}$ for which $\mathbf{P}^{-1} = \mathbf{P}^H$, so that $\mathbf{P}\mathbf{P}^H = \mathbf{P}^H\mathbf{P} = \mathbf{I}$, is said to be *unitary*. An *orthogonal* matrix is a *real* unitary matrix $\mathbf{P}$, so that $\mathbf{P}^{-1} = \mathbf{P}^T$ and $\mathbf{P}\mathbf{P}^T = \mathbf{P}^T\mathbf{P} = \mathbf{I}$.

(7.30) ***Example.*** If

$$\mathbf{P}_1 = \frac{1}{2}\begin{bmatrix} 1+i & -1+i \\ 1+i & 1-i \end{bmatrix} \quad \text{and} \quad \mathbf{P}_2 = \frac{\sqrt{2}}{2}\begin{bmatrix} 1 & 1 \\ 1 & -1 \end{bmatrix},$$

then $\mathbf{P}_1$ and $\mathbf{P}_2$ are both unitary, while $\mathbf{P}_2$ is also orthogonal, as direct computation of $\mathbf{P}_1^H\mathbf{P}_1$, $\mathbf{P}_2^H\mathbf{P}_2$, and $\mathbf{P}_2^T\mathbf{P}_2$ shows. Note that $\mathbf{P}_1$ is not orthogonal (not being real, for example).

Note that every orthogonal matrix is unitary, but not every unitary matrix is orthogonal; *any statement true for all unitary matrices is also true for all orthogonal matrices.* If you do not need to deal with complex numbers, then for simplicity interpret every statement about unitary matrices as being made about (real) orthogonal matrices.

Unitary matrices have remarkable properties.

(7.31) ***Theorem*** (unitary matrices)

(a) A $p \times p$ matrix $\mathbf{P}$ is unitary (or orthogonal) if and only if its columns form an orthonormal set.

(b) A $p \times p$ matrix $\mathbf{P}$ is unitary (or orthogonal) if and only if its rows form an orthonormal set.

(c) If $\mathbf{P}$ is unitary (or orthogonal), then $|\det \mathbf{P}| = 1$ (or $\det \mathbf{P} = \pm 1$).

(d) If $\mathbf{P}$ and $\mathbf{Q}$ are both unitary (or both orthogonal), then so is $\mathbf{PQ}$.

(e) If $\mathbf{P}$ is unitary (or orthogonal) and $(\cdot, \cdot)$ is the standard inner product, then:

1. $(\mathbf{Px}, \mathbf{Py}) = (\mathbf{x}, \mathbf{y})$ for all $\mathbf{x}$ and $\mathbf{y}$, so the angle between $\mathbf{Px}$ and $\mathbf{Py}$ equals that between $\mathbf{x}$ and $\mathbf{y}$.
2. $\|\mathbf{Px}\|_2 = \|\mathbf{x}\|_2$ for all $\mathbf{x}$, so the length of $\mathbf{Px}$ equals that of $\mathbf{x}$.
3. $\|\mathbf{P}\|_2 = 1$.

(f) If λ is an eigenvalue of the unitary (or orthogonal) matrix $\mathbf{P}$, then $|\lambda| = 1$.

(g) If $\mathbf{P}$ is $p \times p$ and unitary (or orthogonal) while $\mathbf{A}$ is $p \times r$ and $\mathbf{B}$ is $r \times p$, then

$$\|\mathbf{PA}\|_2 = \|\mathbf{A}\|_2 \text{ and } \|\mathbf{BP}\|_2 = \|\mathbf{B}\|_2.$$

PROOF

(a) This follows from the material leading up to Definition 7.29.

(b) This follows by applying (a) to $\mathbf{P}^H$ (or to $\mathbf{P}^T$).

(c) Note that

$$1 = \det \mathbf{I} = \det \mathbf{P}^H\mathbf{P} = (\det \mathbf{P}^H)(\det \mathbf{P}) = |\det \mathbf{P}|^2.$$

For orthogonal $\mathbf{P}$, observe that $\det \mathbf{P}$ is real since $\mathbf{P}$ is real.

(d) $(\mathbf{PQ})^H(\mathbf{PQ}) = \mathbf{Q}^H\mathbf{P}^H\mathbf{PQ} = \mathbf{Q}^H\mathbf{IQ} = \mathbf{I}$.

(e) 1. $(\mathbf{Px}, \mathbf{Py}) = (\mathbf{Px})^H(\mathbf{Py}) = \mathbf{x}^H\mathbf{P}^H\mathbf{Py} = \mathbf{x}^H\mathbf{y} = (\mathbf{x}, \mathbf{y})$, and the angle between vectors is defined by inner products.

2. This follows from (e)(1) with $\mathbf{x} = \mathbf{y}$.

3. $\|\mathbf{P}\|_2$ equals the supremum of $\|\mathbf{Px}\|_2/\|\mathbf{x}\|_2$, which equals 1 for all $\mathbf{x} \neq \mathbf{0}$.

(f) If $\mathbf{Px} = \lambda\mathbf{x}$, then from (e)(2),

$$\|\mathbf{x}\|_2 = \|\mathbf{Px}\|_2 = \|\lambda\mathbf{x}\|_2 = |\lambda|\,\|\mathbf{x}\|_2;$$

since $\mathbf{x} \neq \mathbf{0}$, this gives $|\lambda| = 1$.

(g) $\|\mathbf{PA}\|_2$ equals the supremum of $\|\mathbf{PAx}\|_2/\|\mathbf{x}\|_2$, and by (e)(2) this ratio equals $\|\mathbf{Ax}\|_2/\|\mathbf{x}\|_2$, whose supremum equals $\|\mathbf{A}\|_2$. $\|\mathbf{BP}\|_2$ equals the supremum of $\|\mathbf{BPx}\|_2/\|\mathbf{x}\|_2$, and this ratio equals

$$\frac{\|\mathbf{BPx}\|_2}{\|\mathbf{Px}\|_2} = \frac{\|\mathbf{By}\|_2}{\|\mathbf{y}\|_2};$$

since $\mathbf{P}$ is nonsingular, $\mathbf{y}$ assumes all nonzero vectors as $\mathbf{x}$ assumes all nonzero vectors, so the supremum of this last ratio equals $\|\mathbf{B}\|_2$. ∎

Part (e) is responsible for much of the practical importance of unitary matrices computationally. Intuitively interpreted, this part says that multiplication by a unitary matrix does not make small vectors large (or make large vectors small) and does not make "nearly dependent" vectors "more independent" (or make "quite independent" vectors "nearly dependent"). Thus small measuring or modeling errors remain small when multiplied by unitary matrices, for example. And if data appear "quite independent" after being transformed by unitary matrices, then the original data were also "quite independent." Another reason for the practical importance of *real* unitary matrices (that is, orthogonal matrices) is that every such matrix can be computed as a product of special orthogonal matrices ("Householder matrices") that are especially convenient for computational work; we return to this shortly.

Unitary matrices were introduced for the purpose of performing similarity transformations; we now consider this use.

(7.32) ***Definition.*** If there exists a unitary matrix $\mathbf{P}$ such that $\mathbf{P}^H\mathbf{AP} = \mathbf{B}$, then $\mathbf{B}$ is said to be *unitarily similar* to $\mathbf{A}$ and to be obtained from $\mathbf{A}$ by means of a *unitary similarity transformation.* If $\mathbf{P}$ is *real* and hence *orthogonal,* then $\mathbf{B}$ is said to be *orthogonally similar* to $\mathbf{A}$ and to be obtained from $\mathbf{A}$ by means of an *orthogonal similarity transformation.*

Since these new concepts are special cases of general similarity, all our results on similar matrices hold here as well.

(7.33) ***Theorem*** (unitary similarity)

(a) The results of Theorems 7.22, 7.23, and 7.25 hold with "similar" replaced by "unitarily similar" (or "orthogonally similar") and $\mathbf{P}^{-1}$ replaced by $\mathbf{P}^H$ (or $\mathbf{P}^T$).

(b) If $\mathbf{A}$ and $\mathbf{B}$ are unitarily or orthogonally similar, then $\|\mathbf{A}\|_2 = \|\mathbf{B}\|_2$.

PROOF

(a) This is immediate since unitary similarity is a similarity.

(b) If $\mathbf{P}$ is unitary, then so is $\mathbf{P}^H$ since $(\mathbf{P}^H)^H\mathbf{P}^H = \mathbf{P}\mathbf{P}^H = \mathbf{I}$; Theorem 7.31(g) then gives

$$\|\mathbf{B}\|_2 = \|\mathbf{P}^H\mathbf{A}\mathbf{P}\|_2 = \|\mathbf{P}^H\mathbf{A}\|_2 = \|\mathbf{A}\|_2. \quad \blacksquare$$

You should now be able to solve Problems 1 to 9.

Householder Reflections

As with any matrix, a unitary matrix $\mathbf{P}$ can be used to define a linear transformation $\mathcal{T}$ via $\mathcal{T}(\mathbf{x}) = \mathbf{P}\mathbf{x}$. Theorem 7.31(e) states that $\mathcal{T}$ preserves lengths and angles. We now consider a geometrically simple transformation in the plane that preserves lengths and angles and show that it is, in fact, described by a *real* unitary matrix (that is, an orthogonal matrix) of an especially important type.

(7.34) ***Example.*** Consider the fixed straight line ℓ perpendicular to a given vector $\mathbf{w}$ in the plane. Suppose that we transform each vector $\mathbf{x}$ in the plane by *reflection about* ℓ: That is, $\mathbf{x}$ is transformed to $\mathcal{T}(\mathbf{x}) = \mathbf{y}$, the mirror image of

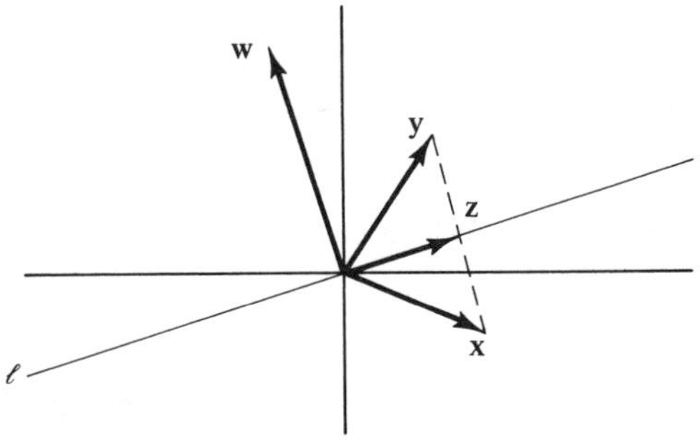

$\mathbf{x}$ on the opposite side of ℓ. (See the diagram.) The vector from $\mathbf{x}$ to $\mathbf{y}$—that is, $\mathbf{y} - \mathbf{x}$—is clearly perpendicular to ℓ and so must be a multiple of $\mathbf{w}$:

$$\mathbf{y} - \mathbf{x} = \alpha\mathbf{w}.$$

The point $\mathbf{z}$ halfway along this vector must lie on ℓ and thus be perpendicular to $\mathbf{w}$:

$$\mathbf{w}^T(\mathbf{x} + 0.5\alpha\mathbf{w}) = 0,$$

which gives $\alpha = -2\mathbf{w}^T\mathbf{x}/\mathbf{w}^T\mathbf{w}$. Therefore,

$$\begin{aligned}\mathbf{y} &= \mathbf{x} + \alpha\mathbf{w} = \mathbf{x} - (2\mathbf{w}^T\mathbf{x}/\mathbf{w}^T\mathbf{w})\mathbf{w} = \mathbf{x} - (2/\mathbf{w}^T\mathbf{w})\mathbf{w}\mathbf{w}^T\mathbf{x} \\ &= \{\mathbf{I} - (2/\mathbf{w}^T\mathbf{w})\mathbf{w}\mathbf{w}^T\}\,\mathbf{x} = \mathbf{H}_\mathbf{w}\mathbf{x}\end{aligned}$$

where

$$\mathbf{H}_\mathbf{w} = \mathbf{I} - (2/\mathbf{w}^T\mathbf{w})\mathbf{w}\mathbf{w}^T$$

is a 2×2 matrix. That is, the transformation $\mathcal{T}$ is, in fact, the linear transformation $\mathcal{T}(\mathbf{x}) = \mathbf{H}_\mathbf{w}\mathbf{x}$ defined by multiplication by the 2×2 matrix $\mathbf{H}_\mathbf{w}$. If, for example, $\mathbf{w} = [1 \quad 2]^T$, then $\mathbf{H}_\mathbf{w} = \mathbf{I} - \frac{2}{5}\mathbf{w}\mathbf{w}^T$, so

$$\mathbf{H}_\mathbf{w} = \begin{bmatrix} 1 & 0 \\ 0 & 1 \end{bmatrix} - \frac{2}{5}\begin{bmatrix} 1 & 2 \\ 2 & 4 \end{bmatrix} = \begin{bmatrix} \frac{3}{5} & -\frac{4}{5} \\ -\frac{4}{5} & -\frac{3}{5} \end{bmatrix}.$$

Note that $\mathbf{H}_\mathbf{w}$ is symmetric and orthogonal.

The same sort of transformation makes sense in higher-dimensional spaces; the reflection is in the direction of some vector $\mathbf{w}$, reflecting about the set of vectors orthogonal to $\mathbf{w}$. Essentially the same argument as in Example 7.34 shows that such elementary reflections are still represented by matrices $\mathbf{H}_\mathbf{w}$.

(7.35) ***Definition.*** For nonzero $\mathbf{w}$ in $\mathbb{R}^p$, the $p \times p$ *Householder matrix* $\mathbf{H}_\mathbf{w}$ is defined as

$$\mathbf{H}_\mathbf{w} = \mathbf{I}_p - \left(\frac{2}{\mathbf{w}^T\mathbf{w}}\right)\mathbf{w}\mathbf{w}^T.$$

In the special case of Example 7.34, $\mathbf{H}_\mathbf{w}$ was symmetric and orthogonal. This is true generally, since

$$\begin{aligned}\mathbf{H}_\mathbf{w}^T\mathbf{H}_\mathbf{w} &= \{\mathbf{I} - (2/\mathbf{w}^T\mathbf{w})\mathbf{w}\mathbf{w}^T\}^T\{\mathbf{I} - (2/\mathbf{w}^T\mathbf{w})\mathbf{w}\mathbf{w}^T\} \\ &= \{\mathbf{I} - (2/\mathbf{w}^T\mathbf{w})\mathbf{w}\mathbf{w}^T\}\{\mathbf{I} - (2/\mathbf{w}^T\mathbf{w})\mathbf{w}\mathbf{w}^T\} \\ &= \mathbf{I} - (4/\mathbf{w}^T\mathbf{w})\mathbf{w}\mathbf{w}^T + (2/\mathbf{w}^T\mathbf{w})(2/\mathbf{w}^T\mathbf{w})\mathbf{w}\mathbf{w}^T\mathbf{w}\mathbf{w}^T \\ &= \mathbf{I} - (4/\mathbf{w}^T\mathbf{w})\mathbf{w}\mathbf{w}^T + (4/\mathbf{w}^T\mathbf{w})\mathbf{w}\mathbf{w}^T \\ &= \mathbf{I}.\end{aligned}$$

We summarize.

(7.36) ***Theorem*** (Householder matrices). Let $\mathbf{H}_\mathbf{w}$ be the Householder matrix defined by a nonzero $\mathbf{w}$ in $\mathbb{R}^p$. Then:
(a) $\mathbf{H}_\mathbf{w}$ is symmetric.
(b) $\mathbf{H}_\mathbf{w}$ is orthogonal.

(c) For each $\mathbf{x}$, $\mathbf{H_w x}$ equals the reflection of $\mathbf{x}$ about the subspace of all $\mathbf{v}$ orthogonal to $\mathbf{w}$.
(d) $\det \mathbf{H_w} = -1$.
(e) For any nonzero $\mathbf{x}$ and nonzero $\mathbf{y}$ in $\mathbb{R}^p$ with $\mathbf{x} \neq \mathbf{y}$, there exists an $\mathbf{H_w}$ such that $\mathbf{H_w x} = \alpha \mathbf{y}$ for some real number α. More precisely:
1. Unless $\mathbf{x}$ equals a positive multiple of $\mathbf{y}$, $\mathbf{w}$ can be taken as

$$\mathbf{w} = \mathbf{x} - \left(\frac{\|\mathbf{x}\|_2}{\|\mathbf{y}\|_2}\right)\mathbf{y} \quad \text{and then} \quad \mathbf{H_w x} = \left(\frac{\|\mathbf{x}\|_2}{\|\mathbf{y}\|_2}\right)\mathbf{y}.$$

2. Unless $\mathbf{x}$ equals a negative multiple of $\mathbf{y}$, $\mathbf{w}$ can be taken as

$$\mathbf{w} = \mathbf{x} + \left(\frac{\|\mathbf{x}\|_2}{\|\mathbf{y}\|_2}\right)\mathbf{y} \quad \text{and then} \quad \mathbf{H_w x} = -\left(\frac{\|\mathbf{x}\|_2}{\|\mathbf{y}\|_2}\right)\mathbf{y}.$$

PROOF

(a), (b), and (c) follow from the preceding discussion.
(d) Let $\mathbf{A}$ be a nonsingular $p \times p$ matrix whose first column equals $\mathbf{w}$ and whose remaining columns are orthogonal to $\mathbf{w}$; we can find such columns by choosing them as a basis for the subspace of vectors orthogonal to $\mathbf{w}$. It is easy to see that $\mathbf{H_w A}$ has $-\mathbf{w}$ as its first column and that its remaining columns are identical with those of $\mathbf{A}$, in order. Therefore,

$$\det(\mathbf{H_w A}) = -\det \mathbf{A};$$

but it also equals $\det(\mathbf{H_w})\det(\mathbf{A})$. Division by $\det \mathbf{A}$ gives $\det \mathbf{H_w} = -1$.
(e) follows by direct computation of $\mathbf{H_w x}$ using the given vectors $\mathbf{w}$; the conditions on the relationship of $\mathbf{x}$ and $\mathbf{y}$ guarantee $\mathbf{w} \neq \mathbf{0}$. ■

Theorem 7.36(e) is heavily used in the applications of Householder matrices, especially to transform a given $\mathbf{x}$ to a multiple of $\mathbf{e}_1$: $\mathbf{H_w x} = \|\mathbf{x}\|_2 \mathbf{e}_1$. The Householder matrix in effect "zeros out" the entries below the first in $\mathbf{x}$.

(7.37) ***Example.*** Consider the vector $\mathbf{x} = [3 \quad 4]^T$ and the problem of transforming it by a Householder transformation into a multiple of $\mathbf{e}_1$. By Theorem 7.36(e) with $\mathbf{y} = \mathbf{e}_1$, we can for instance take

$$\mathbf{w} = \mathbf{x} - \|\mathbf{x}\|_2 \mathbf{e}_1 = \mathbf{x} - 5\mathbf{e}_1 = [-2 \quad 4]^T.$$

This gives

$$\mathbf{H_w} = \mathbf{I}_2 - \tfrac{2}{20}\mathbf{w}\mathbf{w}^T = \begin{bmatrix} 0.6 & 0.8 \\ 0.8 & -0.6 \end{bmatrix} \quad \text{and} \quad \mathbf{H_w x} = 5\mathbf{e}_1.$$

Problem 13 in Section 5.9 described how to use the "zeroing out" power of Householder matrices to transform a matrix $\mathbf{A}$ into an upper-triangular matrix by premultiplying **A** by a sequence of Householder matrices; a slight modification in that process (Problem 18 in this Section) based on Theorem 7.36(e)(1) gives part (a) of the following result, which is of great computational importance.

(7.38) **Key Theorem** (Householder matrices and QR)

(a) Suppose that $\mathbf{A}$ is $p \times q$ and real. A sequence $\mathbf{H}_1, \ldots, \mathbf{H}_q$ of at most q Householder matrices can be easily computed (as in Problem 13 of Section 5.9) so that

$$\mathbf{H}_q\mathbf{H}_{q-1}\cdots\mathbf{H}_1\mathbf{A} = \mathbf{R},$$

where $\mathbf{R}$ is upper-triangular and has nonnegative entries on the main diagonal; equivalently,

$$\mathbf{A} = \mathbf{QR},$$

where $\mathbf{Q} = \mathbf{H}_1\mathbf{H}_2\cdots\mathbf{H}_q$ is $p \times p$ and orthogonal.

(b) Every $p \times p$ orthogonal matrix $\mathbf{Q}$ can be written as the product $\mathbf{Q} = \mathbf{H}_1\mathbf{H}_2\cdots\mathbf{H}_p$ of at most p Householder matrices.

PROOF

(a) Problem 13 of Section 5.9 and Problem 18.

(b) Apply (a) to $\mathbf{Q}$, so that $\mathbf{H}_p\cdots\mathbf{H}_1\mathbf{Q} = \mathbf{R}$; $\mathbf{R}$ is orthogonal (since its factors are orthogonal), it is upper-triangular, and its main-diagonal entries are nonnegative. The first column of $\mathbf{R}$ is some multiple of $\mathbf{e}_1$, and this column is orthogonal to the remaining columns of $\mathbf{R}$; therefore, the first row of $\mathbf{R}$ has zeros in all but the first entry. Arguing similarly with the second column/row, and so on, we see that $\mathbf{R}$ is diagonal. Since $\mathbf{R}^T\mathbf{R} = \mathbf{I}$, the diagonal entries are $+1$ or -1; since they are nonnegative, they equal $+1$—that is, $\mathbf{R} = \mathbf{I}$. Thus $\mathbf{Q} = \mathbf{H}_1\mathbf{H}_2\cdots\mathbf{H}_p$. ■

The first part of **Key Theorem 7.38** is crucial to state-of-the-art computational procedures in such areas as least-squares computations and eigensystem computations. The second part tells us that (a sequence of) Householder matrices can accomplish anything that general orthogonal matrices can, and thus helps explain why they have become indispensible in matrix computations.

You should now be able to solve Problems 1 to 20.

Elementary Rotations

If you try to visualize a transformation $\mathscr{T}(\mathbf{v}) = \mathbf{Pv}$ defined by a unitary or orthogonal matrix (so that it preserves length and angle), you are probably more likely to picture some sort of rotation of vectors than to picture a reflection (as described by a Householder matrix). Although—unlike the case with reflections—not every orthogonal matrix can be written as a product of especially simple rotations (Problem 25), these so-called *elementary rotations* are useful in applications.

(7.39) ***Example.*** In $\mathbb{R}^2$, consider the transformation $\mathscr{T}$ that transforms each vector $\mathbf{x}$ by rotating it through an angle θ measured in the direction from the x_1-axis toward the x_2-axis (that is, counterclockwise) so as to obtain $\mathbf{x}' = \mathscr{T}(\mathbf{x})$.

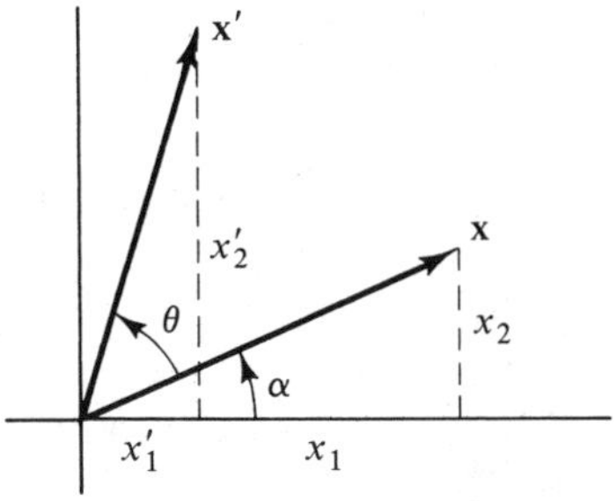

$$\|\mathbf{x}'\|_2 = \|\mathbf{x}\|_2$$

$$\begin{aligned} x_1' &= \|\mathbf{x}'\|_2 \cos(\alpha + \theta) \\ &= \|\mathbf{x}\|_2 \cos\alpha \cos\theta \\ &\quad - \|\mathbf{x}\|_2 \sin\alpha \sin\theta \\ &= x_1 \cos\theta - x_2 \sin\theta \end{aligned}$$

$$x_2' = x_1 \sin\theta + x_2 \cos\theta$$

A little trigonometry produces the formulas above for $\mathbf{x}'$ in terms of $\mathbf{x}$. Because the formulas are linear, we can write them in matrix notation as $\mathbf{x}' = \mathscr{T}(\mathbf{x}) = \mathbf{P}\mathbf{x}$, where

$$\mathbf{P} = \begin{bmatrix} \cos\theta & -\sin\theta \\ \sin\theta & \cos\theta \end{bmatrix}.$$

It is simple to verify that $\mathbf{P}^T\mathbf{P} = \mathbf{I}$, so $\mathbf{P}$ is orthogonal.

We can consider the same sort of simple planar rotations and the matrices that generate them in higher dimensions.

(7.40) ***Definition.*** The $p \times p$ *elementary rotation matrix* $\mathbf{R} = \mathbf{R}_{kn}(\theta)$ is defined by:

(a) $\langle \mathbf{R} \rangle_{kk} = \langle \mathbf{R} \rangle_{nn} = \cos\theta$.
(b) $\langle \mathbf{R} \rangle_{kn} = -\sin\theta$ and $\langle \mathbf{R} \rangle_{nk} = \sin\theta$.
(c) $\langle \mathbf{R} \rangle_{ii} = 1$ for $i \neq k$ and $i \neq n$.
(d) $\langle \mathbf{R} \rangle_{ij} = 0$ for all entries other than those defined in (a), (b), and (c).

The properties of elementary rotation matrices are straightforward to derive.

(7.41) ***Theorem*** (rotation matrices). Let $\mathbf{R} = \mathbf{R}_{kn}(\theta)$ be a $p \times p$ elementary rotation matrix. Then:

(a) $\mathbf{R}$ is orthogonal.
(b) For each $\mathbf{x}$, $\mathbf{R}\mathbf{x}$ equals the vector that results from rotating $\mathbf{x}$ through an angle θ in the x_k-x_n-plane in the direction from the x_k-axis toward the x_n-axis.
(c) $\det \mathbf{R} = 1$.

PROOF. Problem 40. ■

Schur, Diagonal, and Hessenberg Forms

In Section 7.4, after discussing similarity transformations we previewed the result from Chapter 9 on the special Jordan form to which every $p \times p$ matrix $\mathbf{A}$ can be reduced by a similarity transformation; this special form provides information on the structure of the eigensystem of $\mathbf{A}$. We paused to discuss unitary and orthogonal similarities and matrices because of computational problems with general similarities that these special similarities avoid. We now turn to previews of special forms that can be produced using unitary or orthogonal similarities.

(7.42) **Preview of Theorem 8.2 and Key Theorem 8.6**
(a) Every $p \times p$ matrix $\mathbf{A}$ is unitarily similar to a matrix $\mathbf{T}$ in *Schur form* defined as follows: $\mathbf{T}$ is upper-triangular and the eigenvalues of $\mathbf{A}$, repeated according to their algebraic multiplicities, are the main-diagonal entries of $\mathbf{T}$. If $\mathbf{A}$ and its eigenvalues are real, then $\mathbf{A}$ is *orthogonally* similar to such a $\mathbf{T}$.
(b) The Schur form of every *normal* matrix $\mathbf{A}$ (a matrix satisfying $\mathbf{A}^H\mathbf{A} = \mathbf{A}\mathbf{A}^H$—hermitian and real symmetric matrices, for example) is diagonal.

Part (b) of this result gives information on the eigensystem of a normal matrix.

(7.43) **Preview of Key Theorem 8.8.** A $p \times p$ matrix is normal (for example, real symmetrix or hermitian) if and only if $\mathbf{A}$ has a linearly independent set of p eigenvectors that may be chosen to form an orthonormal set.

Result (7.43) says that $p \times p$ normal matrices $\mathbf{A}$ are precisely those (1) that have a full set of p eigenvectors forming an easy-to-use basis (that is, an orthonormal one), (2) whose transformation $\mathscr{T}(\mathbf{v}) = \mathbf{A}\mathbf{v}$ can be represented by a simple diagonal matrix with respect to some orthonormal basis, and (3) that can be decomposed into a simple product $\mathbf{A} = \mathbf{P}\mathbf{\Lambda}\mathbf{P}^H$ with diagonal $\mathbf{\Lambda}$ and unitary $\mathbf{P}$.

Note. As in the preview (7.26), this only tells us that such forms *exist*; it does *not* describe a constructive approach for finding them in practice.

There is one computationally useful "simplified form" of a matrix $\mathbf{A}$ that can be produced directly by using a sequence of Householder matrices.

(7.44) ***Theorem*** (Hessenberg form). Suppose that $\mathbf{A}$ is $p \times p$ and real. There is an easily computible sequence of at most $p - 2$ Householder matrices $\mathbf{H}_1, \ldots,$ $\mathbf{H}_{p-2}$ implementing an orthogonal similarity

$$(\mathbf{H}_{p-2} \cdots \mathbf{H}_2\mathbf{H}_1)\mathbf{A}(\mathbf{H}_1\mathbf{H}_2 \cdots \mathbf{H}_{p-2}) = \mathbf{H}$$

such that the transformed matrix $\mathbf{H}$ is in (upper) *Hessenberg form* defined as follows: $\mathbf{H}$ is almost upper-triangular in that only its first subdiagonal can contain nonzero entries, that is $\langle \mathbf{H} \rangle_{ij} = 0$ for $i \geq j + 2$.

PROOF. Problems 35 and 36. ■

PROBLEMS 7.5

1. Show that the following matrix is orthogonal, and find its determinant.

$$\begin{bmatrix} \frac{1}{3} & \frac{2}{3} & -\frac{2}{3} \\ \frac{2}{3} & -\frac{2}{3} & -\frac{1}{3} \\ \frac{2}{3} & \frac{1}{3} & \frac{2}{3} \end{bmatrix}$$

▷ 2. Find numbers a, b, and c so that $\mathbf{P}$ is orthogonal, where

$$\mathbf{P} = \begin{bmatrix} \frac{\sqrt{6}}{3} & -\frac{\sqrt{3}}{3} & a \\ \frac{\sqrt{6}}{6} & \frac{\sqrt{3}}{3} & b \\ \frac{\sqrt{6}}{6} & \frac{\sqrt{3}}{3} & c \end{bmatrix}.$$

3. Suppose that $\mathbf{P}$ is real. Prove that $\mathbf{P}$ is orthogonal if and only if $\imath\mathbf{P}$ is unitary.
4. Suppose that $\mathbf{R}$ and $\mathbf{S}$ are real. Prove that $\mathbf{P} = \mathbf{R} + \imath\mathbf{S}$ is unitary if and only if $\mathbf{R}^T\mathbf{R} + \mathbf{S}^T\mathbf{S} = \mathbf{I}$ and $\mathbf{R}^T\mathbf{S} = -\mathbf{S}^T\mathbf{R}$.

▷ 5. Show that $\mathbf{P}$ is unitary if and only if $\mathbf{P}^H$ is unitary.
6. Show that $\mathbf{P}$ is unitary if and only if $\mathbf{P}^T$ is unitary.
7. Show that $\mathbf{P}$ is orthogonal if and only if $\mathbf{P}^T$ is orthogonal.

▷ 8. Suppose that $\mathbf{P}$ is $p \times p$ and unitary. Show that $1/\sqrt{p} \leq \|\mathbf{P}\| \leq \sqrt{p}$ for both the 1-norm and the ∞-norm.
9. Find the eigensystem of the orthogonal matrix

$$\mathbf{A} = \begin{bmatrix} \cos\theta & -\sin\theta \\ \sin\theta & \cos\theta \end{bmatrix}.$$

10. Find the 2×2 matrix that reflects vectors in $\mathbb{R}^2$ across the line $x_1 = x_2$.
11. If $\mathbf{w}$ is complex, show that $\mathbf{I} - (2/\mathbf{w}^H\mathbf{w})\mathbf{w}\mathbf{w}^H$ is unitary.

▷ 12. Find a Householder matrix $\mathbf{H}_\mathbf{w}$ so that $\mathbf{H}_\mathbf{w}\mathbf{e}_1 = \mathbf{e}_2$ in $\mathbb{R}^2$.
13. Find a Householder matrix $\mathbf{H}_\mathbf{w}$ so that $\mathbf{H}_\mathbf{w}[1 \quad 2]^T$ is a multiple of $[2 \quad 1]^T$.

▷ 14. The work of multiplying a $p \times p$ matrix $\mathbf{P}$ times a $p \times p$ matrix $\mathbf{A}$ is usually about p^3 multiplications and p^3 additions. Show that $\mathbf{PA}$ can be computed at a cost of at most $2p^2 + p + 1$ multiplications and $2p^2 - 1$ additions if $\mathbf{P}$ is a Householder matrix $\mathbf{H}_\mathbf{w}$ and $\mathbf{w}$ is available.

15. To be able to access the p^2 entries of a $p \times p$ matrix $\mathbf{A}$ on demand usually requires p^2 storage locations. Show that only p locations are needed if $\mathbf{A}$ is a Householder matrix $\mathbf{H_w}$ and $\mathbf{w}$ is available.

16. Prove Theorem 7.36(e).

▷ **17.** Show that if the first r entries of $\mathbf{w}$ equal zero, then the first r rows of $\mathbf{H_w A}$ equal the first r rows of $\mathbf{A}$ in order.

18. Modify Problem 13 of Section 5.9 so as to obtain *nonnegative* entries on the main diagonal of the matrices $\mathbf{A}_i$ by using Theorem 7.36(e)(1) with $\mathbf{y} = \mathbf{e}_1$.

19. As in **Key Theorem 7.38**, premultiply the matrix below by a sequence of Householder matrices to produce an upper-triangular matrix $\mathbf{R}$.

$$\mathbf{A} = \begin{bmatrix} 1 & 2 & 4 \\ 0 & 1 & 1 \\ 1 & 4 & 6 \end{bmatrix}$$

▷ **20.** Do as in Problem 19 but with

$$\mathbf{A} = \begin{bmatrix} 1 & 2 & 3 \\ 0 & 1 & 1 \\ 1 & 4 & 6 \end{bmatrix}.$$

21. Find the 3×3 rotation matrix $\mathbf{R}_{13}(45^\circ)$.

▷ **22.** Find the 3×3 rotation matrix $\mathbf{R}_{21}(90^\circ)$.

23. Find the single 3×3 orthogonal matrix $\mathbf{P}$ so that multiplication by $\mathbf{P}$ is equivalent to *first* rotating the x_1-axis 45° toward the x_3-axis (see Problem 21) and *then* rotating the new x_2-axis 90° toward the new x_1-axis (see Problem 22).

24. Write the general 2×2 rotation $\mathbf{R}_{12}(\theta)$ as a product of Householder matrices.

▷ **25.** Use $\det \mathbf{R}_{kn}(\theta) = 1$ to show that not every orthogonal matrix equals a product of elementary rotation matrices.

26. Show that the Schur form of a matrix is not unique by considering $\mathbf{A}$ and $\mathbf{P}^H\mathbf{AP}$, where

$$\mathbf{A} = \begin{bmatrix} 1 & 0 & 3 \\ 0 & 1 & 4 \\ 0 & 0 & 2 \end{bmatrix} \quad \text{and} \quad \mathbf{P} = \begin{bmatrix} 0 & -1 & 0 \\ 1 & 0 & 0 \\ 0 & 0 & 1 \end{bmatrix}.$$

27. Let the $p \times p$ matrix $\mathbf{A}$ define a linear transformation $\mathcal{T}$ from $\mathbb{C}^p$ to $\mathbb{C}^p$ by $\mathcal{T}(\mathbf{v}) = \mathbf{Av}$. Show that there is an orthonormal basis for $\mathbb{C}^p$ with respect to which $\mathcal{T}$ is represented by a Schur form of $\mathbf{A}$.

▷ **28.** Show that every real symmetric matrix is normal.

29. Show that every hermitian matrix is normal.

30. Show that every unitary (or orthogonal) matrix is normal.

▷ **31.** Show that if $\mathbf{A}$ is unitarily similar to a diagonal matrix, then $\mathbf{A}$ is normal.

32. Verify that each of the following matrices is normal and show that its eigenvectors form an orthogonal set.

(a) $\begin{bmatrix} 2 & 1 \\ 1 & 2 \end{bmatrix}$ (b) $\begin{bmatrix} 2 & -1 \\ 1 & 2 \end{bmatrix}$ (c) $\begin{bmatrix} 2 & 1+i \\ 1-i & 3 \end{bmatrix}$

33. Show that if **A** is real and normal and has real eigenvalues, then **A** is symmetric.

34. Show that the matrix

$$\begin{bmatrix} 2i & 1 \\ 1 & 0 \end{bmatrix}$$

has only one eigenvector in a linearly independent set, showing that *nonreal* symmetric matrices need not have a full set of eigenvectors as real symmetric matrices do.

▷ **35.** Suppose that **A** is real and $p \times p$.

(a) Show that you can pick $p \times 1$ $\mathbf{w}$ with $\langle \mathbf{w} \rangle_1 = 0$ so that the entries in the first column of $\mathbf{H_w A}$ equal zero in the third row and below (but not necessarily in the first two rows).

(b) For this choice of $\mathbf{w}$, show that $\mathbf{H_w A H_w}$ has the same structure as asserted for $\mathbf{H_w A}$.

36. Use Problem 35 to show that you can treat one column of **A** at a time and obtain a Hessenberg form for **A** as asserted in Theorem 7.44.

▷ **37.** Use the method in Problem 36 to reduce to Hessenberg form

$$\begin{bmatrix} 1 & 1 & -1 \\ 0 & 1 & 3 \\ 2 & 2 & 1 \end{bmatrix}.$$

𝔐 **38.** Use MATLAB or similar software to find a Hessenberg form of the transition matrix **A** in the dairy-market model of Example 2.6.

𝔐 **39.** Use MATLAB or similar software to find a Schur form of the matrix in Problem 38.

40. Prove Theorem 7.41.

7.6 *COMPUTER SOFTWARE FOR FINDING EIGENSYSTEMS*

Unlike the situation with solving linear equations, we have not discussed the practical computation of eigenvalues and eigenvectors; the topic is much too detailed and specialized for this book to treat in detail—see Golub and Van Loan (46), for example. Although the problem of computing eigensystems is appreciably more complicated than that of solving linear systems, over the years experts have developed excellent general-purpose computer programs for the efficient and accurate determination of eigensystems. As with linear equations, such programs are generally available either free or at low cost.

Method: the QR Algorithm

The algorithm of choice—analogous to Gauss elimination for linear equations—for computing eigensystems is the QR algorithm for real matrices and its extension, the QZ algorithm, for complex matrices; we indicate the outline of the basic QR method (leaving out the important refinements that make it so effective) in order to communicate the idea behind the method.

(7.45) ***Outline of the Basic QR Algorithm.***

1. Given the real $p \times p$ matrix $\mathbf{A}$, use $p - 2$ Householder matrices to compute the orthogonally similar matrix $\mathbf{H}$ in Hessenberg form as in Theorem 7.44. Define $\mathbf{A}_1 = \mathbf{H}$ and set $i = 1$.
2. Use p Householder matrices as in **Key Theorem 7.38** to compute a QR-decomposition of $\mathbf{A}_i$ as $\mathbf{A}_i = \mathbf{Q}_i\mathbf{R}_i$, where $\mathbf{Q}_i$ is orthogonal and $\mathbf{R}_i$ is upper-triangular.
3. Define $\mathbf{A}_{i+1} = \mathbf{R}_i\mathbf{Q}_i$. If $\mathbf{A}_{i+1}$ has converged to a Schur form $\mathbf{T}$ as in (7.42), then stop; otherwise, increase i by 1 and return to step 2.

On first glance, this method seems mysterious: Why define $\mathbf{A}_{i+1}$ as $\mathbf{R}_i\mathbf{Q}_i$? Note that

$$\mathbf{R}_i\mathbf{Q}_i = \mathbf{Q}_i^T\mathbf{Q}_i\mathbf{R}_i\mathbf{Q}_i = \mathbf{Q}_i^T\mathbf{A}_i\mathbf{Q}_i,$$

so $\mathbf{A}_{i+1}$ is in fact orthogonally similar to $\mathbf{A}_i$. (It also turns out that $\mathbf{A}_{i+1}$ inherits the Hessenberg shape of $\mathbf{A}_i$, which reduces computational effort in obtaining the QR-decompositions.) Thus the matrices $\mathbf{A}_i$ are all orthogonally similar to $\mathbf{A}$; when the basic method is properly "souped up," the sequence of $\mathbf{A}_i$ converges rapidly to a Schur form $\mathbf{T}$ of $\mathbf{A}$ as in (7.42). This gives the eigenvalues of $\mathbf{A}$; if desired, the eigenvectors (or invariant subspaces, or selected eigenvectors, or . . .) are then found from $\mathbf{T}$ by special procedures (see Golub and Van Loan). In practice, the eigenvalues of a $p \times p$ matrix can usually be accurately approximated in at most about $15p^3$—and often considerably fewer—arithmetic operations by this method. As a simple example, we used MATLAB to generate a random 20×20 matrix and a random 20×20 symmetric matrix; for each matrix, the eigenvalues were found by MATLAB with about $3p^3 = 24{,}000$ MATLAB *flops* (floating-point operations) requiring about 6 seconds on our microcomputer, compared with about 1 second to find the inverse of the same matrix.

What Software to Get

Just as LINPACK sets the standard for software for matrix computations involving equations, EISPACK does so for eigensystem computations. Developed in the same cooperative fashion as LINPACK but earlier, the EISPACK software is available in various forms and from various sources commercial and public—the same ones as for LINPACK, in fact (see Section 3.9). Documentation on the

EISPACK routines is available in *Matrix Eigensystem Routines: EISPACK Guide, 2nd ed.* (52) and in *Matrix Eigensystem Routines: EISPACK Guide Extension* (44). Anyone needing to compute eigensystems should consider obtaining the EISPACK system.

PROBLEMS 7.6

1. Determine whether the EISPACK software is available to you.

2. Determine what software is available to you for computing the eigensystems of:
(a) real symmetric matrices
(b) general real matrices
(c) complex hermitian matrices
(d) general complex matrices

𝔐 **3.** Use MATLAB or similar software to find the QR-decompositions so as to implement the straightforward QR-algorithm (7.45) on the matrix below.

$$\begin{bmatrix} 1 & 2 & 0 & 0 \\ 2 & 3 & 4 & 0 \\ 0 & 4 & 5 & 6 \\ 0 & 0 & 6 & 7 \end{bmatrix}.$$

Note that this matrix is in Hessenberg form and observe that each new $\mathbf{A}_i$ is also in Hessenberg form; iterate until $\mathbf{A}_i$ is approximately diagonal. Compare with the eigenvalues found directly by MATLAB or similar software.

7.7 CONDITION OF EIGENSYSTEMS

Recall from Section 6.4 the general notion of the *condition* of the answer to a problem: how the answer changes in comparison to changes in the data—such as those caused by errors in measuring or modeling. This topic is fundamental in any area of applied mathematics; we consider it briefly with respect to eigenvalues and eigenvectors of $\mathbf{A}$ in terms of changes in the data $\mathbf{A}$.

Gerschgorin Circles

The following result is fundamental to the study of the condition of eigensystems.

(7.46) ***Theorem*** (Gerschgorin circles)

(a) Each (real or complex) eigenvalue λ of $p \times p$ $\mathbf{B}$ with $\langle \mathbf{B} \rangle_{ij} = b_{ij}$ satisfies at least one of the inequalities

$$|\lambda - b_{ii}| \leq r_i, \quad \text{where} \quad r_i = \sum_{\substack{j=1 \\ j \neq i}}^{p} |b_{ij}| \qquad (i = 1, \ldots, p).$$

That is, each eigenvalue lies in at least one of the discs with center $b_{ii} = \langle \mathbf{B} \rangle_{ii}$ and radius r_i in the complex plane.

(b) If the union of n of the discs is disjoint from the remainder, then there are precisely n eigenvalues of $\mathbf{B}$ in that union.

PROOF

(a) If λ and $\mathbf{x} \neq \mathbf{0}$ with $\langle \mathbf{x} \rangle_i = x_i$ satisfy $\mathbf{Bx} = \lambda\mathbf{x}$, then

$$(\lambda - b_{ii})x_i = \sum_{j=1}^{p}{}' \, b_{ij}x_j \qquad (i = 1, \ldots, p),$$

where the prime indicates that the term for $j = i$ has been omitted. Suppose that x_k has $|x_k|$ largest for all x_j, so that $|x_j/x_k| \leq 1$ for all j. Then

$$|\lambda - b_{kk}| \leq \sum_{j=1}^{p}{}' \, |b_{kj}| \left| \frac{x_j}{x_k} \right| \leq \sum_{j=1}^{p}{}' \, |b_{kj}|$$

holds for every eigenvalue for some value of k, proving the first part of the theorem.

(b) Problems 9 and 10. ■

(7.47) ***Example.*** Consider the matrix

$$\mathbf{B} = \begin{bmatrix} 1 & 0.1 & 2 \\ 0.01 & 10 & 10 \\ -0.01 & 1 & 100 \end{bmatrix}.$$

Direct application of the Gerschgorin circle theorem to $\mathbf{B}$ shows that the three eigenvalues satisfy

$$|\lambda - 1| \leq 2.1, \qquad |\lambda - 10| \leq 10.01, \qquad |\lambda - 100| \leq 1.01.$$

Since the third disc is disjoint from the others there is precisely one eigenvalue λ_3 with $|\lambda_3 - 100| \leq 1.01$. Since the first two discs overlap, however, we can only conclude that two eigenvalues lie somewhere in the union of the two discs. We also know that the eigenvalues of $\mathbf{B}$ and of $\mathbf{B}^T$ are equal. Gerschgorin's theorem applied to $\mathbf{B}^T$ yields the existence of precisely one eigenvalue in each of the three disjoint discs

$$|\lambda_1 - 1| \leq 0.02, \qquad |\lambda_2 - 10| \leq 1.1, \qquad |\lambda_3 - 100| \leq 12.$$

Eigenvalue Condition

Consider now perturbations that change a matrix $\mathbf{A}$ to $\mathbf{A} + \delta\mathbf{A}$; suppose that $p \times p$ $\mathbf{A}$ has a linearly independent set of p eigenvectors, so that—see **Key Theorem 7.14**—there is a nonsingular matrix $\mathbf{P}$ of eigenvectors of $\mathbf{A}$ such that $\mathbf{P}^{-1}\mathbf{AP} = \mathbf{\Lambda}$, where $\mathbf{\Lambda}$ is diagonal and contains the eigenvalues of $\mathbf{A}$. Then

$$\mathbf{P}^{-1}(\mathbf{A} + \delta\mathbf{A})\mathbf{P} = \mathbf{P}^{-1}\mathbf{AP} + \mathbf{P}^{-1}\delta\mathbf{AP} = \mathbf{\Lambda} + \delta\mathbf{\Lambda},$$

where $\delta\Lambda = \mathbf{P}^{-1}\,\delta\mathbf{A}\mathbf{P}$ is not in general diagonal. Since the eigenvalues of similar matrices are identical, we can study the eigenvalues of $\Lambda + \delta\Lambda$ instead of those of $\mathbf{A} + \delta\mathbf{A}$. Since $\langle\Lambda\rangle_{ii} = \lambda_i$, the ith eigenvalue of $\mathbf{A}$, the Gerschgorin circle theorem immediately says that each eigenvalue λ of $\Lambda + \delta\Lambda$ lies in at least one of the discs centered at λ_i with radius given by the sum of the magnitudes of the off-diagonal entries in the ith row of $\delta\Lambda$—which is at most

$$\|\delta\Lambda\|_\infty \le \|\mathbf{P}^{-1}\|_\infty \|\delta\mathbf{A}\|_\infty \|\mathbf{P}\|_\infty.$$

This proves the ∞-norm version of the following theorem; for the other norms, see the 2nd edition of this book, for example.

(7.48) ***Theorem*** (eigenvalue condition). Suppose that $p \times p$ $\mathbf{A}$ has a linearly independent set of p eigenvectors $\mathbf{x}_1, \ldots, \mathbf{x}_p$ with associated eigenvalues λ_i, and that $\mathbf{P}$ is the (nonsingular) matrix

$$\mathbf{P} = [\mathbf{x}_1 \quad \cdots \quad \mathbf{x}_p].$$

If λ is an eigenvalue of the perturbed matrix $\mathbf{A} + \delta\mathbf{A}$, then λ satisfies at least one of the inequalities

$$|\lambda - \lambda_i| \le \|\mathbf{P}^{-1}\|\,\|\mathbf{P}\|\,\|\delta\mathbf{A}\| \qquad \text{for } i = 1, \ldots, p,$$

where $\|\cdot\|$ is any of the 1-norm, 2-norm, or ∞-norm. In particular, if $\mathbf{P}$ is unitary or orthogonal—for example, if $\mathbf{A}$ is real symmetric or hermitian or more generally normal—so that $\|\mathbf{P}\|_2 = \|\mathbf{P}^{-1}\|_2 = 1$, then λ satisfies at least one of $|\lambda - \lambda_i| \le \|\delta\mathbf{A}\|_2$.

This theorem states that the condition of the eigenvalues of $\mathbf{A}$—how large $|\lambda - \lambda_i|$ can be in comparison to $\|\delta\mathbf{A}\|$—depends on the factor $\|\mathbf{P}\|\,\|\mathbf{P}^{-1}\|$, which is the condition number $c(\mathbf{P})$ of the matrix $\mathbf{P}$ of eigenvectors of $\mathbf{A}$.

(7.49) ***Example.*** Consider how the eigenvalues change when

$$\mathbf{A} = \begin{bmatrix} 1 & 10^6 \\ 0 & 2 \end{bmatrix}$$

is perturbed to

$$\mathbf{A}' = \mathbf{A} + \delta\mathbf{A} = \begin{bmatrix} 1 & 10^6 \\ \epsilon & 2 \end{bmatrix}.$$

The matrix $\mathbf{A}$ has eigenvalues $\lambda_1 = 1$, $\lambda_2 = 2$, and associated eigenvectors $\mathbf{x}_1 = \mathbf{e}_1$, $\mathbf{x}_2 = \mathbf{e}_1 + 10^{-6}\mathbf{e}_2$. Thus we have

$$\mathbf{P} = \begin{bmatrix} 1 & 1 \\ 0 & 10^{-6} \end{bmatrix}, \qquad \mathbf{P}^{-1} = \begin{bmatrix} 1 & -10^6 \\ 0 & 10^6 \end{bmatrix},$$

so that—by Theorem 7.48—the eigenvalues λ' of $\mathbf{A}'$ lie in the discs

$$|\lambda' - 1| \le 2(1 + 10^6)\epsilon, \qquad |\lambda' - 2| \le 2(1 + 10^6)\epsilon,$$

indicating very large perturbations (compared to ϵ). In fact, for $\epsilon = 0.75 \times 10^{-6}$ the perturbed eigenvalues are precisely $\lambda_1' = 0.5$ and $\lambda_2' = 2.5$, compared with $\lambda_1 = 1$ and $\lambda_2 = 2$, so the eigenvalues have changed by about $10^6\epsilon$.

Eigenvector Condition

A detailed discussion of the condition of the eigen*vectors* is beyond the scope of this book. We merely examine three examples of the condition of eigenvectors of *diagonal* matrices to provide an indication of the nature of the difficulties involved.

(7.50) ***Example.*** Consider the matrix

$$\mathbf{A} = \begin{bmatrix} 1 & 0 \\ 0 & 2 \end{bmatrix},$$

having eigenvectors $\mathbf{e}_1$ and $\mathbf{e}_2$, and the perturbed matrix

$$\mathbf{A} + \delta\mathbf{A} = \begin{bmatrix} 1 & \epsilon \\ 0 & 2 \end{bmatrix},$$

having eigenvectors $\mathbf{e}_1$ and $\epsilon\mathbf{e}_1 + \mathbf{e}_2$. The eigenvectors of $\mathbf{A}$ are thus perturbed by $\|\mathbf{e}_1 - \mathbf{e}_1\| = 0$ and by $\|\mathbf{e}_2 - (\mathbf{e}_2 + \epsilon\mathbf{e}_1)\| = \|\epsilon\mathbf{e}_1\|$, which is of order ϵ; we see that in this case the eigenvectors are well-conditioned.

(7.51) ***Example.*** Consider the matrix

$$\mathbf{A} = \begin{bmatrix} 2 + \delta & 0 \\ 0 & 2 \end{bmatrix},$$

where $\delta \neq 0$, having the eigenvectors $\mathbf{e}_1$ and $\mathbf{e}_2$. Consider also the perturbed matrix

$$\mathbf{A} + \delta\mathbf{A} = \begin{bmatrix} 2 + \delta & \epsilon \\ 0 & 2 \end{bmatrix}$$

having eigenvectors $\mathbf{e}_1$ and $\mathbf{e}_2 - (\epsilon/\delta)\mathbf{e}_1$. Perturbing $\mathbf{A}$ by ϵ thus perturbs its eigenvectors by 0 and $|\epsilon/\delta|$, which is $1/|\delta|$ times the size of the perturbation in $\mathbf{A}$. Thus *the condition of the eigenvectors depends on the separation δ of the two eigenvalues.*

(7.52) ***Example.*** Consider the matrices $\mathbf{A}$ and $\mathbf{A} + \delta\mathbf{A}$ given by

$$\mathbf{A} = \begin{bmatrix} 2 & 0 \\ 0 & 2 \end{bmatrix} \quad \text{and} \quad \mathbf{A} + \delta\mathbf{A} = \begin{bmatrix} 2 & \epsilon \\ 0 & 2 \end{bmatrix}.$$

The matrix $\mathbf{A} + \delta\mathbf{A}$ has only *one* eigenvector in a linearly independent set, say $\mathbf{e}_1$, as long as $\epsilon \neq 0$. The matrix $\mathbf{A}$, however, has a linearly independent set of *two* eigenvectors that can be chosen as *any* linearly independent set

$\{\mathbf{x}_1, \mathbf{x}_2\}$. If, for example, we have $\mathbf{x}_1 = [1 \quad 1]^T$ and $\mathbf{x}_2 = [1 \quad -1]^T$, then we have not only "lost" an eigenvector in going from $\mathbf{A}$ to $\mathbf{A} + \delta\mathbf{A}$, but also the perturbation in either eigenvector of $\mathbf{A}$ is at least of magnitude 1. Had we instead chosen $\mathbf{x}_1 = \mathbf{e}_1$, say, then the perturbation in the first eigenvector of $\mathbf{A}$ would have been $\mathbf{0}$.

These three examples indicate that we might expect a matrix with *distinct and well-separated* eigenvalues to have well-conditioned eigenvectors, but matrices with repeated or clustered eigenvalues to possibly have ill-conditioned eigenvectors. This indeed is generally the case. For more detail on this subtle issue, see Golub and Van Loan (46), Stewart (53), or Wilkinson (56).

PROBLEMS 7.7

▷ **1.** Use the Gerschgorin circle theorem to bound the eigenvalues of

$$\mathbf{B} = \begin{bmatrix} 1 & -10^{-5} & 2\cdot 10^{-5} \\ 4\cdot 10^{-5} & 0.5 & -3\cdot 10^{-5} \\ -10^{-5} & 3\cdot 10^{-5} & 0.1 \end{bmatrix}.$$

▷ **2.** Let $\mathbf{S}$ be the 3×3 diagonal matrix with $\langle\mathbf{S}\rangle_{11} = \alpha$, $\langle\mathbf{S}\rangle_{22} = \langle\mathbf{S}\rangle_{33} = 1$.
(a) Find $\mathbf{S}^{-1}\mathbf{BS}$ for the matrix $\mathbf{B}$ in Problem 1, and write down the Gerschgorin circles for the eigenvalues of $\mathbf{S}^{-1}\mathbf{BS}$ (which equal the eigenvalues of $\mathbf{B}$).
(b) Pick α to make the radius r_1 of the circle centered at $\langle\mathbf{S}^{-1}\mathbf{BS}\rangle_{11}$ as small as possible without that circle overlapping the other two, and thus locate λ_1 in as small a circle as you can.
(c) Use analogous matrices $\mathbf{S}$ to estimate λ_2 and λ_3.

3. Describe the perturbations in the eigenvalues of $\mathbf{A}$ when $\mathbf{A}$ is perturbed by $\delta\mathbf{A}$, where

$$\mathbf{A} = \begin{bmatrix} 1 & \alpha \\ 0 & 2 \end{bmatrix} \quad \text{and} \quad \delta\mathbf{A} = \begin{bmatrix} 0 & 0 \\ \epsilon & 0 \end{bmatrix}.$$

4. Use the Gerschgorin circle theorem to bound the eigenvalues of

$$\mathbf{B} = \begin{bmatrix} 2 & 0.1 & 0.2 \\ -0.1 & 2 & -0.1 \\ 1 & -1 & 10 \end{bmatrix}.$$

5. Use the approach of Problem 2 to refine the bounds in Problem 4.

6. If λ is an eigenvalue of $\mathbf{B}$ so that $\mathbf{B} - \lambda\mathbf{I}$ is singular, there is a (*left-eigenvector*) $\mathbf{y} \neq \mathbf{0}$ with $\mathbf{y}^T\mathbf{B} = \lambda\mathbf{y}^T$. Use this $\mathbf{y}$ to give a proof analogous to that of Theorem 7.46 to show that the eigenvalues λ of $\mathbf{B}$ satisfy $|\lambda - b_{ii}| \leq r_i'$, where r_i' is the column sum analogous to the row sum r_i.

▷ **7.** Suppose that $\mathbf{P}^{-1}\mathbf{AP} = \mathbf{\Lambda}$ is diagonal and that μ and $\mathbf{v}$ are an *approximate* eigenvalue/eigenvector in the sense that $\mathbf{r} = \mathbf{Av} - \mu\mathbf{v}$ is viewed as "small."

(a) Prove that there is an eigenvalue λ of $\mathbf{A}$ satisfying

$$|\lambda - \mu| \leq \|\mathbf{P}\| \, \|\mathbf{P}^{-1}\| \frac{\|\mathbf{r}\|}{\|\mathbf{v}\|},$$

where $\|\cdot\|$ denotes any of the 1-norm, 2-norm, or ∞-norm.

(b) If $\mathbf{A}$ is normal, show that

$$|\lambda - \mu| \leq \frac{\|\mathbf{r}\|_2}{\|\mathbf{v}\|_2}.$$

8. Apply the estimate in Problem 7(b) with $\mu = 1$ and compare with the true eigenvalues when

$$\mathbf{A} = \begin{bmatrix} 1.9 & 1 \\ 1 & 2.1 \end{bmatrix} \quad \text{and} \quad \mathbf{v} = \begin{bmatrix} 1 \\ -1 \end{bmatrix}.$$

9. It is known that the roots of a polynomial are continuous functions of its coefficients. Prove that the eigenvalues of a matrix are continuous functions of its entries.

10. Prove Theorem 7.46(b) as follows. Let $\mathbf{D}$ be the diagonal matrix whose main diagonal equals that of $\mathbf{B}$ and define $\mathbf{E} = \mathbf{B} - \mathbf{D}$; define G_t as the union of the n Gerschgorin discs in question but for the matrix $\mathbf{D} + t\mathbf{E}$ for $0 \leq t \leq 1$. Show that G_0 contains exactly n eigenvalues, and use Problem 9 to show that G_t always contains exactly n eigenvalues; G_1 is the region of Theorem 7.46(b).

▷ **11.** Show that if each of the p Gerschgorin discs of a $p \times p$ matrix is disjoint from the others, then each contains exactly one eigenvalue; conclude that if in addition $\mathbf{B}$ is real then the eigenvalues must be real and satisfy $b_{ii} - r_i \leq \lambda \leq b_{ii} + r_i$ in the notation of Theorem 7.46.

12. Assume as in Problem 11 that $\mathbf{B}$ is real and has disjoint Gerschgorin discs. Show that if $b_{ii} > r_i$ for all i, then the eigenvalues of $\mathbf{B}$ are strictly positive.

13. Use Problem 11 to give as much information as you can about the eigenvalues of

$$\mathbf{B} = \begin{bmatrix} 6 & 1 & 1 & -2 \\ 1 & 14 & 1 & 1 \\ 2 & 2 & -9 & -1 \\ 1 & -1 & 1 & -20 \end{bmatrix}.$$

7.8 *MISCELLANEOUS PROBLEMS*

PROBLEMS 7.8

1. If $\mathbf{A}$ is $p \times p$ and real, show that its eigenvalues are real or complex-conjugate pairs; if in addition p is odd, show that there is at least one real eigenvalue.

▷ **2.** Show that the $n \times n$ matrix

$$\begin{bmatrix} k & 1 & 0 & \cdots & 0 \\ 1 & k & 1 & \cdots & 0 \\ 0 & 1 & k & \cdots & 0 \\ & & \cdots & & \\ 0 & 0 & 0 & \cdots & k \end{bmatrix}$$

has eigenvectors $\mathbf{x}_i$ whose jth entry is given by $\sin\{ij\pi/(n+1)\}$. Deduce the corresponding eigenvalues.

3. Prove that if $\mathbf{A}$ is real and $p \times p$ with p even and det $\mathbf{A}$ strictly negative, then $\mathbf{A}$ has at least two real eigenvalues.

4. Prove that the characteristic polynomial of

$$\mathbf{A} = \begin{bmatrix} \mathbf{B} & \mathbf{0} \\ \mathbf{0} & \mathbf{C} \end{bmatrix},$$

where $\mathbf{B}$ and $\mathbf{C}$ are square, is the product of the characteristic polynomials of $\mathbf{B}$ and $\mathbf{C}$. Investigate the relationship among the eigenvectors of $\mathbf{A}$, $\mathbf{B}$, and $\mathbf{C}$.

▷ **5.** Show that every polynomial equation is the characteristic equation of some matrix. More specifically, show that

$$(-1)^n \det \begin{bmatrix} -a_1-\lambda & -a_2 & \cdots & -a_{n-1} & -a_n \\ 1 & -\lambda & \cdots & 0 & 0 \\ 0 & 1 & \cdots & 0 & 0 \\ & & \cdots & & \\ 0 & 0 & \cdots & 1 & -\lambda \end{bmatrix} = \lambda^n + a_1\lambda^{n-1} + \cdots + a_n.$$

6. If $\mathbf{A}$ is $p \times p$ with rank k, prove that $\lambda = 0$ is an eigenvalue whose algebraic multiplicity is at least $p - k$, and give an example to show that it can exceed $p - k$.

7. Let $\mathbf{P}$ denote the $n \times n$ permutation matrix (Definition 3.53) for which $\langle\mathbf{P}\rangle_{1n} = 1$ and $\langle\mathbf{P}\rangle_{i,i-1} = 1$ for $2 \le i \le n$. Prove that its characteristic polynomial is $\lambda^n - 1$. If we define

$$\mu = \exp\left(\frac{2\pi i}{n}\right) = \cos\left(\frac{2\pi}{n}\right) + i \sin\left(\frac{2\pi}{n}\right),$$

show that the eigenvectors of $\mathbf{P}$ are

$$\mathbf{x}_r = [\mu^r \quad \mu^{2r} \quad \cdots \quad \mu^{nr}]^T \qquad (r = 1, \ldots, n)$$

corresponding to eigenvalues μ^{n-r}, respectively. Prove that these are orthogonal. Prove that

$$\mathbf{Pe}_i = \mathbf{e}_{i+1} \qquad (i = 1, \ldots, n-1); \qquad \mathbf{Pe}_n = \mathbf{e}_1.$$

Deduce that if **A** is the *circulant*

$$\mathbf{A} = \begin{bmatrix} c_0 & c_{n-1} & \cdots & c_1 \\ c_1 & c_0 & \cdots & c_2 \\ & \cdots & & \\ c_{n-1} & c_{n-2} & \cdots & c_0 \end{bmatrix},$$

then $\mathbf{A} = f(\mathbf{P})$ where $f(x) = c_0 + c_1 x + \cdots + c_{n-1}x^{n-1}$. Deduce that the eigenvalues of **A** are given by $f(\mu^{n-r})$, $r = 1, \ldots, n$, with corresponding eigenvectors $\mathbf{x}_r$, defined above.

8. Generalize **Key Theorem 7.10(c)** by proving the following. Suppose that $\lambda_1, \ldots, \lambda_r$ is a collection of distinct eigenvalues of **A** and that for each i the set $S_i = \{\mathbf{x}_{i1}, \mathbf{x}_{i2}, \ldots, \mathbf{x}_{ir_i}\}$ is a linearly independent set of eigenvectors associated with λ_i. Then the union of all the S_i is linearly independent.

▷ **9.** Either prove or give a counter example: If **A** is similar to **A**′ and **B** to **B**′ (and all are $p \times p$), then **AB** is similar to **A**′**B**′.

▷ **10.** Suppose that f is a polynomial.

(a) Use Theorem 7.25(e) to show that if **A** and **B** are similar and $f(\mathbf{A}) = \mathbf{0}$, then $f(\mathbf{B}) = \mathbf{0}$.

(b) Prove that if **A** is similar to a diagonal matrix and f is the characteristic polynomial of **A**, then $f(\mathbf{A}) = \mathbf{0}$.

▷ **11.** Suppose that **P** is a $p \times p$ matrix for which $\|\mathbf{Px}\|_2 = \|\mathbf{x}\|_2$ for all **x**. Show that $(\mathbf{Px}, \mathbf{Py}) = (\mathbf{x}, \mathbf{y})$ for all **x** and **y** using the standard inner product for either $\mathbb{R}^p$ or $\mathbb{C}^p$.

12. Use Problem 11 to prove that $p \times p$ **P** is unitary if and only if $\|\mathbf{Px}\|_2 = \|\mathbf{x}\|_2$ for all **x** in $\mathbb{C}^p$.

13. Suppose that **B** is unitarily similar to **A**. Prove that $\mathbf{B}^H\mathbf{B}$ is unitarily similar to $\mathbf{A}^H\mathbf{A}$.

8

Eigensystems of symmetric, hermitian, and normal matrices, with applications

The theory outlined in Chapter 7 concerning the eigensystems of normal matrices is explained in more detail in this chapter. That theory is then exploited to develop some powerful tools for applied work. Fundamental in this development are ***Key Theorems 8.6, 8.8, 8.19****, and* ***8.26****, together with* ***Key Corollaries 8.9*** *and* ***8.20****.*

8.1 INTRODUCTION

Perhaps reflecting an inherent symmetry in nature, mathematical models of real-world problems often produce matrices that are *symmetric* (or, more generally, hermitian or normal). Consider the oscillating-masses problem modeled in Section 2.5, for example. Although the problem itself does not seem symmetric—the system is fixed at the top but free at the bottom, for example—the model leads to matrices $\mathbf{K}$ and $\mathbf{M}$ in (2.29) that *are* symmetric.

Problem 9 in Section 7.2 extended that model to p masses coupled by p springs; once again symmetric matrices $\mathbf{K}$ and $\mathbf{M}$ appeared. The natural frequencies of oscillation of the system are $\omega/2\pi$, where $\mathbf{K} - \omega^2\mathbf{M}$ is singular. We have earlier remarked that this means that $\lambda = \omega^2$ is an eigenvalue of (nonsymmetric) $\mathbf{M}^{-1}\mathbf{K}$; we can in fact reformulate this as an eigenvalue problem for—once again—*symmetric* matrices. Since $\mathbf{M} = \mathbf{diag}(m_1, \ldots, m_p)$ with $m_i > 0$ we can write $\mathbf{M} = \mathbf{D}^2$, where

$$\mathbf{D} = \mathbf{diag}(\sqrt{m_1}, \ldots, \sqrt{m_p}).$$

Then $\mathbf{D}$ is nonsingular, so $\mathbf{K} - \omega^2\mathbf{M}$ is singular if and only if $\mathbf{D}^{-1}(\mathbf{K} - \omega^2\mathbf{M})\mathbf{D}^{-1}$ is singular—that is, if and only if $\lambda = \omega^2$ is an eigenvalue of $\mathbf{B} = \mathbf{D}^{-1}\mathbf{K}\mathbf{D}^{-1}$. And $\mathbf{B}$ is *symmetric*:

$$\mathbf{B}^T = (\mathbf{D}^{-1}\mathbf{K}\mathbf{D}^{-1})^T = (\mathbf{D}^{-1})^T\mathbf{K}^T(\mathbf{D}^{-1})^T = (\mathbf{D}^{-1})\mathbf{K}(\mathbf{D}^{-1}) = \mathbf{B},$$

since $\mathbf{D}^{-1}$ is diagonal (hence symmetric) and $\mathbf{K}$ is symmetric. Thus the physically important fundamental frequencies of the oscillating system are found from the eigenvalues of the symmetric matrix $\mathbf{B}$ even though the system itself seems nonsymmetric.

(8.1) ***Example.*** Consider the concrete model of two coupled masses in Example 2.31, where the matrices $\mathbf{K}$ and $\mathbf{M}$ are given. It is easy to find the matrix $\mathbf{B} = \mathbf{D}^{-1}\mathbf{K}\mathbf{D}^{-1}$, where $\mathbf{M} = \mathbf{D}^2$:

$$\mathbf{B} = \begin{bmatrix} 12 & -4\sqrt{2} \\ -4\sqrt{2} & 8 \end{bmatrix},$$

which is indeed symmetric.

The theory of the eigensystems of symmetric matrices is important for another reason as well: It gives a powerful tool—the singular value decomposition—for use with *general* matrices. The results of this chapter reflect both motivations for the study of symmetric matrices and their generalizations.

PROBLEMS 8.1

▷ **1.** Find the relationship between vectors ξ solving $(\mathbf{K} - \omega^2\mathbf{M})\xi = \mathbf{0}$ and vectors $\mathbf{x}$ solving $(\mathbf{B} - \omega^2\mathbf{I})\mathbf{x} = \mathbf{0}$, where $\mathbf{B} = \mathbf{D}^{-1}\mathbf{K}\mathbf{D}^{-1}$ and $\mathbf{D}^2 = \mathbf{M}$.

2. Find the matrix $\mathbf{B} = \mathbf{D}^{-1}\mathbf{K}\mathbf{D}^{-1}$, where $\mathbf{D}^2 = \mathbf{M}$, for the system of 10 coupled masses in Problem 10 of Section 7.2, and show that $\mathbf{B}$ is symmetric.

3. Find the general formula for the matrix $\mathbf{B} = \mathbf{D}^{-1}\mathbf{K}\mathbf{D}^{-1}$, where $\mathbf{D}^2 = \mathbf{M}$, for the system of p coupled masses in Problem 9 of Section 7.2, and show that $\mathbf{B}$ is symmetric.

8.2 SCHUR FORM AND DECOMPOSITION; NORMAL MATRICES

We saw with the Gauss-reduced form and the row-echelon form that special forms to which every matrix can be transformed may be extremely useful in analyzing the properties of matrices. This is a useful approach for the study of eigensystems as well.

Schur Form

(8.2) ***Theorem*** (Schur form and decomposition). Let $\mathbf{A}$ be $p \times p$.

(a) $\mathbf{A}$ is unitarily similar to an upper-triangular matrix $\mathbf{T} = \mathbf{P}^H\mathbf{A}\mathbf{P}$ with $\mathbf{P}$ unitary and with the eigenvalues of $\mathbf{A}$ (repeated according to their algebraic multiplicities) on the main diagonal of $\mathbf{T}$. $\mathbf{T}$ is called a *Schur form* of $\mathbf{A}$ and the decomposition $\mathbf{A} = \mathbf{PTP}^H$ a *Schur decomposition* of $\mathbf{A}$.

(b) If $\mathbf{A}$ and its eigenvalues are real, then $\mathbf{P}$ may be taken real and hence orthogonal.

PROOF. Both parts of the theorem clearly are true if $p = 1$. We proceed by induction, assuming the theorem true for $p = k$ and seeking to prove it for $p = k + 1$. So suppose that $\mathbf{A}$ is $(k + 1) \times (k + 1)$. Let λ_1 be an eigenvalue of $\mathbf{A}$ with associated eigenvector $\mathbf{x}_1$ normalized so that $\|\mathbf{x}_1\|_2 = 1$; if $\mathbf{A}$ and λ_1 are real, then by **Key Theorem 7.10(a)** we can take $\mathbf{x}_1$ real. Since we can extend $\{\mathbf{x}_1\}$ to form a basis for $\mathbb{C}^{k+1}$ (or $\mathbb{R}^{k+1}$ if $\mathbf{x}_1$ is real) and then use the Gram-Schmidt process to produce from it an orthonormal basis, there is a set of vectors (real if $\mathbf{x}_1$ is real) $\mathbf{w}_1, \ldots, \mathbf{w}_k$ such that $\{\mathbf{x}_1, \mathbf{w}_1, \ldots, \mathbf{w}_k\}$ is orthonormal; thus the matrix

$$\mathbf{U} = [\mathbf{x}_1 \quad \mathbf{w}_1 \quad \mathbf{w}_2 \quad \cdots \quad \mathbf{w}_k] = [\mathbf{x}_1 \quad \mathbf{W}]$$

is unitary (orthogonal if $\mathbf{x}_1$ is real). We now compute $\mathbf{A}' = \mathbf{U}^H\mathbf{A}\mathbf{U}$:

$$\begin{aligned}\mathbf{A}' = \mathbf{U}^H\mathbf{A}\mathbf{U} &= [\mathbf{x}_1 \quad \mathbf{W}]^H\mathbf{A}[\mathbf{x}_1 \quad \mathbf{W}] = [\mathbf{x}_1 \quad \mathbf{W}]^H[\mathbf{A}\mathbf{x}_1 \quad \mathbf{A}\mathbf{W}] \\ &= [\mathbf{x}_1 \quad \mathbf{W}]^H[\lambda_1\mathbf{x}_1 \quad \mathbf{A}\mathbf{W}] \\ &= \begin{bmatrix} \lambda_1 & \mathbf{x}_1^H\mathbf{A}\mathbf{W} \\ \mathbf{0} & \mathbf{W}^H\mathbf{A}\mathbf{W} \end{bmatrix} = \begin{bmatrix} \lambda_1 & \mathbf{b}^H \\ \mathbf{0} & \mathbf{C} \end{bmatrix},\end{aligned}$$

since $\|\mathbf{x}_1\|_2 = 1$ and since $\mathbf{W}^H\mathbf{x}_1 = \mathbf{0}$ because $\mathbf{U}$ is unitary. Since $\mathbf{A}'$ is similar to $\mathbf{A}$, the eigenvalues of $\mathbf{A}$ and $\mathbf{A}'$ are identical, including multiplicities. By expanding $\det(\mathbf{A}' - \lambda\mathbf{I})$ by its first column, we see that the characteristic polynomial of $\mathbf{A}'$ equals $\lambda_1 - \lambda$ times that of $\mathbf{C}$; thus the eigenvalues of $\mathbf{A}$ in addition to λ_1 are just those of $\mathbf{C}$. But $\mathbf{C}$ is $k \times k$ (and real if $\mathbf{A}$ and λ_1 are), so our inductive hypothesis holds and we can find a unitary $\mathbf{V}$ (orthogonal if $\mathbf{A}$ and its eigenvalues are real) so that $\mathbf{V}^H\mathbf{C}\mathbf{V} = \mathbf{T}'$, with $\mathbf{T}'$ upper-triangular and with the eigenvalues of $\mathbf{C}$ (hence of $\mathbf{A}$) on its main diagonal. If we now define $\mathbf{P}$ as

$$\mathbf{P} = \mathbf{U}\begin{bmatrix} 1 & \mathbf{0} \\ \mathbf{0} & \mathbf{V} \end{bmatrix}, \quad \text{then} \quad \mathbf{P}^H\mathbf{A}\mathbf{P} = \begin{bmatrix} \lambda_1 & \mathbf{b}^H\mathbf{V} \\ \mathbf{0} & \mathbf{T}' \end{bmatrix},$$

which has the proper form and thus proves that the inductive hypothesis holds for $p = k + 1$ also. Therefore, it is true for all p. ∎

Note that this is an existence theorem, but does not say that $\mathbf{T}$ is easily computed; in practice one needs to know the eigenvalues to obtain $\mathbf{T}$. It is a powerful theoretical tool, however. We now express this theorem in terms of transformation representations rather than decompositions or similarities {see (7.19)}.

(8.3) ***Corollary*** (Schur representations). If $\mathscr{T}$ is the linear transformation from $\mathbb{C}^p$ to $\mathbb{C}^p$ defined by $\mathscr{T}(\mathbf{v}) = \mathbf{A}\mathbf{v}$, then there is an orthonormal basis for $\mathbb{C}^p$ with respect to which $\mathscr{T}$ is represented by a Schur form of $\mathbf{A}$.

(8.4) ***Example.*** Consider the transition matrix **A** of the dairy-competition model in Example 2.6 of Section 2.2; the eigenvalues and eigenvectors of **A** were found in Example 7.2 of Section 7.1. To construct a Schur form, we start with **A**, λ_1, and a normalized $\mathbf{x}_1$:

$$\mathbf{A} = \begin{bmatrix} 0.8 & 0.2 & 0.1 \\ 0.1 & 0.7 & 0.3 \\ 0.1 & 0.1 & 0.6 \end{bmatrix}, \qquad \lambda_1 = 0.6, \quad \text{and} \quad \mathbf{x}_1 = \frac{\sqrt{2}}{2}\begin{bmatrix} 1 \\ -1 \\ 0 \end{bmatrix}.$$

We can use $\mathbf{x}_1$ as the first column of a unitary matrix, say

$$\mathbf{Q} = \begin{bmatrix} \frac{\sqrt{2}}{2} & \frac{\sqrt{2}}{2} & 0 \\ -\frac{\sqrt{2}}{2} & \frac{\sqrt{2}}{2} & 0 \\ 0 & 0 & 1 \end{bmatrix},$$

where we have chosen the last two columns of **Q** as simply as we could to make **Q** unitary; we then calculate

$$\mathbf{A}_1 = \mathbf{Q}^H\mathbf{A}\mathbf{Q} = \begin{bmatrix} 0.6 & 0.1 & -0.1\sqrt{2} \\ 0 & 0.9 & 0.2\sqrt{2} \\ 0 & 0.1\sqrt{2} & 0.6 \end{bmatrix}.$$

We next consider the 2×2 submatrix indicated above, namely

$$\begin{bmatrix} 0.9 & 0.2\sqrt{2} \\ 0.1\sqrt{2} & 0.6 \end{bmatrix},$$

which is easily found to have an eigenvalue $\lambda = 1$ with an associated normalized eigenvector

$$\begin{bmatrix} \frac{2\sqrt{2}}{3} \\ \frac{1}{3} \end{bmatrix}.$$

We add, as a second column, the simplest normalized orthogonal vector we can think of to get a unitary matrix

$$\begin{bmatrix} \frac{2\sqrt{2}}{3} & \frac{1}{3} \\ \frac{1}{3} & \frac{-2\sqrt{2}}{3} \end{bmatrix};$$

finally, we use the matrix

$$\begin{bmatrix} 1 & 0 & 0 \\ 0 & \frac{2\sqrt{2}}{3} & \frac{1}{3} \\ 0 & \frac{1}{3} & \frac{-2\sqrt{2}}{3} \end{bmatrix}$$

as the basis of a further unitary similarity to be applied to $\mathbf{A}_1$, yielding for $\mathbf{A}$ the desired triangular form

$$\begin{bmatrix} 0.6 & \frac{0.1\sqrt{2}}{3} & \frac{1}{6} \\ 0 & 1 & -0.1\sqrt{2} \\ 0 & 0 & 0.5 \end{bmatrix}.$$

Normal Matrices

Theorem 8.2 is surprisingly powerful: It contains striking information on the eigensystem of a symmetric matrix (**Key Corollary 8.9**). In fact, it allows us to analyze the eigensystem of a broader class of matrices than merely symmetric ones.

(8.5) ***Definition.*** A *normal* matrix is a $p \times p$ matrix $\mathbf{A}$ satisfying $\mathbf{A}^H\mathbf{A} = \mathbf{A}\mathbf{A}^H$.

Certainly, hermitian matrices ($\mathbf{A}^H = \mathbf{A}$) and their special case of real symmetric matrices ($\mathbf{A}^T = \mathbf{A}$) are normal, since $\mathbf{A}^H\mathbf{A}$ and $\mathbf{A}\mathbf{A}^H$ both equal $\mathbf{A}^2$. Similarly, unitary matrices are normal ($\mathbf{A}^H\mathbf{A}$ and $\mathbf{A}\mathbf{A}^H$ both equal $\mathbf{I}$). We now use the Schur form of normal matrices to obtain a striking result.

(8.6) **Key Theorem** (diagonal form). A $p \times p$ matrix $\mathbf{A}$ is normal (for example, real symmetric, hermitian, or unitary) if and only if $\mathbf{A}$ is unitarily similar to a diagonal (Schur form) matrix $\mathbf{D} = \mathbf{P}^H\mathbf{A}\mathbf{P}$, where $\mathbf{P}$ is unitary and $\mathbf{D}$ is diagonal with the eigenvalues of $\mathbf{A}$ (repeated according to their algebraic multiplicities) on the main diagonal of $\mathbf{D}$; if $\mathbf{A}$ and its eigenvalues are real, then $\mathbf{P}$ can be taken real and hence orthogonal.

PROOF. (diagonal form $\Rightarrow$ normal) If there exists unitary $\mathbf{P}$ with $\mathbf{P}^H\mathbf{A}\mathbf{P} = \mathbf{D}$ diagonal, then $\mathbf{A}$ is certainly normal since diagonal matrices commute:

$$\begin{aligned}\mathbf{A}^H\mathbf{A} &= (\mathbf{PDP}^H)^H(\mathbf{PDP}^H) = \mathbf{PD}^H\mathbf{P}^H\mathbf{PDP}^H = \mathbf{PD}^H\mathbf{DP}^H = \mathbf{PDD}^H\mathbf{P}^H \\ &= \mathbf{PDP}^H\mathbf{PD}^H\mathbf{P}^H = (\mathbf{PDP}^H)(\mathbf{PDP}^H)^H = \mathbf{AA}^H.\end{aligned}$$

(normal $\Rightarrow$ diagonal form) Suppose that $\mathbf{A}$ is normal, and let $\mathbf{T}$ be a Schur form with $\mathbf{P}^H\mathbf{A}\mathbf{P} = \mathbf{T}$. $\mathbf{T}$ is also normal:

$$\mathbf{T}^H\mathbf{T} = (\mathbf{P}^H\mathbf{AP})^H(\mathbf{P}^H\mathbf{AP}) = \mathbf{P}^H\mathbf{A}^H\mathbf{AP} = \mathbf{P}^H\mathbf{AA}^H\mathbf{P} = (\mathbf{P}^H\mathbf{AP})(\mathbf{P}^H\mathbf{AP})^H = \mathbf{TT}^H,$$

as required. Let t_{ij} denote $\langle \mathbf{T} \rangle_{ij}$; we have $t_{ij} = 0$ for $i > j$. Since $\mathbf{TT}^H = \mathbf{T}^H\mathbf{T}$, their (i, i)-entries are equal. We have

$$\langle \mathbf{TT}^H \rangle_{ii} = |t_{ii}|^2 + |t_{i,i+1}|^2 + \cdots + |t_{ip}|^2,$$

while also

$$\langle \mathbf{T}^H\mathbf{T} \rangle_{ii} = |t_{1i}|^2 + |t_{2i}|^2 + \cdots + |t_{ii}|^2.$$

Equating these two expressions and subtracting the common term $|t_{ii}|^2$ from each side gives

$$|t_{i,i+1}|^2 + \cdots + |t_{ip}|^2 = |t_{1i}|^2 + \cdots + |t_{i-1,i}|^2$$

for all i. If we take $i = 1$ in this equality, we get 0 on the right-hand side since it has no terms; so on the left side $t_{12} = t_{13} = \cdots = t_{1p} = 0$. If we then take $i = 2$, we get $|t_{1i}|^2$ on the right-hand side, which equals zero; thus $t_{23} = t_{24} = \cdots = t_{2p} = 0$. Continuing in this fashion shows that $\mathbf{T}$ is in fact diagonal, so we may take $\mathbf{D} = \mathbf{T}$. The remark on the reality of $\mathbf{P}$ follows from the same remark for the Schur form. ■

As usual, we can immediately interpret any such decomposition or similarity result in terms of transformation representations.

(8.7) ***Corollary.*** Suppose that $\mathscr{T}$ is the linear transformation from $\mathbb{C}^p$ to $\mathbb{C}^p$ defined by $\mathscr{T}(\mathbf{v}) = \mathbf{Av}$. Then $\mathscr{T}$ is represented by a diagonal matrix with respect to some orthonormal basis if and only if $\mathbf{A}$ is normal.

Of course, we can also immediately interpret this result in terms of eigensystems of normal matrices. This form of the theorem is so important that we place it in a special section for emphasis.

PROBLEMS 8.2

1. Show that a $p \times p$ upper-triangular matrix is already in Schur form.
2. ▷ As in Example 8.4, find a Schur form of the matrix in Problem 2 of Section 7.2.
3. As in Example 8.4, find a Schur form of the matrix in Problem 1 of Section 7.2.
4. As in Example 8.4, find a Schur form of the matrix $\mathbf{A}^T$, where the matrix $\mathbf{A}$ is as in that example.
5. Show that Schur form is not unique by showing that $\mathbf{A}$ and $\mathbf{P}^H\mathbf{AP}$ ($\neq \mathbf{A}$) are Schur forms of $\mathbf{A}$, where

$$\mathbf{A} = \begin{bmatrix} 1 & 0 & 3 \\ 0 & 1 & 4 \\ 0 & 0 & 2 \end{bmatrix} \quad \text{and} \quad \mathbf{P} = \begin{bmatrix} 0 & -1 & 0 \\ 1 & 0 & 0 \\ 0 & 0 & 1 \end{bmatrix}.$$

▷ **6.** Show that skew-hermitian matrices ($\mathbf{A}^H = -\mathbf{A}$) are normal.

7. Suppose that $p \times p$ $\mathbf{A}$ is upper-triangular but definitely not diagonal; show that $\mathbf{A}$ is not normal.

8. Given that the eigenvalues of the normal matrix $\mathbf{A}$ below are 9, 9, -3, find a Schur form.

$$\mathbf{A} = \begin{bmatrix} 5 & 4 & -4 \\ 4 & 5 & 4 \\ -4 & 4 & 5 \end{bmatrix}$$

▷ **9.** Determine whether $\mathbf{A}$ is normal and find a Schur form.

$$\mathbf{A} = \begin{bmatrix} 2 & 1+i \\ 1-i & 3 \end{bmatrix}.$$

▷ **10.** Show that $\mathbf{A}$ is normal (but neither symmetric, hermitian, unitary, nor skew-hermitian) and find a Schur form.

$$\mathbf{A} = \begin{bmatrix} 1 & -1 \\ 1 & 1 \end{bmatrix}.$$

11. For each of the matrices $\mathbf{A}$ in Problems 8, 9, and 10, find an orthonormal basis and a diagonal matrix $\mathbf{D}$ such that the linear transformation $\mathscr{T}$ defined by $\mathscr{T}(\mathbf{v}) = \mathbf{A}\mathbf{v}$ is represented by $\mathbf{D}$ with respect to that basis (used in both domain and range).

8.3 *EIGENSYSTEMS OF NORMAL MATRICES*

Recall from (7.19) that statements about decompositions/similarities, about transformation representations, and about eigensystems are all equivalent. We now interpret **Key Theorem 8.6** and its corollary in the remaining version: eigensystems.

(8.8) **Key Theorem** (normality and eigensystems). A $p \times p$ matrix $\mathbf{A}$ is normal (for example, real symmetric, hermitian, or unitary) if and only if $\mathbf{A}$ has a linearly independent set of p eigenvectors that may be chosen so as to form an orthonormal set.

PROOF. This follows immediately from **Key Theorems 8.6** and **7.14** (on decompositions and eigenvectors). ■

(8.9) **Key Corollary** (hermitian matrix eigensystems). The eigenvalues of a $p \times p$ hermitian (or real symmetric) matrix $\mathbf{A}$ are real, and the associated eigenvectors may be chosen so as to form an orthonormal set of p vectors.

PROOF. The second part follows from the normality of **A** by **Key Theorem 8.8**. Since we can write $\mathbf{A} = \mathbf{PDP}^H$ for diagonal **D** containing the eigenvalues on the main diagonal and since $\mathbf{A}^H = \mathbf{A}$, we have

$$\mathbf{D}^H = (\mathbf{P}^H\mathbf{A}\mathbf{P})^H = \mathbf{P}^H\mathbf{A}^H\mathbf{P} = \mathbf{P}^H\mathbf{A}\mathbf{P} = \mathbf{D},$$

and thus **D** is real. ∎

(8.10) ***Example.*** Consider the 2×2 matrix **B** from Example 8.1, whose eigenvalues give us the fundamental frequencies of the oscillating masses in Section 2.5. We easily find the eigenvalues to be $\lambda_1 = 4$ and $\lambda_2 = 16$ and eigenvectors to be $[\sqrt{2} \quad 2]^T$ and $[\sqrt{2} \quad -1]^T$, respectively. Normalizing these gives the orthonormal set $\{\mathbf{x}_1 \quad \mathbf{x}_2\}$ of eigenvectors

$$\mathbf{x}_1 = \begin{bmatrix} \frac{\sqrt{3}}{3} & \frac{\sqrt{6}}{3} \end{bmatrix}^T \quad \text{and} \quad \mathbf{x}_2 = \begin{bmatrix} \frac{\sqrt{6}}{3} & \frac{-\sqrt{3}}{3} \end{bmatrix}^T.$$

Some comment is necessary concerning the phrase "*may be chosen* so as to form an orthonormal set" of eigenvectors in **Key Theorem 8.8** and **Key Corollary 8.9**. Since $c\mathbf{x}$ is an eigenvector for $c \neq 0$ if $\mathbf{x}$ is an eigenvector, we certainly can force the eigenvectors to be normalized to 2-norm 1. Eigenvectors associated with distinct eigenvalues are automatically orthogonal (Problem 8), but the eigenvectors associated with a multiple eigenvalue need not be orthogonal unless we choose them that way. An example should clarify this point.

(8.11) ***Example.*** Consider the real symmetric matrix

$$\mathbf{A} = \begin{bmatrix} 7 & -16 & -8 \\ -16 & 7 & 8 \\ -8 & 8 & -5 \end{bmatrix}.$$

The characteristic polynomial is found to be

$$\lambda^3 - 9\lambda^2 - 405\lambda - 2187,$$

with roots $\lambda_1 = 27$, $\lambda_2 = \lambda_3 = -9$. The equations $(\mathbf{A} - \lambda_1\mathbf{I})\mathbf{x} = \mathbf{0}$ readily yield precisely one eigenvector, as we should expect, since λ_1 is a simple root:

$$\mathbf{x}_1 = \alpha \begin{bmatrix} -2 \\ 2 \\ 1 \end{bmatrix}$$

where α is an arbitrary constant. The equations $(\mathbf{A} - \lambda_2\mathbf{I})\mathbf{x} = \mathbf{0}$ reduce to a single equation

$$2x_1 - 2x_2 - x_3 = 0.$$

The general solution of this is $x_3 = \beta$ arbitrary, $x_2 = \gamma$ arbitrary, and then $x_1 = \gamma + \beta/2$. Thus all eigenvectors associated with $\lambda_2 = \lambda_3$ have the form

$$\mathbf{x} = \gamma \begin{bmatrix} 1 \\ 1 \\ 0 \end{bmatrix} + \beta \begin{bmatrix} \frac{1}{2} \\ 0 \\ 1 \end{bmatrix}.$$

The two eigenvectors in this linear combination are linearly independent *but not orthogonal* to one another (note that they *are* orthogonal to $\mathbf{x}_1$). We therefore must *choose* two different linear combinations of the above form so as to be mutually orthogonal. For the first, we simply take, say, $\gamma = 1$ and $\beta = 0$, getting $[1 \quad 1 \quad 0]^T$. For the second, we must choose γ and β so that the resulting $[\gamma + \beta/2 \quad \gamma \quad \beta]^T$ is orthogonal to the first, namely $[1 \quad 1 \quad 0]^T$. That is, we require $\gamma + \beta/2 + \gamma = 0$, which gives $\beta = -4\gamma$; taking $\gamma = 1$ and thus $\beta = -4$ gives the other eigenvector as $[-1 \quad 1 \quad -4]^T$. Normalizing our eigenvectors finally gives an orthonormal set of eigenvectors

$$\begin{bmatrix} \frac{\sqrt{2}}{2} & \frac{\sqrt{2}}{2} & 0 \end{bmatrix}^T \quad \text{and} \quad \begin{bmatrix} \frac{-\sqrt{2}}{6} & \frac{\sqrt{2}}{6} & \frac{-2\sqrt{2}}{3} \end{bmatrix}^T \quad \text{for} \quad \lambda_2 = \lambda_3 = -9,$$

together with

$$\begin{bmatrix} \frac{-2}{3} & \frac{2}{3} & \frac{1}{3} \end{bmatrix}^T \quad \text{for} \quad \lambda_1 = 27.$$

You should now be able to solve Problems 1 to 9.

Solvability of Equations

Suppose that $\mathbf{A}$ is $p \times p$. We know that we can only expect to be able to solve $\mathbf{Ax} = \mathbf{b}$ given arbitrary $\mathbf{b}$ when $\mathbf{A}$ is nonsingular. However, we know that even with $\mathbf{A}$ singular there are *some* $\mathbf{b}$ so that $\mathbf{Ax} = \mathbf{b}$ is solvable. The question is: How can we characterize such $\mathbf{b}$ (other than by saying it is in the column space of $\mathbf{A}$, a tautology)? For normal matrices $\mathbf{A}$, there is an elegant characterization.

(8.12) ***Theorem*** (normal matrices and solvability). Suppose that $\mathbf{A}$ is a $p \times p$ normal matrix having eigenvalues λ_i and an orthonormal set of associated eigenvectors $\{\mathbf{x}_1, \ldots, \mathbf{x}_p\}$. Consider the question of the existence of a solution $\mathbf{x}$ to $(\mathbf{A} - \lambda\mathbf{I})\mathbf{x} = \mathbf{b}$, where $\mathbf{b}$ is a given $p \times 1$ matrix and λ is a given number. Then:

(a) If λ is not one of the eigenvalues, then the equation has a unique solution, namely

$$\mathbf{x} = \sum_{i=1}^{p} \frac{(\mathbf{x}_i, \mathbf{b})}{\lambda_i - \lambda} \mathbf{x}_i.$$

(b) If λ equals one of the eigenvalues λ_i, then the equation has a solution if and only if **b** is orthogonal to all the eigenvectors associated with λ_i. When there are solutions, there are infinitely many, and each can be obtained by adding to one solution **x** an arbitrary linear combination of the eigenvectors associated with λ_i.

PROOF. Since $\{\mathbf{x}_1, \ldots, \mathbf{x}_p\}$ is an orthonormal basis, we can write **b** and possible solutions **x** in terms of these vector as

$$\mathbf{b} = (\mathbf{x}_1, \mathbf{b})\mathbf{x}_1 + \cdots + (\mathbf{x}_p, \mathbf{b})\mathbf{x}_p$$

$$\mathbf{x} = \alpha_1\mathbf{x}_1 + \alpha_2\mathbf{x}_2 + \cdots + \alpha_p\mathbf{x}_p \qquad \text{(where the } \alpha_i \text{ are unknown)}$$

according to Theorem 5.74 on orthogonal bases. Thus the equation $\mathbf{b} = (\mathbf{A} - \lambda\mathbf{I})\mathbf{x}$ is equivalent to

$$\mathbf{b} = \sum_{j=1}^{p} (\mathbf{x}_j, \mathbf{b})\mathbf{x}_j = (\mathbf{A} - \lambda\mathbf{I}) \sum_{j=1}^{p} \alpha_j\mathbf{x}_j = \sum_{j=1}^{p} \alpha_j(\lambda_j - \lambda)\mathbf{x}_j;$$

hence a solution $\mathbf{x} = \alpha_1\mathbf{x}_1 + \cdots + \alpha_p\mathbf{x}_p$ exists if and only if

$$\alpha_j(\lambda_j - \lambda) = (\mathbf{x}_j, \mathbf{b})$$

for $j = 1, \ldots, p$. The theorem now follows easily. ■

(8.13) ***Example.*** Consider applying Theorem 8.12 when

$$\mathbf{A} = \begin{bmatrix} 12 & -4\sqrt{2} \\ -4\sqrt{2} & 8 \end{bmatrix},$$

whose eigenvalues $\lambda_1 = 4$ and $\lambda_2 = 16$ and associated eigenvectors are given in Example 8.10. We consider conditions in order that $(\mathbf{A} - \lambda_1\mathbf{I})\mathbf{x} = \mathbf{b}$ be solvable. According to Theorem 8.12, **b** must be orthogonal to $\mathbf{x}_1 = [\sqrt{3}/3 \quad \sqrt{6}/3]^T$. That is,

$$\begin{bmatrix} 8 & -4\sqrt{2} \\ -4\sqrt{2} & 4 \end{bmatrix}\begin{bmatrix} x \\ y \end{bmatrix} = \begin{bmatrix} c \\ d \end{bmatrix}$$

has a solution if and only if $c\sqrt{3}/3 + d\sqrt{6}/3 = 0$—that is, $c = -d\sqrt{2}$.

Application: Forced Oscillations and Resonance

As an example of how this theorem is used in applications, we apply it to studying the oscillating masses from Section 2.5 when those masses are subject to external forces. To be specific, we suppose that an oscillating force is applied to the lower of the two masses, and we ask whether there will still be an oscillating solution of fixed amplitude as in the case without such an external force.

To get a feel for what to expect, consider what happens when you start pushing a child in a swing. The swing has a natural frequency, and if the frequency of your pushes is the same as that natural frequency, you can make the swing go higher and higher (that is, the amplitude is *not* fixed); this is the phenomenon of *resonance.* If you push at a different frequency, however, the amplitude of the swinging will not build up. We want to see whether our model can indicate any of this behavior in the case of the coupled masses.

Suppose that an oscillating downward force $F\cos(\mu t + \alpha)$ of frequency $\mu/(2\pi)$ with $\mu \neq 0$ is applied to the lower mass in the specific case of Example 2.31. Then the second equation in (2.26) is changed by the addition of this term to its right-hand side, giving

$$\begin{aligned} m_1 X_1'' &= -k_1 X_1 + k_2(X_2 - X_1) \\ m_2 X_2'' &= -k_2(X_2 - X_1) + F\cos(\mu t + \alpha). \end{aligned}$$

Since these equations are linear, their general solution is the sum of two terms: (1) the general solution of the homogeneous system (when $F = 0$); and (2) a particular solution for particular F, μ, and α. Section 2.5 developed (1) as linear combinations of $\sin \omega t$ and $\cos \omega t$ for special values of ω; see (2.32).

To find a particular solution (2), we still seek oscillating solutions of fixed magnitude and of unknown frequency $\omega/(2\pi)$ as in (2.27), where $\omega \neq 0$ is to be determined. Substituting those expressions for X_1 and X_2 into the preceding equations gives equations like those just before (2.28), except that the second equation there has the zero on the right-hand side replaced by $F\cos(\mu t + \alpha)$, which equals

$$(F\cos\alpha)\cos\mu t - (F\sin\alpha)\sin\mu t.$$

A little calculation shows that the only way a nontrivial linear combination of $\sin \omega t$ and $\cos \omega t$ can be identically equal to a nontrivial linear combination of $\sin \mu t$ and $\cos \mu t$ for all t is to have $\omega = \mu$ and the coefficients of corresponding functions equal. Equating corresponding coefficients leads to (2.28), except that the second equation there has $-F\sin\alpha$ on the right-hand side and the fourth has $F\cos\alpha$ there. Writing in matrix terms, we finally obtain for a particular solution the analogue of (2.30), using the notation of (2.29):

(8.14)
$$\begin{aligned} &(\mathbf{K} - \mu^2\mathbf{M})\boldsymbol{\xi} = \mathbf{b}_1 \quad \text{and} \quad (\mathbf{K} - \mu^2\mathbf{M})\boldsymbol{\eta} = \mathbf{b}_2, \quad \text{where} \\ &\mathbf{b}_1 = [0 \quad -F\sin\alpha]^T \quad \text{and} \quad \mathbf{b}_2 = [0 \quad F\cos\alpha]^T. \end{aligned}$$

To be specific, we revert to the matrices $\mathbf{K}$ and $\mathbf{M}$ given in Example 2.31; Example 2.34 showed that $\mathbf{K} - \mu^2\mathbf{M}$ is nonsingular except when $\mu^2 = 4$ or $\mu^2 = 16$. Since $\mathbf{K} - \mu^2\mathbf{M}$ is symmetric, it is certainly normal and Theorem 8.12 applies. If $\mu^2 \neq 4$ and $\mu^2 \neq 16$, there is a unique solution $\boldsymbol{\xi}$, $\boldsymbol{\eta}$ to (8.14), which means that $\mathbf{X}(t) = \boldsymbol{\xi}\sin\mu t + \boldsymbol{\eta}\cos\mu t$ is a particular solution of the differential equations. Adding this to the general solution (2.35) of the homogeneous system shows the general solution of the differential equations to be a linear combination of $\sin \mu t$, cos μt, sin $4t$, cos $4t$, sin $2t$, and cos $2t$. That is,

> *if the frequency $\mu/(2\pi)$ of the applied force is different from each of the natural frequencies $1/\pi$ and $2/\pi$, then the motion of the coupled masses consists of a linear combination of three fixed-amplitude oscillations at those three distinct frequencies.*

But what if $\mu^2 = 4$ or $\mu^2 = 16$ (the numbers determining the natural frequencies $1/\pi$ and $2/\pi$)? According to Theorem 8.12, the system (8.14) will have a solution if and only if $\mathbf{b}_1$ and $\mathbf{b}_2$ are orthogonal to the vectors $\mathbf{x}$ for which $(\mathbf{K} - \mu^2\mathbf{M})\mathbf{x} = \mathbf{0}$. It is easy to find all such vectors $\mathbf{x}$ from the results in Example 8.10; we discover that $\mathbf{b}_1$ and $\mathbf{b}_2$ can be orthogonal to such $\mathbf{x}$ if and only if $F = 0$ (that is, there is no applied force). Since we assume that there is an applied force $F \neq 0$, $\mathbf{b}_1$ and $\mathbf{b}_2$ do not meet the conditions needed for (8.14) to have solutions. Therefore,

> *if the frequency $\mu/(2\pi)$ of the applied force equals either natural frequency $1/\pi$ or $2/\pi$, then the motion of the coupled masses cannot consist of a linear combination of fixed-amplitude oscillations.*

Just as in the case of the child's swing, this is the phenomenon of *resonance*: The external force is applied at the natural frequency of the system. Further knowledge of differential equations would allow us to show that in this case there are solutions that are linear combinations of $t(\cos \mu t)$ and $t(\sin \mu t)$—oscillatory solutions *with amplitudes that grow without bound.*

PROBLEMS 8.3

▷ **1.** (a) If $\mathbf{A}$ is $p \times p$, show that λ is an eigenvalue of $\mathbf{A}$ if and only if $\bar{\lambda}$ is an eigenvalue of $\mathbf{A}^H$.
(b) If $\mathbf{A}$ is normal, show that $\mathbf{x}$ is an eigenvector of $\mathbf{A}$ associated with the eigenvalue λ of $\mathbf{A}$ if and only if $\mathbf{x}$ is an eigenvector of $\mathbf{A}^H$ associated with the eigenvalue $\bar{\lambda}$ of $\mathbf{A}^H$.

2. For the matrix $\mathbf{A}$ below, find a pair of eigenvectors that are *not* orthogonal, and also find a pair that *are* ortho*normal.*

$$\mathbf{A} = \begin{bmatrix} 2 & 0 \\ 0 & 2 \end{bmatrix}.$$

3. Find the eigenvalues and eigenvectors of $\mathbf{A}$ so that the eigenvectors form an orthonormal set, where

$$\mathbf{A} = \begin{bmatrix} 2 & -1 & 0 \\ -1 & 3 & -1 \\ 0 & -1 & 2 \end{bmatrix}.$$

▷ **4.** Given that the eigenvalues of $\mathbf{A}$ below are 9, 9, and -3, find an orthonormal set of associated eigenvectors.

$$\mathbf{A} = \begin{bmatrix} 5 & 4 & -4 \\ 4 & 5 & 4 \\ -4 & 4 & 5 \end{bmatrix}$$

5. Show that the matrix below is normal (but neither real symmetric, hermitian, unitary, nor skew-hermitian), and find its eigenvalues and an orthonormal set of eigenvectors.

$$\mathbf{A} = \begin{bmatrix} 1 & -1 \\ 1 & 1 \end{bmatrix}$$

6. Suppose that $\mathbf{A}$ is skew-hermitian ($\mathbf{A}^H = -\mathbf{A}$).

(a) Prove that $i\mathbf{A}$ is hermitian.

(b) Prove that the eigenvalues of $\mathbf{A}$ are pure imaginary.

▷ **7.** Prove that a normal matrix is hermitian if and only if its eigenvalues are real.

8. By using Problem 1 and considering $(\lambda_1 - \lambda_2)\mathbf{x}_1^H\mathbf{x}_2$, show that if $\lambda_1 \neq \lambda_2$ are eigenvalues associated with the eigenvectors $\mathbf{x}_1$ and $\mathbf{x}_2$ of a normal matrix, then $\mathbf{x}_1$ and $\mathbf{x}_2$ are orthogonal.

𝔐 **9.** Use MATLAB or similar software to find the eigenvalues and eigenvectors of the matrix $\mathbf{B} = \mathbf{D}^{-1}\mathbf{K}\mathbf{D}^{-1}$, where $\mathbf{D}^2 = \mathbf{M}$ (see Problem 2 of Section 8.1), for $\mathbf{K}$ and $\mathbf{M}$ as in Problem 10 of Section 7.2; verify that the eigenvectors are (approximately) orthogonal.

▷ **10.** Use Theorem 8.12 on normal matrices and solvability to find necessary and sufficient conditions on b_1 and b_2 for the system

$$\begin{bmatrix} 2-\lambda & 1 \\ 1 & 2-\lambda \end{bmatrix}\begin{bmatrix} x_1 \\ x_2 \end{bmatrix} = \begin{bmatrix} b_1 \\ b_2 \end{bmatrix}$$

to have a solution when:

(a) $\lambda = 1$

(b) $\lambda = 3$

11. Use Theorem 8.12 on normal matrices and solvability to find necessary and sufficient conditions on b_1 and b_2 for there to be a solution to

$$\begin{bmatrix} 3 & -9 \\ -9 & 27 \end{bmatrix}\begin{bmatrix} x_1 \\ x_2 \end{bmatrix} = \begin{bmatrix} b_1 \\ b_2 \end{bmatrix}.$$

▷ **12.** Suppose that $\mathbf{A}$ is normal. Use Theorem 8.12 on normal matrices and solvability to prove that there is at most one solution to $\mathbf{Ax} = \mathbf{b}$ for arbitrary $\mathbf{b}$ if and only if there is at least one solution to $\mathbf{Ax} = \mathbf{b}$ for arbitrary $\mathbf{b}$.

13. Suppose that $\mathbf{A}$ is $p \times p$ and normal. Use orthogonal projection and Theorem 8.12 to prove that each vector $\mathbf{v}$ in $\mathbb{C}^p$ can be written as $\mathbf{v} = \mathbf{Ax} + \mathbf{h}$ for some $\mathbf{x}$ in $\mathbb{C}^p$ and some $\mathbf{h}$ satisfying $\mathbf{Ah} = \mathbf{0}$.

14. Provide the details to derive (8.14).

15. Suppose that in our forced-oscillation model (with **K** and **M** as in Example 2.31) we take $F = 2$, $\alpha = \pi/4$, and $\mu = 3$. Find ξ and η solving (8.14).

𝔐 **16.** Use MATLAB or similar software to find the frequencies at which resonance would occur were forced oscillations applied to the 10 coupled masses of Problem 10 in Section 7.2.

8.4 APPLICATION: SINGULAR VALUE DECOMPOSITION

We have thus far explored at least two useful decompositions that use unitary or orthogonal matrices: (1) the QR-decompositions $\mathbf{A} = \mathbf{QR}$ for $p \times q$ $\mathbf{A}$ (there are various versions of this; compare **Key Theorems 5.82** and **7.38**); and (2) the Schur decomposition $\mathbf{A} = \mathbf{PTP}^H$ for $p \times p$ $\mathbf{A}$.

According to Theorem 6.17 on change of basis and representations, these can be interpreted as statements about representations of the linear transformation $\mathscr{T}$ defined by $\mathscr{T}(\mathbf{v}) = \mathbf{Av}$ with respect to various orthonormal bases: (1) $\mathscr{T}$ is represented by $\mathbf{R}$ with respect to the orthonormal ordered basis $\{\mathbf{e}_1; \ldots; \mathbf{e}_q\}$ for the domain and the orthonormal ordered basis consisting of the columns of $\mathbf{Q}$ for the range of $\mathscr{T}$; and (2) $\mathscr{T}$ is represented by $\mathbf{T}$ with respect to the orthonormal ordered basis consisting of the columns of $\mathbf{P}$ used for both domain and range. The first version (QR) prescribes the basis for the domain but allows that for the range to be chosen to produce a simple representation. The second (Schur) prescribes that the same basis be used for both domain and range. We now examine what can be accomplished by careful choice of unrestricted and possibly different orthonormal bases for both the domain and range. The result is of great practical importance.

Singular Value Decomposition

This section will show that every $p \times q$ matrix $\mathbf{A}$ can be decomposed as

(8.15) $\mathbf{A} = \mathbf{U\Sigma V}^H$, where $\mathbf{U}$ is $p \times p$ and unitary, $\mathbf{V}$ is $q \times q$ and unitary, and $\mathbf{\Sigma}$ is $p \times q$ and "diagonal" in that $\langle\mathbf{\Sigma}\rangle_{ij} = 0$ unless $i = j$, in which case we write $\langle\mathbf{\Sigma}\rangle_{ii} = \sigma_i$ with σ_i real and nonnegative.

(8.16) ***Example.*** Each matrix below is a $\mathbf{\Sigma}$ of the type described in (8.15):

$$\begin{bmatrix} 2 & 0 & 0 \\ 0 & 0 & 0 \end{bmatrix}, \quad \begin{bmatrix} 3 & 0 \\ 0 & 0 \\ 0 & 0 \end{bmatrix}, \quad \begin{bmatrix} 3 & 0 & 0 \\ 0 & 2 & 0 \end{bmatrix},$$

$$\begin{bmatrix} 4 & 0 \\ 0 & 6 \\ 0 & 0 \end{bmatrix}, \quad \begin{bmatrix} 2 & 0 & 0 \\ 0 & 3 & 0 \\ 0 & 0 & 5 \end{bmatrix}.$$

Before we prove that (8.15) can be obtained, we consider first what it would mean if true. We note first that $\mathbf{AV} = \mathbf{U\Sigma}$ means $\mathbf{Av}_i = \sigma_i\mathbf{u}_i$ for $1 \le i \le \min\{p, q\}$, where the columns of $\mathbf{U}$ and $\mathbf{V}$ are the $\mathbf{u}_i$ and $\mathbf{v}_i$. Also, we would have

$$\begin{aligned}\mathbf{A}^H\mathbf{A} &= (\mathbf{U\Sigma V}^H)^H(\mathbf{U\Sigma V}^H) = \mathbf{V\Sigma}^H\mathbf{U}^H\mathbf{U\Sigma V}^H \\ &= \mathbf{V}(\mathbf{\Sigma}^H\mathbf{\Sigma})\mathbf{V}^H \qquad \text{(since } \mathbf{U} \text{ is unitary)},\end{aligned}$$

where $\mathbf{\Sigma}^H\mathbf{\Sigma} = \mathbf{D} = \mathbf{V}^H(\mathbf{A}^H\mathbf{A})\mathbf{V}$ is $q \times q$ and diagonal with real nonnegative entries on its main diagonal. According to **Key Theorem 8.6**, this means that

> *the eigenvectors of* $\mathbf{A}^H\mathbf{A}$ *make up* $\mathbf{V}$*, with the associated (real nonnegative) eigenvalues on the diagonal of* $\mathbf{D} = \mathbf{\Sigma}^H\mathbf{\Sigma}$.

Similarly, we find

$$\mathbf{AA}^H = \mathbf{U}(\mathbf{\Sigma\Sigma}^H)\mathbf{U}^H,$$

where $\mathbf{\Sigma\Sigma}^H = \mathbf{D}' = \mathbf{U}^H(\mathbf{AA}^H)\mathbf{U}$ is $p \times p$ and diagonal with real nonnegative entries on its main diagonal, which in turn means that

> *the eigenvectors of* $\mathbf{AA}^H$ *make up* $\mathbf{U}$*, with the associated (real nonnegative) eigenvalues on the diagonal of* $\mathbf{D}' = \mathbf{\Sigma\Sigma}^H$.

The diagonal entries of both $\mathbf{\Sigma\Sigma}^H$ and $\mathbf{\Sigma}^H\mathbf{\Sigma}$ are just the σ_i^2 (perhaps with some zeros added), where $\langle\mathbf{\Sigma}\rangle_{ii} = \sigma_i$.

We can use these results to construct an approach to prove that (8.15) can be accomplished: We can *define* $\mathbf{V}$ to contain the eigenvectors of $\mathbf{A}^H\mathbf{A}$, *define* the σ_i so that the eigenvalues of $\mathbf{A}^H\mathbf{A}$ are the σ_i^2, and *define* $\mathbf{U}$ to contain the eigenvectors of $\mathbf{AA}^H$. We would then need to *prove* that indeed $\mathbf{A} = \mathbf{U\Sigma V}^H$. Since the proofs are rather technical, we give an illustrative example of how this approach could work.

(8.17) ***Example.*** Consider

$$\mathbf{A} = \begin{bmatrix} 1 & 1 \\ 2 & 2 \\ 2 & 2 \end{bmatrix}.$$

Since

$$\mathbf{A}^H\mathbf{A} = \begin{bmatrix} 9 & 9 \\ 9 & 9 \end{bmatrix}$$

has (nonnegative!) eigenvalues 18 and 0, we can take $\sigma_1 = \sqrt{18} = 3\sqrt{2}$ and $\sigma_2 = 0$. A normalized pair of eigenvectors of $\mathbf{A}^H\mathbf{A}$ can easily be found as

$$\mathbf{v}_1 = [\sqrt{2}/2 \quad \sqrt{2}/2]^T \text{ and } \mathbf{v}_2 = [\sqrt{2}/2 \quad -\sqrt{2}/2]^T,$$

from which we will form $\mathbf{V} = [\mathbf{v}_1 \quad \mathbf{v}_2]$ in the hope of obtaining (8.15). Similarly, we find that

$$\mathbf{A}\mathbf{A}^H = \begin{bmatrix} 2 & 4 & 4 \\ 4 & 8 & 8 \\ 4 & 8 & 8 \end{bmatrix}$$

has eigenvalues 18 (which does equal σ_1^2), 0 (which does equal σ_2^2), and 0. Associated normalized eigenvectors are easily found as

$$\mathbf{u}_1 = \begin{bmatrix} \frac{1}{3} & \frac{2}{3} & \frac{2}{3} \end{bmatrix}^T, \qquad \mathbf{u}_2 = \begin{bmatrix} -\frac{2\sqrt{5}}{5} & \frac{\sqrt{5}}{5} & 0 \end{bmatrix}^T,$$

$$\mathbf{u}_3 = \begin{bmatrix} \frac{2\sqrt{5}}{15} & \frac{4\sqrt{5}}{15} & -\frac{\sqrt{5}}{3} \end{bmatrix}^T,$$

from which we will form $\mathbf{U} = [\mathbf{u}_1 \quad \mathbf{u}_2 \quad \mathbf{u}_3]$ in the hope of obtaining (8.15). We now check whether $\mathbf{A} = \mathbf{U}\mathbf{\Sigma}\mathbf{V}^H$ and find that indeed it does:

$$\mathbf{U}\mathbf{\Sigma}\mathbf{V}^H = \begin{bmatrix} \frac{1}{3} & -\frac{2\sqrt{5}}{5} & \frac{2\sqrt{5}}{15} \\ \frac{2}{3} & \frac{\sqrt{5}}{5} & \frac{4\sqrt{5}}{15} \\ \frac{2}{3} & 0 & -\frac{\sqrt{5}}{3} \end{bmatrix} \begin{bmatrix} 3\sqrt{2} & 0 \\ 0 & 0 \\ 0 & 0 \end{bmatrix} \begin{bmatrix} \frac{\sqrt{2}}{2} & \frac{\sqrt{2}}{2} \\ \frac{\sqrt{2}}{2} & -\frac{\sqrt{2}}{2} \end{bmatrix} = \mathbf{A}.$$

Observe also that $\mathbf{A}\mathbf{v}_i = \sigma_i \mathbf{u}_i$ and $\mathbf{A}^H \mathbf{u}_i = \sigma_i \mathbf{v}_i$ for $i = 1, 2$.

Note that we can easily construct the decomposition (8.15) in the simplest cases:

(8.18)

1. For $\mathbf{A}$ a $p \times q$ zero matrix, we have $\mathbf{0} = \mathbf{I}_p \mathbf{0} \mathbf{I}_q^H$, so $\mathbf{U} = \mathbf{I}_p$, $\mathbf{\Sigma} = \mathbf{0}$, and $\mathbf{V} = \mathbf{I}_q$.
2. For $\mathbf{A}$ a nonzero $p \times 1$ column matrix $\mathbf{a}$, we can use the Gram-Schmidt process to produce an orthonormal basis $\mathbf{a}/\|\mathbf{a}\|_2, \mathbf{u}_2, \ldots, \mathbf{u}_p$ for $\mathbb{C}^p$ and write
$$\mathbf{a} = \mathbf{U}(\|\mathbf{a}\|_2 \mathbf{e}_1)[1]$$
with $\mathbf{U} = [\mathbf{a}/\|\mathbf{a}\|_2 \quad \mathbf{u}_2 \quad \cdots \quad \mathbf{u}_p]$, $\mathbf{\Sigma} = \|\mathbf{a}\|_2 \mathbf{e}_1$, $\mathbf{V} = [1]$.
3. If we can obtain a decomposition (8.15) for $\mathbf{A}^H = \mathbf{U}_0 \mathbf{\Sigma}_0 \mathbf{V}_0^H$, say, then
$$\mathbf{A} = \mathbf{U}\mathbf{\Sigma}\mathbf{V}^H$$
with $\mathbf{U} = \mathbf{V}_0$, $\mathbf{\Sigma} = \mathbf{\Sigma}_0^T$, $\mathbf{V} = \mathbf{U}_0$.
4. For $\mathbf{A}$ a nonzero row matrix, we can apply (2) to $\mathbf{A}^H$ and then use (3).

We can now proceed more formally.

(8.19) **Key Theorem** (singular value decomposition). Let $\mathbf{A}$ be $p \times q$.

(a) There exists a $p \times p$ unitary matrix $\mathbf{U}$ (orthogonal if $\mathbf{A}$ is real), a $q \times q$ unitary matrix $\mathbf{V}$ (orthogonal if $\mathbf{A}$ is real), and a $p \times q$ "diagonal" matrix $\mathbf{\Sigma}$ with $\langle\mathbf{\Sigma}\rangle_{ij} = 0$ for $i \neq j$ and $\langle\mathbf{\Sigma}\rangle_{ii} = \sigma_i \geq 0$ with $\sigma_1 \geq \sigma_2 \geq \cdots \geq \sigma_s$, where $s = \min\{p, q\}$, such that the *singular value decomposition*

$$\mathbf{A} = \mathbf{U\Sigma V}^H \qquad (\mathbf{A} = \mathbf{U\Sigma V}^T \text{ if } \mathbf{A} \text{ is real})$$

is valid.

(b) The numbers σ_i^2 *make up* the eigenvalues of $\mathbf{A}^H\mathbf{A}$ (perhaps with some zeros added), and the associated eigenvectors are the columns $\mathbf{v}_i$ of $\mathbf{V}$; similarly, the σ_i^2 *make up* the eigenvalues of $\mathbf{AA}^H$ (perhaps with some zeros added), and the associated eigenvectors are the columns $\mathbf{u}_i$ of $\mathbf{U}$. The σ_i are called the *singular values* of $\mathbf{A}$, the $\mathbf{u}_i$ the associated *left singular vectors* of $\mathbf{A}$, and the $\mathbf{v}_i$ the associated *right singular vectors* of $\mathbf{A}$; they are related by $\mathbf{Av}_i = \sigma_i\mathbf{u}_i$ for $1 \leq i \leq s$.

PROOF. Under the assumption (8.15)—which is just (a) here—we already derived (b); thus we need only prove (a). Because of (8.18)(3) and (1), we may assume $p \geq q$ and $\mathbf{A} \neq \mathbf{0}$. Our proof uses induction: (1) by (8.15)(2) the result is true for $(p - q + 1) \times 1$ matrices; (2) we suppose that it is true for $(p - q + k) \times k$ matrices and prove that it is true for $p' \times q'$ matrices where $p' = p - q + k + 1$ and $q' = k + 1$.

Suppose that $\mathbf{A}$ is $p' \times q'$ with $p' \geq q'$, that $\mathbf{A} \neq \mathbf{0}$, and that the theorem is true for $(p' - 1) \times (q' - 1)$ matrices. $\mathbf{A}^H\mathbf{A}$ is hermitian (the real case will not be mentioned explicitly hereafter), nonzero, and has at least one nonzero eigenvalue (otherwise, $\mathbf{A} = \mathbf{0}$). If $\mathbf{A}^H\mathbf{Av} = \lambda\mathbf{v}$, then

$$\lambda\|\mathbf{v}\|_2^2 = \mathbf{v}^H\lambda\mathbf{v} = \mathbf{v}^H\mathbf{A}^H\mathbf{Av} = \|\mathbf{Av}\|_2^2,$$

so all eigenvalues of $\mathbf{A}^H\mathbf{A}$ are nonnegative; let λ_1 be the largest positive eigenvalue, let $\sigma_1 = +\sqrt{\lambda_1}$, let $\mathbf{v}_1$ be an associated normalized eigenvector, and let $\mathbf{V}_1 = [\mathbf{v}_1 \quad \mathbf{V}_0]$ be a unitary matrix formed from an orthonormal set of eigenvectors of $\mathbf{A}^H\mathbf{A}$. Define $\mathbf{u}_1 = \mathbf{Av}_1/\sigma_1$, so that $\|\mathbf{u}_1\|_2 = 1$, and let $\mathbf{U}_1 = [\mathbf{u}_1 \quad \mathbf{U}_0]$ be a unitary matrix (built by the Gram-Schmidt process, say). We have

$$\mathbf{Av}_1 = \sigma_1\mathbf{u}_1, \qquad \mathbf{u}_1^H\mathbf{u}_1 = 1, \qquad \mathbf{U}_0^H\mathbf{u}_1 = \mathbf{0}, \qquad \mathbf{v}_1^H\mathbf{V}_0 = \mathbf{0},$$

$$\mathbf{u}_1^H\mathbf{AV}_0 = \left(\frac{\mathbf{Av}_1}{\sigma_1}\right)^H \mathbf{AV}_0 = \left(\frac{\mathbf{A}^H\mathbf{Av}_1}{\sigma_1}\right)^H \mathbf{V}_0 = \sigma_1\mathbf{v}_1^H\mathbf{V}_0 = \mathbf{0}.$$

This gives

$$\mathbf{A}' = \mathbf{U}_1^H\mathbf{AV}_1 = [\mathbf{u}_1 \quad \mathbf{U}_0]^H\mathbf{A}[\mathbf{v}_1 \quad \mathbf{V}_0] = \begin{bmatrix} \sigma_1 & \mathbf{0} \\ \mathbf{0} & \mathbf{U}_0^H\mathbf{AV}_0 \end{bmatrix} = \begin{bmatrix} \sigma_1 & \mathbf{0} \\ \mathbf{0} & \mathbf{A}_0 \end{bmatrix}$$

with $\mathbf{A}_0$ $(p' - 1) \times (q' - 1)$. By the inductive hypothesis,

$$\mathbf{A}_0 = \mathbf{P}_0\boldsymbol{\Sigma}_0\mathbf{Q}_0^H \quad \text{and} \quad \mathbf{P}_0^H\mathbf{A}_0\mathbf{Q}_0 = \boldsymbol{\Sigma}_0$$

with $\mathbf{P}_0$ and $\mathbf{Q}_0$ unitary. But then

$$\begin{bmatrix} 1 & \mathbf{0} \\ \mathbf{0} & \mathbf{P}_0 \end{bmatrix}^H \mathbf{U}_1^H\mathbf{A}\mathbf{V}_1 \begin{bmatrix} 1 & \mathbf{0} \\ \mathbf{0} & \mathbf{Q}_0 \end{bmatrix} = \begin{bmatrix} \sigma_1 & \mathbf{0} \\ \mathbf{0} & \boldsymbol{\Sigma}_0 \end{bmatrix} = \boldsymbol{\Sigma},$$

as required to complete the inductive argument. ■

In practice, the computation of a singular value decomposition of $p \times q$ $\mathbf{A}$ is performed by computer software that makes use of efficient techniques—see Section 7.6—for finding the eigensystem of a symmetric or hermitian matrix; the total effort involved is usually a small multiple of $p^2q + pq^2 + \{\min(p, q)\}^3$—roughly comparable to several matrix inversions when $p = q$.

You should now be able to solve Problems 1 to 12.

Singular Values and Rank

The singular value decomposition contains a great deal of information about the matrix $\mathbf{A}$.

(8.20) **Key Corollary.** Let $\mathbf{A}$ be $p \times q$.
(a) The rank k of $\mathbf{A}$ equals the number of nonzero singular values of $\mathbf{A}$.
(b) The first k left singular vectors $\mathbf{u}_1, \ldots, \mathbf{u}_k$ form an orthonormal basis for the column space of $\mathbf{A}$, that is, for the range of $\mathscr{T}$ defined by $\mathscr{T}(\mathbf{v}) = \mathbf{A}\mathbf{v}$.
(c) The last $q - k$ right singular vectors $\mathbf{v}_{k+1}, \ldots, \mathbf{v}_q$ form an orthonormal basis for the null space of $\mathscr{T}$.

PROOF. Problems 13 and 14. ■

Thus the singular value decomposition can be used to determine the rank of a matrix. More important, it can be used to analyze the various ranks that $\mathbf{A}$ might possibly have if its entries are subject to errors (such as measuring or modeling errors). We saw in Example 6.31 that the number of independent factors in a chemical reaction can be determined as the rank of a certain matrix of measured concentrations, and we saw how errors in these measurements could cause us to believe that the rank is greater than it probably really is; we dealt with this by seeking to find the least rank of all matrices $\mathbf{C} + \delta\mathbf{C}$, where the perturbations $\delta\mathbf{C}$ in the concentration matrix $\mathbf{C}$ were to be smaller than the errors we believed to be inherent in our measurements. The following theorem, whose proof is omitted, is the basis for accomplishing this.

(8.21) ***Theorem*** (approximate rank and singular values). Let the $p \times q$ matrix $\mathbf{A}$ of rank k have a singular value decomposition using the notation in **Key Theorem 8.19**.

(a) For given $r < k$, the matrix $\mathbf{A}_r$ of rank r that minimizes $\|\mathbf{A} - \mathbf{A}'\|_2$ over all $p \times q$ matrices $\mathbf{A}'$ of rank r is given by

$$\mathbf{A}_r = \sigma_1 \mathbf{u}_1 \mathbf{v}_1^H + \sigma_2 \mathbf{u}_2 \mathbf{v}_2^H + \cdots + \sigma_r \mathbf{u}_r \mathbf{v}_r^H,$$

and the minimum is $\|\mathbf{A} - \mathbf{A}_r\|_2 = \sigma_{r+1}$.

(b) For given $\epsilon > 0$, the matrix $\mathbf{A}_{\text{min}}$ of least rank of all matrices $\mathbf{A}'$ satisfying $\|\mathbf{A} - \mathbf{A}'\|_2 \le \epsilon$ equals $\mathbf{A}_r$ from (a), where $\sigma_{r+1} \le \epsilon$ but $\sigma_r > \epsilon$.

𝔐 (8.22) ***Example.*** Consider the concentration matrix $\mathbf{C}$ of Example 6.31 again, where its entries are subject to experimental errors of size 0.015 and where the problem is to determine the number of independent factors in the reaction studied; this number is known to equal the rank of $\mathbf{C}$ if perfect measurements are made. MATLAB computed the singular values of $\mathbf{C}$ to be $\sigma_1 = 9.5213$, $\sigma_2 = 0.0071$, and $\sigma_3 = 0.0023$. By **Key Corollary 8.20**, the measured $\mathbf{C}$ has rank $k = 3$; however, by Theorem 8.21, there is a matrix of rank two only 0.0023 away and a matrix of rank one only 0.0071 away, both distances measured in the 2-norm. Since no entry of a matrix $\mathbf{A}$ can exceed $\|\mathbf{A}\|$ in any of the 1-norm, 2-norm, or ∞-norm (since $\|\mathbf{A}\mathbf{e}_i\| \le \|\mathbf{A}\|$), the entries of these rank 2 and rank 1 matrices are within experimental error of those measured in $\mathbf{C}$. We can only safely say that the rank of the perfectly measured $\mathbf{C}$ is at least one, so we can only be confident of there being one independent factor in the chemical reaction. Note that it might even be reasonable to *replace* $\mathbf{C}$ by the nearest rank 1 approximation $\mathbf{C}_1 = \sigma_1 \mathbf{u}_1 \mathbf{v}_1^T$:

$$\mathbf{C}_1 = \sigma_1 \mathbf{u}_1 \mathbf{v}_1^T = \begin{bmatrix} 1.0193 & 2.0280 & 4.2011 \\ 0.2567 & 0.5107 & 1.0580 \\ 1.7395 & 3.4610 & 7.1696 \end{bmatrix},$$

which is indeed within measurement error of the measured $\mathbf{C}$.

The ability of the singular value decomposition to tell us how to obtain low-rank approximations to a given matrix can be useful in data compression—for example, in sending photographs from space. Consider a camera on a satellite far from Earth; how can thousands of photographs be relayed home? Each photo can be *discretized* or *digitized* by breaking the image into many tiny squares and assigning a blackness level to that square. For example, we could impose a 1000×1000 grid on the photo and assign a blackness level of from 0 to 10 for each tiny square, relay to Earth the 1,000,000 integers for each photo, and then reconstruct the photo on Earth.

The problem with this notion is the relaying of 1,000,000 pieces of data for each photo, when there may easily be thousands of photos along with other data.

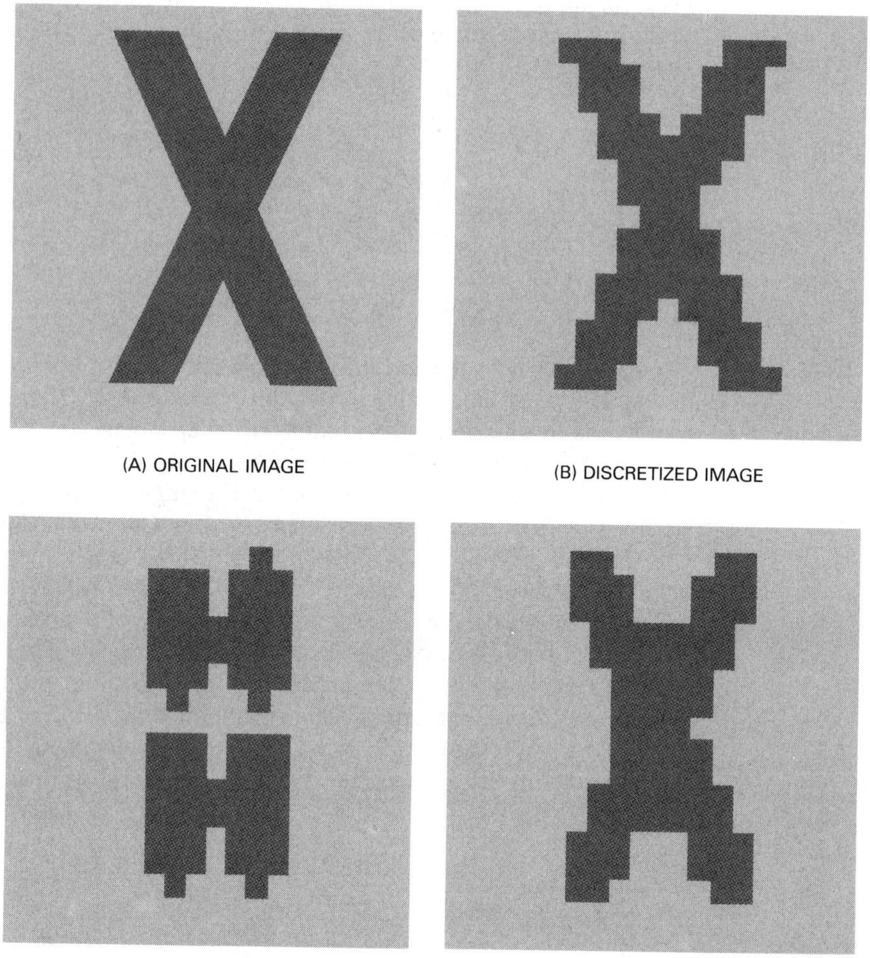

(A) ORIGINAL IMAGE (B) DISCRETIZED IMAGE

(C) RANK-1 RECONSTRUCTION (D) RANK-2 RECONSTRUCTION

A method is needed to compress the data so as to transmit a smaller amount. One approach is to find the singular value decomposition of the 1000×1000 matrix $\mathbf{B}$ of blackness levels and see whether a low-rank approximation

$$\mathbf{B}_r = \sigma_1 \mathbf{u}_1 \mathbf{v}_1^T + \cdots + \sigma_r \mathbf{u}_r \mathbf{v}_r^T$$

might adequately represent the image; this could be transmitted to Earth *not as* $\mathbf{B}_r$ *explicitly but rather in the form of the* $2r$ *vectors* $\mathbf{u}_i$ and $\mathbf{v}_i$ *and the* r *numbers* σ_i. If $r = 5$ is sufficient to reconstruct the photo, for example, only $2 \times 5 \times 1000 + 5 = 10{,}005$ pieces of data are needed per photo rather than 1,000,000—a savings of almost 99%.

𝔐 (8.23) ***Example.*** We consider a crude and simple example of the approach just discussed. Suppose that the first **X** in the figure on page 344 is a photo of an X-shaped object, where only black and white are allowed as blackness levels. We impose a crude 20×20 grid on the photo, and assign the value 1 (black) to each square that is at least half black and assign the value 0 otherwise. This gives a discretized version of the photo, with $20(20) = 400$ entries. If we perfectly transmit these 400 entries and then construct an approximating photo by making an entire square black if its value equals 1, we obtain the second **X** in the figure. We want to send a smaller amount of data, however; the question is whether we can do so and obtain a reconstructed photo that is about as close to the first **X** (the perfect photo) *as is the second* **X**. MATLAB computed the singular values of the 20×20 blackness matrix $\mathbf{B}$ as approximately $\sigma_1 = 7.3$, $\sigma_2 = 4.6$, $\sigma_3 = 2.2$, $\sigma_4 = 1.7$, $\sigma_5 = 0.8$, $\sigma_6 = 0.7$, $\sigma_7 = 0.5$, and the other $\sigma_i \approx 0.0000$. Thus the rank of $\mathbf{B}$ is computed by MATLAB to equal 7. The closest rank 1 approximation to $\mathbf{B}$ is

$$\mathbf{B}_1 = \sigma_1 \mathbf{u}_1 \mathbf{v}_1^T, \quad \text{and} \quad \|\mathbf{B} - \mathbf{B}_1\|_2 = \sigma_2 = 4.6;$$

the information in $\mathbf{B}_1$ is contained in the 41 numbers of σ_1, $\mathbf{u}_1$, and $\mathbf{v}_1$. If we compute $\mathbf{B}_1$ and color any square black that has a corresponding entry in $\mathbf{B}_1$ of at least $\frac{1}{2}$, we get the split shape in the figure; it is unsatisfactory. However, there is a rank 2 matrix $\mathbf{B}_2 = \sigma_1 \mathbf{u}_1 \mathbf{v}_1^T + \sigma_2 \mathbf{u}_2 \mathbf{v}_2^T$ requiring only 82 numbers and having $\|\mathbf{B} - \mathbf{B}_2\|_2 = \sigma_3 = 2.2$. If we similarly reconstruct the photo from $\mathbf{B}_2$, we get the final shape on page 344. It clearly is X-shaped and seems fairly close to the discretized version of the true picture (see the second **X** again); in fact, this rank 2 reconstruction omits only four squares it should have blackened and blackens only four that it should not have. Eight errors in 400 squares—98% correct—seems adequate performance.

PROBLEMS 8.4

1. Verify that $\mathbf{A}\mathbf{v}_i = \sigma_i \mathbf{u}_i$ in Example 8.17.

▷ 2. Find a singular value decomposition of:
   (a) $[1 \quad 2 \quad -2]^T$ (b) $[-4 \quad 12 \quad 3]$

3. "Right" in "right singular vectors $\mathbf{v}_i$" refers to $\mathbf{A}\mathbf{v}_i = \sigma_i \mathbf{u}_i$. Show that $\mathbf{u}_i^H \mathbf{A} = \sigma_i \mathbf{v}_i^H$, as explanation of "left."

▷ 4. The proof of **Key Theorem 8.19** with nonzero $\mathbf{A}$ asserts that $\mathbf{A}^H\mathbf{A} \neq \mathbf{0}$ and that $\mathbf{A}^H\mathbf{A}$ has at least one nonzero eigenvalue. Prove these assertions.

5. Suppose that $\mathbf{A}$ is $p \times q$. Show that there are orthonormal bases for $\mathbb{C}^q$ and $\mathbb{C}^p$ with respect to which the linear transformation $\mathscr{T}$ defined by $\mathscr{T}(\mathbf{v}) = \mathbf{A}\mathbf{v}$ is represented by the matrix $\mathbf{\Sigma}$ in a singular value decomposition of $\mathbf{A}$.

6. Find a singular value decomposition of

$$\mathbf{A} = \begin{bmatrix} 1 & -1 & 2 \\ -1 & 1 & -2 \end{bmatrix}.$$

▷ 7. Use the singular value decomposition to prove that $\|\mathbf{A}\|_2 = \sigma_1$.

8. The *Frobenius norm* $\|\mathbf{A}\|_F$ of a $p \times q$ matrix $\mathbf{A}$ is defined as the square root of the sum of the squares of the magnitudes of all the entries of $\mathbf{A}$.
   (a) Show that $\|\mathbf{A}\|_F \geq 0$ and $\|\mathbf{A}\|_F = 0$ if and only if $\mathbf{A} = \mathbf{0}$, that $\|\alpha\mathbf{A}\|_F = |\alpha|\,\|\mathbf{A}\|_F$, and that $\|\mathbf{A} + \mathbf{B}\|_F \leq \|\mathbf{A}\|_F + \|\mathbf{B}\|_F$.
   (b) Compute $\|\mathbf{I}_p\|_F$ and deduce that $\|\cdot\|_F$ cannot be a transformation norm induced by some vector norms.

9. Let $\mathbf{A}$ be $p \times q$ and let $\mathbf{U}$ and $\mathbf{V}$ be unitary. Show that

$$\|\mathbf{A}\|_F = \|\mathbf{UA}\|_F = \|\mathbf{AV}\|_F = \|\mathbf{UAV}\|_F,$$

with $\|\cdot\|_F$ as in Problem 8.

▷ 10. Use Problem 9 to prove that $\|\mathbf{A}\|_F^2$ equals the sum of the squares of the singular values of $\mathbf{A}$.

11. Suppose that for theoretical reasons it is known that a certain matrix $\mathbf{Q}_0$ should be unitary, but that measurement errors have produced a nonunitary $\mathbf{Q}_0$. Prove that the unitary matrix $\mathbf{Q}_1$ that minimizes $\|\mathbf{Q} - \mathbf{Q}_0\|_F$ over all unitary matrices $\mathbf{Q}$ equals $\mathbf{UV}^H$ if $\mathbf{Q}_0 = \mathbf{U\Sigma V}^H$ is a singular value decomposition of $\mathbf{Q}_0$.

▷ 12. As in Problem 11, find the closest unitary matrices $\mathbf{Q}_1$ to each of the matrices $\mathbf{Q}_0$ below.

(a) $\mathbf{Q}_0 = \begin{bmatrix} 1 & -0.1 \\ 0.1 & 1 \end{bmatrix}$ (b) $\mathbf{Q}_0 = \begin{bmatrix} \frac{3}{5} & 1 \\ \frac{4}{5} & 0 \end{bmatrix}$.

13. Prove **Key Corollary 8.20(a)** on rank and singular values.

14. Prove **Key Corollary 8.20(b)** and **(c)** on singular vectors and bases.

𝔐 15. Use MATLAB or similar software to compute the rank 3 approximation $\mathbf{B}_3$ to the blackness matrix $\mathbf{B}$ in Example 8.24 and reconstruct the photo using $\mathbf{B}_3$.

𝔐 16. Make a 40×40 discretization of the $\mathbf{X}$ in Example 8.23 and use MATLAB or similar software to compute low-rank approximations to the photo to see how good an approximation you can obtain using little information.

𝔐 17. Use MATLAB or similar software to find the closest rank 5 and closest rank 2 matrices to the matrix $\mathbf{H}_6$ in Example 5.84.

8.5 APPLICATION: LEAST SQUARES AND THE PSEUDOINVERSE

Least-squares problems are central to mathematical modeling. We saw as early as Chapter 2, for example, that a good way to find a straight line that comes as close as possible to passing through some given points is to pose the problem in terms of least squares; see Section 2.6. This was one instance of a general phenomenon: One has a model depending on various parameters to be determined so as to model as accurately as possible how some system has actually behaved in the past. Problem 12 in Section 2.6 is of this type: It asks you to find the entries in the transition matrix for the dairy-competition model of Section 2.2 so that the model corresponds to data on how the market shares have actually changed. Problem 22 in Section 5.9 is also of this type.

More generally, the problem is to determine the parameters $\mathbf{x}$ so that the model's prediction $\mathbf{Ax}$ is close to the measured data $\mathbf{b}$. With least *squares*, we seek to find $\mathbf{x}$ to minimize $\|\mathbf{Ax} - \mathbf{b}\|_2$ over all possible $\mathbf{x}$, where $\mathbf{A}$ is $p \times q$ and given, $\mathbf{b}$ is $p \times 1$ and given, and $\mathbf{x}$ is $q \times 1$ and unknown. Problem 26 in Section 5.9 stated that $\mathbf{x}$ solves this if and only if $\mathbf{A}^H\mathbf{Ax} = \mathbf{A}^H\mathbf{b}$; as was shown in Problem 24 of Section 5.9, however, this can be a poor approach computationally. That section showed that a much better method can be based on the normalized QR-decomposition of $\mathbf{A}$: Write $p \times q$ $\mathbf{A}$ of rank k as $\mathbf{A} = \mathbf{QR}$ where $\mathbf{Q}$ is $p \times k$ and has orthonormal columns and $\mathbf{R}$ is $k \times q$ upper-triangular of rank k, and then $\mathbf{x}$ solves $\mathbf{Rx} = \mathbf{Q}^T\mathbf{b}$—which can be solved by simple back-substitution. We now consider another method for solving least-squares problems that is equally effective.

Singular Value Decompositions and Least Squares

Suppose that $\mathbf{A} = \mathbf{U\Sigma V}^H$ is the singular value decomposition from **Key Theorem 8.19** for the $p \times q$ matrix $\mathbf{A}$ of rank k. We consider the problem of minimizing $\|\mathbf{Ax} - \mathbf{b}\|_2$ with respect to $\mathbf{x}$. By Theorem 7.31 on unitary matrices, we can write

$$\|\mathbf{Ax} - \mathbf{b}\|_2 = \|\mathbf{U\Sigma V}^H\mathbf{x} - \mathbf{b}\|_2 = \|\mathbf{\Sigma y} - \mathbf{U}^H\mathbf{b}\|_2,$$

where $\mathbf{y} = \mathbf{V}^H\mathbf{x}$ is the new variable with respect to which we are minimizing. Thus $\mathbf{x}$ minimizes $\|\mathbf{Ax} - \mathbf{b}\|_2$ if and only if $\mathbf{y}$ $(= \mathbf{V}^H\mathbf{x})$ minimizes $\|\mathbf{\Sigma y} - \mathbf{b}'\|_2$, where we use $\mathbf{b}'$ to denote $\mathbf{U}^H\mathbf{b}$. But

$$\begin{aligned}\|\mathbf{\Sigma y} - \mathbf{b}'\|_2^2 &= |\sigma_1 y_1 - b_1'|^2 + \cdots + |\sigma_k y_k - b_k'|^2 \\ &\quad + |b_{k+1}'|^2 + \cdots + |b_p'|^2,\end{aligned}$$

where $y_i = \langle \mathbf{y} \rangle_i$ and $b_i' = \langle \mathbf{b}' \rangle_i$. This expression is minimized by making as many terms zero as possible: $y_i = b_i'/\sigma_i$ for $1 \le i \le k$ and y_i arbitrary for $k + 1 \le i \le q$. Since $\|\mathbf{x}\|_2 = \|\mathbf{Vy}\|_2 = \|\mathbf{y}\|_2$, the $\mathbf{x}$ that has least norm from among all solutions to the least-squares problem comes from $y_i = 0$ for $k + 1 \le i \le q$; all other solutions can be obtained by adding to that $\mathbf{x}$ an arbitrary linear combination of $\mathbf{v}_{k+1}, \ldots, \mathbf{v}_q$, the last $q - k$ columns of $\mathbf{V}$.

This gives a method for solving least-squares problems:

(8.24) To find $\mathbf{x}$ minimizing $\|\mathbf{Ax} - \mathbf{b}\|_2$ with $\mathbf{A}$ having rank k:

1. Find the singular value decomposition $\mathbf{A} = \mathbf{U\Sigma V}^H$.
2. Compute $\mathbf{b}' = \mathbf{U}^H\mathbf{b}$.

3. Compute $\mathbf{y}$ with $y_i = b'_i/\sigma_i$ for $1 \le i \le k$, $y_i = 0$ otherwise.
4. Compute $\mathbf{x}_0 = \mathbf{V}\mathbf{y}$.
5. $\mathbf{x}_0$ solves the least-squares problem, and among all such solutions $\mathbf{x}_0$ has the smallest 2-norm; any other $\mathbf{x}'$ is a solution if and only if $\mathbf{x}'$ equals $\mathbf{x}_0$ plus a linear combination of the last $q - k$ columns of $\mathbf{V}$.

The process (8.24) can be more compactly described by the simple device of defining a matrix that handles step 3.

(8.25) ***Definition.*** Suppose that $\mathbf{\Sigma}$ is a $p \times q$ matrix with $\langle\mathbf{\Sigma}\rangle_{ij} = 0$ for $i \neq j$ and with $\langle\mathbf{\Sigma}\rangle_{ii} = \sigma_i$ for all i, with $\sigma_i \neq 0$ for $1 \le i \le k$ and $\sigma_i = 0$ for $k + 1 \le i \le \min\{p, q\}$. Then $\mathbf{\Sigma}^+$ is that $q \times p$ matrix (*note reversal of p and q*) whose only nonzero entries are $\langle\mathbf{\Sigma}^+\rangle_{ii} = 1/\sigma_i$ for $1 \le i \le k$.

In this notation, $\mathbf{x}_0$ in (8.24) is just $\mathbf{x}_0 = \mathbf{V}\mathbf{\Sigma}^+\mathbf{U}^H\mathbf{b}$; we have proved the following important result.

(8.26) **Key Theorem** (least squares and singular values). Suppose that $\mathbf{A} = \mathbf{U}\mathbf{\Sigma}\mathbf{V}^H$ is the singular value decomposition of the $p \times q$ matrix $\mathbf{A}$ of rank k, and that $\mathbf{A}^+ = \mathbf{V}\mathbf{\Sigma}^+\mathbf{U}^H$ is the so-called *pseudoinverse* of $\mathbf{A}$, where $\mathbf{\Sigma}^+$ is as in Definition 8.25. Then:
(a) $\mathbf{x}_0 = \mathbf{A}^+\mathbf{b}$ minimizes $\|\mathbf{Ax} - \mathbf{b}\|_2$ with respect to $\mathbf{x}$.
(b) Among all minimizers $\mathbf{x}'$ of $\|\mathbf{Ax} - \mathbf{b}\|_2$, $\mathbf{x}_0 = \mathbf{A}^+\mathbf{b}$ has least 2-norm.
(c) $\mathbf{x}'$ minimizes $\|\mathbf{Ax} - \mathbf{b}\|_2$ if and only if $\mathbf{x}' = \mathbf{x}_0 + \mathbf{v}$, where $\mathbf{v}$ is an arbitrary linear combination of the final $q - k$ columns of $\mathbf{V}$ and $\mathbf{x}_0 = \mathbf{A}^+\mathbf{b}$.

(8.27) ***Example.*** Consider the least-squares problem $\mathbf{Ax} \approx \mathbf{b}$:

$$\begin{bmatrix} 1 & 1 \\ 2 & 2 \\ 2 & 2 \end{bmatrix} \begin{bmatrix} x_1 \\ x_2 \end{bmatrix} \approx \begin{bmatrix} 15 \\ 15 \\ -30 \end{bmatrix}.$$

The singular value decomposition of $\mathbf{A}$ was obtained in Example 8.17. Following the procedure in (8.24), we compute $\mathbf{b}' = \mathbf{U}^H\mathbf{b}$, $\mathbf{y} = \mathbf{\Sigma}^+\mathbf{b}'$, and then $\mathbf{x}_0 = \mathbf{V}\mathbf{y}$:

$$\mathbf{b}' = \begin{bmatrix} \frac{1}{3} & \frac{2}{3} & \frac{2}{3} \\ -\frac{2\sqrt{5}}{5} & \frac{\sqrt{5}}{5} & \frac{4\sqrt{5}}{15} \\ \frac{2\sqrt{5}}{15} & \frac{4\sqrt{5}}{15} & -\frac{\sqrt{5}}{3} \end{bmatrix} \begin{bmatrix} 15 \\ 15 \\ -30 \end{bmatrix} = \begin{bmatrix} -5 \\ -11\sqrt{5} \\ 16\sqrt{5} \end{bmatrix},$$

$$\mathbf{y} = \begin{bmatrix} \frac{1}{3\sqrt{2}} & 0 & 0 \\ 0 & 0 & 0 \end{bmatrix} \begin{bmatrix} -5 \\ -11\sqrt{5} \\ 16\sqrt{5} \end{bmatrix} = \begin{bmatrix} -\frac{5\sqrt{2}}{6} \\ 0 \end{bmatrix},$$

$$\mathbf{x}_0 = \begin{bmatrix} \frac{\sqrt{2}}{2} & \frac{\sqrt{2}}{2} \\ \frac{\sqrt{2}}{2} & -\frac{\sqrt{2}}{2} \end{bmatrix} \begin{bmatrix} -\frac{5\sqrt{2}}{6} \\ 0 \end{bmatrix} = \begin{bmatrix} -\frac{5}{6} \\ -\frac{5}{6} \end{bmatrix}.$$

Alternatively, we can compute the pseudoinverse $\mathbf{A}^+ = \mathbf{V}\boldsymbol{\Sigma}^+\mathbf{U}^H$:

$$\mathbf{A}^+ = \begin{bmatrix} \frac{\sqrt{2}}{2} & \frac{\sqrt{2}}{2} \\ \frac{\sqrt{2}}{2} & -\frac{\sqrt{2}}{2} \end{bmatrix} \begin{bmatrix} \frac{1}{3\sqrt{2}} & 0 & 0 \\ 0 & 0 & 0 \end{bmatrix} \begin{bmatrix} \frac{1}{3} & \frac{2}{3} & \frac{2}{3} \\ -\frac{2\sqrt{5}}{5} & \frac{\sqrt{5}}{5} & 0 \\ \frac{2\sqrt{5}}{15} & \frac{4\sqrt{5}}{15} & -\frac{\sqrt{5}}{15} \end{bmatrix}$$

$$= \begin{bmatrix} \frac{1}{18} & \frac{1}{9} & \frac{1}{9} \\ \frac{1}{18} & \frac{1}{9} & \frac{1}{9} \end{bmatrix}.$$

We then get $\mathbf{x}_0$ directly as $\mathbf{x}_0 = \mathbf{A}^+\mathbf{b}$:

$$\mathbf{x}_0 = \begin{bmatrix} \frac{1}{18} & \frac{1}{9} & \frac{1}{9} \\ \frac{1}{18} & \frac{1}{9} & \frac{1}{9} \end{bmatrix} \begin{bmatrix} 15 \\ 15 \\ -30 \end{bmatrix} = \begin{bmatrix} -\frac{5}{6} \\ -\frac{5}{6} \end{bmatrix},$$

as before. All other solutions $\mathbf{x}'$ are of the form $\mathbf{x}' = \mathbf{x}_0 + \alpha\mathbf{v}_2$:

$$\mathbf{x}' = \mathbf{x}_0 + \alpha\mathbf{v}_2 = \begin{bmatrix} -\frac{5}{6} \\ -\frac{5}{6} \end{bmatrix} + \alpha \begin{bmatrix} \frac{\sqrt{2}}{2} \\ -\frac{\sqrt{2}}{2} \end{bmatrix}.$$

When solving a least-squares problem in practice, it is often useful to modify the matrix $\boldsymbol{\Sigma}$ of singular values. Suppose, as is so often the case, that the entries in $\mathbf{A}$ are subject to measurement errors; then the smallest nonzero singular values computed for $\mathbf{A}$ may well have been zero if the measurements had been perfect. Using the reciprocal of such a singular value to solve a least-squares problem may well prove disastrous. It is often better to replace $\boldsymbol{\Sigma}$ by a matrix $\boldsymbol{\Sigma}_0$ obtained by replacing the smallest singular values in $\boldsymbol{\Sigma}$ by exact zeros, which is the effect of the construction in Theorem 8.21; those singular values that are of the magnitude of the errors inherent in the data should usually be treated this way. See Problem 6.

You should now be able to solve Problems 1 to 6.

The Pseudoinverse

The term "pseudoinverse" for the matrix $\mathbf{A}^+$ in **Key Theorem 8.26** is a reasonable one to use: $\mathbf{A}^+$ behaves in several ways as an inverse of sorts. For one thing, $\mathbf{x}_0 = \mathbf{A}^+\mathbf{b}$ solves the least-squares problem $\mathbf{Ax} \approx \mathbf{b}$ just as $\mathbf{x}_0 = \mathbf{A}^{-1}\mathbf{b}$ solves $\mathbf{Ax} = \mathbf{b}$ with nonsingular $\mathbf{A}$. Also $(\mathbf{U\Sigma V}^H)^+ = \mathbf{V\Sigma}^+\mathbf{U}^H$ just as $(\mathbf{U\Sigma V}^H)^{-1} = \mathbf{V\Sigma}^{-1}\mathbf{U}^H$ when $\mathbf{\Sigma}$ is square and nonsingular. And, of course, $\mathbf{A}^+ = \mathbf{A}^{-1}$ when $\mathbf{A}$ is nonsingular (Problem 9). $\mathbf{A}^+$ is one example of a variety of generalized inverses that have been developed; this particular member of that class is also called the *Moore-Penrose generalized inverse* in honor of those who developed its theory.

Although we derived the pseudoinverse through the singular value decomposition, it can be characterized independently of that; this result, which we state without proof, can aid in computing $\mathbf{A}^+$.

(8.28) ***Theorem*** (pseudoinverse characterization). A $p \times q$ matrix $\mathbf{A}$ has exactly one pseudoinverse $\mathbf{A}^+$, and a $q \times p$ matrix $\mathbf{X}$ equals the pseudoinverse $\mathbf{A}^+$ if and only if $\mathbf{X}$ satisfies the following three conditions:
(a) $\mathbf{AXA} = \mathbf{A}$.
(b) $\mathbf{XAX} = \mathbf{X}$.
(c) $\mathbf{AX}$ and $\mathbf{XA}$ are hermitian.

(8.29) ***Theorem*** ($\{\mathbf{BC}\}^+$). Suppose that the $p \times q$ matrix $\mathbf{A}$ has rank k and that $\mathbf{A} = \mathbf{BC}$, where $\mathbf{B}$ is $p \times k$ and of rank k while $\mathbf{C}$ is $k \times q$ and of rank k. Then

$$\mathbf{A}^+ = \mathbf{C}^H(\mathbf{CC}^H)^{-1}(\mathbf{B}^H\mathbf{B})^{-1}\mathbf{B}^H.$$

PROOF. $\mathbf{CC}^H$ and $\mathbf{B}^H\mathbf{B}$ are nonsingular; for example, $\mathbf{B}^H\mathbf{Bx} = \mathbf{0}$ implies $\mathbf{0} = \mathbf{x}^H\mathbf{B}^H\mathbf{Bx} = \|\mathbf{Bx}\|_2^2$, so $\mathbf{Bx} = \mathbf{0}$ and hence $\mathbf{x} = \mathbf{0}$ since $\mathbf{B}$ has rank k. The theorem follows by direct verification of the conditions in Theorem 8.28. ■

An example of how one obtains $\mathbf{A} = \mathbf{BC}$ for such $\mathbf{B}$ and $\mathbf{C}$ is in Problem 11. Another important case comes from the normalized QR-decomposition, which is precisely of the form required in Theorem 8.29 with $\mathbf{B} = \mathbf{Q}$ having orthonormal columns and $\mathbf{C} = \mathbf{R}$ being upper-triangular.

(8.30) ***Corollary*** (*QR and pseudoinverses*). Suppose that $\mathbf{A} = \mathbf{QR}$ is a normalized QR-decomposition of the $p \times q$ matrix $\mathbf{A}$ of rank k, so that $\mathbf{Q}$ is $p \times k$ with orthonormal columns and $\mathbf{R}$ is $k \times q$, upper-triangular, and of rank k. Then

$$\mathbf{A}^+ = \mathbf{R}^H(\mathbf{RR}^H)^{-1}\mathbf{Q}^H.$$

Note that it is simpler to solve the least-squares problem $\mathbf{Ax} \approx \mathbf{b}$ with the normalized QR-decomposition by solving $\mathbf{Rx}_0 = \mathbf{Q}^H\mathbf{b}$ than by writing $\mathbf{x}_0 = \mathbf{A}^+\mathbf{b}$ with $\mathbf{A}^+$ as in Corollary 8.30; as with Gauss elimination for systems of linear equations, it is easier to solve systems directly than to find inverses (pseudo or real) and use them to get the solution.

PROBLEMS 8.5

1. Find Σ^+ for each Σ in Example 8.16.
▷ 2. Find $\mathbf{0}^+$.
3. Use (8.24) to solve the least-squares problem

$$\begin{bmatrix} 1 & 1 \\ 2 & 2 \end{bmatrix} \begin{bmatrix} a \\ b \end{bmatrix} \approx \begin{bmatrix} 2 \\ 8 \end{bmatrix}.$$

𝔐 4. Use MATLAB or similar software to implement (8.24) and solve the least-squares problem in Problem 22 of Section 5.9.
𝔐 5. Use MATLAB or similar software to implement (8.24) and solve the least-squares problem in Problem 12 of Section 2.6.
𝔐 6. Consider the problem of predicting the U.S. population in 1990 as described in Problem 3 of Section 2.6, but use a quadratic to fit the data:

$$(\text{population in millions}) \approx a + b(\text{year}) + c(\text{year})^2.$$

Use MATLAB or similar software to find a singular value decomposition of the matrix $\mathbf{A}$ in that model, and note the variation in size among the singular values.
(a) Use the full singular value decomposition to find the least-squares solution and then predict the population in 1990.
(b) Replace the smallest singular value by zero and use the resulting singular value decomposition (of a slightly different rank 2 matrix) to get a least-squares solution and to predict the 1990 population.
(c) Replace the two smallest singular values by zero and again solve and predict as in (b).
(d) Based on this experiment, what is your personal prediction for the 1990 population?
𝔐 7. Let ϵ be a number such that $1 + \epsilon$ is calculated accurately on your computer but $1 + \epsilon^2$ is calculated as 1; for example, $\epsilon = 10^{-10}$ does this on most microcomputers with math coprocessors. Use MATLAB or similar software to find the singular value decomposition of the matrix $\mathbf{A}$ and solve the corresponding least-squares problem in Problem 24 of Section 5.9; compare with the solution similarly obtained when you replace the smallest singular value by an exact zero.
▷ 8. Use the pseudoinverse to solve the least-squares problem

$$\begin{bmatrix} 1 \\ 2 \\ -2 \end{bmatrix} x \approx \begin{bmatrix} 2 \\ 3 \\ -3 \end{bmatrix}.$$

▷ 9. Prove that $\mathbf{A}^+ = \mathbf{A}^{-1}$ when $\mathbf{A}$ is nonsingular.
10. Prove that the pseudoinverse $\mathbf{X} = \mathbf{A}^+$ satisfies conditions (a)–(c) in Theorem 8.28.

11. Suppose that $p \times q$ $\mathbf{A}$ has rank k and that $\mathbf{A}$ is partitioned as

$$\mathbf{A} = \begin{bmatrix} \mathbf{A}_{11} & \mathbf{A}_{12} \\ \mathbf{A}_{21} & \mathbf{A}_{22} \end{bmatrix},$$

where $\mathbf{A}_{11}$ is $k \times k$ and nonsingular. Prove that $\mathbf{A} = \mathbf{B}_1\mathbf{C}_1$ and $\mathbf{A} = \mathbf{B}_2\mathbf{C}_2$, where

$$\mathbf{B}_1 = \begin{bmatrix} \mathbf{I} \\ \mathbf{A}_{21}\mathbf{A}_{11}^{-1} \end{bmatrix}, \qquad \mathbf{C}_1 = [\mathbf{A}_{11} \quad \mathbf{A}_{12}],$$

$$\mathbf{B}_2 = \begin{bmatrix} \mathbf{A}_{11} \\ \mathbf{A}_{21} \end{bmatrix}, \quad \text{and} \quad \mathbf{C}_2 = [\mathbf{I} \quad \mathbf{A}_{11}^{-1}\mathbf{A}_{12}].$$

▷ **12.** Use Problem 11 and Theorem 8.29 to find the pseudoinverse of

$$\mathbf{A} = \begin{bmatrix} -1 & 0 & 1 & 2 \\ -1 & 1 & 0 & -1 \\ 0 & -1 & 1 & 3 \\ 0 & 1 & -1 & -3 \\ 1 & -1 & 0 & 1 \\ 1 & 0 & -1 & -2 \end{bmatrix}.$$

13. Show that a singular value decomposition is not necessarily unique by finding all singular value decompositions of $\mathbf{I}$ with $\mathbf{U} = \mathbf{V}$; show, however, that all of these decompositions lead to the same pseudoinverse.

14. Prove Theorem 8.29 in detail.

15. Prove:

(a) The rank of $\mathbf{A}^+$ is the same as the rank of $\mathbf{A}$.

(b) If $\mathbf{A}$ is symmetric, then $\mathbf{A}^+$ is symmetric.

(c) $(c\mathbf{A})^+ = (1/c)\mathbf{A}^+$ for $c \neq 0$.

(d) $(\mathbf{A}^+)^T = (\mathbf{A}^T)^+$

(e) $(\mathbf{A}^+)^+ = \mathbf{A}$.

(f) Show by a counterexample that in general $(\mathbf{AB})^+ \neq \mathbf{B}^+\mathbf{A}^+$.

(g) If $\mathbf{A}$ is $m \times r$, $\mathbf{B}$ is $r \times n$, and both matrices are of rank r, then $(\mathbf{AB})^+ = \mathbf{B}^+\mathbf{A}^+$.

▷ **16.** Suppose that $\mathbf{c}$ is $p \times 1$ and $\mathbf{r}$ is $1 \times p$. Find:

(a) $\mathbf{r}^+$ (b) $\mathbf{c}^+$ (c) $(\mathbf{cr})^+$

17. Use the normalized QR-decomposition $\mathbf{A} = \mathbf{QR}$ in Example 5.83 to find $\mathbf{A}^+$ via Corollary 8.30.

18. Use the normalized QR-decomposition $\mathbf{A} = \mathbf{QR}$ in Problem 15 of Section 5.9 to find $\mathbf{A}^+$ via Corollary 8.30.

𝔐 **19.** Use MATLAB or similar software to find the pseudoinverse $\mathbf{A}^+$ for the rank 3, rank 2, and rank 1 matrices in Problem 6.

8.6 MISCELLANEOUS PROBLEMS

PROBLEMS 8.6

1. Schur form of a matrix is not uniquely determined. Analyze the variation possible for a normal matrix.
2. The *Lanczos method* that follows reduces the eigenvalue problem of a general real symmetric $n \times n$ matrix to that for a much simpler matrix. Let $\mathbf{A}$ be a real symmetric matrix and $\mathbf{u}_1$ an arbitrary real vector. Set

$$\mathbf{v}_1 = \mathbf{A}\mathbf{u}_1, \qquad \mathbf{u}_2 = \mathbf{v}_1 - \alpha_1\mathbf{u}_1,$$

where α_1 is determined so as to make $\mathbf{u}_2$ and $\mathbf{u}_1$ orthogonal, which gives $\alpha_1 = (\mathbf{u}_1, \mathbf{v}_1)/(\mathbf{u}_1, \mathbf{u}_1)$. Next, form

$$\mathbf{v}_2 = \mathbf{A}\mathbf{u}_2, \qquad \mathbf{u}_3 = \mathbf{v}_2 - \alpha_2\mathbf{u}_2 - \beta_1\mathbf{u}_1,$$

where α_2, β_1 are determined so that $\mathbf{u}_3$ is orthogonal to $\mathbf{u}_2$ and $\mathbf{u}_1$. This gives $\alpha_2 = (\mathbf{u}_2, \mathbf{v}_2)/(\mathbf{u}_2, \mathbf{u}_2)$, $\beta_1 = (\mathbf{u}_1, \mathbf{v}_2)/(\mathbf{u}_1, \mathbf{u}_1)$. Next, form

$$\mathbf{v}_3 = \mathbf{A}\mathbf{u}_3, \qquad \mathbf{u}_4 = \mathbf{v}_3 - \alpha_3\mathbf{u}_3 - \beta_2\mathbf{u}_2 - \gamma_1\mathbf{u}_1,$$

where α_3, β_2, γ_1 are determined so that $\mathbf{u}_4$ is orthogonal to $\mathbf{u}_3$, $\mathbf{u}_2$, $\mathbf{u}_1$. Prove that this gives $\gamma_1 = 0$. Show that at the general step we have

$$\mathbf{v}_r = \mathbf{A}\mathbf{u}_r, \qquad \mathbf{u}_{r+1} = \mathbf{v}_r - \alpha_r\mathbf{u}_r - \beta_{r-1}\mathbf{u}_{r-1},$$

where $\mathbf{u}_{r+1}$ is orthogonal to $\mathbf{u}_1, \ldots, \mathbf{u}_r$. Prove also that $\mathbf{u}_{n+1}$ must be identically zero. Let $\mathbf{x}$ be an eigenvector of $\mathbf{A}$ and set

$$\mathbf{x} = c_1\mathbf{u}_1 + \cdots + c_n\mathbf{u}_n.$$

Form $\mathbf{A}\mathbf{x}$ and express $\mathbf{A}\mathbf{u}_r = \mathbf{v}_r$ in terms of the $\mathbf{u}_s$ by the formulas above. Deduce that the eigenvalues of $\mathbf{A}$ coincide with those of the tridiagonal matrix

$$\begin{bmatrix} \alpha_1 & \beta_1 & 0 & \cdots & 0 \\ 1 & \alpha_2 & \beta_2 & \cdots & 0 \\ 0 & 1 & \alpha_3 & \cdots & 0 \\ & & \cdots & & \\ 0 & 0 & 0 & \cdots & \alpha_n \end{bmatrix}.$$

(The orthogonality conditions used to determine the α_i, β_i also minimize the $\|\mathbf{u}_i\|_2$. This is why the method, due to C. Lanczos, is also known as the *method of minimized iterations.*)

▷ 3. Let $\mathbf{A}$ be $p \times p$, and let $\mathbf{D} = \mathbf{diag}(\epsilon, \epsilon^2, \ldots, \epsilon^p)$, where $\epsilon \neq 0$. Describe $\mathbf{D}^{-1}\mathbf{A}\mathbf{D}$.

4. Let $\mathbf{T} = \mathbf{P}^H\mathbf{A}\mathbf{P}$ be a Schur form of $p \times p$ $\mathbf{A}$. Use $\mathbf{T}$ and Problem 3 to show that there is a nonsingular matrix $\mathbf{S}$ such that $\mathbf{S}^{-1}\mathbf{A}\mathbf{S}$ is upper-triangular and all its non-main-diagonal entries are as small in magnitude as we choose.

5. Given a nonsingular matrix $\mathbf{S}$, we can define a vector norm $\|\cdot\|_{\mathbf{S}}$ by $\|\mathbf{x}\|_{\mathbf{S}} = \|\mathbf{S}^{-1}\mathbf{x}\|_{\infty}$; this induces the transformation norm $\|\mathbf{A}\|_{\mathbf{S}} = \|\mathbf{S}^{-1}\mathbf{A}\mathbf{S}\|_{\infty}$. Use Problem 4 to show that there is an induced norm $\|\cdot\|_{\mathbf{S}}$, depending on $\mathbf{A}$, so that $\|\mathbf{A}\|_{\mathbf{S}}$ is as close as we choose to the maximum magnitude of the eigenvalues of $\mathbf{A}$.

▷ 6. Prove that the eigenvalues of the normal matrix $\mathbf{A}$ are all equal if and only if $\mathbf{A} = c\mathbf{I}$ for some number c.

7. Show that $(\mathbf{I} - \mathbf{A}^{+}\mathbf{A})\mathbf{x}$ is the orthogonal projection of $\mathbf{x}$ onto the subspace consisting of all $\mathbf{v}$ with $\mathbf{A}\mathbf{v} = \mathbf{0}$, that is, onto the null space of $\mathbf{A}$.

▷ 8. Show that $\mathbf{A}\mathbf{A}^{+}\mathbf{x}$ is the orthogonal projection of $\mathbf{x}$ onto the column space of $\mathbf{A}$.

9

Eigensystems of general matrices, with applications

The theory outlined in Chapter 7 concerning the eigensystems of general matrices is more fully treated in this chapter. That theory is then exploited to analyze the behavior over time of systems whose evolution is described by discrete or continuous models and to study iterative methods for the solution of linear equations. Fundamental in these developments are ***Key Theorems 9.4**, **9.11**, **9.27**, and **9.57**.*

9.1 *INTRODUCTION*

Chapter 7 presented the basic facts about eigensystems of square matrices, stressing the equivalence of various viewpoints: special forms and decompositions, transformation representations, and eigensystems. From the eigensystem perspective, the main points were:

1. Every $p \times p$ matrix $\mathbf{A}$ has p eigenvalues if each distinct eigenvalue λ_i is counted according to its algebraic multiplicity m_i as a root of the characteristic polynomial of $\mathbf{A}$.
2. Associated with each distinct eigenvalue λ_i is at least one eigenvector, and any collection of such eigenvectors—each associated with a different distinct eigenvalue—forms a linearly independent set.
3. Associated with each distinct eigenvalue λ_i is some maximum number μ_i of eigenvectors in a linearly independent set of such eigenvectors; that is, the subspace of eigenvectors associated with λ_i has dimension μ_i—the geometric multiplicity of λ_i.

One important question concerned the relationship between μ_i and m_i: How many eigenvectors in an independent set could be associated with a repeated eigenvalue? Some simple examples revealed that we could have μ_i as any number from 1 to m_i. For example, it is easy to show that $\lambda_1 = 4$ is an eigenvalue of algebraic multiplicity $m_1 = 3$ in each of the matrices below, but the geometric multiplicities are $\mu_1 = 3$, $\mu_1 = 2$, and $\mu_1 = 1$, respectively:

$$\begin{bmatrix} 4 & 0 & 0 \\ 0 & 4 & 0 \\ 0 & 0 & 4 \end{bmatrix} \quad \text{has} \quad \mu_1 = 3; \qquad \begin{bmatrix} 4 & 1 & 0 \\ 0 & 4 & 0 \\ 0 & 0 & 4 \end{bmatrix} \quad \text{has} \quad \mu_1 = 2;$$

$$\begin{bmatrix} 4 & 1 & 0 \\ 0 & 4 & 1 \\ 0 & 0 & 4 \end{bmatrix} \quad \text{has} \quad \mu_1 = 1.$$

From the viewpoint of decompositions and similarities, we developed **Key Theorem 7.14**:

> $p \times p$ **A** is similar to a diagonal matrix $\mathbf{\Lambda}$—equivalently, **A** can be decomposed as $\mathbf{A} = \mathbf{P\Lambda P}^{-1}$ with diagonal $\mathbf{\Lambda}$—if and only if **A** has a linearly independent set of p eigenvectors; the eigenvectors are the columns of **P**, and the eigenvalues the main diagonal of $\mathbf{\Lambda}$.

This important result concerns the situation where the geometric and algebraic multiplicities are equal for all eigenvalues: $\mu_i = m_i$ for all i. What remains to study is the more complicated case when at least one μ_i is less than its m_i, so that we total fewer than p eigenvectors in a linearly independent set.

(9.1) ***Definition.*** A $p \times p$ matrix that fails to have a linearly independent set of p eigenvectors is said to be *defective*; otherwise, it is *nondefective*.

Thus the problem that remains is to understand in detail the structure of the eigensystem of defective matrices. When we sought to study the structure of eigensystems of normal matrices in Chapter 8, we found that it was effective to exploit the viewpoint of decompositions or simple forms first, and then to shift perspectives and interpret the results in terms of eigensystems. The same approach is followed here.

PROBLEMS 9.1

1. Verify the assertions above concerning the eigensystems of the three explicitly given matrices.
2. Solve Problem 8 of Section 7.4, showing that $\mu_i \le m_i$.

9.2 JORDAN FORM

This section develops a special form to which each matrix can be reduced by a similarity transformation. We develop the special form in three stages: Schur form, block-diagonal upper-triangular form, and finally Jordan form. Once we know what to seek as the Jordan form, we will be able to find that form (in simple cases) directly rather than through the three stages.

Block-Diagonal Upper-Triangular Forms

Suppose that $\mathbf{A}$ is $p \times p$; we know from Theorem 8.2 that $\mathbf{A}$ is orthogonally similar to its Schur form $\mathbf{T}$: an upper-triangular matrix with the eigenvalues of $\mathbf{A}$ on its main diagonal. From the way in which $\mathbf{T}$ was constructed in that theorem, it is clear that we can build $\mathbf{T}$ in such a way that each distinct eigenvalue λ_i is repeated (m_i times, of course) in consecutive positions along the main diagonal of $\mathbf{T}$. To reach the second stage in developing the Jordan form, we show that $\mathbf{T}$ in turn is similar to another upper-triangular matrix with the same structure and the additional property of being block-diagonal; we call this a *block-diagonal upper-triangular form.*

(9.2) ***Lemma*** (block-diagonal upper-triangular form). Suppose that $\mathbf{T}$ is upper-triangular and, moreover, has the form

$$\mathbf{T} = \begin{bmatrix} \mathbf{T}_{11} & \mathbf{T}_{12} & \cdots & \mathbf{T}_{1s} \\ \mathbf{0} & \mathbf{T}_{22} & \cdots & \mathbf{T}_{2s} \\ & \cdots & & \\ \mathbf{0} & \mathbf{0} & \cdots & \mathbf{T}_{ss} \end{bmatrix},$$

where each $\mathbf{T}_{ii}$ is $m_i \times m_i$ and upper-triangular, all the main-diagonal entries of $\mathbf{T}_{ii}$ equal λ_i, and the λ_i are distinct for $1 \le i \le s$. Then $\mathbf{T}$ is similar to a block-diagonal upper-triangular matrix

$$\mathbf{V} = \begin{bmatrix} \mathbf{V}_1 & \mathbf{0} & \cdots & \mathbf{0} \\ \mathbf{0} & \mathbf{V}_2 & \cdots & \mathbf{0} \\ & \cdots & & \\ \mathbf{0} & \mathbf{0} & \cdots & \mathbf{V}_s \end{bmatrix},$$

where each $\mathbf{V}_i$ is $m_i \times m_i$ and upper-triangular, all the main-diagonal entries of $\mathbf{V}_i$ equal λ_i above, and the λ_i are of course distinct.

PROOF. The basic idea is to use similarity transformations based on the elementary matrices $\mathbf{E}_{ij}(c)$ of Definition 3.31; recall that $\mathbf{E}_{ij}(c)$ is nonsingular and that its inverse equals $\mathbf{E}_{ij}(-c)$. Suppose that $i < j$, and consider the similarity transformation

$$\mathbf{E}_{ij}(-c)\mathbf{T}\mathbf{E}_{ij}(c) = \mathbf{T}'.$$

This replaces the (i, j)-entry of $\mathbf{T}$ by

$$\langle \mathbf{T} \rangle_{ij} + c(\langle \mathbf{T} \rangle_{ii} - \langle \mathbf{T} \rangle_{jj})$$

and otherwise modifies only entries in the ith row (to the right of the (i, j)-entry) and in the jth column (above that entry). If we choose i and j to correspond to rows in different blocks $\mathbf{T}_{mm}$ and $\mathbf{T}_{nn}$, then

$$\langle \mathbf{T} \rangle_{ii} - \langle \mathbf{T} \rangle_{jj} = \lambda_m - \lambda_n \neq 0,$$

so we can pick $c = -\langle \mathbf{T} \rangle_{ij}/(\lambda_m - \lambda_n)$ and the (i, j)-entry of the similar matrix $\mathbf{T}'$ will equal zero. A sequence of such similarity transformations can be performed that will replace the blocks

$$\mathbf{T}_{s-1,s}, \ \mathbf{T}_{s-2,s-1}, \ \mathbf{T}_{s-2,s}, \ \ldots, \ \mathbf{T}_{12}, \ \mathbf{T}_{13}, \ \ldots \ \mathbf{T}_{1s}$$

in that order by zero blocks (proceeding from bottom to top and left to right within each block). This gives $\mathbf{V}$. ■

Jordan Form

Recall that the main diagonal of each $\mathbf{V}_i$ has all entries equal to λ_i: $\langle \mathbf{V}_i \rangle_{kk} = \lambda_i$ for all k. By some technically rather complex arguments that we omit, it is possible to show that each such matrix $\mathbf{V}_i$ is similar to a block-diagonal upper-triangular matrix $\mathbf{J}_i$, each of whose blocks is a *Jordan block* $\mathbf{J}(\lambda_i)$:

(9.3) ***Definition.*** A *Jordan block* is a square upper-triangular matrix $\mathbf{J}(\lambda)$ such that:
(a) All its main-diagonal entries equal λ: $\langle \mathbf{J}(\lambda) \rangle_{ii} = \lambda$.
(b) All its entries on the first superdiagonal equal 1: $\langle \mathbf{J}(\lambda) \rangle_{i,i+1} = 1$.
(c) All its other entries equal 0.
Thus

$$\mathbf{J}(\lambda) = \begin{bmatrix} \lambda & 1 & 0 & \cdots & 0 \\ 0 & \lambda & 1 & \cdots & 0 \\ & & \cdots & & \\ 0 & 0 & 0 & \cdots & \lambda \end{bmatrix}.$$

Since the combination of successive similarities is also a similarity, the transformations from $\mathbf{A}$ to $\mathbf{T}$ to $\mathbf{V}$ and then to Jordan blocks can be expressed as a single similarity transformation.

(9.4) **Key Theorem** (Jordan form). Each $p \times p$ $\mathbf{A}$ is similar to a matrix $\mathbf{J}$ in *Jordan form*: $\mathbf{J} = \mathbf{Q}^{-1}\mathbf{A}\mathbf{Q}$, and $\mathbf{A} = \mathbf{Q}\mathbf{J}\mathbf{Q}^{-1}$, with

$$\mathbf{Q}^{-1}\mathbf{A}\mathbf{Q} = \begin{bmatrix} \mathbf{J}_1 & \mathbf{0} & \cdots & \mathbf{0} \\ \mathbf{0} & \mathbf{J}_2 & \cdots & \mathbf{0} \\ & & \cdots & \\ \mathbf{0} & \mathbf{0} & \cdots & \mathbf{J}_\mu \end{bmatrix} = \mathbf{J},$$

where each $\mathbf{J}_r$ is an $n_r \times n_r$ Jordan block and $\mu = \mu_1 + \cdots + \mu_s$ equals the sum of the geometric multiplicities of the distinct eigenvalues of $\mathbf{A}$. The same distinct eigenvalue may occur in different Jordan blocks $\mathbf{J}_r$, but the total number of *blocks* with that eigenvalue equals its geometric multiplicity μ_i while the total number of main-diagonal *entries* with that eigenvalue equals its algebraic multiplicity m_i. The numbers n_r and the total number of blocks are uniquely determined by $\mathbf{A}$.

Note that this is an existence theorem, but not a constructive one: no method is provided for finding $\mathbf{J}$ in practice. However the very existence of a $\mathbf{J}$ in such a special form enables us to compute it. Since $\mathbf{Q}^{-1}\mathbf{A}\mathbf{Q} = \mathbf{J}$, we have $\mathbf{A}\mathbf{Q} = \mathbf{Q}\mathbf{J}$. Writing $\mathbf{Q}$ in terms of its columns as

$$\mathbf{Q} = [\mathbf{q}_1 \quad \mathbf{q}_2 \quad \cdots \quad \mathbf{q}_p]$$

shows that $\mathbf{A}\mathbf{Q} = \mathbf{Q}\mathbf{J}$ is equivalent to $\mathbf{A}\mathbf{q}_i = \lambda\mathbf{q}_i + v_i\mathbf{q}_{i-1}$, where λ is the eigenvalue in the Jordan block affecting $\mathbf{q}_i$ and v_i equals 0 or 1. More precisely, since the $\mathbf{J}_r$ are $n_r \times n_r$, the columns of $\mathbf{Q}$ affected by the block $\mathbf{J}_r$ in the product $\mathbf{Q}\mathbf{J} = \mathbf{A}\mathbf{Q}$ are just the n_r numbered

$$n_1 + n_2 + \cdots + n_{r-1} + 1 \text{ to } n_1 + n_2 + \cdots + n_r.$$

For convenience, denote these columns of $\mathbf{Q}$ by $\mathbf{v}_{r1}, \ldots, \mathbf{v}_{rn_r}$. It then follows from $\mathbf{A}\mathbf{Q} = \mathbf{Q}\mathbf{J}$ that

(9.5) $\mathbf{A}\mathbf{v}_{r1} = \lambda_r\mathbf{v}_{r1}$ and $\mathbf{A}\mathbf{v}_{rj} = \lambda_r\mathbf{v}_{rj} + \mathbf{v}_{r,j-1}$ for $j = 2, \ldots, n_r$, where $\mathbf{v}_{rj}$ is the column of $\mathbf{Q}$ numbered $n_1 + \cdots + n_{r-1} + j$ and where $\mathbf{Q}^{-1}\mathbf{A}\mathbf{Q} = \mathbf{J}$ and $\mathbf{A}\mathbf{Q} = \mathbf{Q}\mathbf{J}$ and where the rth Jordan block of $\mathbf{J}$ is $n_r \times n_r$.

This complicated relationship (9.5) is the key to understanding the eigensystem of $\mathbf{A}$; we return to this later. For now, we show how to use (9.5) as a guide to finding a Jordan form of $\mathbf{A}$ when $\mathbf{A}$ is small enough for hand calculations. The crucial observation is that (9.5) means that $\mathbf{v}_{r1}$ is an eigenvector of $\mathbf{A}$; we therefore find the eigenvectors $\mathbf{v}_{r1}$ and then use them to help find the remaining $\mathbf{v}_{rj}$.

(9.6) ***Example.*** Consider finding a Jordan form of

$$\mathbf{A} = \begin{bmatrix} 5 & 4 & 3 \\ -1 & 0 & -3 \\ 1 & -2 & 1 \end{bmatrix}.$$

We easily find that $\det(\mathbf{A} - \lambda\mathbf{I}) = (-2 - \lambda)(4 - \lambda)^2$, and that $\lambda_1 = -2$ is a simple eigenvalue while $\lambda_2 = 4$ has algebraic multiplicity $m_2 = 2$ but only one eigenvector ($\mu_2 = 1$); an eigenvector for λ_1 is $[1 \quad -1 \quad -1]^T$, which we take as $\mathbf{v}_{11}$ for (9.5), while one for λ_2 is $[1 \quad -1 \quad 1]$, which we take as $\mathbf{v}_{21}$ for (9.5). Since $\mu_1 = \mu_2 = 1$, there is only one Jordan block with each eigenvalue; since $m_2 = 2$, the eigenvalue $\lambda_2 = 4$ must appear twice on the diagonal and so must be in a 2 × 2 block. Thus, according to (9.5), we seek $\mathbf{v}_{22}$ so that $\mathbf{A}\mathbf{v}_{22} - \lambda_2\mathbf{v}_{22} = \mathbf{v}_{21}$, that is,

$$\begin{bmatrix} 1 & 4 & 3 \\ -1 & -4 & -3 \\ 1 & -2 & -3 \end{bmatrix}\begin{bmatrix} x \\ y \\ z \end{bmatrix} = \begin{bmatrix} 1 \\ -1 \\ 1 \end{bmatrix}, \quad \text{with solution} \quad \begin{bmatrix} 0 \\ 1 \\ -1 \end{bmatrix} + \alpha\mathbf{v}_{21}$$

for arbitrary α (since $\mathbf{A}\mathbf{v}_{21} - \lambda_2\mathbf{v}_{21} = \mathbf{0}$). We take $\alpha = 0$, giving the third column of $\mathbf{Q}$ as $\mathbf{v}_{22} = [0 \quad 1 \quad -1]^T$. (9.5) in this case is

$$\mathbf{A}\mathbf{v}_{11} = \lambda_1\mathbf{v}_{11}, \quad \mathbf{A}\mathbf{v}_{21} = \lambda_2\mathbf{v}_{21}, \quad \mathbf{A}\mathbf{v}_{22} = \lambda_2\mathbf{v}_{22} + \mathbf{v}_{21}.$$

Forming the matrix $\mathbf{Q}$ from the columns $\mathbf{v}_{11}, \mathbf{v}_{21}, \mathbf{v}_{31}$ produces the Jordan form $\mathbf{J}$:

$$\mathbf{Q} = \begin{bmatrix} 1 & 1 & 0 \\ -1 & -1 & 1 \\ -1 & 1 & -1 \end{bmatrix}, \qquad \mathbf{Q}^{-1} = \frac{1}{2}\begin{bmatrix} 0 & -1 & -1 \\ 2 & 1 & 1 \\ 2 & 2 & 0 \end{bmatrix},$$

$$\mathbf{Q}^{-1}\mathbf{A}\mathbf{Q} = \begin{bmatrix} -2 & 0 & 0 \\ 0 & 4 & 1 \\ 0 & 0 & 4 \end{bmatrix} = \mathbf{J}.$$

(9.7) ***Example.*** Consider finding a Jordan form for

$$\mathbf{A} = \begin{bmatrix} 2 & 2 & -1 \\ -1 & -1 & 1 \\ -1 & -2 & 2 \end{bmatrix}.$$

We easily find $\det(\mathbf{A} - \lambda\mathbf{I}) = (1 - \lambda)^3$; there is one distinct eigenvalue $\lambda_1 = 1$ with algebraic multiplicity $m_1 = 3$. We can find only two eigenvectors, for example $\mathbf{x}_1 = [1 \quad 0 \quad 1]^T$ and $\mathbf{x}_2 = [0 \quad 1 \quad 2]^T$; thus $\mu_1 = 2$. There must be two Jordan blocks, so one is 1 × 1 and one is 2 × 2. We can take $\mathbf{v}_{11} = \mathbf{x}_1$ as one eigenvector column of $\mathbf{Q}$; the second eigenvector column $\mathbf{v}_{21}$ of $\mathbf{Q}$ must be some combination of $\mathbf{x}_1$ and $\mathbf{x}_2$ such that we can solve (9.5) for a third column $\mathbf{v}_{22}$. We let $\mathbf{v}_{21} = \alpha\mathbf{x}_1 + \beta\mathbf{x}_2$ and seek to solve for $\mathbf{v}_{22}$ from $(\mathbf{A} - \mathbf{I})\mathbf{v}_{22} = \alpha\mathbf{x}_1 + \beta\mathbf{x}_2$:

$$\begin{bmatrix} 1 & 2 & -1 \\ -1 & -2 & 1 \\ -1 & -2 & 1 \end{bmatrix}\begin{bmatrix} x \\ y \\ z \end{bmatrix} = \begin{bmatrix} \alpha \\ \beta \\ \alpha + 2\beta \end{bmatrix}.$$

Trying to solve these shows that we need $\beta = -\alpha$ for there to be a solution, namely $x = \alpha + \gamma - 2\delta$, $y = \delta$, $z = \gamma$ for arbitrary γ and δ. Now we must take $\alpha \neq 0$ in order that $\mathbf{v}_{21} \neq \mathbf{0}$, but we can certainly take $\gamma = \delta = 0$ for simplicity; we choose $\alpha = 1$, so $\beta = -1$. This gives

$$\mathbf{v}_{11} = [1 \quad 0 \quad 1]^T, \quad \mathbf{v}_{21} = [1 \quad -1 \quad -1]^T, \quad \mathbf{v}_{22} = [1 \quad 0 \quad 0]^T,$$

with

$$\mathbf{A}\mathbf{v}_{11} = \lambda_1 \mathbf{v}_{11}, \quad \mathbf{A}\mathbf{v}_{21} = \lambda_1 \mathbf{v}_{21}, \quad \text{and} \quad \mathbf{A}\mathbf{v}_{22} = \lambda_1 \mathbf{v}_{22} + \mathbf{v}_{21}.$$

Forming $\mathbf{Q}$ from these columns $\mathbf{v}_{ij}$, we find

$$\mathbf{Q} = \begin{bmatrix} 1 & 1 & 1 \\ 0 & -1 & 0 \\ 1 & -1 & 0 \end{bmatrix}, \qquad \mathbf{Q}^{-1} = \begin{bmatrix} 0 & -1 & 1 \\ 0 & -1 & 0 \\ 1 & 2 & -1 \end{bmatrix},$$

$$\mathbf{Q}^{-1}\mathbf{A}\mathbf{Q} = \begin{bmatrix} 1 & 0 & 0 \\ 0 & 1 & 1 \\ 0 & 0 & 1 \end{bmatrix} = \mathbf{J}.$$

You should now be able to solve Problems 1 to 10.

Cayley-Hamilton Theorem

In going from a general block-diagonal upper-triangular form to the special Jordan form, we claimed that it was useful to have this simpler form. We present now the first of several instances of Jordan form for purposes of analysis.

For a typical $k \times k$ Jordan block $\mathbf{J}(\lambda)$ as in Definition 9.3, consider the simple matrix $\mathbf{E}_k = \mathbf{J}(\lambda) - \lambda\mathbf{I}$; $\mathbf{E}_k$ has ones on its first superdiagonal and zeros elsewhere. It is easy to see that $\mathbf{E}_k^2$ has ones on its *second* superdiagonal and zeros elsewhere, that $\mathbf{E}_k^3$ has ones on its *third* superdiagonal and zeros elsewhere, and so on until $\mathbf{E}_k^k = \mathbf{0}$.

Next, suppose that $\mathbf{J}$ is a Jordan form of a $p \times p$ matrix $\mathbf{A}$ as in **Key Theorem 9.4**, and consider any one of the Jordan blocks $\mathbf{J}_n$ in which a particular eigenvalue λ_i appears; since λ_i appears at most m_i times in all of $\mathbf{J}$, the block $\mathbf{J}_n$ can be at most $m_i \times m_i$. Therefore, by the preceding paragraph, $(\lambda_i\mathbf{I} - \mathbf{J}_n)^{m_i} = \mathbf{0}$.

Now, consider the characteristic polynomial f of $\mathbf{A}$:

$$f(\lambda) = \det(\mathbf{A} - \lambda\mathbf{I}) = (\lambda_1 - \lambda)^{m_1}(\lambda_2 - \lambda)^{m_2} \cdots (\lambda_s - \lambda)^{m_s}.$$

If we consider $f(\mathbf{J})$, since powers of $\mathbf{J}$ commute with $\mathbf{I}$ and with powers of $\mathbf{J}$, we have

$$f(\mathbf{J}) = (\lambda_1\mathbf{I} - \mathbf{J})^{m_1}(\lambda_2\mathbf{I} - \mathbf{J})^{m_2} \cdots (\lambda_s\mathbf{I} - \mathbf{J})^{m_s}.$$

By what we have already shown, every block $(\lambda_i\mathbf{I} - \mathbf{J}_n)^{m_i} = \mathbf{0}$ whenever $\mathbf{J}_n$ involves λ_i, so the block corresponding to $\mathbf{J}_n$ in $f(\mathbf{J})$ must equal zero; but every block in $f(\mathbf{J})$ is such a block, so $f(\mathbf{J}) = \mathbf{0}$. This leads to the following remarkable result.

(9.8) ***Theorem*** (Cayley-Hamilton theorem). If $f(\lambda)$ is the characteristic polynomial of a matrix $\mathbf{A}$, then $f(\mathbf{A}) = \mathbf{0}$.

PROOF. Let $\mathbf{Q}^{-1}\mathbf{AQ} = \mathbf{J}$ be a Jordan form of $\mathbf{A}$. We have already proved that $f(\mathbf{J}) = \mathbf{0}$. But $f(\mathbf{A}) = \mathbf{Q}f(\mathbf{J})\mathbf{Q}^{-1}$ by Theorem 7.25(e), so $f(\mathbf{A}) = \mathbf{Q0Q}^{-1} = \mathbf{0}$. ■

(9.9) ***Example.*** Consider the matrix $\mathbf{A}$ in Example 9.6, where $f(\lambda) = -\lambda^3 + 6\lambda^2 - 32$; then

$$f(\mathbf{A}) = -\mathbf{A}^3 + 6\mathbf{A}^2 - 32\mathbf{I}$$

$$= -\begin{bmatrix} 112 & 84 & 36 \\ -48 & -20 & -36 \\ 48 & 12 & 28 \end{bmatrix} + 6\begin{bmatrix} 24 & 14 & 6 \\ -8 & 2 & -6 \\ 8 & 2 & 10 \end{bmatrix} - 32\begin{bmatrix} 1 & 0 & 0 \\ 0 & 1 & 0 \\ 0 & 0 & 1 \end{bmatrix} = \mathbf{0},$$

as asserted by the Cayley-Hamilton theorem.

PROBLEMS 9.2

1. Reduce to block-diagonal upper-triangular form the matrix

$$\begin{bmatrix} -2 & 4 & 3 \\ 0 & 4 & 2 \\ 0 & 0 & 4 \end{bmatrix}.$$

2. Reduce to block-diagonal upper-triangular form the matrix

$$\begin{bmatrix} 1 & -3 & 2 \\ 0 & 1 & -1 \\ 0 & 0 & 2 \end{bmatrix}.$$

▷ 3. Reduce to block-diagonal upper-triangular form the matrix

$$\begin{bmatrix} 1 & -2 & 3 & -4 \\ 0 & 1 & -1 & -2 \\ 0 & 0 & 1 & 4 \\ 0 & 0 & 0 & -3 \end{bmatrix}.$$

4. Write out (9.5) in detail for $\mathbf{J} =$

(a) $\begin{bmatrix} 4 & 0 & 0 & 0 \\ 0 & 4 & 1 & 0 \\ 0 & 0 & 4 & 0 \\ 0 & 0 & 0 & 2 \end{bmatrix}$ (b) $\begin{bmatrix} 3 & 0 & 0 & 0 & 0 \\ 0 & 4 & 1 & 0 & 0 \\ 0 & 0 & 4 & 0 & 0 \\ 0 & 0 & 0 & 4 & 1 \\ 0 & 0 & 0 & 0 & 4 \end{bmatrix}$

(c) $\begin{bmatrix} 4 & 1 & 0 & 0 & 0 \\ 0 & 4 & 1 & 0 & 0 \\ 0 & 0 & 4 & 0 & 0 \\ 0 & 0 & 0 & 4 & 1 \\ 0 & 0 & 0 & 0 & 4 \end{bmatrix}$

5. Consider the matrix

$$\mathbf{A} = \begin{bmatrix} 2+\epsilon & \epsilon \\ -\epsilon & 2-\epsilon \end{bmatrix}.$$

(a) Assuming perfect arithmetic, find the Jordan form of **A** when $\epsilon \neq 0$ and when $\epsilon = 0$.

(b) Assuming computer arithmetic with ϵ so small that $2 \pm \epsilon$ is evaluated as 2, find the Jordan form of the resulting computed **A**.

(c) Use the results of (a) and (b) to discuss the sensitivity of Jordan form to perturbations in the entries of the matrix.

▷ **6.** Reduce to Jordan form the matrix

$$\begin{bmatrix} 0 & 1 & 0 \\ 0 & 0 & 1 \\ 6 & -1 & -4 \end{bmatrix}.$$

7. Reduce to Jordan form the matrix in Problem 3.

8. Prove that **A** is similar to **B** if and only if **A** and **B** have a common Jordan form.

9. Suppose that **A** and **B** are $p \times p$, that each has a linearly independent set of p eigenvectors, and that $\mathbf{AB} = \mathbf{BA}$. Suppose that $\mathbf{P}^{-1}\mathbf{BP} = \mathbf{\Lambda}$, where $\mathbf{\Lambda}$ is diagonal and has as its main-diagonal entries the eigenvalues λ_i, each of algebraic and geometric multiplicity m_i.

(a) Show that $\mathbf{P}^{-1}\mathbf{AP}$ must be block-diagonal with main-diagonal blocks of size $m_i \times m_i$.

(b) Prove that there exists **Q** with $\mathbf{Q}^{-1}\mathbf{AQ}$ and $\mathbf{Q}^{-1}\mathbf{BQ}$ both diagonal.

(c) Prove that (b) holds if and only if **A** and **B** commute: $\mathbf{AB} = \mathbf{BA}$.

▷ **10.** Reduce to Jordan form the matrix

$$\mathbf{A} = \begin{bmatrix} 2 & -1 & 0 & 0 \\ 1 & 2 & 0 & 0 \\ 0 & 0 & 2 & 1 \\ 0 & 0 & -1 & 2 \end{bmatrix}.$$

11. Verify the Cayley-Hamilton theorem for the matrix in Example 9.7.

12. For the matrix **A** in Example 9.7, show that $g(\mathbf{A}) = \mathbf{0}$ if

$$g(x) = (1-x)^2 = 1 - 2x + x^2.$$

13. Suppose that $h(x)$ is a polynomial such that $h(\mathbf{J}) = \mathbf{0}$, where **J** is the Jordan form in Example 9.7. Show that $h(1) = h'(1) = 0$, so that

$$h(x) = (1-x)^2 q(x)$$

for some polynomial q; conclude that $(1-x)^2$ is the polynomial $m(x)$ of least degree (and leading coefficient 1) for which $m(\mathbf{A}) = \mathbf{0}$—$m$ is the so-called *minimal polynomial* of **A**.

▷ **14.** As in Problem 13, find the minimal polynomial of

$$\mathbf{A} = \begin{bmatrix} 4 & 1 & 0 & 0 & 0 \\ 0 & 4 & 0 & 0 & 0 \\ 0 & 0 & 4 & 1 & 0 \\ 0 & 0 & 0 & 4 & 1 \\ 0 & 0 & 0 & 0 & 4 \end{bmatrix}.$$

▷ **15.** Generalize Problems 13 and 14 to describe the minimal polynomial of a general Jordan form **J**.

9.3 *EIGENSYSTEMS FOR GENERAL MATRICES*

The Jordan form developed in the preceding section provides detailed information on the eigensystem of a matrix. The key is to interpret the relationship $\mathbf{J} = \mathbf{Q}^{-1}\mathbf{A}\mathbf{Q}$ in the form $\mathbf{AQ} = \mathbf{QJ}$ as represented in (9.5):

Review of (9.5)

$$\mathbf{A}\mathbf{v}_{r1} = \lambda_r \mathbf{v}_{r1} \quad \text{and} \quad \mathbf{A}\mathbf{v}_{rj} = \lambda_r \mathbf{v}_{rj} + \mathbf{v}_{r,j-1} \qquad \text{for } j = 2, \ldots, n_r.$$

Generalized Eigenvectors and Invariant Subspaces

To make the general relationship (9.5) more concrete, consider first the specific case in Example 9.6. In that case the relations are

$$\mathbf{A}\mathbf{v}_{11} = \lambda_1 \mathbf{v}_{11}, \qquad \mathbf{A}\mathbf{v}_{21} = \lambda_2 \mathbf{v}_{21}, \quad \text{and} \quad \mathbf{A}\mathbf{v}_{22} = \lambda_2 \mathbf{v}_{22} + \mathbf{v}_{21}, \tag{9.10}$$

with $\lambda_1 = -2$ a simple eigenvalue and $\lambda_2 = 4$ an eigenvalue of algebraic multiplicity $m_2 = 2$ but geometric multiplicity $\mu_2 = 1$. The vectors $\mathbf{v}_{11}$ and $\mathbf{v}_{21}$ are the two eigenvectors, while $\mathbf{v}_{22}$ is said to be a *generalized eigenvector*. Since $m_2 = 2$ we might hope to find two eigenvectors associated with λ_2, but this is impossible (if they are to form an independent set); instead, we associate the generalized eigenvector $\mathbf{v}_{22}$ with λ_2 so as to provide the second of two ($= m_2$) special vectors associated with λ_2. In what way is $\mathbf{v}_{22}$ "special"? Recall that, if we had two eigenvectors associated with λ_2, they would span a two-dimensional invariant subspace. $\mathbf{v}_{22}$ substitutes in this capacity: $\mathbf{v}_{21}$ *and* $\mathbf{v}_{22}$ *together span a two-dimensional invariant subspace.* To verify this, we check whether **A** times any linear combination of $\mathbf{v}_{21}$ and $\mathbf{v}_{22}$ is another such combination:

$$\begin{aligned}\mathbf{A}(\alpha\mathbf{v}_{21} + \beta\mathbf{v}_{22}) &= \alpha\mathbf{A}\mathbf{v}_{21} + \beta\mathbf{A}\mathbf{v}_{22} = \alpha(\lambda_2\mathbf{v}_{21}) + \beta(\lambda_2\mathbf{v}_{22} + \mathbf{v}_{21}) \\ &= (\alpha\lambda_2 + \beta)\mathbf{v}_{21} + (\beta\lambda_2)\mathbf{v}_{22},\end{aligned}$$

as asserted.

This happens in general as well: The vectors $\mathbf{v}_{rj}$ in (9.5) for $1 \le j \le n_r$ span an n_r-dimensional invariant subspace. The uniqueness of the numbers n_r and of the number of blocks in the Jordan form implies that no such invariant subspace can be split into invariant subspaces of smaller dimension. This proves the following interpretation in eigensystem terms for the Jordan form theorem.

(9.11) **Key Theorem** (eigensystem structure). Let $p \times p$ $\mathbf{A}$ have a linearly independent set of $\mu = \mu_1 + \cdots + \mu_s$ eigenvectors associated with distinct eigenvalues $\lambda_1, \ldots, \lambda_s$, where λ_i is of algebraic multiplicity m_i and geometric multiplicity μ_i. Then there exist positive integers $n_1, \ldots, n_\mu$ with $n_1 + \cdots + n_\mu = p$ and a linearly independent set of p vectors $\mathbf{v}_{rj}$ for $1 \le r \le \mu$ and $1 \le j \le n_r$ such that:

(a) $\{\mathbf{v}_{11}, \mathbf{v}_{21}, \ldots, \mathbf{v}_{\mu 1}\}$ is a linearly independent set of eigenvectors of $\mathbf{A}$, and no such set can contain more vectors.

(b) For each r with $1 \le r \le \mu$, the set of vectors $\mathbf{v}_{rj}$ for $1 \le j \le n_r$ is a basis for an invariant subspace $\mathscr{V}_r$ for $\mathbf{A}$ that cannot be further decomposed into two nontrivial invariant subspaces with no nonzero vectors in common.

(c) If λ is the eigenvalue associated with $\mathbf{v}_{r1}$, then the $\mathbf{v}_{rj}$ for $2 \le j \le n_r$ are called *generalized eigenvectors* associated with λ and they satisfy

$$\mathbf{A}\mathbf{v}_{rj} = \lambda \mathbf{v}_{rj} + \mathbf{v}_{r,j-1}.$$

The description above is complicated, but then so also is the situation it describes. The following schematic may clarify matters.

(9.12)

	Bases for Invariant Subspaces $\mathscr{V}_1, \ldots, \mathscr{V}_\mu$			
	For subspace $\mathscr{V}_1$ ↓	For subspace $\mathscr{V}_2$ ↓	$\cdots$	For subspace $\mathscr{V}_\mu$ ↓
Eigensystem →	$\mathbf{v}_{11}$	$\mathbf{v}_{21}$	$\cdots$	$\mathbf{v}_{\mu 1}$
	$\mathbf{v}_{12}$ $\vdots$ $\mathbf{v}_{1n_1}$	$\mathbf{v}_{22}$ $\vdots$ $\mathbf{v}_{2n_2}$	$\cdots$	$\mathbf{v}_{\mu 2}$ $\vdots$ $\mathbf{v}_{\mu n_\mu}$

(9.13) ***Example.*** The situation found in Example 9.7 is described in the language of **Key Theorem 9.11** as follows:

(a) $\{\mathbf{v}_{11}, \mathbf{v}_{21}\}$ is a linearly independent set of eigenvectors of $\mathbf{A}$ containing the maximum possible number, and $\mathbf{v}_{11}$ and $\mathbf{v}_{21}$ are both associated with $\lambda_1 = 1$.

(b) $\{\mathbf{v}_{11}\}$ and $\{\mathbf{v}_{21}, \mathbf{v}_{22}\}$ each spans an invariant subspace of $\mathbf{A}$.
(c) $\mathbf{A}\mathbf{v}_{22} = \lambda_1\mathbf{v}_{22} + \mathbf{v}_{21}$.

You should now be able to solve Problems 1 to 3.

Left Eigenvectors: Solvability of Equations

In the case of normal matrices, the fact that a unitary similarity produced a diagonal form demonstrated the mutual orthogonality of the eigenvectors and the validity of Theorem 8.12 characterizing the solvability of systems of equations. An analogous result holds more generally, but a new concept must be introduced to substitute for the mutual orthogonality of eigenvectors.

To develop this idea we start once again with the key relationship $\mathbf{Q}^{-1}\mathbf{A}\mathbf{Q} = \mathbf{J}$ for the Jordan form $\mathbf{J}$. Postmultiplication by $\mathbf{Q}^{-1}$ gives $\mathbf{Q}^{-1}\mathbf{A} = \mathbf{J}\mathbf{Q}^{-1}$; let $\mathbf{y}_i^H$ denote the ith row of $\mathbf{Q}^{-1}$, so that

$$\mathbf{Q}^{-1} = \begin{bmatrix} \mathbf{y}_1^H \\ \mathbf{y}_2^H \\ \cdots \\ \mathbf{y}_p^H \end{bmatrix}.$$

Similar to the relation we found using the columns of $\mathbf{Q}$, we find the relation $\mathbf{y}_i^H\mathbf{A} = \lambda\mathbf{y}_i^H + \xi_i\mathbf{y}_{i+1}^H$, where λ is the eigenvalue in the Jordan block affecting $\mathbf{y}_i^H$ in the product $\mathbf{J}\mathbf{Q}^{-1}$ and ξ_i equals 0 or 1. More precisely, ξ_i will equal 0 and thus $\mathbf{y}_i^H\mathbf{A} = \lambda\mathbf{y}_i^H$ precisely when i is the row number of a *bottom* row in some Jordan block of $\mathbf{J}$. Thus there is one such vector $\mathbf{y}_i^H$ for each of the μ Jordan blocks in $\mathbf{A}$.

(9.14) ***Definition.*** A nonzero vector $\mathbf{y}$ for which $\mathbf{y}^H\mathbf{A} = \lambda\mathbf{y}^H$ is called a *left-eigenvector* associated with the eigenvalue λ of $\mathbf{A}$.

Since this relationship is equivalent to $\mathbf{A}^T\bar{\mathbf{y}} = \lambda\bar{\mathbf{y}}$, where $\bar{\mathbf{y}}$ is the complex conjugate of $\mathbf{y}$, it follows that λ is an eigenvalue of $\mathbf{A}^T$; since det $\mathbf{B}$ = det $\mathbf{B}^T$ for all $\mathbf{B}$, we know that $\det(\mathbf{A}^T - \lambda\mathbf{I}) = 0$ if and only if $\det(\mathbf{A} - \lambda\mathbf{I}) = 0$, and therefore left-eigenvectors associated with a number λ exist if and only if λ is an eigenvalue of $\mathbf{A}$, as was implicit in Definition 9.14.

Recall that the left-eigenvectors introduced above were selected from the rows $\mathbf{y}_i^H$ of $\mathbf{Q}^{-1}$. If the columns of $\mathbf{Q}$ are denoted by $\mathbf{q}_j$, then the relation $\mathbf{Q}^{-1}\mathbf{Q} = \mathbf{I}$ is equivalent to

$$\mathbf{y}_i^H\mathbf{q}_i = 1 \quad \text{and} \quad \mathbf{y}_i^H\mathbf{q}_j = 0 \quad \text{if} \quad i \neq j:$$

the $\mathbf{y}_i$ and the $\mathbf{q}_j$ form what is called a *bi-orthogonal* system. Recalling that μ of the $\mathbf{y}_i$ are left-eigenvectors of $\mathbf{A}$ immediately gives the following result.

(9.15) ***Theorem*** (left-eigenvectors). Let $\mathbf{A}$, λ_i, m_i, μ_i, μ, and $\mathbf{v}_{ri}$ be as in **Key Theorem 9.11**. Then:

(a) There is a linearly independent set of μ left-eigenvectors $\mathbf{u}_1, \ldots, \mathbf{u}_\mu$ of $\mathbf{A}$, and no such set can contain more vectors.

(b) $\mathbf{u}_r$ is associated with the same eigenvalue λ as is the (right) eigenvector $\mathbf{v}_{r1}$.

(c) $(\mathbf{u}_r, \mathbf{v}_{rn_r}) = \mathbf{u}_r^H\mathbf{v}_{rn_r} = 1$.

(d) $(\mathbf{u}_r, \mathbf{v}_{ij}) = \mathbf{u}_r^H\mathbf{v}_{ij} = 0$ unless $i = r$, $j = n_r$ as in (c).

(9.16) ***Example.*** Consider the matrix $\mathbf{A}$ and its Jordan form $\mathbf{J}$ found in Example 9.7. The $\mathbf{u}_i$ are to be obtained from the rows of $\mathbf{Q}^{-1}$ corresponding to the bottom row of each Jordan block in $\mathbf{J}$. Thus

$$\mathbf{u}_1^H = [0 \quad -1 \quad 1]$$

is the first row of $\mathbf{Q}^{-1}$ and

$$\mathbf{u}_2^H = [1 \quad 2 \quad -1]$$

is the last row of $\mathbf{Q}^{-1}$. It is easy to verify that

$$\mathbf{u}_1^H\mathbf{A} = \lambda\mathbf{u}_1^H, \qquad \mathbf{u}_2^H\mathbf{A} = \lambda\mathbf{u}_2^H, \qquad \text{where } \lambda = 1.$$

Moreover, $\mathbf{u}_1^H\mathbf{v}_{11} = \mathbf{u}_2^H\mathbf{v}_{22} = 1$ as required; the orthogonality conditions also hold.

The solvability of equations $(\mathbf{A} - \lambda\mathbf{I})\mathbf{x} = \mathbf{b}$ can now be treated. If λ is not an eigenvalue, $\mathbf{A} - \lambda\mathbf{I}$ is of course nonsingular and there is a unique solution. What if λ is an eigenvalue, however? Writing $\mathbf{A} = \mathbf{QJQ}^{-1}$ with its Jordan form $\mathbf{J}$ and then premultiplying by $\mathbf{Q}^{-1}$ converts the equation to $(\mathbf{J} - \lambda\mathbf{I})\boldsymbol{\xi} = \boldsymbol{\beta}$, where $\boldsymbol{\xi} = \mathbf{Q}^{-1}\mathbf{x}$ and $\boldsymbol{\beta} = \mathbf{Q}^{-1}\mathbf{b}$. Any Jordan block involving the eigenvalue λ will have a 1 in the $(i, i + 1)$-entry of the ith row except in the bottom row of that block, which will in fact be a zero row all across $\mathbf{J} - \lambda\mathbf{I}$. Thus there will be a solution $\boldsymbol{\xi}$ if and only if the entry in that row of $\boldsymbol{\beta}$ equals 0 for each such row; note that this is the very row that produces the left eigenvector $\mathbf{u}_i$ in $\mathbf{Q}^{-1}$, so $\langle\boldsymbol{\beta}\rangle_i = \mathbf{u}_i^H\mathbf{b}$. This generalizes Theorem 8.12 on normal matrices and solvability.

(9.17) ***Theorem*** (general matrices and solvability). Suppose that $\mathbf{A}$ is a $p \times p$ matrix having eigenvalues λ_i of geometric multiplicity μ_i for $1 \le i \le s$ and a linearly independent set of $\mu = \mu_1 + \cdots + \mu_s$ left-eigenvectors $\mathbf{u}_j$.

(a) If λ is not an eigenvalue of $\mathbf{A}$, then $(\mathbf{A} - \lambda\mathbf{I})\mathbf{x} = \mathbf{b}$ has a unique solution for each $\mathbf{b}$.

(b) If λ equals an eigenvalue of $\mathbf{A}$, then $(\mathbf{A} - \lambda\mathbf{I})\mathbf{x} = \mathbf{b}$ has a solution if and only if $\mathbf{b}$ is orthogonal to all the left-eigenvectors $\mathbf{u}_j$ associated with λ. When there are solutions, there are infinitely many, and each can be obtained by adding to one solution $\mathbf{x}$ an arbitrary linear combination of (right-) eigenvectors of $\mathbf{A}$ associated with λ.

PROOF. Problem 6. ∎

(9.18) ***Example.*** The matrix **A** of the equations

$$6x_1 + 2x_2 = b_1$$
$$3x_1 + x_2 = b_2$$

is singular (so $\lambda = 0$ is an eigenvalue); left-eigenvectors are multiples of $\mathbf{u}_1 = [1 \quad -2]^T$. The system of equations has a solution if and only if

$$0 = [1 \quad -2][b_1 \quad b_2]^T, \quad \text{that is,} \quad b_1 = 2b_2.$$

PROBLEMS 9.3

1. Show that the $\mathbf{v}_{rj}$ for $1 \le j \le n_r$ in (9.5) do indeed span a n_r-dimensional invariant subspace.

▷ **2.** Find bases for the invariant subspaces of **Key Theorem 9.11** for the matrix:

(a) in Example 9.7 (b) in Example 9.8

(c) $\begin{bmatrix} 2 & 2 & -2 \\ 0 & 2 & -3 \\ 0 & 0 & -1 \end{bmatrix}$ (d) $\begin{bmatrix} 5 & 4 & 3 \\ -1 & 0 & -3 \\ 1 & -2 & 1 \end{bmatrix}$ (e) $\begin{bmatrix} 2 & 2 & -1 \\ -1 & -1 & 1 \\ -1 & -2 & 4 \end{bmatrix}$

(f) in Problem 3 of Section 9.2
(g) in Problem 6 of Section 9.2
(h) in Problem 10 of Section 9.2

3. Show that if one of the invariant subspaces of **Key Theorem 9.11** splits into two nontrivial invariant subspaces intersecting only in $\{\mathbf{0}\}$, each would contain an eigenvector of a Jordan block, which would contradict the block's structure.

4. Verify the bi-orthogonality asserted for the vectors in Example 9.16.

▷ **5.** Find the left-eigenvectors and verify the bi-orthogonality with the (right-) eigenvectors of the matrix:

(a) in Example 9.6
(b) in Example 9.7
(c) in Problem 3 of Section 9.2
(d) in Problem 6 of Section 9.2
(e) in Problem 10 of Section 9.2

6. Prove Theorem 9.17(b) on solvability of equations.

▷ **7.** Find necessary and sufficient conditions on the b_i in order that there be a solution to

$$2x_1 + 2x_2 + 4x_3 + x_4 = b_1$$
$$-3x_1 - 3x_2 - 6x_3 + 2x_4 = b_2$$
$$-6x_1 - 6x_2 - 12x_3 + 3x_4 = b_3$$
$$x_1 + x_2 + 2x_3 + x_4 = b_4.$$

8. Let $\mathscr{T}$ be the linear transformation from $\mathbb{C}^p$ to $\mathbb{C}^p$ defined by $\mathscr{T}(\mathbf{v}) = \mathbf{A}\mathbf{v}$, and let $\mathbf{A}^H$ define its adjoint transformation $\mathscr{T}^*$ (see Definition 6.10 and Problem 24 of Section 6.1). Prove that $\mathscr{T}(\mathbf{v}) = \mathbf{w}$ is solvable, given $\mathbf{w}$, if and only if $\mathbf{w}$ is orthogonal to the null space of $\mathscr{T}^*$.

9.4 *APPLICATION: DISCRETE SYSTEM EVOLUTION AND MATRIX POWERS*

The Jordan form is extremely important in applied mathematics because it provides the answer key for a set of questions concerning a broad class of applications.

Section 2.2 modeled how the market shares controlled by three dairies evolved from month to month; the result was (2.4)—$\mathbf{x}_{r+1} = \mathbf{A}\mathbf{x}_r$—where the three components of $\mathbf{x}_r$ represent the fractions of the market held by each dairy in month r and where the transition matrix $\mathbf{A}$ describes the effects of market forces.

Similarly, Section 2.3 modeled the evolution of competing populations of foxes and chickens by (2.14)—$\mathbf{x}_{i+1} = \mathbf{A}\mathbf{x}_i$—where the two components of $\mathbf{x}_i$ represent the numbers of foxes and chickens at time i and the matrix $\mathbf{A}$ describes their competition.

More generally, mathematical models often lead to a $p \times 1$ matrix $\mathbf{x}_i$ that describes the state of some complicated system at time i and to a $p \times p$ matrix $\mathbf{A}$ that represents the internal workings of the system in such a way that

(9.19)
$$\mathbf{x}_{i+1} = \mathbf{A}\mathbf{x}_i$$

models the evolution of the system over time. Denoting the initial state by $\mathbf{x}_0$ allows us to write this equivalently as

(9.20)
$$\mathbf{x}_i = \mathbf{A}^i\mathbf{x}_0 \qquad \text{for } i \geq 0.$$

Important questions addressed in Chapter 2 in those specific examples and of central importance generally are: What happens as time passes? Does the system tend to equilibrium? Do the states $\mathbf{x}_i$ become arbitrarily large? Do the states tend to zero? Do they oscillate?

Thanks to (9.20), the Jordan form of $\mathbf{A}$ can handle such questions. Writing $\mathbf{A} = \mathbf{Q}\mathbf{J}\mathbf{Q}^{-1}$ for $\mathbf{J}$ a Jordan form of $\mathbf{A}$ allows the application of Theorem 7.25(a), so that

(9.21)
$$\mathbf{A}^i = \mathbf{Q}\mathbf{J}^i\mathbf{Q}^{-1} \text{ and therefore}$$
$$\mathbf{x}_i = \mathbf{Q}\mathbf{J}^i(\mathbf{Q}^{-1}\mathbf{x}_0).$$

This allows us to study the behavior of $\mathbf{J}^i$—a much simpler problem than for $\mathbf{A}^i$ directly. It becomes even simpler with the observation from **Key Theorem 9.4** that $\mathbf{J}$ itself is a block-diagonal matrix with Jordan blocks $\mathbf{J}_r$ on the diagonal; we then need only study powers $\mathbf{J}_r^i$, since

$$\mathbf{J}^i = \begin{bmatrix} \mathbf{J}_1^i & \mathbf{0} & \cdots & \mathbf{0} \\ \mathbf{0} & \mathbf{J}_2^i & \cdots & \mathbf{0} \\ & \cdots & & \\ \mathbf{0} & \mathbf{0} & \cdots & \mathbf{J}_\mu^i \end{bmatrix}.$$

Nondefective Matrices

To start simply, we suppose that each Jordan block $\mathbf{J}_r$—and hence $\mathbf{J}$ itself—is *diagonal*; by **Key Theorem 7.14**, this is equivalent to $\mathbf{A}$'s having a linearly independent set of p eigenvectors—that is, $\mathbf{A}$ is nondefective. As usual in this case we write $\mathbf{\Lambda}$ in place of $\mathbf{J}$, with the eigenvalues $\lambda_1, \ldots, \lambda_p$ as the diagonal entries of $\mathbf{\Lambda}$. Thus the matrix whose behavior is to be studied is simply

(9.22)
$$\mathbf{\Lambda}^i = \begin{bmatrix} \lambda_1^i & 0 & \cdots & 0 \\ 0 & \lambda_2^i & \cdots & 0 \\ & \cdots & & \\ 0 & 0 & \cdots & \lambda_p^i \end{bmatrix}.$$

For any number λ, it is clear how λ^i behaves as i tends to infinity:

(9.23)
(a) λ^i tends to 0 if and only if $|\lambda| < 1$.
(b) $|\lambda^i|$ tends to infinity if and only if $|\lambda| > 1$.
(c) λ^i is bounded—that is, $|\lambda^i| \le c$ for some constant c and for all i—if and only if $|\lambda| \le 1$.

To extend (9.23) to apply to $\mathbf{\Lambda}^i$ and $\mathbf{A}^i$ rather than to λ^i requires notions of convergence—"tending to"—and boundedness for matrices; the matrix norms of Definition 6.22 and **Key Theorem 6.23** provide just what is needed, since we already have discussed convergence in terms of norms in Definition 5.58 and Theorem 5.59. For completeness and ease of reference we restate here the result of combining all those ideas.

(9.24) ***Summary On Matrix Sequences.***
(a) A sequence $\{\mathbf{A}_i\}$ of $p \times q$ matrices is said to *converge* to the $p \times q$ matrix $\mathbf{A}_\infty$ if and only if the sequences of corresponding entries converge: $\langle \mathbf{A}_i \rangle_{jk}$ converges to $\langle \mathbf{A}_\infty \rangle_{jk}$ for all j and k.
(b) A sequence $\{\mathbf{A}_i\}$ of $p \times q$ matrices is said to be *bounded* if and only if the sequences of entries are bounded: $|\langle \mathbf{A}_i \rangle_{jk}| \le c$ for some constant c and all i, j, and k.

(c) The sequence of matrices $\mathbf{A}_i$ converges to $\mathbf{A}_\infty$ if and only if the sequence of numbers $\|\mathbf{A}_i - \mathbf{A}_\infty\|$ converges to zero, where $\|\cdot\|$ is any matrix norm; thus, if the sequence converges as determined in one norm, it converges as determined in all norms.

(d) The sequence of matrices $\mathbf{A}_i$ is bounded if and only if the sequence of numbers $\|\mathbf{A}_i\|$ is bounded, where $\|\cdot\|$ denotes any matrix norm; thus, if the sequence is bounded as determined in one norm, it is bounded as determined in all norms.

(9.25) ***Example.*** The 2×2 matrix

$$\mathbf{\Lambda} = \begin{bmatrix} \lambda_1 & 0 \\ 0 & \lambda_2 \end{bmatrix} \quad \text{with} \quad \mathbf{\Lambda}^i = \begin{bmatrix} \lambda_1^i & 0 \\ 0 & \lambda_2^i \end{bmatrix}$$

typifies the general case in (9.22). Using the ∞-norm gives

$$\|\mathbf{\Lambda}^i\|_\infty = \max(|\lambda_1^i|, |\lambda_2^i|) = \{\max(|\lambda_1|, |\lambda_2|)\}^i.$$

If we let $\rho = \max(|\lambda_1|, |\lambda_2|)$, then clearly $\|\mathbf{\Lambda}^i\|_\infty$—and hence $\mathbf{\Lambda}^i$ itself—tends to zero if and only if $\rho < 1$ and is bounded if and only if $\rho \le 1$. When $\rho > 1$, certainly some entries of $\mathbf{\Lambda}^i$ tend to infinity in magnitude, but not all entries do so.

The number ρ introduced in Example 9.25 is useful generally.

(9.26) ***Definition.*** The *spectral radius* $\rho(\mathbf{A})$ of a $p \times p$ matrix $\mathbf{A}$ equals the magnitude of the eigenvalue having largest magnitude:

$$\rho(\mathbf{A}) = \max(|\lambda_1|, \ldots, |\lambda_p|).$$

Note from **Key Theorem 7.10(e)** that

$$\rho(\mathbf{A}) \le \|\mathbf{A}\|$$

for all matrix norms.

The analysis in Example 9.25 clearly extends to $p \times p$ diagonal matrices $\mathbf{\Lambda}$. To extend it to $p \times p$ nondefective matrices $\mathbf{A} = \mathbf{Q\Lambda Q}^{-1}$ and related sequences $\mathbf{A}^i\mathbf{x}_0$ requires only the following observations that follow from the fact that matrix norms are defined as transformation norms and the facts that $\mathbf{A}^i = \mathbf{Q\Lambda}^i\mathbf{Q}^{-1}$ and $\mathbf{\Lambda}^i = \mathbf{Q}^{-1}\mathbf{A}^i\mathbf{Q}$:

$$\|\mathbf{A}^i\| \le \|\mathbf{Q}\|\,\|\mathbf{Q}^{-1}\|\,\|\mathbf{\Lambda}^i\|$$

$$\|\mathbf{A}^i\| \ge \|\mathbf{\Lambda}^i\|/(\|\mathbf{Q}\|\,\|\mathbf{Q}^{-1}\|)$$

$$\|\mathbf{A}^i\mathbf{x}_0\| \le \|\mathbf{A}^i\|\,\|\mathbf{x}_0\| \qquad \text{for all } \mathbf{x}_0$$

$\|\mathbf{A}^i\mathbf{v}\| = \{\rho(\mathbf{A})\}^i\|\mathbf{v}\|$ if $\mathbf{v}$ is an eigenvector associated with an eigenvalue of magnitude $\rho(\mathbf{A})$.

(9.27) **Key Theorem** (nondefective matrix powers). Let $p \times p$ $\mathbf{A}$ be nondefective. Then, as i tends to plus infinity:

(a) $\mathbf{A}^i$ converges to zero if and only if the spectral radius $\rho(\mathbf{A}) < 1$.

(b) $\mathbf{A}^i\mathbf{x}_0$ converges to zero for every $\mathbf{x}_0$ if and only if $\rho(\mathbf{A}) < 1$.

(c) $\mathbf{A}^i$ is bounded if and only if $\rho(\mathbf{A}) \leq 1$.

(d) $\mathbf{A}^i\mathbf{x}_0$ is bounded for every $\mathbf{x}_0$ if and only if $\rho(\mathbf{A}) \leq 1$.

(e) $\|\mathbf{A}^i\|$ tends to infinity in some matrix norm (and therefore in *all* matrix norms) if and only if $\rho(\mathbf{A}) > 1$.

(f) $\|\mathbf{A}^i\mathbf{x}_0\|$ tends to infinity for *some* $\mathbf{x}_0$ in some vector norm (and therefore in *all* vector norms) if and only if $\rho(\mathbf{A}) > 1$.

(9.28) ***Example.*** The transition matrix $\mathbf{A}$ for the dairy-competition model in Example 2.6 was shown in Example 7.2 to have eigenvalues $\lambda_1 = 0.5$, $\lambda_2 = 0.6$, and $\lambda_3 = 1.0$ together with a linearly independent set of three associated eigenvectors $\mathbf{v}_1$, $\mathbf{v}_2$, and $\mathbf{v}_3$. Since, then, $\rho(\mathbf{A}) = 1$, it follows from **Key Theorem 9.27** that $\mathbf{A}^i$ and $\mathbf{A}^i\mathbf{x}_0$ are bounded for all $\mathbf{x}_0$. In fact, since $\mathbf{v}_1$, $\mathbf{v}_2$, and $\mathbf{v}_3$ form a basis for $\mathbb{R}^3$, $\mathbf{x}_0$ can be written as

$$\mathbf{x}_0 = a_1\mathbf{v}_1 + a_2\mathbf{v}_2 + a_3\mathbf{v}_3.$$

Therefore,

$$\mathbf{A}^i\mathbf{x}_0 = a_1(0.5)^i\mathbf{v}_1 + a_2(0.6)^i\mathbf{v}_2 + a_3(1.0)^i\mathbf{v}_3,$$

which shows that $\mathbf{A}^i\mathbf{x}_0$ converges to a multiple $a_3\mathbf{v}_3$ for every $\mathbf{x}_0$. In that model, the sum of the entries of each state $\mathbf{x}_i$ ($= \mathbf{A}^i\mathbf{x}_0$) equals 1, so this must be true in the limit $a_3\mathbf{v}_3$ as well; the limit must then equal

$$[0.45 \quad 0.35 \quad 0.20]^T$$

for all $\mathbf{x}_0$, as observed experimentally in Section 2.2. This finally answers one of the fundamental questions raised there about that model.

You should now be able to solve Problems 1 to 5.

Defective Matrices

The more difficult matter of the behavior of $\mathbf{A}^i$ for defective $\mathbf{A}$ remains. As before, a 2×2 example illuminates the possibilities.

(9.29) ***Example.*** A 2×2 Jordan block has the form

$$\mathbf{J} = \begin{bmatrix} \lambda & 1 \\ 0 & \lambda \end{bmatrix},$$

from which follows

$$\mathbf{J}^2 = \begin{bmatrix} \lambda^2 & 2\lambda \\ 0 & \lambda^2 \end{bmatrix}, \qquad \mathbf{J}^3 = \begin{bmatrix} \lambda^3 & 3\lambda^2 \\ 0 & \lambda^3 \end{bmatrix}, \qquad \mathbf{J}^4 = \begin{bmatrix} \lambda^4 & 4\lambda^3 \\ 0 & \lambda^4 \end{bmatrix},$$

$$\mathbf{J}^i = \begin{bmatrix} \lambda^i & i\lambda^{i-1} \\ 0 & \lambda^i \end{bmatrix} = \lambda^i \begin{bmatrix} 1 & \dfrac{i}{\lambda} \\ 0 & 1 \end{bmatrix}.$$

Since $i\lambda^{i-1}$ tends to zero if and only if $|\lambda| < 1$, $\mathbf{J}^i$ tends to zero if and only if $|\lambda| < 1$, while $\|\mathbf{J}^i\|$ tends to inifinity *if* $|\lambda| > 1$—just as for nondefective matrices. The "only if" for $\|\mathbf{J}^i\|$ tending to infinity and the situation when $|\lambda| = 1$ is different, however. The formula for $\mathbf{J}^i$ shows that

$$\|\mathbf{J}^i\|_\infty = |\lambda|^i(1 + i/|\lambda|);$$

for $|\lambda| = 1$, this gives $\|\mathbf{J}^i\|_\infty = 1 + i$. Thus in this case $\|\mathbf{J}^i\|$ is bounded if and only if $|\lambda| < 1$, in which case $\mathbf{J}^i$ actually tends to $\mathbf{0}$; and $\|\mathbf{J}^i\|$ tends to infinity if and only if $|\lambda| \geq 1$.

That the results of Example 9.29 extend to $n \times n$ Jordan blocks is clear. The extension to general Jordan form—and thus to general $\mathbf{A}$ by $\mathbf{A}^i = \mathbf{QJ}^i\mathbf{Q}^{-1}$—is then straightforward; the only delicate point involves eigenvalues λ with $|\lambda| = 1$: if they appear only in 1×1 Jordan blocks then the powers of those blocks are bounded, but if they appear in $n \times n$ blocks with $n > 1$, the powers of those blocks blow up as in Example 9.29.

(9.30) **Key Theorem** (matrix powers). Let $\mathbf{A}$ be $p \times p$ and possibly defective. Then, as i tends to plus infinity:

(a) $\mathbf{A}^i$ converges to zero if and only if the spectral radius $\rho(\mathbf{A}) < 1$.
(b) $\mathbf{A}^i\mathbf{x}_0$ tends to zero for every $\mathbf{x}_0$ if and only if $\rho(\mathbf{A}) < 1$.
(c) $\mathbf{A}^i$ is bounded if and only if:
   1. $\rho(\mathbf{A}) \leq 1$; *and*
   2. if λ is an eigenvalue with $|\lambda| = 1$, then its algebraic multiplicity equals its geometric multiplicity.
(d) $\mathbf{A}^i\mathbf{x}_0$ is bounded for every $\mathbf{x}_0$ if and only if the conditions (1) *and* (2) of (c) above *both* hold.
(e) $\|\mathbf{A}^i\|$ tends to infinity if and only if *either*:
   1. $\rho(\mathbf{A}) > 1$; *or*
   2. $\rho(\mathbf{A}) = 1$ and there is an eigenvalue λ of $\mathbf{A}$ with $|\lambda| = 1$ and whose geometric multiplicity is strictly less than its algebraic multiplicity.
(f) $\|\mathbf{A}^i\mathbf{x}_0\|$ tends to infinity for *some* $\mathbf{x}_0$ if and only if *either* (1) *or* (2) of (e) above holds.

(9.31) ***Example.*** The matrix $\mathbf{A}$ for the competing-populations model with $k = 0.18$ in Example 2.17 was shown in Example 6.18 to have a double eigenvalue

$\lambda_1 = \lambda_2 = 0.9$ (although such terminology was unavailable then). The matrix is defective, since $(\mathbf{A} - 0.9\mathbf{I})\mathbf{x} = \mathbf{0}$ requires

$$\begin{bmatrix} -0.30 & 0.5 \\ -0.18 & 0.3 \end{bmatrix} \begin{bmatrix} x \\ y \end{bmatrix} = \begin{bmatrix} 0 \\ 0 \end{bmatrix},$$

for which every solution is just a multiple of $[5 \quad 3]^T$. Since $\rho(\mathbf{A}) = 0.9 < 1$, we still know that $\mathbf{A}^i$ tends to zero, as does $\mathbf{A}^i\mathbf{x}_0$ for every $\mathbf{x}_0$. This proves what was indicated experimentally in Example 2.17: the populations in this case die off, no matter what the starting distribution between foxes and chickens.

You should now be able to solve Problems 1 to 9.

Nonnegative and Markov Matrices

Example 9.28 analyzed the dairy-competition model of Example 2.6 and showed, in that specific case, that $\mathbf{A}^i\mathbf{x}_0$ converges to a multiple of the eigenvector $\mathbf{v}_3$ associated with $\lambda_3 = 1$ no matter what initial $\mathbf{x}_0$ is taken. Example 2.6 was included in Chapter 2 as an illustration of more general models that lead to *Markov matrices.*

(9.32) ***Definition.*** A $p \times q$ matrix $\mathbf{A}$ is said to be *nonnegative* if and only if all its entries are nonnegative: $\langle \mathbf{A} \rangle_{ij} \geq 0$. A *Markov matrix* is a $p \times p$ nonnegative matrix such that the sums of its entries in each column equal 1.

Much of the behavior seen in Example 2.6 is typical of Markov matrices.

If we define the $p \times 1$ matrix $\mathbf{1} = [1 \quad 1 \quad \cdots \quad 1]^T$, then $\mathbf{1}^T\mathbf{A} = \mathbf{1}^T$, since the entries of each column of a Markov matrix $\mathbf{A}$ sum to 1; thus $\lambda = 1$ is an eigenvalue of $\mathbf{A}$. Since $\|\mathbf{A}\|_1 = 1$, $|\lambda| \leq 1$ for every eigenvalue λ of $\mathbf{A}$ by **Key Theorem 7.10(e)**; these two facts together imply that $\rho(\mathbf{A}) = 1$. Moreover, since $\|\mathbf{A}^i\|_1 \leq \|\mathbf{A}\|_1^i = 1$, we know that $\mathbf{A}^i$ is bounded; **Key Theorem 9.30** implies then that each eigenvalue of magnitude 1 has equal algebraic and geometric multiplicities—that is, $\mathbf{A}$ has a "full set" of eigenvectors associated with such eigenvalues. This is not, however, sufficient to prove that $\mathbf{A}^i\mathbf{x}_0$ converges for all $\mathbf{x}_0$, as was true in Example 2.6. The Markov matrix

$$\mathbf{A} = \begin{bmatrix} 0 & 1 \\ 1 & 0 \end{bmatrix},$$

for example, has eigenvalues $+1$ and -1 and associated eigenvectors $\mathbf{v}_1 = [1 \quad 1]^T$ and $\mathbf{v}_2 = [1 \quad -1]^T$; if we write $\mathbf{x}_0 = a_1\mathbf{v}_1 + a_2\mathbf{v}_2$, then $\mathbf{A}^i\mathbf{x}_0$ equals either $\mathbf{x}_0$ (for even i) or $\mathbf{x}_0 - 2a_2\mathbf{v}_2$ (for odd i), so $\mathbf{A}^i\mathbf{x}_0$ converges (to $\mathbf{x}_0$, in fact) if and only if $a_2 = 0$—that is, when $\mathbf{x}_0$ is $\mathbf{0}$ or an eigenvector associated with $\lambda = 1$.

If we could somehow be certain that all eigenvalues of **A** other than $\lambda = 1$ had magnitude strictly less than 1, the situation just illustrated could not arise; this requires, however, some additional conditions on the Markov matrix **A**, as the 2×2 example just presented demonstrates.

The theory developed to treat this issue applies more generally than just to Markov matrices. Some terminology is necessary before proceeding.

(9.33) ***Definition.*** A $p \times p$ matrix is said to be *reducible* if and only if there exists a permutation matrix **P** such that

$$\mathbf{PAP}^T = \begin{bmatrix} \mathbf{B} & \mathbf{C} \\ \mathbf{0} & \mathbf{D} \end{bmatrix}$$

with **B** $r \times r$ and **D** $(p - r) \times (p - r)$ for $1 \le r \le p - 1$. A matrix is *irreducible* if and only if it is not reducible.

The term "reducible" refers to the fact that, for a reducible matrix **A** with its rows and columns permuted as in the definition, the p equations $\mathbf{Ax} = \mathbf{b}$ in p unknowns reduce to solving $p - r$ equations in $p - r$ unknowns with coefficient matrix **D**, followed by solving r equations in r unknowns with coefficient matrix **B** (but note the third matrix following). The matrices

$$\begin{bmatrix} 2 & 5 \\ 0 & 3 \end{bmatrix} \quad \text{and} \quad \begin{bmatrix} 7 & 0 \\ 4 & 2 \end{bmatrix} \quad \text{are reducible,}$$

while

$$\begin{bmatrix} 0 & 1 \\ 1 & 0 \end{bmatrix} \quad \text{and} \quad \begin{bmatrix} 3 & 4 \\ 1 & 2 \end{bmatrix} \quad \text{are irreducible.}$$

The fourth example illustrates the fact that every square matrix of strictly positive entries is irreducible; the third example shows that our intuitive description of the concept "reducible" is not *equivalent* to the definition.

Much of what has been demonstrated for Markov matrices holds for irreducible nonnegative matrices; this elegant theory is named after its primary developers, O. Perron and G. Frobenius. We state the results without proof.

(9.34) ***Theorem*** (Perron-Frobenius). Let **A** be a $p \times p$ irreducible nonnegative matrix. Then:

(a) **A** has a positive real eigenvalue λ_1 with $\lambda_1 = \rho(\mathbf{A})$.

(b) λ_1 above has an associated eigenvector **x** with strictly positive entries.

(c) λ_1 has algebraic multiplicity $m_1 = 1$.

(d) All eigenvalues λ of **A** other than λ_1 satisfy $|\lambda| < |\lambda_1|$ if and only if there is a positive integer k with all entries of $\mathbf{A}^k$ strictly positive.

(e) If the main-diagonal entries of $\mathbf{A}$ are strictly positive, then all entries of $\mathbf{A}^{p-1}$ are strictly positive and—by (d)—all eigenvalues λ of $\mathbf{A}$ other than λ_1 satisfy $|\lambda| < |\lambda_1|$.
(f) $\rho(\mathbf{A})$ strictly increases if any entry of $\mathbf{A}$ strictly increases.

Parts (a)–(c) say that irreducible nonnegative matrices are much like Markov matrices. Part (d) says that the condition we needed for a Markov matrix in order to show that $\mathbf{A}^i\mathbf{x}_0$ converges for all $\mathbf{x}_0$ is equivalent to having all entries of $\mathbf{A}^k$ positive for some k; part (e) says that this will happen if the diagonal entries of $\mathbf{A}$ are positive. A straight-forward corollary of this theorem handles the generalization of Example 2.6.

(9.35) ***Corollary*** (Markov-matrix powers). Suppose that $\mathbf{A}$ is an irreducible Markov matrix with strictly positive diagonal entries (a Markov matrix is nonnegative and its entries in each column sum to 1). Then:
(a) For any $\mathbf{x}_0$, $\mathbf{A}^i\mathbf{x}_0$ converges to $\alpha\mathbf{x}^*$, where $\mathbf{x}^*$ is an eigenvector associated with the eigenvalue $\lambda_1 = 1$, where $\mathbf{x}^*$ has strictly positive entries that sum to 1, and where α equals the sum of the entries in $\mathbf{x}_0$.
(b) $\mathbf{A}^i$ converges to $[\mathbf{x}^* \quad \mathbf{x}^* \quad \cdots \quad \mathbf{x}^*]$.

PROOF. Problem 11. ■

An example of this theorem is the 3×3 matrix from Example 2.6; since all its entries are strictly positive, it satisfies the corollary's hypotheses, while the computations in Examples 2.6 and 2.9 illustrate the conclusions with

$$\mathbf{x}^* = [0.45 \quad 0.35 \quad 0.20]^T.$$

PROBLEMS 9.4

1. Prove **Key Theorem 9.27(b), (d),** and **(f)** by writing $\mathbf{x}_0$ as a linear combination of the eigenvectors and then computing $\mathbf{A}^i\mathbf{x}_0$.
2. Prove **Key Theorem 9.27(a), (c),** and **(e)**.
3. ▷ Analyze the behavior of $\mathbf{x}_i$ in the competing-populations model in Example 2.13 as i tends to infinity if $k = 0.10$.
4. Analyze the behavior of $\mathbf{x}_i$ in the competing-populations model in Example 2.13 as i tends to infinity if $k = 0.16$.
5. ▷ Determine the behavior of $\mathbf{A}^i$ and of $\mathbf{A}^i\mathbf{x}_0$ for arbitrary $\mathbf{x}_0$ if

$$\mathbf{A} = \begin{bmatrix} \frac{1}{6} & \frac{\sqrt{6}}{3} & \frac{-\sqrt{2}}{6} \\ \frac{\sqrt{6}}{3} & 0 & \frac{\sqrt{3}}{3} \\ \frac{-\sqrt{2}}{6} & \frac{\sqrt{3}}{3} & \frac{1}{3} \end{bmatrix}.$$

6. Prove **Key Theorem 9.30**.

7. Determine the behavior of $\mathbf{A}^i$ if

$$\mathbf{A} = \begin{bmatrix} 0.5 & -4 & 0.5 \\ 0 & 1 & 0 \\ 0 & -1 & 1 \end{bmatrix}.$$

8. When x is a real number with $|x| < 1$, we show that $(1-x)^{-1} = 1 + x + x^2 + \cdots$ by *first* showing that the series converges and *second* writing $(1-x)(1+x+\cdots+x^n) = 1 - x^{n+1}$ and taking the limit as n tends to infinity. Use a similar argument to show that if $\mathbf{A}$ is a $p \times p$ matrix with $\rho(\mathbf{A}) < 1$ then $\mathbf{I} - \mathbf{A}$ is nonsingular and its inverse is given by the convergent series

$$(\mathbf{I} - \mathbf{A})^{-1} = \mathbf{I} + \mathbf{A} + \mathbf{A}^2 + \cdots.$$

▷ 9. Use Problem 8 to prove that if $\alpha > \rho(\mathbf{A})$ then $\alpha\mathbf{I} - \mathbf{A}$ is nonsingular and

$$(\alpha\mathbf{I} - \mathbf{A})^{-1} = \alpha^{-1}\left(\mathbf{I} + \frac{\mathbf{A}}{\alpha} + \frac{\mathbf{A}^2}{\alpha^2} + \cdots\right).$$

▷ 10. Show that:

(a) $\begin{bmatrix} 0 & 1 \\ 1 & 0 \end{bmatrix}$ is irreducible

(b) $\begin{bmatrix} 7 & 0 & 6 \\ 3 & 1 & 2 \\ 5 & 0 & 4 \end{bmatrix}$ is reducible

(c) $\begin{bmatrix} 2 & 0 \\ 3 & 4 \end{bmatrix}$ is reducible

11. Use the Perron-Frobenius theorem (9.34) to prove Corollary 9.35.

𝔐 12. Verify (f) in the Perron-Frobenius theorem (9.34) by using MATLAB or similar software for explicitly examining the spectral radius of the matrix $\mathbf{A}(\alpha)$ as $\alpha \geq 0$ increases, where

$$\mathbf{A}(\alpha) = \begin{bmatrix} 1 & 4 & 1 + 0.9\alpha \\ 3 & 0.8\alpha & 3 \\ \alpha & 5 & 1 \end{bmatrix}.$$

▷ 13. Characterize the $\mathbf{x}_0$ such that $\mathbf{A}^i\mathbf{x}_0$ converges to zero for a Markov matrix $\mathbf{A}$.

14. Suppose that $\mathbf{A}$ is nonnegative. Use Problem 9 to show that if $\alpha > \rho(\mathbf{A})$ then $(\alpha\mathbf{I} - \mathbf{A})^{-1}$ is nonnegative.

15. Consider the irreducible nonnegative matrix $\mathbf{B}$ with $\langle\mathbf{B}\rangle_{ij} = 0$ except when $|i - j| = 1$, in which case $\langle\mathbf{B}\rangle_{ij} = 1$. By comparing $\mathbf{B}$ to the matrix $\mathbf{B}'$ obtained from $\mathbf{B}$ by increasing to 2 the 1 in the top and bottom rows and using (e) of the Perron-Frobenius theorem, show that $\rho(\mathbf{B}) < 2$.

▷ **16.** Use Problems 14 and 15 to show that $(2\mathbf{I} - \mathbf{B})^{-1}$ is nonnegative, where $\mathbf{B}$ is the matrix from Problem 15.

𝔐 **17.** Use MATLAB or similar software to verify (a)–(d) of the Perron-Frobenius theorem and Corollary 9.35 for:
(a) the transition matrix of the model in Problem 4 of Section 2.2
(b) the transition matrix of the model in Problem 6 of Section 2.2

𝔐 **18.** Use MATLAB or similar software to verify (a)–(d) of the Perron-Frobenius theorem for the 6×6 Hilbert matrix $\mathbf{H}_6$, where $\langle \mathbf{H}_6 \rangle_{ij} = 1/(i + j - 1)$.

▷ **19.** Suppose that $\mathbf{A}$ is a square matrix with strictly positive entries. Prove that $\rho(\mathbf{A}) \le \|\mathbf{A}\|_1$ and that $\rho(\mathbf{A}) = \|\mathbf{A}\|_1$ if and only if $\mathbf{A}/\rho(\mathbf{A})$ is a Markov matrix.

9.5 *APPLICATION: CONTINUOUS SYSTEM EVOLUTION AND MATRIX EXPONENTIALS*

This section provides another illustration of the analytical value of the Jordan form. In the preceding section we extended our examples from Sections 2.2 and 2.3 to the evolution of general systems whose state at one specific point in time is given as a linear transformation of the state at a specific earlier time. If the time interval involved is extremely small, or if the change in the state is extremely small relative to the state, it is often useful to assume that the state $\mathbf{x}$ is defined for *all* time t by the function $\mathbf{x}(t)$ and that information on state changes is given in terms of the derivative $\dot{\mathbf{x}}$, where $\dot{\mathbf{x}}$ denotes

$$\dot{\mathbf{x}}(t) = \begin{bmatrix} \dfrac{dx_1}{dt}(t) \\ \dfrac{dx_2}{dt}(t) \\ \vdots \\ \dfrac{dx_p}{dt}(t) \end{bmatrix}$$

if $\mathbf{x}$ is $p \times 1$. In other instances, particularly in the engineering and physical sciences, time is most naturally considered to be a continuously sampled variable, so that the natural descriptions of how the state $\mathbf{x}$ of a system evolves use the derivative $\dot{\mathbf{x}}$ and ordinary differential equations. We now give three examples to illustrate how some ordinary differential equations arising in applications can be expressed in a standard form $\dot{\mathbf{x}} = \mathbf{A}\mathbf{x} + \mathbf{f}$.

(9.36) ***Example.*** The population-growth models of Section 2.3 arose essentially from the assumption that growth is proportional to present population:

$$p_{i+1} - p_i = (b - d)p_i$$

in the notation of that section. For small time intervals or rapidly growing populations the analogous continuous model would use the time derivative $\dot{p}$:

$$\dot{p}(t) = (\beta - \delta)p(t),$$

with β and δ reflecting the birth and death rates. For competing populations such as the fox-chicken model of that section, we might replace (2.13) with

$$\dot{F}(t) = -0.4F(t) + 0.5C(t), \qquad F(t_0) = 100$$

$$\dot{C}(t) = -kF(t) + 0.2C(t), \qquad C(t_0) = 1000$$

or, in matrix notation,

$$\dot{\mathbf{x}} = \mathbf{A}\mathbf{x}, \qquad \mathbf{x}(t_0) = \mathbf{x}_0,$$

where

$$\mathbf{x} = \begin{bmatrix} F \\ C \end{bmatrix}, \qquad \mathbf{A} = \begin{bmatrix} -0.4 & 0.5 \\ -k & 0.2 \end{bmatrix}, \qquad \mathbf{x}_0 = \begin{bmatrix} 100 \\ 1000 \end{bmatrix}.$$

(9.37) ***Example.*** Suppose that $v(t)$ denotes the total income at time t of a given business enterprise (or perhaps the GNP of an economy), and suppose that, if $i(t)$ of this income is reinvested, the rate of increase of v will be proportional to i, so that

$$\dot{v}(t) = gi(t)$$

for some growth rate g. We propose to reinvest a fixed fraction r of income, so that $i(t) = rv(t)$, and we suppose that the remaining income, $(1 - r)v(t)$, is to be distributed as profits to the shareholders after taxes and other expenses are deducted, so that the rate of increase of profit p is proportional to $(1 - r)v$, that is,

$$\dot{p} = s(1 - r)v.$$

Altogether then, we have the model

$$\dot{v} = grv,$$

$$\dot{p} = s(1 - r)v,$$

or, in matrix notation,

$$\dot{\mathbf{x}} = \mathbf{A}\mathbf{x},$$

where

$$\mathbf{x} = \begin{bmatrix} v \\ p \end{bmatrix}, \qquad \mathbf{A} = \begin{bmatrix} gr & 0 \\ s(1 - r) & 0 \end{bmatrix}.$$

Here g, r, and s are constants.

(9.38) ***Example.*** Section 2.5 derived the differential equations (2.26) for the motion of a pair of masses coupled by springs; suppose in addition that downward external forces $F_1(t)$ and $F_2(t)$ are applied to the masses m_1 and m_2. This adds F_1 to the right-hand side of the first equation in (2.26) and F_2 to that of the second. To eliminate second derivatives, we introduce auxiliary variables: Let $x_1 = X_1$, $x_2 = \dot{X}_1$, $x_3 = X_2$, and $x_4 = \dot{X}_2$. This requires additional equations $\dot{x}_1 = x_2$ and $\dot{x}_3 = x_4$. The result is

$$\begin{aligned} \dot{x}_1 &= x_2 \\ m_1\dot{x}_2 &= -k_1x_1 + k_2(x_3 - x_1) + F_1 \\ \dot{x}_3 &= x_4 \\ m_2\dot{x}_4 &= -k_2(x_3 - x_1) + F_2. \end{aligned}$$

With $\mathbf{x} = [x_1 \quad x_2 \quad x_3 \quad x_4]^T$, this system can be written as $\dot{\mathbf{x}} = \mathbf{A}\mathbf{x} + \mathbf{f}$, where

$$\mathbf{A} = \begin{bmatrix} 0 & 1 & 0 & 0 \\ -\dfrac{k_1 + k_2}{m_1} & 0 & \dfrac{k_2}{m_1} & 0 \\ 0 & 0 & 0 & 1 \\ \dfrac{k_2}{m_2} & 0 & -\dfrac{k_2}{m_2} & 0 \end{bmatrix} \quad \text{and} \quad \mathbf{f} = \begin{bmatrix} 0 \\ \dfrac{F_1}{m_1} \\ 0 \\ \dfrac{F_2}{m_2} \end{bmatrix}.$$

Each of the three preceding examples led to a differential equation

(9.39) $$\dot{\mathbf{x}} = \mathbf{A}\mathbf{x} + \mathbf{f},$$

where $\mathbf{A}$ is a given $p \times p$ matrix of constants, $\mathbf{f}$ is a given $p \times 1$ matrix of functions of t, and $\mathbf{x}$ is a $p \times 1$ matrix of unknown functions of t to be found so as to satisfy (9.39) identically in t. In such models of evolving systems it is typical to be given the state at some time t_0, so that

(9.40) $$\mathbf{x}(t_0) = \mathbf{x}_0$$

for some given t_0 and given constant vector $\mathbf{x}_0$. For simplicity we suppose that $\mathbf{t}_0 = 0$ and—for the moment—that $\mathbf{f} = \mathbf{0}$, so that we treat

(9.41) $$\dot{\mathbf{x}} = \mathbf{A}\mathbf{x}, \qquad \mathbf{x}(0) = \mathbf{x}_0.$$

Differential Equations with Nondefective Matrices

As usual, the analysis is simpler when $\mathbf{A}$ has a linearly independent set of p eigenvectors, so for now we assume this; for such nondefective $\mathbf{A}$, we can write

$$\mathbf{P}^{-1}\mathbf{A}\mathbf{P} = \mathbf{\Lambda} \quad \text{and} \quad \mathbf{A} = \mathbf{P}\mathbf{\Lambda}\mathbf{P}^{-1}$$

with diagonal $\mathbf{\Lambda} = \mathbf{diag}(\lambda_1, \ldots, \lambda_p)$. Since $\mathbf{P}^{-1}$ is a matrix of constants, if we premultiply $\dot{\mathbf{x}} = \mathbf{A}\mathbf{x}$ by $\mathbf{P}^{-1}$ we get $d(\mathbf{P}^{-1}\mathbf{x})/dt = (\mathbf{P}^{-1}\mathbf{A}\mathbf{P})(\mathbf{P}^{-1}\mathbf{x})$, or $\dot{\mathbf{y}} = \mathbf{\Lambda}\mathbf{y}$ if we let $\mathbf{y} = \mathbf{P}^{-1}\mathbf{x}$. Certainly, $\mathbf{y}(0) = \mathbf{P}^{-1}\mathbf{x}(0) = \mathbf{P}^{-1}\mathbf{x}_0$, so $\mathbf{y}(0)$ is known also. Denoting $\mathbf{P}^{-1}\mathbf{x}_0$ by $\mathbf{y}_0$, we can write our system as

(9.42)
$$\dot{\mathbf{y}} = \mathbf{\Lambda}\mathbf{y}, \qquad \mathbf{y}(0) = \mathbf{y}_0.$$

With $\mathbf{y} = [y_1 \quad y_2 \quad \cdots \quad y_p]^T$, this is nothing other than

(9.43)
$$\dot{y}_i = \lambda_i y_i, \qquad y_i(0) = \langle \mathbf{y}_0 \rangle_i \qquad \text{for } 1 \le i \le p.$$

The scalar equations (9.43) are solved by $y_i(t) = \langle \mathbf{y}_0 \rangle_i \exp(\lambda_i t)$, where exp denotes the exponential function: $\exp(s) = e^s$ for all s. In matrix notation, this is

(9.44) $\mathbf{y}(t) = \mathbf{L}(t)\mathbf{y}_0$, where

$$\mathbf{L}(t) = \begin{bmatrix} \exp(\lambda_1 t) & 0 & \cdots & 0 \\ 0 & \exp(\lambda_2 t) & \cdots & 0 \\ & \cdots & & \\ 0 & 0 & \cdots & \exp(\lambda_p t) \end{bmatrix}$$

Note that $\mathbf{L}(t)$ is diagonal. Since $\mathbf{x} = \mathbf{P}\mathbf{y}$ and $\mathbf{y}_0 = \mathbf{P}^{-1}\mathbf{x}_0$, (9.44) yields

(9.45)
$$\mathbf{x}(t) = \mathbf{P}\mathbf{L}(t)\mathbf{P}^{-1}\mathbf{x}_0$$

as the solution to (9.41). This can also be written as

$$\mathbf{x}(t) = a_1\mathbf{p}_1 \exp(\lambda_1 t) + a_2\mathbf{p}_2 \exp(\lambda_2 t) + \cdots + a_p\mathbf{p}_p \exp(\lambda_p t),$$

where $\mathbf{p}_i$ denotes the ith column of $\mathbf{P}$ and $a_i = \langle \mathbf{P}^{-1}\mathbf{x}_0 \rangle_i$. We summarize what has been proved.

(9.46) ***Theorem*** (nondefective matrix solutions). Let $p \times p$ $\mathbf{A}$ be nondefective with $\mathbf{P}^{-1}\mathbf{A}\mathbf{P} = \mathbf{\Lambda}$ diagonal with eigenvalues λ_i of $\mathbf{A}$ on the main diagonal of $\mathbf{\Lambda}$. Then the solution $\mathbf{x}(t)$ to (9.41) is

$$\mathbf{x}(t) = \mathbf{P}\mathbf{L}(t)\mathbf{P}^{-1}\mathbf{x}_0,$$

where $\mathbf{L}(t)$ is the diagonal matrix in (9.44).

(9.47) ***Example.*** If $k = 0.16$ in the competing-populations model of Example 9.36, then

$$\mathbf{A} = \begin{bmatrix} -0.4 & 0.5 \\ -0.16 & 0.2 \end{bmatrix}$$

has eigenvalues $\lambda_1 = 0$, $\lambda_2 = -0.2$ and is reduced to diagonal Jordan form by

$$\mathbf{P} = \begin{bmatrix} 5 & 5 \\ 4 & 2 \end{bmatrix}, \qquad \mathbf{P}^{-1} = \begin{bmatrix} -0.2 & 0.5 \\ 0.4 & -0.5 \end{bmatrix}.$$

The diagonal entries of $\mathbf{L}(t)$ are then $\exp(0t) = 1$ and $\exp(-0.2t)$; Theorem 9.46 easily gives $\mathbf{x}(t) = [F(t) \quad C(t)]^T$ where

$$F(t) = 2400 - 2300 \exp(-0.2t) \quad \text{and} \quad C(t) = 1920 - 920 \exp(-0.2t).$$

As time passes, the populations tend to stable values of 2400 and 1920.

Differential Equations with Defective Matrices

The situation is almost as simple for defective matrices; the preceding analysis has to be repeated, but with $\mathbf{J}$ in place of $\mathbf{\Lambda}$ in (9.42):

(9.48)
$$\dot{\mathbf{y}} = \mathbf{J}\mathbf{y}, \qquad \mathbf{y}(0) = \mathbf{y}_0.$$

Since $\mathbf{J}$ is composed of μ Jordan blocks, we can treat each block independently—just as we treated each y_i independently for nondefective matrices. Denote by $z_1, \ldots, z_n$ for the moment the variables affected by some $n \times n$ Jordan block associated with the eigenvalue λ. Then the equations for this block in (9.48) are

(9.49)
$$\begin{aligned}
\dot{z}_1 &= \lambda z_1 + z_2, && \text{so that} \quad d\{z_1 \exp(-\lambda t)\}/dt = z_2 \exp(-\lambda t) \\
&\ \vdots \\
\dot{z}_{n-1} &= \lambda z_{n-1} + z_n, && \text{so that} \quad d\{z_{n-1} \exp(-\lambda t)\}/dt = z_n \exp(-\lambda t) \\
\dot{z}_n &= \lambda z_n, && \text{so that} \quad d\{z_n \exp(-\lambda t)\}/dt = 0.
\end{aligned}$$

These can be solved successively, starting with the bottom equation; the result is that z_r equals $\exp(\lambda t)$ times a polynomial of degree $n - r$ in t with coefficients involving the values of $z_i(0)$ (Problem 4). In matrix form, this is expressed as

(9.50) $\mathbf{z}(t) = \mathbf{k}_\lambda(t)\mathbf{z}(0)$, where

$$\mathbf{k}_\lambda(t) = \exp(\lambda t) \begin{bmatrix} 1 & t & \dfrac{t^2}{2} & \cdots & \dfrac{t^{n-1}}{(n-1)!} \\ 0 & 1 & t & \cdots & \dfrac{t^{n-2}}{(n-2)!} \\ & & \cdots & & \\ 0 & 0 & 0 & \cdots & 1 \end{bmatrix}$$

The full Jordan form $\mathbf{J}$ in (9.48) consists of μ blocks like that producing this solution (9.50). Each such block yields a $\mathbf{k}_\lambda(t)$ with λ replaced by the appropriate eigenvalue λ_i in that block. Reconstructing $\mathbf{x}$ from $\mathbf{y}$ as in (9.45) finally gives

(9.51) $\mathbf{x}(t) = \mathbf{P}\mathbf{K}(t)\mathbf{P}^{-1}\mathbf{x}_0$, where

$$\mathbf{K}(t) = \begin{bmatrix} \mathbf{k}_1(t) & \mathbf{0} & \cdots & \mathbf{0} \\ \mathbf{0} & \mathbf{k}_2(t) & \cdots & \mathbf{0} \\ & \cdots & & \\ \mathbf{0} & \mathbf{0} & \cdots & \mathbf{k}_\mu(t) \end{bmatrix},$$

and where $\mathbf{k}_i(t)$ denotes $\mathbf{k}_\lambda(t)$ in (9.50) with λ replaced by the eigenvalue in the ith Jordan block $\mathbf{J}_i$ in a Jordan form $\mathbf{J}$ of $\mathbf{A}$.

We summarize.

(9.52) ***Theorem*** (general solutions). Let $p \times p$ $\mathbf{A}$ have Jordan form $\mathbf{J} = \mathbf{P}^{-1}\mathbf{A}\mathbf{P}$ with the eigenvalues λ_i of $\mathbf{A}$ on the main diagonal of $\mathbf{J}$. Then the solution $\mathbf{x}(t)$ to (9.41) is given by

$$\mathbf{x}(t) = \mathbf{P}\mathbf{K}(t)\mathbf{P}^{-1}\mathbf{x}_0,$$

where $\mathbf{K}(t)$ is the block-diagonal matrix in (9.51).

(9.53) ***Example.*** If $k = 0.18$ in the competing-population model in Example 9.36, then

$$\mathbf{A} = \begin{bmatrix} -0.40 & 0.5 \\ -0.18 & 0.2 \end{bmatrix}$$

has a double eigenvalue $\lambda_1 = -0.1$, and every eigenvector is a multiple of $[5 \quad 3]^T$. This means that $\mathbf{A}$ has a 2×2 Jordan block in its Jordan form, which is easily found to be

$$\mathbf{J} = \mathbf{P}^{-1}\mathbf{A}\mathbf{P} = \begin{bmatrix} 0.02 & 0.3 \\ -0.06 & 0.1 \end{bmatrix}\begin{bmatrix} -0.40 & 0.5 \\ -0.18 & 0.2 \end{bmatrix}\begin{bmatrix} 5 & -15 \\ 3 & 1 \end{bmatrix}$$
$$= \begin{bmatrix} -0.1 & 1 \\ 0 & -0.1 \end{bmatrix}.$$

The formula (9.51) then gives $\mathbf{x}(t) = [F(t) \quad C(t)]^T$, where

$$F(t) = (100 + 470t)\exp(-0.1t), \qquad C(t) = (1000 + 282t)\exp(-0.1t).$$

Both populations die out as t tends to infinity.

To this point we have assumed that $\mathbf{f}(t) = \mathbf{0}$ in (9.39). Instead of treating the nonzero case now, we take another approach. We could have produced everything above by introducing *matrix exponentials* $\exp(t\mathbf{A})$ instead of reducing the problem to scalar equations as in (9.43) and (9.49). In order to illustrate both approaches, we next develop matrix exponentials and then use them to handle (9.39) for $\mathbf{f} \neq \mathbf{0}$.

You should now be able to solve Problems 1 to 7.

Matrix Exponentials

It should be clear from the preceding that exponentials play a fundamental role in solving differential equations: $\dot{x} = ax$ with $x(0) = x_0$ is solved by $x(t) = \exp(at)x_0$, for example. It seems natural to consider the possibility of solving $\dot{\mathbf{x}} = \mathbf{A}\mathbf{x}$ with $\mathbf{x}(0) = \mathbf{x}_0$ by $\exp(\mathbf{A}t)\mathbf{x}_0$, if $\exp(\mathbf{A}t)$ is appropriately defined. Letting $\mathbf{B}$ denote $\mathbf{A}t$, we consider the expression $\exp(\mathbf{B})$.

For *numbers b*, $\exp(b)$ is defined by the power series

$$\exp(b) = 1 + b + \frac{b^2}{2!} + \frac{b^3}{3!} + \cdots,$$

which converges to $\exp(b)$ for all real or complex numbers b. By using matrix norms and the concept of convergence as determined by norms, it can be shown that

$$\mathbf{I} + \mathbf{B} + \frac{\mathbf{B}^2}{2!} + \frac{\mathbf{B}^3}{3!} + \cdots$$

converges for every square real or complex matrix $\mathbf{B}$ and can thus be used as a definition of $\exp(\mathbf{B})$. It is even easy to see what this series converges to. If we find the Jordan form $\mathbf{J} = \mathbf{P}^{-1}\mathbf{B}\mathbf{P}$, then by substituting $\mathbf{P}\mathbf{J}\mathbf{P}^{-1}$ for $\mathbf{B}$ in the series we get

$$\exp(\mathbf{B}) = \mathbf{P}\left(\mathbf{I} + \mathbf{J} + \frac{\mathbf{J}^2}{2!} + \cdots\right)\mathbf{P}^{-1} = \mathbf{P}\exp(\mathbf{J})\mathbf{P}^{-1}.$$

Moreover, $\mathbf{J}$ and its powers are so simple that we can explicitly evaluate this series in $\mathbf{J}$. From the formula that results, the following can be proved directly (Problems 8–11).

(9.54) ***Theorem*** (matrix exponentials)

(a) The infinite series

$$\exp(\mathbf{A}) = \mathbf{I} + \mathbf{A} + \frac{\mathbf{A}^2}{2!} + \frac{\mathbf{A}^3}{3!} + \cdots$$

converges for every square real or complex matrix $\mathbf{A}$.

(b) $\exp(r\mathbf{A})\exp(s\mathbf{A}) = \exp(s\mathbf{A})\exp(r\mathbf{A}) = \exp\{(r+s)\mathbf{A}\}$ for all numbers r, s, and all square matrices $\mathbf{A}$.

(c) $\mathbf{A}\exp(\mathbf{A}) = \exp(\mathbf{A})\mathbf{A}$ for all square $\mathbf{A}$.

(d) The derivative $d\{\exp(t\mathbf{A})\}/dt = \mathbf{A}\exp(t\mathbf{A})$.

(e) Suppose that $\mathbf{J} = \mathbf{Q}^{-1}\mathbf{A}\mathbf{Q}$ is a Jordan form of $\mathbf{A}$—using the notation of **Key Theorem 9.4**—with $n_r \times n_r$ Jordan blocks $\mathbf{J}_r$ for $1 \le r \le \mu$. Then $\exp(\mathbf{A})$ can be found by setting $t = 1$ in the following formula for $\exp(t\mathbf{A})$: For each number t, $\exp(t\mathbf{A}) = \mathbf{Q}\exp(t\mathbf{J})\mathbf{Q}^{-1}$, where

$$\exp(t\mathbf{J}) = \begin{bmatrix} \exp(t\mathbf{J}_1) & \mathbf{0} & \cdots & \mathbf{0} \\ & & \cdots & \\ \mathbf{0} & \mathbf{0} & \cdots & \exp(t\mathbf{J}_\mu) \end{bmatrix},$$

and where $\exp(t\mathbf{J}_r)$ for the $n_r \times n_r$ Jordan block $\mathbf{J}_r$ associated with the eigenvalue λ is $\exp(t\mathbf{J}_r) = \mathbf{k}_\lambda(t)$ with $\mathbf{k}_\lambda(t)$ as given in (9.50), replacing n there by n_r.

(f) $\exp(\mathbf{A})$ is nonsingular and $\{\exp(\mathbf{A})\}^{-1} = \exp(-\mathbf{A})$.

(g) $\exp(\mathbf{0}) = \mathbf{I}$.

With this information, Theorem 9.52 on solving $\dot{\mathbf{x}} = \mathbf{A}\mathbf{x}$ can be restated.

(9.55) ***Corollary.*** The solution to $\dot{\mathbf{x}} = \mathbf{A}\mathbf{x}$ with $\mathbf{x}(0) = \mathbf{x}_0$ is

$$\mathbf{x}(t) = \exp(t\mathbf{A})\mathbf{x}_0.$$

(9.56) ***Example.*** Example 9.53 presented 2×2 $\mathbf{A}$ and its Jordan form

$$\mathbf{J} = \begin{bmatrix} -0.1 & 1 \\ 0 & -0.1 \end{bmatrix} \quad \text{for} \quad \mathbf{A} = \begin{bmatrix} -0.40 & 0.5 \\ -0.18 & 0.2 \end{bmatrix}.$$

Using $\mathbf{P}$ and $\mathbf{P}^{-1}$ from Example 9.53 gives

$$\exp(t\mathbf{A}) = \exp(-0.1t)\begin{bmatrix} 1 - 0.3t & 0.5t \\ -0.18t & 1 + 0.3t \end{bmatrix};$$

setting $t = 1$ gives

$$\exp(\mathbf{A}) = \exp(1\mathbf{A}) = \exp(-0.1)\begin{bmatrix} 0.7 & 0.5 \\ -0.18 & 1.3 \end{bmatrix}.$$

Once we found the solutions to the differential equations studied in Examples 9.47 and 9.53, we could immediately see how they behaved as t tended to infinity. Since those solutions can be written as $\mathbf{x}(t) = \exp(t\mathbf{A})\mathbf{x}_0$, those results should follow from properties of $\mathbf{A}$ and of $\exp(t\mathbf{A})$. The explicit formula for $\exp(t\mathbf{A})$ contained in Theorem 9.54 provides an immediate proof of such results, once you understand the behavior of $\exp(\lambda t)$ for possibly complex λ. If $\lambda = \alpha + \beta i$ with α and β real, then the infinite series for $\exp(\lambda t)$ splits into real and imaginary parts and yields $\exp\{(\alpha + \beta i)t\} = \exp(\alpha t)\{\cos \beta t + i \sin \beta t\}$. If α, the real part of λ, is negative, then $\exp(\lambda t)$ tends to zero as t tends to infinity; if the real part of λ is positive, $|\exp(\lambda t)|$ tends to infinity; and if the real part equals zero, $\exp(\lambda t)$ is bounded.

(9.57) **Key Theorem** (exponential behavior). As t tends to plus infinity:

(a) $\exp(t\mathbf{A})$ tends to zero if and only if all the eigenvalues of $\mathbf{A}$ have strictly negative real parts.

(b) $\exp(t\mathbf{A})$ is bounded—that is, $\|\exp(t\mathbf{A})\| \le c$ for some constant c and all $t \ge 0$ for one norm and hence for every norm—if and only if:

1. the real part of each eigenvalue of $\mathbf{A}$ is less than or equal to zero; *and*
2. every eigenvalue of $\mathbf{A}$ with zero real part has its algebraic and geometric multiplicities equal.

(c) $\|\exp(t\mathbf{A})\|$ tends to infinity—in one norm and hence in every norm—if and only if *either*:

1. some eigenvalue of $\mathbf{A}$ has a strictly positive real part; *or*
2. some eigenvalue of $\mathbf{A}$ has zero real part and has its geometric multiplicity unequal to its algebraic.

(9.58) ***Example.*** The problems in Examples 9.47 and 9.53 illustrate this theorem. The 2×2 matrix $\mathbf{A}$ in Example 9.47 has $\lambda_1 = 0$ and $\lambda_2 = -0.2$; all solutions are bounded, as was the one there. The 2×2 matrix $\mathbf{A}$ in Example 9.53 has $\lambda_1 = \lambda_2 = -0.2$; all solutions tend to zero, as did the one there.

Inhomogeneous Equations

We return finally to the *inhomogeneous* equation (9.39):

(9.59)
$$\dot{\mathbf{x}} = \mathbf{A}\mathbf{x} + \mathbf{f}, \qquad \mathbf{x}(0) = \mathbf{x}_0,$$

where $\mathbf{f}$ is a function of t. The approach we now use, based on matrix exponentials, could also have been used when $\mathbf{f} = \mathbf{0}$. Premultiplying both sides of (9.59) by $\exp(-t\mathbf{A})$ gives

(9.60)
$$\exp(-t\mathbf{A})\dot{\mathbf{x}} - \exp(-t\mathbf{A})\mathbf{A}\mathbf{x} = \exp(-t\mathbf{A})\mathbf{f}.$$

The left-hand side of (9.60) is just the derivative of $Z(t)$ defined as $\exp(-t\mathbf{A})\mathbf{x}(t)$, so the right-hand side also equals $\dot{Z}(t)$. Since $Z(t) - Z(0)$ equals the integral of $\dot{Z}(\tau)$ from 0 to t, we get

$$\exp(-t\mathbf{A})\mathbf{x}(t) - \exp(0\mathbf{A})\mathbf{x}(0) = \int_0^t \exp(-\tau\mathbf{A})\mathbf{f}(\tau)\, d\tau.$$

Since $\exp(0\mathbf{A}) = \mathbf{I}$ and $\exp(t\mathbf{A}) = \{\exp(-t\mathbf{A})\}^{-1}$, premultiplication of both sides by $\exp(t\mathbf{A})$ completes the derivation of the solution to the inhomogeneous equation.

(9.61) ***Theorem*** (inhomogeneous differential equations). The solution $\mathbf{x}(t)$ to

$$\dot{\mathbf{x}} = \mathbf{A}\mathbf{x} + \mathbf{f}, \qquad \mathbf{x}(0) = \mathbf{x}_0,$$

where $\mathbf{A}$ and $\mathbf{x}_0$ are constant and $\mathbf{f}$ may depend on t, is given by

$$\mathbf{x}(t) = \exp(t\mathbf{A})\mathbf{x}_0 + \int_0^t \exp\{(t-\tau)\mathbf{A}\}\mathbf{f}(\tau)\,d\tau.$$

Note that if $\mathbf{f} = \mathbf{0}$, this reduces to Corollary 9.55.

PROBLEMS 9.5

▷ **1.** Write the differential equations of the triatomic molecule in Problem 4 of Section 2.5 in the form $\dot{\mathbf{x}} = \mathbf{A}\mathbf{x}$.

2. Solve $\dot{\mathbf{x}} = \mathbf{A}\mathbf{x}$, $\mathbf{x}(0) = \mathbf{x}_0$, with

$$\mathbf{A} = \begin{bmatrix} 0 & 1 & 0 \\ 0 & 0 & 1 \\ 6 & -1 & -4 \end{bmatrix}, \qquad \mathbf{x}_0 = \begin{bmatrix} 12 \\ -12 \\ 12 \end{bmatrix}.$$

3. Solve the differential equations for the competing-populations model of Example 9.36 for $k = 0.1$.

4. Prove that (9.50) follows from (9.49).

5. Solve $\dot{\mathbf{x}} = \mathbf{A}\mathbf{x}$, $\mathbf{x}(0) = [2 \quad -2 \quad 2]^T$, with the matrix $\mathbf{A}$ from Example 9.6.

▷ **6.** Solve

$$\begin{aligned} \dot{x}_1 &= 2x_1 + x_2, & x_1(0) &= 1 \\ \dot{x}_2 &= 2x_2, & x_2(0) &= 1. \end{aligned}$$

7. Solve the differential equations of Example 9.37 with $g = 0.08$, $r = 0.2$, $s = 0.7$, $v(0) = 100$, $p(0) = 5$.

▷ **8.** Prove Theorem 9.54(g), and find $\exp(\mathbf{I})$.

9. Use the formula for $\exp(\mathbf{A})$ in Theorem 9.54(e) to prove Theorem 9.54(b) and (c).

10. Prove that the series $\mathbf{I} + \mathbf{A} + \mathbf{A}^2/2! + \mathbf{A}^3/3! + \cdots$ in Theorem 9.54(a) converges.

▷ **11.** Prove Theorem 9.54(f).

12. Find $\exp(t\mathbf{A})$ and then $\exp(\mathbf{A})$ for $\mathbf{A}$ in Example 9.47.

13. Solve $\dot{\mathbf{x}} = \mathbf{A}\mathbf{x} + \mathbf{f}$, $\mathbf{x}(0) = \mathbf{x}_0$, with $\mathbf{A}$ and $\mathbf{x}_0$ as in Example 9.47 but with $\mathbf{f}(t) = [1 \quad 1]^T$.

14. Solve $\dot{\mathbf{x}} = \mathbf{A}\mathbf{x} + \mathbf{f}$, $\mathbf{x}(0) = \mathbf{x}_0$, with $\mathbf{A}$ and $\mathbf{x}_0$ as in Problem 5, but with $\mathbf{f}(t) = [1 \quad 0 \quad t]^T$.

15. Verify that $\{\exp(\mathbf{A})\}^{-1} = \exp(-\mathbf{A})$ for the matrix $\mathbf{A}$ in Example 9.47.

▷ **16.** Find $\exp(\mathbf{A})$ for the $\mathbf{A}$ in Example 9.6.

17. Without solving the differential equations, determine the behavior as t tends to plus infinity of the solution $\mathbf{x}(t)$ to the competing-populations model of Example 9.36 with $k = 0.1$.

▷ **18.** Find a value of k in the competing-populations model of Example 9.36 such that, as t tends to plus infinity, all solutions are bounded but not all solutions tend to zero.

𝔐 **19.** Use MATLAB or similar software to sum several terms in the series for exp(**A**) to verify experimentally that the series converges to exp(**A**) for the matrix **A** in Example 9.56.

9.6 *APPLICATION: ITERATIVE SOLUTION OF LINEAR EQUATIONS*

Mathematical models of very complex physical or social systems often involve systems of linear equations involving thousands of variables, systems that may have to be solved many times in the model. Even modern supercomputers may be unable to store and solve the equations using Gauss elimination rapidly enough (if at all) to be useful or cost effective.

It is often the case in such problems that each of the thousands of equations involves only a few of the thousands of variables; this arises from the fact that quite often each parameter in a large model only interacts *directly* with a few of the other parameters. In an economic model, for example, each individual conducts business with only a few others; in a complicated physical structure, for another example, each member is connected to only a few others. The matrix representing such equations is therefore *sparse*: It contains only a fraction of a percent of nonzero entries.

(9.62) ***Example.*** Suppose that we have a 10×10 matrix **A** with the following structure, where "x" represents a nonzero entry in the original **A**, "0" represents a zero in the original **A** whose presence can be used to advantage during Gauss elimination, and "$\otimes$" represents a zero in the original **A** which is replaced by a nonzero number during elimination before elimination can proceed far enough to take advantage of the zero.

$$\mathbf{A} = \begin{bmatrix} x & x & 0 & 0 & 0 & 0 & 0 & 0 & 0 & x \\ x & x & x & 0 & 0 & 0 & 0 & 0 & 0 & x \\ x & x & x & x & 0 & 0 & 0 & 0 & 0 & x \\ x & \otimes & x & x & x & 0 & 0 & 0 & 0 & x \\ x & \otimes & \otimes & x & x & x & 0 & 0 & 0 & x \\ x & \otimes & \otimes & \otimes & x & x & x & 0 & 0 & x \\ x & \otimes & \otimes & \otimes & \otimes & x & x & x & 0 & x \\ x & \otimes & \otimes & \otimes & \otimes & \otimes & x & x & x & x \\ x & \otimes & \otimes & \otimes & \otimes & \otimes & \otimes & x & x & x \\ x & \otimes & \otimes & \otimes & \otimes & \otimes & \otimes & \otimes & x & x \end{bmatrix}$$

Although only 44 of the 100 entries of **A** are nonzero, fully half of the 56 zero entries have to be treated as nonzero during elimination; and in solving a system $\mathbf{Ax} = \mathbf{b}$, back-substitution will involve multiplying by zero 28 times. If possible, we would like to avoid the loss of sparsity and the wasted work of multiplying by zeros.

One approach is to develop special Gauss-elimination procedures for sparse matrices; just noting the location of the many zeros in the Gauss-reduced form above, for example, avoids the unnecessary multiplications. And the simple device of reordering the unknowns as $x_2, x_3, \ldots, x_{10}, x_1$ preserves much of the sparsity during elimination: in effect it moves the first column to the far right of the matrix, after which Gauss elimination with interchanges can take advantage of every zero in the lower triangle of **A** while replacing at most 13 zeros in the upper triangle.

This example indicates the potentially enormous savings possible by designing special Gauss-elimination programs for sparse matrices. This important topic is beyond the scope of this book, however—see Bjorck, Plemmons, and Schneider (38); Bunch and Rose (39); Duff and Stewart (41); George and Liu (45); and Rose and Willoughby (51).

There are alternative approaches to taking advantage of sparsity, ones that can easily exploit the sparsity of **A** and maintain low storage and computational requirements. Among the most powerful of these methods are *iterative* methods: Rather than compute the solution directly as in Gauss elimination, they compute a (theoretically infinite) sequence of approximate solutions $\mathbf{x}_r$ intended to converge rapidly to the solution. The remainder of this section examines such methods.

As a model problem for illustrations, we take $\mathbf{Ax} = \mathbf{b}$, where

(9.63) **A** is 20×20, symmetric, and tridiagonal:

$\langle \mathbf{A} \rangle_{ii} = 2$ for $1 \leq i \leq 20$, $\langle \mathbf{A} \rangle_{i,i-1} = \langle \mathbf{A} \rangle_{i-1,i} = -1$ for $2 \leq i \leq 20$, and all other $\langle \mathbf{A} \rangle_{ij} = 0$.

b is 20×1: $\langle \mathbf{b} \rangle_i = 0.01i$ for $1 \leq i \leq 20$.

Note that only 58 of **A**'s 400 entries are nonzero; this figure of 15% is high compared to what occurs in practice. Note too that Gauss elimination on **A** is simple and iterative methods are not needed; we merely use **A** to illustrate the formulas below.

Each of the iterative methods to be introduced solves for the ith entry of a new approximation $\mathbf{x}'$ by using the ith equation and the values of the old approximation $\mathbf{x}$ (and perhaps some of the already computed entries of $\mathbf{x}'$); they differ in how they do this. A sequence of approximations $\mathbf{x}_r$ is produced by starting with $\mathbf{x}$ as $\mathbf{x}_0$, computing $\mathbf{x}'$ from $\mathbf{x}$ and letting $\mathbf{x}_1$ equal $\mathbf{x}'$, setting $\mathbf{x}$ to $\mathbf{x}_1$, computing $\mathbf{x}'$ from $\mathbf{x}$ and letting $\mathbf{x}_2$ equal $\mathbf{x}'$, and so on. (We employ this notation in order to reduce the number of subscripts and superscripts involved.)

Throughout this section, $\mathbf{A}$ *is* $p \times p$, a_{ij} *denotes* $\langle \mathbf{A} \rangle_{ij}$, x_i *denotes* $\langle \mathbf{x} \rangle_i$, x_i' *denotes* $\langle \mathbf{x}' \rangle_i$, *and* b_i *denotes* $\langle \mathbf{b} \rangle_i$.

(9.64) ***Example.*** The *Jacobi* iterative method substitutes the old approximation $\mathbf{x}$ into the ith equation for all variables but the ith and solves for that variable as x_i':

$$a_{ii}x_i' = -a_{i1}x_1 - \cdots - a_{i,i-1}x_{i-1} - a_{i,i+1}x_{i+1} - \cdots - a_{ip}x_p + b_i \quad \text{for } 1 \le i \le p.$$

In the sample problem (9.63), this becomes simply

$$2x_i' = x_{i-1} + x_{i+1} + 0.01i \qquad \text{for } 1 \le i \le 20.$$

(9.65) ***Example.*** The *Gauss–Seidel* iterative method substitutes the new approximation $\mathbf{x}'$ into the ith equation for those variables $x_1', \ldots, x_{i-1}'$ that have been updated and substitutes the old approximation $\mathbf{x}$ for all other variables but the ith and solves for that variable as x_i':

$$a_{ii}x_i' = -a_{i1}x_1' - \cdots - a_{i,i-1}x_{i-1}' - a_{i,i+1}x_{i+1} - \cdots - a_{ip}x_p + b_i \quad \text{for } 1 \le i \le p.$$

In the sample problem (9.63), this becomes simply

$$2x_i' = x_{i-1}' + x_{i+1} + 0.01i \qquad \text{for } 1 \le i \le 20.$$

(9.66) ***Example.*** The *successive over-relaxation* or *SOR* iterative method computes x_i' as follows: It first computes a temporary approximation x^* to the ith variable as in the Gauss–Seidel process, and then finds $x_i' = x_i + \omega(x^* - x_i)$ for a fixed *over-relaxation parameter* ω, usually greater than 1, whose use is intended to move the approximation more rapidly toward the solution. This gives

$$a_{ii}x^* = -a_{i1}x_1' - \cdots - a_{i,i-1}x_{i-1}' - a_{i,i+1}x_{i+1} - \cdots - a_{ip}x_p + b_i,$$
$$x_i' = x_i + \omega(x^* - x_i) \qquad \text{for } 1 \le i \le p.$$

In the sample problem (9.63), this becomes simply

$$2x^* = x_{i-1}' + x_{i+1} + 0.01i, \qquad x_i' = x_i + \omega(x^* - x_i) \qquad \text{for } 1 \le i \le 20.$$

Not every iterative method we might create will produce approximations that converge to a solution, nor will the three methods above produce such approximations for every problem. These three methods—and especially SOR with proper choice of ω—are extremely effective on a large class of matrices that arise in solving boundary-value problems for ordinary and partial differential equations; the sample problem (9.63) is, in fact, just such a matrix. To determine whether a specific method will be effective on a specific problem requires some analytical tools.

Matrix Splittings

The application of these methods to the sample problem (9.63) reveals that they are simple computationally. In order to analyze the behavior of these methods, we introduce matrix notation; *it is important to realize that the matrix versions are **not** used in the actual computations, but only in the analysis.* Write **A** as

(9.67) $\mathbf{A} = \mathbf{L} + \mathbf{D} + \mathbf{U}$, where **D** is a diagonal matrix, **L** is a lower-triangular matrix with zeros on its main diagonal, and **U** is an upper-triangular matrix with zeros on its main diagonal.

Obviously, **L** equals **A**'s lower triangle, **D** equals **A**'s main diagonal, and **U** equals **A**'s upper triangle. Examination of the formulas for the Jacobi, Gauss–Seidel, and SOR methods shows that they can be restated as follows (Problems 5 and 6):

(9.68) (a) The Jacobi computation is

$$\mathbf{D}\mathbf{x}' = -(\mathbf{L} + \mathbf{U})\mathbf{x} + \mathbf{b}.$$

(b) The Gauss–Seidel computation is

$$(\mathbf{L} + \mathbf{D})\mathbf{x}' = -\mathbf{U}\mathbf{x} + \mathbf{b}.$$

(c) The SOR computation is

$$(\mathbf{L} + \omega^{-1}\mathbf{D})\mathbf{x}' = -\{\mathbf{U} + (1 - \omega^{-1})\mathbf{D}\}\mathbf{x} + \mathbf{b}.$$

Each of these is in fact a special case of a class of methods based on matrix splittings $\mathbf{A} = \mathbf{M} + \mathbf{N}$; we write

(9.69)
$$\mathbf{A} = \mathbf{M} + \mathbf{N} \quad \text{and} \quad \mathbf{M}\mathbf{x}' = -\mathbf{N}\mathbf{x} + \mathbf{b}.$$

Each of the methods in (9.68) is of this type:

Jacobi uses $\mathbf{M} = \mathbf{D}$, $\quad \mathbf{N} = \mathbf{L} + \mathbf{U}$.

Gauss–Seidel uses $\mathbf{M} = \mathbf{L} + \mathbf{D}$, $\quad \mathbf{N} = \mathbf{U}$.

SOR uses $\mathbf{M} = \mathbf{L} + \omega^{-1}\mathbf{D}$, $\quad \mathbf{N} = \mathbf{U} + (1 - \omega^{-1})\mathbf{D}$.

The sequence of approximations $\mathbf{x}_r$ computed by such methods, the true solution $\tilde{\mathbf{x}}$, and the errors $\boldsymbol{\delta}_r = \mathbf{x}_r - \tilde{\mathbf{x}}$ then satisfy

(9.70)
$$\begin{aligned}\mathbf{M}\mathbf{x}_{r+1} &= -\mathbf{N}\mathbf{x}_r + \mathbf{b}\\ \mathbf{M}\tilde{\mathbf{x}} &= -\mathbf{N}\tilde{\mathbf{x}} + \mathbf{b}\\ \mathbf{M}\boldsymbol{\delta}_{r+1} &= -\mathbf{N}\boldsymbol{\delta}_r,\end{aligned}$$

where the third equation comes from subtracting the second from the first. If $\mathbf{M}$ is nonsingular, this becomes

(9.71)
$$\begin{aligned}\boldsymbol{\delta}_{r+1} &= \mathbf{H}\boldsymbol{\delta}_r, \text{ where } \mathbf{H} = -\mathbf{M}^{-1}\mathbf{N}, \text{ and thus}\\ \boldsymbol{\delta}_r &= \mathbf{H}^r\boldsymbol{\delta}_0.\end{aligned}$$

Since we want $\mathbf{x}_r$ to converge to $\tilde{\mathbf{x}}$, we want $\boldsymbol{\delta}_r$ to converge to zero no matter what the starting error $\boldsymbol{\delta}_0$. Since $\boldsymbol{\delta}_r = \mathbf{H}^r\boldsymbol{\delta}_0$, the question of the behavior of $\boldsymbol{\delta}_r$ is covered by **Key Theorem 9.30**: To ensure that the errors $\boldsymbol{\delta}_r$ converge to zero, we need to have the spectral radius $\rho(\mathbf{H}) < 1$. Clearly, the smaller $\rho(\mathbf{H})$, the more rapidly $\boldsymbol{\delta}_r$ should be expected in general to converge to zero. The spectral radius $\rho(\mathbf{H})$ is the analytical tool required for the analysis of the performance of particular iterative methods on particular problems.

𝔐 (9.72) ***Example.*** We used MATLAB to compute the eigenvalues and then the spectral radius of the appropriate matrices $\mathbf{H}$ for the Jacobi, Gauss–Seidel, and SOR (with various values of ω) methods in Examples 9.64–9.66 on the sample problem (9.63); denote these matrices by $\mathbf{H}_J$, $\mathbf{H}_{GS}$, and $\mathbf{H}_\omega$, respectively (note that SOR with $\omega = 1$ is just Gauss–Seidel, so $\mathbf{H}_1 = \mathbf{H}_{GS}$). MATLAB calculated $\rho(\mathbf{H}_J) = 0.9888$ and $\rho(\mathbf{H}_{GS}) = 0.9778$, both of which are rather close to 1; since the errors $\|\boldsymbol{\delta}_r\|$ are expected to behave like $\{\rho(\mathbf{H})\}^r$, this indicates rather slow convergence. By solving for r in $\{\rho(\mathbf{H})\}^r = \frac{1}{10}$ to see how many steps are required to reduce the error by a factor of 10, we find that about 204 steps are required for the Jacobi method and 103 steps for the Gauss–Seidel method. (This factor of about 2 between the steps in these methods is typical.) We now examine how SOR performs for various values of ω. We used MATLAB to calculate $\rho(\mathbf{H}_\omega)$ for ω from 0.0 to 2.0 in steps of 0.2 and observed that it was smallest between 1.6 and 2.0. We then calculated it in steps of 0.05 in that region, observing it to be least from 1.7 to 1.8. We then calculated it there in steps of 0.01 and found it to be least at $\omega = 1.75$. The table below shows representative values.

ω	0.6	1.2	1.6	1.7	1.74	1.75	1.76	1.8
$\rho(\mathbf{H}_\omega)$	0.9905	0.9666	0.9056	0.8479	0.7562	0.7500	0.7600	0.8000

The table reveals that proper choice of ω can make a dramatic difference in the speed of convergence. Choosing the optimum ω of about 1.75 gives $\rho(\mathbf{H}_{1.75}) = 0.75$, which means that only about 8 steps—compared to 204 for Jacobi and 103 for Gauss–Seidel—of SOR with $\omega = 1.75$ are needed to reduce the error by a factor of 10.

The comparisons illustrated in Example 9.72 are typical for a large class of matrices arising in the numerical solution of differential equations; the speed of Gauss–Seidel is typically twice that of Jacobi, with SOR (*with an optimal choice of* ω) dramatically faster. The following theorem, stated without proof (see the 2nd edition of this book, for example), illustrates this.

(9.73) ***Theorem*** (Jacobi, Gauss–Seidel, and SOR). Suppose that $\mathbf{A}$ is a tridiagonal matrix with nonzero main-diagonal entries:

$$\langle \mathbf{A} \rangle_{ii} \neq 0, \qquad \langle \mathbf{A} \rangle_{ij} = 0 \qquad \text{if } |i - j| > 1.$$

Let $\mathbf{H}_\mathrm{J}$, $\mathbf{H}_\mathrm{GS}$, and $\mathbf{H}_\omega$ be the matrices whose spectral radii govern the convergence of these methods. Then:

(a) $\rho(\mathbf{H}_\mathrm{GS}) = \{\rho(\mathbf{H}_\mathrm{J})\}^2$—if Jacobi converges, then so does Gauss–Seidel and at twice the speed; if Jacobi diverges, then so does Gauss–Seidel and at twice the speed.

(b) If $\mathbf{A}$ has real eigenvalues, then the value of ω that minimizes $\rho(\mathbf{H}_\omega)$ is

$$\omega^* = 2/\{1 + (1 - \rho^2)^{1/2}\},$$

where $\rho = \rho(\mathbf{H}_\mathrm{J})$; for this optimal ω^*, $\rho(\mathbf{H}_{\omega^*}) = \omega^* - 1$.

In order to interpret Theorem 9.73, suppose for the moment that $\rho(\mathbf{H}_\mathrm{J}) = 1 - \epsilon$ for some small ϵ, so that the Jacobi method converges slowly. Then

$$\rho(\mathbf{H}_\mathrm{GS}) = (1 - \epsilon)^2 = 1 - 2\epsilon + \epsilon^2 \approx 1 - 2\epsilon,$$

somewhat smaller. Also,

$$\omega^* = \frac{2}{\{1 + (2\epsilon - \epsilon^2)^{1/2}\}} \approx 2 - 2(2\epsilon)^{1/2},$$

and so $\rho(\mathbf{H}_{\omega^*}) \approx 1 - 2(2\epsilon)^{1/2}$, which is appreciably smaller. If $\epsilon = 0.00005$, for example, then

$$\rho(\mathbf{H}_\mathrm{J}) = 0.99995, \qquad \rho(\mathbf{H}_\mathrm{GS}) \approx 0.99990, \quad \text{and} \quad \rho(\mathbf{H}_{\omega^*}) \approx 0.98.$$

An interesting and important aspect of this theorem is that it requires information only on $\rho(\mathbf{H}_\mathrm{J})$ in order to determine the optimal ω for SOR; this is useful since $\mathbf{H}_\mathrm{J}$ is relatively simple. Writing $\mathbf{A} = \mathbf{L} + \mathbf{D} + \mathbf{U}$ as in (9.67) and finding

$\mathbf{H}_J = -\mathbf{D}^{-1}(\mathbf{L} + \mathbf{U})$ shows that the Jacobi method will be convergent if, for example, $\|\mathbf{H}_J\|_\infty < 1$; this inequality holds if and only if **A** is strictly row-diagonally dominant—the magnitude of $\langle\mathbf{A}\rangle_{ii}$ is strictly greater than the sum of the magnitudes of $\langle\mathbf{A}\rangle_{ij}$ for $j \neq i$—which often holds in applications {or a slightly modified version of this condition holds, in which the "strictly" is relaxed, that is sufficient to guarantee $\rho(\mathbf{H}_J) < 1$}.

PROBLEMS 9.6

1. Write out the equations for applying the Jacobi method to

$$2u + v = 4$$
$$u + 2v = 5.$$

▷ 2. Write out the equations for applying the Gauss–Seidel method in Problem 1.

3. Write out the equations for applying the SOR method in Problem 1.

4. Verify (9.68)(a), (b), and (c) in Problems 1, 2, and 3.

5. Derive (9.68)(a) and (b).

6. Derive (9.68)(c).

▷ 7. Find the matrices $\mathbf{H}_J$, $\mathbf{H}_{GS}$, and $\mathbf{H}_\omega$ that serve as **H** in (9.71) in Problems 1, 2, and 3.

8. Use the general formulas for **M** and **N** to find the iteration matrices **H** of (9.71) in general for the Jacobi, Gauss–Seidel, and SOR methods.

▷ 9. Find $\rho(\mathbf{H}_J)$, $\rho(\mathbf{H}_{GS})$, and $\rho(\mathbf{H}_{1.1})$ in Problems 1, 2, and 3.

10. In Problems 1, 2, and 3 (with $\omega = 1.1$), compare the actual number of steps necessary to reduce the error by a factor of 10 (starting with $u_0 = v_0 = 0$) with the number $n = -\log_{10}\{\rho(\mathbf{H})\}$ predicted theoretically as the solution to $\{\rho(\mathbf{H})\}^n = \frac{1}{10}$.

▷ 11. Interchange the two equations to be solved in Problem 1.
   (a) Show that the Jacobi method does not converge if applied to this new system.
   (b) Determine whether the Gauss–Seidel method converges.
   (c) Explore whether you can find ω so that SOR will converge.

12. Describe some matrix splittings $\mathbf{A} = \mathbf{M} + \mathbf{N}$ for the matrix in Example 9.62 that might be effective for solving $\mathbf{Ax} = \mathbf{b}$; remember that you need to be able to solve $\mathbf{Mx}_{r+1} = -\mathbf{Nx}_r + \mathbf{b}$ easily at each step.

𝔐 13. Modify the main-diagonal entries in the matrix **A** of the sample problem (9.63) by setting $\langle\mathbf{A}\rangle_{ii} = 2 + 0.1i$. Use MATLAB or similar software to find $\rho(\mathbf{H}_J)$, $\rho(\mathbf{H}_{GS})$, and $\rho(\mathbf{H}_\omega)$ for enough of a range of ω values to allow you to approximate an optimal $\omega = \omega^*$.

𝔐 14. (a) Use $\rho(\mathbf{H}_J)$ for the sample problem (9.63) to determine the optimal $\omega = \omega^*$ for SOR from Theorem 9.73.
   (b) Compare this value with the experimentally determined value 1.75.

(c) Use MATLAB or similar software to compute $\rho(\mathbf{H}_{\omega*})$ and then compare it with the theoretical value $\omega^* - 1$.

𝔐 **15.** Do as requested in Problem 14, but for the modified sample problem of Problem 13; for (b), compare with your experimentally determined optimal ω from Problem 13.

▷ **16.** (a) Find the eigenvalues of the matrix $\mathbf{H}_\omega$ for Problem 3, and then determine ω so as to minimize $\rho(\mathbf{H}_\omega)$.

(b) Compare your answer with that obtained by using Theorem 9.73 and $\rho(\mathbf{H}_J) = 0.5$.

9.7 *MISCELLANEOUS PROBLEMS*

PROBLEMS 9.7

1. Prove that two $p \times p$ nondefective matrices are similar if and only if they have the same characteristic polynomial.

▷ **2.** Suppose that $\mathbf{A}$ is a real 2×2 matrix with complex conjugate eigenvalues $\alpha \pm \beta i$. Show that there exists a *real* $\mathbf{P}$ such that

$$\mathbf{P}^{-1}\mathbf{A}\mathbf{P} = \begin{bmatrix} \alpha & \beta \\ -\beta & \alpha \end{bmatrix}.$$

3. Show that there are nondefective matrices, with repeated eigenvalues, that are not normal by considering

$$\mathbf{A} = \begin{bmatrix} 2 & 0 & 0 \\ 0 & 3 & 1 \\ 0 & 0 & 2 \end{bmatrix}.$$

4. Problem 9 in Section 9.2 shows that if $\mathbf{A}$ and $\mathbf{B}$ are both nondefective and $\mathbf{AB} = \mathbf{BA}$, then $\mathbf{P}^{-1}\mathbf{AP}$ will be block diagonal whenever $\mathbf{P}^{-1}\mathbf{BP}$ is diagonal. Suppose that we are given nondefective $\mathbf{A}$; if we can find a $\mathbf{B}$ that commutes with $\mathbf{A}$ and whose eigensystem is easy to find, then we can produce a simple $\mathbf{P}^{-1}\mathbf{AP}$ from which we may be able to find $\mathbf{A}$'s eigensystem easily. As a simple example, consider the system of resistors and capacitors on page 396. If q_i is the charge on the ith capacitor, remembering that the voltage across a capacitor C containing charge q is q/C, and the current through it is $i = dq/dt$, prove that the equations for the circuit are, assuming all $R_{ij} = R$ and $C_i = C$,

$$RC\frac{d\mathbf{q}}{dt} = \mathbf{A}\mathbf{q}, \qquad \mathbf{A} = \begin{bmatrix} -2 & 1 & 1 \\ 1 & -2 & 1 \\ 1 & 1 & -2 \end{bmatrix}, \qquad \mathbf{q} = \begin{bmatrix} q_1 \\ q_2 \\ q_3 \end{bmatrix}.$$

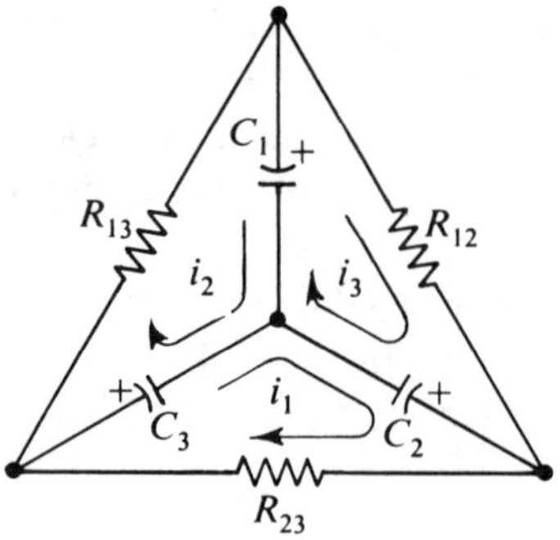

Because of symmetry, the circuit is unchanged if capacitor 1 is called 2, 2 is called 3, and 3 is called 1. This suggests that we try the following permutation matrix for **B**. The eigenvectors of **B**, which give the columns of **P**, are easily found. Letting $\omega = \frac{1}{2}(-1 + i\sqrt{3})$, so that $\omega^3 = 1$ and $\omega^2 + \omega + 1 = 0$, we find

$$\mathbf{B} = \begin{bmatrix} 0 & 1 & 0 \\ 0 & 0 & 1 \\ 1 & 0 & 0 \end{bmatrix}$$

$$\mathbf{P} = \begin{bmatrix} 1 & 1 & 1 \\ 1 & \omega & \omega^2 \\ 1 & \omega^2 & \omega \end{bmatrix}$$

$$\mathbf{P}^{-1} = \frac{1}{3}\begin{bmatrix} 1 & 1 & 1 \\ 1 & \omega^2 & \omega \\ 1 & \omega & \omega^2 \end{bmatrix}.$$

We readily find $\mathbf{AB} = \mathbf{BA}$ so that we can use the method above. This gives

$$\mathbf{P}^{-1}\mathbf{AP} = \begin{bmatrix} 0 & 0 & 0 \\ 0 & -3 & 0 \\ 0 & 0 & -3 \end{bmatrix},$$

so that the eigenvalues of **A** are 0, -3, -3 and the eigenvectors are the columns of **P**.

▷ **5.** Suppose that $p \times p$ **A** has a simple eigenvalue λ_1 with $|\lambda_1| = \rho(\mathbf{A})$, and that all other eigenvalues λ satisfy $|\lambda| < \rho(\mathbf{A})$. Show how to use the values $\mathbf{A}^i\mathbf{x}_0$ computed from an initial $\mathbf{x}_0$ so as to obtain an approximation to λ_1.

6. Provide the details for the following outline of a proof that an $n \times n$ Markov matrix with strictly positive entries has no eigenvalues λ with $|\lambda| = 1$ other than the simple eigenvalue $\lambda = 1$: If $|\lambda| = 1$ and λ is an eigenvalue of **A**, then let $\mathbf{x} \neq \mathbf{0}$

satisfy $\mathbf{x}^T\mathbf{A} = \lambda\mathbf{x}^T$ and

$$\max\{|x_1|, |x_2|, \ldots, |x_n|\} = |x_{i_0}| = 1$$

for some i_0. From the i_0th equation in $\mathbf{x}^T\mathbf{A} = \lambda\mathbf{x}^T$ deduce that

$$|x_{i_0}| \leq \sum_{j=1}^{n} a_{ji_0}|x_j| = |x_{i_0}| + \sum_{j=1}^{n} a_{ji_0}[|x_j| - |x_{i_0}|],$$

that therefore $\sum_{j=1}^{n} a_{ji_0}[|x_j| - |x_{i_0}|] = 0$, and hence that $|x_j| = |x_{i_0}|$ for all j. From

$$\lambda x_1 = \sum_{j=1}^{n} a_{j1}x_j \quad \text{and} \quad |\lambda x_1| = |x_1| = \cdots = |x_n| = 1$$

deduce that $x_1 = \cdots = x_n$ so that $\mathbf{x} = c\mathbf{1}$, $\mathbf{1}^T = [1, \ldots, 1]$. From

$$\lambda c\mathbf{1}^T = \lambda\mathbf{x}^T = \mathbf{x}^T\mathbf{A} = c\mathbf{1}^T\mathbf{A} = c\mathbf{1}^T,$$

deduce that $\lambda = 1$. Conclude that $\lambda = 1$ is a simple eigenvalue since $\|\mathbf{A}^i\|_1 \leq 1$. This proves the result.

▷ **7.** Suppose that $\mathbf{A}$ and $\mathbf{B}$ are $p \times p$ and that $\mathbf{A}$ and $\mathbf{B}$ commute: $\mathbf{AB} = \mathbf{BA}$. Prove that

$$\exp(\mathbf{A} + \mathbf{B}) = \exp(\mathbf{A})\exp(\mathbf{B}) = \exp(\mathbf{B})\exp(\mathbf{A}),$$

and show by 2×2 example that this need not hold when $\mathbf{A}$ and $\mathbf{B}$ do not commute.

10

Quadratic forms and variational characterizations of eigenvalues

This chapter continues the development of the properties of eigensystems by considering some of their extremal properties related to special quadratic functions; many of the results have a distinctly geometric flavor. ***Key Theorems*** *are* ***10.14***, ***10.18***, ***10.25***, ***10.28***, *and* ***10.32***.

10.1 INTRODUCTION

After constant functions and linear functions (already studied extensively through linear equations and linear transformations), the quadratic functions are next in level of complexity. Such functions arise in diverse application areas, but matrix methods allow a unified study of their properties; conversely, quadratic functions provide important insights into matrix concepts, particularly eigensystems.

We will study special quadratics called *quadratic forms*: $(\mathbf{x}, \mathbf{Ax}) = \mathbf{x}^H\mathbf{Ax}$, where $\mathbf{A}$ is a $p \times p$ hermitian matrix, the variable $\mathbf{x}$ ranges over $\mathbb{R}^p$ or $\mathbb{C}^p$, and $(\cdot, \cdot)$ denotes the standard inner product in these spaces. We first illustrate how such quadratic forms arise naturally.

(10.1) ***Example.*** In two dimensions, the so-called *conic sections* are the simplest curves and play a fundamental role in two-dimensional geometry. Classically, the ellipse (hyperbola) was defined as the locus of points such that the sum (difference) of its distances from two fixed points, called the *foci*, equals a constant α. Analytic geometry allows the description of such curves by equations such as $x_1^2/25 + x_2^2/16 = 1$—an ellipse with foci at $(\pm 3, 0)$ and constant $\alpha = 10$—and $x_1^2/16 - x_2^2/9 = 1$—a hyperbola with foci at

$(\pm 5, 0)$ and constant $\alpha = 8$. We can as well describe these via quadratic forms: $(\mathbf{x}, \mathbf{Ax}) = 1$ with

$$\mathbf{A} = \begin{bmatrix} \frac{1}{25} & 0 \\ 0 & \frac{1}{16} \end{bmatrix}$$

for the ellipse above, for example. In p dimensions, the equation $(\mathbf{x}, \mathbf{Ax}) = 1$ for $p \times p$ $\mathbf{A}$ yields curves analogous to conic sections and fundamental in the geometry of $\mathbb{R}^p$ and $\mathbb{C}^p$.

(10.2) ***Example.*** Many applications involve optimization: reaching some goal in the best (in some sense) possible way. Thus we seek to minimize cost, to minimize idle time, to maximize efficiency, and so on. Mathematically this involves minimizing or maximizing some function $f(\mathbf{x})$ of p variables $x_1, \ldots, x_p$ (written as the components of the column matrix $\mathbf{x}$). If f is extremized at $\mathbf{x} = \mathbf{0}$, for simplicity, then we can get an idea of the behavior of f nearby from the Taylor series

$$f(x_1, \ldots, x_p) = (f)_0 + \sum_{i=1}^{p} (f_i)_0 x_i + \tfrac{1}{2} \sum_{i=1}^{p} \sum_{j=1}^{p} (f_{ij})_0 x_i x_j + \cdots,$$

where

$$f_i = \frac{\partial f}{\partial x_i}, \qquad f_{ij} = \frac{\partial^2 f}{\partial x_i \, \partial x_j},$$

and $(\cdot)_0$ indicates evaluation at $\mathbf{0}$. At an extremizing point the first derivatives f_i of course equal zero, so the local behavior of f is approximately described by the quadratic form $(\mathbf{x}, \mathbf{Ax})$ with $\langle \mathbf{A} \rangle_{ij} = (f_{ij})_0$. This generalizes the theory from $p = 1$, where the nature of the extremum—minimum, maximum, or saddle point—is determined by the second derivative of f.

(10.3) ***Example.*** In the study of dynamics of physical systems, important physical quantities such as kinetic energy and potential energy are often approximated by quadratic forms (via Taylor series) near an equilibrium state of the system. The theory of small vibrations about equilibrium in many-coordinate physical dynamical systems therefore naturally involves quadratic forms.

(10.4) ***Example.*** In the statistical analysis of data described by random variables $x_1, \ldots, x_p$, the *expected value* of x_i is the mean value of x_i over a large number of trials and is denoted by $E(x_i)$. E is linear in that $E(a_1x_1 + a_2x_2) = a_1E(x_1) + a_2E(x_2)$ for all constants a_i. Suppose that the random variables have zero means, so that $E(x_i) = 0$ for all i. The *variance* of x_i is defined as $E(x_i^2)$, and the *total variance* V of the set of x_i is defined as $V = E(x_1^2 + \cdots + x_p^2)$. The *covariance matrix* $\mathbf{S}$ is the $p \times p$ matrix with $\langle \mathbf{S} \rangle_{ij} = s_{ij} = E(x_ix_j)$. Thus the variance $E(x_i^2)$ equals s_{ii} for all i, and the total variance V equals $s_{11} + \cdots + s_{pp}$.

It is often important in the statistical analysis of data to seek some small set of new variables or factors that explain the experimental results. A common approach is to seek a new variable

$$y_1 = a_1x_1 + \cdots + a_px_p$$

whose variations in some sense reflect as much as possible of the variations in the x_i. Technically, this requires that $\mathbf{a} = [a_1 \quad \cdots \quad a_p]^T$ be chosen so as to maximize the variance $E(y_1^2)$ in y_1 subject to the constraint that $\|\mathbf{a}\|_2 = 1$. Since $\mathbf{S} = E(\mathbf{x}\mathbf{x}^T)$ if $\mathbf{x} = [x_1 \quad \cdots \quad x_p]^T$ and $y_1 = \mathbf{a}^T\mathbf{x} = \mathbf{x}^T\mathbf{a}$, we have

$$E(y_1^2) = E(\mathbf{a}^T\mathbf{x}\mathbf{x}^T\mathbf{a}) = \mathbf{a}^T\mathbf{S}\mathbf{a}.$$

That is, $\mathbf{a}$ is sought to maximize $(\mathbf{a}, \mathbf{S}\mathbf{a})$ subject to the constraint $(\mathbf{a}, \mathbf{a}) = 1$. Again, quadratic forms are at the heart of the issue.

Having seen some examples of the varied ways in which quadratic forms arise, we are ready to begin their study. We start with two variables, where intuition and geometric insight can guide us.

PROBLEMS 10.1

▷ **1.** Write out explicitly in terms of x_1 and x_2 the quadratic forms generated by:
(a) $\begin{bmatrix} 1 & 0 \\ 0 & -2 \end{bmatrix}$ (b) $\begin{bmatrix} 2 & 1 \\ 1 & 3 \end{bmatrix}$ (c) $\begin{bmatrix} a & b \\ b & c \end{bmatrix}$

2. Graph the ellipse with equation $x_1^2/25 + x_2^2/16 = 1$.

3. Graph the ellipse with equation $x_1^2/16 + x_2^2/25 = 1$.

▷ **4.** Find an equation for and then graph the ellipse with foci at $(0, \pm 4)$ and with the constant α for the sum of the distances being 10.

5. Graph the hyperbola with equation $x_1^2/16 - x_2^2/9 = 1$.

6. Graph the hyperbola with equation $x_1^2/9 - x_2^2/16 = 1$.

▷ **7.** Find an equation for and then graph the hyperbola with foci at $(\pm 13, 0)$ and with the constant α for the difference of the distances being 24.

8. Find the Taylor series through quadratic terms for $\exp(2x_1^2 - 4x_1x_2 + 4x_2^2)$ expanded about the minimizing point $x_1 = x_2 = 0$.

10.2 QUADRATIC FORMS IN $\mathbb{R}^2$

Before exploring quadratic forms in depth, we should explain why hermitian (or real symmetric) matrices are involved naturally. Consider a general quadratic in two real variables, $Q = ax_1^2 + bx_1x_2 + cx_2^2$, where a, b, and c are real. Various matrices **A** produce this quadratic form as $(\mathbf{x}, \mathbf{Ax}) = \mathbf{x}^T\mathbf{Ax}$:

$$\begin{bmatrix} a & 0 \\ b & c \end{bmatrix}, \quad \begin{bmatrix} a & b \\ 0 & c \end{bmatrix}, \quad \begin{bmatrix} a & b+1 \\ b-1 & c \end{bmatrix}, \quad \begin{bmatrix} a & \frac{b}{2} \\ \frac{b}{2} & c \end{bmatrix},$$

to list but a few. Only one of these (the last) is real symmetric (hence hermitian); because we were able to develop such strong results for hermitian matrices in Chapter 8, we seek to take advantage of this by *choosing* a hermitian **A** to generate the quadratic form.

The question now is how to make use of the fact that the matrix

$$\mathbf{A} = \begin{bmatrix} a & \frac{b}{2} \\ \frac{b}{2} & c \end{bmatrix}$$

defining the quadratic form

$$Q(\mathbf{x}) = (\mathbf{x}, \mathbf{Ax}) = \mathbf{x}^T\mathbf{Ax} = ax_1^2 + bx_1x_2 + cx_2^2$$

is real symmetric (hence hermitian, hence normal). By **Key Corollary 8.9** and **Key Theorem 8.6**, there is an orthogonal matrix **P** and a real diagonal matrix $\mathbf{\Lambda}$ such that $\mathbf{P}^T\mathbf{AP} = \mathbf{\Lambda}$; the main diagonal entries of $\mathbf{\Lambda}$ are the eigenvalues λ_1 and λ_2 of **A**, while the columns $\mathbf{v}_1$ and $\mathbf{v}_2$ of **P** form an orthonormal set of associated eigenvectors. Substitution of $\mathbf{A} = \mathbf{P\Lambda P}^T$ into the definition of Q gives

$$Q(\mathbf{x}) = \mathbf{x}^T(\mathbf{P\Lambda P}^T)\mathbf{x} = (\mathbf{P}^T\mathbf{x})^T\mathbf{\Lambda}(\mathbf{P}^T\mathbf{x}).$$

That is,

(10.5) $$Q(\mathbf{x}) = (\mathbf{x}, \mathbf{Ax}) = (\xi, \mathbf{\Lambda}\xi) = \lambda_1\xi_1^2 + \lambda_2\xi_2^2, \quad \text{where} \quad \xi = \mathbf{P}^T\mathbf{x}.$$

The quadratic form Q takes an extremely simple form in terms of the new variables ξ. For this to be useful, two questions need answers:

1. What is the nature of the quadratic form in ξ?
2. How can information on the quadratic form in ξ be converted to information on the quadratic form in **x**?

A standard device for understanding a function of two variables is to examine its *level curves*: Curves C_t consisting of all those points at which the function equals t, for various values of t. This is commonly done with weather maps that show curves of constant temperature and with topographic maps that show curves of constant altitude; such curves easily let us see the temperature patterns in the one case and identify hills and other terrain features in the other. We first therefore examine level curves $\tilde{C}_t$ of all those points ξ at which $\tilde{Q}(\xi) = t$, where $\tilde{Q}(\xi) = \lambda_1\xi_1^2 + \lambda_2\xi_2^2$. Second, we show how to obtain information about the corresponding level curves C_t for $Q(\mathbf{x})$.

Graphs in ξ Variables

The level curve $\tilde{C}_t$ is the graph on a ξ_1–ξ_2-coordinate plane of

(10.6) $$\tilde{Q}(\xi) = t, \quad \text{where} \quad \tilde{Q}(\xi) = \lambda_1\xi_1^2 + \lambda_2\xi_2^2;$$

this is merely one of the conic sections familiar from analytic geometry and calculus, which we briefly review. First, the "typical" cases.

(10.7) ***Ellipse.*** If λ_1, λ_2, and t are all strictly positive (or all strictly negative), then $\tilde{C}_t$ is an ellipse whose major and minor axes lie on the ξ_1-axis and ξ_2-axis. Dividing (10.6) by t and rearranging gives the standard form of an ellipse:

$$\frac{\xi_1^2}{A^2} + \frac{\xi_2^2}{B^2} = 1, \quad \text{where} \quad A = \left(\frac{t}{\lambda_1}\right)^{1/2} \quad \text{and} \quad B = \left(\frac{t}{\lambda_2}\right)^{1/2}.$$

If $A \geq B$, the foci are at $(\pm C, 0)$, where $C^2 = A^2 - B^2$; the ellipse is the set of all points the sum of whose distances from the two foci equals $2A$.

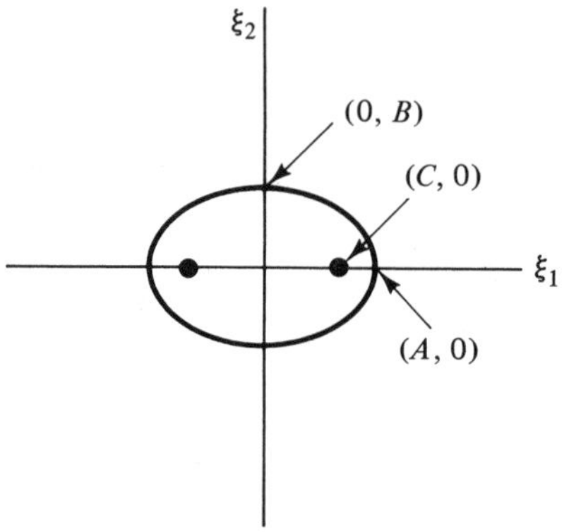

$$\frac{\xi_1^2}{A^2} + \frac{\xi_2^2}{B^2} = 1, \qquad A \geq B, \qquad C^2 = A^2 - B^2$$

(10.8) ***Hyperbola.*** If t is nonzero and if λ_1 and λ_2 are nonzero and of opposite sign, $\tilde{C}_t$ is a hyperbola; suppose, for definiteness, that $t/\lambda_1 > 0$ and $t/\lambda_2 < 0$. Dividing (10.6) by t and rearranging gives the standard form of a hyperbola:

$$\frac{\xi_1^2}{A^2} - \frac{\xi_2^2}{B^2} = 1, \quad \text{where} \quad A = \left(\frac{t}{\lambda_1}\right)^{1/2} \quad \text{and} \quad B = \left(-\frac{t}{\lambda_2}\right)^{1/2}.$$

The foci are at $(\pm C, 0)$, where $C^2 = A^2 + B^2$; the hyperbola is the set of all points the difference of whose distances to the two foci equals $2A$ in magnitude.

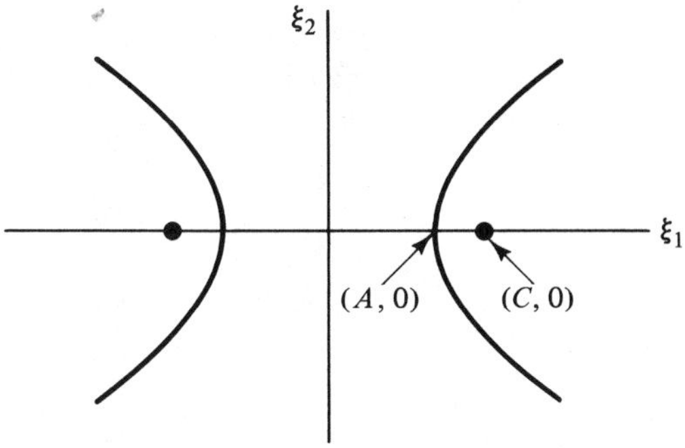

$$\frac{\xi_1^2}{A^2} - \frac{\xi_2^2}{B^2} = 1, \qquad C^2 = A^2 + B^2$$

(10.9) ***Parallel Lines.*** If exactly one of the λ_i equals zero while the other has the same sign as (nonzero) t, then $\tilde{C}_t$ is a pair of horizontal or vertical parallel lines; suppose, for definiteness, that $\lambda_1 > 0$, $\lambda_2 = 0$, and $t > 0$. Then (10.6) becomes $\xi_1 = \pm(t/\lambda_1)^{1/2}$—a pair of lines parallel to the ξ_2-axis. $\tilde{C}_t$ can be described as the set of all points whose distance from the line $\xi_1 = 0$ equals $|(t/\lambda_1)^{1/2}|$.

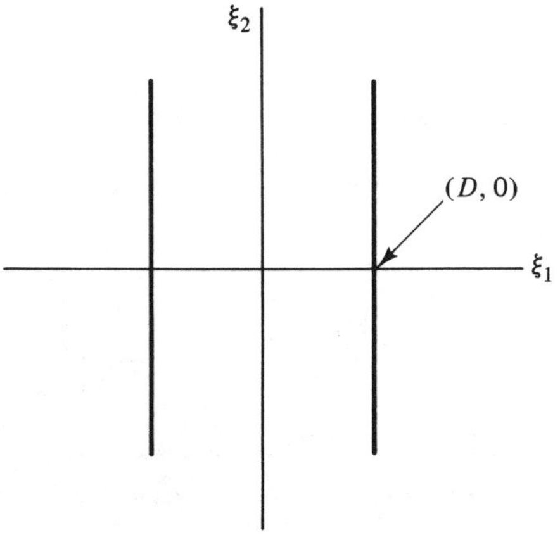

$$\xi_1 = \pm D = (t/\lambda_1)^{1/2}$$

There are also some "exceptional" cases—such as $\lambda_1 > 0$, $\lambda_2 > 0$, $t < 0$, in which case $\tilde{C}_t$ is empty; see Problems 2 and 3.

(10.10) ***Example.*** The quadratic form $Q(\mathbf{x}) = 13x_1^2 + 6\sqrt{3}x_1x_2 + 7x_2^2$ is generated by the real symmetric matrix

$$\mathbf{A} = \begin{bmatrix} 13 & 3\sqrt{3} \\ 3\sqrt{3} & 7 \end{bmatrix}.$$

The eigenvalues are easily found to be $\lambda_1 = 16$ and $\lambda_2 = 4$ with associated orthonormal eigenvectors

$$\mathbf{v}_1 = \begin{bmatrix} \frac{\sqrt{3}}{2} & \frac{1}{2} \end{bmatrix}^T \quad \text{and} \quad \mathbf{v}_2 = \begin{bmatrix} \frac{1}{2} & -\frac{\sqrt{3}}{2} \end{bmatrix}^T.$$

With $\mathbf{P} = [\mathbf{v}_1 \quad \mathbf{v}_2]$, we obtain $\mathbf{P}^H\mathbf{A}\mathbf{P} = \mathbf{\Lambda}$ diagonal with entries 16 and 4. We have

$$\tilde{Q}(\xi) = 16\xi_1^2 + 4\xi_2^2,$$

whose typical level curves $\tilde{C}_t$ are ellipses: $\tilde{C}_{64}$, for example, has as its graph

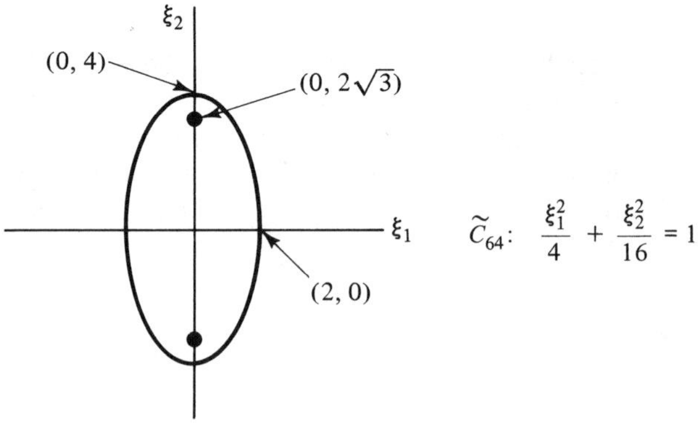

The relation between $\mathbf{x}$ and ξ is

$$\xi = \mathbf{P}^T\mathbf{x} = \begin{bmatrix} \frac{\sqrt{3}}{2} & \frac{1}{2} \\ \frac{1}{2} & -\frac{\sqrt{3}}{2} \end{bmatrix}\mathbf{x}, \qquad \mathbf{x} = \mathbf{P}\xi = \begin{bmatrix} \frac{\sqrt{3}}{2} & \frac{1}{2} \\ \frac{1}{2} & -\frac{\sqrt{3}}{2} \end{bmatrix}\xi.$$

You should now be able to solve Problems 1 to 4.

Graphs in x Variables

We now convert the detailed information in (10.7–9) on the level curves $\tilde{C}_t$ in ξ variables to information on the level curves C_t in $\mathbf{x}$ variables. For two variables, this is traditionally approached through considering the $\mathbf{x}$-to-ξ transformation as an elementary rotation; this does not generalize satisfactorily to p variables, however. We take instead a viewpoint that generalizes.

The first step is to note that ξ is on $\tilde{C}_t$ if and only if $\mathbf{x} = \mathbf{P}\xi$ is on C_t (note the same value of t). ξ on $\tilde{C}_t$ means $t = (\xi, \mathbf{\Lambda}\xi)$, which equals $(\mathbf{x}, \mathbf{Ax})$ according to (10.5), and conversely. In other words,

> C_t is precisely the image of $\tilde{C}_t$ under the linear transformation $\mathscr{T}$ defined by $\mathscr{T}(\xi) = \mathbf{P}\xi$.

The second step is to recall the crucial fact about orthogonal matrices: They preserve length—$\|\mathscr{T}(\xi)\|_2 = \|\mathbf{P}\xi\|_2 = \|\xi\|_2$ for all ξ.

Finally, note that each level curve $\tilde{C}_t$ in ξ variables was defined in terms of lengths and points: in (10.7) $\tilde{C}_t$ was the set of points $\xi = [\xi_1 \quad \xi_2]^T$ the sum of whose distances from the foci

$$\tilde{\mathbf{f}}_1 = [C \quad 0]^T \quad \text{and} \quad \tilde{\mathbf{f}}_2 = [-C \quad 0]^T$$

equals $2A$. This means that C_t is the set of points $\mathbf{x} = \mathbf{P}\xi$ the sum of whose distances from $\mathbf{P}\tilde{\mathbf{f}}_1$ and $\mathbf{P}\tilde{\mathbf{f}}_2$ equals $2A$. This is just an ellipse; since the distance $2C$ between the foci and the sum $2A$ is the same for $\tilde{C}_t$ and for C_t, the ellipses are congruent—C_t could be placed on top of $\tilde{C}_t$ as a perfect match.

Thus we can conclude generally that

> the level curves C_t in $\mathbf{x}$ variables are congruent to the level curves $\tilde{C}_t$ in ξ variables and are therefore ellipses, hyperbolas, parallel straight lines, or one of the exceptional cases. (Note that this depends crucially on the fact that $\mathbf{P}$ is unitary; see Problem 5.)

Another generalizable way to visualize the transformation from $\tilde{C}_t$ to C_t by $\mathbf{P}$ is to recall that $\mathbf{P}$ can be written as the product of at most two elementary reflections about a line {**Key Theorem 7.38(b)**}. If you visualize a reflection in a plane about a line as *a rotation of the plane through three dimensional space through* 180° *using the given line as the axis of rotation*, then it is easy to see that this transforms any geometric figure into a figure congruent to the original. Thus every orthogonal transformation—as a product of such reflections—does likewise.

(10.11) ***Example.*** The curves $\tilde{C}_t$ were found in Example 10.10 to be ellipses, and in particular $\tilde{C}_{64}$ has ξ foci

$$\tilde{\mathbf{f}}_1 = [0 \quad C]^T \quad \text{and} \quad \tilde{\mathbf{f}}_2 = [0 \quad -C]^T,$$

with the sum of the relevant distances equaling $2A$, where

$$A^2 = 16 \quad \text{and} \quad C^2 = A^2 - B^2 = 16 - 4 = 12.$$

For C_t, therefore, the **x** foci are

$$\mathbf{f}_1 = \mathbf{P}\tilde{\mathbf{f}}_1 = \begin{bmatrix} \dfrac{C}{2} & -\dfrac{C\sqrt{3}}{2} \end{bmatrix}^T = [\sqrt{3} \quad -3]^T$$

and

$$\mathbf{f}_2 = \mathbf{P}\tilde{\mathbf{f}}_2 = \begin{bmatrix} -\dfrac{C}{2} & \dfrac{C\sqrt{3}}{2} \end{bmatrix}^T = [-\sqrt{3} \quad 3]^T,$$

and the sum of the relevant distances equals $2A = 8$. The graph below is of C_{64}: $(\mathbf{x}, \mathbf{A}\mathbf{x}) = 64$.

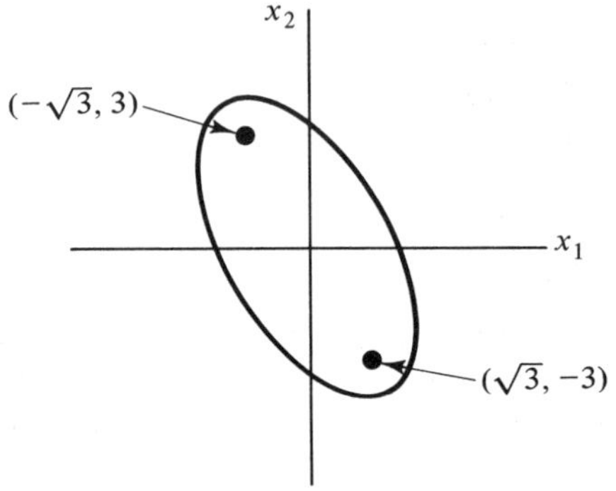

C_{64}: $13x_1^2 + 6\sqrt{3}\,x_1x_2 + 7x_2^2 = 64$

PROBLEMS 10.2

1. (a) Show that each of the four matrices at the very start of this section generates the same quadratic form on $\mathbb{R}^2$.
   (b) Find all 2×2 real matrices **A** that generate the same quadratic form on $\mathbb{R}^2$.
   (c) Find all 2×2 real symmetric matrices that generate it.
   (d) Find all (possibly complex) 2×2 hermitian matrices **A** that generate it on $\mathbb{R}^2$.

▷ 2. Find all conditions on λ_i and t for which the level curves $\tilde{C}_t$ of (10.6) are empty.

3. Find all conditions on λ_i and t for which the level curves $\tilde{C}_t$ of (10.6) are single points.

▷ 4. For each quadratic form in **x** below, find the simple form (10.5) in ξ and graph the level curves $\tilde{C}_t$ for various t so as to see how $\tilde{C}_t$ varies with t:
   (a) $x_1^2 + 4x_1x_2 - 2x_2^2$ (b) $x_1^2 + 12x_1x_2 + 4x_2^2$
   (c) $2x_1^2 + 4x_1x_2 + 4x_2^2$ (d) $-5x_1^2 - 8x_1x_2 - 5x_2^2$
   (e) $11x_1^2 + 2x_1x_2 + 3x_2^2$

▷ 5. Consider the circle $\tilde{C}$ in ξ given by $\xi_1^2 + \xi_2^2 = 9$. We know that the set of $\mathbf{x} = \mathbf{P}\xi$ as ξ varies over $\tilde{C}$ is congruent to $\tilde{C}$ and is thus a circle if **P** is unitary.

Show that the curve C of all $\mathbf{x} = \mathbf{P}\xi$ as ξ varies over $\tilde{C}$ is *not* congruent to $\tilde{C}$ if $\mathbf{P} = \mathbf{diag}(8, 1)$ (which is *not* unitary).

6. Graph the level curve C_t in $\mathbf{x}$ corresponding to each level curve $\tilde{C}_t$ in ξ for each of the quadratic forms in Problem 4(a), (b), (c), (d), and (e).

10.3 QUADRATIC FORMS IN $\mathbb{R}^p$ AND $\mathbb{C}^p$

Except for omitting graphs because of our inability to draw in p dimensions, we follow here the same route that served us well in the preceding section on $\mathbb{R}^2$. The first issue is the restriction to hermitian or real symmetric matrices.

In practice, one is usually only concerned with quadratic forms $Q(\mathbf{x}) = (\mathbf{x}, \mathbf{Ax})$ that assume only *real* values; in cases with complex Q, one usually can study only the real part of Q. So we assume that $Q(\mathbf{x})$ is real, *but not that* $\mathbf{x}$ *and* $\mathbf{A}$ *are real.* Even when $\mathbf{A}$ and $\mathbf{x}$ are possibly complex, $Q(\mathbf{x})$ will necessarily be real if, for example, $\mathbf{A}$ is hermitian ($\mathbf{A}^H = \mathbf{A}$):

$$\mathbf{x}^H(\mathbf{Ax}) = \mathbf{x}^H\mathbf{Ax} = \mathbf{x}^H\mathbf{A}^H\mathbf{x} = (\mathbf{Ax})^H\mathbf{x};$$

since the far-right term also equals the complex conjugate of the far-left term, their common value must therefore be real. But might there be *non*-hermitian matrices whose quadratic forms are real? The theorem below answers this.

(10.12) ***Theorem*** (real quadratic forms). Suppose that the quadratic form $Q(\mathbf{x}) = (\mathbf{x}, \mathbf{Ax})$ is generated by a $p \times p$ matrix $\mathbf{A}$. Then:

(a) $Q(\mathbf{x})$ is real for all complex $\mathbf{x}$ if and only if $\mathbf{A}$ is hermitian.

(b) $Q(\mathbf{x})$ is real for all real $\mathbf{x}$ if and only if $\mathbf{A} = \mathbf{R} + \mathbf{H}$ where $\mathbf{R}$ is real and $\mathbf{H}$ is hermitian; in such case with $\mathbf{A} = \mathbf{R} + \mathbf{H}$, the quadratic form generated by $\mathbf{A}$ is identical with that generated by a real symmetric matrix $\mathbf{A}'$, where $\mathbf{A}' = (\mathbf{A} + \mathbf{A}^T)/2$.

PROOF

(a) (hermitian $\Rightarrow$ real) This was demonstrated above.

(real $\Rightarrow$ hermitian) Since $Q(\mathbf{e}_r) = \langle \mathbf{A} \rangle_{rr}$, the diagonal entries $a_{11}, \ldots, a_{pp}$ of $\mathbf{A}$ must be real; let $\mathbf{D} = \mathbf{diag}(a_{11}, \ldots, a_{pp})$. Since

$$(\mathbf{x}, \mathbf{Dx}) = a_{11}|x_1|^2 + \cdots + a_{pp}|x_p|^2$$

is real, as is $(\mathbf{x}, \mathbf{Ax})$, so is the quadratic form generated by $\mathbf{A}_0 = \mathbf{A} - \mathbf{D}$. For $r > s$, let α denote $\langle \mathbf{A}_0 \rangle_{rs} = \langle \mathbf{A} \rangle_{rs}$, β denote $\langle \mathbf{A}_0 \rangle_{sr} = \langle \mathbf{A} \rangle_{sr}$, $\mathbf{x}_{rs}$ denote $\mathbf{e}_r + \mathbf{e}_s$, and $\mathbf{y}_{rs}$ denote $\mathbf{e}_r + i\mathbf{e}_s$. Then c defined as $(\mathbf{x}_{rs}, \mathbf{A}_0\mathbf{x}_{rs}) = \beta + \alpha$ and d defined as $(\mathbf{y}_{rs}, \mathbf{A}_0\mathbf{y}_{rs}) = i(\beta - \alpha)$ must both be real. Solving for α and β in terms of the real c and d gives

$$\alpha = \frac{c}{2} + \frac{id}{2} \quad \text{and} \quad \beta = \frac{c}{2} - \frac{id}{2},$$

that is, $\beta = \bar{\alpha}$, so $\mathbf{A}_0$ (and also $\mathbf{A}$) is hermitian since r and s are arbitrary.

(b) ($\mathbf{R} + \mathbf{H} \Rightarrow$ real) The quadratic form on $\mathbb{R}^p$ generated by real $\mathbf{R}$ is real, and that generated by hermitian $\mathbf{H}$ is real by (a); thus that generated by $\mathbf{A} = \mathbf{R} + \mathbf{H}$ is real.

(real $\Rightarrow \mathbf{R} + \mathbf{H}$) Let $\mathbf{D}$, $\mathbf{A}_0$, r, s, α, β, $\mathbf{x}_{rs}$, and c be as in the proof of (a). $\mathbf{D}$ and $c = \alpha + \beta$ are again real. Let $\mathbf{R}$ be the real $p \times p$ matrix formed from the real parts of the entries of $\mathbf{A}$, and $\mathbf{H}$ be the pure imaginary $p \times p$ matrix formed from the imaginary parts of the entries of $\mathbf{A}$; clearly $\mathbf{A} = \mathbf{R} + \mathbf{H}$. Then

$$\text{real } c = \langle \mathbf{A} \rangle_{rs} + \langle \mathbf{A} \rangle_{sr} = \langle \mathbf{R} \rangle_{rs} + \langle \mathbf{R} \rangle_{sr} + \langle \mathbf{H} \rangle_{rs} + \langle \mathbf{H} \rangle_{sr},$$

so the pure imaginary number $\langle \mathbf{H} \rangle_{rs} + \langle \mathbf{H} \rangle_{sr}$ is real, hence zero. This means that $\langle \mathbf{H} \rangle_{rs} = -\langle \mathbf{H} \rangle_{sr}$, and thus $\mathbf{H}$ is hermitian as claimed since r and s are arbitrary.

($\mathbf{A}$ and $\mathbf{A}'$)

$$\begin{aligned}(\mathbf{x}, \mathbf{A}'\mathbf{x}) &= \mathbf{x}^T\mathbf{A}'\mathbf{x} = \mathbf{x}^T(\mathbf{A} + \mathbf{A}^T)\mathbf{x}/2 = \mathbf{x}^T\mathbf{A}\mathbf{x}/2 + (\mathbf{A}\mathbf{x})^T\mathbf{x}/2 \\ &= \mathbf{x}^T\mathbf{A}\mathbf{x}/2 + \mathbf{x}^T(\mathbf{A}\mathbf{x})/2 = \mathbf{x}^T\mathbf{A}\mathbf{x},\end{aligned}$$

so the quadratic forms are certainly identical. To see that $\mathbf{A}'$ is real and symmetric, note first that

$$(\mathbf{A} + \mathbf{A}^T)^T = \mathbf{A}^T + (\mathbf{A}^T)^T = \mathbf{A}^T + \mathbf{A} = \mathbf{A} + \mathbf{A}^T,$$

so $\mathbf{A}'$ is symmetric. For the reality of $\mathbf{A}'$, write

$$\begin{aligned}\mathbf{A} + \mathbf{A}^T &= \mathbf{R} + \mathbf{H} + \mathbf{R}^T + \mathbf{H}^T = \mathbf{R} + \mathbf{R}^T + \mathbf{H} + \bar{\mathbf{H}}^H \\ &= \mathbf{R} + \mathbf{R}^T + \mathbf{H} + \bar{\mathbf{H}}\end{aligned}$$

(since $\mathbf{H}$ is hermitian), which is real. ■

This justifies the restriction to quadratic forms generated by hermitian and real symmetric matrices.

(10.13) ***Definition***

(a) A *quadratic form on* $\mathbb{C}^p$ is a real-valued function

$$Q(\mathbf{x}) = (\mathbf{x}, \mathbf{A}\mathbf{x}) = \mathbf{x}^H\mathbf{A}\mathbf{x}$$

defined by a $p \times p$ hermitian matrix $\mathbf{A}$ for all $\mathbf{x}$ in $\mathbb{C}^p$.

(b) A *quadratic form on* $\mathbb{R}^p$ is a real-valued function

$$Q(\mathbf{x}) = (\mathbf{x}, \mathbf{A}\mathbf{x}) = \mathbf{x}^T\mathbf{A}\mathbf{x}$$

defined by a $p \times p$ real symmetric matrix $\mathbf{A}$ for all $\mathbf{x}$ in $\mathbb{R}^p$.

There is no difference in the theories for quadratic forms on $\mathbb{C}^p$ and $\mathbb{R}^p$ except for the appearance of ^{H} in place of ^{T}. We treat the two cases simultaneously by using $(\mathbf{x}, \mathbf{A}\mathbf{x})$ to denote the quadratic form, where $(\cdot, \cdot)$ as usual denotes the standard inner product on $\mathbb{R}^p$ or $\mathbb{C}^p$; we generally refer simply to the quadratic form $Q(\mathbf{x}) = (\mathbf{x}, \mathbf{A}\mathbf{x})$ without explicit reference to $\mathbb{R}$ or $\mathbb{C}$.

We proceed as in the previous section.

(10.14) **Key Theorem** (diagonal quadratic forms). Let the $p \times p$ hermitian matrix $\mathbf{A}$ define the quadratic form $Q(\mathbf{x}) = (\mathbf{x}, \mathbf{A}\mathbf{x})$. Then:

(a) There exist p real numbers $\lambda_1, \ldots, \lambda_p$ and a unitary matrix $\mathbf{P}$ such that

$$Q(\mathbf{x}) = \tilde{Q}(\boldsymbol{\xi}) = \lambda_1|\xi_1|^2 + \cdots + \lambda_p|\xi_p|^2 \qquad \text{with } \boldsymbol{\xi} = \mathbf{P}^H\mathbf{x} \text{ and } \mathbf{x} = \mathbf{P}\boldsymbol{\xi}.$$

$\tilde{Q}$ is said to be a *diagonal form* of Q, and Q is said to have been *diagonalized* by the transformation to $\boldsymbol{\xi}$, since

$$\tilde{Q}(\boldsymbol{\xi}) = (\boldsymbol{\xi}, \boldsymbol{\Lambda}\boldsymbol{\xi}) \qquad \text{with } \boldsymbol{\Lambda} = \mathbf{diag}(\lambda_1, \ldots, \lambda_p).$$

(b) The numbers λ_i are the eigenvalues of $\mathbf{A}$, the columns of $\mathbf{P}$ form an orthonormal set of associated eigenvectors, and $\mathbf{P}^H\mathbf{A}\mathbf{P} = \boldsymbol{\Lambda}$.

(c) If $\mathbf{A}$ is real symmetric and $\mathbf{x}$ is real, then $\mathbf{P}$ may be taken to be orthogonal and $|\xi_i|^2$ may be replaced by ξ_i^2.

PROOF. This follows directly from **Key Corollary 8.9** and **Key Theorem 8.6** by substituting $\mathbf{x} = \mathbf{P}\boldsymbol{\xi}$ into $Q(\mathbf{x})$ as in the preceding section. ∎

(10.15) ***Example.*** Consider the quadratic form defined by the 3×3 real symmetric matrix $\mathbf{A}$ in Example 8.11:

$$Q(\mathbf{x}) = 7x_1^2 + 7x_2^2 - 5x_3^2 - 32x_1x_2 - 16x_1x_3 + 16x_2x_3.$$

The eigensystem of $\mathbf{A}$ was found in that example to be

$$\lambda_1 = 27 \qquad \text{with } \mathbf{v}_1 = [-\tfrac{2}{3} \quad \tfrac{2}{3} \quad \tfrac{1}{3}]^T$$

and

$$\lambda_2 = \lambda_3 = -9 \qquad \text{with } \mathbf{v}_2 = \begin{bmatrix} \frac{\sqrt{2}}{2} & \frac{\sqrt{2}}{2} & 0 \end{bmatrix}^T \text{ and}$$

$$\mathbf{v}_3 = \begin{bmatrix} -\frac{\sqrt{2}}{6} & \frac{\sqrt{2}}{6} & -\frac{2\sqrt{2}}{3} \end{bmatrix}^T.$$

If we form $\mathbf{P} = [\mathbf{v}_1 \quad \mathbf{v}_2 \quad \mathbf{v}_3]$ and then define new variables $\boldsymbol{\xi} = \mathbf{P}^H\mathbf{x}$ so that $\mathbf{x} = \mathbf{P}\boldsymbol{\xi}$, we have

$$Q(\mathbf{x}) = \tilde{Q}(\boldsymbol{\xi}) = 27\xi_1^2 - 9\xi_2^2 - 9\xi_3^2,$$

where

$$\xi_1 = \left(-\frac{2}{3}\right)x_1 + \left(\frac{2}{3}\right)x_2 + \left(\frac{1}{3}\right)x_3$$

$$\xi_2 = \left(\frac{\sqrt{2}}{2}\right)x_1 + \left(\frac{\sqrt{2}}{2}\right)x_2$$

$$\xi_3 = \left(-\frac{\sqrt{2}}{6}\right)x_1 + \left(\frac{\sqrt{2}}{6}\right)x_2 + \left(-\frac{2\sqrt{2}}{3}\right)x_3$$

and

$$x_1 = \left(-\frac{2}{3}\right)\xi_1 + \left(\frac{\sqrt{2}}{2}\right)\xi_2 + \left(-\frac{\sqrt{2}}{6}\right)\xi_3$$
$$x_2 = \left(\frac{2}{3}\right)\xi_1 + \left(\frac{\sqrt{2}}{2}\right)\xi_2 + \left(\frac{\sqrt{2}}{6}\right)\xi_3$$
$$x_3 = \left(\frac{1}{3}\right)\xi_1 + \left(-\frac{2\sqrt{2}}{3}\right)\xi_3.$$

The simple form of $\tilde{Q}(\xi)$ involving only three terms shows the advantage of diagonalizing by an appropriate change of variables.

You should now be able to solve Problem 1.

Level Surfaces in $\mathbb{R}^p$

The analogues of the level curves we studied in Section 10.2 are the *level surfaces* S_t: the set of all $\mathbf{x}$ satisfying $Q(\mathbf{x}) = t$. Suppose that $p \times p$ $\mathbf{A}$ defining the quadratic form is real symmetric and that $\mathbf{x}$ varies over $\mathbb{R}^p$. According to **Key Theorem 10.14**, Q can be diagonalized by an orthogonal transformation $\mathbf{P}$ leading to the diagonal quadratic form $\tilde{Q}(\xi)$ with $\mathbf{x} = \mathbf{P}\xi$. As in Section 10.2, it is clear that $\mathbf{x}$ is on the level surface S_t if and only if ξ is on the level surface $\tilde{S}_t$ of those ξ satisfying $\tilde{Q}(\xi) = t$. **Key Theorem 7.38(b)** then says that $\mathbf{P}$ can be written as the product of at most p elementary reflections, each of which clearly produces surfaces congruent to $\tilde{S}_t$; this implies that

the level surfaces S_t of $Q(\mathbf{x}) = t$ are congruent to those $\tilde{S}_t$ of $\tilde{Q}(\xi) = t$.

The nature of these level surfaces depends, as in the case of $p = 2$, on the signs of the terms λ_i in the diagonal form $\tilde{Q}$—that is, on the signs of the eigenvalues of $\mathbf{A}$. In two dimensions, for example, we saw that λ_1 and λ_2 of the same nonzero sign produce ellipses, that λ_1 and λ_2 of opposite nonzero signs produce hyperbolas, and that one zero λ_i produces parallel straight lines—each a fundamentally different geometric shape. In three dimensions, where

$$\tilde{Q}(\xi) = \lambda_1\xi_1^2 + \lambda_2\xi_2^2 + \lambda_3\xi_3^2,$$

the typical cases of $\tilde{S}_t$ *for* t *positive*, say, can be classified and visualized in physical space as follows:

1. *Ellipsoid* if λ_1, λ_2, and λ_3 are strictly positive. If two of the λ_i are equal, the surface is an ellipsoid of revolution (cross sections in one direction are circles); if three λ_i are equal, the surface is a sphere.

2. *Hyperboloid of one sheet* if two λ_i are positive and one is negative. If the positive λ_i are equal, the surface is a hyperboloid of revolution.
3. *Hyperboloid of two sheets* if one λ_i is positive and two are negative. If the negative λ_i are equal, the surface is a hyperboloid of revolution.
4. *Elliptic or hyperbolic cylinder* if exactly one of the λ_i equals zero. The cylinder is elliptic if the other λ_i are both positive, hyperbolic if they have opposite nonzero signs.
5. *Two parallel planes* if exactly two of the λ_i equal zero and the third is positive.

Similar classifications are valid in p dimensions, although the results are difficult to visualize. Nonetheless, such classification can be performed for the level surfaces $\tilde{S}_t$ of the diagonal form $\tilde{Q}(\xi)$ with the knowledge that the classification holds for the general level surface S_t of $Q(\mathbf{x})$ since S_t is congruent to $\tilde{S}_t$.

(10.16) ***Example.*** Consider the level surface S_1 for the quadratic form $Q(\mathbf{x})$ of Example 10.15:

$$7x_1^2 + 7x_2^2 - 5x_3^2 - 32x_1x_2 - 16x_1x_3 + 16x_2x_3 = 1.$$

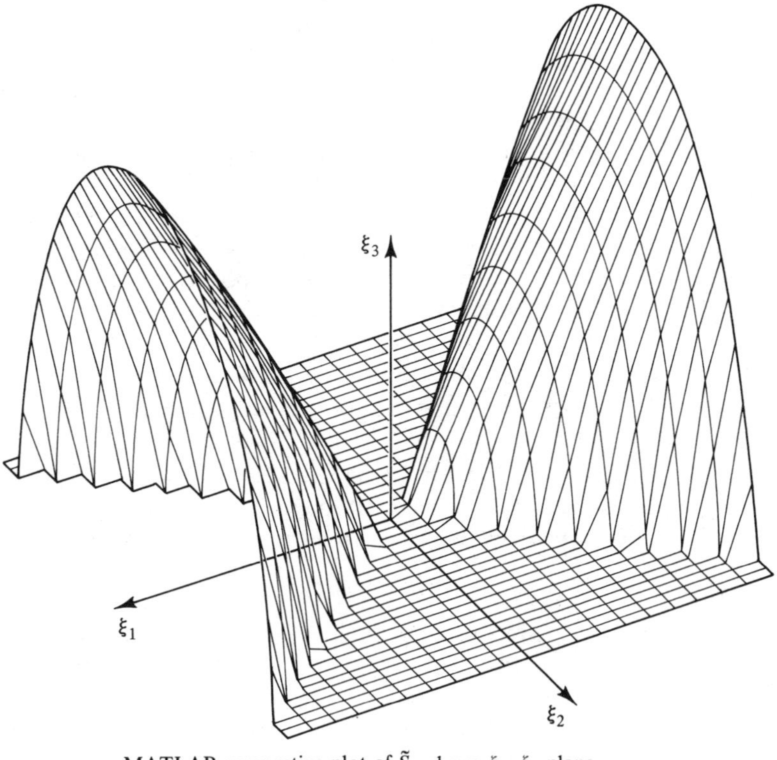

MATLAB perspective plot of $\tilde{S}_1$ above ξ_1–ξ_2-plane

In this form the nature of the surface S_1 is unrecognizable. From Example 10.15, however, we know that

$$Q(\mathbf{x}) = \tilde{Q}(\xi) = 27\xi_1^2 - 9\xi_2^2 - 9\xi_3^2;$$

by (3) above, the level surface $\tilde{S}_1$ is a two-sheeted hyperboloid of revolution. Since S_1 is congruent to $\tilde{S}_1$, we know that the level surface S_1 of $Q(\mathbf{x}) = 1$ is a two-sheeted hyperboloid of revolution—a fact rather difficult to determine without the diagonalization to $\tilde{Q}(\xi)$.

You should now be able to solve Problems 1 to 3.

Definite Quadratic Forms

In general, a quadratic form $Q(\mathbf{x})$ will assume positive, negative, and zero values for various values of $\mathbf{x}$. Sometimes, however, quadratic forms represent physical phenomena for which particular signed values are meaningless, and in such cases we expect Q not to assume values with the meaningless sign. A simple example of a quadratic form that is never negative is $(x_1 - x_2)^2$ on $\mathbb{R}^2$, although it does equal zero for some nonzero $\mathbf{x}$. An example of a quadratic form that is never negative and, in addition, is zero only when $\mathbf{x} = \mathbf{0}$ is $x_1^2 + x_2^2$ on $\mathbb{R}^2$. Special terminology is used to distinguish such quadratic forms.

(10.17) ***Definition***

(a) A quadratic form $Q(\mathbf{x})$ is said to be *positive definite* if $Q(\mathbf{x}) > 0$ for all $\mathbf{x} \neq \mathbf{0}$.

(b) $Q(\mathbf{x})$ is said to be *positive semidefinite* if $Q(\mathbf{x}) \geq 0$ for all $\mathbf{x}$.

(c) $Q(\mathbf{x})$ is said to be *negative definite* or *negative semidefinite* if and only if $-Q(\mathbf{x})$ is positive definite or positive semidefinite, respectively.

(d) $Q(\mathbf{x})$ is said to be *indefinite* if and only if there are $\mathbf{x}_-$ and $\mathbf{x}_+$ such that $Q(\mathbf{x}_-) < 0 < Q(\mathbf{x}_+)$.

(e) A *matrix* $\mathbf{A}$ is said to be positive definite (et cetera) when the quadratic form $(\mathbf{x}, \mathbf{Ax})$ it defines is positive definite (et cetera).

Note that a positive definite matrix is also positive semidefinite, but that the converse need not be true (although it might be).

Key Theorem 10.14 says that each quadratic form can be diagonalized, so that

$$Q(\mathbf{x}) = \tilde{Q}(\xi) = \lambda_1|\xi_1|^2 + \cdots + \lambda_p|\xi_p|^2,$$

where the λ_i are the eigenvalues of the $p \times p$ hermitian matrix $\mathbf{A}$ defining Q and $\xi = \mathbf{P}^H\mathbf{x}$ with $\mathbf{P}$ unitary. Since $\mathbf{P}$ and $\mathbf{P}^H$ are nonsingular, $\mathbf{x} = \mathbf{0}$ if and only if $\xi = \mathbf{0}$, where $\xi = \mathbf{P}^H\mathbf{x}$ and $\mathbf{x} = \mathbf{P}\xi$; and, clearly, $\tilde{Q}$ is positive definite if and only if all the λ_i are strictly positive. This type of argument immediately yields the following characterization of definite matrices.

(10.18) **Key Theorem** (definite matrices). Suppose that $\mathbf{A}$ is hermitian. Then:

(a) $\mathbf{A}$ is positive (negative) definite if and only if all its eigenvalues are strictly positive (negative).

(b) $\mathbf{A}$ is positive (negative) semidefinite if and only if all its eigenvalues are nonnegative (nonpositive).

(c) $\mathbf{A}$ is indefinite if and only if $\mathbf{A}$ has both strictly positive and strictly negative eigenvalues.

PROOF. Problem 4. ■

To use this theorem to check whether a matrix is positive definite requires knowledge of its eigenvalues; it seems reasonable to believe that there should be a characterization requiring only information that might be more directly checked. For example, since every $p \times p$ $\mathbf{A}$ has $\det \mathbf{A} = \lambda_1 \cdots \lambda_p$, it is clear that every positive definite $\mathbf{A}$ must have a positive determinant. Also, since $(\mathbf{e}_i, \mathbf{A}\mathbf{e}_i) = \langle \mathbf{A} \rangle_{ii}$, every positive definite matrix must have positive main-diagonal entries. If we consider $(\mathbf{x}, \mathbf{A}\mathbf{x})$ only for those $\mathbf{x}$ that are nonzero only in the ith and jth entries, then $(\mathbf{x}, \mathbf{A}\mathbf{x})$ equals $(\mathbf{x}_0, \mathbf{A}_0\mathbf{x}_0)$, where $\mathbf{A}_0$ is 2×2 and $\mathbf{x}_0$ is 2×1:

$$\mathbf{A}_0 = \begin{bmatrix} \langle \mathbf{A} \rangle_{ii} & \langle \mathbf{A} \rangle_{ij} \\ \langle \mathbf{A} \rangle_{ji} & \langle \mathbf{A} \rangle_{jj} \end{bmatrix} \quad \text{and} \quad \mathbf{x}_0 = \begin{bmatrix} \langle \mathbf{x} \rangle_i \\ \langle \mathbf{x} \rangle_j \end{bmatrix}.$$

Therefore, the quadratic form defined by $\mathbf{A}_0$ must also be positive definite, so $\det \mathbf{A}_0$ must be positive, for example. By applying the same argument with $\mathbf{x}$ nonzero only in entries i, j, and k, we find that the determinants of various 3×3 submatrices of $\mathbf{A}$ must be positive also. And so on with 4×4, $5 \times 5, \ldots,$ and $(p-1) \times (p-1)$ submatrices. Among all these necessary conditions on the positivity of determinants is sufficient information to guarantee that the original matrix is positive definite.

(10.19) ***Theorem*** (definiteness and determinants). Suppose that $p \times p$ $\mathbf{A}$ is hermitian. Then:

(a) $\mathbf{A}$ is positive definite if and only if each of its $k \times k$ *principal submatrices* $\mathbf{A}_k$, for $1 \le k \le p$, has a strictly positive determinant, where $\mathbf{A}_k$ is the $k \times k$ matrix formed from the entries in the upper-left corner of $\mathbf{A}$:

$$\langle \mathbf{A}_k \rangle_{ij} = \langle \mathbf{A} \rangle_{ij} \qquad \text{for } 1 \le i \le k \quad \text{and} \quad 1 \le j \le k.$$

(b) $\mathbf{A}$ is positive definite if and only if each pivot is strictly positive when $\mathbf{A}$ is reduced to Gauss-reduced or row-echelon form *without interchanging rows.*

PROOF. Problems 8, 9, and 10. ■

(10.20) ***Example.*** To determine whether the quadratic form

$$2x_1^2 + x_2^2 + 6x_3^2 + 2x_1x_2 + x_1x_3 + 4x_2x_3$$

is positive definite we form a hermitian matrix defining it and transform this to Gauss-reduced form, keeping track of pivots:

$$\begin{bmatrix} 2 & 1 & \frac{1}{2} \\ 1 & 1 & 2 \\ \frac{1}{2} & 2 & 6 \end{bmatrix} \xrightarrow{\text{pivot} = 2} \begin{bmatrix} 1 & \frac{1}{2} & \frac{1}{4} \\ 0 & \frac{1}{2} & \frac{7}{4} \\ 0 & \frac{7}{4} & \frac{47}{8} \end{bmatrix}$$

$$\xrightarrow{\text{pivot} = 1/2} \begin{bmatrix} 1 & \frac{1}{2} & \frac{1}{4} \\ 0 & 1 & \frac{7}{2} \\ 0 & 0 & -\frac{1}{4} \end{bmatrix} \xrightarrow{\text{pivot} = -1/4} \begin{bmatrix} 1 & 0 & \frac{1}{4} \\ 0 & 1 & \frac{7}{2} \\ 0 & 0 & 1 \end{bmatrix}.$$

Since one of the pivots is negative, the quadratic form is not positive definite.

PROBLEMS 10.3

▷ **1.** For each quadratic form $Q(\mathbf{x})$ below, find a diagonal quadratic form $\tilde{Q}(\xi) = Q(\mathbf{x})$.
(a) $3x_1^2 + 2x_2^2 + 3x_3^2 - 2x_1x_2 - 2x_2x_3$
(b) Q generated by $\mathbf{A}$ in Problem 8 of Section 8.2
(c) Q generated by $\mathbf{A}$ in Problem 3 of Section 8.3

2. Describe the nature of each of the level surfaces $\tilde{S}_t$ in Problem 1(a), (b), and (c).

$\mathfrak{M}$ **3.** Use MATLAB or similar software to produce a perspective plot of one nontrivial level surface $\tilde{S}_t$ for each $\tilde{Q}$ in Problem 1(a), (b), and (c).

4. Prove **Key Theorem 10.18.**

▷ **5.** By considering $-\mathbf{A}$, use Theorem 10.19 to state and prove a theorem characterizing negative definite matrices.

6. By using Theorem 10.19 on $\mathbf{A} + \epsilon\mathbf{I}$, state and prove a theorem characterizing positive semidefinite matrices.

▷ **7.** Use Theorem 10.19 to determine which of the matrices are positive definite in Problem 1(a), (b), and (c).

8. Section 3.7 presented the LU-decomposition of a matrix. Prove that $p \times p$ $\mathbf{A}$ can be written $\mathbf{A} = \mathbf{LU}$ with $\mathbf{U}$ unit-upper-triangular and $\mathbf{L}$ lower-triangular with $\langle \mathbf{L} \rangle_{ii} \neq 0$ for all i if and only if the determinants of the principal submatrices $\mathbf{A}_k$ are nonzero for $1 \leq k \leq p$.

▷ **9.** Suppose that a $p \times p$ hermitian nonsingular matrix $\mathbf{A}$ can be decomposed as $\mathbf{A} = \mathbf{L}_1\mathbf{D}_0\mathbf{U}_1$ with $\mathbf{L}_1$ unit-lower-triangular, $\mathbf{D}_0$ diagonal, and $\mathbf{U}_1$ unit-upper-triangular. Prove that $\mathbf{L}_1 = \mathbf{U}_1^H$, so that $\mathbf{A} = \mathbf{L}_1\mathbf{D}_0\mathbf{L}_1^H$.

10. Use Problems 8 and 9 to prove Theorem 10.19.

▷ **11.** Suppose that $\mathbf{A}$ is hermitian and positive semidefinite. Prove that $\mathbf{A}$ is positive definite if and only if $\mathbf{A}$ is nonsingular, and deduce that such $\mathbf{A}$ is singular if and only if there exists an $\mathbf{x}$ with $(\mathbf{x}, \mathbf{Ax}) = 0$.

12. Suppose that **A**, **B**, and **S** are $p \times p$, that **S** is nonsingular, and that $\mathbf{B} = \mathbf{S}^H\mathbf{A}\mathbf{S}$; **A** and **B** are said to be *hermitian congruent*. Prove that **A** is positive definite (or positive semidefinite, or negative definite, or negative semidefinite, or indefinite) if and only if **B** is likewise.

13. Prove that every $p \times p$ hermitian matrix **A** is hermitian congruent—see Problem 12—to a matrix of the form

$$\mathbf{diag}(1, \ldots, 1, -1, \ldots, -1, 0, \ldots, 0)$$

where there are r entries $+1$, s entries -1, and $p - r - s$ entries 0, and that the numbers r and s equal the numbers of positive and of negative (respectively) eigenvalues of **A** and so will be the same in every such form. (This is Sylvester's law of inertia.)

▷ **14.** (a) Suppose that **A** is hermitian and positive definite; prove that **A** is nonsingular and that $\mathbf{A}^{-1}$ is positive definite.

(b) Suppose that **A** is hermitian and nonsingular; prove that **A** is positive definite (or negative definite, or indefinite) if and only if $\mathbf{A}^{-1}$ is likewise.

15. Use Problems 8 and 9 and Theorem 10.19 to prove that **A** is hermitian (or real symmetric) and positive definite if and only if **A** can be written in the *Cholesky decomposition*

$$\mathbf{A} = \mathbf{C}\mathbf{C}^H \qquad (\text{or } \mathbf{A} = \mathbf{C}\mathbf{C}^T),$$

where **C** is lower-triangular and $\langle \mathbf{C} \rangle_{ii} \neq 0$ for all i.

10.4 EXTREMIZING QUADRATIC FORMS: RAYLEIGH'S PRINCIPLE

Applied problems involving quadratic forms often lead to the task of finding the extreme values—maxima and minima—of a quadratic form $(\mathbf{x}, \mathbf{A}\mathbf{x})$ subject to the constraint $(\mathbf{x}, \mathbf{x}) = 1$. The statistical problem in Example 10.4, for example, required such a maximization in order to determine a new variable that contained as much as possible of the variance in an experiment. The diagonalization of quadratic forms makes the study of such extremum problems rather straightforward.

(10.21) ***Example.*** Consider maximizing the diagonalized quadratic form

$$\tilde{Q}(\xi) = \lambda_1\xi_1^2 + \lambda_2\xi_2^2 \quad \text{subject to} \quad \xi_1^2 + \xi_2^2 = 1$$

where $0 < \lambda_1 \leq \lambda_2$. The level curves $\tilde{C}_t$ of all ξ satisfying $\tilde{Q}(\xi) = t$ are ellipses (for $t > 0$) with major axes along the ξ_1-axis, with semimajor axes equal to $(t/\lambda_1)^{1/2}$, and with semiminor axes equal to $(t/\lambda_2)^{1/2}$. These ellipses "expand" as t increases, and thus $t = \tilde{Q}(\xi)$ is greatest with $(\xi, \xi) = 1$ when $\tilde{C}_t$ circumscribes the circle $(\xi, \xi) = 1$ and is tangent to it at the tip of the semiminor axis—that is, when $t = \lambda_2$. Even without the geometry, of course, it is clear that $\lambda_1\xi_1^2 + \lambda_2\xi_2^2$ is maximized by making ξ_2 as large as possible (since

$0 < \lambda_1 \le \lambda_2$) and that this occurs with $\xi_2 = \pm 1$ (since $\xi_1^2 + \xi_2^2 = 1$), giving $\tilde{Q}(\xi) = \lambda_2$.

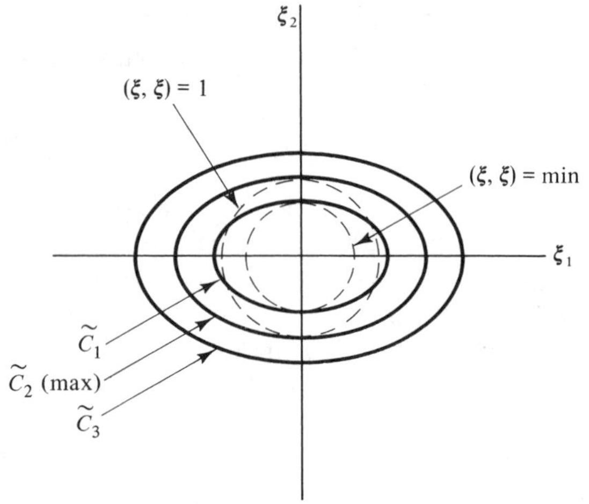

It is interesting to observe that the same sketch shows that, if we seek the point ξ on $\tilde{C}_1$ with *least* (ξ, ξ), this occurs again with an inscribed circle tangent at the tip of the semiminor axis with $(\xi, \xi) = 1/\lambda_2$. Such *dual* problems arise often in mathematics generally and in extremum problems especially: The solution to a particular *maximum* problem is equivalent to the solution of a related *minimum* problem.

The fact that the largest eigenvalue equals the maximum of $\tilde{Q}(\xi)$ subject to the constraint $(\xi, \xi) = 1$ holds more generally than just in Example 10.21. Before proving this, we introduce an important device that simplifies the constraint.

The Rayleigh Quotient

(10.22) ***Theorem*** (equivalent extremum problems). Suppose that $\mathbf{A}$ is hermitian. Then $\mathbf{x}_0$ maximizes $(\mathbf{x}, \mathbf{Ax})/(\mathbf{x}, \mathbf{x})$ subject to $\mathbf{x} \neq \mathbf{0}$ and yields the maximum M if and only if $\mathbf{x}_1 = \mathbf{x}_0/\|\mathbf{x}_0\|_2$ maximizes $Q(\mathbf{x})$ subject to $(\mathbf{x}, \mathbf{x}) = 1$ and yields $Q(\mathbf{x}_1) = M$.

PROOF. ($\mathbf{x}_0$ max $\Rightarrow$ $\mathbf{x}_1$ max) Suppose that $\mathbf{x}_0$ is as stated and define

$$\mathbf{x}_1 = \mathbf{x}_0/\|\mathbf{x}_0\|_2.$$

For any $\mathbf{x}$ with $(\mathbf{x}, \mathbf{x}) = 1$, we have

$$Q(\mathbf{x}) = (\mathbf{x}, \mathbf{Ax}) = \frac{(\mathbf{x}, \mathbf{Ax})}{(\mathbf{x}, \mathbf{x})} \le M = \frac{(\mathbf{x}_0, \mathbf{Ax}_0)}{(\mathbf{x}_0, \mathbf{x}_0)} = (\mathbf{x}_1, \mathbf{Ax}_1) = Q(\mathbf{x}_1),$$

and thus $\mathbf{x}_1$ is as claimed.

($\mathbf{x}_1$ max $\Rightarrow$ $\mathbf{x}_0$ max) Suppose that $\mathbf{x}_1$ maximizes $Q(\mathbf{x})$ subject to $(\mathbf{x}, \mathbf{x}) = 1$. Let $\mathbf{x}_0 = \alpha\mathbf{x}_1$ for arbitrary real $\alpha \neq 0$. For any nonzero $\mathbf{x}$, $\tilde{\mathbf{x}}$ defined as $\tilde{\mathbf{x}} = \mathbf{x}/\|\mathbf{x}\|_2$ has $(\tilde{\mathbf{x}}, \tilde{\mathbf{x}}) = 1$ and thus

$$Q(\tilde{\mathbf{x}}) \leq M = Q(\mathbf{x}_1);$$

but

$$Q(\tilde{\mathbf{x}}) = \frac{(\mathbf{x}, \mathbf{A}\mathbf{x})}{(\mathbf{x}, \mathbf{x})} \quad \text{and} \quad Q(\mathbf{x}_1) = \frac{(\mathbf{x}_1, \mathbf{A}\mathbf{x}_1)}{(\mathbf{x}_1, \mathbf{x}_1)}$$

since $(\mathbf{x}_1, \mathbf{x}_1) = 1$, so the preceding inequality becomes

$$\frac{(\mathbf{x}, \mathbf{A}\mathbf{x})}{(\mathbf{x}, \mathbf{x})} \leq M = \frac{(\mathbf{x}_1, \mathbf{A}\mathbf{x}_1)}{(\mathbf{x}_1, \mathbf{x}_1)} = \frac{(\mathbf{x}_0, \mathbf{A}\mathbf{x}_0)}{(\mathbf{x}_0, \mathbf{x}_0)}$$

and so $\mathbf{x}_0$ is as claimed. ■

This theorem shows that we can replace the constraint $(\mathbf{x}, \mathbf{x}) = 1$ by $\mathbf{x} \neq \mathbf{0}$ if we deal with the special quotient $Q(\mathbf{x})/(\mathbf{x}, \mathbf{x})$ rather than just $Q(\mathbf{x})$.

(10.23) ***Definition.*** The *Rayleigh quotient* of a hermitian matrix $\mathbf{A}$ is the function $\rho_{\mathbf{A}}$ defined for $\mathbf{x} \neq \mathbf{0}$ by

$$\rho_{\mathbf{A}}(\mathbf{x}) = \frac{(\mathbf{x}, \mathbf{A}\mathbf{x})}{(\mathbf{x}, \mathbf{x})} \qquad \text{for } \mathbf{x} \neq \mathbf{0}.$$

The diagonalization of Q by the introduction of ξ in place of $\mathbf{x}$ is equally helpful in dealing with the Rayleigh quotient. If, in the notation of **Key Theorem 10.14**, we have

$$\mathbf{P}^H\mathbf{A}\mathbf{P} = \mathbf{\Lambda} = \mathbf{diag}(\lambda_1, \ldots, \lambda_p) \quad \text{and} \quad \xi = \mathbf{P}^H\mathbf{x},\ \mathbf{x} = \mathbf{P}\xi$$

with unitary $\mathbf{P}$, then not only does

$$Q(\mathbf{x}) = \tilde{Q}(\xi), \quad \text{but also} \quad (\mathbf{x}, \mathbf{x}) = (\xi, \xi)$$

since $\mathbf{P}$ is unitary. Since $\rho_{\mathbf{A}}(\mathbf{x}) = Q(\mathbf{x})/(\mathbf{x}, \mathbf{x})$, this says that $\rho_{\mathbf{A}}(\mathbf{x}) = \rho_{\mathbf{\Lambda}}(\xi)$.

(10.24) ***Theorem*** (diagonalized Rayleigh quotient). If $\mathbf{A}$ is hermitian, then $\rho_{\mathbf{A}}(\mathbf{x}) = \rho_{\mathbf{\Lambda}}(\xi)$, where $\mathbf{\Lambda}$ and ξ are as in **Key Theorem 10.14:**

$$\mathbf{P}^H\mathbf{A}\mathbf{P} = \mathbf{\Lambda} = \mathbf{diag}(\lambda_1, \ldots, \lambda_p), \qquad \mathbf{x} = \mathbf{P}\xi,$$

and $\mathbf{P}$ is unitary.

We now turn to showing that the extreme eigenvalues λ_1 and λ_p solve extremum problems involving $\rho_{\mathbf{A}}(\mathbf{x})$.

Extremum Problems and Extremal Eigenvalues

(10.25) **Key Theorem** ($\lambda_{\min}$, $\lambda_{\max}$, and $\rho_{\mathbf{A}}$). Let $p \times p$ $\mathbf{A}$ be hermitian and have eigenvalues

$$\lambda_1 \le \lambda_2 \le \cdots \le \lambda_p$$

and an orthonormal set of associated eigenvectors $\mathbf{v}_1, \ldots, \mathbf{v}_p$. Then:

(a) $\lambda_1 \le \rho_{\mathbf{A}}(\mathbf{x}) \le \lambda_p$ for all $\mathbf{x} \ne \mathbf{0}$.

(b) λ_1 is the minimum value of $\rho_{\mathbf{A}}(\mathbf{x})$ for $\mathbf{x} \ne \mathbf{0}$, and $\rho_{\mathbf{A}}(\mathbf{x}) = \lambda_1$ if and only if $\mathbf{x}$ is an eigenvector associated with λ_1.

(c) λ_p is the maximum value of $\rho_{\mathbf{A}}(\mathbf{x})$ for $\mathbf{x} \ne \mathbf{0}$, and $\rho_{\mathbf{A}}(\mathbf{x}) = \lambda_p$ if and only if $\mathbf{x}$ is an eigenvector associated with λ_p.

PROOF. By Theorem 10.24 we can treat the simpler $\rho_{\Lambda}(\xi) = \rho_{\mathbf{A}}(\mathbf{x})$. Since $\mathbf{x} = \mathbf{v}_i$ if and only if $\xi = \mathbf{P}^H\mathbf{v}_i = \mathbf{e}_i$, the ith unit column matrix, we need only prove that $\lambda_1 \le \rho_{\Lambda}(\xi) \le \lambda_p$, that $\mathbf{e}_1$ minimizes $\rho_{\Lambda}(\xi)$ with $\rho_{\Lambda}(\mathbf{e}_1) = \lambda_1$; and that $\mathbf{e}_p$ maximizes $\rho_{\Lambda}(\xi)$ with $\rho_{\Lambda}(\mathbf{e}_p) = \lambda_p$.

(a) This follows immediately from (b) and (c) below.

(b) Clearly, $\rho_{\Lambda}(\mathbf{e}_1) = \lambda_1 1^2/1^2 = \lambda_1$, so $\rho_{\Lambda}(\mathbf{e}_1) = \lambda_1$; therefore, $\rho_{\mathbf{A}}(\mathbf{v}_1) = \lambda_1$. Also

$$\begin{aligned}\lambda_1 &= (\lambda_1|\xi_1|^2 + \lambda_1|\xi_2|^2 + \cdots + \lambda_1|\xi_p|^2)/(|\xi_1|^2 + \cdots + |\xi_p|^2)\\ &\le (\lambda_1|\xi_1|^2 + \lambda_2|\xi_2|^2 + \cdots + \lambda_p|\xi_p|^2)/(|\xi_1|^2 + \cdots + |\xi_p|^2)\\ &= \rho_{\Lambda}(\xi)\end{aligned}$$

with equality holding if and only if $\xi_i = 0$ for all i for which $\lambda_i > \lambda_1$—that is, if and only if $\Lambda\xi = \lambda_1\xi$. Replacing ξ by $\mathbf{P}^H\mathbf{x}$ completes the proof of (b).

(c) follows by applying (b) to $-\mathbf{A}$ (see Problem 3). ■

(10.26) ***Corollary*** (matrix 2-norm). For any $p \times q$ matrix $\mathbf{A}$, the matrix 2-norm equals the largest singular value of $\mathbf{A}$: $\|\mathbf{A}\|_2 = \sigma_1$.

PROOF

$$\begin{aligned}\|\mathbf{A}\|_2^2 &= \left\{\max \frac{\|\mathbf{Ax}\|_2}{\|\mathbf{x}\|_2}\right\}^2 = \max \frac{\|\mathbf{Ax}\|_2^2}{\|\mathbf{x}\|_2^2}\\ &= \max \frac{(\mathbf{Ax}, \mathbf{Ax})}{(\mathbf{x}, \mathbf{x})} = \max \frac{(\mathbf{x}, \mathbf{A}^H\mathbf{Ax})}{(\mathbf{x}, \mathbf{x})}\\ &= \lambda_q,\end{aligned}$$

the largest eigenvalue of the hermitian matrix $\mathbf{A}^H\mathbf{A}$. But σ_1 is defined by $\sigma_1^2 = \lambda_q$. ■

Since $\rho_{\mathbf{A}}(\mathbf{x}) = \lambda_1$ if and only if $\mathbf{x}$ is an eigenvector of $\mathbf{A}$ associated with λ_1, we might reasonably expect that $\rho_{\mathbf{A}}(\mathbf{x})$ will be near λ_1 when $\mathbf{x}$ is near an eigenvector associated with λ_1. This should hold for other eigenvalues as well, since

$$\rho_{\mathbf{A}}(\mathbf{v}_i) = \frac{(\mathbf{v}_i, \mathbf{A}\mathbf{v}_i)}{(\mathbf{v}_i, \mathbf{v}_i)} = \frac{(\mathbf{v}_i, \lambda_i \mathbf{v}_i)}{(\mathbf{v}_i, \mathbf{v}_i)} = \lambda_i$$

for $\mathbf{v}_i$ an eigenvector associated with λ_i. Suppose that $\mathbf{x} = \mathbf{v}_i + \boldsymbol{\epsilon}$; then direct calculation shows that

$$\rho_{\mathbf{A}}(\mathbf{x}) = \lambda_i + \frac{\|\boldsymbol{\epsilon}\|_2^2\{\rho_{\mathbf{A}}(\boldsymbol{\epsilon}) - \lambda_i\}}{\|\mathbf{x}\|_2^2};$$

that is, if $\|\mathbf{x} - \mathbf{v}_i\|_2$ is of size ϵ, then $\rho_{\mathbf{A}}(\mathbf{x}) - \lambda_i$ is of size ϵ^2. When using this to approximate the largest eigenvalue λ_p, we know of course that $\rho_{\mathbf{A}}(\mathbf{x}) \leq \lambda_p$ and $\rho_{\mathbf{A}}(\boldsymbol{\epsilon}) \geq \lambda_1$, so we obtain

$$\lambda_p \geq \rho_{\mathbf{A}}(\mathbf{x}) \geq \lambda_p - \frac{(\lambda_p - \lambda_1)\|\mathbf{x} - \mathbf{v}_p\|_2^2}{\|\mathbf{x}\|_2^2}$$

and in particular certainly we have a guaranteed lower bound $\rho_{\mathbf{A}}(\mathbf{x})$ on λ_p. Similarly $\rho_{\mathbf{A}}(\mathbf{x})$ is an upper bound on λ_1, and

$$\lambda_1 \leq \rho_{\mathbf{A}}(\mathbf{x}) \leq \lambda_1 + \frac{(\lambda_p - \lambda_1)\|\mathbf{x} - \mathbf{v}_1\|_2^2}{\|\mathbf{x}\|_2^2}.$$

In vibration problems, the smallest eigenvalue corresponds to the lowest natural frequency of the vibrating system, and it is often possible on physical grounds to guess the shape of such a vibration and thereby obtain information allowing a rough approximation to $\mathbf{v}_1$; the analysis above shows that such a rough approximation to $\mathbf{v}_1$ can in fact produce a good estimate of λ_1. See also Problem 1 in Section 10.5.

(10.27) ***Example.*** The matrix **A** below arose from a vibration problem in which, on physical grounds, it is reasonable to expect the eigenvector $\mathbf{v}_1$ associated with the smallest eigenvalue λ_1 to have nonnegative entries.

$$\mathbf{A} = \begin{bmatrix} 1.7 & -1 & 0 \\ -1 & 2 & -1 \\ 0 & -1 & 2 \end{bmatrix}$$

(From the Gerschgorin circle theorem and the symmetry of **A** we can see that the eigenvalues of **A** are nonnegative. Problem 14 of Section 9.4 shows that $\mathbf{A}^{-1}$ exists and is nonnegative, and the Perron-Frobenius theorem then states that in fact $\mathbf{v}_1$ is positive. So the physical grounds have solid theoretical support.) Any $\mathbf{x}$ gives a $\rho_{\mathbf{A}}(\mathbf{x})$ that is an upper bound on λ_1; for example,

$$\rho_{\mathbf{A}}([1 \quad 1 \quad 1]^T) = 0.57$$

must be greater than or equal to λ_1. Exact calculation shows that in fact $\lambda_1 = 0.5$ and that $[1.0 \quad 1.2 \quad 0.8]^T$ is an associated eigenvector, so that $\rho_{\mathbf{A}}(\mathbf{x})$ differed from λ_1 by 0.07.

You should now be able to solve Problems 1 to 5.

Extremum Problems and Intermediate Eigenvalues

Example 10.4 introduced a quadratic form whose maximizing vector **a** (subject to $\|\mathbf{a}\|_2 = 1$) provides a new variable $y_1 = a_1x_1 + \cdots + a_px_p$ in a statistical problem; this new variable contains as much of the experimental variance as possible and is called the first *principal component* of the covariance matrix **S** defining the quadratic form $Q(\mathbf{a}) = (\mathbf{a}, \mathbf{Sa})$. We now know that this **a** maximizes $\rho_S(\mathbf{a})$ subject to $\mathbf{a} \neq \mathbf{0}$ and will in fact be an eigenvector of **S** associated with the largest eigenvalue of **S**.

If the new variable y_1 does not adequately represent the total variance V, a statistician may seek a second principal component

$$y_2 = b_1x_1 + \cdots + b_px_p$$

that is statistically independent of the new variable y_1 already found; it is known from statistics that this statistical independence is equivalent to the orthogonality condition $(\mathbf{a}, \mathbf{b}) = 0$. Thus **b** is sought to maximize $\rho_S(\mathbf{b})$ subject to both $\mathbf{b} \neq \mathbf{0}$ and $(\mathbf{a}, \mathbf{b}) = 0$, where **a** is the given solution to the original problem of maximizing $\rho_S(\mathbf{a})$ subject to $\mathbf{a} \neq \mathbf{0}$.

The foregoing problem in statistics is only one instance of how such extremum problems arise in applications. The general problem is to maximize $\rho_A(\mathbf{x})$ subject to both

$$\mathbf{x} \neq \mathbf{0} \quad \text{and} \quad (\mathbf{x}, \mathbf{v}_p) = 0,$$

where $\mathbf{v}_p$ is an eigenvector associated with the largest eigenvalue λ_p of **A** and thus maximizes $\rho_A(\mathbf{x})$ subject just to $\mathbf{x} \neq \mathbf{0}$. A solution to this problem turns out to be $\mathbf{v}_{p-1}$, an eigenvector associated with the second largest eigenvalue λ_{p-1}; this in turn leads to a further extremum problem of maximizing $\rho_A(\mathbf{x})$ subject to

$$\mathbf{x} \neq \mathbf{0}, \qquad (\mathbf{x}, \mathbf{v}_p) = 0, \quad \text{and} \quad (\mathbf{x}, \mathbf{v}_{p-1}) = 0.$$

The following classic theorem characterizes the solutions of all such problems in terms of the eigensystem of **A**.

(10.28) **Key Theorem** (Rayleigh's principle). Let $p \times p$ **A** be hermitian and have eigenvalues $\lambda_1 \leq \lambda_2 \leq \cdots \leq \lambda_p$ and an orthonormal set of associated eigenvectors $\mathbf{v}_1, \ldots, \mathbf{v}_p$. Let

S_j = the set of all $\mathbf{x} \neq \mathbf{0}$ that are orthogonal to the j eigenvectors $\mathbf{v}_p, \mathbf{v}_{p-1}, \ldots, \mathbf{v}_{p-j+1}$ associated with the j largest eigenvalues, and let

T_j = the set of all $\mathbf{x} \neq \mathbf{0}$ that are orthogonal to the j eigenvectors $\mathbf{v}_1, \mathbf{v}_2, \ldots, \mathbf{v}_j$ associated with the j smallest eigenvalues.

Then:

(a) $\rho_{\mathbf{A}}(\mathbf{x}) \geq \lambda_{j+1}$ for all $\mathbf{x}$ in T_j, and $\rho_{\mathbf{A}}(\mathbf{x}) = \lambda_{j+1}$ with $\mathbf{x}$ in T_j if and only if $\mathbf{x}$ is an eigenvector associated with λ_{j+1}. Thus $\mathbf{v}_{j+1}$ minimizes $\rho_{\mathbf{A}}(\mathbf{x})$ over T_j with minimum value λ_{j+1}.

(b) $\rho_{\mathbf{A}}(\mathbf{x}) \leq \lambda_{p-j}$ for all $\mathbf{x}$ in S_j, and $\rho_{\mathbf{A}}(\mathbf{x}) = \lambda_{p-j}$ with $\mathbf{x}$ in S_j if and only if $\mathbf{x}$ is an eigenvector associated with λ_{p-j}. Thus $\mathbf{v}_{p-j}$ maximizes $\rho_{\mathbf{A}}(\mathbf{x})$ over S_j with maximum value λ_{p-j}.

PROOF. We as usual work with $\rho_{\Lambda}(\boldsymbol{\xi}) = \rho_{\mathbf{A}}(\mathbf{x})$, where

$$\mathbf{P}^H\mathbf{A}\mathbf{P} = \boldsymbol{\Lambda} = \mathbf{diag}(\lambda_1, \ldots, \lambda_p),$$

$\mathbf{P} = [\mathbf{v}_1 \quad \cdots \quad \mathbf{v}_p]$ is unitary, and $\mathbf{x} = \mathbf{P}\boldsymbol{\xi}$, $\boldsymbol{\xi} = \mathbf{P}^H\mathbf{x}$. Recall that $\mathbf{x} = \mathbf{v}_i$ if and only if $\boldsymbol{\xi} = \mathbf{e}_i$; thus the condition $(\mathbf{x}, \mathbf{v}_i) = 0$ is equivalent to $(\boldsymbol{\xi}, \mathbf{e}_i) = 0$—that is, to $\xi_i = 0$.

(a) Clearly, $\mathbf{x}$ is in T_j if and only if $\boldsymbol{\xi} = \mathbf{P}^H\mathbf{x}$ is in $\tilde{T}_j$, the set of $\boldsymbol{\xi} \neq \mathbf{0}$ with $\xi_1 = \xi_2 = \cdots = \xi_j = 0$. For such $\boldsymbol{\xi}$, an argument like that in the proof of (b) in **Key Theorem 10.25** easily shows that $\rho_{\Lambda}(\boldsymbol{\xi})$ satisfies $\rho_{\Lambda}(\boldsymbol{\xi}) \geq \lambda_{j+1}$ and $\rho_{\Lambda}(\boldsymbol{\xi}) = \lambda_{j+1}$ if and only if $\xi_r = 0$ for all r such that $\lambda_r > \lambda_{j+1}$. Substituting $\boldsymbol{\xi} = \mathbf{P}^H\mathbf{x}$ completes the proof for (a).

(b) follows by applying (a) to $-\mathbf{A}$—Problem 8. ■

(10.29) ***Example.*** Suppose that the covariance matrix $\mathbf{S}$ of an experiment (see Example 10.4) in three variables x_1, x_2, x_3 is

$$\mathbf{S} = \begin{bmatrix} 0.4 & 0.1 & 0.1 \\ 0.1 & 0.3 & 0.2 \\ 0.1 & 0.2 & 0.3 \end{bmatrix};$$

we seek a new variable whose variance equals at least 55% of the total variance

$$V = 0.4 + 0.3 + 0.3 = 1.0.$$

According to Example 10.4 and the preceding, we take

$$y_1 = a_1x_1 + a_2x_2 + a_3x_3,$$

where $\mathbf{a}$ maximizes $\rho_{\mathbf{S}}(\mathbf{a})$ for $\mathbf{a} \neq \mathbf{0}$—that is,

$$\mathbf{a} = \frac{[1 \quad 1 \quad 1]^T}{\sqrt{3}},$$

a normalized eigenvector associated with the largest eigenvalue $\lambda_3 = 0.6$ of $\mathbf{S}$. Also, $\rho_{\mathbf{S}}(\mathbf{a}) = \lambda_3 = 0.6 > 55\% \times 1.0$, so the first principal component

$$y_1 = \frac{x_1 + x_2 + x_3}{\sqrt{3}}$$

is sufficient. Had we wanted to account for, say, 80% of the total variance V, y_1 would have been inadequate and we would have sought $\mathbf{b}$ to maximize $\rho_{\mathbf{S}}(\mathbf{b})$ subject to $\mathbf{b} \neq \mathbf{0}$ and $(\mathbf{b}, \mathbf{a}) = 0$. That is, $\mathbf{b}$ would be the eigenvector

$$\mathbf{b} = \frac{[2 \quad -1 \quad -1]^T}{\sqrt{6}}$$

associated with the second largest eigenvalue $\lambda_2 = 0.3$. The second principal component

$$y_2 = \frac{2x_1 - x_2 - x_3}{\sqrt{6}}$$

accounts for an additional $\rho_{\mathbf{S}}(\mathbf{b}) = 0.3$ of the variance, giving a total of

$$0.6 + 0.3 > 80\% \times 1.0,$$

so the first two principal components are sufficient.

PROBLEMS 10.4

▷ **1.** Use $\rho_{\mathbf{A}}$ to find lower bounds for the largest eigenvalue and upper bounds for the smallest eigenvalue of

$$\mathbf{A} = \begin{bmatrix} 0 & -1 & 0 \\ -1 & -1 & 1 \\ 0 & 1 & 0 \end{bmatrix}.$$

▷ **2.** An eigenvector associated with the lowest eigenvalue of the matrix below has the form $\mathbf{x}_\alpha = [1 \quad \alpha \quad 1]^T$. Find the exact value of α by defining the function $\mathbf{f}(\alpha) = \rho_{\mathbf{A}}(\mathbf{x}_\alpha)$ and using calculus to minimize $f(\alpha)$.

$$\mathbf{A} = \begin{bmatrix} 3 & -1 & 0 \\ -1 & 2 & -1 \\ 0 & -1 & 3 \end{bmatrix}$$

3. Prove **Key Theorem 10.25(c)** by applying **(b)** to $-\mathbf{A}$.

4. Suppose that $\mathbf{A}$ is hermitian and positive definite. Prove that $\mathbf{x}^*$ maximizes $Q(\mathbf{x}) = (\mathbf{x}, \mathbf{A}\mathbf{x})$ subject to the constraint $(\mathbf{x}, \mathbf{x}) = 1$ if and only if $\alpha\mathbf{x}^*$ minimizes $(\mathbf{x}, \mathbf{x})$ subject to the constraint $Q(\mathbf{x}) = 1$, where $\alpha = 1/(\mathbf{x}^*, \mathbf{A}\mathbf{x}^*)^{1/2}$.

▷ **5.** For each matrix $\mathbf{A}$ below, use $\rho_{\mathbf{A}}$ to obtain lower bounds on the greatest eigenvalue and upper bounds on the least eigenvalue.

(a) $\begin{bmatrix} 3 & -1 & 0 \\ -1 & 2 & -1 \\ 0 & -1 & 3 \end{bmatrix}$ (b) $\begin{bmatrix} 7 & -16 & -8 \\ -16 & 7 & 8 \\ -8 & 8 & -5 \end{bmatrix}$ (c) $\begin{bmatrix} 2 & -1 & 0 \\ -1 & 3 & -1 \\ 0 & -1 & 2 \end{bmatrix}$

6. Using $\mathbf{v}_3 = [1 \quad -1 \quad -1]^T$ as an eigenvector associated with the largest eigenvalue λ_3 of the matrix **A** in Problem 1, use ρ_{A} to obtain lower bounds on the second largest eigenvalue λ_2.

▷ **7.** Using $\mathbf{v}_1 = [1 \quad 2 \quad -1]^T$ as an eigenvector associated with the least eigenvalue λ_1 of the matrix **A** in Problem 1, use ρ_{A} to obtain upper bounds on the second smallest eigenvalue λ_2.

8. Prove **Key Theorem 10.28(b)** by applying **(a)** to $-\mathbf{A}$.

9. Much as in Example 10.29, suppose that the covariance matrix **S** of an experiment in three variables x_1, x_2, x_3 is

$$\mathbf{S} = \begin{bmatrix} 0.6 & 0.3 & 0.3 \\ 0.3 & 0.5 & 0.4 \\ 0.3 & 0.4 & 0.5 \end{bmatrix};$$

find principal components to account for (a) 80% and (b) 95% of the total variance 1.6.

▷ **10.** For each of the three matrices in Problem 5(a), (b), and (c), use the eigenvectors $\mathbf{v}_1$ and $\mathbf{v}_3$ given below as associated with the least eigenvalue λ_1 and largest eigenvalue λ_3, respectively, in order to obtain upper and lower bounds on the middle eigenvalue λ_2:
(a) $\mathbf{v}_1 = [1 \quad 2 \quad 1]^T, \quad \mathbf{v}_3 = [1 \quad -1 \quad 1]^T$
(b) $\mathbf{v}_1 = [1 \quad 1 \quad 0]^T, \quad \mathbf{v}_3 = [-2 \quad 2 \quad 1]^T$
(c) $\mathbf{v}_1 = [1 \quad 1 \quad 1]^T, \quad \mathbf{v}_3 = [1 \quad -2 \quad 1]^T$

10.5 EXTREMIZING QUADRATIC FORMS: THE MIN-MAX PRINCIPLE

Rayleigh's principle in the preceding section characterizes each eigenvalue and eigenvector of a $p \times p$ hermitian matrix **A** in terms of an extremum problem. Note, however, that this characterization of the eigenvalues/eigenvectors other than the largest and smallest requires knowledge of eigenvectors other than the one being characterized; to characterize or estimate λ_j with $1 < j < p$ we need the eigenvectors $\mathbf{v}_{j+1}, \ldots, \mathbf{v}_p$ or the eigenvectors $\mathbf{v}_1, \ldots, \mathbf{v}_{j-1}$. The object of this section is to present a characterization that allows estimation of each eigenvalue entirely independently of the rest of the eigensystem.

The Second-Largest Eigenvalue

For simplicity we consider first λ_{p-1}, the second largest of the eigenvalues

$$\lambda_1 \leq \lambda_2 \leq \cdots \leq \lambda_{p-1} \leq \lambda_p.$$

Recall that the Rayleigh quotient $\rho_{\mathbf{A}}$ satisfies $\rho_{\mathbf{A}}(\mathbf{x}) \le \lambda_p$ with equality when $\mathbf{x}$ is an eigenvector associated with λ_p. Suppose now that we restrict $\mathbf{x}$ to lie in the plane $(\mathbf{c}, \mathbf{x}) = 0$ for some arbitrary fixed $\mathbf{c}$; this constraint is of the type used to characterize λ_{p-1} in Rayleigh's principle: if $\mathbf{c} = \mathbf{v}_p$, then the maximum $M(\mathbf{c})$ of $\rho_{\mathbf{A}}(\mathbf{x})$ in this plane equals λ_{p-1} and occurs with $\mathbf{x} = \mathbf{v}_{p-1}$: $M(\mathbf{v}_p) = \lambda_{p-1}$. What happens, however, if $\mathbf{c}$ is not a multiple of $\mathbf{v}_p$?

We claim that $M(\mathbf{c})$, the maximum of $\rho_{\mathbf{A}}(\mathbf{x})$ as $\mathbf{x}$ varies subject to $\mathbf{x}$ being in the plane $\pi_{\mathbf{c}}$ of $\mathbf{x}$ satisfying $(\mathbf{c}, \mathbf{x}) = 0$, for a given fixed $\mathbf{c}$, always satisfies $M(\mathbf{c}) \ge \lambda_{p-1}$. Since $M(\mathbf{c})$ is the *maximum* of $\rho_{\mathbf{A}}$ as $\mathbf{x}$ varies, to show $M(\mathbf{c}) \ge \lambda_{p-1}$ only requires us to produce one $\mathbf{x}_0$ in $\pi_{\mathbf{c}}$ with $\rho_{\mathbf{A}}(\mathbf{x}_0) \ge \lambda_{p-1}$. To do this, consider three cases: (a) $\mathbf{v}_p$ is orthogonal to $\mathbf{c}$; (b) $\mathbf{v}_{p-1}$ is orthogonal to $\mathbf{c}$; and (c) neither $\mathbf{v}_p$ nor $\mathbf{v}_{p-1}$ is orthogonal to $\mathbf{c}$.

(a) If $\mathbf{v}_p$ is orthogonal to $\mathbf{c}$, then $\mathbf{x}_0 = \mathbf{v}_p$ is in $\pi_{\mathbf{c}}$, so

$$\rho_{\mathbf{A}}(\mathbf{x}_0) = \rho_{\mathbf{A}}(\mathbf{v}_p) = \lambda_p \ge \lambda_{p-1},$$

as asserted.

(b) If $\mathbf{v}_{p-1}$ is orthogonal to $\mathbf{c}$, then $\mathbf{x}_0 = \mathbf{v}_{p-1}$ is in $\pi_{\mathbf{c}}$, so

$$\rho_{\mathbf{A}}(\mathbf{x}_0) = \rho_{\mathbf{A}}(\mathbf{v}_{p-1}) = \lambda_{p-1},$$

as asserted.

(c) If neither $\mathbf{v}_p$ nor $\mathbf{v}_{p-1}$ is orthogonal to $\mathbf{c}$, then define

$$\mathbf{x}_0 = \mathbf{v}_{p-1} - \alpha\mathbf{v}_p \quad \text{with} \quad \alpha = \frac{(\mathbf{c}, \mathbf{v}_{p-1})}{(\mathbf{c}, \mathbf{v}_p)} \ne 0;$$

this makes $(\mathbf{c}, \mathbf{x}_0) = 0$ and $\mathbf{x}_0 \ne \mathbf{0}$. Direct calculation gives

$$\rho_{\mathbf{A}}(\mathbf{x}_0) = \lambda_{p-1} + \frac{\|\alpha\mathbf{v}_p\|_2^2(\lambda_p - \lambda_{p-1})}{\|\mathbf{x}_0\|_2^2} \ge \lambda_{p-1},$$

as asserted.

Thus $M(\mathbf{c}) \ge \lambda_{p-1}$ for all $\mathbf{c}$.

The next step is to minimize $M(\mathbf{c})$ over all possible vectors $\mathbf{c}$. We know already that $M(\mathbf{c}) \ge \lambda_{p-1}$ for all $\mathbf{c}$. If we take $\mathbf{v}_p$ as $\mathbf{c}$, then $M(\mathbf{v}_p) = \lambda_{p-1}$ as noted above. In other words,

$$\lambda_{p-1} = \min M(\mathbf{c}) \qquad \text{as } \mathbf{c} \text{ varies.}$$

Remembering the definition of $M(\mathbf{c})$ gives

$$\lambda_{p-1} = \min_{\mathbf{c}} \max_{(\mathbf{c},\mathbf{x})=0} \rho_{\mathbf{A}}(\mathbf{x}). \tag{10.30}$$

The important fact about (10.30) is that it characterizes λ_{p-1} without reference to the rest of the eigensystem of $\mathbf{A}$.

(10.31) ***Example.*** The 3×3 matrix $\mathbf{S}$ in Example 10.29 has strictly positive entries and is irreducible, so the Perron-Frobenius theorem (9.34) tells us that the largest (in this case, largest in magnitude as well as largest including sign) eigenvalue λ_3 is positive and has an associated eigenvector of strictly positive entries. Suppose that we want information on λ_2 rather than λ_3; we know that $\lambda_2 = \min M(\mathbf{c})$, where $M(\mathbf{c})$ is the maximum of $\rho_S(\mathbf{x})$ subject to $(\mathbf{c}, \mathbf{x}) = 0$. If we take $\mathbf{c} = [1 \quad 0 \quad 1]^T$, for example, then $\mathbf{x}$ is restricted to the form $[\alpha \quad \beta \quad -\alpha]^T$ from which

$$\rho_S(\mathbf{x}) = \frac{(0.5\alpha^2 + 0.3\beta^2 - 0.2\alpha\beta)}{(2\alpha^2 + \beta^2)};$$

if we maximize this in terms of α and β we find that

$$M([1 \quad 0 \quad 1]^T) = 0.35.$$

Since $\lambda_2 = \min M(\mathbf{c})$, we have $\lambda_2 \leq 0.35$ as an estimate of and bound on λ_2. In fact, $\lambda_2 = 0.3$.

The Min-Max Principle in General

The preceding treated just the characterization of λ_{p-1}. The general case is similar.

(10.32) **Key Theorem** (min-max principle). Let $p \times p$ $\mathbf{A}$ be hermitian and have eigenvalues $\lambda_1 \leq \lambda_2 \leq \cdots \leq \lambda_p$ and an orthonormal set of associated eigenvectors $\mathbf{v}_1, \ldots, \mathbf{v}_p$. Then for $0 \leq r \leq p - 1$:

(a)
$$\lambda_{p-r} = \min_{\mathbf{c}_1, \ldots, \mathbf{c}_r} \max_{(\mathbf{c}_i, \mathbf{x}) = 0} \rho_A(\mathbf{x}),$$

where the minimum is with respect to all sets of exactly r arbitrary vectors $\mathbf{c}_1, \ldots, \mathbf{c}_r$, while the maximum is with respect to all $\mathbf{x} \neq \mathbf{0}$ satisfying $(\mathbf{c}_i, \mathbf{x}) = 0$ for $1 \leq i \leq r$.

(b)
$$\lambda_{r+1} = \max_{\mathbf{c}_1, \ldots, \mathbf{c}_r} \min_{(\mathbf{c}_i, \mathbf{x}) = 0} \rho_A(\mathbf{x}),$$

where the maximum is with respect to all sets of exactly r arbitrary vectors $\mathbf{c}_1, \ldots, \mathbf{c}_r$, while the minimum is with respect to all $\mathbf{x} \neq \mathbf{0}$ satisfying $(\mathbf{c}_i, \mathbf{x}) = 0$ for $1 \leq i \leq r$.

PROOF. As usual, we diagonalize and treat $\tilde{Q}(\xi) = Q(\mathbf{x})$ and $\rho_\Lambda(\xi) = \rho_A(\mathbf{x})$ with

$$\mathbf{P}^H\mathbf{A}\mathbf{P} = \mathbf{\Lambda} = \mathbf{diag}(\lambda_1, \ldots, \lambda_p),$$

$$\mathbf{P} = [\mathbf{v}_1 \quad \cdots \quad \mathbf{v}_p] \text{ unitary}, \quad \mathbf{P}\xi = \mathbf{x}, \quad \text{and} \quad \xi = \mathbf{P}^H\mathbf{x}.$$

Since $\mathbf{P}$ is unitary, the condition $(\mathbf{c}_i, \mathbf{x}) = 0$ is equivalent to the condition $(\tilde{\mathbf{c}}_i, \xi) = 0$ with $\tilde{\mathbf{c}}_i = \mathbf{P}^H\mathbf{c}_i$.

(a) The proof follows that leading to (10.30) for λ_{p-1}. For any set of $\tilde{\mathbf{c}}_1, \ldots, \tilde{\mathbf{c}}_r$, denote by $\tilde{\Pi}_\mathbf{c}$ the set of all ξ satisfying $(\tilde{\mathbf{c}}_i, \xi) = 0$ for $1 \leq i \leq r$

and denote by $M(\mathbf{c})$ the maximum as ξ varies of $\rho_\Lambda(\xi)$ for ξ in $\tilde{\Pi}_\mathbf{c}$ where the $\tilde{\mathbf{c}}_i = \mathbf{P}^H\mathbf{c}_i$ are fixed. We are done if we can show that the minimum of $M(\mathbf{c})$ over all possible choices of the $\tilde{\mathbf{c}}_i = \mathbf{P}^H\mathbf{c}_i$ equals λ_{p-r}. We first show that $M(\mathbf{c}) \geq \lambda_{p-r}$; since $M(\mathbf{c})$ is the *maximum* of $\rho_\Lambda(\xi)$ over $\tilde{\Pi}_\mathbf{c}$, to do this we need only find ξ in $\tilde{\Pi}_\mathbf{c}$ with $\rho_\Lambda(\xi) \geq \lambda_{p-r}$, which we now do. We claim that there is a ξ in $\tilde{\Pi}_\mathbf{c}$ with $\xi_1 = \xi_2 = \cdots = \xi_{p-r-1} = 0$, since being in $\tilde{\Pi}_\mathbf{c}$ is equivalent to satisfying r homogeneous equations in the remaining $r+1$ variables $\xi_{p-r}, \ldots, \xi_p$; having more unknowns than homogeneous equations means that there are some nonleading variables and hence some nonzero solutions. For any such nonzero solution we have

$$\begin{aligned}\rho_\Lambda(\xi) &= \frac{\lambda_1|\xi_1|^2 + \cdots + \lambda_p|\xi_p|^2}{|\xi_1|^2 + \cdots + |\xi_p|^2}\\ &= \frac{\lambda_{p-r}|\xi_{p-r}|^2 + \cdots + \lambda_p|\xi_p|^2}{|\xi_{p-r}|^2 + \cdots + |\xi_p|^2}\\ &\geq \lambda_{p-r},\end{aligned}$$

where the final inequality follows by replacing each λ_i for $i \geq p-r$ by the smaller λ_{p-r}. This completes the argument that $M(\mathbf{c}) \geq \lambda_{p-r}$. By taking $\tilde{\mathbf{c}}_i = \mathbf{e}_{p-i+1}$ for $1 \leq i \leq r$ we force

$$\xi_{p-r+1} = \cdots = \xi_p = 0$$

and obtain $M(\mathbf{c}) = \lambda_{p-r}$ from Rayleigh's principle. Thus λ_{p-r} equals the minimum of $M(\mathbf{c})$, as claimed, completing the proof of (a).

(b) follows by applying (a) to $-\mathbf{A}$—Problem 2. ■

You should now be able to solve Problems 1 to 5.

Comparing Eigenvalues of Different Matrices

The min-max characterization of eigenvalues independent of the rest of the eigensystem leads to a fundamental result on the relationship of eigenvalues of different matrices; this result provides information on how the frequencies of vibrating systems change when parameters in the system change—for example, we can conclude mathematically that the pitch of a guitar string rises when the string is tightened.

(10.33) ***Theorem*** (comparing eigenvalues). Suppose that the $p \times p$ hermitian matrices $\mathbf{A}$ and $\mathbf{B}$ satisfy $(\mathbf{x}, \mathbf{Ax}) \leq (\mathbf{x}, \mathbf{Bx})$ for all $\mathbf{x}$, and let $\lambda_1 \leq \cdots \leq \lambda_p$ and $\mu_1 \leq \cdots \leq \mu_p$ be the eigenvalues of $\mathbf{A}$ and $\mathbf{B}$, respectively. Then $\lambda_i \leq \mu_i$ for $1 \leq i \leq p$.

PROOF. Clearly $\rho_A(\mathbf{x}) \le \rho_B(\mathbf{x})$ for all $\mathbf{x}$. Therefore, the maximum of $\rho_A(\mathbf{x})$ with $\mathbf{x}$ restricted to any set Π is less than or equal to the maximum of $\rho_B(\mathbf{x})$ with $\mathbf{x}$ restricted to that same set. Similarly, the minimum of these maxima as Π varies will be less for $\mathbf{A}$ than for $\mathbf{B}$. That is,

$$\min_{\mathbf{c}_1,\ldots,\mathbf{c}_r} \max_{(\mathbf{c}_i,\mathbf{x})=0} \rho_A(\mathbf{x}) \le \min_{\mathbf{c}_1,\ldots,\mathbf{c}_r} \max_{(\mathbf{c}_i,\mathbf{x})=0} \rho_B(\mathbf{x}).$$

The left-hand term above equals λ_{p-r} while the right-hand equals μ_{p-r}. ∎

(10.34) ***Example.*** In Section 8.1 we showed that the fundamental frequencies $\omega/2\pi$ of the oscillating-masses example of Section 2.5 can be obtained from $\omega^2 = \lambda$, where λ is any eigenvalue of

$$\mathbf{B} = \mathbf{D}^{-1}\mathbf{K}\mathbf{D}^{-1} \quad \text{and} \quad \mathbf{D}^2 = \mathbf{M}$$

with $\mathbf{M}$ and $\mathbf{K}$ as in (2.29). The theorem above can be used to see how these frequencies vary with the parameters in the model; suppose, for example, that the spring constant k_1 for the first spring in the model increases from k_1 to $k_1' = k_1 + k$. Then the matrix changes from $\mathbf{B}$ to $\mathbf{B}'$:

$$\mathbf{B}' = \begin{bmatrix} \dfrac{k_1 + k + k_2}{m_1} & \dfrac{-k_2}{(m_1 m_2)^{1/2}} \\ \dfrac{-k_2}{(m_1 m_2)^{1/2}} & \dfrac{k_2}{m_2} \end{bmatrix} = \mathbf{B} + \begin{bmatrix} \dfrac{k}{m_1} & 0 \\ 0 & 0 \end{bmatrix}.$$

Clearly, the quadratic form for $\mathbf{B}'$ exceeds that for $\mathbf{B}$ by $(k/m_1)x_1^2$. According to the theorem comparing eigenvalues, those of $\mathbf{B}'$ exceed those of $\mathbf{B}$. That is, the natural frequencies of the oscillating masses increase if the first spring becomes stiffer.

PROBLEMS 10.5

▷ **1.** Use the min-max principle to obtain an upper bound for the second largest eigenvalue λ_3 of $\mathbf{A}$ below, given that the largest eigenvalue λ_4 has an associated eigenvector with strictly positive entries (by the Perron-Frobenius theorem):

$$\mathbf{A} = \begin{bmatrix} 1 & 3 & 1 & 2 \\ 3 & 2 & 4 & 1 \\ 1 & 4 & 2 & 1 \\ 2 & 1 & 1 & 3 \end{bmatrix}.$$

2. Prove **Key Theorem 10.32(b)** by applying **(a)** to $-\mathbf{A}$.

3. Use $\mathbf{c} = [1 \quad 1 \quad 1]^T$ in the min-max principle to obtain an upper bound on the second largest eigenvalue of the matrix $\mathbf{A}$ in Example 10.29.

▷ **4.** Do as asked in Problem 3, except use $\mathbf{c} = [1 \quad 2 \quad 3]^T$.

5. Using $[1 \quad 1 \quad -1]^T$ as near $\mathbf{v}_1$ and $[1 \quad 0 \quad 0]^T$ as near $\mathbf{v}_3$, use the min-max (and max-min) principle to obtain upper and lower bounds on the middle eigenvalue of the matrix in Problem 1 of Section 10.4.

6. As in Example 10.34, determine what happens to the frequencies of the oscillating-masses system when the second spring constant k_2 increases.

▷ 7. Without finding the eigenvalues $\lambda_1 \le \lambda_2 \le \lambda_3$ of **A** or $\mu_1 \le \mu_2 \le \mu_3$ of **B**, show that $\lambda_i \le \mu_i$ for $i = 1, 2, 3$, if

$$\mathbf{A} = \begin{bmatrix} -1 & 2 & -1 \\ 2 & 3 & 2 \\ -1 & 2 & -5 \end{bmatrix} \quad \text{and} \quad \mathbf{B} = \begin{bmatrix} 2 & 3 & -2 \\ 3 & 9 & 6 \\ -2 & 6 & 0 \end{bmatrix}.$$

𝔐 8. Use MATLAB or similar software to find the eigenvalues in Problem 7 and verify the assertion.

9. Prove that each of the eigenvalues of a hermitian matrix is nondecreasing as the main-diagonal entries of the matrix increase.

▷ 10. For each of the four eigenvalues of the matrix **A** below, find an interval in which the eigenvalue must lie by using the fact that the eigenvalues of the matrix **E** below are (approximately) ± 0.3 and ± 0.8:

$$\mathbf{A} = \begin{bmatrix} 1 & -1 & 0 & 0 \\ -1 & 1.4 & -1 & 0 \\ 0 & -1 & 2 & -1 \\ 0 & 0 & -1 & 1.8 \end{bmatrix} \quad \text{and} \quad \mathbf{E} = \begin{bmatrix} 0 & 1 & 0 & 0 \\ 1 & 0 & 1 & 0 \\ 0 & 1 & 0 & 1 \\ 0 & 0 & 1 & 0 \end{bmatrix}.$$

𝔐 11. Use MATLAB or similar software to verify that the intervals found in Problem 10 actually contain the eigenvalues of **A**.

12. Under the hypotheses of Theorem 10.33, *use Rayleigh's principle* rather than the min-max principle to show that $\lambda_1 \le \mu_1$ and $\lambda_p \le \mu_p$, and explain why Rayleigh's principle cannot be used to prove the similar results for the intermediate eigenvalues λ_i and μ_i.

10.6 *MISCELLANEOUS PROBLEMS*

PROBLEMS 10.6

1. Consider a curve (a conic section) defined by a quadratic function including linear and constant terms:

$$ax_1^2 + 2kx_1x_2 + bx_2^2 + 2cx_1 + 2dx_2 + e = 0.$$

(a) Show that the substitution $x_1 = \eta_1 + x_{10}$, $x_2 = \eta_2 + x_{20}$, replaces this equation by

$$a\eta_1^2 + 2k\eta_1\eta_2 + b\eta_2^2 + 2C\eta_1 + 2D\eta_2 + E = 0,$$

where

$$C = ax_{10} + kx_{20} + c, \qquad D = kx_{10} + bx_{20} + d,$$

and E is for you to determine.

(b) Show that if either $ab - k^2 \neq 0$ or $ab - k^2 = 0$ with $ad = bc$, then x_{10} and x_{20} can be chosen so that all first-degree terms vanish and we are left with

$$a\eta_1^2 + 2k\eta_1\eta_2 + b\eta_2^2 + E = 0,$$

whose graph can be analyzed by the method of Section 10.2.

(c) Show that, if $a \neq 0$ and $ab = k^2$ and $ad \neq bc$, then x_{10} and x_{20} can be chosen so that the equation becomes

$$a\eta_1^2 + 2k\eta_1\eta_2 + b\eta_2^2 + 2D\eta_2 + E = 0.$$

(d) Show that this last equation can be reduced by a further change of variables to $\alpha\xi_1^2 + 2\gamma\xi_2 = f$, an equation for a parabola.

▷ **2.** Use Problem 1 to analyze the nature of the curves defined by the equations below, and sketch the curves:

(a) $5x_1^2 - 8x_1x_2 + 5x_2^2 - 18x_1 + 18x_2 + 8 = 0$

(b) $9x_1^2 - 12x_1x_2 + 4x_2^2 - 42x_1 - 2x_2 + 7 = 0$

3. In some applications it is important to find a single nonsingular matrix $\mathbf{S}$ such that $\mathbf{S}^H\mathbf{A}\mathbf{S}$ and $\mathbf{S}^H\mathbf{B}\mathbf{S}$ are both diagonal, given $p \times p$ $\mathbf{A}$ and $\mathbf{B}$. Suppose that $\mathbf{A}$ and $\mathbf{B}$ are hermitian and $\mathbf{B}$ is positive definite.

(a) Show that there exists a unitary matrix $\mathbf{P}$ and

$$\mathbf{D} = \mathbf{diag}(d_1, \ldots, d_p)$$

with all $d_i > 0$ such that

$$\mathbf{B} = (\mathbf{PD})(\mathbf{PD})^H = \mathbf{PD}^2\mathbf{P}^H \quad \text{and} \quad \mathbf{Q}^H\mathbf{B}\mathbf{Q} = \mathbf{I},$$

where $\mathbf{Q} = \mathbf{PD}^{-1}$.

(b) Show that there exists a unitary matrix $\mathbf{R}$ such that

$$\mathbf{R}^H(\mathbf{Q}^H\mathbf{A}\mathbf{Q})\mathbf{R} = \mathbf{\Lambda} \text{ diagonal}.$$

(c) Let $\mathbf{S} = \mathbf{QR}$ and show that

$$\mathbf{S}^H\mathbf{A}\mathbf{S} = \mathbf{\Lambda} \quad \text{while} \quad \mathbf{S}^H\mathbf{B}\mathbf{S} = \mathbf{I},$$

as desired.

4. In Problem 3, let

$$\mathbf{S} = [\mathbf{s}_1 \quad \cdots \quad \mathbf{s}_p] \quad \text{and} \quad \mathbf{\Lambda} = \mathbf{diag}(\lambda_1, \ldots, \lambda_p).$$

Show that λ_i and $\mathbf{s}_i$ solve the *generalized eigenvalue problem* $\mathbf{A}\mathbf{s}_i = \lambda_i\mathbf{B}\mathbf{s}_i$.

5. As in Problem 3, simultaneously reduce $\mathbf{A}$ and $\mathbf{B}$ to diagonal form, where

$$\mathbf{A} = \begin{bmatrix} 75 & 35 \\ 35 & -117 \end{bmatrix}, \qquad \mathbf{B} = \begin{bmatrix} 5 & -3 \\ -3 & 5 \end{bmatrix}.$$

▷ **6.** Suppose that $\mathbf{A}$ is $p \times p$, hermitian, and positive definite and that $\mathbf{B}$ is $p \times q$.
(a) Prove that $\mathbf{A}' = \mathbf{B}^H\mathbf{A}\mathbf{B}$ is positive semidefinite.
(b) Prove that $\mathbf{A}'$ is positive definite if and only if the rank of $\mathbf{B}$ equals q.

7. Prove that every principal submatrix of a hermitian positive definite matrix is hermitian positive definite.

▷ **8.** Suppose that $\mathbf{A}$, $\mathbf{B}$, and $\mathbf{C}$ are $p \times p$, hermitian, and positive definite. Prove that if

$$\det(\mathbf{A} + \lambda\mathbf{B} + \lambda^2\mathbf{C}) = 0,$$

then the real part of λ is negative.

▷ **9.** Suppose that $\mathbf{A}$ and $\mathbf{B}$ are $p \times p$ and hermitian, and that $\mathbf{B}$ is positive definite. Prove that the eigenvalues of $\mathbf{BA}$ are real.

▷ **10.** Suppose that $\mathbf{B}$ and $\mathbf{C}$ are $p \times p$ and hermitian, that $\mathbf{B}$ is positive definite, and that $\mathbf{C}$ is positive semidefinite. Show that:
(a) $\mathbf{B} + \mathbf{C}$ is positive definite.
(b) $\det \mathbf{B} \leq \det(\mathbf{B} + \mathbf{C})$.
(c) $\mathbf{B}^{-1} - (\mathbf{B} + \mathbf{C})^{-1}$ is positive semidefinite.

11. Find a diagonal quadratic form $\tilde{Q}(\xi) = Q(\mathbf{x})$ for each quadratic form $Q(\mathbf{x})$ below and sketch the level surface $\tilde{Q}(\xi) = 1$.
(a) $5x_1^2 + 6x_2^2 + 7x_3^2 - 4x_1x_2 - 4x_2x_3$
(b) $3x_1^2 + 2x_2^2 + 2x_3^2 + 2x_1x_2 + 4x_2x_3 + 2x_3x_1$
(c) $x_1^2 + x_2^2 + x_3^2 + 4x_1x_2 + 4x_2x_3 + 4x_3x_1$

▷ **12.** (a) If $\mathbf{A}$ is $p \times p$, hermitian, and positive definite, prove that

$$\det \mathbf{A} \leq a_{11}a_{22} \cdots a_{pp},$$

where $a_{ii} = \langle\mathbf{A}\rangle_{ii}$.
(b) If $\mathbf{B}$ is $p \times p$ with $\langle\mathbf{B}\rangle_{ij} = b_{ij}$, prove that

$$|\det \mathbf{B}|^2 \leq \prod_{i=1}^{p}\left(\sum_{j=1}^{p} |b_{ij}|^2\right).$$

This is known as *Hadamard's inequality*. The absolute value of the determinant of a matrix can be interpreted as the volume in p-dimensional space of the solid whose edges are described by the row vectors forming the rows of the matrix. Hadamard's inequality then says that this volume is less than or equal to that of the p-dimensional *rectangular* solid whose sides have the same lengths. Verify this for $p = 2, 3$ from your knowledge of the geometry of parallelograms and parallelepipeds in two and three dimensions. Under what conditions is equality attained in Hadamard's inequality? Give an independent proof of the equality in this case. Show that if $b = \max|b_{ij}|$, then

$$|\det \mathbf{B}| \leq b^p p^{p/2}.$$

13. Prove that $\mathbf{x}(t)$ tends to zero as t tends to plus infinity if $\dot{\mathbf{x}} = \mathbf{Ax}$ and $\mathbf{A}$ is negative definite.

▷ **14.** $\mathbf{A}$ is said to be *idempotent* if $\mathbf{A}^2 = \mathbf{A}$. Prove:
  (a) Each eigenvalue of an idempotent matrix is either 0 or 1.
  (b) The only nonsingular idempotent matrix is the identity matrix.
  (c) A necessary and sufficient condition that a *hermitian* $p \times p$ matrix be idempotent is that k of its eigenvalues equal 1, and the remaining $p - k$ equal zero, where k is the rank of the matrix.
  (d) The trace of a hermitian idempotent matrix is equal to its rank.

15. Suppose that n measurements are represented by a vector $\mathbf{x} = [x_1, \ldots, x_n]^T$. The mean and variance of the measurements are given by

$$\bar{x} = \frac{1}{n}\sum_{i=1}^{n} x_i, \qquad s^2 = \frac{1}{n}\sum_{i=1}^{n} (x_i - \bar{x})^2.$$

Show that

$$s^2 = (\mathbf{x}, \{\mathbf{I} - n^{-1}\mathbf{J}\}\mathbf{x}),$$

where $\mathbf{J}$ is a square matrix, all of whose elements are unity, and prove that $\mathbf{I} - n^{-1}\mathbf{J}$ is idempotent (see Problem 14). (Applications of idempotent matrices occur in statistics and regression theory.)

16. Extend Theorem 10.19(b) on definiteness and elimination by showing that, for a real symmetric matrix, the numbers of positive, of zero, and of negative pivots equal the numbers of positive, zero, and negative eigenvalues.

▷ **17.** Prove that:
  (a) If $\mathbf{A}$ is hermitian and positive definite, then there is a hermitian and positive definite $\mathbf{B}$ with $\mathbf{B}^2 = \mathbf{A}$.
  (b) If $\mathbf{A}$ is hermitian and positive semidefinite, then there is a hermitian and positive semidefinite $\mathbf{B}$ with $\mathbf{B}^2 = \mathbf{A}$.
  (c) $\mathbf{B}$ in both (a) and (b) is unique.

▷ **18.** Prove that:
  (a) $\{\exp(\mathbf{B})\}^H = \exp(\mathbf{B}^H)$.
  (b) $\exp(i\mathbf{A})$ is unitary if $\mathbf{A}$ is hermitian.
  (c) If $\mathbf{U}$ is unitary, then there exists a hermitian $\mathbf{A}$ with $\mathbf{U} = \exp(i\mathbf{A})$.
  (d) If $\mathbf{A}$ is square, then there exist hermitian positive semidefinite $\mathbf{M}$ and $\mathbf{N}$ and unitary $\mathbf{U}$ and $\mathbf{V}$ such that $\mathbf{A} = \mathbf{UM} = \mathbf{NV}$. {Part (d) is known as the *polar representation* of $\mathbf{A}$ by analogy with $a = r\exp(i\theta)$ for each complex number a. $\mathbf{M}$ and $\mathbf{N}$ are unique; if $\mathbf{A}$ is nonsingular, then $\mathbf{U} = \mathbf{V}$ and this matrix is unique also.}

19. Suppose that $p \times p$ $\mathbf{A}$ is hermitian, that $\mathbf{B}$ is the real matrix formed from the real parts of the entries of $\mathbf{A}$, and that $i\mathbf{C}$ is the pure imaginary matrix formed from the imaginary parts of the entries of $\mathbf{A}$—that is, $\mathbf{A} = \mathbf{B} + i\mathbf{C}$. Define the

$2p \times 2p$ matrix $\mathscr{A}$ as

$$\mathscr{A} = \begin{bmatrix} \mathbf{B} & \mathbf{C} \\ -\mathbf{C} & \mathbf{B} \end{bmatrix}.$$

(a) Show that $\mathscr{A}$ is real symmetric.

(b) Show that the quadratic form generated by $\mathbf{A}$ on $\mathbb{C}^p$ and the quadratic form generated by $\mathscr{A}$ on $\mathbb{R}^{2p}$ are equivalent in the sense that, for $\mathbf{u}$ and $\mathbf{v}$ in $\mathbb{R}^p$,

$$(\{\mathbf{u} + i\mathbf{v}\}, \mathbf{A}\{\mathbf{u} + i\mathbf{v}\}) = (\mathbf{x}, \mathscr{A}\mathbf{x}) \qquad \text{if} \quad \mathbf{x} = [\mathbf{u}^T \quad \mathbf{v}^T]^T.$$

(c) Describe how this could be used to find the eigenvalues and eigenvectors of complex hermitian $\mathbf{A}$ without computing with complex numbers.

20. Rayleigh's principle is stated for hermitian matrices $\mathbf{A}$.

(a) Show that for $\alpha \neq 0$ it is false for the non-hermitian matrix

$$\mathbf{A} = \begin{bmatrix} 0 & \alpha \\ 0 & 0 \end{bmatrix}.$$

(b) Although their quadratic forms over $\mathbb{R}^2$ are identical, show that $\mathbf{A}$ above and $(\mathbf{A} + \mathbf{A}^T)/2$ do not have the same eigenvalues.

21. Suppose that $\mathbf{A}$ and $\mathbf{B}$ are hermitian and $\mathbf{B}$ is positive definite; define

$$_{\mathbf{B}}\rho_{\mathbf{A}}(\mathbf{x}) = \frac{(\mathbf{x}, \mathbf{Ax})}{(\mathbf{x}, \mathbf{Bx})}.$$

Extend Rayleigh's principle to apply to this generalization of the Rayleigh quotient, where now the eigenvalues characterized are generalized eigenvalues λ_i with $\mathbf{Av}_i = \lambda_i\mathbf{Bv}_i$ (see Problems 3 and 4).

22. Suppose that $\mathbf{A}$ is hermitian; prove that the 2-norm of $\mathbf{A}$ equals the spectral radius of $\mathbf{A}$.

11

Linear programming

This chapter applies matrix methods to the solution of linear programs, one of the largest areas where matrices are used in business as well as science and engineering. Information on both theory and methods is presented, including the famous simplex method and its recent challenge by Karmarkar. ***Key Theorems 11.37****, **11.44**, **11.50**, and **11.53** are fundamental.*

11.1 ANALYSIS OF A SIMPLE EXAMPLE

In a wide variety of economic, political, social, and scientific operations, situations arise in which one wants to minimize or maximize some quantity that measures the efficiency or some other important aspect of an activity: total output, cost, profit, and the like. Many optimization problems of this type are known as *mathematical programming* problems or *mathematical programs.* This chapter examines a special but important subclass of mathematical programs involving only *linear* equations and inequalities; this section begins with a model problem from Section 2.7.

A Model Problem

Recall the production-planning model from Section 2.7 and its problem of how to allocate a plant's resources (limited available time on three types of machines) between two products so as to maximize profits. The analysis there produced a linear program (2.54), (2.55):

(11.1)

$$\text{maximize } M = 40x_1 + 60x_2$$

subject to the constraints

$$\begin{aligned} 2x_1 + x_2 &\le 70 \\ x_1 + x_2 &\le 40 \\ x_1 + 3x_2 &\le 90 \\ x_1 &\ge 0 \\ x_2 &\ge 0. \end{aligned}$$

This was rewritten in matrix notation as (2.56), (2.57), (2.58):

(11.2) maximize $M = \mathbf{c}^T\mathbf{x}$

subject to the constraints

$\mathbf{Ax} \le \mathbf{b}$, $\mathbf{x} \ge \mathbf{0}$, where

$$\mathbf{A} = \begin{bmatrix} 2 & 1 \\ 1 & 1 \\ 1 & 3 \end{bmatrix}, \quad \mathbf{b} = \begin{bmatrix} 70 \\ 40 \\ 90 \end{bmatrix}, \quad \mathbf{c} = \begin{bmatrix} 40 \\ 60 \end{bmatrix}, \quad \mathbf{x} = \begin{bmatrix} x_1 \\ x_2 \end{bmatrix}.$$

The optimal solution $\mathbf{x} = [15 \quad 25]^T$ was obtained geometrically in (2.59) by examining the constraint set (those $\mathbf{x}$ satisfying the constraints) and observing that the problem was to find the greatest value of M for which the straight-line graph of $\mathbf{c}^T\mathbf{x} = M$ intersected the constraint set:

(11.3)

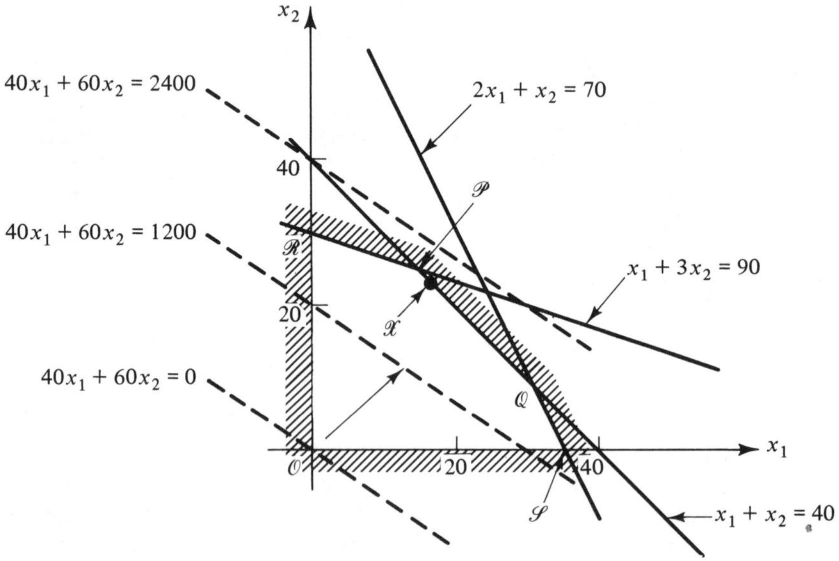

$\mathscr{O}$, $\mathscr{R}$, $\mathscr{P}$, $\mathscr{Q}$, $\mathscr{S}$ are vertices relevant to the simplex method; $\mathscr{X}$ appears in Karmarkar's method.

It is geometrically obvious that solutions will always occur at vertices of the constraint set, no matter what $\mathbf{c}$ might equal: a last point at which the graph of $\mathbf{c}^T\mathbf{x} = M$ intersects the constraint set will always be a vertex. *Thus a possible approach to solving linear programs is simply to evaluate* $\mathbf{c}^T\mathbf{x}$ *at each vertex in the knowledge that vertices giving the largest value provide a solution.* There are at least two problems with this: (1) It may be that $\mathbf{c}^T\mathbf{x}$ can be made arbitrarily large while satisfying the constraints, and this might not be detected; and (2) when one generalizes from two variables to q, the number of vertices grows rapidly with the number of variables and constraints, making a comprehensive examination of the vertices far too expensive computationally.

The famous *simplex method* uses the basic idea of examining the vertices but deals with the two difficulties above by (1) implementing a guaranteed approach to detecting unboundedness; and (2) avoiding examining vertices with smaller values of $\mathbf{c}^T\mathbf{x}$ than have already been found. Thus, while the random method above would test all five vertices $\mathcal{O}$, $\mathcal{S}$, $\mathcal{Q}$, $\mathcal{P}$, $\mathcal{R}$ in (11.3), the simplex method starting at $\mathcal{O}$ might examine $\mathcal{O}$, $\mathcal{R}$, $\mathcal{P}$ in order or $\mathcal{O}$, $\mathcal{S}$, $\mathcal{Q}$, $\mathcal{P}$ in order before detecting that $\mathcal{P}$ is the solution; in fact, it uses the first order $\mathcal{O}$, $\mathcal{R}$, $\mathcal{P}$.

For approximately 40 years, the simplex method was almost unquestionably the method of choice for solving linear programs. This changed dramatically in the middle 1980s with the startling invention of *Karmarkar's method* and other *interior methods* that grew from it. As of this writing, careful implementations of Karmarkar's method and its cousins are claimed by some to be several times faster than the simplex method *on special classes of highly structured problems.* Research on these interior methods and testing of their implementations is being carried out throughout the world; what is learned from this work may totally change this chapter in the next edition. For the present, we attempt only to indicate what is involved in the new methods. The basic difference from what we have so far described is that the interior methods do not simply look at vertices and move from one to another along edges; rather, they move through the interior of the constraint set as will be illustrated later in this section.

Introducing Slack Variables

Before demonstrating the simplex and interior methods, we transform the linear program into a form more amenable to attack by matrix methods. Most of what has been treated in this book involves *equalities* rather than the *inequalities* of (11.2); our powerful computational tools—row operations, elimination, back-substitution, and the like—have been developed for equations. To take advantage of these tools, we reformulate the linear program (11.2) in such a way that equalities appear except for the extremely simple inequalities $\mathbf{x} \geq \mathbf{0}$.

Introduce nonnegative *slack variables* x_3, x_4, x_5 into the first three inequalities of (11.1) so that (11.1) becomes

(11.4)
$$\begin{aligned} &\text{maximize } M = 40x_1 + 60x_2 \\ &\text{subject to the constraints} \\ &2x_1 + x_2 + x_3 = 70 \\ &x_1 + x_2 + x_4 = 40 \\ &x_1 + 3x_2 + x_5 = 90 \\ &x_j \geq 0 \quad (j = 1, 2, \ldots, 5). \end{aligned}$$

The nonnegativity of x_3, x_4, x_5 makes (11.4) equivalent to (11.1). In matrix notation, we *extend* the matrices **A**, **x**, and **c** to allow for five x_i:

(11.5)
$$\text{maximize } M = \mathbf{c}_e^T \mathbf{x}_e$$

subject to the constraints

$$\mathbf{A}_e \mathbf{x}_e = \mathbf{b}, \quad \mathbf{x}_e \geq \mathbf{0}, \quad \text{where}$$

$$\mathbf{A}_e = \begin{bmatrix} 2 & 1 & 1 & 0 & 0 \\ 1 & 1 & 0 & 1 & 0 \\ 1 & 3 & 0 & 0 & 1 \end{bmatrix}, \qquad \mathbf{c}_e = [40 \quad 60 \quad 0 \quad 0 \quad 0]^T,$$

$$\mathbf{x}_e = [x_1 \quad x_2 \quad x_3 \quad x_4 \quad x_5]^T, \quad \text{and} \quad \mathbf{b} = [70 \quad 40 \quad 90]^T.$$

Important points in the original formulation are the vertices $\mathscr{O}$, $\mathscr{S}$, $\mathscr{Q}$, $\mathscr{P}$, $\mathscr{R}$; these points remain important and are easily identified in the new setting. The values x_i for each point are

$\mathscr{O}$: 0, 0, 70, 40, 90 $\qquad$ $\mathscr{S}$: 35, 0, 0, 5, 55

$\mathscr{Q}$: 30, 10, 0, 0, 30 $\qquad$ $\mathscr{P}$: 15, 25, 15, 0, 0.

$\mathscr{R}$: 0, 30, 40, 10, 0

Note that each vertex corresponds to $\mathbf{x}_e$ *having exactly two zero entries and exactly three nonzero entries.* The nonzero variables in such a solution are called *basic variables*, the zero variables are called *nonbasic variables*, and the solution $\mathbf{x}_e$ itself is called a *basic solution.* Solutions to $\mathbf{A}_e\mathbf{x}_e = \mathbf{b}$ that also satisfy $\mathbf{x}_e \geq \mathbf{0}$ are called *feasible.* Since vertices correspond to basic feasible solutions, the simplex method will move from basic feasible solution to basic feasible solution, since we expect an optimal solution to occur as a basic feasible one (that is, a vertex).

The Simplex Method

The simplex method moves from vertex to vertex and so must start at a vertex, that is, at a basic feasible solution. One such basic feasible solution—corresponding to $\mathcal{O}$ in (11.3)—is obvious from (11.4):

(11.6) $$x_1 = 0, \quad x_2 = 0, \quad x_3 = 70, \quad x_4 = 40, \quad x_5 = 90.$$

This is easily spotted because each of x_3, x_4, and x_5 (the nonzero and hence basic variables) appears in only one equation and so its value is easily obtained when x_1 and x_2 (the zero and hence nonbasic variables) are set to zero. Note that $M = 40x_1 + 60x_2$ is defined in terms of the nonbasic variables only.

Now we are ready to move to another vertex with a larger value of M. Since the value of M changes, we treat M as a variable and rewrite the equations in (11.4) as

(11.7)
$$\begin{aligned}
0M + 2x_1 + 1x_2 + 1x_3 \qquad\qquad &= 70.\\
0M + 1x_1 + 1x_2 + \qquad + 1x_4 \qquad &= 40\\
0M + 1x_1 + 3x_2 + \qquad\qquad + 1x_5 &= 90\\
1M - 40x_1 - 60x_2 \qquad\qquad\qquad &= 0.
\end{aligned}$$

As usual, we find it advantageous to describe equations using an augmented matrix:

(11.8)
$$\left[\begin{array}{cccccc|c}
0 & 2 & 1 & 1 & 0 & 0 & 70\\
0 & 1 & 1 & 0 & 1 & 0 & 40\\
0 & 1 & 3 & 0 & 0 & 1 & 90\\
1 & -40 & -60 & 0 & 0 & 0 & 0
\end{array}\right].$$

Since $x_1 = x_2 = 0$ at this point, $M = 0$. To increase $M = 40x_1 + 60x_2$, we can make either x_1 or x_2 (or both) positive; *the simplex method always changes one variable at a time*, and we choose to increase x_2 since a unit increase there increases M by 60 rather than by 40 as for x_1. Keeping $x_1 = 0$ but changing x_2 requires changes in x_3, x_4, and x_5 in order to satisfy (11.7):

$$x_3 = 70 - x_2, \qquad x_4 = 40 - x_2, \quad \text{and} \quad x_5 = 90 - 3x_2.$$

Since we require $x_i \geq 0$, the first equation immediately above requires $x_2 \leq 70$, the second requires $x_2 \leq 40$, and the third requires $x_2 \leq 30$. To meet all three restrictions requires that x_2 be increased only to the smallest bound, 30; this gives

(11.9) $$x_1 = 0, \quad x_2 = 30, \quad x_3 = 40, \quad x_4 = 10, \quad x_5 = 0, \quad \text{with} \quad M = 1800$$

as a new *basic feasible solution*. Note that this point corresponds to $\mathcal{R}$ in (11.3).

The next step is to repeat the process just completed. Recall that two useful conditions were met as we started with (11.7), (11.8): (1) each basic (nonzero) variable appeared in only one equation—each basic variable has only one nonzero entry in its column in (11.8); and (2) M is expressed in terms of the nonbasic (zero) variables only—zeros appear in the basic variable columns in the bottom row of (11.8). These two conditions coalesce:

(11.10) Each basic variable has only one nonzero entry in its column in the augmented matrix, and this is not in the bottom row.

Since the basic variables are now x_2, x_3, and x_4, the standard format described by (11.10) no longer is true for the augmented matrix (11.8)—the x_2-column violates (11.10). This can be rectified by Gauss-Jordan elimination: using the third row (so as not to disturb the zeros in the x_3- and x_4-columns) to eliminate both above and below in the x_2-column transforms (11.8) to

(11.11)
$$\left[\begin{array}{cccccc|c} 0 & \frac{5}{3} & 0 & 1 & 0 & -\frac{1}{3} & 40 \\ 0 & \frac{2}{3} & 0 & 0 & 1 & -\frac{1}{3} & 10 \\ 0 & \frac{1}{3} & 1 & 0 & 0 & \frac{1}{3} & 30 \\ 1 & -20 & 0 & 0 & 0 & 20 & 1800 \end{array}\right].$$

This has the form (11.10) for the x_2-, x_3-, and x_4-columns; note too that the bottom row says $M - 20x_1 + 20x_5 = 1800$, where $x_1 = x_5 = 0$ are the nonbasic variables.

We are ready to move again to another basic feasible solution with an increased value of M. Since now $M = 1800 + 20x_1 - 20x_5$ and $x_i \geq 0$, we can increase M by increasing x_1 to be positive while keeping $x_5 = 0$. This of course forces changes in the basic variables:

$$x_3 = 40 - \tfrac{5}{3}x_1, \quad x_4 = 10 - \tfrac{2}{3}x_1, \quad x_2 = 30 - \tfrac{1}{3}x_1.$$

Keeping $x_i \geq 0$ requires x_1 no larger than 24, 15, and 90, respectively; thus x_1 can only increase to 15, the least of these. This yields a new basic feasible solution:

(11.12)
$$x_1 = 15, \quad x_2 = 25, \quad x_3 = 15, \quad x_4 = 0, \quad x_5 = 0, \quad \text{with} \quad M = 2100.$$

Note that this point corresponds to $\mathscr{P}$ in (11.3). Since the basic variables are now x_1, x_2, x_3, (11.11) is no longer in the standard form (11.10). Using the second row (so as not to disturb the zeros in the x_2- and x_3-columns) to eliminate both above and below in the x_1-column produces

(11.13)
$$\left[\begin{array}{cccccc|c} 0 & 0 & 0 & 1 & -\frac{5}{2} & \frac{1}{2} & 15 \\ 0 & 1 & 0 & 0 & \frac{3}{2} & -\frac{1}{2} & 15 \\ 0 & 0 & 1 & 0 & -\frac{1}{2} & \frac{1}{2} & 25 \\ 1 & 0 & 0 & 0 & 30 & 10 & 2100 \end{array}\right].$$

This has the form (11.10) for the x_1-, x_2-, and x_3-columns; note, too, that the bottom row says

$$M + 30x_4 + 10x_5 = 2100,$$

where $x_4 = x_5 = 0$ are the nonbasic variables.

We are ready to move to another basic feasible solution. However, from

$$M = 2100 - 30x_4 - 10x_5$$

we see that increases in x_4 or x_5 will *decrease* M; *this is the signal in the simplex method that an optimal solution has been found.* Indeed,

$$x_1 = 15, x_2 = 25, \quad \text{and} \quad M = 2100$$

is the optimal solution found graphically.

The simplex method as implemented with augmented matrices is conceptually simple:

(11.14) Starting with an augmented matrix in the form (11.10):

1. Determine which nonbasic (zero) variable to increase.
2. Determine how far it can be increased, and do so.
3. Identify the new basic and nonbasic variables.
4. Use Gauss-Jordan elimination to put the augmented matrix in the form (11.10) with respect to the new basic and nonbasic variables.

The process (11.14) is repeated until an optimal solution is found (more will be said about this later). The tasks above are in fact rather mechanical:

(11.15) Starting with an augmented matrix in the form (11.10):

1. Increase the nonbasic variable with the largest negative entry in the bottom row.
2. Divide the right-hand-side entries by the positive entries in the column of the increasing nonbasic variable, and increase it to the least of these quotients.
3. The basic variables are the nonzero variables, while the nonbasic variables are the zero variables.
4. Use Gauss-Jordan elimination to put the augmented matrix in the form (11.10) with respect to the new basic and nonbasic variables.

Note, too, that the first column of the augmented matrix never entered the computation in any central manner; it served solely to let us interpret the relationship between M and the variables x_i. As long as we remember how to interpret the bottom row, we can omit the first column from the augmented matrices (11.8), (11.11), and (11.13).

You should now be able to solve Problems 1 to 5.

Projected Gradients: An Interior Method

One key idea in *interior methods* generally and in *Karmarkar-like methods* particularly is to avoid the restriction to moving only along the edges of the constraint set. Assume that slack variables have been introduced to put the problem in the form (11.5). Suppose that our present approximate solution is $\mathbf{x}_e$ and that we are considering directions of possible movement from $\mathbf{x}_e$. *We seek that direction that gives the most rapid increase in the function to be maximized.* Since derivatives measure rate of change, we seek that direction $\mathbf{n}$ (that is, a unit vector $\mathbf{n}$) from $\mathbf{x}_e$ in which—if we are maximizing some function $f(\mathbf{x}_e)$—the derivative of f in the direction $\mathbf{n}$ is greatest.

In our case, the function being maximized is $f(\mathbf{x}_e) = (\mathbf{c}_e, \mathbf{x}_e)$. The derivative is just the derivative of $f(\mathbf{x}_e + t\mathbf{n})$ at $t = 0$; in our case

$$f(\mathbf{x}_e + t\mathbf{n}) = f(\mathbf{x}_e) + t(\mathbf{c}_e, \mathbf{n}),$$

so this directional derivative is simply $(\mathbf{c}_e, \mathbf{n})$. Thus our problem of choosing a direction becomes:

find a unit vector $\mathbf{n}$—so $(\mathbf{n}, \mathbf{n}) = 1$—to maximize $(\mathbf{c}_e, \mathbf{n})$.

Recalling that

$$(\mathbf{c}_e, \mathbf{n}) = \|\mathbf{c}_e\|_2 \|\mathbf{n}\|_2 \cos\theta = \|\mathbf{c}_e\|_2 \cos\theta,$$

where θ is the angle between $\mathbf{c}_e$ and $\mathbf{n}$, we see that $(\mathbf{c}_e, \mathbf{n})$ is maximized with $\theta = 0$—that is, with $\mathbf{n}$ in the direction $\mathbf{c}_e$.

The analysis above has neglected an important aspect of our problem: the *constraints* $\mathbf{A}_e\mathbf{x}_e = \mathbf{b}$. There is no guarantee that $\mathbf{x}_e + t\mathbf{n}$ will satisfy the constraints unless we force that condition on $\mathbf{n}$, namely:

$$\mathbf{b} = \mathbf{A}_e(\mathbf{x}_e + t\mathbf{n}) = \mathbf{A}_e\mathbf{x}_e + t\mathbf{A}\mathbf{n} = \mathbf{b} + t\mathbf{A}\mathbf{n}.$$

Therefore,

we should restrict the direction $\mathbf{n}$ *to lie in the null space* $\mathcal{N}$ *of* $\mathbf{A}_e$: $\mathbf{A}_e\mathbf{n} = \mathbf{0}$.

We must look again at maximizing $(\mathbf{c}_e, \mathbf{n})$ but now with $\mathbf{n}$ restricted to the subspace $\mathcal{N}$ of vectors satisfying $\mathbf{A}_e\mathbf{n} = \mathbf{0}$. Let P_0 denote orthogonal projection onto this subspace as in Section 5.8; recall that $\mathbf{c}_e - P_0\mathbf{c}_e$ is therefore orthogonal to all vectors in $\mathcal{N}$. Writing

$$\begin{aligned}(\mathbf{c}_e, \mathbf{n}) &= (\mathbf{c}_e - P_0\mathbf{c}_e + P_0\mathbf{c}_e, \mathbf{n}) = (\mathbf{c}_e - P_0\mathbf{c}_e, \mathbf{n}) + (P_0\mathbf{c}_e, \mathbf{n})\\ &= 0 + (P_0\mathbf{c}_e, \mathbf{n})\\ &= (P_0\mathbf{c}_e, \mathbf{n})\end{aligned}$$

shows that we must choose $\mathbf{n}$ of length 1 in $\mathcal{N}$ to maximize $(P_0\mathbf{c}_e, \mathbf{n})$, where $P_0\mathbf{c}_e$ is itself in $\mathcal{N}$. The solution is therefore for $\mathbf{n}$ to be a unit vector in the direction $P_0\mathbf{c}_e$.

In summary:

(11.16) The unit vector $\mathbf{n}$ in the direction of most rapid increase of $(\mathbf{c}_e, \mathbf{x})$, constrained by $\mathbf{A}_e\mathbf{n} = \mathbf{0}$ so that the constraints are satisfied, is given by the direction of the orthogonal projection of $\mathbf{c}_e$ onto the set $\mathcal{N}$ on $\mathbf{n}$ satisfying $\mathbf{A}_e\mathbf{n} = \mathbf{0}$. This direction is called the *projected gradient*; see Problem 8.

Finding this direction $\mathbf{n}$ in practice is not the formidable problem it might seem at first glance. Recall **Key Corollary 8.20(c)** from Section 8.4 on singular value decompositions of matrices: If $\mathbf{A}_e = \mathbf{U\Sigma V}^H$ is a singular value decomposition of $p \times q$ $\mathbf{A}_e$ of rank k, then the last $q - k$ columns of $\mathbf{V}$ form an orthonormal basis for the space of $\mathbf{n}$ satisfying $\mathbf{A}_e\mathbf{n} = \mathbf{0}$. If

$$\mathbf{V}_0 = [\mathbf{v}_{k+1} \quad \cdots \quad \mathbf{v}_q]$$

is formed from these columns, the $q \times q$ matrix $\mathbf{V}_0\mathbf{V}_0^H$ accomplishes this orthogonal projection:

$$P_0\mathbf{c}_e = \mathbf{V}_0\mathbf{V}_0^H\mathbf{c}_e \qquad \text{(see Theorem 5.79).}$$

There are other ways to compute $P_0\mathbf{c}_e$, but this is one of the best (at least in this case of few variables and few constraints) because of the availability of good software for finding $\mathbf{V}$.

In the specific case of our sample linear program (11.4), (11.5), we have to compute the orthogonal projection of

$$\mathbf{c}_e = [40 \quad 60 \quad 0 \quad 0 \quad 0]^T \quad \text{onto } \mathcal{N}.$$

MATLAB produces the direction

(11.17) $$[2.5 \quad 3.75 \quad -8.75 \quad -6.25 \quad -13.75]^T$$

for movement from our initial feasible vector $[0 \quad 0 \quad 70 \quad 40 \quad 90]^T$. Thus we are considering $\mathbf{x}_e$ of the form

$$\mathbf{x}_e = [0 \quad 0 \quad 70 \quad 40 \quad 90]^T + t[2.5 \quad 3.75 \quad -8.75 \quad -6.25 \quad -13.75]^T,$$

all of which will satisfy $\mathbf{A}_e\mathbf{x}_e = \mathbf{b}$ no matter the value of t. However, we also need $\mathbf{x}_e \geq \mathbf{0}$, so t cannot exceed 70/8.75, 40/6.25, or 90/13.75—the least of which equals 6.4. Using this value for t gives a new $\mathbf{x}_e$ equal to

$$[16 \quad 24 \quad 14 \quad 0 \quad 2]^T.$$

In summary: We started with the same initial feasible point as in the simple method:

(11.18) $$x_1 = 0, \qquad x_2 = 0, \qquad x_3 = 70, \qquad x_4 = 40, \qquad x_5 = 90, \quad \text{with} \quad M = 0$$

and in one step moved *across the interior of the constraint set* to reach

(11.19) $\quad x_1 = 16, \quad x_2 = 24, \quad x_3 = 14, \quad x_4 = 0, \quad x_5 = 2, \quad \text{with} \quad M = 2080;$

this is quite near $x_1 = 15$, $x_2 = 25$, with $M = 2100$—the optimal solution. Note that we did not arrive at a vertex of the constraint set; this new point corresponds to that marked $\mathscr{X}$ on (11.3).

The discussion above indicates some of the attraction of interior methods generally and of this *projected-gradient* method particularly. But it neglected one difficulty: we cannot repeat the process as described and expect to move away from (11.19)! Why? Because the choice of the direction $\mathbf{n}$ in which to move did not depend on the $\mathbf{x}_e$ from which we move; $\mathbf{n}$ is simply defined by $\mathbf{c}_e$ and the null space $\mathscr{N}$, neither of which change. Thus if we apply our method again at (11.19), we again find the direction (11.17) and discover that we cannot increase M in this direction; we stop at (11.19), which is not optimal.

The basic idea of the projected-gradient method above has been known for years; although ways had been found to avoid the breakdown of the method, none were competitive with the simplex method. Karmarkar's contribution was to find a powerful way around the difficulty that may well compete with the simplex approach.

A Karmarkar-Like Method

Karmarkar's approach can be seen as one of transforming the linear program after each step to a *new problem for which the projected-gradient direction is different from that for the original problem.* Several ingenious transformations have been used and are being studied by researchers as to their impact. Here we describe just one of the ideas: *rescaling.*

Suppose that each variable is *rescaled*—that is, replaced by a positive multiple of itself: $x_i' = x_i/d_i$, say. In matrix notation, this is $\mathbf{x}_e' = \mathbf{D}^{-1}\mathbf{x}_e$ or $\mathbf{x}_e = \mathbf{D}\mathbf{x}_e'$, where $\mathbf{D} = \mathbf{diag}(d_1, \ldots, d_5)$ in our case. In the new variables $\mathbf{x}_e'$, the constraints become

$$\mathbf{b} = \mathbf{A}_e\mathbf{x}_e = \mathbf{A}_e\mathbf{D}\mathbf{x}_e' \quad \text{and} \quad \mathbf{0} \le \mathbf{x}_e = \mathbf{D}\mathbf{x}_e';$$

the function to maximize is

$$\mathbf{c}_e^T\mathbf{x}_e = \mathbf{c}_e^T\mathbf{D}\mathbf{x}_e' = (\mathbf{D}\mathbf{c}_e)^T\mathbf{x}_e'.$$

Thus the rescaled program (11.5) is now

(11.20)
$$\begin{aligned}&\text{maximize } (\mathbf{D}\mathbf{c}_e)^T\mathbf{x}_e'\\ &\text{subject to the constraints}\\ &(\mathbf{A}_e\mathbf{D})\mathbf{x}_e' = \mathbf{b}, \qquad \mathbf{x}_e' \ge \mathbf{0}.\end{aligned}$$

If we use the projected-gradient idea from above, we need to find the orthogonal projection of $\mathbf{D}\mathbf{c}_e$ onto the null space of $\mathbf{A}\mathbf{D}$; once we find this direction

$\mathbf{n}'$, we will move to a new $\mathbf{x}'_e + t\mathbf{n}'$. Since $\mathbf{x}_e = \mathbf{D}\mathbf{x}'_e$, this corresponds to moving to a new $\mathbf{x}_e + t\mathbf{D}\mathbf{n}'$. Thus:

(11.21) Rescaling by $\mathbf{x}_e = \mathbf{D}\mathbf{x}'_e$ has the following effect—the new direction of movement from $\mathbf{x}_e$ is $\mathbf{n} = \mathbf{D}\mathbf{n}'$, where $\mathbf{n}'$ is in the direction of the orthogonal projection $P'_0(\mathbf{D}\mathbf{c}_e)$ of $\mathbf{D}\mathbf{c}_e$ onto the null space of $\mathbf{A}_e\mathbf{D}$.

This can be interpreted another way. Recall from Theorem 5.73 on best approximation that the orthogonal projection of $\mathbf{v}$ onto $\mathscr{V}_0$ is the closest point to $\mathbf{v}$ in $\mathscr{V}_0$. Thus $P'_0(\mathbf{D}\mathbf{c}_e)$ is the closest point to $\mathbf{D}\mathbf{c}_e$ in the null space of $\mathbf{A}_e\mathbf{D}$. Now $\mathbf{v}'$ in the null space of $\mathbf{A}_e\mathbf{D}$ means

$$\mathbf{0} = (\mathbf{A}_e\mathbf{D})\mathbf{v}' = \mathbf{A}_e(\mathbf{D}\mathbf{v}') = \mathbf{A}_e\mathbf{v},$$

so $\mathbf{v} = \mathbf{D}\mathbf{v}'$ is in the null space of $\mathbf{A}_e$. This says that $P'_0(\mathbf{D}\mathbf{c}_e)$ equals $\mathbf{D}\mathbf{v}$, where $\mathbf{v}$ minimizes

$$(\mathbf{D}\mathbf{c}_e - \mathbf{D}^{-1}\mathbf{v}, \mathbf{D}\mathbf{c}_e - \mathbf{D}^{-1}\mathbf{v})$$

over $\mathbf{v}$ in the null space of $\mathbf{A}_e$. And it is in this direction $\mathbf{v}$—or, rather, $\mathbf{n} = \mathbf{v}/\|\mathbf{v}\|_2$—in which we move the $\mathbf{x}_e$ variables. We rewrite the above expression as

$$(\mathbf{D}^{-1}\{\mathbf{D}^2\mathbf{c}_e - \mathbf{v}\}, \mathbf{D}^{-1}\{\mathbf{D}^2\mathbf{c}_e - \mathbf{v}\})$$

and introduce a new nonstandard inner product $\langle\cdot,\cdot\rangle$ defined by

$$\langle\mathbf{u}, \mathbf{w}\rangle = (\mathbf{D}^{-1}\mathbf{u}, \mathbf{D}^{-1}\mathbf{w});$$

then $\mathbf{v}$ minimizes

$$\langle\mathbf{D}^2\mathbf{c}_e - \mathbf{v}, \mathbf{D}^2\mathbf{c}_e - \mathbf{v}\rangle.$$

That is, the direction of movement from $\mathbf{x}_e$ is the closest to $\mathbf{D}^2\mathbf{c}_e$ as measured with this new inner product. And $\mathbf{D}^2\mathbf{c}_e$ has a simple interpretation: the function we are maximizing in the linear program equals

$$\mathbf{c}_e^T\mathbf{x}_e = (\mathbf{c}_e, \mathbf{x}_e) = \langle\mathbf{D}^2\mathbf{c}_e, \mathbf{x}_e\rangle.$$

That is, $\mathbf{D}^2\mathbf{c}_e$ is in the direction of most rapid increase when the new inner product is used.

We summarize:

(11.22) Rescaling by $\mathbf{x}_e = \mathbf{D}\mathbf{x}'_e$ has the following effect—the new direction of movement from $\mathbf{x}_e$ is the projected-gradient direction with respect to the new nonstandard inner product

$$\langle\mathbf{u}, \mathbf{w}\rangle = (\mathbf{D}^{-1}\mathbf{u}, \mathbf{D}^{-1}\mathbf{w}) = \mathbf{u}^T\mathbf{D}^{-2}\mathbf{w}.$$

That is, rescaling simply introduces a new way of measuring length and angle when finding the projected-gradient direction.

Nothing said above explains *how* the rescaling **D** is chosen. One of Karmarker's ideas is that being on the edge of the constraint set causes difficulty, so one should never move in a particular direction all the way to the edge, and one should rescale the variables so that the present approximation $\mathbf{x}_e'$ is in some sense centered in the constraint set. The effectiveness of Karmarkar-like methods depends strongly on precisely how such scaling and other transformations are designed and implemented. Research and experimentation indicate that proper resolution of these subtle issues allows the method not only rapidly to get close to the optimal solution but also rapidly to identify variables that are basic at an optimal solution and then to leap directly to that optimal solution. The details of how to do this are presently the subject of intense research and debate, and are beyond the scope of this book.

You should now be able to solve Problems 1 to 11.

Duality

Recall that (11.1) is serving as an example to illustrate some general concepts of linear programming. We next illustrate the important concept of *duality*, presenting it from two different viewpoints.

First we examine the idea in the setting of the application from which (11.1) derived. Recall that the problem is to determine the numbers x_1 and x_2 of two types of products to produce in order to maximize profits, given that the profit equals 40 on each unit of the first product and 60 on each unit of the second; the constraints reflect the fact that the machines used to manufacture the product have only 70, 40, and 90 hours available on them, and that one unit of each product requires certain amounts of each machine's time. See Section 2.7 and especially (2.53).

Suppose that another division (division 2, say) of the company wants to use those same machines to manufacture some other products; division 2 needs to figure out how much it should pay division 1 per hour on each machine in order that division 1 will turn the machine over to division 2, and of course division 2 wants to do this as cheaply as possible. We model this mathematically by letting y_1, y_2, y_3 be the (as yet unknown) prices that division 2 will pay per hour of time on each of the three types of machines. The total payment to division 1, which division 2 wants to minimize, will thus be $m = 70y_1 + 40y_2 + 90y_3$. In order for division 1 to agree to the deal, division 1 must do as well this way as by using the machine. When division 1 declines to produce one unit of its first product, it loses \$40 from its profits; however, it frees 2, 1, and 1 hours, respectively, from the three machines, for which it can receive $2y_1 + 1y_2 + 1y_3$ from division 2. Thus the swap appeals to division 1 as far as the first product is concerned as long as its gain exceeds its loss: $2y_1 + y_2 + y_3 \geq 40$. Analyzing the second product similarly leads to $y_1 + y_2 + 3y_3 \geq 60$. Thus the problem for division 2 is to determine prices y_1, y_2, and y_3 so as to

(11.23)
$$\begin{aligned}&\text{minimize } m = 70y_1 + 40y_2 + 90y_3\\&\text{subject to the constraints}\\&2y_1 + y_2 + y_3 \geq 40\\&y_1 + y_2 + 3y_3 \geq 60\\&y_1 \geq 0, \quad y_2 \geq 0, \quad y_3 \geq 0.\end{aligned}$$

This linear program (11.23) is called the *dual* of (11.1). In matrix notation, its relation to (11.2) is striking:

(11.24)
$$\begin{aligned}&\text{minimize } m = \mathbf{b}^T\mathbf{y}\\&\text{subject to the constraints}\\&\mathbf{A}^T\mathbf{y} \geq \mathbf{c}, \quad \mathbf{y} \geq \mathbf{0},\\&\text{where } \mathbf{A}, \mathbf{b}, \text{ and } \mathbf{c} \text{ are as in (11.2) and}\\&\mathbf{y} = [y_1 \quad y_2 \quad y_3]^T.\end{aligned}$$

Comparing (11.2) and (11.24) shows that "minimize" replaces "maximize," "$\mathbf{A}^T$" replaces "$\mathbf{A}$," "$\geq$" replaces "$\leq$," "$\mathbf{c}$" replaces "$\mathbf{b}$," and "$\mathbf{b}$" replaces "$\mathbf{c}$."

This dual program (11.24) can be derived another way, using geometry rather than the application setting. Examine the figure (11.3) and especially the optimal point $\mathscr{P}$. The "northeasterly" perpendicular to the graph of $40x_1 + 60x_2 = 2100$ through $\mathscr{P}$ lies in the angle between the similar perpendiculars to $\mathscr{RP}$ and $\mathscr{PQ}$. This says that the first perpendicular is a nonnegative linear combination of the other two. Since these perpendiculars are just $[40 \quad 60]^T$, $[1 \quad 3]^T$, and $[1 \quad 1]^T$, this combination is easily found: $[40 \quad 60]^T = 30[1 \quad 1]^T + 10[1 \quad 3]^T$. Note that $[40 \quad 60]^T$ is just $\mathbf{c}$, while the other two vectors are transposes of those rows of $\mathbf{A}$ corresponding to the inequalities that in fact hold as equalities at the optimal $\mathscr{P}$. In matrix notation we can write this as

(11.25)
$$\begin{aligned}&\mathbf{A}^T\mathbf{y}^* = \mathbf{c}\\&\text{where } \mathbf{y}^* = [0 \quad 30 \quad 10]^T \geq \mathbf{0}.\end{aligned}$$

The surprising truth is that this $\mathbf{y}^*$ solves the linear program (11.24). To see this, suppose that $\mathbf{y}$ satisfies the dual constraints $\mathbf{A}^T\mathbf{y} \geq \mathbf{c}$ and $\mathbf{y} \geq \mathbf{0}$, and let $\mathbf{x}^*$ be the optimal solution $[15 \quad 25]^T$ to the original—the so-called *primal*—program (11.2).

Since $\mathbf{Ax}^* \le \mathbf{b}$ while $\mathbf{y} \ge \mathbf{0}$, we have

$$\mathbf{y}^T\mathbf{Ax}^* \le \mathbf{y}^T\mathbf{b}.$$

Since also $\mathbf{A}^T\mathbf{y} \ge \mathbf{c}$ while $\mathbf{x}^* \ge \mathbf{0}$, we have

$$\mathbf{x}^{*T}\mathbf{A}^T\mathbf{y} \ge \mathbf{x}^{*T}\mathbf{c}.$$

Combining these two inequalities and using $\mathbf{y}^T\mathbf{Ax}^* = (\mathbf{Ax}^*)^T\mathbf{y} = \mathbf{x}^{*T}\mathbf{A}^T\mathbf{y}$ gives

(11.26) $$\mathbf{b}^T\mathbf{y} \ge \mathbf{c}^T\mathbf{x}^* \qquad \text{for all } \mathbf{y} \text{ satisfying } \mathbf{A}^T\mathbf{y} \ge \mathbf{c}, \quad \mathbf{y} \ge \mathbf{0}.$$

Remember that $\mathbf{y}^*$ has zero entries where $\mathbf{Ax}^* - \mathbf{b}$ has nonzero entries, while the nonzero entries of $\mathbf{y}^*$ appear where $\mathbf{Ax}^* - \mathbf{b}$ has zero entries. Therefore,

$$\mathbf{y}^{*T}(\mathbf{Ax}^* - \mathbf{b}) = 0, \quad \text{that is,} \quad \mathbf{y}^{*T}\mathbf{Ax}^* = \mathbf{y}^{*T}\mathbf{b}.$$

Since $\mathbf{y}^{*T}\mathbf{A} = (\mathbf{A}^T\mathbf{y}^*)^T \sim \mathbf{c}^T$ from (11.25), we obtain $\mathbf{c}^T\mathbf{x}^* = \mathbf{y}^{*T}\mathbf{b}$. Combining this with (11.26) yields

(11.27) $$\mathbf{b}^T\mathbf{y} \ge \mathbf{b}^T\mathbf{y}^* = \mathbf{c}^T\mathbf{x}^* \qquad \text{for all } \mathbf{y} \text{ satisfying } \mathbf{A}^T\mathbf{y} \ge \mathbf{c}, \quad \mathbf{y} \ge \mathbf{0}.$$

In other words, $\mathbf{y}^*$ minimizes $\mathbf{b}^T\mathbf{y}$ subject to the constraints $\mathbf{A}^T\mathbf{y} \ge \mathbf{c}$, $\mathbf{y} \ge \mathbf{0}$—so $\mathbf{y}^*$ solves the dual linear program (11.24). Note, too, that $\mathbf{b}^T\mathbf{y}^* = \mathbf{c}^T\mathbf{x}^*$: The optimal (minimum) value m of the dual equals the optimal (maximum) value M of the primal. In our sample (11.2), (11.24), we had $M = 2100$; now

$$m = \mathbf{b}^T\mathbf{y}^* = [70 \quad 40 \quad 90][0 \quad 30 \quad 10]^T = 1200 + 900 = 2100 = M$$

as claimed. In our application, this says the division 2 can offer payment for the machines that will give division 1 exactly the same profit as from using the machines (which seems intuitively what should be offered).

We will see in Section 11.4 that the dual is of great practical importance.

PROBLEMS 11.1

▷ **1.** Solve the linear program below (a) geometrically and (b) by the simplex method.

Maximize $x_1 + 2x_2$

subject to the constraints

$$\begin{aligned} x_1 \qquad &\le 25 \\ x_1 + x_2 &\le 30 \\ -x_1 + x_2 &\le 10 \\ x_1 \ge 0 \qquad & x_2 \ge 0. \end{aligned}$$

▷ **2.** Solve the linear program below (a) geometrically and (b) by the simplex method.

$$\text{Maximize } x_1 + 2x_2$$

subject to the constraints

$$\begin{aligned} -x_1 + x_2 &\le 10 \\ x_2 &\le 20 \\ x_1 + x_2 &\le 60 \\ x_1 \qquad &\le 50 \\ x_1 \ge 0 \qquad x_2 &\ge 0. \end{aligned}$$

3. Solve the linear program below (a) geometrically and (b) by the simplex method.

$$\text{Maximize } 3x_1 + 2x_2$$

subject to the constraints

$$\begin{aligned} x_1 + 2x_2 &\le 70 \\ x_1 + x_2 &\le 40 \\ 3x_1 + x_2 &\le 90 \\ x_1 \ge 0, \qquad x_2 &\ge 0. \end{aligned}$$

4. Explain why, in the simplex method, the present value of M corresponding to the latest basic feasible solution $\mathbf{x}_e$ appears as the (4, 7)-entry in the augmented matrices in our examples.

5. *Subtract* a slack variable x_4 from the inequality $x_2 - x_1 \ge -5$ so that it becomes $x_2 - x_1 - x_4 = -5$ in order to use the simplex method to solve the linear program

$$\text{maximize } 2x_1 + x_2$$

subject to the constraints

$$\begin{aligned} x_1 + x_2 &\le 10 \\ -x_1 + x_2 &\ge -5 \\ x_1 \ge 0, \qquad x_2 &\ge 0. \end{aligned}$$

▷ **6.** Show that as far as only the variables x_1 and x_2 are concerned, the projected-gradient direction (11.17) is just a multiple of $\mathbf{c}$, that is, of $[40 \quad 60]^T$.

7. Show that no movement increasing M is possible in the projected-gradient direction (11.17) from the basic feasible vectors corresponding to $\mathcal{O}$, $\mathcal{S}$, $\mathcal{Q}$, $\mathcal{P}$, $\mathcal{R}$ other than $\mathcal{O}$.

▷ **8.** Suppose that $f(x_1, x_2, \ldots, x_p) = \mathbf{c}^T\mathbf{x}$, where $\langle \mathbf{x} \rangle_i = x_i$, and $\mathbf{c}$ and $\mathbf{x}$ are both $p \times 1$. Show that ∇f, the $p \times 1$ matrix formed from the first partial derivatives of f—$\langle \nabla f \rangle_i = \partial f / \partial x_i$—satisfies $\nabla f = \mathbf{c}$. (This explains the term "projected *gradient*.")

𝔐 **9.** Use MATLAB or similar software to calculate the projected-gradient direction for the linear program in Problem 1, and then move as far as possible in that direction from your initial basic feasible vector there.

𝔐 **10.** Do as in Problem 9 but for the program in Problem 2.

𝔐 **11.** Consider the model linear program of this section.
(a) Find the feasible vector that results from moving only 95% of the way to the edge of the constraint set from the basic feasible point (11.18) toward (11.19) in the direction (11.17).
(b) Use MATLAB or similar software to find the scaled projected-gradient direction obtained by using $\mathbf{D} = \mathbf{diag}(0.95, 1.75, 0.675, 0.18, 0.18)$.
(c) Move from the point reached in (a) in the direction found in (b) to reach the edge of the constraint set.

▷ **12.** Find the dual for the linear program in Problem 1; use the geometric argument at the optimal point of the primal program to find an optimal vector for the dual.

13. Do as in Problem 12 but for the program in Problem 2.

14. Do as in Problem 12 but for the program in Problem 3.

▷ **15.** Rewrite the dual (11.23) as a maximization problem (that is, maximize $-m = -70y_1 - 40y_2 - 90y_3$) and use the simplex method to solve the resulting linear program.

16. Do as in Problem 15 but for the dual of the program of Problem 1.

17. Do as in Problem 15 but for the dual of the program of Problem 2.

18. Do as in Problem 15 but for the dual of the program of Problem 3.

11.2 *A GENERAL LINEAR PROGRAM*

To define a general linear program precisely requires some notation.

Terminology

(11.28) ***Definition.*** Suppose that $\mathbf{A}$ and $\mathbf{B}$ are both $p \times q$ real matrices. We write $\mathbf{A} \geq \mathbf{B}$ if and only if $\langle \mathbf{A} \rangle_{ij} \geq \langle \mathbf{B} \rangle_{ij}$ for $1 \leq i \leq p$ and $1 \leq j \leq q$; similar definitions hold for $>$, $\leq$, and $<$. If $\mathbf{A} \geq \mathbf{0}$, then $\mathbf{A}$ is called *nonnegative*; if $\mathbf{A} > \mathbf{0}$, then $\mathbf{A}$ is called (strictly) *positive.*

This lays the groundwork for a definition of a general linear program.

(11.29) ***Definition***
(a) Any problem of extremizing (minimizing or maximizing) a function $\mathbf{c}^T\mathbf{x}$ with $\mathbf{x}$ in $\mathbb{R}^q$ subject to a finite number of linear equality and inequality constraints using $\leq$ or $\geq$ or both is called a *linear program.*

(b) A linear program is said to be in *standard form* if and only if it is written as

$$\text{maximize } \mathbf{c}^T\mathbf{x}$$

$$\text{subject to the constraints}$$

$$\mathbf{A}\mathbf{x} \leq \mathbf{b}, \qquad \mathbf{x} \geq \mathbf{0}.$$

There is no universally agreed standard form for linear programs; the one above is one of many possibilities. Any linear program can be rewritten in this standard form. Any variable z that is not originally constrained to be nonnegative can be replaced by letting $z = z_+ - z_-$ with $z_+ \geq 0$ and $z_- \geq 0$; any linear inequality $\boldsymbol{\alpha}^T\mathbf{x} \geq \beta$ can be replaced by $(-\boldsymbol{\alpha}^T)\mathbf{x} \leq (-\beta)$; and any linear equality $\boldsymbol{\alpha}^T\mathbf{x} = \beta$ can be replaced by the pair $\boldsymbol{\alpha}^T\mathbf{x} \leq \beta$ and $(-\boldsymbol{\alpha}^T\mathbf{x}) \leq (-\beta)$.

In order to apply the powerful methods developed for working with equations, we convert the standard form to one involving equalities and simple inequalities, just as in Section 11.1. If $\mathbf{A}$ is $p \times q$, we introduce *slack variables* $x_{q+1}, \ldots, x_{q+p}$—one for each of the p inequalities in $\mathbf{A}\mathbf{x} \leq \mathbf{b}$. We define *extended* matrices $\mathbf{A}_e$, $\mathbf{x}_e$, and $\mathbf{c}_e$ just as in Section 11.1:

(11.30)
$$\begin{aligned} \mathbf{A}_e &= [\mathbf{A} \quad \mathbf{I}_p] \text{ is } p \times (p + q) \\ \mathbf{x}_e &= [\mathbf{x}^T \quad x_{q+1} \quad \cdots \quad x_{q+p}]^T \text{ is } (q + p) \times 1 \\ \mathbf{c}_e &= [\mathbf{c}^T \quad \mathbf{0}^T]^T \text{ is } (q + p) \times 1. \end{aligned}$$

These allow an equivalent *slack-variable form* of our standard form in Definition 11.29:

(11.31)
$$\text{maximize } \mathbf{c}_e^T\mathbf{x}_e$$

$$\text{subject to the constraints}$$

$$\mathbf{A}_e\mathbf{x}_e = \mathbf{b}, \qquad \mathbf{x}_e \geq \mathbf{0}, \quad \text{where} \quad \mathbf{A}_e \text{ is } p \times (q + p).$$

Note that a linear program might be posed with p equality constraints in $q + p$ variables as in (11.31) without their arising from the addition of slack variables to a problem in standard form; we nonetheless refer to a linear program in the form (11.31) as in slack-variable form regardless of its origin or the precise structure of $\mathbf{A}_e$—other than being $p \times (q + p)$.

(11.32) ***Definition***

(a) Any vector that satisfies all the constraints of a linear program is said to be *feasible* for that linear program.

(b) Any feasible vector that extremizes the function in a linear program is said to be *optimal* (feasible).

The next concept is that of *basic* feasible vectors; in Section 11.1 these corresponded to the vertices of the constraint set in $\mathbb{R}^2$ for the linear program in standard form. With two variables a vertex was identified as the intersection of exactly two of the lines defined by making equalities out of exactly two of the inequalities. In the slack-variable form with five variables and three equations, the equivalent procedure was to set two variables to zero (thereby selecting those two inequalities to be satisfied as equalities) and solve the three equations for the three remaining variables.

For a linear program in standard form in $\mathbb{R}^q$, the analogous process would be to identify vertices by demanding that some q of the p inequalities be solved as equalities. In the slack-variable form (11.31), the equivalent procedure is to set q of the variables in $\mathbf{x}_e$ to zero and then solve the p equations $\mathbf{A}_e\mathbf{x}_e = \mathbf{b}$ for the p remaining variables. Such a solution is termed *basic*.

(11.33) ***Definition.*** A feasible vector $\mathbf{x}_e$ for the slack-variable form (11.31) is said to be a *basic* feasible vector if and only if (at least) q of the entries of $\mathbf{x}_e$ equal zero.

Computationally, the process of *finding* basic feasible vectors involves setting some q of the entries of $\mathbf{x}_e$ to zero and then *solving the p equations for the remaining p entries of* $\mathbf{x}_e$. To be certain of being able to solve such equations requires the relevant $p \times p$ submatrix of $\mathbf{A}_e$ to be nonsingular—which need not be the case in practice. Dealing with this situation is one of the complications for linear programming that are beyond the scope of this book; we therefore restrict ourselves to situations in which this cannot arise by treating programs that are *nondegenerate* (a slightly stronger condition).

(11.34) ***Definition.*** A linear program in the slack-variable form (11.31) is said to be *nondegenerate* if and only if every $p \times p$ submatrix of the augmented matrix $[\mathbf{A}_e \quad \mathbf{b}]$ is nonsingular; otherwise, the program is said to be *degenerate*.

Fundamental Theory

In Section 11.1 the "nonbasic variables" were those we set to zero, and then the variables for which we solved always turned out to be nonzero (we called these "basic variables"). Had some of these calculated variables turned out to equal zero, it would have been difficult for an observer to distinguish them from those variables initially set to zero. We consider whether this can occur.

Suppose that we set to zero q of the $q + p$ entries of $\mathbf{x}_e$ in the slack-variable form of a nondegenerate linear program; this leaves p entries and p equations to determine them. Since the coefficient matrix of this system of p equations in p

unknowns is a submatrix of $\mathbf{A}_e$, it is nonsingular by the assumption of nondegeneracy. Thus there is a solution to $\mathbf{A}_e\mathbf{x}_e = \mathbf{b}$, which of course writes $\mathbf{b}$ as a linear combination (using the entries of $\mathbf{x}_e$ as coefficients) of the columns of $\mathbf{A}_e$. If any of the entries of $\mathbf{x}_e$ are zero in addition to the q entries set to zero, then at least $q + 1$ entries of $\mathbf{x}_e$ are zero; this leaves at most $p - 1$ entries of $\mathbf{x}_e$ nonzero. Therefore, $\mathbf{b}$ would be written as a linear combination of $p - 1$ columns of $\mathbf{A}_e$, and hence the $p \times p$ submatrix of $[\mathbf{A}_e \quad \mathbf{b}]$ formed from those $p - 1$ columns of $\mathbf{A}_e$ and $\mathbf{b}$ would be singular—in contradiction to the assumption of nondegeneracy. This completes a proof of the following.

(11.35) ***Theorem*** (basic vectors). If the linear program in slack-variable form (11.31) is nondegenerate, then each basic feasible vector has exactly q zero entries; these zero variables are called *nonbasic*, while the p nonzero variables are called *basic*.

Our approach to solving linear programs was based on the idea that they *have* solutions and that a solution can always be found *at a vertex* (that is, as a *basic* feasible solution in the slack-variable setting). Of course, linear programs need not have solutions: The problem of maximizing $x_1 + x_2$ subject to $x_1 \geq 0$ and $x_2 \geq 0$ has no solution because $x_1 + x_2$ can be made arbitrarily large while satisfying the constraints; and the problem of maximizing x_1 subject to $x_1 \leq 1$ and $x_1 \geq 2$ has no solution because no such numbers x_1 exist. But otherwise—when the constraint set is nonempty and the function to be maximized is bounded above on the constraint set—it seems geometrically clear that an optimal feasible vector exists and that an optimal *basic* feasible vector exists. Geometrically clear or not, this must be proved.

(11.36) ***Lemma.*** Suppose that the slack-variable form (11.31) of a linear program is nondegenerate, that $M = \mathbf{c}_e^T\mathbf{x}_e$ cannot be arbitrarily large on the constraint set, and that $\tilde{\mathbf{x}}_e$ is a feasible point. Then there exists a *basic* feasible point $\mathbf{x}_e^*$ with at least as large a value of M:

$$\mathbf{c}_e^T\mathbf{x}_e^* \geq \mathbf{c}_e^T\tilde{\mathbf{x}}_e.$$

PROOF. If $\tilde{\mathbf{x}}_e$ is itself a basic feasible vector, we can take $\mathbf{x}_e^* = \tilde{\mathbf{x}}_e$ and we are done. We therefore assume that $\tilde{\mathbf{x}}_e$ is not a basic vector; since the linear program is nondegenerate, Theorem 11.35 implies that $\tilde{\mathbf{x}}_e$ has at most $q - 1$ zero entries. We show that a feasible vector can be found with at least one more zero entry than in $\tilde{\mathbf{x}}_e$ and with at least as large a value of M; by repeating this process we obtain a feasible vector with at least q zero entries, and this must be the required basic feasible vector $\mathbf{x}_e^*$. The only thing for us to prove is that we can indeed find a feasible $\mathbf{x}'_e$ with $\mathbf{c}_e^T\mathbf{x}'_e \geq \mathbf{c}_e^T\tilde{\mathbf{x}}_e$ and with at least one more zero entry than in $\tilde{\mathbf{x}}_e$; we now prove this.

Since every $p \times p$ submatrix of $p \times (q+p)$ $\mathbf{A}_e$ is nonsingular, $\mathbf{A}_e$ has rank p; by Theorem 6.8 on domains, images, and null spaces, the null space of $\mathbf{A}_e$—the set of solutions $\mathbf{n}$ to $\mathbf{A}_e\mathbf{n} = \mathbf{0}$—has dimension q and hence has some basis $\mathbf{n}_1, \ldots, \mathbf{n}_q$. Let $\mathbf{x}'_e = \tilde{\mathbf{x}}_e + \mathbf{n}$, where $\mathbf{n} = \alpha_1\mathbf{n}_1 + \cdots + \alpha_q\mathbf{n}_q$ with the α_i to be determined; $\mathbf{A}_e\mathbf{x}'_e = \mathbf{b}$ for all choices of the α_i. We seek to determine the α_i so that $\mathbf{x}'_e \geq \mathbf{0}$, so that $\mathbf{x}'_e$ has at least one more zero entry than in $\tilde{\mathbf{x}}_e$, and so that $\mathbf{c}_e^T\mathbf{n} \geq 0$ (which implies that $\mathbf{c}_e^T\mathbf{x}'_e \geq \mathbf{c}_e^T\tilde{\mathbf{x}}_e$). Let $\boldsymbol{\alpha} = [\alpha_1, \ \cdots \ \alpha_q]^T$ be a nonzero solution to the $q-1$ homogeneous equations in q unknowns formed by setting to zero some $q-1$ of the entries of $\mathbf{n}$, including all of the at most $q-1$ entries corresponding to the zero entries in $\tilde{\mathbf{x}}_e$; by **Key Theorem 4.13(3)**, such a nonzero $\boldsymbol{\alpha}$ exists since there are fewer (homogeneous) equations than unknowns. Since $\boldsymbol{\alpha}$ is nonzero and the $\mathbf{n}_i$ form a linearly independent set, $\mathbf{n} = \alpha_1\mathbf{n}_1 + \cdots + \alpha_q\mathbf{n}_q$ is nonzero. All the remaining $p+1$ entries of $\tilde{\mathbf{x}}_e$ are nonzero, and at least one of the corresponding entries of $\mathbf{n}$ must be nonzero since $\mathbf{n}$ is nonzero. By replacing $\mathbf{n}$ by $-\mathbf{n}$ if $\mathbf{c}_e^T\mathbf{n} < 0$, we can assume that $\mathbf{c}_e^T\mathbf{n} \geq 0$. If all the nonzero entries of $\mathbf{n}$ are strictly positive and $\mathbf{c}_e^T\mathbf{n} > 0$, then M could be made arbitrarily large at the feasible vectors $\tilde{\mathbf{x}}_e + t\mathbf{n}$ by making t arbitrarily large, in contradiction to the assumption that M is bounded; if all these entries are positive but $\mathbf{c}_e^T\mathbf{n} = 0$, we can again replace $\mathbf{n}$ by $-\mathbf{n}$ and still have $\mathbf{c}_e^T\mathbf{n} \geq 0$ (actually $= 0$) with some entry of the new $\mathbf{n}$ negative. Thus we can assume that at least one entry of $\mathbf{n}$ is strictly negative. Let r number the entry for which $-\langle\tilde{\mathbf{x}}_e\rangle_i/\langle\mathbf{n}\rangle_i$ is least among the strictly negative $\langle\mathbf{n}\rangle_i$ and let t equal this least (positive) ratio. Then $\mathbf{x}'_e = \tilde{\mathbf{x}}_e + t\mathbf{n}$ is feasible, has at least one more zero entry—the rth—than in $\tilde{\mathbf{x}}_e$, and has $\mathbf{c}_e^T\mathbf{x}'_e \geq \mathbf{c}_e^T\tilde{\mathbf{x}}_e$. ■

(11.37) **Key Theorem** (linear program solvability). Suppose that a linear program in slack-variable form (11.31) is nondegenerate, that the constraint set is nonempty, and that the function M being maximized is bounded above on the constraint set. Then there exists a basic feasible vector $\mathbf{x}_e^*$ that maximizes M over the constraint set—that is, an optimal basic feasible vector exists.

PROOF. Since the linear program is nondegenerate, the basic feasible solutions can be obtained by setting any q entries of $\mathbf{x}_e$ to zero, solving for the remaining p entries from the p equations $\mathbf{A}_e\mathbf{x}_e = \mathbf{b}$, and checking whether $\mathbf{x}_e \geq \mathbf{0}$. By Lemma 11.36, at least one such basic feasible vector exists since the constraint set is non-empty. Since there are at most N_{pq} subsets of q entries of the $p+q$ entries that can be chosen to set to zero, where $N_{pq} = (q+p)!/(q!p!)$, the number N of basic feasible vectors satisfies $1 \leq N \leq N_{pq}$. Let $\mathbf{x}_e^*$ maximize $M = \mathbf{c}_e^T\mathbf{x}_e$ as $\mathbf{x}_e$ ranges over the N basic feasible vectors. Then $\mathbf{x}_e^*$ is the required optimal basic feasible vector; no feasible $\tilde{\mathbf{x}}_e$ can give a larger value of M since then, by Lemma 11.36, one of the N basic feasible vectors would give an even larger value of M, in contradiction to $\mathbf{x}_e^*$'s giving the largest of these. ■

PROBLEMS 11.2

▷ **1.** Put into standard form the linear program

$$\text{maximize} \quad -2x_1 + 5x_2$$

subject to the constraints

$$2x_1 - x_2 \leq 7$$

$$3x_1 + x_2 = 6$$

$$x_1 \geq 0, \qquad x_2 \geq 0.$$

2. Put the linear program of Problem 1 into slack-variable form.

3. Put into standard form the linear program

$$\text{minimize} \quad 4x_1 - 3x_2 + 2x_3$$

subject to the constraints

$$x_1 + x_2 \geq 4$$

$$x_1 - x_2 \leq 6$$

$$x_1 + 3x_2 + 6x_3 = 5$$

$$x_1 \geq 0, \qquad x_3 \geq 0.$$

▷ **4.** Put the linear program of Problem 3 into slack-variable form.

5. Put into standard form the linear program

$$3x_1 + 2x_2 \geq 7$$

$$-x_1 - 5x_2 \geq 3$$

$$x_1 \geq 5$$

$$x_2 \geq 0.$$

6. Put the linear program of Problem 5 into standard form.

7. Give an example of a degenerate linear program and a basic feasible solution to it with strictly more than q zero entries.

8. Determine whether the linear programs from the following Problems in Section 11.1 are degenerate:
(a) Problem 1 (b) Problem 2
(c) Problem 3 (d) Problem 5

▷ **9.** (a) Determine whether the linear program below is degenerate.
(b) Solve the linear program by the simplex method.

$$\text{Maximize} \quad x_1 + 2x_2$$

subject to the constraints

$$\begin{aligned} x_1 \qquad &\leq 10 \\ x_2 &\leq 10 \\ -x_1 + x_2 &\leq 10 \\ x_1 \geq 0, \qquad & x_2 \geq 0 \end{aligned}$$

10. Use the construction in Lemma 11.36 to find a feasible vector for the model problem (11.1) having (at least) one more zero entry than in

$$\mathbf{x}_e = [20 \quad 15 \quad 15 \quad 5 \quad 25]^T.$$

▷ **11.** Suppose that you need to maximize the minimum of the two nonnegative numbers u and v subject to the constraints that $2u + v \leq 5$ and $u + 3v \leq 8$. Write a linear program whose solution will let you compute u and v, and put the program in standard form.

11.3 SOLVING A GENERAL LINEAR PROGRAM

As with systems of linear equations and with eigensystem problems, linear programs are in practice solved by sophisticated computer software. In this section we merely extend the basic ideas on methods presented in Section 11.1 to the general linear programs of Section 11.2 without going into detail; such an introduction should be sufficient for most users of linear programming.

Karmarkar-Like Methods

We return to the slack-variable version (11.31) of linear programs:

$$\text{maximize} \quad M = \mathbf{c}_e^T \mathbf{x}_e$$

subject to the constraints

$$\mathbf{A}_e \mathbf{x}_e = \mathbf{b}, \qquad \mathbf{x}_e \geq \mathbf{0}, \quad \text{with} \quad \mathbf{A}_e \; p \times (q + p).$$

The key aspects to remember about the important new Karmarkar-like methods are that (1) they are *interior* methods rather than methods restricted to movement along the edges of the constraint set; and (2) they operate on *iteratively transformed* versions of the linear program—that is, at each step the program is again transformed before a new approximate solution is found.

Although other approaches are possible, the interior-method aspect of the methods is well illustrated by the projected-gradient approach: Move from the present approximate solution $\mathbf{x}_e$ to a new approximate solution $\mathbf{x}_e + t\mathbf{d}$, where $\mathbf{d} = P_0 \mathbf{c}_e$ and P_0 denotes orthogonal projection onto the null space $\mathcal{N}$ of $\mathbf{A}_e$—the set of $\mathbf{n}$ satisfying $\mathbf{A}_e \mathbf{n} = \mathbf{0}$. We mentioned in Section 11.1 that the orthogonal projection $P_0\mathbf{c}_e$ could easily be computed by using the singular value decomposition $\mathbf{A}_e = \mathbf{U}\boldsymbol{\Sigma}\mathbf{V}^H$ of $\mathbf{A}_e$; this is certainly true for moderate p and q, but linear programs often involve enormous p and q (and sparse, highly structured $\mathbf{A}_e$), in which case other approaches may be preferable. Finding the closest point $P_0\mathbf{c}_e$ to $\mathbf{c}_e$ in $\mathscr{N}$ is essentially a least-squares problem, although not one in the form $\mathbf{Ax} \approx \mathbf{y}$ as discussed in Sections 5.9 and 8.5, where methods using the QR or singular value decompositions were presented; the next theorem shows that projection and least squares are the same, however.

(11.38) ***Theorem*** (projection onto null spaces). Let $\mathbf{B}$ be $p \times q$ and let P_0 denote orthogonal projection onto the null space $\mathscr{N}$ of $\mathbf{B}$—all those $\mathbf{n}$ satisfying $\mathbf{Bn} = \mathbf{0}$. Then $P_0\mathbf{y} = \mathbf{y} - \mathbf{B}^H\tilde{\mathbf{x}}$ where $\tilde{\mathbf{x}}$ solves the least-squares problem $\mathbf{B}^H\mathbf{x} \approx \mathbf{y}$.

PROOF. Recall from Problem 26 in Section 5.9 that $\tilde{\mathbf{x}}$ solves $\mathbf{Cx} \approx \mathbf{y}$ if and only if $\mathbf{C}^H\mathbf{C}\tilde{\mathbf{x}} = \mathbf{C}^H\mathbf{y}$; in our case $\mathbf{C} = \mathbf{B}^H$, so $\tilde{\mathbf{x}}$ solves $\mathbf{B}^H\mathbf{x} \approx \mathbf{y}$ if and only if $\mathbf{BB}^H\tilde{\mathbf{x}} = \mathbf{By}$. This in turn is equivalent to $\mathbf{B}(\mathbf{y} - \mathbf{B}^H\tilde{\mathbf{x}}) = \mathbf{0}$—that is, to $\mathbf{y} - \mathbf{B}^H\tilde{\mathbf{x}}$ being in the null space $\mathscr{N}$ of $\mathbf{B}$. Let $\mathbf{n}_0$ denote $\mathbf{y} - \mathbf{B}^H\tilde{\mathbf{x}}$; our problem is to show that $\mathbf{n}_0 = P_0\mathbf{y}$. Let $\mathbf{n}_1 = \mathbf{n}_0 - P_0\mathbf{y}$. Since $\mathbf{n}_1$ is in $\mathscr{N}$, $(\mathbf{y} - P_0\mathbf{y}, \mathbf{n}_1) = 0$ by Theorem 5.71(a). Then

$$\begin{aligned}\|\mathbf{n}_1\|^2 &= (\mathbf{n}_1, \mathbf{n}_1) = (\mathbf{n}_0 - \mathbf{y} + \mathbf{y} - P_0\mathbf{y}, \mathbf{n}_1) = (-\mathbf{B}^H\tilde{\mathbf{x}}, \mathbf{n}_1)\\ &= (-\tilde{\mathbf{x}}, \mathbf{Bn}_1) = (-\tilde{\mathbf{x}}, \mathbf{0}) = 0. \quad \blacksquare\end{aligned}$$

Thus, in our linear programming problem, any method for solving $\mathbf{A}_e^H\tilde{\mathbf{z}} \approx \mathbf{c}_e$ for $\tilde{\mathbf{z}}$ allows the computation of the new direction $\mathbf{d}$ as $\mathbf{d} = P_0\mathbf{c}_e = \mathbf{c}_e - \mathbf{A}_e^H\tilde{\mathbf{z}}$. One method for computing $\tilde{\mathbf{z}}$ is through the singular value decomposition, which of course we could instead use to compute $P_0\mathbf{c}_e$ directly as mentioned in Section 11.1. We might also solve $\mathbf{A}_e^H\tilde{\mathbf{z}} \approx \mathbf{c}_e$ by using the QR-decomposition of $\mathbf{A}_e$. The problem of solving extremely large sparse least-squares problems has been—and is being—investigated extensively for reasons unrelated to linear programming; this work is now being applied in the study of Karmarkar-like methods.

As stressed in Section 11.1, the projected-gradient approach as described cannot be applied repeatedly since the same direction $\mathbf{d}$ would be produced at each step; the simplest of the transformations proposed by Karmarkar—*scaling* the variables $\mathbf{x}_e = \mathbf{Dx}'_e$ with a diagonal matrix $\mathbf{D}$—avoids this difficulty, since the next direction $\mathbf{d}$ is the orthogonal projection of $\mathbf{Dc}_e$ (rather than of $\mathbf{c}_e$) onto the null space of $\mathbf{A}_e\mathbf{D}$ (rather than of $\mathbf{A}_e$). The simple nature of $\mathbf{D}$ may make the least-squares problems that must be solved at each step somewhat more tractable than they would be with just treating $\mathbf{A}_e\mathbf{D}$ from scratch each time $\mathbf{D}$ changes. Additional transformations are proposed by Karmarkar that have even more dramatic effects on the problem—replacing the linear function $\mathbf{c}_e^T\mathbf{x}_e$ by logarithmic expressions such as $\log(\mathbf{c}_e^T\mathbf{x}_e)$, nonlinear changes in the variables $\mathbf{x}_e$, and the like. Although it is the nature of these transformations that appears to be the secret to the power of the methods, this topic and further discussion of Karmarkar-like methods are beyond the scope of this book.

The Simplex Method: the Tableau Implementation

The simplex method works with the augmented matrix for the slack-variable form of a linear program, with the addition of a final equation $M - \mathbf{c}_e^T\mathbf{x}_e = u$ to allow for changing values of M; recall from Section 11.1 that we agreed to omit the column of the augmented matrices corresponding to M since that column never enters the computation. Thus the simplex method treats what is called the *simplex tableau*:

(11.39)
$$\mathbf{T} = \begin{bmatrix} \mathbf{A}_e & \mathbf{b} \\ -\mathbf{c}_e^T & u \end{bmatrix}.$$

The simplex method as described in Section 11.1 primarily involves performing Gauss–Jordan elimination steps on $\mathbf{T}$, producing a new tableau $\mathbf{T}'$; just as with systems of linear equations, the question of whether the solutions of the two problems are identical must be addressed.

(11.40) ***Theorem*** (equivalence of tableaus). Suppose that a sequence of elementary row operations transforms a simplex tableau $\mathbf{T}$ into a simplex tableau $\mathbf{T}'$, where

$$\mathbf{T} = \begin{bmatrix} \mathbf{A}_e & \mathbf{b} \\ -\mathbf{c}_e^T & u \end{bmatrix} \quad \text{and} \quad \mathbf{T}' = \begin{bmatrix} \mathbf{A}'_e & \mathbf{b}' \\ -(\mathbf{c}'_e)^T & u' \end{bmatrix}$$

and where the row operations performed with the bottom row can only be of the form: add to the bottom row a multiple of a higher row. Then $\mathbf{z}^*$ solves the linear program

$$\text{maximize} \quad M = \mathbf{c}_e^T\mathbf{x}_e$$
$$\text{subject to the constraints}$$
$$\mathbf{A}_e\mathbf{x}_e = \mathbf{b}, \qquad \mathbf{x}_e \geq \mathbf{0}$$

if and only if $\mathbf{z}^*$ solves the linear program

$$\text{maximize} \quad M' = \mathbf{c}_e'^T\mathbf{x}'_e$$
$$\text{subject to the constraints}$$
$$\mathbf{A}'_e\mathbf{x}'_e = \mathbf{b}' \qquad \mathbf{x}'_e \geq \mathbf{0}.$$

PROOF. Since the bottom row cannot be used to change the higher rows, $[\mathbf{A}'_e \ \ \mathbf{b}']$ is produced from $[\mathbf{A}_e \ \ \mathbf{b}]$ by elementary row operations; by Theorem 4.12 on Gauss elimination and solution sets, the constraint sets of the two linear programs are therefore identical. Because of the restriction on the operations performed on the bottom row, the bottom rows $[-\mathbf{c}_e^T \ \ u]$ and $[-\mathbf{c}_e'^T \ \ u']$ must be related by $[-\mathbf{c}_e'^T \ \ u'] = [\mathbf{c}_e^T \ \ u] + \mathbf{w}^T[\mathbf{A}_e \ \ \mathbf{b}]$.

Thus $-\mathbf{c}_e'^T = -\mathbf{c}_e^T + \mathbf{w}^T\mathbf{A}_e$ and $u' = u + \mathbf{w}^T\mathbf{b}$. We already know that the two constraint sets are identical; for any $\mathbf{z}$ in either, we have

$$M = \mathbf{c}_e^T\mathbf{z} + u \quad \text{and} \quad M' = \mathbf{c}_e'^T\mathbf{z} + u'.$$

Therefore,

$$\begin{aligned} M - M' &= (\mathbf{c}_e^T - \mathbf{c}_e'^T)\mathbf{z} + (u - u') \\ &= \mathbf{w}^T\mathbf{A}_e\mathbf{z} - \mathbf{w}^T\mathbf{b} \\ &= 0. \end{aligned}$$

Since the constraint sets are identical and the functions M and M' to maximize are identical, the two programs are equivalent. ■

The simplex method generates a sequence of tableaus, each of which describes equivalent programs; analysis of the simplex method then requires knowing:

1. how to tell from a tableau whether a solution has been found;
2. how to tell from a tableau whether a solution even exists;
3. how to produce the next tableau from the present one.

We answer these three "how to"s without proof—see Problems 5–8—by generalizing the method from Section 11.1. Recall that the simplex method is performed in such a way that at each step (for a nondegenerate linear program) precisely q entries—the nonbasic variables—of $\mathbf{x}_e$ equal zero, and precisely p entries—the basic variables—are strictly positive.

(11.41) 1. A solution has been found when all entries (other than possibly u in the last column) in the bottom row of the tableau are greater than or equal to zero.

If (11.41) fails to hold, then another simplex step is performed. The most negative entry in the bottom row (other than possibly u in the last column) is identified (in case of ties, pick any "winner"); this will correspond to a presently nonbasic (zero) variable x_n, say, and it is this variable that is to be made basic (positive) in the next simplex step. To see how large x_n can be made, each *positive* entry in the column of x_n is divided into the corresponding entry of the last column, and the smallest of these positive quotients will be the new value of x_n.

(11.42) 2. It is known that M can be made arbitrarily large in the constraint set, so that there is no optimal solution, when there are no strictly positive entries in the x_n-column of the variable x_n that is to be made basic.

If (11.42) fails to hold, suppose that (one of) the least quotient(s) above occurs in row i of the x_n-column.

(11.43) 3. The next tableau is produced by using the ith row to eliminate above and below that row in the x_n-column.

Assuming that the simplex method starts with a basic feasible solution, the process above increases the formerly zero variable x_n until it forces another variable to equal zero; thus there are still at least q zero variables—that is, another basic feasible vector has been produced. Since the process strictly increases M each time, it can never return to a previous basic feasible vector (of which there are only finitely many). We summarize.

(11.44) **Key Theorem** (simplex solution). If the simplex method is applied to a nondegenerate linear program in slack-variable form, then in a finite number of steps exactly one of the following alternatives will occur:
(a) The condition in (11.41) holds and an optimal basic feasible solution has been found.
(b) The condition in (11.42) holds and there is no optimal solution.

You should now be able to solve Problems 1 to 9.

The Simplex Method: Revised Implementation

One aspect of the simplex method as described appears quite inefficient. To determine whether we have found a solution and, if not, which nonbasic variable x_n to make basic, only the q nonzero entries of the bottom row of the tableau are needed; if a further simplex step is required, then only the $2p$ entries of the last column and the x_n-column are needed in order to determine whether M is unbounded, how large x_n becomes if not, and the next pivot for elimination. That is:

Only $2p + q$ of the $(p + 1)(q + p + 1)$ entries of the tableau are required in order to determine whether a solution has been found, whether a solution is known not to exist, and which are to be the next basic and nonbasic variables.

The so-called *revised simplex method* seeks to implement the simplex method without producing the $qp + p^2 + 1$ extraneous entries.

At the start, recall that we have a basic feasible solution: q entries of $\mathbf{x}_e$ have been set to zero and the p positive entries have been found as the solution to p linear equations in p unknowns. *All that is needed for this step is the inverse (or better yet, an LU-decomposition) of the $p \times p$ **submatrix** of the original $\mathbf{A}_e$ that is the coefficient matrix for this system, together with the right-hand side* **b**. Similarly, *with this inverse and the rows of the original tableau the new bottom row can be computed.* From the new bottom row can be learned as usual whether an optimal basic feasible vector has been found. If not and another simplex step is to be taken, what is needed next are the entries in the one column above the soon-to-be-basic variable;

this column has not yet been computed, but *this column can be found from the inverse obtained earlier and the original tableau's one corresponding column.* With this column in hand, we determine which presently nonzero (basic) variable becomes zero. We once again have q zero variables and p positive variables and can continue the process; it is important to observe that only one of the p nonzero variables has been changed, so that *the new $p \times p$ matrix whose inverse (or LU-decomposition) we require differs from the old $p \times p$ submatrix of $\mathbf{A}_e$ in only one column, and thus can be efficiently updated (rather than be computed from scratch) by methods such as in Theorem 3.63.*

The revised simplex method, as sketched above, essentially manipulates $p \times p$ submatrices of $\mathbf{A}_e$ and a few other columns without ever computing the new versions of full tableaus. When the number p of constraints is modest compared with the number q of variables, the savings in the revised simplex implementation can be quite significant. Most state-of-the-art software for the simplex method uses the revised implementation with modifications to handle numerical difficulties.

The Simplex Method: Getting Started

One issue we have rather consistently avoided until now is that of getting started: how to find an *initial* basic feasible solution. In theory we could just methodically require sets of q entries of $\mathbf{x}_e$ to equal zero, solve for the remaining p entries from the p equations, and then check to see whether $\mathbf{x}_e \geq \mathbf{0}$. Even in our small example in Section 11.1, however, there would be 10 such systems of three equations in three unknowns to solve, only five of which actually produce feasible vectors; as the numbers of variables and constraints grow, this approach becomes ever more unattractive. Most linear programming software includes a so-called *Phase I* stage designed to locate a basic feasible vector for the start of the simplex method (*Phase II*). The trick is to define a new linear program with two properties: (1) the new program has an obvious initial basic feasible vector; and (2) any solution to the new program is a basic feasible vector for the original program. If done properly, this will even detect whether the constraint set is empty for the original program. We consider an example in detail.

(11.45) ***Example.*** Consider the sample program of Section 11.1 with a modification in its second constraint:

(11.46)
$$
\begin{aligned}
&\text{maximize} \quad 40x_1 + 60x_2\\
&\text{subject to the constraints}\\
&2x_1 + x_2 \leq 70\\
&-x_1 - x_2 \leq -40\\
&x_1 + 3x_2 \leq 90\\
&x_1 \geq 0, \qquad x_2 \geq 0.
\end{aligned}
$$

No longer does $x_1 = x_2 = 0$ satisfy the constraints, nor is it particularly obvious that the constraints can be satisfied at all. If we insert a slack variable s in the awkward second inequality, the inequality becomes the equality

$$-x_1 - x_2 + s = -40,$$

and the difficulty shows up in the fact that trying $x_1 = x_2 = 0$ makes the slack variable s negative—which is not allowed. The clever trick is to introduce still another variable, called an *artificial variable*, to handle the difficulty. Specifically, we add slack variables x_3, x_4, and x_5 in the three inequalities as usual, but also we *subtract an artificial variable* (x_6) *in the awkward inequality*:

(11.47)
$$\begin{aligned} 2x_1 + x_2 + x_3 \qquad\qquad &= 70 \\ -x_1 - x_2 \quad + x_4 \quad - x_6 &= -40 \\ x_1 + 3x_2 \qquad + x_5 \qquad &= 90 \\ \text{all } x_i \geq 0. \end{aligned}$$

In this new set of constraints, we can set $x_1 = x_2 = 0$, giving $x_3 = 70$ and $x_5 = 90$ as in Section 11.1, and set $x_4 = 0$ with $x_6 = 40$. Note that we now have $p = 3$ constraints and $q + p = 6$ variables, so a basic feasible vector for these new constraints should have q $(=3)$ zero entries, which the vector above indeed has. Unfortunately, a vector that is feasible for (11.47) need not be feasible for the original set (11.46) of constraints *unless* x_6 *equals zero: If* $x_6 = 0$ in (11.47), then x_1 to x_5 is feasible for (11.46). Since $x_6 \geq 0$, seeking $x_6 = 0$ is equivalent to *minimizing* x_6 subject to (11.47); and if this minimum is strictly positive, then there is no feasible vector for the original program (11.47). That is,

> we can determine whether there are feasible vectors for (11.46), and can find one if one exists, by solving the linear program of minimizing x_6 subject to the constraints (11.47).

Rewriting the problem as a *maximization* problem by seeking to maximize $M = -x_6$, we are ready to begin the simplex method; the initial tableau is

$$\left[\begin{array}{rrrrrr|r} 2 & 1 & 1 & 0 & 0 & 0 & 70 \\ -1 & -1 & 0 & 1 & 0 & -1 & -40 \\ 1 & 3 & 0 & 0 & 1 & 0 & 90 \\ 0 & 0 & 0 & 0 & 0 & 1 & 0 \end{array}\right].$$

and the initial basic feasible vector is as found above:

$x_1 = x_2 = x_4 = 0$ as nonbasic variables
$x_3 = 70$, $x_5 = 90$, $x_6 = 40$ as basic variables.

This is *not* in the standard form (11.10) with each basic variable having only one nonzero entry in its column (and not in the bottom row): the x_6 column violates this, so we use the second row to eliminate the offending entry and obtain

$$\left[\begin{array}{rrrrrr|r} 2 & 1 & 1 & 0 & 0 & 0 & 70 \\ 1 & 1 & 0 & -1 & 0 & 1 & 40 \\ 1 & 3 & 0 & 0 & 1 & 0 & 90 \\ -1 & -1 & 0 & 1 & 0 & 0 & -40 \end{array}\right],$$

which is in the form (11.10). There is a tie for most negative entry in the bottom row; we (randomly) pick the second column. The quotients formed with positive entries in that column are 90/3, 40/1, and 70/1, of which the least—30—occurs with the third row; that (3, 2)-entry therefore is the pivot for Gauss–Jordan elimination, giving as the next tableau

$$\left[\begin{array}{rrrrrr|r} \frac{5}{3} & 0 & 1 & 0 & -\frac{1}{3} & 0 & 40 \\ \frac{2}{3} & 0 & 0 & -1 & \frac{2}{3} & 1 & 10 \\ \frac{1}{3} & 1 & 0 & 0 & \frac{1}{3} & 0 & 30 \\ -\frac{2}{3} & 0 & 0 & 1 & \frac{1}{3} & 0 & -10 \end{array}\right].$$

The (most) negative entry in the bottom row is $-\frac{2}{3}$; the quotients formed with positive entries in that column are

$$30/(\tfrac{1}{3}) = 90,\ 10/(\tfrac{2}{3}) = 15, \quad \text{and} \quad 40/(\tfrac{5}{3}) = 24,$$

of which the least—15—occurs with the second row; that (2, 1)-entry therefore is the pivot for Gauss–Jordan elimination, giving as the next tableau

$$\left[\begin{array}{rrrrrr|r} 0 & 0 & 1 & \frac{5}{2} & -2 & -\frac{5}{2} & 15 \\ 1 & 0 & 0 & -\frac{3}{2} & 1 & \frac{3}{2} & 15 \\ 0 & 1 & 0 & \frac{1}{2} & 0 & -\frac{1}{2} & 25 \\ 0 & 0 & 0 & 0 & 1 & 1 & 0 \end{array}\right].$$

Since there are no negative entries in the bottom row, the simplex method has found a solution to the problem of maximizing $-x_6$ subject to the constraints (11.47). The basic variables are those corresponding to the unit column matrices—x_1, x_2, x_3—and the nonbasic variables are the others—x_4, x_5, x_6; since nonbasic variables equal zero, this gives

$$x_4 = x_5 = x_6 = 0, \qquad x_1 = 15, \qquad x_2 = 25, \qquad x_3 = 15.$$

Note that $x_6 = 0$, and hence x_1 through x_5 give a basic feasible vector for the original linear program (11.46). For the purpose of illustration, we proceed with the solution of the original linear program (11.46) in slack-variable form.

This gives an initial tableau of

$$\left[\begin{array}{rrrrr|r} 2 & 1 & 1 & 0 & 0 & 70 \\ -1 & -1 & 0 & 1 & 0 & -40 \\ 1 & 3 & 0 & 0 & 1 & 90 \\ -40 & -60 & 0 & 0 & 0 & 0 \end{array}\right].$$

The basic feasible vector found by the artificial variable method has x_1, x_2, x_3 as basic variables, and so—according to (11.10)—the corresponding columns should be unit column matrices. Only x_3 satisfies this, so we have to perform Gauss–Jordan elimination in the x_1- and x_2- columns; this produces a tableau in the proper form of (11.10):

$$\left[\begin{array}{rrrrr|r} 0 & 0 & 1 & \frac{5}{2} & \frac{1}{2} & 15 \\ 1 & 0 & 0 & -\frac{3}{2} & -\frac{1}{2} & 15 \\ 0 & 1 & 0 & \frac{1}{2} & \frac{1}{2} & 25 \\ 0 & 0 & 0 & -30 & 10 & 2100 \end{array}\right].$$

The (most) negative entry in the bottom row is -30; the quotients with the positive entries above that are $25/(\frac{1}{2}) = 50$ and $15/(\frac{5}{2}) = 6$, of which the smallest—6—occurs in the top row. Therefore, we use that (1, 4)-entry as the pivot for Gauss–Jordan elimination, producing the tableau

$$\left[\begin{array}{rrrrr|r} 0 & 0 & \frac{2}{5} & 1 & \frac{1}{5} & 6 \\ 1 & 0 & \frac{3}{5} & 0 & -\frac{1}{5} & 24 \\ 0 & 1 & -\frac{1}{2} & 0 & \frac{2}{5} & 22 \\ 0 & 0 & 30 & 0 & 16 & 2280 \end{array}\right].$$

Since there are no negative entries on the bottom row, we have reached an optimal basic feasible vector. The basic (nonzero) variables correspond to the unit column matrices in the tableau and are therefore x_1, x_2, and x_4, leaving x_3 and x_5 as nonbasic (zero) variables. Setting $x_3 = x_5 = 0$ gives $x_1 = 24$, $x_2 = 22$, and $x_4 = 6$ as an optimal solution, with $M = 2280$. This solves the original linear program (11.46) with $x_1 = 24$, $x_2 = 22$, and $M = 2280$.

PROBLEMS 11.3

1. Write out the linear programs represented by each of the tableaus (11.11) and (11.13) produced in solving (11.1).

▷ **2.** Maximize $-4x_1 - x_2 + x_3 - 2x_4$

subject to the constraints

$$\begin{aligned} 3x_1 - 3x_2 + x_3 \qquad &= 3 \\ 6x_2 - 2x_3 + x_4 &= 2 \end{aligned}$$

all $x_i \geq 0$.

3. Maximize $x_1 + 4x_2 + 3x_3$

subject to the constraints

$$3x_1 + 2x_2 + x_3 \leq 4$$

$$x_1 + 5x_2 + 4x_3 \leq 14$$

all $x_i \geq 0$.

4. Maximize $3x_1 - x_2$

subject to the constraints

$$-2x_1 + x_2 \leq 1$$

$$x_1 - 2x_2 \leq 2$$

$$x_1 \geq 0, \qquad x_2 \geq 0.$$

5. Show that (11.41) on detecting a solution in the simplex method is valid as follows:

(a) Show that M cannot be increased by making any of the nonbasic variables positive.

(b) Use nondegeneracy to show that the present feasible vector is the only one with the present nonbasic variables all zero.

(c) Deduce that the present feasible vector is optimal.

6. Show that (11.42) on detecting an unbounded solution is valid.

▷ **7.** Show that, in a nondegenerate problem, if (11.41) and (11.42) fail to hold at a feasible vector, the last column of the present tableau (except possibly for the bottom entry) is strictly positive and a new basic feasible solution can be found with a strictly larger M.

8. If all entries of the bottom row of a tableau (except possibly in the last column) are greater than or equal to zero and if one of the entries corresponding to a *nonbasic* variable equals zero, there is more than one optimal basic feasible vector. Prove this by showing how another can be constructed by increasing that nonbasic variable.

▷ **9.** (a) Maximize $40x_1 + 40x_2$ subject to the constraints in (7.1) by arguing geometrically, and show that there are infinitely many optimal feasible vectors and two optimal basic feasible vectors.

(b) Use the simplex method to solve this problem and verify that the condition described in Problem 8 occurs to signal the nonuniqueness.

10. In the revised simplex method, the main need is not for the inverse of the $p \times p$ submatrix $\mathbf{B}$ of $\mathbf{A}_e$ but for the ability to solve $\mathbf{Bz} = \mathbf{w}$ given various $\mathbf{w}$. $\mathbf{B}$ changes in only one column at each step. Explain how to use an *LU*-decomposition of $\mathbf{B}$ with Theorem 3.63(b) to take advantage of this.

▷ **11.** (a) Use artificial variables to find an initial basic feasible point for the linear program below.

(b) Solve it.

$$\begin{aligned}&\text{Maximize} \quad x_1 + x_2\\&\text{subject to the constraints}\\&-x_1 + x_2 \le 10\\&\quad x_1 + x_2 \ge 5\\&\ 2x_1 + x_2 \le 40\\&\quad x_1 \ge 0, \qquad x_2 \ge 0.\end{aligned}$$

12. Do as in Problem 11(a) and (b) but for the linear program

$$\begin{aligned}&\text{maximize} \quad 2x_1 + 3x_2\\&\text{subject to the constraints}\\&\ x_1 - \quad x_2 \ge 2\\&\ x_1 + 0.2x_2 \ge 4\\&3x_1 - \quad x_2 \ge 14\\&\ x_1 \ge 0, \qquad x_2 \ge 0.\end{aligned}$$

13. Do as in Problem 11(a) and (b) but for the linear program

$$\begin{aligned}&\text{minimize} \quad 2x_1 - x_2\\&\text{subject to the constraints}\\&\quad x_1 + 2x_2 \ge 3\\&10x_1 + \quad x_2 \ge 11\\&\ 4x_1 + 3x_2 \le 33\\&x_1 \ge 0, \qquad x_2 \ge 0.\end{aligned}$$

14. Do as in Problem 11(a) and (b) but for the linear program

$$\begin{aligned}&\text{maximize} \quad x_1 + x_2 + x_3\\&\text{subject to the constraints}\\&\ x_1 + \ x_2 \qquad\quad \ge 3\\&\ x_1 + 2x_2 + x_3 \ge 4\\&2x_1 + \ x_2 + x_3 \le 2\\&\ x_1 \ge 0, \qquad x_2 \ge 0, \qquad x_3 \ge 0.\end{aligned}$$

▷ **15.** Use artificial variables to determine whether there are any vectors satisfying

$$\begin{aligned} 2x_1 + x_2 &\geq 70 \\ x_1 + x_2 &\leq 40 \\ x_1 + 3x_2 &\geq 90 \\ x_1 \geq 0, \qquad x_2 &\geq 0. \end{aligned}$$

16. Do as in Problem 11(a) and (b) but for the linear program

minimize $3x_1 - 5x_2$

subject to the constraints

$$\begin{aligned} 2x_1 + x_2 &\geq 70 \\ x_1 + x_2 &\geq 40 \\ x_1 + 3x_2 &\geq 90 \\ x_1 \geq 0, \qquad x_2 &\geq 0. \end{aligned}$$

17. Do as in Problem 11(a) and (b) but for the linear program

maximize $3x_2 - x_1$

subject to the constraints

$$\begin{aligned} x_1 &\leq 10 \\ x_1 + x_2 &\geq 5 \\ x_1 - x_2 &\leq 5 \\ x_1 \geq 0, \qquad x_2 &\geq 0. \end{aligned}$$

11.4 DUALITY

Section 11.1 showed that two different linear programs—(11.1)/(11.2) and (11.23)/(11.24)—were surprisingly related both geometrically and through the application from which they arose. The present section treats this subject of *duality* more generally and indicates its computational importance.

(11.48) ***Definition.*** A linear program in standard form

maximize $M = \mathbf{c}^T\mathbf{x}$

subject to the constraints

$\mathbf{Ax} \leq \mathbf{b}, \qquad \mathbf{x} \geq \mathbf{0}, \quad$ where $\mathbf{A}$ is $p \times q$, is said to be a *primal* linear program. The *dual* of this primal is defined to be the linear program

$$\text{minimize} \quad m = \mathbf{b}^T\mathbf{y}$$

subject to the constraints

$$\mathbf{A}^T\mathbf{y} \geq \mathbf{c}, \qquad \mathbf{y} \geq \mathbf{0}, \quad \text{where} \quad \mathbf{A}^T \text{ is } q \times p.$$

(11.49) ***Theorem.*** The dual of the dual is the primal.

PROOF. To write the dual of the dual we need the dual in standard form. We can rewrite the dual in Definition 11.48 as

$$\text{maximize} \quad M' = \mathbf{c}'^T\mathbf{y} \quad \text{with} \quad \mathbf{c}' = -\mathbf{b},$$

subject to the constraints

$$\mathbf{A}'\mathbf{y} \leq \mathbf{b}', \qquad \mathbf{y} \geq \mathbf{0} \quad \text{with} \quad \mathbf{A}' = -\mathbf{A}^T \quad \text{and} \quad \mathbf{b}' = -\mathbf{c},$$

which is in standard form. Its dual is

$$\text{minimize} \quad m' = \mathbf{b}'^T\mathbf{z}\,(= -\mathbf{c}^T\mathbf{z})$$

subject to the constraints

$$\mathbf{A}'^T\mathbf{z} \geq \mathbf{c}', \qquad \mathbf{z} \geq \mathbf{0} \quad (\text{that is, } -\mathbf{A}\mathbf{z} \geq -\mathbf{b}, \mathbf{z} \geq \mathbf{0}),$$

which is equivalent to the primal. ■

Duality and the Simplex Method

In the final tableau (11.13) for the simplex method applied to solve the primal linear program (11.2) in Section 11.1, the entries 0, 30 and 10 on its bottom row in the columns corresponding to the (basic) slack variables in the original tableau (11.8) are precisely the numbers used as the entries of $\mathbf{y}^*$ in (11.25) that was shown to solve the dual linear program (11.24). That is, the simplex tableau for the optimal solution to the primal program also contained the optimal solution to the dual program. This is true more generally; the relationship is simplest when $\mathbf{x} = \mathbf{0}$ can be taken as an initial feasible vector (that is, when $\mathbf{b} \geq \mathbf{0}$), so we make this assumption for convenience.

(11.50) **Key Theorem** (primal, dual, and simplex method). Suppose that the simplex method has been applied to the slack-variable form of the nondegenerate primal linear program of Definition 11.48, in which $\mathbf{b} \geq \mathbf{0}$, and has obtained a final tableau with an optimal basic feasible solution for the primal program. Let $[\mathbf{y}^{*T} \quad u]$ denote the last $q + 1$ entries of the bottom row of that final tableau. Then $\mathbf{y}^*$ is an optimal solution of the dual linear program of Definition 11.48 and $\mathbf{b}^T\mathbf{y}^* = \mathbf{c}^T\mathbf{x}^* = u$, where $\mathbf{x}^*$ is optimal for the primal program.

PROOF. Since $\mathbf{b} \geq \mathbf{0}$ we can take $\mathbf{x}$, the first q entries of $\mathbf{x}_e$, to be zero and obtain an initial basic feasible vector. This means that the starting tableau

$$\mathbf{T} = \begin{bmatrix} \mathbf{A} & \mathbf{I} & \mathbf{b} \\ -\mathbf{c}^T & \mathbf{0}^T & 0 \end{bmatrix}$$

has the standard form (11.10) for beginning the simplex method. The final tableau $\mathbf{T}'$ is produced from $\mathbf{T}$ by a sequence of elementary row operations, with the restriction that the only operations involving the bottom row replace that bottom row by itself plus a multiple of a higher row. This means that the bottom row of $\mathbf{T}'$ equals the bottom row of $\mathbf{T}$ plus a linear combination of the higher rows of $\mathbf{T}$. The bottom row of $\mathbf{T}$ is just $[-\mathbf{c}^T \quad \mathbf{0}^T \quad 0]$, and we partition the bottom row of $\mathbf{T}'$ as $[\mathbf{d}^T \quad \mathbf{y}^{*T} \quad u]$ in the same manner. We have shown that there is some $\mathbf{v}$ with

$$[\mathbf{d}^T \quad \mathbf{y}^{*T} \quad u] = [-\mathbf{c}^T \quad \mathbf{0}^T \quad 0] + \mathbf{v}^T[\mathbf{A} \quad \mathbf{I} \quad \mathbf{b}].$$

The rule for partitioned multiplication tells us that

$$\mathbf{d}^T = -\mathbf{c}^T + \mathbf{v}^T\mathbf{A}, \qquad \mathbf{y}^{*T} = \mathbf{v}^T, \quad \text{and} \quad u = \mathbf{v}^T\mathbf{b}.$$

Now, recall what it means for the simplex method to terminate with the tableau $\mathbf{T}'$: the bottom row (except possibly for u) is nonnegative, and u equals the optimal value $\mathbf{c}_e^T\mathbf{x}_e$ since the nonzero entries in the bottom row of $\mathbf{T}'$ correspond to the zero (nonbasic) entries of the $\mathbf{x}_e$ according to (11.10). Thus $\mathbf{d} \geq \mathbf{0}$, $\mathbf{y}^* \geq \mathbf{0}$, and $\mathbf{v}^T = \mathbf{y}^{*T}$, from which follows

$$\mathbf{A}^T\mathbf{y}^* \geq \mathbf{c}, \qquad \mathbf{y}^* \geq \mathbf{0}.$$

That is, $\mathbf{y}^*$ is a feasible vector for the dual linear program. From $u = \mathbf{v}^T\mathbf{b}$ and $\mathbf{c}_e^T\mathbf{x}_e = u$ for the optimal $\mathbf{x}_e$, since $\mathbf{v} = \mathbf{y}^*$ it follows that

$$\mathbf{b}^T\mathbf{y}^* = \mathbf{c}^T\mathbf{x}^* \text{ for the optimal } \mathbf{x}^* \text{ formed from the optimal } \mathbf{x}_e.$$

But then if $\mathbf{y}$ is any feasible vector for the dual program with $\mathbf{y} \geq \mathbf{0}$ and $\mathbf{A}^T\mathbf{y} \geq \mathbf{c}$, since $\mathbf{A}\mathbf{x}^* \leq \mathbf{b}$ we get

$$\mathbf{b}^T\mathbf{y} \geq (\mathbf{A}\mathbf{x}^*)^T\mathbf{y} = \mathbf{x}^{*T}\mathbf{A}^T\mathbf{y} \geq \mathbf{x}^{*T}\mathbf{c} = \mathbf{b}^T\mathbf{y}^*,$$

as above. But if $\mathbf{b}^T\mathbf{y} \geq \mathbf{b}^T\mathbf{y}^*$ for all feasible $\mathbf{y}$, then $\mathbf{y}^*$ solves the dual linear program. ■

(11.51) ***Example.*** Consider the primal linear program

$$\begin{aligned} \text{maximize} \quad & 2x_1 + 2x_2 + 3x_3 \\ \text{subject to} \quad & 2x_1 + x_2 + 6x_3 \leq 1 \\ & x_1 + 2x_2 + x_3 \leq 1 \\ x_1 \geq 0, \quad & x_2 \geq 0, \quad x_3 \geq 0, \end{aligned}$$

and its dual linear program

$$\begin{aligned} \text{minimize} \quad & y_1 + y_2 \\ \text{subject to} \quad & 2y_1 + y_2 \geq 2 \\ & y_1 + 2y_2 \geq 2 \\ & 6y_1 + y_2 \geq 3 \\ y_1 \geq 0, \quad & y_2 \geq 0. \end{aligned}$$

The initial tableau for the slack-variable version of the primal is

$$\left[\begin{array}{rrrrr|r} 2 & 1 & 6 & 1 & 0 & 1 \\ 1 & 2 & 1 & 0 & 1 & 1 \\ -2 & -2 & -3 & 0 & 0 & 0 \end{array}\right].$$

We pivot on the (1, 3)-entry and produce the tableau

$$\left[\begin{array}{rrrrr|r} \frac{1}{3} & \frac{1}{6} & 1 & \frac{1}{6} & 0 & \frac{1}{6} \\ \frac{2}{3} & \frac{11}{6} & 0 & -\frac{1}{6} & 1 & \frac{5}{6} \\ -1 & -\frac{3}{2} & 0 & \frac{1}{2} & 0 & \frac{1}{2} \end{array}\right].$$

We pivot on the (2, 2)-entry and produce the tableau

$$\left[\begin{array}{rrrrr|r} \frac{3}{11} & 0 & 1 & \frac{2}{11} & -\frac{1}{11} & \frac{1}{11} \\ \frac{4}{11} & 1 & 0 & -\frac{1}{11} & \frac{6}{11} & \frac{5}{11} \\ -\frac{5}{11} & 0 & \frac{1}{2} & \frac{4}{11} & \frac{9}{11} & \frac{13}{11} \end{array}\right].$$

We pivot on the (1, 1)-entry and produce the tableau

$$\left[\begin{array}{rrrrr|r} 1 & 0 & \frac{11}{3} & \frac{2}{3} & -\frac{1}{3} & \frac{1}{3} \\ 0 & 1 & -\frac{4}{3} & -\frac{1}{3} & \frac{2}{3} & \frac{1}{3} \\ 0 & 0 & \frac{13}{6} & \frac{2}{3} & \frac{2}{3} & \frac{4}{3} \end{array}\right],$$

so that the solution in the $\mathbf{x}$ variables is $x_1 = \frac{1}{3}$, $x_2 = \frac{1}{3}$, $x_3 = 0$, while the solution in the $\mathbf{y}$ variables comes from the fourth and fifth entries in the last row, namely $y_1 = y_2 = \frac{2}{3}$. Note that, as claimed, the minimum value $y_1 + y_2$ for the dual ($\frac{4}{3}$) equals the maximum value $2x_1 + 2x_2 + 3x_3$ for the primal.

Using Duality Computationally

The primal/dual relationship can be quite useful computationally, since the simplex tableau for the solution to either one provides a solution to the other (since the dual of the dual is the primal). Thus in practice we can solve whichever linear program is more convenient computationally.

Suppose, for example, that we have far more constraints than variables, so that $p \gg q$. Each step of the simplex method requires elimination in a large number

p of rows, and the revised simplex method requires inverses of large $p \times p$ matrices. The dual program, however, is $q \times p$; the revised simplex method, for example, requires only inverses of the much smaller $q \times q$ matrices when applied to this dual. Once the $q \times p$ dual is solved, a solution to the primal can be read off from the final tableau as in **Key Theorem 11.50**. Solving the dual can be much more efficient when $p \gg q$.

Suppose, as another example, that you are to solve several linear programs that differ only by the addition of new constraints, a situation that arises in applications. Once you have found an optimal solution for one primal program, it is not clear how to make use of it to make solving the next program easier since that optimal solution need not satisfy the additional constraint. However, adding a constraint to the primal merely adds a *variable* to the dual, and the optimal solution of the first dual provides an initial feasible vector for the new dual just by setting the new dual variable to zero. The dual programs can handle this situation easily.

(11.52) ***Example.*** Suppose that in our sample primal linear program (11.1) that modeled a production-planning problem, the plant management learns that government safety regulations on their products will require an additional manufacturing step on a fourth type of machine, and that the available machine time and requirements for each product's use of the machine translate into the additional constraint

$$1.5x_1 + 2x_2 \leq 70.$$

The original optimal solution is $x_1 = 15$, $x_2 = 25$, but this does not satisfy the new constraint. Rather than starting the simplex method over from $x_1 = x_2 = 0$ to solve the new linear program formed by joining this new constraint to those in (11.1), we instead treat the dual linear program (restated in maximization language):

$$\text{maximize} \quad -70y_1 - 40y_2 - 90y_3 - 70y_4$$

subject to the constraints

$$\begin{aligned} 2y_1 + y_2 + y_3 + 1.5y_4 &\geq 40 \\ y_1 + y_2 + 3y_3 + 2y_4 &\geq 60 \end{aligned}$$

all $y_i \geq 0$.

This differs from the original program's dual (11.23) only in the presence of an additional variable y_4. The final accomplishment of Section 11.1 was to show that $y_1 = 0$, $y_2 = 30$, $y_3 = 10$ was an optimal feasible vector for the original dual; by joining this with $y_4 = 0$ we obtain a feasible vector (although no longer optimal) for our new dual program. It is, in fact, a *basic* feasible vector since it has two zero entries. Inserting slack variables y_5 and y_6 in the inequalities produces an initial tableau

$$\left[\begin{array}{cccccc|c} 2 & 1 & 1 & \frac{3}{2} & -1 & 0 & 40 \\ 1 & 1 & 3 & 2 & 0 & -1 & 60 \\ 70 & 40 & 90 & 70 & 0 & 0 & 0 \end{array}\right].$$

The basic (nonzero) variables are $y_2 = 30$ and $y_3 = 10$, so the y_2- and y_3-columns should be unit column matrices; Gauss–Jordan elimination produces a tableau that complies with (11.10):

$$\left[\begin{array}{cccccc|c} \frac{5}{2} & 1 & 0 & \frac{5}{4} & -\frac{3}{2} & \frac{1}{2} & 30 \\ -\frac{1}{2} & 0 & 1 & \frac{1}{4} & \frac{1}{2} & -\frac{1}{2} & 10 \\ 15 & 0 & 0 & -\frac{5}{2} & 15 & 25 & -2100 \end{array}\right].$$

The simplex procedure tells us to pivot on the (1, 4)-entry $\frac{5}{4}$, giving the tableau

$$\left[\begin{array}{cccccc|c} 2 & \frac{4}{5} & 0 & 1 & -\frac{6}{5} & \frac{2}{5} & 24 \\ -1 & -\frac{1}{5} & 1 & 0 & \frac{4}{5} & -\frac{3}{5} & 4 \\ 20 & 2 & 0 & 0 & 12 & 26 & -2040 \end{array}\right].$$

Since the bottom entries are nonnegative (other than possibly the last column), we have a solution $y_1 = y_2 = y_5 = y_6 = 0$, $y_3 = 4$, and $y_5 = 24$. This was not what interested us, however; we solved the dual so as to solve the primal (the dual's dual). According to **Key Theorem 11.50**, we can read off the primal solution from the entries in the bottom row beneath the slack variables: $x_1 = 12$, $x_2 = 26$. This production plan gives a new maximum profit of $40(12) + 60(26) = 2040$. Use of the dual required one simplex step on a 3×7 tableau; use of the primal would have required two simplex steps on a 5×7 tableau.

Duality Theory

The theory of duality in linear programming is elegant and deep, with interpretations—as we saw in Section 11.1 in our sample problem—that are algebraic, geometric, and applied. We have already developed and proved some portions of this theory—Theorem 11.49 and **Key Theorem 11.50**, for example. We restate some of that and present some additional facts in the following results.

(11.53) **Key Theorem** (duality theory). Let "the primal" and "the dual" refer to the linear programs of Definition 11.48. Then:

(a) The dual of the dual is the primal.

(b) If $\mathbf{x}$ is feasible for the primal and $\mathbf{y}$ is feasible for the dual, then $M = \mathbf{c}^T\mathbf{x} \leq \mathbf{b}^T\mathbf{y} = m$.

(c) If $\mathbf{x}^*$ is feasible for the primal and $\mathbf{y}^*$ is feasible for the dual and $\mathbf{c}^T\mathbf{x}^* = \mathbf{b}^T\mathbf{y}^*$, then $\mathbf{x}^*$ is optimal for the primal and $\mathbf{y}^*$ is optimal for the dual.

(d) If the dual (or primal) has a feasible vector, then the values of M (or m) on the primal (or dual) constraint set are bounded above (or below).
(e) If both the primal and the dual are nondegenerate and both have feasible vectors, then both have optimal feasible vectors and the maximum value of the primal's M equals the minimum value of the dual's m.
(f) If the primal (or dual) fails to have a feasible vector and the dual (or primal) is nondegenerate, then neither has an optimal feasible vector.

PROOF

(a) See Theorem 11.49.
(b) Simply compute: $\mathbf{c}^T\mathbf{x} \le (\mathbf{y}^T\mathbf{A})\mathbf{x} = \mathbf{y}^T(\mathbf{A}\mathbf{x}) \le \mathbf{y}^T\mathbf{b} = \mathbf{b}^T\mathbf{y}$.
(c) If $\mathbf{x}$ is feasible for the primal, then by (b) $\mathbf{c}^T\mathbf{x} \le \mathbf{b}^T\mathbf{y}^*$, which equals $\mathbf{c}^T\mathbf{x}^*$ by assumption; but $\mathbf{c}^T\mathbf{x} \le \mathbf{c}^T\mathbf{x}^*$ for all $\mathbf{x}$ feasible for the primal means that $\mathbf{x}^*$ is optimal for the primal. Similarly for the dual optimality of $\mathbf{y}^*$.
(d) This follows immediately from (b).
(e) By (d), both programs are bounded; by **Key Theorem 11.44** on simplex solutions, both programs have optimal solutions obtainable by the simplex method. **Key Theorem 11.50** on the primal, dual, and simplex method shows that the minimum ($\mathbf{b}^T\mathbf{y}^*$ in that theorem) for the dual equals the maximum ($\mathbf{c}^T\mathbf{x}^*$ there) for the primal.
(f) If the dual *does* have an optimal vector, then **Key Theorem 11.44** says that it can be found by the simplex method and **Key Theorem 11.50** says that from that final tableau we can construct an (optimal) feasible vector for the primal, which by assumption does not exist. Thus the dual cannot have an optimal vector. Similarly for the alternate statement. ■

(11.54) ***Example.*** Consider the inequalities

$$2x_1 + x_2 \ge 70$$
$$x_1 + x_2 \le 40$$
$$x_1 + 3x_2 \ge 90$$
$$x_1 \ge 0, \qquad x_2 \ge 0$$

and the question of whether there exist any vectors satisfying them. We add a function to maximize, $M = \mathbf{c}^T\mathbf{x}$, with $\mathbf{c} = [1 \quad 1]^T$, say, and consider the dual (multiplying the $\ge$ inequalities by -1 in order to get the proper form for finding the dual):

minimize $\quad -70y_1 + 40y_2 - 90y_3$

subject to the constraints

$$-2y_1 + y_2 - y_3 \ge 1$$
$$-y_1 + y_2 - 3y_3 \ge 1$$
$$y_1 \ge 0, \qquad y_2 \ge 0, \qquad y_3 \ge 0.$$

If we let $y_1 = 2\alpha$, $y_2 = 1 + 5\alpha$, and $y_3 = \alpha$, then the dual constraints are satisfied if $\alpha \geq 0$; also

$$m = -70y_1 + 40y_2 - 90y_3 = 40 - 3\alpha,$$

which is unbounded below since we can take α arbitrarily large and positive. By **Key Theorem 11.53(d)**, there can exist no feasible points for the primal constraints.

There are elegant connections between duality in linear programming and the theory of linear inequalities, especially so-called *theorems of the alternative* (somewhat akin to the Fredholm Alternative for equalities). As an illustration of the connection, suppose that it is impossible to find an $\mathbf{x}$ that satisfies $\mathbf{Ax} \leq \mathbf{0}$, $\mathbf{x} \geq \mathbf{0}$, and $\mathbf{c}^T\mathbf{x} > 0$. Since $\mathbf{x} = \mathbf{0}$ satisfies the first two inequalities, this means that $\mathbf{0}$ solves the linear program of maximizing $\mathbf{c}^T\mathbf{x}$ subject to $\mathbf{Ax} \leq \mathbf{0}$ and $\mathbf{x} \geq \mathbf{0}$. Were this linear program nondegenerate—*which it is not, it is in fact* ***degenerate***—we could use (f) of our duality-theory theorem to deduce that there is a vector $\mathbf{y}$ that is feasible for the dual, so $\mathbf{A}^T\mathbf{y} \geq \mathbf{c}$ and $\mathbf{y} \geq \mathbf{0}$. Although we have not developed here a powerful enough theory to prove this, it is nonetheless true that such a $\mathbf{y}$ exists. It is also easy to see that the converse holds: If there is such a $\mathbf{y}$, then there can be no $\mathbf{x}$ as above since, if there were, we would have $0 < \mathbf{c}^T\mathbf{x} \leq (\mathbf{y}^T\mathbf{A})\mathbf{x} = \mathbf{y}^T(\mathbf{Ax}) \leq 0$, a self-contradiction. So the following theorem of the alternative is in fact valid (although we lack the tools for a short proof).

(11.55) ***Theorem*** (a theorem of the alternative). Let $p \geq q$ $\mathbf{A}$ and $q \geq 1$ $\mathbf{c}$ be real matrices. Then exactly one of the following two alternative holds:
Either
1. there exists an $\mathbf{x}$ solving $\mathbf{Ax} \leq 0$, $\mathbf{x} \geq 0$, $\mathbf{c}^T\mathbf{x} > 0$;

or

2. there exists a $\mathbf{y}$ solving $\mathbf{A}^T\mathbf{y} \geq \mathbf{c}$, $\mathbf{y} \geq \mathbf{0}$.

There are many related such theorems, many of which are equivalent. We conclude with one of the most famous as a corollary of the preceding; geometrically, it states that when no vector $\mathbf{z}$ that makes an obtuse angle with each of a set of vectors $\mathbf{b}_i$ can possibly make an acute angle with $\mathbf{d}$, then $\mathbf{d}$ is a nonnegative linear combination of the $\mathbf{b}_i$.

(11.56) ***Corollary*** (Farkas' theorem of the alternative). Let $p \times q$ $\mathbf{B}$ and $q \times 1$ $\mathbf{d}$ be real matrices. Then exactly one of the following two alternatives holds:
Either
1. there exists a $\mathbf{z}$ solving $\mathbf{Bz} \leq \mathbf{0}$, $\mathbf{d}^T\mathbf{z} > \mathbf{0}$;

or

2. there exists a $\mathbf{y}$ solving $\mathbf{B}^T\mathbf{y} = \mathbf{d}$, $\mathbf{y} \geq \mathbf{0}$.

PROOF. Let $\mathbf{A} = [\mathbf{B} \quad -\mathbf{B}]$, $\mathbf{c}^T = [\mathbf{d}^T \quad -\mathbf{d}^T]$, $\mathbf{x}^T = [\mathbf{u}^T \quad \mathbf{v}^T]$. Then (1) is equivalent to solving $\mathbf{Ax} \leq \mathbf{0}$, $\mathbf{x} \geq \mathbf{0}$, where we write $\mathbf{z} = \mathbf{u} - \mathbf{v}$ with $\mathbf{u} \geq \mathbf{0}$ and $\mathbf{v} \geq \mathbf{0}$. Thus, by Theorem 11.55, not solving (1) is equivalent to solving for $\mathbf{y}$ with $\mathbf{A}^T\mathbf{y} \geq \mathbf{c}$ and $\mathbf{y} \geq \mathbf{0}$; but $\mathbf{A}^T\mathbf{y} \geq \mathbf{c}$ is just

$$\begin{bmatrix} \mathbf{B}^T \\ -\mathbf{B}^T \end{bmatrix} \mathbf{y} \geq \begin{bmatrix} \mathbf{d}^T \\ -\mathbf{d}^T \end{bmatrix},$$

which is equivalent to $\mathbf{B}^T\mathbf{y} = \mathbf{d}$. ■

PROBLEMS 11.4

▷ **1.** By rewriting the program in standard form, find the dual of the linear program

$$\text{maximize} \quad \mathbf{f}^T\mathbf{z}$$

subject to the constraints

$$\mathbf{Bz} = \mathbf{d}, \qquad \mathbf{z} \geq \mathbf{0}.$$

2. By rewriting the program in standard form, find the dual of the linear program (there are *no* nonnegativity constraints)

$$\text{minimize} \quad \mathbf{f}^T\mathbf{z}$$

subject to the constraints

$$\mathbf{Bz} = \mathbf{d}.$$

▷ **3.** Find the dual of each of the following problems from Section 11.2:
(a) Problem 1
(b) Problem 3
(c) Problem 5

4. Find the dual of the dual of (11.1) and show that it is just (11.1) again.

5. Use the simplex method on the dual in order to solve the linear program

$$\text{maximize} \quad x_1 + x_2$$

subject to the constraints

$$\begin{aligned} -x_1 + x_2 &\leq 10 \\ x_1 + 2x_2 &\leq 50 \\ 5x_1 + x_2 &\leq 160 \\ x_2 &\leq 15 \end{aligned}$$

$$x_1 \geq 0, \qquad x_2 \geq 0.$$

6. Suppose that **A** is $p \times q$ in a linear program in standard form, so that the tableau for the slack-variable version of the primal is $(p+1) \times (q+p+1)$ while the tableau for the slack-variable version of the dual is $(q+1) \times (q+p+1)$. Count the arithmetic operations involved in Gauss–Jordan elimination in a column of each tableau and then explain why it is generally better to solve the dual than the primal when $p \gg q$.
7. Join a new constraint $4x_1 + x_2 \leq 80$ to the linear program (11.1) and then use duality as in Example 11.52 to solve the new primal.
8. Use the simplex method to solve the dual in Example 11.54 and thereby discover that the dual is unbounded and that the primal has no feasible vector.
9. Use duality to show that the linear program in Problem 16 of Section 11.3 has no optimal solution.

▷ 10. Use duality to discover the nature of or the solution to the linear program

$$\begin{aligned}&\text{minimize} \quad -x_1 + 2x_2\\&\text{subject to the constraints}\\&-5x_1 + x_2 \geq 2\\&\;\;4x_1 - x_2 \geq 3\\&x_1 \geq 0, \qquad x_2 \geq 0.\end{aligned}$$

11. Use duality to discover the nature of or the solution to the linear program

$$\begin{aligned}&\text{maximize} \quad x_1 - x_2\\&\text{subject to the constraints}\\&-2x_1 + \;\;x_2 \leq 2\\&\;\;\;x_1 - 2x_2 \leq 1\\&\;\;\;x_1 + \;\;x_2 \leq 4\\&x_1 \geq 0, \qquad x_2 \geq 0.\end{aligned}$$

▷ 12. Theorem 11.55 was used to prove the Farkas theorem of the alternative; show that these two theorems are equivalent by using the Farkas theorem to prove Theorem 11.55.

13. Apply the Farkas theorem of the alternative to $[\mathbf{z}^T \;\; \alpha]^T$ in an appropriate way in order to prove the following theorem of the alternative:
either

there is a solution to $\mathbf{Bz} \leq \mathbf{f}$,

or

there is a solution to $\mathbf{B}^T\mathbf{w} = \mathbf{0}$, $\mathbf{w} \geq \mathbf{0}$, $\mathbf{f}^T\mathbf{w} = -1$,

but never both.

11.5 MISCELLANEOUS PROBLEMS

PROBLEMS 11.5

1. Show that the linear program

$$\text{maximize}\quad 0.75x_1 - 150x_2 + 0.02x_3 - 6x_4$$

subject to the constraints

$$\tfrac{1}{4}x_1 - 60x_2 - \tfrac{1}{25}x_3 + 9x_4 \leq 0$$

$$\tfrac{1}{2}x_1 - 90x_2 - \tfrac{1}{50}x_3 + 3x_4 \leq 0$$

$$x_3 \leq 1$$

$$\text{all } x_i \geq 0$$

is degenerate. When the simplex method is applied, there are "ties" in deciding which row to use in elimination; if these ties are decided by choosing the row nearest the top of the tableau, then the simplex method cycles and never converges. Demonstrate this by applying the simplex method. (Degeneracy *often* arises in applications, but in fact cycling *rarely* arises in these situations; the inventor of the simplex method, George Dantzig, recalls only seeing one case in practice since he invented the method.)

2. Consider the linear program

$$\text{minimize}\quad \mathbf{c}_e^T\mathbf{x}_e$$

$$\text{subject to}\quad \mathbf{A}_e\mathbf{x}_e = \mathbf{b}, \qquad \mathbf{x}_e \geq \mathbf{0},$$

where $\mathbf{A}_e$ is $m \times n$ and $(n/2) < m < n$ and $\mathbf{A}_e$ has full rank m. Show how to reduce this to a *smaller* problem of size $(n - m) \times n$ as follows. Show that we can find a nonsingular $n \times n$ matrix $\mathbf{S}$ such that

$$\mathbf{A}_e\mathbf{S} = [\mathbf{B} \quad \mathbf{0}],$$

where $\mathbf{B}$ is $m \times m$ and nonsingular. Let $\mathbf{y} = \mathbf{S}^{-1}\mathbf{x}_e$ and partition the matrices so that

$$\mathbf{y} = \mathbf{S}^{-1}\mathbf{x}_e = \begin{bmatrix}\mathbf{y}_1\\ \mathbf{y}_2\end{bmatrix}, \qquad \mathbf{S} = [\mathbf{S}_1 \quad \mathbf{S}_2],$$

where $\mathbf{y}_1$ is in $\mathbb{R}^m$, $\mathbf{y}_2$ is in $\mathbb{R}^{n-m}$, $\mathbf{S}_1$ is $n \times m$, and $\mathbf{S}_2$ is $n \times (n - m)$. Show then that $\mathbf{A}_e\mathbf{x}_e = \mathbf{b}$ and $\mathbf{x}_e \geq \mathbf{0}$ if and only if $\mathbf{y}_1 = \mathbf{B}^{-1}\mathbf{b}$ and $\mathbf{S}_1\mathbf{y}_1 + \mathbf{S}_2\mathbf{y}_2 \geq \mathbf{0}$. Deduce that the primal linear program is equivalent to

$$\text{minimize}\quad \mathbf{c}^T\mathbf{S}_2\mathbf{y}_2$$

$$\text{subject to}\quad \mathbf{S}_2\mathbf{y}_2 \geq -\mathbf{S}_1\mathbf{B}^{-1}\mathbf{b},$$

whose dual is

$$\text{maximize} \quad -(\mathbf{S}_1\mathbf{B}^{-1}\mathbf{b})^T\mathbf{x}_2$$
$$\text{subject to} \quad \mathbf{S}_2^T\mathbf{x}_2 = \mathbf{S}_2^T\mathbf{c}, \qquad \mathbf{x}_2 \geq \mathbf{0}$$

and the matrix of the equality constraints is now only $(n - m) \times n$. (This technique was communicated to us by David M. Gay.)

3. (a) Use the method of Problem 2 on the slack-variable form of our model problem (11.1) to change from five variables and three equalities to five variables and two equalities.
(b) Solve the linear program in the new form and use it to obtain a solution to the original program.

▷ **4.** A university library is open 24 hours per day, and each librarian works a steady eight-hour shift beginning at 12 midnight, 4 a.m., 8 a.m., 12 noon, 4 p.m., or 8 p.m. To handle the demands for service, the library requires the following numbers of librarians on hand during various time periods: 3 from midnight to 3:59 a.m.; 2 from 4 a.m. to 7:59 a.m.; 10 from 8 a.m. to 11:59 a.m.; 14 from noon to 3:59 p.m.; 8 from 4 p.m. to 7:59 p.m.; and 10 from 8 p.m. to 11:59 p.m. Let $x_1, x_2, \ldots, x_6$ denote the number of persons to start their eight-hour shift at midnight, 4 a.m., . . . , 8 p.m., respectively, and pose as a linear program the problem of minimizing the total number of librarians used to operate the library. Show that an optimal solution is given by $x_1 = 2$, $x_2 = 0$, $x_3 = 14$, $x_4 = 0$, $x_5 = 8$, $x_6 = 2$, once the system is in operation.

5. We have often discussed ways for picking $\mathbf{z}$ so that $\mathbf{Bz}$ is as close as possible to $\mathbf{w}$ *in the 2-norm sense*, when $\mathbf{B}$ and $\mathbf{w}$ are given; this is just our standard least-squares problem that can be solved with the QR- or singular value decompositions, for example. Linear programming can be used to solve this in an ∞-norm sense. To require the maximum of the absolute values of two real variables u and v to be as small as possible, we can minimize m subject to the constraints $u \leq m$, $-u \leq m$, $v \leq m$, $-v \leq m$.
(a) Explain why this minimizes the maximum of $|u|$ and $|v|$.
(b) Use this idea to form a linear program whose solution determines $\mathbf{z}$ that minimizes $\|\mathbf{Bz} - \mathbf{w}\|_\infty$.

▷ **6.** Linear programming can be used in a similar way to that in Problem 5 to determine $\mathbf{z}$ that minimizes $\|\mathbf{Bz} - \mathbf{w}\|_1$. To minimize $|u| + |v|$ one can minimize $m_1 + m_2$ subject to the constraints $u \leq m_1$, $-u \leq m_1$, $v \leq m_2$, $-v \leq m_2$.
(a) Explain why this minimizes $|u| + |v|$.
(b) Use this idea to form a linear program whose solution determines $\mathbf{z}$ that minimizes $\|\mathbf{Bz} - \mathbf{w}\|_1$.

7. Use Problem 5 to solve the following problem. It is assumed that the number N of cell divisions in a certain organism in each time period is approximately $N_0 + bp$, where p is the amount of a certain growth stimulus that is added and where N_0 and b are model parameters to be determined. An experiment is performed using three different amounts $p_1 = 1$, $p_2 = 3$, $p_3 = 6$ of the stimulus, and it is found that the respective number of cell divisions is $N_1 = 40$, $N_2 = 102$, $N_3 = 190$. Determine the model parameters N_0 and b so as to minimize the maximum error in the model over the three experiments, that is to minimize

$$\max\{|N_0 + b - 40|,\quad |N_0 + 3b - 102|,\quad |N_0 + 6b - 190|\}.$$

8. Suppose that a bread company has three bakeries with production capacities of 5000, 7000, and 9000 loaves daily. Suppose also that the firm ships to five warehouses for further distribution, and that the daily demands at these warehouses are 2000, 6000, 8000, 4000, and 1000. Because of the various distances between the three bakeries and the five warehouses, the shipping costs vary depending on who ships to whom. Suppose that the dollar cost of shipping 1000 loaves from bakery i to warehouse j is the (i, j)-entry in the following cost matrix $\mathbf{C}$:

$$\mathbf{C} = \begin{bmatrix} 70 & 30 & 20 & 40 & 20 \\ 60 & 50 & 80 & 30 & 40 \\ 30 & 20 & 50 & 70 & 10 \end{bmatrix}$$

Set up a linear program to determine how many loaves to ship from each bakery to each warehouse each day in order to meet the warehouse needs at least cost. (This is an example of the famous *transportation problem* and is of enormous practical importance to large businesses.) The solution turns out to be the following; bakery i should ship to warehouse j the number of loaves in the (i, j)-entry of the following solution matrix $\mathbf{X}$:

$$\mathbf{X} = \begin{bmatrix} 0 & 0 & 5000 & 0 & 0 \\ 0 & 3000 & 0 & 4000 & 0 \\ 2000 & 3000 & 3000 & 0 & 1000 \end{bmatrix}.$$

The least cost is \$650 daily. Verify that this solution is feasible for your linear program and that its cost is as stated.

▷ **9.** Suppose that $\mathbf{x}^*$ is optimal for the primal and that $\mathbf{y}^*$ is optimal for the dual in Definition 11.48, and let $\mathbf{s}_\mathbf{x}$ and $\mathbf{s}_\mathbf{y}$ be the corresponding slack variables: $\mathbf{A}\mathbf{x}^* + \mathbf{s}_\mathbf{x} = \mathbf{b}$ and $\mathbf{A}^T\mathbf{y}^* - \mathbf{s}_\mathbf{y} = \mathbf{c}$, with $\mathbf{s}_\mathbf{x} \geq \mathbf{0}$ and $\mathbf{s}_\mathbf{y} \geq \mathbf{0}$.

(a) Prove the *complementarity condition* for linear programs:

$$\mathbf{s}_\mathbf{x}^T\mathbf{y}^* = \mathbf{s}_\mathbf{y}^T\mathbf{x}^* = 0.$$

(b) Deduce that $\mathbf{c}^T$ is a nonnegative linear combination of the rows of $\mathbf{A}$ corresponding to those constraints satisfied as equalities at $\mathbf{x}^*$ and of $-\mathbf{e}_i^T$ for those i for which $\langle\mathbf{x}^*\rangle_i = 0$.

Appendix one

Answers and aids to selected problems

Chapter 1

Problems 1.1

1. (a) $7i$. (d) $6 - 4i$. (g) $-46 - 9i$. (k) $-\frac{3}{2} - i$.

4. (b) y. (f) 0.

Problems 1.2

1. $x = 6$, $z = 2$.

4. $\langle \mathbf{A} + \mathbf{A} \rangle_{ij} = \langle \mathbf{A} \rangle_{ij} + \langle \mathbf{A} \rangle_{ij} = 2\langle \mathbf{A} \rangle_{ij} = \langle 2\mathbf{A} \rangle_{ij}$.

8. $\langle \mathbf{A} + \mathbf{0} \rangle_{ij} = \langle \mathbf{A} \rangle_{ij} + \langle \mathbf{0} \rangle_{ij} = \langle \mathbf{A} \rangle_{ij} + 0 = \langle \mathbf{A} \rangle_{ij}$ for half.

11. $\langle \mathbf{A} + (-\mathbf{A}) \rangle_{ij} = \langle \mathbf{A} \rangle_{ij} + \langle -\mathbf{A} \rangle_{ij} = \langle \mathbf{A} \rangle_{ij} + (-\langle \mathbf{A} \rangle_{ij}) = 0 = \langle \mathbf{0} \rangle_{ij}$.

14. $\langle (-r)\mathbf{A} \rangle_{ij} = (-r)\langle \mathbf{A} \rangle_{ij} = -(r\langle \mathbf{A} \rangle_{ij}) = -\langle r\mathbf{A} \rangle_{ij} = \langle -(r\mathbf{A}) \rangle_{ij}$.

Problems 1.3

4. $\langle \mathbf{A} \rangle_{3j} = 4\langle \mathbf{A} \rangle_{1ij}$, and $\langle \mathbf{AB} \rangle_{3i} = \langle \mathbf{A} \rangle_{31}\langle \mathbf{B} \rangle_{1i} + \cdots + \langle \mathbf{A} \rangle_{3q}\langle \mathbf{B} \rangle_{qi}$.

7. **CD** makes sense for $m \times n$ **C** and $k \times s$ **D** if and only if $n = k$.

11. (a) $\langle \mathbf{A} \square \mathbf{B} \rangle_{ij} = \langle \mathbf{A} \rangle_{ij}\langle \mathbf{B} \rangle_{ij} = \langle \mathbf{B} \rangle_{ij}\langle \mathbf{A} \rangle_{ij} = \langle \mathbf{B} \square \mathbf{A} \rangle_{ij}$. (c) $\langle \mathbf{E} \rangle_{ij} = 1$ for all i, j.

14. **A** must commute with

$$\begin{bmatrix} 0 & 0 \\ 1 & 0 \end{bmatrix}, \quad \begin{bmatrix} 0 & 0 \\ 0 & 1 \end{bmatrix}, \quad \begin{bmatrix} 1 & 0 \\ 0 & 0 \end{bmatrix}, \quad \text{and} \quad \begin{bmatrix} 0 & 1 \\ 0 & 0 \end{bmatrix}.$$

17. $\begin{bmatrix} 1 & 0 \\ 0 & 1 \end{bmatrix}\begin{bmatrix} a & b & c \\ d & e & f \end{bmatrix} = \begin{bmatrix} a & b & c \\ d & e & f \end{bmatrix}$.

20. $\mathbf{A}^n = \begin{bmatrix} r^n & nr^{n-1} \\ 0 & r^n \end{bmatrix}$, and work from there.

24. $\langle \mathbf{A0} \rangle_{ij} = \langle \mathbf{A} \rangle_{i1}\langle \mathbf{0} \rangle_{1j} + \cdots + \langle \mathbf{A} \rangle_{iq}\langle \mathbf{0} \rangle_{qj} = \langle \mathbf{A} \rangle_{i1}0 + \cdots + \langle \mathbf{A} \rangle_{iq}0 = 0 = \langle \mathbf{0} \rangle_{ij}$.

26. (b) $\begin{bmatrix} -1 & 3 & 6 \\ -2 & 6 & 12 \\ -3 & 9 & 18 \end{bmatrix}$. (c) $[23]$.

30. $\langle (c\mathbf{A})^H \rangle_{ij} = \overline{\langle c\mathbf{A} \rangle}_{ji} = \bar{c}\overline{\langle \mathbf{A} \rangle}_{ji} = \bar{c}\langle \mathbf{A}^H \rangle_{ij} = \langle \bar{c}\mathbf{A}^H \rangle_{ij}$.

33. Use mathematical induction: (1) true for $k = 2$; (2) if true for $k = i$ then true for $k = i + 1$ by writing $(\mathbf{A}_1 \cdots \mathbf{A}_i)\mathbf{A}_{i+1}$ and using truth for $k = 2$ and for $k = i$.

37. $\begin{bmatrix} a & b + ci \\ b - ci & d \end{bmatrix}$ for real a, b, c, d.

41. (a) $(\mathbf{A}^2)^T = (\mathbf{AA})^T = \mathbf{A}^T\mathbf{A}^T = (\mathbf{A}^T)^2$. (c) Consider $\mathbf{A} = \begin{bmatrix} 0 & 1 \\ 0 & 0 \end{bmatrix}$.

45. (b) $(\mathbf{A} - \mathbf{A}^T)^T = \mathbf{A}^T - \mathbf{A}^{TT} = \mathbf{A}^T - \mathbf{A} = -(\mathbf{A} - \mathbf{A}^T)$.

48. Write $\mathbf{A} = \mathbf{B} + i\mathbf{C}$ with **B** and **C** real, and write out what $\mathbf{A}^H = \mathbf{A}$ means.

51. Since e^{kix} evaluated at x_j equals $z^{k(j-1)}$, the equations become simply

$$f_j = c_0 + c_1 z^{j-1} + c_2(z^{j-1})^2 + \cdots + c_{N-1}(z^{j-1})^{N-1}$$

for $j = 1, 2, \ldots, N$.

Problems 1.4

1. There is none.

4. $\mathbf{L} = \begin{bmatrix} \frac{1}{2} & \frac{1}{2} & 0 \\ -\frac{1}{2} & \frac{1}{2} & 0 \end{bmatrix} + a\begin{bmatrix} \frac{1}{2} & -\frac{5}{2} & 1 \\ 0 & 0 & 0 \end{bmatrix} + b\begin{bmatrix} 0 & 0 & 0 \\ \frac{1}{2} & -\frac{5}{2} & 0 \end{bmatrix}$.

8. (a) **L** is a left-inverse of **A** if and only if $\mathbf{LA} = \mathbf{I}$ if and only if $(\mathbf{LA})^T = \mathbf{I}^T$ if and only if $\mathbf{A}^T\mathbf{L}^T = \mathbf{I}$ if and only if $\mathbf{L}^T$ is a right-inverse for $\mathbf{A}^T$.

10. (a) $\begin{bmatrix} 4 & 5 \\ 3 & 4 \end{bmatrix}$. (d) $\begin{bmatrix} 1 & 1 \\ 3 & 4 \end{bmatrix}$.

12. (a) Use $\mathbf{diag}(d_1, \ldots, d_p)\mathbf{diag}(e_1, \ldots, e_p) = \mathbf{diag}(d_1e_1, \ldots, d_pe_p)$.
(b) See Problem 6 in Section 1.3.

15. Verify $(\mathbf{A}^T)(\mathbf{A}^{-1})^T = (\mathbf{A}^{-1})^T(\mathbf{A}^T) = \mathbf{I}$ using Theorem 1.22(d).

19. Use Theorem 1.35(c).

21. See hint for Problem 33 in Section 1.3.

23. (d) $z = \cos \pi/6 \approx 0.87$.

26. $\mathbf{X} = \begin{bmatrix} 2 & -1 \\ -7 & 6 \end{bmatrix}$.

29. (b) Both methods should produce $[-8 \quad 28 \quad -20]^T$.

32. (b) $s(x) = 2 \sin x + 2 \cos 2x$.

Problems 1.5

2. $\mathbf{A} + \mathbf{B} = \begin{bmatrix} 0 & 8 & 0 \\ 6 & 5 & -8 \\ 9 & 10 & -1 \end{bmatrix}$ either way.

5. $\mathbf{AB} = \begin{bmatrix} 1 & -11 & 35 \\ 9 & 28 & 25 \end{bmatrix}$ either way.

7. You get $[19 \quad -25 \quad 58]^T$ either way.

11. (a) Write out $\begin{bmatrix} \mathbf{A} & \mathbf{0} & \mathbf{0} \\ \mathbf{0} & \mathbf{B} & \mathbf{0} \\ \mathbf{0} & \mathbf{0} & \mathbf{C} \end{bmatrix}\begin{bmatrix} \mathbf{E} & \mathbf{F} & \mathbf{G} \\ \mathbf{H} & \mathbf{J} & \mathbf{K} \\ \mathbf{L} & \mathbf{M} & \mathbf{N} \end{bmatrix} = \begin{bmatrix} \mathbf{I} & \mathbf{0} & \mathbf{0} \\ \mathbf{0} & \mathbf{I} & \mathbf{0} \\ \mathbf{0} & \mathbf{0} & \mathbf{I} \end{bmatrix}$.

13. Write out $\mathbf{AX} = \mathbf{XA} = \mathbf{I}$ in detail and use the nonsingularity of **A** to solve for **B**, **w**, **z**, **a**, where

$$\mathbf{X} = \begin{bmatrix} \mathbf{B} & \mathbf{w} \\ \mathbf{z}^T & \mathbf{a} \end{bmatrix}.$$

16. $\begin{bmatrix} 7 & 8 \\ 1 & 1 \end{bmatrix}$.

19. $[-c - \frac{1}{3} \quad -c + \frac{2}{3} \quad c]$ for arbitrary c.

Problems 1.6

2. Consider the main-diagonal entries of $\mathbf{A}^T\mathbf{A}$ and $\mathbf{B}^H\mathbf{B}$ to show that $\mathbf{A}$ and $\mathbf{B}$ equal $\mathbf{0}$.

3. Use $(\mathbf{I} + \mathbf{K})(\mathbf{I} - \mathbf{K}) = \mathbf{I} - \mathbf{K}^2 = (\mathbf{I} - \mathbf{K})(\mathbf{I} + \mathbf{K})$, Theorem 1.22(d), and Theorem 1.35(c).

4. Write out $\mathbf{AR} = \mathbf{I}$ or $\mathbf{LA} = \mathbf{I}$ and see that the zero row or column gives a contradiction.

6. Use (a) to deduce (b).

Chapter 2

Problems 2.2

4. (a) $\mathbf{A} = \begin{bmatrix} 0.85 & 0.15 & 0.05 \\ 0.05 & 0.75 & 0.05 \\ 0.10 & 0.10 & 0.90 \end{bmatrix}$.

6. (a) $\mathbf{A} = \begin{bmatrix} 0.70 & 0.20 & 0.20 \\ 0.20 & 0.60 & 0.30 \\ 0.10 & 0.20 & 0.50 \end{bmatrix}$.

10. $[\frac{1}{3} \quad \frac{1}{6} \quad \frac{1}{2}]^T$.

13. They do.

17. $\mathbf{A}^{\text{odd}}[x_0 \quad y_0 \quad z_0]^T = [z_0 \quad y_0 \quad x_0]^T$.

21. $\mathbf{A}^{\text{odd}} = \mathbf{A}$, $\mathbf{A}^{\text{even}} = \mathbf{I}_3$.

Problems 2.3

3. "War-like."

5. (a) Compare $1000(1.1)^i$ with $10(1.15)^i$. (b) With i about 104.

9. F_i tends to 2400 and C_i to 1920.

12. At $k = 0.16$.

Problems 2.4

3. $d_1 = -0.0078$, $d_2 = 0.0561$.

7. $d_1 = 9$, $d_2 = 7.5$, $d_3 = 9.5$.

Problems 2.5

4. (b) $(\mathbf{K} - \omega^2\mathbf{M})\boldsymbol{\xi} = \mathbf{0}$ and $(\mathbf{K} - \omega^2\mathbf{M})\boldsymbol{\eta} = \mathbf{0}$, where

$$\mathbf{K} = k\begin{bmatrix} 1 & -1 & 0 \\ -1 & 2 & -1 \\ 0 & -1 & 1 \end{bmatrix} \quad \text{and} \quad \mathbf{M} = \begin{bmatrix} m & 0 & 0 \\ 0 & M & 0 \\ 0 & 0 & m \end{bmatrix}.$$

6. $\omega^2 = 0$, 6, or 10.

Problems 2.6

3. $9x_1 + 17{,}460x_2 = 1288.3,\ 17{,}460x_1 + 3{,}387{,}400x_2 = 2{,}510{,}405.$

6. $y \approx -3.194 + 2.522t.$

10. $\mathbf{x} = [-0.027 \quad 1.216]^T.$

13. $5a + 26b + 198c = 33,\ 26a + 198b + 1664c = 227,\ 198a + 1664b + 14{,}802c = 1871.$

Problems 2.7

3. Minimize $\mathbf{b}^T\mathbf{y}$ with $\mathbf{b} = [70 \quad 40 \quad 90]^T$ and the constraints

$$\begin{bmatrix} 2 & 1 & 1 \\ 1 & 1 & 3 \end{bmatrix} \mathbf{y} \geq \begin{bmatrix} 40 \\ 60 \end{bmatrix}.$$

5. The first inequality becomes $2x_1 + x_2 \leq 55$, still satisfied by 15, 25. This is still optimal since the new allowable plans form a subset of earlier allowable plans.

8. $x_1 = 0,\ x_2 = 30.$

10. Any positive c, d with $1 \leq c/d \leq 2$.

Chapter 3

Problems 3.2

1. (a) $x_1 = -1,\ x_2 = 2,\ x_3 = 4.$

4. $x_1 = 1,\ x_2 = -2,\ x_3 = -1.$

8. (a) $\begin{bmatrix} 1 & -\frac{1}{3} & -\frac{2}{3} & 0 \\ 0 & 1 & \frac{22}{5} & \frac{39}{5} \\ 0 & 0 & 1 & 2 \end{bmatrix}.$

10. No solutions.

12. $\mathbf{b} = [\alpha \quad \beta \quad \alpha + \beta]^T$, α and β arbitrary.

Problems 3.3

1. $x = y = 1.$

4. $x_1 = -1,\ x_2 = 2,\ x_3 = 1,\ x_4 = 0.$

8. (a) Columns 1 and 2; x_1, x_2; the last column is nonleading, all variables are leading, so exactly one solution.

10. $\begin{bmatrix} -3a & -3b & -3c \\ 0 & 0 & 0 \\ a & b & c \end{bmatrix} = \mathbf{X}$ for arbitrary a, b, c.

Problems 3.4

1. $\mathbf{A}^{-1} = \dfrac{1}{12}\begin{bmatrix} -7 & -6 & 5 \\ 2 & 0 & 2 \\ 1 & -6 & 1 \end{bmatrix}.$

4. $\mathbf{x}_1 = [1 \quad 1 \quad 1]^T,\ \mathbf{x}_2 = [4 \quad 1 \quad 2]^T,\ \mathbf{x}_3 = [1 \quad 2 \quad 1]^T.$

Problems 3.5

2. $\mathbf{E}_i(c) = \mathbf{diag}(1, \ldots, 1, c, 1, \ldots, 1)$ with c in the ith position; just calculate $\mathbf{E}_i(c)\mathbf{A}$.

6. $\mathbf{F} = \begin{bmatrix} 0 & \frac{1}{2} & 0 \\ 1 & 0 & 0 \\ 0 & -2 & 1 \end{bmatrix}$.

8. Use $(\mathbf{EA})^{-1} = \mathbf{A}^{-1}\mathbf{E}^{-1}$ and Problem 7.

Problems 3.6

1. (b) 0.57.
4. (a) $x_1 = -0.04$, $x_2 = 2.2$.
6. Residual is $[0.01 \quad 0]^T$.
9. Residual is $[0 \quad -0.0001]^T$.

Problems 3.7

1. $\alpha_{11} = 2$, $m_{21} = -4$, $m_{31} = -2$, $\alpha_{22} = -3$, $m_{32} = 2$, $\alpha_{33} = -1$.
4. Consider solving $\mathbf{L}\mathbf{x}_i = \mathbf{e}_i$ for all i.
6. Use mathematical induction; for two lower-triangular matrices **A** and **B**, write out in detail $\langle \mathbf{AB} \rangle_{ij}$ for $j > i$.
10. Use Problems 4–8 to see that $\mathbf{L}_2^{-1}\mathbf{L}_1$ is lower-triangular while $\mathbf{U}_2\mathbf{U}_1^{-1}$ is unit-upper-triangular, and deduce that each equals **I**.
14. $\mathbf{A} = \begin{bmatrix} 1 & 0 \\ 0 & 1 \end{bmatrix}\begin{bmatrix} 2 & 0 \\ 2 & -3 \end{bmatrix}\begin{bmatrix} 1 & -2 \\ 0 & 1 \end{bmatrix} = \begin{bmatrix} 0 & 1 \\ 1 & 0 \end{bmatrix}\begin{bmatrix} 2 & 0 \\ 2 & 3 \end{bmatrix}\begin{bmatrix} 1 & -\frac{7}{2} \\ 0 & 1 \end{bmatrix}$.
17. (a) $\mathbf{A} = \begin{bmatrix} 1 & 0 \\ 3 & 1 \end{bmatrix}\begin{bmatrix} 2 & 0 \\ 0 & 0 \end{bmatrix}$.

(c) $\mathbf{A} = \begin{bmatrix} 1 & 0 & 0 & 0 \\ -2 & 1 & 0 & 0 \\ -1 & 0 & 1 & 0 \\ 3 & 0 & -1 & 1 \end{bmatrix}\begin{bmatrix} 2 & 2 & 3 & 4 \\ 0 & 0 & 7 & 12 \\ 0 & 0 & 7 & 12 \\ 0 & 0 & 0 & 2 \end{bmatrix}$.

Problems 3.8

1. A row times a column involves s products; summing s terms requires $s - 1$ additions; each of the r rows must multiply each of the t columns.
5. (a) 430. (b) 119,210.
7. For multiplication/divisions, for example, the $\mathbf{P}^T\mathbf{LU}$ method costs $(p^3 - p)/3 + kp^2$, the other costs p^3 for the inverse and then kp^2 for the solutions, and $(p^3 - p)/3 + kp^2 \leq p^3 + kp^2$ for all $k \geq 0$ and all $p \geq 0$.
11. Note that **K** is only 1×1.
15. (a) To divide row i by the pivot costs 1 multiply/divide for $1 \leq i \leq p - 1$. To eliminate in the next row costs 1 multiply/divide and 1 add/subtract for $1 \leq i \leq p - 1$.

Problems 3.10

2. (a) For $k = 0$, $x_1 = 3 - 3\alpha$, $x_2 = 1$, $x_3 = \alpha$ for arbitrary α.
   (b) No solution for $k = -4$.

5. $l_{i1} = a_{i1}$ for $1 \le i \le p$; then $u_{1j} = a_{1j}/l_{11}$ for $2 \le j \le p$; then $l_{i2} = a_{i2} - l_{i1}u_{12}$ for $2 \le i \le p$; then $u_{2j} = (a_{2j} - l_{21}u_{1j})/l_{22}$; and so on.

7. $\mathbf{L} = \begin{bmatrix} 1 & 0 \\ 2 & 1 \end{bmatrix}$, $\mathbf{U} = \begin{bmatrix} 2 & 2 \\ 0 & -5 \end{bmatrix}$.

10. (a) First write $\mathbf{A} = \mathbf{LU}$ as in **Key Theorem 3.48**, then let $\mathbf{D}_0 = \mathbf{diag}(\alpha_{11}, \ldots, \alpha_{pp})$, $\mathbf{U}_1 = \mathbf{U}$, $\mathbf{L}_1 = \mathbf{LD}_0^{-1}$.

13. If $\mathbf{L}_1\mathbf{D}_0\mathbf{U}_1 = \mathbf{L}_1'\mathbf{D}_0\mathbf{U}_1'$, then $\mathbf{L}_1^{-1}\mathbf{L}_1'$ is unit-lower-triangular and equals the upper-triangular $\mathbf{D}_0\mathbf{U}_1\mathbf{U}_1'^{-1}\mathbf{D}_0'^{-1}$, both of which must then equal $\mathbf{I}$. So $\mathbf{L}_1 = \mathbf{L}_1'$ and unit-upper-triangular $\mathbf{U}_1\mathbf{U}_1'^{-1}$ equals diagonal $\mathbf{D}_0^{-1}\mathbf{D}_0'$, so both equal $\mathbf{I}$.

15. $\mathbf{L}_1\mathbf{D}_0\mathbf{U}_1 = \mathbf{A} = \mathbf{A}^T = \mathbf{U}_1^T\mathbf{D}_0^T\mathbf{L}_1^T$, and use Problem 13 or its proof.

Chapter 4 *Problems 4.2*

3. (a) $\begin{bmatrix} 1 & 1 & -8 & -14 \\ 0 & 1 & -3 & -6 \\ 0 & 0 & 0 & 0 \end{bmatrix}$. (d) $\begin{bmatrix} 1 & -\frac{1}{4} & \frac{1}{2} & \frac{3}{2} \\ 0 & 1 & -\frac{2}{19} & -\frac{6}{19} \\ 0 & 0 & 0 & 0 \end{bmatrix}$.

5. The first p columns serve to make the matrix in the right form.

8. Apply to $[\mathbf{A} \quad \mathbf{B}]$ the row operations that put $\mathbf{A}$ in Gauss-reduced form $\mathbf{G}$ and explain why $[\mathbf{G} \quad \mathbf{B}']$ that you get is in Gauss-reduced form by using the fact that $\mathbf{A}$ and $\mathbf{G}$ have rank p.

10. Interchange rows so that the nonzero rows are in the top $p - n$ rows, reduce the top $(p - n) \times p$ matrix to its Gauss-reduced form, and explain why this puts $\mathbf{A}$ in Gauss-reduced form.

14. The row-echelon form equals $\mathbf{I}_2$

Problems 4.3

1. (a) Infinitely many. (c) No solution. (f) One solution.

3. Solutions when $\mathbf{b} = [\alpha \quad \beta \quad \alpha + \beta]^T$.

7. The common point is (x, y) if and only if $[\mathbf{a} \quad \mathbf{b}][x \quad y]^T = \mathbf{c}$.

9. (a) $\mathbf{h} = \alpha[-\frac{1}{2} \quad 1 \quad 0 \quad 0]^T$ for arbitrary α.

13. Each $\mathbf{R} = \mathbf{R}_0 + \mathbf{Y}$ where $\mathbf{R}_0$ is any fixed right-inverse and $\mathbf{Y}$ is any solution to $\mathbf{AY} = \mathbf{0}$.

16. $\mathbf{Ax}_\infty = \mathbf{x}_\infty$ means $\mathbf{0} = \mathbf{Ax}_\infty - \mathbf{x}_\infty = (\mathbf{A} - \mathbf{I})\mathbf{x}_\infty$.

Problems 4.4

2. (a) Let $\mathbf{F}$ be a nonsingular matrix that reduces $\mathbf{A}$ to Gauss-reduced form $\mathbf{R}$, and argue that rank p means $\mathbf{R} = \mathbf{I}_p$. Show then that the same row operations on $[\mathbf{A} \quad \mathbf{I}_p]$ produce $\mathbf{F}[\mathbf{A} \quad \mathbf{I}_p] = [\mathbf{FA} \quad \mathbf{F}] = [\mathbf{I}_p \quad \mathbf{F}]$. Since $\mathbf{FA} = \mathbf{I}_p$, $\mathbf{F} = \mathbf{A}^{-1}$.

6. Examine the solvability of $\mathbf{Ax}_i = \mathbf{e}_i$ using **Key Theorem 4.13.**

8. (a) Show that there is an $\mathbf{x} \neq \mathbf{0}$ with $(\mathbf{AB})\mathbf{x} = \mathbf{0}$ by finding $\mathbf{x} \neq \mathbf{0}$ with $\mathbf{Bx} = \mathbf{0}$.

11. Let $\mathbf{F}$ be nonsingular with $\mathbf{FA} = \mathbf{R}_1$ in Gauss-reduced form with rank k_1, and let $\mathbf{G}$ be nonsingular with $\mathbf{G}(\mathbf{AA}^T) = \mathbf{R}_2$ in Gauss-reduced form with rank k_2. Show that $\mathbf{F}(\mathbf{AA}^T)$ has $p - k_1$ zero rows and deduce that $k_2 \le k_1$. Show that $\mathbf{GAA}^T\mathbf{G}^T = \mathbf{R}_2\mathbf{G}^T$ has $p - k_2$ zero rows, and deduce that $\mathbf{GA}$ has $p - k_2$ zero rows by examining $\langle \mathbf{GAA}^T\mathbf{G}^T \rangle_{ii}$, and then deduce that $k_1 \le k_2$.

Problems 4.5

1. (c) det $\mathbf{A} = 30$.
4. Expand repeatedly along the first row or column.
8. det $\mathbf{A} = 60$.
10. Use det $\mathbf{A}^T =$ det $\mathbf{A}$.
12. If $\mathbf{T}$ is lower-triangular, expand repeatedly along the top row.
15. Problem 14 says that the line passes through the three points (x, y), (x_1, y_1), (x_2, y_2) if and only if this holds.
18. Use mathematical induction on n and Theorems 4.35 and 4.36.

Problems 4.6

1. $\mathbf{A}(\mathbf{adj\ A}) = \mathbf{0}$.
4. (b) $\frac{1}{11}\begin{bmatrix} 6 & -2 & -3 \\ -3 & 1 & 7 \\ 2 & 3 & -1 \end{bmatrix}$.
7. $\mathbf{A}(\mathbf{adj\ A}) = (\det \mathbf{A})\mathbf{I}_p$, so $(\det \mathbf{A}) \det(\mathbf{adj\ A}) = (\det \mathbf{A})^p$.
9. (b) The determinants equal 1 and $10^8 + 1$.
12. $x_1 = 2$, $x_2 = -1$.
14. (a) $\mathbf{x}' = \mathbf{A}^{-1}\mathbf{b}' = \mathbf{A}^{-1}(\mathbf{b} + \epsilon\mathbf{e}_i) = \mathbf{A}^{-1}\mathbf{b} + \epsilon\mathbf{A}^{-1}\mathbf{e}_i = \mathbf{x}_0 + \epsilon\mathbf{A}^{-1}\mathbf{e}_i$.
    (b) $\langle \mathbf{x}_0 \rangle_j$ changes by $\epsilon A_{ij}/(\det \mathbf{A})$ with A_{ij} the (i, j)-cofactor.

Problems 4.7

2. If $\mathbf{F}$ is nonsingular with $\mathbf{FR}$ in Gauss-reduced form while $\mathbf{G}$ is nonsingular with $\mathbf{GS}$ in Gauss-reduced form, then $\mathbf{diag}(\mathbf{F}, \mathbf{G})$ is nonsingular and puts $\mathbf{A}$ into a form that a few interchanges easily make Gauss-reduced.
5. There is a nonsingular $\mathbf{F}$ with $\mathbf{FA} = \mathbf{G}$ in Gauss-reduced form, with $\mathbf{G} = [\mathbf{g} \quad \mathbf{0} \quad \cdots \quad \mathbf{0}]^T$. Then $\mathbf{A} = \mathbf{F}^{-1}\mathbf{G} = \mathbf{cr}$ with $\mathbf{c}$ the first column of $\mathbf{F}^{-1}$ and $\mathbf{r}$ the first row $\mathbf{g}^T$ of $\mathbf{G}$.
7. Partition $\mathbf{R}$ into $\mathbf{R}_{ij}$, write $\mathbf{AR} = \mathbf{diag}(\mathbf{I}, \ldots, \mathbf{I})$, and observe that this requires $\mathbf{A}_{ii}\mathbf{R}_{ii} = \mathbf{I}$.
10. (a) Expand by the last row or column.
11. Starting with the next-to-last column, subtract from each column x_1 times the column to its left. Then expand along the top row and factor out $(x_p - x_1) \cdots (x_2 - x_1)$ from the result. Finally, use induction to handle the $(p - 1) \times (p - 1)$ Vandermonde determinant that results.
14. $\det\begin{bmatrix} x_1^2 & x_1y_1 & y_1^2 & x_1 & y_1 & 1 \\ x_2^2 & x_2y_2 & y_2^2 & x_2 & y_2 & 1 \\ x_3^2 & x_3y_3 & y_3^2 & x_3 & y_3 & 1 \\ x_4^2 & x_4y_4 & y_4^2 & x_4 & y_4 & 1 \\ x_5^2 & x_5y_5 & y_5^2 & x_5 & y_5 & 1 \\ x^2 & xy & y^2 & x & y & 1 \end{bmatrix} = 0.$
18. If $\mathbf{A}^T\mathbf{Ax} = \mathbf{0}$, then $0 = \mathbf{x}^T\mathbf{A}^T\mathbf{Ax} = \|\mathbf{Ax}\|_2^2$, so $\mathbf{Ax} = \mathbf{0}$ and thus $\mathbf{x} = \mathbf{0}$ since $\mathbf{A}$ has rank q.

Chapter 5 *Problems 5.1*

4. About at 5.8 mph at an angle of 30° downstream of across-river.

7. Airspeed about 141 mph; wind about 28 mph to the northeast.

9. The magnitude of $\alpha\vec{u}$, is $|\alpha|$ times that of $\vec{u}$, in the same (or opposite) direction as $\vec{u}$ if $\alpha \geq 0\ (<0)$.

Problems 5.2

3. $0 + 0 = 0$, $r0 = 0$, $-0 = 0$, so all in Definition 5.8 hold.

5. (a) No. (b) Yes. (c) No.

7. If $\mathscr{V}_0$ is empty, then it is not a subspace. If it contains some vector $\mathbf{v}_0$, then $\mathbf{v}_0 + \mathbf{v}_0$ is not in $\mathscr{V}_0$ since $\mathbf{b} + \mathbf{b} \neq \mathbf{b}$ for $\mathbf{b} \neq \mathbf{0}$.

10. It must contain $\mathbf{0}$.

14. Use Example 5.14 on $\mathbf{A} - \mathbf{I}_q$.

Problems 5.3

4. $\mathbf{u}_1$ and $\mathbf{u}_2$.

6. Show that $a + bt + ct^2 = \alpha(2) + \beta(3 + t) + \gamma(2 - t^2)$ is solved by $\gamma = -c$, $\beta = b$, $\alpha = a/2 - c$.

9. The only solution to $a(\mathbf{u} + \mathbf{v}) + b(\mathbf{v} + \mathbf{w}) + c(\mathbf{w} + \mathbf{u}) = \mathbf{0}$, that is, to $a + c = 0$, $a + b = 0$, $b + c = 0$, is $a = b = c = 0$.

12. $\mathbf{v}_3 = \mathbf{v}_1 + \mathbf{v}_2$. $\mathbf{v}_1$ is dependent on $\{\mathbf{v}_2, \mathbf{v}_3\}$, $\mathbf{v}_1$ on $\{\mathbf{v}_2, \mathbf{v}_3, \mathbf{v}_4\}$, $\mathbf{v}_2$ on $\{\mathbf{v}_1, \mathbf{v}_3\}$, $\mathbf{v}_2$ on $\{\mathbf{v}_1, \mathbf{v}_3, \mathbf{v}_4\}$, $\mathbf{v}_3$ on $\{\mathbf{v}_1, \mathbf{v}_2\}$, and $\mathbf{v}_3$ on $\{\mathbf{v}_1, \mathbf{v}_2, \mathbf{v}_4\}$.

15. The second one.

19. An infinite set F of vectors is linearly dependent if and only if some linear combination of a *finite* set of vectors from F equals $\mathbf{0}$ without all coefficients being 0.

20. (b) A finite subset will be some finite collection of $\mathbf{x}_j$; let N be the largest subscript used. By adding $0\mathbf{x}_i$ if $\mathbf{x}_i$ is not one of the vectors in the combination, we can assume the set to be $\mathbf{x}_1, \ldots, \mathbf{x}_N$. The jth term in the sequence defined by the linear combination $c_1\mathbf{x}_1 + \cdots + c_N\mathbf{x}_N$ is c_j if $j \leq N$, so $c_j = 0$.

Problems 5.4

1. $\mathbf{v}_1, \mathbf{v}_2, \mathbf{v}_4$.

5. The first three vectors, for example.

7. The first three vectors, for example.

8. Each polynomial by definition is a linear combination of the given vectors; the set is linearly independent by Problem 20(a) of Section 5.3.

11. $-6\mathbf{v}_1 + 2\mathbf{v}_2 + \mathbf{v}_3 = \mathbf{0}$.

14. No; there are only two vectors in the set, $\mathbb{R}^3$ has dimension 3, so three vectors are required to span.

16. (a) $[6 \quad -9 \quad 5]^T$.

19. $\mathbf{v}$ is linearly dependent on $\mathbf{v}_1, \ldots, \mathbf{v}_r$ if and only if $\mathbf{v} = c_1\mathbf{v}_1 + \cdots + c_r\mathbf{v}_r$ if and only if $c_B(\mathbf{v}) = c_B(c_1\mathbf{v}_1 + \cdots + c_r\mathbf{v}_r) = c_1c_B(\mathbf{v}_1) + \cdots + c_rc_B(\mathbf{v}_r)$ if and only if $c_B(\mathbf{v})$ is linearly dependent on $c_B(\mathbf{v}_1), \ldots, c_B(\mathbf{v}_r)$.

21. Use Problems 18 and 19.

24. $\mathbf{M} = [\mathbf{v}_1 \quad \mathbf{v}_2 \quad \mathbf{v}_3]$.

28. $\mathbf{M} = \begin{bmatrix} 1 & \frac{1}{2} & 1 \\ -1 & \frac{5}{2} & -2 \\ 0 & -\frac{1}{2} & 1 \end{bmatrix}$.

Problems 5.5

1. Use the first two vectors in S.

5. $k = 1, 3,$ or 4.

7. Join $[1 \quad -1 \quad 0 \quad 1]^T$, for example.

11. Use the first two columns.

13. Use the first three columns.

14. Dimension $= 3$. (a) $[1 \quad 0 \quad 0 \quad 2]^T, [0 \quad 1 \quad 0 \quad 0]^T, [0 \quad 0 \quad 1 \quad 1]^T$.

16. The common row-echelon form is

$$\begin{bmatrix} 1 & 0 & -\frac{1}{5} \\ 0 & 1 & -\frac{7}{5} \end{bmatrix}.$$

19. $[1 \quad 0 \quad 0 \quad 0 \quad 2 \quad 1]^T, [0 \quad 1 \quad 0 \quad 0 \quad 3 \quad 1]^T, [0 \quad 0 \quad 1 \quad 0 \quad 1 \quad 2]^T,$
$[0 \quad 0 \quad 0 \quad 1 \quad -1 \quad 1]^T$.

22. (b) $\mathbf{A} = \mathbf{B} = \mathbf{I}$, for example; $\mathbf{A} = \mathbf{B} = \mathbf{e}_1\mathbf{e}_2^T$, for example.

Problems 5.6

1. (a) $\|\mathbf{x}\|_1 = 7$. (c) $\|\mathbf{x}\|_2 = \sqrt{54}$. (f) $\|\mathbf{x}\|_\infty = 5$.

5. The $\|\cdot\|_1$-unit-circle is a square with vertices $(1, 0), (0, 1), (-1, 0), (0, -1)$.

8. (a) $\|\mathbf{u} + \mathbf{v}\|_2^2 = \|\mathbf{u}\|_2^2 + \mathbf{u}^H\mathbf{v} + \mathbf{v}^H\mathbf{u} + \|\mathbf{v}\|_2^2 \le \|\mathbf{u}\|_2^2 + 2\|\mathbf{u}\|_2\|\mathbf{v}\|_2 + \|\mathbf{v}\|_2^2 = (\|\mathbf{u}\|_2 + \|\mathbf{v}\|_2)^2$.
(b) $\|\mathbf{u}\| + t\mathbf{v}\|_2^2 \le (\|\mathbf{u}\|_2 + \|t\mathbf{v}\|_2)^2$ gives $t\mathbf{u}^H\mathbf{v} + \bar{t}\mathbf{v}^H\mathbf{u} \le 2|t|\,\|\mathbf{u}\|_2\|\mathbf{v}\|_2$. Let $t = \mathbf{v}^H\mathbf{u}$, the complex conjugate of $\mathbf{u}^H\mathbf{v}$.

12. $\mathbf{x}_\infty = [2 \quad 0 \quad 1]^T$.

13. $\|\mathbf{u}\| = \|\mathbf{v} + (\mathbf{u} - \mathbf{v})\| \le \|\mathbf{v}\| + \|\mathbf{u} - \mathbf{v}\|$, so $\|\mathbf{u}\| - \|\mathbf{v}\| \le \|\mathbf{u} - \mathbf{v}\|$; interchanging $\mathbf{u}$ and $\mathbf{v}$ gives the result.

15. $|\mathbf{u}_i^H\mathbf{v}_i - \mathbf{u}_\infty^H\mathbf{v}_\infty| = |\mathbf{u}_i^H\mathbf{v}_i - \mathbf{u}_i^H\mathbf{v}_\infty + \mathbf{u}_i^H\mathbf{v}_\infty - \mathbf{u}_\infty^H\mathbf{v}_\infty| \le |\mathbf{u}_i^H(\mathbf{v}_i - \mathbf{v}_\infty)| + |(\mathbf{u}_i - \mathbf{u}_\infty)^H\mathbf{v}_\infty| \le$ $\|\mathbf{u}_i\|\,\|\mathbf{v}_i - \mathbf{v}_\infty\| + \|\mathbf{u}_i - \mathbf{u}_\infty\|\,\|\mathbf{v}_\infty\|$ which tends to zero since $\|\mathbf{u}_i\|$ is bounded—$\|\mathbf{u}_i\| \le$ $\|\mathbf{u}_\infty\| + \|\mathbf{u}_i - \mathbf{u}_\infty\|$.

18. $\|[a \quad b \quad c]^T\| = (a^2/2 + 11b^2/6 + 47c^2/60 + 2ab + 5ac/3 + 31bc/12)^{1/2}$.

Problems 5.7

1. $30°$.

4. $[c - 2b \quad b \quad c]^T$.

7. The equation $(\mathbf{n}, \mathbf{x}) = 0$ has one leading variable; the other $p - 1$ variables may be given arbitrary values, and this can be used to construct a basis containing $p - 1$ vectors.

9. Use $\|\mathbf{u} + \mathbf{v}\|^2 = \|\mathbf{u}\|^2 + \|\mathbf{v}\|^2 + 2(\mathbf{u}, \mathbf{v})$.

12. (a) Show that the properties of $(\cdot, \cdot)$ on the space are inherited in the subspace.
(b) 30.4°, or 0.53 radian.

15. Verify the conditions of Definition 5.63 using the properties of c_B.
(a) For example, $(\mathbf{u}, \mathbf{u})_{\mathscr{V}} = (c_B(\mathbf{u}), c_B(\mathbf{u})) \geq 0$, and this equals 0 if and only if $c_B(\mathbf{u}) = \mathbf{0}$ if and only if $\mathbf{u} = \mathbf{0}$, as needed.

18. Use induction on p. For $p = 2$, $\|\alpha_1\mathbf{v}_1 + \alpha_2\mathbf{v}_2\|^2 = |\alpha_1|^2\|\mathbf{v}_1\|^2 + |\alpha_2|^2\|\mathbf{v}_2\|^2 + \bar{\alpha}_1\alpha_2(\mathbf{v}_1, \mathbf{v}_2) + \alpha_1\bar{\alpha}_2(\mathbf{v}_2, \mathbf{v}_1)$.

Problems 5.8

1. (a) $[0 \quad 2 \quad 3]^T$

4. Note $\|\mathbf{v} - \mathbf{v}\| = 0$ and use Theorem 5.73 on best approximation.

7. If the original set is orthogonal with nonzero vectors, then the process reproduces the set.

10. $1 + t$, $-\frac{1}{2} + t/2 + t^2$, $\frac{2}{3} - 2t/3 + 2t^2/3$.

12. $\mathbf{u}_1 = [1 \quad 1+\epsilon \quad 1 \quad 1]^T$, $\mathbf{u}_2 = [0 \quad -\epsilon \quad \epsilon \quad 0]^T$, $\mathbf{u}_3 = [0 \quad -\epsilon/2 \quad -\epsilon/2 \quad \epsilon]^T$; angles within about ϵ of 90°.

16. $\mathbf{H}_+\mathbf{x} = \mathbf{x} - 2(\mathbf{x} + \mathbf{y}\|\mathbf{x}\|_2/\|\mathbf{y}\|_2)(\|\mathbf{x}\|_2^2 + \mathbf{x}^T\mathbf{y}\|\mathbf{x}\|_2/\|\mathbf{y}\|_2)/(2\|\mathbf{x}\|_2^2 + 2\mathbf{x}^T\mathbf{y}\|\mathbf{x}\|_2/\|\mathbf{y}\|_2) = -\mathbf{y}\|\mathbf{x}\|_2/\|\mathbf{y}\|_2$.

17. (b) Take $\mathbf{y} = \mathbf{e}_1$ in Problem 16 and pick the + or − in $\mathbf{w}$ so that $\mathbf{w} \neq \mathbf{0}$.

Problems 5.9

1. $\mathbf{Q}^T\mathbf{Q}$ is diagonal with nonnegative entries.

4. $$\frac{1}{9}\begin{bmatrix} 5 & -2 & 4 \\ -2 & 8 & 2 \\ 4 & 2 & 5 \end{bmatrix}.$$

8. $$\mathbf{A} = \begin{bmatrix} \frac{1}{\sqrt{2}} & \frac{-1}{\sqrt{3}} & \frac{1}{\sqrt{6}} \\ 0 & \frac{1}{\sqrt{3}} & \frac{2}{\sqrt{6}} \\ \frac{1}{\sqrt{2}} & \frac{1}{\sqrt{3}} & \frac{-1}{\sqrt{6}} \end{bmatrix}\begin{bmatrix} \sqrt{2} & 3\sqrt{2} & \frac{9}{2}\sqrt{2} \\ 0 & \sqrt{3} & \frac{4}{3}\sqrt{3} \\ 0 & 0 & -\frac{1}{6}\sqrt{6} \end{bmatrix}.$$

11. If $\mathbf{y}$ is in the column space of $\mathbf{A}$, then $\mathbf{y} = \mathbf{A}\mathbf{x} = \mathbf{Q}_0(\mathbf{R}_0\mathbf{x})$, which writes $\mathbf{y}$ as a linear combination of the orthogonal set of columns of $\mathbf{Q}_0$. Apply the same argument to the matrix formed from the first i columns of $\mathbf{A}$.

16. The first Householder matrix, for example, is

$$\begin{bmatrix} \pm\frac{\sqrt{2}}{2} & 0 & \pm\frac{\sqrt{2}}{2} \\ 0 & 1 & 0 \\ \pm\frac{\sqrt{2}}{2} & 0 & \mp\frac{\sqrt{2}}{2} \end{bmatrix}.$$

20. $\mathbf{A} = [\sqrt{5}/5 \quad 2\sqrt{5}/5]^T[\sqrt{5} \quad \sqrt{5}]$, $\mathbf{x} = [3.6 - \alpha \quad \alpha]^T$ for arbitrary α.

23. $\mathbf{A} = [\mathbf{q}_1 \quad \mathbf{q}_2][\mathbf{r}_1 \quad \mathbf{r}_2]$ with $\mathbf{q}_1 = [1 \quad 1 \quad 1 \quad 1 \quad 1]^T/\sqrt{5}$, $\mathbf{r}_1 = [\sqrt{5} \quad 0]^T$, $\mathbf{q}_2 = [-26 \quad -11 \quad -1 \quad 14 \quad 24]^T/\sqrt{1570}$, $\mathbf{r}_2 = [26\sqrt{5}/5 \quad \sqrt{1570}/5]^T$.

Problems 5.10

1. Each complex number equals $a + bi$ for real a and b; verify the conditions in Definition 5.8.

4. $1 - \beta + \alpha\beta \neq 0$.

5. (b) Basis: $\{[2 \quad -7 \quad -1]^T\}$.

9. Let $(\mathbf{v}, \mathbf{u}) = s\|\mathbf{u}\|_2\|\mathbf{v}\|_2$ with $|s| = 1$. For real t, show that $\|\mathbf{u} - st\mathbf{v}\|_2^2 = (\|\mathbf{u}\|_2 - t\|\mathbf{v}\|_2)^2$ and let $t = \|\mathbf{u}\|_2/\|\mathbf{v}\|_2$ for $\mathbf{v} \neq \mathbf{0}$; treat $\mathbf{v} = \mathbf{0}$ separately.

11. (b) The new set is linearly dependent since $\mathbf{v}_0$ is a linear combination of the others.

Chapter 6 *Problems 6.1*

1. If $\mathscr{T}(\alpha_1\mathbf{v}_1 + \alpha_2\mathbf{v}_2) = \alpha_1\mathscr{T}(\mathbf{v}_1) + \alpha_2\mathscr{T}(\mathbf{v}_2)$, first take $\alpha_1 = \alpha_2 = 1$, then take $\alpha_2 = 0$. Conversely, $\mathscr{T}(\alpha_1\mathbf{v}_1 + \alpha_2\mathbf{v}_2) = \mathscr{T}(\alpha_1\mathbf{v}_1) + \mathscr{T}(\alpha_2\mathbf{v}_2) = \alpha_1\mathscr{T}(\mathbf{v}_1) + \alpha_2\mathscr{T}(\mathbf{v}_2)$.

2. $\mathscr{I}(\mathbf{v}_1 + \mathbf{v}_2) = \mathbf{v}_1 + \mathbf{v}_2 = \mathscr{I}(\mathbf{v}_1) + \mathscr{I}(\mathbf{v}_2)$, $\mathscr{I}(\alpha\mathbf{v}) = \alpha\mathbf{v} = \alpha\mathscr{I}(\mathbf{v})$.

8. $(\mathscr{T}\mathscr{S})(\mathbf{v}_1 + \mathbf{v}_2) = \mathscr{T}\{\mathscr{S}(\mathbf{v}_1 + \mathbf{v}_2)\} = \mathscr{T}\{\mathscr{S}(\mathbf{v}_1) + \mathscr{S}(\mathbf{v}_2)\} = \mathscr{T}\{\mathscr{S}(\mathbf{v}_1)\} + \mathscr{T}\{\mathscr{S}(\mathbf{v}_2)\} = (\mathscr{T}\mathscr{S})(\mathbf{v}_1) + (\mathscr{T}\mathscr{S})(\mathbf{v}_2)$, for example.

11. $\mathbf{y}$ is in the image space if and only if $\mathbf{y} = \mathscr{T}(\mathbf{x}) = \mathbf{A}\mathbf{x} = x_1\mathbf{a}_1 + \cdots + x_q\mathbf{a}_q$, where $\mathbf{A} = [\mathbf{a}_1 \cdots \mathbf{a}_q]$.

14. Use the Subspace Theorem: $\mathscr{T}(\mathbf{0}) = \mathscr{T}(0\mathbf{0}) = 0\mathscr{T}(\mathbf{0}) = \mathbf{0}$, so the null space $\mathscr{N}$ is nonempty; $\mathbf{v}_1$ and $\mathbf{v}_2$ in $\mathscr{N}$ gives $\mathscr{T}(\mathbf{v}_1 + \mathbf{v}_2) = \mathscr{T}(\mathbf{v}_1) + \mathscr{T}(\mathbf{v}_2) = \mathbf{0} + \mathbf{0} = \mathbf{0}$, so $\mathbf{v}_1 + \mathbf{v}_2$ is in $\mathscr{N}$; similarly for $\alpha\mathbf{v}_1$.

18. Use Theorem 6.8.

21. No; it's not one-to-one: $\mathscr{T}(e^{-t}) = \mathbf{0}$, for example.

24. $(\mathscr{T}(\mathbf{u}), \mathbf{v}) = (\mathbf{A}\mathbf{u}, \mathbf{v}) = (\mathbf{u}, \mathbf{A}^H\mathbf{v})$, so $\mathscr{T}^*(\mathbf{v}) = \mathbf{A}^H\mathbf{v}$.

27. (a) $\mathbf{n}$ is in $(\mathscr{R}^*)^\perp$ if and only if $(\mathbf{n}, \mathbf{x}) = 0$ for all $\mathbf{x}$ in $\mathscr{R}^*$ if and only if $(\mathbf{n}, \mathscr{T}^*(\mathbf{y})) = 0$ for all $\mathbf{y}$ in $\mathscr{W}$ if and only if $(\mathscr{T}(\mathbf{n}), \mathbf{y}) = 0$ for all $\mathbf{y}$ in $\mathscr{W}$; using $\mathbf{y} = \mathscr{T}(\mathbf{n})$ helps show that this holds if and only if $\mathbf{n}$ is in the null space $\mathscr{N}$ of $\mathscr{T}$.

(b) Let P_0 denote orthogonal projection onto $\mathscr{R}^*$ and let $\mathbf{x}$ be in $\mathscr{N}^\perp$. Show that $\mathbf{x} - P_0\mathbf{x}$ is in $\mathscr{R}^{*\perp}$ and so—by (a)—is in $\mathscr{N}$, which means that $\mathbf{x}$ is orthogonal to it; write $\|\mathbf{x} - P_0\mathbf{x}\|^2 = (\mathbf{x}, \mathbf{x} - P_0\mathbf{x}) - (P_0\mathbf{x}, \mathbf{x} - P_0\mathbf{x})$ and show this to be zero; conclude that $\mathbf{x}$ is in $\mathscr{R}^*$. The other direction is easier: $\mathbf{x}$ in $\mathscr{R}^*$ means $\mathbf{x} = \mathscr{T}^*(\mathbf{y})$ and $(\mathbf{n}, \mathbf{x}) = (\mathbf{n}, \mathscr{T}^*(\mathbf{y})) = (\mathscr{T}(\mathbf{n}), \mathbf{y}) = (\mathbf{0}, \mathbf{y}) = 0$ for all $\mathbf{n}$ in $\mathscr{N}$.

Problems 6.2

1. $\mathbf{I}$.

3.
$$\begin{bmatrix} 1 & -1 & -11 \\ -1 & 1 & 9 \\ 0 & 1 & 4 \\ 0 & 0 & -1 \end{bmatrix}.$$

6. $\mathbf{A} = \mathbf{diag}(1, 2, 3, 4)$.

9. $\mathcal{T}^{-1}$ exists if and only if $\mathcal{T}$ is one-to-one and onto. $\mathcal{T}$ is one-to-one, for example, if and only if $\mathcal{T}(\mathbf{v}) = \mathbf{0}$ implies $\mathbf{v} = \mathbf{0}$; $\mathcal{T}(\mathbf{v}) = \mathbf{0}$ if and only if $\mathbf{A}\mathbf{v}_B = \mathbf{0}_C = \mathbf{0}$ since C is linearly independent, and $\mathbf{A}\mathbf{v}_B = \mathbf{0}$ implies $\mathbf{v}_B = \mathbf{0}$ if and only if $\mathbf{A}$ is nonsingular.

11. $\mathcal{T}(\mathbf{v}) = \mathbf{0}$ if and only if $\mathbf{A}\mathbf{v}_B = \mathbf{0}_C$, and $\mathbf{0}_C = \mathbf{0}$ since C is linearly independent.

13. $\begin{bmatrix} 1 & -1 \\ 2 & 1 \\ 1 & -2 \end{bmatrix}$.

16. $\begin{bmatrix} 1 & 0 \\ 0 & 1 \\ 0 & 0 \end{bmatrix}$.

Problems 6.3

1. For $\mathbf{v} \neq \mathbf{0}$, $\|\mathcal{T}(\mathbf{v})\|_{\mathcal{W}}/\|\mathbf{v}\|_{\mathcal{V}} \leq k$, and so the supremum is also at most k.

4. $\|\mathcal{I}\|_{\mathcal{V},\mathcal{W}} = 1$.

7. $\|\mathcal{T}\|_{\mathcal{V},\mathcal{W}} = \|f\|$.

9. $\|\mathbf{A}\|_1 = 1.7$, $\|\mathbf{A}\|_\infty = 1.3$.

12. (a) $\|\mathbf{A}\|_\infty = 6$. (d) $\|\mathbf{A}\|_1 = 16$.

15. (b) $\|\mathbf{I}\|$ equals the supremum of $\|\mathbf{Ix}\|/\|\mathbf{x}\| = \|\mathbf{x}\|/\|\mathbf{x}\| = 1$.

Problems 6.4

1. Use **Key Lemma 6.26** and $\|\cdot\|_\infty$.

4. Use **Key Theorem 6.28** with one matrix the main diagonal of **A** and the other the rest of **A**.

6. $1 = \|\mathbf{I}\| = \|\mathbf{A}\mathbf{A}^{-1}\| \leq \|\mathbf{A}\|\,\|\mathbf{A}^{-1}\| = c(\mathbf{A})$.

11. Gauss elimination shows that row 2 $\approx$ 1.1(row 1) $-$ row 3; use (6.30).

14. The magnitude of the entry of largest magnitude in a column matrix equals the ∞-norm of that matrix; the magnitude of the entry of largest magnitude in a $p \times q$ matrix is at most the ∞-norm of the matrix and is at least $(1/q)$ times that ∞-norm.

16. (c) $c(\mathbf{A}) = 19{,}312$.

Problems 6.5

1. (a) $\{[-1 \quad 1 \quad 0 \quad 0]^T, [1 \quad 0 \quad 1 \quad 0]^T, [1 \quad 0 \quad 0 \quad 1]^T\}$, for example, from the general solution to $\mathbf{Ax} = \mathbf{0}$.

4. You must show that $[\cdot]_{B,C}$ is a linear transformation, that $[\mathcal{T}]_{B,C} = \mathbf{0}$ if and only if $\mathcal{T}$ equals the zero transformation $\mathcal{O}$ with $\mathcal{O}(\mathbf{v}) = \mathbf{0}$ for all $\mathbf{v}$, and that for every $p \times q$ **A** there is a $\mathcal{T}$ with $[\mathcal{T}]_{B,C} = \mathbf{A}$. For the last, for example, just define $\mathcal{T}(\mathbf{v}) = \mathbf{w}$ such that $\mathbf{w}_C = \mathbf{A}\mathbf{v}_B$.

7. Represent $\mathcal{T}$ by a $p \times p$ matrix **A**, and use **Key Theorem 4.18** by showing that each of the conditions on $\mathcal{T}$ is equivalent to a similar condition on **A**.

10. Simply use Problems 8 and 9.

14. $((\mathcal{S}\mathcal{T})\mathbf{u}, \mathbf{v}) = (\mathcal{S}\{\mathcal{T}(\mathbf{u})\}, \mathbf{v}) = (\mathcal{T}(\mathbf{u}), \mathcal{S}^*(\mathbf{v})) = (\mathbf{u}, \mathcal{T}^*\{\mathcal{S}^*(\mathbf{v})\}) = (\mathbf{u}, (\mathcal{T}^*\mathcal{S}^*)\mathbf{v})$.

Chapter 7 *Problems 7.1*

1. $\lambda = 1$ or 9.
3. $\lambda_1 = 0.5$, $\lambda_2 = -0.5$.
6. $\mathbf{A}' = \begin{bmatrix} 0.5 & 0 \\ 0 & -0.5 \end{bmatrix}$.

Problems 7.2

2. $\{[1 \quad -2 \quad 0]^T, [0 \quad 3 \quad 1]^T\}$, for example; $\lambda_3 = 6$, $[2 \quad 1 \quad -1]^T$.
4. $\lambda^2 - (a + d)\lambda + (ad - bc)$ for $\begin{bmatrix} a & b \\ c & d \end{bmatrix}$.
7. $-\lambda^3 + 1.8\lambda^2 - 0.95\lambda + 0.15$.
12. For any $i \geq 0$, $\mathbf{A}^i\mathbf{x} = \mathbf{A}^{i-1}(\mathbf{A}\mathbf{x}) = \mathbf{A}^{i-1}\lambda\mathbf{x} = \lambda\mathbf{A}^{i-1}\mathbf{x} = \cdots = \lambda^i\mathbf{x}$. Then $f(\mathbf{A})\mathbf{x} = (c_n\mathbf{A}^n + \cdots + c_1\mathbf{A} + c_0\mathbf{I})\mathbf{x} = c_n\mathbf{A}^n\mathbf{x} + \cdots + c_1\mathbf{A}\mathbf{x} + c_0\mathbf{I}\mathbf{x} = c_n\lambda^n\mathbf{x} + \cdots + c_1\lambda\mathbf{x} + c_0\mathbf{x} = f(\lambda)\mathbf{x}$.
17. $\lambda_1 = 2$, $m_1 = 2$, $\mu_1 = 1$, $\mathbf{v}_1 = [1 \quad 0 \quad 0]^T$; $\lambda_2 = 3$, $m_2 = \mu_2 = 1$, $\mathbf{v}_2 = [0 \quad 1 \quad 1]^T$.

Problems 7.3

1. $\mathbf{P} = \begin{bmatrix} 1 & 1 \\ -1 & 1 \end{bmatrix}$, $\mathbf{\Lambda} = \mathbf{diag}(1, 3)$.
4. No.
7. $B = \{[5 \quad 4]^T; [5 \quad 2]^T\}$, $\mathbf{\Lambda} = \mathbf{diag}(1.0, 0.8)$.
11. No. It fails to have a linearly independent set of two eigenvectors.

Problems 7.4

1. 4, -2, 3.
3. $\lambda_i = \langle \mathbf{U} \rangle_{ii}$.
6. (b) If $\mathbf{P} = \begin{bmatrix} 5 & 5 \\ 4 & 2 \end{bmatrix}$, then $\mathbf{A} = \mathbf{P\Lambda P}^{-1}$ with $\mathbf{\Lambda} = \mathbf{diag}(1.0, 0.8)$. So $\mathbf{x}_i = \mathbf{A}^i\mathbf{x}_0 = \mathbf{P\Lambda}^i\mathbf{P}^{-1}\mathbf{x}_0 = \mathbf{P}\,\mathbf{diag}(1, (0.8)^i)\mathbf{P}^{-1}\mathbf{x}_0$, which tends to $\mathbf{P}\,\mathbf{diag}(1, 0)\mathbf{P}^{-1}\mathbf{x}_0 = [2400 \quad 1920]^T$ if $\mathbf{x}_0 = [100 \quad 1000]^T$.
10. Their eigenvalues are not the same. Also, det $\mathbf{A} \neq$ det $\mathbf{B}$.
13. $\lambda_1 = 3$, $m_1 = 4$, $\mu_1 = 3$; $\lambda_2 = 5$, $m_2 = 3$, $\mu_2 = 1$.

Problems 7.5

2. $a = 0$, $b = \pm\sqrt{2}/2$, $c = \mp\sqrt{2}/2$.
5. $\mathbf{P}$ is unitary if and only if $\mathbf{PP}^H = \mathbf{P}^H\mathbf{P} = \mathbf{I}$; $\mathbf{P}^H$ is unitary if and only if $\mathbf{P}^H\mathbf{P}^{HH} = \mathbf{P}^{HH}\mathbf{P}^H = \mathbf{I}$; these are identical, since $\mathbf{P}^{HH} = \mathbf{P}$.
8. Use Theorem 6.25.
12. $\mathbf{H}_\mathbf{w} = \begin{bmatrix} 0 & 1 \\ 1 & 0 \end{bmatrix}$.

14. $\mathbf{PA} = \mathbf{H_w A} = \mathbf{A} - (2/\mathbf{w}^T\mathbf{w})\mathbf{w}(\mathbf{w}^T\mathbf{A})$. For multiplication/divisions: $\mathbf{w}^T\mathbf{A}$ costs p^2, $\mathbf{w}^T\mathbf{w}$ costs p more, $(2/\mathbf{w}^T\mathbf{w})$ costs 1 more, $(2/\mathbf{w}^T\mathbf{w})\mathbf{w}$ costs p more, $(2/\mathbf{w}^T\mathbf{w})\mathbf{w}(\mathbf{w}^T\mathbf{A})$ costs p^2 more, for a total of $2p^2 + 2p + 1$.

17. Let ω be the $(p - r) \times 1$ matrix formed from the lower $p - r$ entries of $\mathbf{w}$. Then

$$\mathbf{H_w} = \begin{bmatrix} \mathbf{I}_r & \mathbf{0} \\ \mathbf{0} & \mathbf{H}_\omega \end{bmatrix}.$$

20. $\mathbf{R} = \begin{bmatrix} \sqrt{2} & 3\sqrt{2} & \dfrac{9\sqrt{2}}{2} \\ 0 & \sqrt{3} & \dfrac{4\sqrt{3}}{3} \\ 0 & 0 & \dfrac{\sqrt{6}}{6} \end{bmatrix}$, for example.

22. $\mathbf{R}_{21}(90°) = \begin{bmatrix} 0 & 1 & 0 \\ -1 & 0 & 0 \\ 0 & 0 & 1 \end{bmatrix}$.

25. If $\mathbf{Q}$ is a product of elementary rotation matrices, then $\det \mathbf{Q} = 1$. Not all orthogonal matrices satisfy $\det \mathbf{Q} = 1$;

$$\det \begin{bmatrix} 0 & 1 \\ 1 & 0 \end{bmatrix} = -1, \text{ for example.}$$

28. $\mathbf{A}^H\mathbf{A} = \mathbf{A}^T\mathbf{A} = \mathbf{AA} = \mathbf{AA}^T = \mathbf{AA}^H$.

31. If $\mathbf{A} = \mathbf{P}^H\mathbf{DP}$ with unitary $\mathbf{P}$ and diagonal $\mathbf{D}$, then $\mathbf{A}^H\mathbf{A} = (\mathbf{P}^H\mathbf{DP})^H(\mathbf{P}^H\mathbf{DP}) = \mathbf{P}^H\mathbf{D}^H\mathbf{PP}^H\mathbf{DP} = \mathbf{P}^H\mathbf{D}^H\mathbf{DP} = \mathbf{P}^H\mathbf{DD}^H\mathbf{P} = \mathbf{P}^H\mathbf{DPP}^H\mathbf{D}^H\mathbf{P} = \mathbf{AA}^H$.

35. First pick a $(p - 1) \times 1$ ω so that $\mathbf{H}_\omega$ times the lower $p - 1$ entries of the first column $[\alpha \quad \mathbf{c}^T]^T$ equals $\lambda\mathbf{e}'_1$, a multiple of the first $(p - 1) \times 1$ unit column matrix, and then form $\mathbf{w} = [0 \quad \omega^T]^T$. Then

$$\mathbf{H_w A H_w} = \begin{bmatrix} 1 & \mathbf{0}^T \\ \mathbf{0} & \mathbf{H}_\omega \end{bmatrix} \begin{bmatrix} \alpha & \mathbf{r}^T \\ \mathbf{c} & \mathbf{A}_0 \end{bmatrix} \begin{bmatrix} 1 & \mathbf{0}^T \\ \mathbf{0} & \mathbf{H}_\omega \end{bmatrix} = \begin{bmatrix} 1 & \mathbf{0}^T \\ \mathbf{0} & \mathbf{H}_\omega \end{bmatrix} \begin{bmatrix} \alpha & \mathbf{r}^T\mathbf{H}_\omega \\ \mathbf{c} & \mathbf{A}_0\mathbf{H}_\omega \end{bmatrix}$$

$$= \begin{bmatrix} \alpha & \mathbf{r}^T\mathbf{H}_\omega \\ \lambda\mathbf{e}'_1 & \mathbf{H}_\omega\mathbf{A}_0\mathbf{H}_\omega \end{bmatrix}.$$

37. $\begin{bmatrix} 1 & 0 & 0 \\ 0 & 0 & 1 \\ 0 & 1 & 0 \end{bmatrix} \mathbf{A} \begin{bmatrix} 1 & 0 & 0 \\ 0 & 0 & 1 \\ 0 & 1 & 0 \end{bmatrix} = \begin{bmatrix} 1 & -1 & 1 \\ 2 & 1 & 2 \\ 0 & 3 & 1 \end{bmatrix}$ is in Hessenberg form.

Problems 7.7

1. $|\lambda_1 - 1| \leq 3 \times 10^{-5}$, $|\lambda_2 - 0.5| \leq 7 \times 10^{-5}$, $|\lambda_3 - 0.1| \leq 4 \times 10^{-5}$.

2. Roughly $|\lambda_1 - 1| \leq 2 \times 10^{-9}$.

7. (a) Show that $\mathbf{v} = \mathbf{P}(\mathbf{\Lambda} - \mu\mathbf{I})^{-1}\mathbf{P}^{-1}\mathbf{r}$ if μ is not an exact eigenvalue (when the inequality clearly holds), so $\|\mathbf{v}\| \leq \|\mathbf{P}\| \, \|(\mathbf{\Lambda} - \mu\mathbf{I})^{-1}\| \, \|\mathbf{P}^{-1}\| \, \|\mathbf{r}\|$. Then show that $\|(\mathbf{\Lambda} - \mu\mathbf{I})^{-1}\|$ is the reciprocal of the least distance of μ from an eigenvalue of $\mathbf{A}$, say $1/|\mu - \lambda_s|$. Finally, rewrite the inequality as $|\mu - \lambda_s| \leq \|\mathbf{P}\| \, \|\mathbf{P}^{-1}\| \, \|\mathbf{r}\|/\|\mathbf{v}\|$.

11. The first part is $n = 1$ in (b) of the Gerschgorin circle theorem. The eigenvalues of real **B** are the roots of a polynomial with real coefficients, whose roots must therefore appear as real numbers or as complex-conjugate pairs (in which case two would fall in the same Gerschgorin disc). So the eigenvalues are real.

Problems 7.8

2. $\lambda_r = k + 2\cos\{r\pi/(n+1)\}$; just verify that the $\mathbf{x}_i$ are eigenvectors.
5. Use mathematical induction on n, and expand along the last column.
9. The statement is false.
10. (b) Use (a).
11. Use an identity for $(\mathbf{Px}, \mathbf{Py})$ and for $(\mathbf{P}i\mathbf{x}, \mathbf{Py})$ analogous to $ab = (1/4)\{(a+b)^2 - (a-b)^2\}$.

Chapter 8

Problems 8.1

1. $\mathbf{x} = \mathbf{D}\boldsymbol{\xi}$.

Problems 8.2

2. $\begin{bmatrix} -2 & 0 & 0 \\ 0 & -2 & -2\sqrt{6} \\ 0 & 0 & 6 \end{bmatrix}$, for example.
6. $\mathbf{A}^H\mathbf{A} = (-\mathbf{A})\mathbf{A} = \mathbf{A}(-\mathbf{A}) = \mathbf{A}\mathbf{A}^H$.
9. Yes; $\begin{bmatrix} 1 & 0 \\ 0 & 4 \end{bmatrix}$, for example.
10. $\begin{bmatrix} 1+i & 0 \\ 0 & 1-i \end{bmatrix}$, for example.

Problems 8.3

1. (a) For $\mathbf{x} \neq \mathbf{0}$: $\mathbf{Ax} = \lambda\mathbf{x}$ if and only if $\bar{\mathbf{A}}\bar{\mathbf{x}} = \bar{\lambda}\bar{\mathbf{x}}$ if and only if $0 = \det(\bar{\mathbf{A}} - \bar{\lambda}\mathbf{I}) = \det(\bar{\mathbf{A}} - \bar{\lambda}\mathbf{I})^T = \det(\mathbf{A}^H - \bar{\lambda}\mathbf{I})$ if and only if $\bar{\lambda}$ is an eigenvalue of $\mathbf{A}^H$.
   (b) $\mathbf{A} = \mathbf{PDP}^H$ if and only if $\mathbf{A}^H = \mathbf{PD}^H\mathbf{P}^H$, and then use **Key Theorem 7.14.**
4. $\{[1 \quad 1 \quad 0]^T/\sqrt{2}, [-1 \quad 1 \quad 2]^T/\sqrt{6}, [1 \quad -1 \quad 1]^T/\sqrt{3}\}$, for example.
7. Use $\mathbf{A} = \mathbf{PDP}^H$ and $\mathbf{D} = \mathbf{D}^H$ if and only if $\mathbf{D}$ is real.
10. (a) $b_1 - b_2 = 0$. (b) $b_1 + b_2 = 0$.
12. There is at least one solution to $\mathbf{Ax} = \mathbf{b}$ if and only if *either* 0 is not an eigenvalue *or* 0 is an eigenvalue and all $\mathbf{b}$ are orthogonal to all eigenvectors $\mathbf{x}_0$ associated with $\lambda = 0$, but taking $\mathbf{b} = \mathbf{x}_0$ in this latter case shows that $\mathbf{x}_0$ is not an eigenvector and so $\mathbf{A}$ is nonsingular and hence $\mathbf{Ax} = \mathbf{b}$ is uniquely solvable by $\mathbf{x} = \mathbf{A}^{-1}\mathbf{b}$.

Problems 8.4

2. (a) $\begin{bmatrix} \frac{1}{3} & \frac{2}{3} & -\frac{2}{3} \\ \frac{2}{3} & \frac{1}{3} & \frac{2}{3} \\ -\frac{2}{3} & \frac{2}{3} & \frac{1}{3} \end{bmatrix}\begin{bmatrix} 3 \\ 0 \\ 0 \end{bmatrix}[1]$.

(b) $[1][13\quad 0\quad 0]\begin{bmatrix} -\frac{4}{13} & \frac{12}{13} & \frac{3}{13} \\ \frac{3}{13} & \frac{4}{13} & -\frac{12}{13} \\ \frac{12}{13} & \frac{3}{13} & \frac{4}{13} \end{bmatrix}$.

4. The main-diagonal entries of $\mathbf{A}^H\mathbf{A}$ equal the squared 2-norms of the columns of $\mathbf{A} \neq \mathbf{0}$, so at least one such entry is nonzero. $\mathbf{A}^H\mathbf{A}$ is hermitian and so $\mathbf{A}^H\mathbf{A} = \mathbf{P}^H\mathbf{\Lambda}\mathbf{P}$ for diagonal $\mathbf{\Lambda}$, and $\mathbf{\Lambda} = \mathbf{0}$ would imply $\mathbf{A}^H\mathbf{A} = \mathbf{0}$.

7. $\|\mathbf{A}\|_2 = \|\mathbf{U\Sigma V}^H\|_2 = \|\mathbf{\Sigma}\|_2$, which can be evaluated from the definition of a matrix norm.

10. $\|\mathbf{A}\|_F^2 = \|\mathbf{U\Sigma V}^H\|_F^2 = \|\mathbf{\Sigma}\|_F^2 = \sigma_1^2 + \cdots + \sigma_s^2$.

12. (b) $\begin{bmatrix} \frac{2}{\sqrt{5}} & \frac{1}{\sqrt{5}} \\ \frac{1}{\sqrt{5}} & -\frac{2}{\sqrt{5}} \end{bmatrix}\begin{bmatrix} \frac{1}{\sqrt{2}} & \frac{1}{\sqrt{2}} \\ \frac{1}{\sqrt{2}} & -\frac{1}{\sqrt{2}} \end{bmatrix} = \begin{bmatrix} \frac{3}{\sqrt{10}} & \frac{1}{\sqrt{10}} \\ -\frac{1}{\sqrt{10}} & \frac{3}{\sqrt{10}} \end{bmatrix} \approx \begin{bmatrix} 0.95 & 0.32 \\ -0.32 & 0.95 \end{bmatrix}$.

Problems 8.5

2. $(\mathbf{0}_{p\times q})^+ = \mathbf{0}_{q\times p}$.

8. $x = 14/9$.

9. $\mathbf{A} = \mathbf{U\Sigma V}^H$ and $\mathbf{\Sigma}$ are nonsingular so, by construction, $\mathbf{\Sigma}^{-1} = \mathbf{\Sigma}^+$. Then $\mathbf{A}^{-1} = (\mathbf{U\Sigma V}^H)^{-1} = (\mathbf{V}^H)^{-1}\mathbf{\Sigma}^{-1}\mathbf{U}^{-1} = \mathbf{V\Sigma}^+\mathbf{U}^H = \mathbf{A}^+$.

12. $\mathbf{A}^+ = \dfrac{1}{102}\begin{bmatrix} -15 & -18 & 3 & -3 & 18 & 15 \\ 8 & 13 & -5 & 5 & -13 & -8 \\ 7 & 5 & 2 & -2 & -5 & -7 \\ 6 & -3 & 9 & -9 & 3 & -6 \end{bmatrix}$.

16. (b) $\mathbf{c}^+ = \mathbf{c}^H/\mathbf{c}^H\mathbf{c}$. (c) $(\mathbf{cr})^+ = \mathbf{r}^H\mathbf{c}^H/\{(\mathbf{c}^H\mathbf{c})(\mathbf{rr}^H)\}$.

Problems 8.6

3. $\langle\mathbf{D}^{-1}\mathbf{AD}\rangle_{ij} = \epsilon^{i-j}\langle\mathbf{A}\rangle_{ij}$.

6. All eigenvalues of normal $\mathbf{A}$ are all equal (to c, say) if and only if the matrix of eigenvalues is $\mathbf{\Lambda} = c\mathbf{I}$, which holds if and only if $\mathbf{P\Lambda P}^H = c\mathbf{I}$.

8. If $\mathbf{A} = \mathbf{U\Sigma V}^H$, then $\mathbf{A}^+ = \mathbf{V\Sigma}^+\mathbf{U}^H$, so $\mathbf{AA}^+ = \mathbf{U\Sigma V}^H\mathbf{V\Sigma}^+\mathbf{U}^H = \mathbf{U\Sigma\Sigma}^+\mathbf{U}^H$.

$$\mathbf{\Sigma\Sigma}^+ = \begin{bmatrix} \mathbf{I}_k & \mathbf{0} \\ \mathbf{0} & \mathbf{0} \end{bmatrix},$$

where k is the rank, and so $\mathbf{AA}^+ = \mathbf{U}_0\mathbf{U}_0^H$ where $\mathbf{U}_0$ equals the matrix of the first k columns of $\mathbf{U}$. Use Theorem 5.79 and **Key Corollary 8.20(b)**.

Chapter 9

Problems 9.2

3. $\begin{bmatrix} 1 & -2 & 3 & 0 \\ 0 & 1 & -1 & 0 \\ 0 & 0 & 1 & 0 \\ 0 & 0 & 0 & -3 \end{bmatrix}$.

6. $\mathbf{J} = \begin{bmatrix} 1 & 0 & 0 \\ 0 & -2 & 0 \\ 0 & 0 & -3 \end{bmatrix}$, for example.

10. $\mathbf{J} = \begin{bmatrix} 2+i & 0 & 0 & 0 \\ 0 & 2-i & 0 & 0 \\ 0 & 0 & 2+i & 0 \\ 0 & 0 & 0 & 2-i \end{bmatrix}$, for example.

14. $(\lambda - 4)^3 = \lambda^3 - 12\lambda^2 + 48\lambda - 64$.

15. For each distinct eigenvalue λ_i, let the largest Jordan block containing λ_i be $p_i \times p_i$. Then the minimum polynomial equals

$$(\lambda - \lambda_1)^{p_1}(\lambda - \lambda_2)^{p_2} \cdots (\lambda - \lambda_s)^{p_s}.$$

Problems 9.3

2. (a) $\{[1 \quad 0 \quad 1]^T\}$ and $\{[1 \quad -1 \quad -1]^T, [1 \quad 0 \quad 0]^T\}$, for example.
(c) $\{[1 \quad 0 \quad 0]^T, [0 \quad \frac{1}{2} \quad 0]^T\}$ and $\{[0 \quad 1 \quad 1]^T\}$, for example.
(f) $\{[1 \quad 0 \quad 0 \quad 0]^T, [0 \quad -\frac{1}{2} \quad 0 \quad 0]^T, [0 \quad \frac{3}{4} \quad \frac{1}{2} \quad 0]^T\}$ and $\{[15 \quad 2 \quad -8 \quad 8]^T\}$, for example.

5. (a) $\mathbf{v}_{11} = [1 \quad -1 \quad -1]^T$, $\mathbf{v}_{21} = [1 \quad -1 \quad 1]^T$, $\mathbf{v}_{22} = [0 \quad 1 \quad -1]^T$, $\mathbf{u}_1 = [0 \quad 1 \quad 1]^T$, $\mathbf{u}_2 = [1 \quad 1 \quad 0]^T$. $\mathbf{u}_1^T\mathbf{v}_{21} = \mathbf{u}_1^T\mathbf{v}_{22} = \mathbf{u}_2^T\mathbf{v}_{11} = \mathbf{u}_2^T\mathbf{v}_{21} = 0$.
(d) $\mathbf{v}_{11} = [1 \quad 1 \quad 1]^T$, $\mathbf{v}_{21} = [1 \quad -2 \quad 4]^T$, $\mathbf{v}_{31} = [1 \quad -3 \quad 9]^T$, $\mathbf{u}_1 = [6 \quad 5 \quad 1]^T$, $\mathbf{u}_2 = [3 \quad -2 \quad -1]^T$, $\mathbf{u}_3 = [2 \quad -1 \quad -1]^T$. $\mathbf{u}_1^T\mathbf{v}_{21} = \mathbf{u}_1^T\mathbf{v}_{31} = \mathbf{u}_2^T\mathbf{v}_{11} = \mathbf{u}_2^T\mathbf{v}_{31} = \mathbf{u}_3^T\mathbf{v}_{11} = \mathbf{u}_3^T\mathbf{v}_{21} = 0$.

7. $5b_1 + b_2 - 7b_4 = 0$, $9b_1 + b_3 - 12b_4 = 0$.

Problems 9.4

3. Eigenvalues: 1.1, 0.7; so $\|\mathbf{x}_i\|$ tends to infinity for some $\mathbf{x}_0$ (including $\mathbf{x}_0 = [100 \quad 1000]^T$ in Example 2.13).

5. Eigenvalues: 1, -1, 1/2; all powers $\mathbf{A}^i$ and $\mathbf{x}_i$ remain bounded.

9. Let $\mathbf{A}' = \mathbf{A}/\alpha$, so $\rho(\mathbf{A}') = \rho(\mathbf{A})/\alpha < 1$. Then $\alpha\mathbf{I} - \mathbf{A} = \alpha(\mathbf{I} - \mathbf{A}')$ is nonsingular since $\mathbf{I} - \mathbf{A}'$ is and $\alpha \neq 0$, with $(\mathbf{I} - \mathbf{A}')^{-1} = \mathbf{I} + \mathbf{A}' + \mathbf{A}'^2 + \cdots$ and $(\alpha\mathbf{I} - \mathbf{A})^{-1} = \alpha^{-1}(\mathbf{I} - \mathbf{A}')^{-1}$.

10. (a) The only permutation gives $\mathbf{A}$ itself, and this is not in the form required in Definition 9.33.
(c) Use $\mathbf{P} = [\mathbf{e}_2 \quad \mathbf{e}_1]$.

13. $\mathbf{x}_0$ must be orthogonal to the left-eigenvectors of $\mathbf{A}$ associated with $\lambda = 1$.

16. Problem 15 says that we may let $\alpha = 2$ in Problem 14.

19. Every eigenvalue satisfies $|\lambda| \leq \|\mathbf{A}\|$ for every norm, so certainly $\rho(\mathbf{A}) \leq \|\mathbf{A}\|_1$. If $\mathbf{A}' = \mathbf{A}/\rho(\mathbf{A})$ is a Markov matrix, then $1 = \|\mathbf{A}'\|_1 = \|\mathbf{A}\|_1/\rho(\mathbf{A})$, so $\rho(\mathbf{A}) = \|\mathbf{A}\|_1$. If $\rho(\mathbf{A}) = \|\mathbf{A}\|_1$, then every column of $\mathbf{A}$ must sum to 1; if not, some entry of $\mathbf{A}$ could be strictly increased to give a new matrix $\tilde{\mathbf{A}}$ with $\|\tilde{\mathbf{A}}\|_1 = \|\mathbf{A}\|_1$, but also $\rho(\tilde{\mathbf{A}}) > \rho(\mathbf{A})$ by (f) in the Perron-Frobenius theorem, which gives the contradiction $\|\tilde{\mathbf{A}}\|_1 = \|\mathbf{A}\|_1 = \rho(\mathbf{A}) < \rho(\tilde{\mathbf{A}}) \leq \|\tilde{\mathbf{A}}\|_1$.

Problems 9.5

1. $\mathbf{A} = \begin{bmatrix} 0 & 1 & 0 & 0 & 0 & 0 \\ -\frac{k}{m} & 0 & \frac{k}{m} & 0 & 0 & 0 \\ 0 & 0 & 0 & 1 & 0 & 0 \\ \frac{k}{M} & 0 & -\frac{2k}{M} & 0 & \frac{k}{M} & 0 \\ 0 & 0 & 0 & 0 & 0 & 1 \\ 0 & 0 & \frac{k}{m} & 0 & -\frac{k}{m} & 0 \end{bmatrix}$, for example.

6. $x_1(t) = (1 + t)\exp(2t)$, $x_2(t) = \exp(2t)$.

8. $\exp(\mathbf{I}) = e\mathbf{I}$.

11. Use Theorem 9.54(b).

16. $\exp(\mathbf{A}) = \begin{bmatrix} 2e^4 & \frac{3}{2}e^4 - \frac{1}{2}e^{-2} & \frac{1}{2}e^4 - \frac{1}{2}e^{-2} \\ -e^4 & \frac{1}{2}e^{-2} - \frac{1}{2}e^4 & \frac{1}{2}e^{-2} - \frac{1}{2}e^4 \\ e^4 & \frac{1}{2}e^{-2} + \frac{1}{2}e^4 & \frac{1}{2}e^{-2} + \frac{1}{2}e^4 \end{bmatrix}$.

18. Choose k so that no eigenvalue has positive real part and some eigenvalue has zero real part.

Problems 9.6

2. $u_{r+1} = -0.5v_r + 2$, $v_{r+1} = -0.5u_{r+1} + 2.5$.

7. $\mathbf{H_J} = \begin{bmatrix} 0 & -\frac{1}{2} \\ -\frac{1}{2} & 0 \end{bmatrix}$, $\mathbf{H_{GS}} = \begin{bmatrix} 0 & -\frac{1}{2} \\ 0 & \frac{1}{4} \end{bmatrix}$.

9. $\rho(\mathbf{H_J}) = 0.5$, $\rho(\mathbf{H_{GS}}) = 0.25$, $\rho(\mathbf{H}_{1.1}) = 0.1$.

11. (a) $\rho(\mathbf{H_J}) = 2$. (b) $\rho(\mathbf{H_{GS}}) = 4$.

16. (a) $\omega^* = 8 - 4\sqrt{3} \approx 1.072$. Experiment with various values of ω as in Example 9.72 to minimize $\rho(\mathbf{H}_\omega)$.

Problems 9.7

2. If $\mathbf{Ax} = \lambda\mathbf{x}$, then $\mathbf{A\bar{x}} = \overline{\mathbf{Ax}} = \bar{\lambda}\mathbf{\bar{x}}$, so we can take as a linearly independent set of eigenvectors associated with $\lambda = \alpha + i\beta$ and $\bar{\lambda} = \alpha - i\beta$ a set $\{\mathbf{x}, \mathbf{\bar{x}}\} = \{\mathbf{a} + i\mathbf{b}, \mathbf{a} - i\mathbf{b}\}$ for real $\mathbf{a}$ and $\mathbf{b}$. $\{\mathbf{a}, \mathbf{b}\}$ must be linearly independent since $\{\mathbf{a} + i\mathbf{b}, \mathbf{a} - i\mathbf{b}\}$ is. Now $\mathbf{Ax} = \lambda\mathbf{x}$ means $\mathbf{A}(\mathbf{a} + i\mathbf{b}) = (\alpha + i\beta)(\mathbf{a} + i\mathbf{b}) = (\alpha\mathbf{a} - \beta\mathbf{b}) + i(\alpha\mathbf{b} + \beta\mathbf{a})$; this gives $\mathbf{Aa} = \alpha\mathbf{a} - \beta\mathbf{b}$ and $\mathbf{Ab} = \alpha\mathbf{b} + \beta\mathbf{a}$, that is: $\mathbf{A}[\mathbf{a}\ \ \mathbf{b}] = [\mathbf{a}\ \ \mathbf{b}]\begin{bmatrix} \alpha & \beta \\ -\beta & \alpha \end{bmatrix}$. Let $\mathbf{P} = [\mathbf{a}\ \ \mathbf{b}]$.

5. If $\mathbf{x}_0 = c_1\mathbf{v}_1 + \cdots + c_p\mathbf{v}_p$, then $\mathbf{A}^i\mathbf{x}_0 = \lambda_1^i\{c_1\mathbf{v}_1 + c_2(\lambda_2/\lambda_1)^i\mathbf{v}_2 + \cdots + c_p(\lambda_p/\lambda_1)^i\mathbf{v}_p\}$, so a normalized $\mathbf{A}^i\mathbf{x}_0$ will converge to a multiple of the eigenvector $\mathbf{v}_1$ as long as $c_1 \neq 0$.

7. If $\mathbf{A}$ and $\mathbf{B}$ commute, then the order of the products in the series for $\exp(\mathbf{A} + \mathbf{B})$ can be switched as though $\mathbf{A}$ and $\mathbf{B}$ were numbers, yielding the product of the series for exp(**A**) and for exp(**B**). Counterexample: try

$$\mathbf{A} = \mathbf{B}^T = \begin{bmatrix} 0 & 1 \\ 0 & 0 \end{bmatrix}.$$

Chapter 10 *Problems 10.1*

1. (b) $2x_1^2 + 2x_1x_2 + 3x_2^2$.
4. $x_1^2/9 + x_2^2/25 = 1$.
7. $x_1^2/144 - x_2^2/25 = 1$.

Problems 10.2

2. $t < 0$, $\lambda_1 \geq 0$, $\lambda_2 \geq 0$; $t > 0$, $\lambda_1 \leq 0$, $\lambda_2 \leq 0$.
4. (b) $\{(5 + \sqrt{153})/2\}\xi_1^2 + \{(5 - \sqrt{153})/2\}\xi_2^2 = t$, a hyperbola.
5. The curve C is the noncircular ellipse $x_1^2/64 + x_2^2 = 9$.

Problems 10.3

1. (a) $\xi_1^2 + 3\xi_2^2 + 4\xi_3^2$, for example.
5. $\mathbf{A}$ is negative definite if and only if the principal submatrices $\mathbf{A}_k$ satisfy $(-1)^k \det \mathbf{A}_k > 0$ for $1 \leq k \leq p$.
7. (b) Pivots are 5, $\frac{9}{5}$, -27, so it is indefinite.
9. $\mathbf{L}_1\mathbf{D}_0\mathbf{U}_1 = \mathbf{A} = \mathbf{A}^H = (\mathbf{L}_1\mathbf{D}_0\mathbf{U}_1)^H = \mathbf{U}_1^H\mathbf{D}_0^H\mathbf{L}_1^H$; use Problem 13 of Section 3.10.
11. Use **Key Theorem 10.18** relating definiteness and eigenvalues.
14. Use **Key Theorem 10.18** relating definiteness and eigenvalues.

Problems 10.4

1. The eigenvalues are actually -2, 0, 1.
2. The eigenvalues are actually 1, 3, 4.
5. (b) The eigenvalues are actually -5, 23, -9.
7. The eigenvalues are actually -2, 0, 1.
10. (c) The eigenvalues are actually 1, 2, 4.

Problems 10.5

1. The eigenvalues are actually about 8.2057, -2.8089, 0, 2.6032.
4. The eigenvalues are actually 0.6, 0.3, 0.1.
7. Show that $\mathbf{B} - \mathbf{A}$ is positive semidefinite to use Theorem 10.33.
10. Find matrices $\mathbf{A}_l$ and $\mathbf{A}_u$ for which you can find the eigenvalues (for example, from $\alpha\mathbf{I} - \mathbf{E}$) so that $\mathbf{x}^H\mathbf{A}_l\mathbf{x} \leq \mathbf{x}^H\mathbf{A}\mathbf{x} \leq \mathbf{x}^H\mathbf{A}_u\mathbf{x}$, and use Theorem 10.33. $0.6 \leq \lambda_1 \leq 1.2$, $1.1 \leq \lambda_2 \leq 1.7$, $1.7 \leq \lambda_3 \leq 2.3$, $2.2 \leq \lambda_4 \leq 2.8$.

Problems 10.6

2. (a) The curve is similar to the graph of $\xi_1^2 + 9\xi_2^2 = 1$.
6. (a) $\mathbf{x}^H\mathbf{A}'\mathbf{x} = \mathbf{x}^H\mathbf{B}^H\mathbf{A}\mathbf{B}\mathbf{x} = (\mathbf{B}\mathbf{x})^H\mathbf{A}(\mathbf{B}\mathbf{x})$.
(b) $\mathbf{B}\mathbf{x} = \mathbf{0}$ with $\mathbf{x} \neq \mathbf{0}$ if and only if $\mathbf{B}$ has rank less than q.

8. There is an $\mathbf{x} \neq \mathbf{0}$ with $(\mathbf{A} + \lambda\mathbf{B} + \lambda^2\mathbf{C})\mathbf{x} = \mathbf{0}$, so $0 = \mathbf{x}^H\mathbf{A}\mathbf{x} + \lambda\mathbf{x}^H\mathbf{B}\mathbf{x} + \lambda^2\mathbf{x}^H\mathbf{C}\mathbf{x} = a + b\lambda + c\lambda^2$ with a, b, and c positive; use the formula for the roots of a quadratic.

9. From Problem 15 of Section 10.3 or from $\mathbf{B} = \mathbf{P\Lambda P}^H$, write $\mathbf{B} = \mathbf{MM}^H$ with $\mathbf{M}$ nonsingular. Then $\mathbf{C} = \mathbf{M}^{-1}(\mathbf{BA})\mathbf{M}$ and $\mathbf{BA}$ are similar and so have the same eigenvalues. But $\mathbf{C} = \mathbf{M}^{-1}(\mathbf{MM}^H\mathbf{A})\mathbf{M} = \mathbf{M}^H\mathbf{AM}$, which is hermitian since $\mathbf{A}$ is hermitian; thus the eigenvalues are real.

10. From Problem 15 of Section 10.3 or from $\mathbf{B} = \mathbf{P\Lambda P}^H$, write $\mathbf{B} = \mathbf{MM}^H$ for nonsingular $\mathbf{M}$ and let $\mathbf{N} = \mathbf{M}^{-1}$. Define $\mathbf{B}' = \mathbf{NBN}^H = \mathbf{I}$ and $\mathbf{C}' = \mathbf{NCN}^H$. Use eigenvalues to show that (a)–(c) hold with $\mathbf{B}'$ and $\mathbf{C}'$ in the place of $\mathbf{B}$ and $\mathbf{C}$, and then deduce that they also hold for $\mathbf{B}$ and $\mathbf{C}$.

12. (a) Partition $\mathbf{A}$ and define a matrix $\mathbf{E}$ as below:

$$\mathbf{A} = \begin{bmatrix} \mathbf{A}_1 & \mathbf{c} \\ \mathbf{c}^H & a_{pp} \end{bmatrix}, \qquad \mathbf{E} = \begin{bmatrix} \mathbf{I}_{p-1} & \mathbf{0} \\ -\mathbf{c}^H\mathbf{A}_1^{-1} & 1 \end{bmatrix},$$

with $\mathbf{A}_1$ also positive definite and $\det \mathbf{E} = 1$. Then $\det \mathbf{A} = \det \mathbf{EA}$, and $\mathbf{EA}$ equals

$$\mathbf{EA} = \begin{bmatrix} \mathbf{A}_1 & \mathbf{c} \\ \mathbf{0}^H & a_{pp} - \mathbf{c}^H\mathbf{A}_1^{-1}\mathbf{c} \end{bmatrix}.$$

Because of the structure of $\mathbf{EA}$, its determinant equals $(a_{pp} - \mathbf{c}^H\mathbf{A}_1^{-1}\mathbf{c}) \det \mathbf{A}_1$. Since $\mathbf{A}_1$ and thus $\mathbf{A}_1^{-1}$ are positive definite, we get $\det \mathbf{A} \le a_{pp} \det \mathbf{A}_1$. Then use mathematical induction.

(b) If $\det \mathbf{B} = 0$, the result is clearly true. For $\det \mathbf{B} \neq 0$, the matrix $\mathbf{A} = \mathbf{B}^H\mathbf{B}$ is positive definite; apply (a) to this $\mathbf{A}$.

14. (a) $(\mathbf{A} - \lambda\mathbf{I})\mathbf{x} = \mathbf{0}$ implies that $(\mathbf{A}^2 - \lambda\mathbf{A})\mathbf{x} = \mathbf{0}$ or $(1 - \lambda)\mathbf{A}\mathbf{x} = \mathbf{0}$. Hence $\lambda = 1$, or $\mathbf{A}\mathbf{x} = \mathbf{0}$ which implies that $\lambda = 0$.

(b) If $\mathbf{A}$ is nonsingular, $\mathbf{A}^2 = \mathbf{A}$ implies $\mathbf{A} = \mathbf{I}$.

(d) The trace equals the sum of the eigenvalues.

17. For (a) and (b), let $\mathbf{A} = \mathbf{P}^H\mathbf{\Lambda P}$ with nonnegative diagonal $\mathbf{\Lambda}$, and define nonnegative diagonal $\mathbf{L}$ with $\mathbf{L}^2 = \mathbf{\Lambda}$; then $\mathbf{B} = \mathbf{P}^H\mathbf{LP}$ gives $\mathbf{B}^2 = \mathbf{A}$ as needed. (c) Suppose that $\mathbf{A} = \mathbf{B}^2$ for some $\mathbf{B}$; the problem is to show that $\mathbf{B} = \mathbf{P}^H\mathbf{LP}$ from (a) and (b). Let $\mathbf{B} = \mathbf{Q}^H\mathbf{DQ}$ with nonnegative diagonal $\mathbf{D}$, so that $\mathbf{A} = \mathbf{B}^2 = \mathbf{Q}^H\mathbf{D}^2\mathbf{Q}$ and still $\mathbf{A} = \mathbf{P}^H\mathbf{L}^2\mathbf{P}$. Thus $\mathbf{D}^2 = \mathbf{S}^H\mathbf{L}^2\mathbf{S}$, where $\mathbf{S} = \mathbf{PQ}^H$ is unitary, so $\mathbf{SD}^2 = \mathbf{L}^2\mathbf{S}$; that is, $\langle\mathbf{S}\rangle_{ij}d_j^2 = \langle\mathbf{S}\rangle_{ij}l_i^2$, where $\mathbf{D} = \mathbf{diag}(d_1, \ldots, d_p)$ and $\mathbf{L} = \mathbf{diag}(l_1, \ldots, l_p)$. Thus $\langle\mathbf{S}\rangle_{ij}(d_j - l_i)(d_j + l_i) = 0$ for all i and j. If either d_j or l_i is positive, then this gives $\langle\mathbf{S}\rangle_{ij}(d_j - l_i) = 0$—which certainly also holds if both d_j and l_i equal 0. But $\langle\mathbf{S}\rangle_{ij}d_j = \langle\mathbf{S}\rangle_{ij}l_i$ means that $\mathbf{SD} = \mathbf{LS}$, so $\mathbf{D} = \mathbf{S}^H\mathbf{LS}$. Therefore, $\mathbf{B} = \mathbf{Q}^H\mathbf{DQ} = \mathbf{Q}^H\mathbf{S}^H\mathbf{LSQ} = \mathbf{Q}^H\mathbf{QP}^H\mathbf{LPQ}^H\mathbf{Q} = \mathbf{P}^H\mathbf{LP}$, as needed.

18. (d) Let $\mathbf{A} = \mathbf{U}_0\mathbf{\Sigma}_0\mathbf{V}_0^H$ be a singular value decomposition of $\mathbf{A}$. We can rewrite this as $\mathbf{A} = (\mathbf{U}_0\mathbf{V}_0^H)(\mathbf{V}_0\mathbf{\Sigma}_0\mathbf{V}_0^H)$ and $\mathbf{A} = (\mathbf{U}_0\mathbf{\Sigma}_0\mathbf{U}_0^H)(\mathbf{U}_0\mathbf{V}_0^H)$, which is just $\mathbf{A} = \mathbf{UM} = \mathbf{NV}$ with $\mathbf{U} = \mathbf{V} = \mathbf{U}_0\mathbf{V}_0^H$, $\mathbf{M} = \mathbf{V}_0\mathbf{\Sigma}_0\mathbf{V}_0^H$, and $\mathbf{N} = \mathbf{U}_0\mathbf{\Sigma}_0\mathbf{U}_0^H$.

Chapter 11 *Problems 11.1*

1. $x_1 = 10$, $x_2 = 20$.

2. $x_1 = 40$, $x_2 = 20$.

6. $16[2.5 \quad 3.75 \quad -8.75 \quad -6.25 \quad -13.75]^T = [40 \quad 60 \quad -140 \quad -100 \quad -220]^T$.

8. $f(\mathbf{x}) = c_1x_1 + \cdots + c_px_p$, so $\partial f/\partial x_i = c_i$ and $\nabla f = \mathbf{c}$.

12. Minimize $24y_1 + 30y_2 + 10y_3$
subject to the constraints
$y_1 + y_2 - y_3 \geq 1$
$y_2 + y_3 \geq 2$
all $y_i \geq 0$.
$[1 \;\; 2]^T = 0[1 \;\; 0]^T + (3/2)[1 \;\; 1]^T + (1/2)[-1 \;\; 1]^T$, and the optimal solution is $y_1 = 0$, $y_2 = 3/2$, $y_3 = 1/2$.

15. Maximize $-70y_1 - 40y_2 - 90y_3$
subject to the constraints
$-2y_1 - y_2 - y_3 \leq -40$
$-y_1 - y_2 - 3y_3 \leq -60$
all $y_i \geq 0$.
Optimal solution $y_1 = 0$, $y_2 = 30$, $y_3 = 10$.

Problems 11.2

1. Maximize $-2x_1 + 5x_2$
subject to the constraints
$2x_1 - x_2 \leq 7$
$3x_1 + x_2 \leq 6$
$-3x_1 - x_2 \leq -6$
$x_1 \geq 0$, $x_2 \geq 0$.

4. (Using $z_1 = x_1$, $z_2 - z_3 = x_2$, $z_4 = x_3$, with slack variables z_5, z_6, z_7, and z_8):
Maximize $4z_1 + 3z_2 - 3z_3 - 2z_4$
subject to the constraints

$$
\begin{aligned}
-z_1 - z_2 + z_3 \quad + z_5 \quad &= -4\\
z_1 - z_2 + z_3 \quad + z_6 \quad &= 6\\
z_1 + 3z_2 - 3z_3 + 6z_4 \quad + z_7 \quad &= 5\\
-z_1 - 3z_2 + 3z_3 - 6z_4 \quad + z_8 &= -5
\end{aligned}
$$

all $z_i \geq 0$.

9. (a) It is degenerate (columns 1, 3, and 5 from $[\mathbf{A}_e \;\; \mathbf{b}]$ are linearly dependent, for example).

11. Maximize m
subject to the constraints
$m - u \leq 0$
$m - v \leq 0$
$2u + v \leq 5$
$u + 3v \leq 8$
$m \geq 0$, $u \geq 0$, $v \geq 0$.

Problems 11.3

2. The program values are unbounded above, so there is no optimal solution.

7. Since the problem is nondegenerate, exactly q variables equal zero at the present basic feasible vector and the p basic variables are strictly positive; use the structure of the tableau to show that the values of these p variables form the present righthand side. Since the bottom row has a negative entry, say in the x_n-column, with at least one positive entry in the column above, there are quotients of positive right-hand-side entries by positive x_n-column entries to be examined; the least of these will be strictly positive, providing a strict increase in x_n and thus in M.

9. (a) $x_1 = 25 + \alpha$, $x_2 = 15 - \alpha$ is optimal for $0 \le \alpha \le 10$.

11. (b) $x_1 = 10$, $x_2 = 20$.

15. There are none satisfying the constraints.

Problems 11.4

1. Minimize $\mathbf{d}^T\mathbf{u} - \mathbf{d}^T\mathbf{v}$
subject to the constraints
$\mathbf{B}^T\mathbf{u} - \mathbf{B}^T\mathbf{v} \ge \mathbf{f}$
$\mathbf{u} \ge \mathbf{0}$, $\mathbf{v} \ge \mathbf{0}$
(or $\mathbf{u} - \mathbf{v}$ can be replaced by $\mathbf{w}$ without sign constraints on $\mathbf{w}$).

3. (a) Minimize $7y_1 + 6y_2 - 6y_3$
subject to the constraints

$$\begin{aligned} 2y_1 + 3y_2 - 3y_3 &\ge -2 \\ -y_1 + y_2 - y_3 &\ge 5 \end{aligned}$$

all $y_i \ge 0$.

10. There are no feasible vectors for this primal (the dual values are unbounded).

12. $\mathbf{Ax} \le \mathbf{0}$, $\mathbf{x} \ge \mathbf{0}$, $\mathbf{c}^T\mathbf{x} > 0$ is equivalent to $\mathbf{Bx} \le \mathbf{0}$, $\mathbf{c}^T\mathbf{x} > 0$ with $\mathbf{B} = \begin{bmatrix} \mathbf{A} \\ -\mathbf{I} \end{bmatrix}$. This is equivalent to the *unsolvability* of $\mathbf{B}^T\mathbf{y} = \mathbf{c}$, $\mathbf{y} \ge \mathbf{0}$ by Farkas' theorem. But this is just $[\mathbf{A}^T \quad -\mathbf{I}]\mathbf{y} = \mathbf{c}$; partitioning $\mathbf{y}$ by $\mathbf{y}^T = [\mathbf{y}_1^T \quad \mathbf{y}_2^T]$, this becomes $\mathbf{A}^T\mathbf{y}_1 = \mathbf{c} + \mathbf{y}_2 \ge \mathbf{c}$, with $\mathbf{y}_1 \ge \mathbf{0}$.

Problems 11.5

4. Minimize $x_1 + x_2 + x_3 + x_4 + x_5 + x_6$
subject to the constraints

$$\begin{aligned} x_1 \qquad\qquad\qquad\qquad + x_6 &\ge 3 \\ x_1 + x_2 \qquad\qquad\qquad\quad &\ge 2 \\ x_2 + x_3 \qquad\qquad\quad &\ge 10 \\ x_3 + x_4 \qquad\quad &\ge 14 \\ x_4 + x_5 \quad &\ge 8 \\ x_5 + x_6 &\ge 10 \end{aligned}$$

all $x_i \ge 0$.

To show the given feasible solution to be optimal, form the tableau and put it into the standard format for the present basic variables $(x_1, x_3, x_5, x_6, x_7, x_9)$; observe that there are no negative entries in the bottom row (except in the irrelevant final column).

6. (b) Minimize $[\mathbf{0}^T \quad \mathbf{1}^T][\mathbf{z}^T \quad \mathbf{m}^T]^T$—where $\mathbf{1} = [1 \quad 1 \quad \cdots \quad 1]^T$—
subject to the constraints

$$\begin{aligned} \mathbf{Bz} - \mathbf{m} &\le \mathbf{w} \\ -\mathbf{Bz} - \mathbf{m} &\le -\mathbf{w}. \end{aligned}$$

9. (a) By **Key Theorem 11.53(e)**, $\mathbf{c}^T\mathbf{x}^* = \mathbf{b}^T\mathbf{y}^*$. Therefore, $\mathbf{c}^T\mathbf{x}^* = (\mathbf{A}^T\mathbf{y}^* - \mathbf{s}_\mathbf{y})^T\mathbf{x}^* = \mathbf{y}^{*T}\mathbf{Ax}^* - \mathbf{s}_\mathbf{y}^T\mathbf{x}^* = \mathbf{y}^{*T}(\mathbf{b} - \mathbf{s}_\mathbf{x}) - \mathbf{s}_\mathbf{y}^T\mathbf{x}^* = \mathbf{b}^T\mathbf{y}^* - \mathbf{s}_\mathbf{x}^T\mathbf{y}^* - \mathbf{s}_\mathbf{y}^T\mathbf{x}^*$, so $\mathbf{s}_\mathbf{x}^T\mathbf{y}^* + \mathbf{s}_\mathbf{y}^T\mathbf{x}^* = 0$. But $\mathbf{s}_\mathbf{x}$, $\mathbf{s}_\mathbf{y}$, $\mathbf{x}^*$, and $\mathbf{y}^*$ are all nonnegative; thus $\mathbf{s}_\mathbf{x}^T\mathbf{y}^* = \mathbf{s}_\mathbf{y}^T\mathbf{x}^* = 0$.

(b) The equation $\mathbf{A}^T\mathbf{y}^* - \mathbf{s}_\mathbf{y} = \mathbf{c}$ gives $\mathbf{c}^T = \mathbf{y}^{*T}\mathbf{A} - \mathbf{s}_\mathbf{y}^T\mathbf{I}$, which explicitly writes $\mathbf{c}^T$ as a nonnegative linear combination of *all* the rows of $\mathbf{A}$ and of *all* the $-\mathbf{e}_i^T$. From $\mathbf{s}_\mathbf{x}^T\mathbf{y}^* = 0$ with $\mathbf{s}_\mathbf{x} \geq \mathbf{0}$ and $\mathbf{y}^* \geq \mathbf{0}$, it follows that $\langle \mathbf{y}^* \rangle_i = 0$ whenever $\langle \mathbf{s}_\mathbf{x} \rangle_i > 0$; that is, the ith row of $\mathbf{A}$ is not actually involved in the representation of $\mathbf{c}^T$ above if the constraint corresponding to that row is not an equality at $\mathbf{x}^*$. Similarly, $\mathbf{s}_\mathbf{y}^T\mathbf{x}^* = 0$ gives $\langle \mathbf{s}_\mathbf{y} \rangle_i = 0$ whenever $\langle \mathbf{x}^* \rangle_i > 0$; that is, those $-\mathbf{e}_i^T$ for which $\langle \mathbf{x}^* \rangle_i > 0$ are not actually involved in the above representation of $\mathbf{c}^T$.

Appendix two

Bibliography

Theory of Linear Algebra and Matrices

1. F. R. Gantmacher, *Theory of Matrices*, Vols. I. II, Chelsea (1959).
2. P. R. Halmos, *Finite-Dimensional Vector Spaces*, Van Nostrand (1958).
3. M. Marcus and H. Minc, *A Survey of Matrix Theory and Matrix Inequalities*, Allyn and Bacon (1964).
4. T. Muir, *Determinants*, Dover (1960).

Applications of Linear Algebra and Matrices

5. A. Albert, *Regression and the Moore-Penrose Pseudoinverse*, Academic Press (1972).
6. N. R. Amundsen, *Mathematical Models in Chemical Engineering; Matrices and Their Applications*, Prentice-Hall (1966).
7. R. Bellman and K. R. Cooke, *Modern Elementary Differential Equations*, 2nd edition, Addison-Wesley (1971).
8. M. Braun, *Differential Equations and Their Applications*, Springer Verlag (1975).
9. G. B. Dantzig, *Linear Programming and Extensions*, Princeton University Press (1963).
10. R. A. Frazer, W. J. Duncan, and A. R. Collar, *Elementary Matrices and Some Applications to Dynamics and Differential Equations*, Cambridge University Press (1938).
11. F. R. Gantmacher, *Applications of the Theory of Matrices*, Interscience (1959).
12. F. A. Graybill, *An Introduction to Linear Statistical Models*, Vol. I, McGraw-Hill (1961).
13. R. Haberman, *Mathematical Models*, Prentice-Hall (1977).
14. G. Hadley, *Linear Programming*, Addison-Wesley (1962).
15. J. Heading, *Matrix Theory for Physicists*, Wiley (1960).
16. W. C. Hurty and M. F. Rubinstein, *Dynamics of Structures*, Prentice-Hall (1964).
17. S. Karlin, *Mathematical Models and Theory in Games, Programming, and Economics*, Vols. I, II, Addison-Wesley (1959).
18. N. Karmarkar, "A new polynomial-time algorithm for linear programming," *Combinatorica*, Vol. 4, No. 4, 373–395 (1984).
19. J. G. Kemeny and J. L. Snell, *Finite Markov Chains*, Van Nostrand (1960).
20. J. G. Kemeny and J. L. Snell, *Mathematical Models in the Social Sciences*, MIT Press (1972).

21. J. G. Kemeny, J. L. Snell, and G. L. Thompson, *Introduction to Finite Mathematics*, Prentice-Hall (1957).
22. I. Linnik, *Method of Least Squares and Principles of the Theory of Observations*, Pergamon Press (1961).
23. D. Maki and M. Thompson, *Mathematical Models and Applications*, Prentice-Hall (1973).
24. O. L. Mangasarian, *Nonlinear Programming*, McGraw-Hill (1969).
25. H. C. Martin, *Introduction to Matrix Methods of Structural Analysis*, McGraw-Hill (1966).
26. M. Z. Nashed, *Generalized Inverses and Applications*, Academic Press (1976).
27. L. A. Pipes, *Matrix Methods in Engineering*, Prentice-Hall (1963).
28. C. R. Rao, *Advanced Statistical Methods in Biometric Research*, Wiley (1952).
29. J. Robinson, *Structural Matrix Analysis for the Engineer*, Wiley (1966).
30. J. T. Schwartz, *Lectures on the Mathematical Method in Analytical Economics*, Gordon and Breach (1961).
31. S. R. Searle, *Matrix Algebra for the Biological Sciences (Including Applications in Statistics)*, Wiley (1966).
32. S. Senturia and B. Wedlock, *Electronic Circuits and Applications*, Wiley (1975).
33. M. Simonnard, *Linear Programming*, translated by W. S. Jewell, Prentice-Hall (1966).
34. G. Strang, *Introduction to Applied Mathematics*, Wellesley-Cambridge Press (1986).
35. A. M. Tropper, *Matrix Theory for Electrical Engineers*, Addison-Wesley and Harrap (1962).
36. R. E. Walpole and R. H. Myers, *Probability and Statistics for Engineers and Scientists*, 2nd edition, Macmillan (1978).
37. A. V. Weiss, *Matrix Analysis for Electrical Engineers*, Van Nostrand (1964).

Computations for Matrices and Linear Algebra

38. A. Bjorck, R. J. Plemmons, and H. Schneider, *Large-Scale Matrix Problems*, North Holland (1981).
39. J. R. Bunch and D. J. Rose (eds.), *Sparse Matrix Computations*, Academic Press (1976).
40. J. J. Dongarra, C. B. Moler, J. R. Bunch, and G. W. Stewart, *LINPACK User's Guide*, SIAM Publications (1979).
41. I. S. Duff and G. W. Stewart (eds.), *Sparse Matrix Proceedings*, SIAM Publications (1979).
42. G. E. Forsythe, M. A. Malcolm, and C. B. Moler, *Computer Methods for Mathematical Computation*, Prentice-Hall (1977).
43. G. E. Forsythe and C. B. Moler, *Computer Solution of Linear Algebraic Systems*, Prentice-Hall (1967).
44. B. S. Garbow, J. M. Boyle, J. J. Dongarra, and C. B. Moler, *Matrix Eigensystem Routines: EISPACK Guide Extension*, Springer Verlag (1972).
45. J. A. George and J. W. Liu, *Computer Solution of Large Sparse Positive Definite Systems*, Prentice-Hall (1981).
46. G. H. Golub and C. F. Van Loan, *Matrix Computations*, Johns Hopkins University Press (1983).
47. L. A. Hageman and D. M. Young, Jr., *Applied Iterative Methods*, Academic Press (1981).
48. A. S. Householder, *The Theory of Matrices in Numerical Analysis*, Blaisdell (1964).

49. C. L. Lawson and R. J. Hanson, *Solving Least Squares Problems*, Prentice-Hall (1974).

50. B. N. Parlett, *The Symmetric Eigenvalue Problem*, Prentice-Hall (1980).

51. D. Rose and R. Willoughby (eds.), *Sparse Matrices and Their Applications*, Plenum Press (1972).

52. B. T. Smith, J. M. Boyle, J. Dongarra, B. Garbow, Y. Ikebe, V. C. Klema, and C. B. Moler, *Matrix Eigensystem Routines: EISPACK Guide*, 2nd edition, Springer Verlag (1976).

53. G. W. Stewart, *Introduction to Matrix Computations*, Academic Press (1973).

54. R. S. Varga, *Matrix Iterative Analysis*, Prentice-Hall (1962).

55. J. H. Wilkinson, *Rounding Errors in Algebraic Processes*, Prentice-Hall (1963).

56. J. H. Wilkinson, *The Algebraic Eigenvalue Problem*, Oxford (1965).

57. (Unauthored), IMSL mathematical software, International Mathematical and Statistical Laboratories, Inc., 7500 Bellaire Blvd., Houston, TX 77036.

58. (Unauthored), MATLAB matrix software, The MathWorks, Inc., 158 Woodland St., Sherborn, MA 01770.

59. (Unauthored), NAG mathematical software, Numerical Algorithms Group, Inc., 1101 31st St., Downers Grove, IL 60515.

Index of notation

$\mathbf{H}_J$, $\mathbf{H}_{GS}$, $\mathbf{H}_\omega$	iteration matrices for Jacobi, Gauss-Seidel, and SOR methods—Example 9.72, Theorem 9.73
i	pure imaginary $\sqrt{-1}$—Section 1.1
i, j	integers
$\mathbf{I}_r$	$r \times r$ identity matrix—Definition 1.18
$\mathbf{I}$	identity matrix—Definition 1.18
$\mathscr{I}$	identity linear transformation—Problem 2 in Section 6.1
$\mathbf{J}$	Jordan form—**Key Theorem 9.4**, (7.26)
$\mathbf{J}_i$, $\mathbf{J}(\lambda)$	Jordan block—Definition 9.3
$\mathbf{L}$, $\mathbf{L}_0$ in $\mathbf{LU}$ and $\mathbf{L}_0\mathbf{U}_0$	lower-triangular matrix in LU-decompositions—Section 3.7, Definition 1.3
m_i	algebraic multiplicity of an eigenvalue—Definition 7.8
$\mathfrak{M}$	symbol for a computer-based example or problem
$\mathscr{N}$	null space of a linear transformation—Definition 6.5
$\mathbf{P}$ in $\mathbf{P}^H\mathbf{AP}$ and $\mathbf{PDP}^H$	usually a unitary matrix—Definitions 7.29 and 7.32, (7.42), and **Key Theorem 8.6**
$\mathbf{P}$ in $\mathbf{P}^{-1}\mathbf{AP}$ and $\mathbf{P\Lambda P}^{-1}$	nonsingular matrix—Definitions 1.34 and 7.20, **Key Theorem 7.14**
$\mathbf{P}$ in $\mathbf{P}^T\mathbf{LU}$	permutation matrix—Definition 3.51
P_0	orthogonal projection onto a subspace—Theorem 5.71
$\mathscr{P}$	real vector space of real polynomials of arbitrary degree—Example 5.9(f)
$\mathscr{P}^k$	real vector space of real polynomials of degree strictly less than k—Example 5.9(g)
$\mathbf{Q}$, $\mathbf{Q}_0$ in $\mathbf{QR}$ and $\mathbf{Q}_0\mathbf{R}_0$	matrix with orthogonal or orthonormal columns in QR-decompositions—**Key Theorem 5.82**
$\mathbf{Q}$ in $\mathbf{QJQ}^{-1}$ and $\mathbf{Q}^{-1}\mathbf{AQ}$	nonsingular matrix—Definition 1.34, **Key Theorem 9.4**
$Q(\mathbf{x})$, $\tilde{Q}(\xi)$	quadratic forms—Definition 10.13, **Key Theorem 10.14**, (10.5)
$\mathbf{R}$, $\mathbf{R}_0$ in $\mathbf{QR}$ and $\mathbf{Q}_0\mathbf{R}_0$	upper-triangular matrix in QR-decompositions—Definition 1.3, **Key Thoerem 5.82**
$\mathbb{R}$	set of real numbers
$\mathbb{R}^p$	real vector space of $p \times 1$ real column matrices—Example 5.9(b), Definition 5.15
$\mathbf{R}_{kn}(\theta)$	elementary rotation matrix—Definition 7.40
S_t	level surface of a quadratic form in $\mathbf{x}$—Section 10.3
$\tilde{S}_t$	level surface of a quadratic form in ξ—Section 10.3
$\mathscr{T}, \mathscr{S}, \ldots$	script uppercase letters that usually denote linear transformations—Definition 6.1
$\mathscr{T}^*$	adjoint transformation of $\mathscr{T}$—Definition 6.10
$\mathscr{T}^{-1}$	inverse transformation of $\mathscr{T}$—Theorem 6.9
tr $\mathbf{A}$, tr($\mathbf{A}$)	trace of $\mathbf{A}$—Problem 6 in Section 1.2, Theorem 7.6(c)
$\mathbf{U}$ in $\mathbf{U\Sigma V}^H$ and $\mathbf{V\Sigma}^H\mathbf{U}^H$	unitary matrix in singular value decompositions—Definition 7.29, (8.15), **Key Theorem 8.19**
$\mathbf{V}$ in $\mathbf{U\Sigma V}^H$ and $\mathbf{V\Sigma}^H\mathbf{U}^H$	unitary matrix in singular value decompositions—Definition 7.29, (8.15), **Key Theorem 8.19**

$\mathscr{V}, \mathscr{W}, \ldots$	script uppercase letters that usually denote vector spaces—Definition 5.8
$\mathscr{V}_0, \mathscr{W}_0$	usually subspaces—Definition 5.11
$\mathbf{0}$	zero matrix or zero vector—Definitions 1.9 and 5.8
$\mathscr{O}$	zero linear transformation—Problem 3 in Section 6.1
λ	eigenvalue—Definition 7.4
$\mathbf{\Lambda}$	diagonal matrix of eigenvalues—Definition 1.3, **Key Theorem 7.14**
μ_i	geometric multiplicity of an eigenvalue—Definition 7.12
$\rho(\mathbf{A})$	spectral radius of **A**—Definition 9.26
$\rho_{\mathbf{A}}(\mathbf{x})$	Rayleigh quotient for **A** at **x**—Definition 10.23
σ_i	singular value—**Key Theorem 8.19**
$\mathbf{\Sigma}$ in $\mathbf{U\Sigma V}^H$	"diagonal" matrix of singular values—(8.15), **Key Theorem 8.19**
$\langle\cdot\rangle_{ij}$ in $\langle\mathbf{A}\rangle_{ij}$	(i, j)-entry of a matrix—Definition 1.1
$\langle\cdot\rangle_i$ in $\langle\mathbf{x}\rangle_i$	$(i, 1)$- or $(1, i)$-entry of a column- or row-matrix—Definition 1.1
$(\cdot, \cdot)$ in $(\mathbf{u}, \mathbf{v})$	inner product—Definition 5.63, Example 5.64
$\prec\!\langle\cdot\rangle\!\succ$ in $\prec\!\langle x\rangle\!\succ$	value of x after rounding to t floating-point digits—Definition 3.35
$\lVert\cdot\rVert$ in $\lVert\mathbf{x}\rVert$ and $\lVert\mathbf{A}\rVert$	norm of a vector or a matrix—Definitions 5.54 and 6.22, Theorem 5.68
$\lVert\cdot\rVert_{\mathscr{V}}$ in $\lVert\mathbf{v}\rVert_{\mathscr{V}}$	norm of a vector in $\mathscr{V}$—Definition 5.54
$\lVert\cdot\rVert_1, \lVert\cdot\rVert_2, \lVert\cdot\rVert_\infty$ in $\lVert\mathbf{x}\rVert_1, \lVert\mathbf{A}\rVert_2, \lVert\mathbf{b}\rVert_\infty$	1-, 2-, and ∞-norm of a matrix—Definitions 5.55 and 6.22, Problem 7 in Section 8.4, Corollary 10.26
$\lVert\cdot\rVert_F$ in $\lVert\mathbf{A}\rVert_F$	Frobenius norm of a matrix—Problem 8 in Section 8.4
$\lVert\cdot\rVert_{\mathscr{V},\mathscr{W}}$ in $\lVert\mathscr{T}\rVert_{\mathscr{V},\mathscr{W}}$	transformation norm of a linear transformation—Definition 6.19
$[\cdot]$ in $[1 \quad 2 \quad 3]$	brackets defining a matrix—Definition 1.1
$[\cdot]$ in $[a, b]$	closed interval of numbers t satisfying $a \le t \le b$
$<, \le, >, \ge$ with matrices	inequalities for matrices—Definitions 11.28 and 9.32
^{T} in $\mathbf{A}^T$	transpose of a matrix—Definition 1.20
^{H} in $\mathbf{A}^H$	hermitian transpose of a matrix—Definition 1.20
$^{-1}$ in $\mathbf{A}^{-1}$ and $\mathscr{T}^{-1}$	inverse of a matrix or linear transformation—Definition 1.34, Theorem 6.9
$^+$ in $\mathbf{A}^+$	pseudoinverse of a matrix—Definition 8.25, **Key Theorem 8.26**
$*$ in $\mathscr{T}^*$	adjoint transformation of a linear transformation—Definition 6.10
$\bar{\ }$ in $\bar{z}$ and $\bar{\mathbf{A}}$	complex conjugate of a number or matrix—Section 1.1
$\vec{\ }$ in $\vec{u}$	geometrical vector—Section 5.1
$\approx$	approximate equality—(2.37)
■	denotes the end of a proof
▷	denotes a Problem that has an answer or aid in Appendix one

Subject Index

D

M

N

O

P

W

Z